Gas-Phase and Surface Chemistry in Electronic Materials Processing

MATERIALS RESEARCH SOCIETY SYMPOSIUM PROCEEDINGS VOLUME 334

Gas-Phase and Surface Chemistry in Electronic Materials Processing

Symposium held November 29-December 2, 1993, Boston, Massachusetts, U.S.A.

EDITORS:

T.J. Mountziaris

State University of New York
Buffalo, New York, U.S.A.

G.R. Paz-Pujalt

Eastman Kodak Company
Rochester, New York, U.S.A.

F.T.J. Smith

LORAL Infrared and Imaging Systems
Lexington, Massachusetts, U.S.A.

P.R. Westmoreland

University of Massachusetts
Amherst, Massachusetts, U.S.A.

MATERIALS RESEARCH SOCIETY
Pittsburgh, Pennsylvania

This work was supported in part by the Army Research Office under Grant Number DAAH04-93-G-0516. The views, opinions, and/or findings contained in this report are those of the author(s) and should not be construed as an official Department of the Army position, policy, or decision, unless so designated by other documentation.

This work was supported in part by the Office of Naval Research under Grant Number N0014-94-1-0511. The United States Government has a royalty-free license throughout the world in all copyrightable material contained herein.

Single article reprints from this publication are available through
University Microfilms Inc., 300 North Zeeb Road, Ann Arbor, Michigan 48106

CODEN: MRSPDH

Copyright 1994 by Materials Research Society.
All rights reserved.

This book has been registered with Copyright Clearance Center, Inc. For further information, please contact the Copyright Clearance Center, Salem, Massachusetts.

Published by:

Materials Research Society
9800 McKnight Road
Pittsburgh, Pennsylvania 15237
Telephone (412) 367-3003
Fax (412) 367-4373

Library of Congress Cataloging in Publication Data

Gas-phase and surface chemistry in electronic materials processing / editors,
    T.J. Mountziaris, G.R. Paz-Pujalt, F.T.J. Smith, P.R. Westmoreland
        p. cm.—(Materials Research Society symposium proceedings ; v. 334).
        Sponsors: Army Research Office, Office of Naval Research.
        Includes bibliographical references and index.
        ISBN 1-55899-233-2
        1. Electronic materials—Surfaces—Congresses. 2. Surface chemistry—Congresses.
   I. Mountziaris, T.J. II. Paz-Pujalt, G.R. III. Smith, F.T.J. IV. Westmoreland,
   P.R. V. United States. Army Research Office. VI. United States. Office of Naval
   Research. VII. Series: Materials Research Society Symposium Proceedings ; v. 334.
TK7871.G37 1994   94-9632
621.3815'2—dc20   CIP

Manufactured in the United States of America

Contents

PREFACE xiii

ACKNOWLEDGMENTS xv

MATERIALS RESEARCH SOCIETY SYMPOSIUM PROCEEDINGS xvi

PART I: SILICON AND CARBON SYSTEMS

*GAS-PHASE SILICON ATOM DENSITIES IN THE CHEMICAL
VAPOR DEPOSITION OF SILICON FROM SILANE 3
   Michael E. Coltrin, William G. Breiland, and Pauline Ho

*GAS PHASE AND GAS SURFACE KINETICS OF TRANSIENT
SILICON HYDRIDE SPECIES 11
   Joseph M. Jasinski

THEORETICAL PREDICTION OF GAS-PHASE NUCLEATION
KINETICS OF SiO 19
   Michael R. Zachariah and Wing Tsang

ATOMIC LAYER GROWTH OF SiO_2 ON Si(100) USING THE
SEQUENTIAL DEPOSITION OF $SiCl_4$ AND H_2O 25
   Ofer Sneh, Michael L. Wise, Lynne A. Okada, Andrew W. Ott,
   and Steven M. George

INFRARED STUDIES OF OZONE-ORGANOSILICON CHEMISTRIES
FOR SiO_2 DEPOSITION 31
   J.A. Mucha and J. Washington

DIETHYLDIETHOXYSILANE AS A NEW PRECURSOR FOR SiO_2
GROWTH ON SILICON 37
   Michael L. Wise, Ofer Sneh, Lynne A. Okada, Anne C. Dillon,
   and Steven M. George

DECOMPOSITION STUDY OF TEOS IN THERMAL CVD 45
   Hideki Takeuchi, Hirohiko Izumi, and Atsushi Kawasaki

DIRECT MEASUREMENT OF THE REACTIVITY OF NH AND OH
ON A SILICON NITRIDE SURFACE 51
   R.J. Buss, P. Ho, E.R. Fisher, and William G. Breiland

SURFACE CHEMISTRY OF BORON-DOPED SiO_2 CVD: ENHANCED
UPTAKE OF TETRAETHYL ORTHOSILICATE BY HYDROXYL GROUPS
BONDED TO BORON 57
   Michael E. Bartram and Harry K. Moffat

PREDICTIVE SURFACE KINETIC ANALYSIS: THE CASE OF
$TiSi_2$ CVD 63
   M.A. Mendicino, R.P. Southwell, and E.G. Seebauer

HYDROGEN PAIRING ON Si(100)-(2×1): A SITE-BLOCKING
STUDY 69
   Wolf Widdra and W. Henry Weinberg

MOLECULAR DYNAMICS SIMULATION OF LARGE CLUSTER GROWTH 75
   Michael R. Zachariah, Michael J. Carrier, and Estela Blaisten-Barojas

*Invited Paper

ROOM TEMPERATURE FABRICATION OF MICRO-CRYSTALLINE
SILICON FILMS FOR TFT'S BY ECR PECVD 81
    Yoo-Chan Jeon, Seok-Woon Lee, and Seung-Ki Joo

GROWTH OF SILICON FILMS ONTO FLUORORESIN SURFACE
BY ArF EXCIMER LASER 87
    T. Miyokawa, M. Okoshi, K. Toyoda, and M. Murahara

PYROLYTIC LCVD OF SILICON USING A PULSED VISIBLE LASER—
EXPERIMENT AND MODELLING 93
    B. Ivanov, C. Popov, V. Shanov, and D. Filipov

EFFECT OF POST EXPOSED HYDROGEN ON CHEMISORBED
ETHYLENE ON Si(100)-(2×1) 99
    Wolf Widdra, Chen Huang, and W. Henry Weinberg

THE DECOMPOSITION OF METHYLTRICHLOROSILANE:
STUDIES IN A HIGH-TEMPERATURE FLOW REACTOR 105
    Mark D. Allendorf, Thomas H. Osterheld, and Carl F. Melius

HETEROGENEOUS KINETICS OF THE CHEMICAL VAPOR DEPOSITION
OF SILICON CARBIDE FROM METHYLTRICHLOROSILANE 111
    George D. Papasouliotis and Stratis V. Sotirchos

REACTION KINETICS ON DIAMOND: MEASUREMENT OF H
ATOM DESTRUCTION RATES 117
    Stephen J. Harris and Anita M. Weiner

A NOVEL MOLECULAR BEAM REACTOR FOR THE STUDY
OF DIAMOND SURFACE CHEMISTRY 123
    C.A. Wolden, G. Zau, W.T. Conner, H.H. Sawin, and K.K. Gleason

USING MOLECULAR-BEAM MASS SPECTROMETRY TO STUDY THE
PECVD OF DIAMONDLIKE CARBON FILMS 129
    I.B. Graff, R.A. Pugliese, Jr., and P.R. Westmoreland

EFFECT OF OXYGEN ON DIAMOND DEPOSITION IN $CH_4/O_2/H_2$
GAS MIXTURES 135
    H. Matsuyama, N. Sato, and H. Kawakami

MASS SPECTROMETRIC ANALYSIS OF A HIGH PRESSURE,
INDUCTIVELY COUPLED PLASMA DURING DIAMOND FILM GROWTH ... 141
    Peter G. Greuel, Hyun J. Yoon, Doulgas W. Ernie, and
    Jeffrey T. Roberts

SELECTIVE GROWTH OF DIAMOND FILMS ON MIRROR-
POLISHED SILICON SUBSTRATES 147
    M.Y. Mao, S.S. Tan, X.K. Zhang, and W.Y. Wang

PART II: COMPOUND SEMICONDUCTORS

*REAL TIME CONTROL OF III-V SEMICONDUCTOR SURFACES
DURING MOVPE GROWTH BY REFLECTANCE ANISOTROPY
SPECTROSCOPY 155
    K. Ploska, W. Richter, F. Reinhardt, J. Jönsson, J. Rumberg,
    and M. Zorn

GAS-PHASE AND SURFACE DECOMPOSITION OF *TRIS*-DIMETHYLAMINO
ARSENIC 169
    Ming Xi, Sateria Salim, Klavs F. Jensen, and David A. Bohling

*Invited Paper

KINETICS OF LOW-PRESSURE MOCVD OF GaAs FROM
TRIETHYL-GALLIUM AND ARSINE 177
    N.K. Ingle, C. Theodoropoulos, T.J. Mountziaris, R.M. Wexler,
    and F.T.J. Smith

THE ROLE OF UNPRECRACKED HYDRIDE SPECIES ADSORBED
ON THE GaAs(100) IN THE GROWTH OF GaAs BY ULTRAHIGH
VACUUM CHEMICAL VAPOR DEPOSITION USING TRIMETHYLGALLIUM
AND TRIETHYLGALLIUM 183
    Seong-Ju Park, Jeong-Rae Ro, Jae-Ki Sim, Jeong Sook Ha, and
    El-Hang Lee

*CONTROLLED IMPURITY INTRODUCTION IN CVD: CHEMICAL,
ELECTRICAL, AND MORPHOLOGICAL INFLUENCES 189
    T.F. Kuech, J.M. Redwing, J.-W. Huang, and S. Nayak

STUDY OF SILICON INCORPORATION IN GaAs MOVPE LAYERS
GROWN WITH TERTIARYBUTYLARSINE 201
    J.M. Redwing, T.F. Kuech, H. Simka, and K.F. Jensen

PREPARATION AND EVALUATION OF ERBIUM *TRIS*(AMIDE)
SOURCE COMPOUNDS FOR ERBIUM DOPING OF SEMICONDUCTING
MATERIALS 207
    William S. Rees, Jr., Uwe W. Lay, and Anton C. Greenwald

TETRAALKYLDIARSINES AS POTENTIAL PRECURSORS FOR
ELECTRONIC MATERIALS: SYNTHESIS AND CHARACTERIZATION
OF VARIOUS *ISO*-PROPYL ARSENIC COMPOUNDS 213
    Lawrence F. Brough, Liu Gang, Matthew A. Lipkovich,
    Thomas J. Colacot, Virgil L. Goedken, and William S. Rees, Jr.

NEW ZINC-*bis*(DIALKYLAMIDES) POTENTIALLY USABLE AS
SITE-SELECTIVE DOPANTS FOR p-TYPE ZnSe 219
    William S. Rees, Jr. and Oliver Just

GROWTH OF InAs AND $(InAs)_1(GaAs)_1$ SUPERLATTICE QUANTUM
WELL STRUCTURES ON GaAs BY ATOMIC LAYER EPITAXY
USING TRIMETHYLINDIUM-DIMETHYLETHYLAMINE ADDUCT 225
    Nobuyuki Ohtsuka and Osamu Ueda

NEW ORGANOMETALLIC SELENIUM REAGENTS FOR LOW
TEMPERATURE OMCVD OF ZnSe 231
    M. Danek, J-S. Huh, K.F. Jensen, D.C. Gordon, and W.P. Kosar

THE SURFACE CHEMISTRY OF CdTe MOCVD 239
    Wen-Shryang Liu and Gregory B. Raupp

NANOSECOND AND FEMTOSECOND UV LASER ABLATION
OF CdTe (100): TIME-OF-FLIGHT AND ANGULAR
DISTRIBUTIONS 245
    P.D. Brewer, M. Späth, and M. Stuke

PICOSECOND OPTICAL SECOND HARMONIC STUDIES OF
ADSORBATE REDUCTION KINETICS ON CADMIUM SULFIDE 251
    T.W. Scott, J. Martorell, and Y.J. Chang

SELECTIVE DEPOSITION OF ZnS THIN FILMS BY
KrF EXCIMER LASER ON PATTERNED Zn SEEDS 257
    K. Yamane and M. Murahara

*Invited Paper

VAPOR DEPOSITED Ag_2S FILMS AS HIGH RESOLUTION
PHOTOREGISTERING MATERIALS IN THE INFRARED REGION 265
   J. Eneva, S. Kitova, A. Panov, and H. Haefke

PART III: METALLIZATION

*PROBING SELECTIVE DEPOSITION OF ALUMINUM USING
IN SITU INFRARED SPECTROSCOPY 273
   Wayne L. Gladfelter, Michael G. Simmonds, Larry A. Zazzera,
   and John F. Evans

INFLUENCE OF FLOW DYNAMICS ON THE MORPHOLOGY OF
CVD ALUMINUM THIN FILMS 283
   Donna M. Speckman, Denise L. Leung, and Jerry P. Wendt

FIRST PRINCIPLES STUDY OF ALUMINUM DEPOSITION ON
HYDROGEN-TERMINATED Si(100) SURFACE 289
   Carlos Sosa

PROPERTIES OF A NEW LIQUID ORGANO GOLD COMPOUND
FOR MOCVD 293
   Hiroto Uchida, Norimichi Saitou, Masamitu Satou, Masayuki Tebakari,
   and Katsumi Ogi

DEPOSITION OF TUNGSTEN FILMS BY PULSED EXCIMER
LASER ABLATION TECHNIQUE 299
   A.M. Dhote and S.B. Ogale

PART IV: DIELECTRICS AND TRANSITIONAL LAYERS

*GASDYNAMICS AND CHEMISTRY IN THE PULSED LASER
DEPOSITION OF OXIDE DIELECTRIC THIN FILMS 305
   John W. Hastie, David W. Bonnell, Albert J. Paul,
   and Peter K. Schenck

ADSORPTION AND REACTION OF $TiCl_4$ ON W(100) 317
   Wei Chen and Jeffrey T. Roberts

ECR PLASMA ENHANCED MOCVD OF TITANIUM NITRIDE 323
   M.E. Gross, A. Weber, R. Nikulski, C.-P. Klages, R.M. Charatan,
   W.L. Brown, E. Dons, and D.J. Eaglesham

INVESTIGATIONS OF TiN AND Ti FILM DEPOSITION BY
PLASMA ACTIVATED CVD USING CYCLOPENTADIENYL
CYCLOHEPTATRIENYL TITANIUM, A LOW OXIDATION
STATE PRECURSOR 329
   Robert M. Charatan, Mihal E. Gross, and David J. Eaglesham

VAPORIZATION CHEMISTRY OF ORGANOALUMINUM PRECURSORS
TO AlN 335
   Paul D. Crocco, John B. Hudson, Yi Wang, and Leonard V. Interrante

CHARACTERIZATION OF SILICON-NITRIDE FILM GROWTH BY
REMOTE PLASMA-ENHANCED CHEMICAL-VAPOR DEPOSITION
(RPECVD) 341
   Zhong Lu, Yi Ma, Scott Habermehl, and Gerry Lucovsky

SPECTROSCOPIC INVESTIGATIONS OF LASER ABLATED
GERMANIUM OXIDE 347
   Paul J. Wolf, Brian M. Patterson, and Sarath Witanachchi

*Invited Paper

INDIUM TIN OXIDE FILMS FORMATION BY LASER ABLATION 353
    J.P. Zheng and H.S. Kwok

EXCIMER LASER INTERACTIONS WITH PTFE RELEVANT TO
THIN FILM GROWTH 359
    J.T. Dickinson, M.G. Norton, J.-J. Shin, W. Jiang, and
    S.C. Langford

PHOTOCHEMICAL MODIFICATION OF FLUOROCARBON RESIN
SURFACE TO ADHERE WITH EPOXY RESIN 365
    M. Okoshi, T. Miyokawa, H. Kashiura, and M. Murahara

THE EFFECTS OF DEPOSITION PARAMETERS ON THE PROPERTIES
OF SiO_2 FILMS DEPOSITED BY MICROWAVE ECR PLASMAS 373
    T.T. Chau, P.M. Lam, and K.C. Kao

FLOW TUBE KINETICS OF GAS-PHASE CVD REACTIONS 379
    Bruce H. Weiller

HIGH RATE DEPOSITION OF SiO_2 BY THE REMOTE PECVD
TECHNIQUE 385
    Arichika Ishida, Masato Hiramatsu, and Yoshito Kawakyu

KrF EXCIMER LASER INDUCED PHOTOCHEMICAL MODIFICATION
OF POLYPHENYLENESULFIDE SURFACE INTO HYDROPHILIC
PROPERTY 391
    H. Kashiura, M. Okoshi, and M. Murahara

PART V: ETCHING

*NEW DRY-ETCH CHEMISTRIES FOR III-V SEMICONDUCTORS 399
    S.J. Pearton, U.K. Chakrabarti, F. Ren, C.R. Abernathy,
    A. Katz, W.S. Hobson, and C. Constantine

GaAs ETCHING BY Cl_2 AND HCl: Ga- vs. As-LIMITED
ETCHING 413
    Chaochin Su, Zi-Guo Dai, Hui-Qi Hou, Ming Xi, Matthew F. Vernon,
    and Brian E. Bent

SCANNING TUNNELING MICROSCOPY OBSERVATION OF THE
REACTION OF $AlCl_3$ ON Si(111)-7×7 SURFACE 419
    Katsuhiro Uesugi, Takaharu Takiguchi, Michiyoshi Izawa,
    Masamichi Yoshimura, and Takafumi Yao

*CHEMICAL TOPOGRAPHY OF Si ETCHING IN A Cl_2 PLASMA,
STUDIED BY X-RAY PHOTOELECTRON SPECTROSCOPY AND
LASER-INDUCED THERMAL DESORPTION 425
    V.M. Donnelly, K.V. Guinn, C.C. Cheng, and I.P. Herman

KINETICS OF REACTIVE ION ETCHING OF POLYMERS IN AN
OXYGEN PLASMA: THE IMPORTANCE OF DIRECT REACTIVE
ION ETCHING 433
    Sandra W. Graham and Christoph Steinbrüchel

HIGH SELECTIVE ETCHING OF SiO_2/Si BY ArF
EXCIMER LASER 439
    K. Kitamura and M. Murahara

*Invited Paper

PLASMA ETCHING AND SURFACE ANALYSIS OF a-SiC:H FILMS
DEPOSITED BY LOW TEMPERATURE PLASMA ENHANCED
CHEMICAL VAPOR DEPOSITION 445
   J.H. Thomas III, M.J. Loboda, and J.A. Seifferly

CONFINEMENT AND LOW-ENERGY EXTRACTION OF
PHOTO-FRAGMENT IONS USING RF ION TRAPPING 451
   Seiji Yamamoto, Kozo Mochiji, Isao Ochiai, and Naohiko Mikami

SURFACE PERTURBATION BY ENERGETIC PARTICLE BEAMS 457
   Che-Chen Chang, Jung-Yen Yang, and Jaw-Chang Shieh

CLEANING DURING INITIAL STAGES OF EPITAXIAL GROWTH
IN AN ULTRA-HIGH VACUUM RAPID THERMAL CHEMICAL
VAPOR DEPOSITION REACTOR 463
   Mahesh K. Sanganeria, Katherine E. Violette, Mehmet C. Öztürk,
   Gari Harris, C. Archie Lee, and Dennis M. Maher

PART VI: SPECIAL TOPICS: DEPOSITION ON PATTERNED SUBSTRATES, REACTOR DESIGN, HETEROEPITAXY, SELECTIVE EPITAXY

*THE IMPACT OF GAS PHASE AND SURFACE CHEMICAL
REACTIONS ON STEP COVERAGE IN LPCVD 471
   Gregory B. Raupp and Timothy S. Cale

THE INFLUENCE OF TEMPERATURE GRADIENTS ON PARTIAL
PRESSURES IN A CVD REACTOR 483
   T.G.M. Oosterlaken, G.J. Leusink, G.C.A.M. Janssen,
   S. Radelaar, K.J. Kuijlaars, C.R. Kleijn, and H.E.A. Van Den Akker

SURFACE REACTION INTERMEDIATES IN Ge CHEMICAL VAPOR
DEPOSITION ON SILICON 489
   C. Michael Greenlief and Lori A. Keeling

GROWTH AND CHARACTERIZATION OF Si-GaP AND Si-GaP-Si
HETEROSTRUCTURES 495
   N. Dietz, S. Habermehl, J.T. Kelliher, G. Lucovsky, and K.J. Bachmann

THE ROLE OF BARIUM IN THE HETEROEPITAXIAL GROWTH OF
INSULATOR AND SEMICONDUCTORS ON SILICON 501
   T.K. Chu, F. Santiago, M. Stumborg, and C.A. Huber

TOPOGRAPHICAL EFFECTS REGARDING TRENCH STRUCTURES
COVERED WITH RTCVD SiGe THIN FILMS 507
   G. Ritter, J. Schlote, and B. Tillack

NEW DAMAGE-LESS PATTERNING METHOD OF A GaAs OXIDE
MASK AND ITS APPLICATION TO SELECTIVE GROWTH BY MOMBE 513
   Seikoh Yoshida and Masahiro Sasaki

SILICON NUCLEATION ON SILICON DIOXIDE AND SELECTIVE
EPITAXY IN AN ULTRA-HIGH VACUUM RAPID THERMAL
CHEMICAL VAPOR DEPOSITION REACTOR USING DISILANE
IN HYDROGEN 519
   Katherine E. Violette, Mahesh K. Sanganeria, Mehmet C. Öztürk,
   Gari Harris, and Dennis M. Maher

*Invited Paper

PERFECT SELECTIVE Si EPITAXIAL GROWTH REALIZED BY
SYNCHROTRON RADIATION IRRADIATION DURING DISILANE
MOLECULAR BEAM EPITAXY 525
 Yuichi Utsumi, Housei Akazawa, and Masao Nagase

GATE QUALITY OXIDES PREPARED BY RAPID THERMAL
CHEMICAL VAPOR DEPOSITION 531
 R.T. Kuehn, X. Xu, D.J. Holcombe, V. Misra, J.J. Wortman,
 J.R. Hauser, Q.-F. Wang, and D.M. Maher

GROWTH KINETICS AND CORRESPONDING LUMINESCENCE
CHARACTERISTICS OF AMORPHOUS C:H DEPOSITED
FROM CH_4 BY DC SADDLE-FIELD PLASMA-ASSISTED CVD 537
 Roman V. Kruzelecky, Chun Wang, and Stefan Zukotynski

MOMBE GROWTH OF YBCO SUPERCONDUCTING FILMS WITH
IN-SITU MONITORING OF RHEED OSCILLATIONS 543
 K. Endo, F. Hosseini Teherani, S. Yoshida, K. Kajimura,
 and Y. Moriyasu

AUTHOR INDEX 549

SUBJECT INDEX 553

Preface

This volume contains the Proceedings of the symposium on "Gas-Phase and Surface Chemistry in Electronic Materials Processing," held during the 1993 Fall Meeting of the Materials Research Society in Boston, Massachusetts. The invited and contributed papers included in this volume address the current state of the art in vapor-phase thin film processing for microelectronics with emphasis on chemistry and its effects on film quality. As is clearly manifested by the variety of topics and areas of specialization of the authors, the symposium provided a forum for effective interdisciplinary communication among chemists, physicists, materials scientists and engineers. This exchange of ideas is essential for rapid advancement of the many critical technologies that depend on thin films processing.

The papers are organized in six sections. Part I contains papers on silicon and carbon systems. In addition to papers discussing silicon and silicon oxide deposition it includes several papers on diamond film growth. The chemistry underlying the growth and doping of III-V and II-VI compound semiconductors is covered in Part II together with several studies discussing new precursors. Part III contains papers on metallization. The deposition of dielectric and transitional layers is the topic of Part IV. Part V contains papers on etching of thin films. Finally, Part VI covers special topics, such as deposition on patterned substrates, reactor design, heteroepitaxy and selective epitaxy.

The symposium consisted of four days of invited and contributed oral presentations and included two evening poster sessions. All the papers appearing in this volume have undergone peer review, and the authors had the opportunity to revise their manuscripts based on the reviewers' comments. We believe that this volume will be useful to scientists and engineers currently working in, or interested in entering, the challenging and technologically important field of electronic materials processing.

T.J. Mountziaris
G.R. Paz-Pujalt
F.T.J. Smith
P.R. Westmoreland

January 1994

Acknowledgments

We would like to thank the invited speakers, M.E. Coltrin, J.R. Creighton, P.D. Dapkus, V.M. Donnelly, L.H. Dubois, S.M. Gates, W. Gladfelter, J.W. Hastie, J.M. Jasinski, K.F. Jensen, D.W. Kisker, T.F. Kuech, U. Memmert, S.J. Pearton, G.B. Raupp and W. Richter, who provided a framework for organizing the various sessions, and all the participants, many of whom served as reviewers. We would also like to thank the staff of the Materials Research Society for their superb organization of the symposium and assistance in the preparation of this volume, Mr. Costas Theodoropoulos of SUNY at Buffalo for his assistance in the editorial process, Ms. Darlene Innes of SUNY at Buffalo for her assistance in the preparation of this volume, and Mrs. Anne Yeager of Eastman Kodak Co. for secretarial support.

Finally, we would like to thank the Chemistry Division of the Office of Naval Research and the Materials Science Division of the Army Research Office for their generous financial support, which made possible the symposium on which this volume is based.

MATERIALS RESEARCH SOCIETY SYMPOSIUM PROCEEDINGS

Volume 297— Amorphous Silicon Technology—1993, E.A. Schiff, M.J. Thompson, P.G. LeComber, A. Madan, K. Tanaka, 1993, ISBN: 1-55899-193-X
Volume 298— Silicon-Based Optoelectronic Materials, R.T. Collins, M.A. Tischler, G. Abstreiter, M.L. Thewalt, 1993, ISBN: 1-55899-194-8
Volume 299— Infrared Detectors—Materials, Processing, and Devices, A. Appelbaum, L.R. Dawson, 1993, ISBN: 1-55899-195-6
Volume 300— III-V Electronic and Photonic Device Fabrication and Performance, K.S. Jones, S.J. Pearton, H. Kanber, 1993, ISBN: 1-55899-196-4
Volume 301— Rare-Earth Doped Semiconductors, G.S. Pomrenke, P.B. Klein, D.W. Langer, 1993, ISBN: 1-55899-197-2
Volume 302— Semiconductors for Room-Temperature Radiation Detector Applications, R.B. James, P. Siffert, T.E. Schlesinger, L. Franks, 1993, ISBN: 1-55899-198-0
Volume 303— Rapid Thermal and Integrated Processing II, J.C. Gelpey, J.K. Elliott, J.J. Wortman, A. Ajmera, 1993, ISBN: 1-55899-199-9
Volume 304— Polymer/Inorganic Interfaces, R.L. Opila, A.W. Czanderna, F.J. Boerio, 1993, ISBN: 1-55899-200-6
Volume 305— High-Performance Polymers and Polymer Matrix Composites, R.K. Eby, R.C. Evers, D. Wilson, M.A. Meador, 1993, ISBN: 1-55899-201-4
Volume 306— Materials Aspects of X-Ray Lithography, G.K. Celler, J.R. Maldonado, 1993, ISBN: 1-55899-202-2
Volume 307— Applications of Synchrotron Radiation Techniques to Materials Science, D.L. Perry, R. Stockbauer, N. Shinn, K. D'Amico, L. Terminello, 1993, ISBN: 1-55899-203-0
Volume 308— Thin Films—Stresses and Mechanical Properties IV, P.H. Townsend, J. Sanchez, C-Y. Li, T.P. Weihs, 1993, ISBN: 1-55899-204-9
Volume 309— Materials Reliability in Microelectronics III, K. Rodbell, B. Filter, P. Ho, H. Frost, 1993, ISBN: 1-55899-205-7
Volume 310— Ferroelectric Thin Films III, E.R. Myers, B.A. Tuttle, S.B. Desu, P.K. Larsen, 1993, ISBN: 1-55899-206-5
Volume 311— Phase Transformations in Thin Films—Thermodynamics and Kinetics, M. Atzmon, J.M.E. Harper, A.L. Greer, M.R. Libera, 1993, ISBN: 1-55899-207-3
Volume 312— Common Themes and Mechanisms of Epitaxial Growth, P. Fuoss, J. Tsao, D.W. Kisker, A. Zangwill, T.F. Kuech, 1993, ISBN: 1-55899-208-1
Volume 313— Magnetic Ultrathin Films, Multilayers and Surfaces/Magnetic Interfaces— Physics and Characterization (2 Volume Set), C. Chappert, R.F.C. Farrow, B.T. Jonker, R. Clarke, P. Grünberg, K.M. Krishnan, S. Tsunashima/ E.E. Marinero, T. Egami, C. Rau, S.A. Chambers, 1993, ISBN: 1-55899-211-1
Volume 314— Joining and Adhesion of Advanced Inorganic Materials, A.H. Carim, D.S. Schwartz, R.S. Silberglitt, R.E. Loehman, 1993, ISBN: 1-55899-212-X
Volume 315— Surface Chemical Cleaning and Passivation for Semiconductor Processing, G.S. Higashi, E.A. Irene, T. Ohmi, 1993, ISBN: 1-55899-213-8

MATERIALS RESEARCH SOCIETY SYMPOSIUM PROCEEDINGS

Volume 316— Materials Synthesis and Processing Using Ion Beams, R.J. Culbertson, K.S. Jones, O.W. Holland, K. Maex, 1994, ISBN: 1-55899-215-4

Volume 317— Mechanisms of Thin Film Evolution, S.M. Yalisove, C.V. Thompson, D.J. Eaglesham, 1994, ISBN: 1-55899-216-2

Volume 318— Interface Control of Electrical, Chemical, and Mechanical Properties, S.P. Murarka, T. Ohmi, K. Rose, T. Seidel, 1994, ISBN: 1-55899-217-0

Volume 319— Defect-Interface Interactions, E.P. Kvam, A.H. King, M.J. Mills, T.D. Sands, V. Vitek, 1994, ISBN: 1-55899-218-9

Volume 320— Silicides, Germanides, and Their Interfaces, R.W. Fathauer, L. Schowalter, S. Mantl, K.N. Tu, 1994, ISBN: 1-55899-219-7

Volume 321— Crystallization and Related Phenomena in Amorphous Materials, M. Libera, T.E. Haynes, P. Cebe, J. Dickinson, 1994, ISBN: 1-55899-220-0

Volume 322— High-Temperature Silicides and Refractory Alloys, B.P. Bewlay, J.J. Petrovic, C.L. Briant, A.K. Vasudevan, H.A. Lipsitt, 1994, ISBN: 1-55899-221-9

Volume 323— Electronic Packaging Materials Science VII, R. Pollak, P. Børgesen, H. Yamada, K.F. Jensen, 1994, ISBN: 1-55899-222-7

Volume 324— Diagnostic Techniques for Semiconductor Materials Processing, O.J. Glembocki, F.H. Pollak, S.W. Pang, G. Larrabee, G.M. Crean, 1994, ISBN: 1-55899-223-5

Volume 325— Physics and Applications of Defects in Advanced Semiconductors, M.O. Manasreh, M. Lannoo, H.J. von Bardeleben, E.L. Hu, G.S. Pomrenke, D.N. Talwar, 1994, ISBN: 1-55899-224-3

Volume 326— Growth, Processing, and Characterization of Semiconductor Heterostructures, G. Gumbs, S. Luryi, B. Weiss, G.W. Wicks, 1994, ISBN: 1-55899-225-1

Volume 327— Covalent Ceramics II: Non-Oxides, A.R. Barron, G.S. Fischman, M.A. Fury, A.F. Hepp, 1994, ISBN: 1-55899-226-X

Volume 328— Electrical, Optical, and Magnetic Properties of Organic Solid State Materials, A.F. Garito, A. K-Y. Jen, C. Y-C. Lee, L.R. Dalton, 1994, ISBN: 1-55899-227-8

Volume 329— New Materials for Advanced Solid State Lasers, B.H.T. Chai, T.Y. Fan, S.A. Payne, A. Cassanho, T.H. Allik, 1994, ISBN: 1-55899-228-6

Volume 330— Biomolecular Materials By Design, H. Bayley, D. Kaplan, M. Navia, 1994, ISBN: 1-55899-229-4

Volume 331— Biomaterials for Drug and Cell Delivery, A.G. Mikos, R. Murphy, H. Bernstein, N.A. Peppas, 1994, ISBN: 1-55899-230-8

Volume 332— Determining Nanoscale Physical Properties of Materials by Microscopy and Spectroscopy, M. Sarikaya, M. Isaacson, H.K. Wickramasighe, 1994, ISBN: 1-55899-231-6

Volume 333— Scientific Basis for Nuclear Waste Management XVII, A. Barkatt, R. Van Konynenburg, 1994, ISBN: 1-55899-232-4

Volume 334— Gas-Phase and Surface Chemistry in Electronic Materials Processing, T.J. Mountziaris, P.R. Westmoreland, F.T.J. Smith, G.R. Paz-Pujalt, 1994, ISBN: 1-55899-233-2

Volume 335— Metal-Organic Chemical Vapor Deposition of Electronic Ceramics, S.B. Desu, D.B. Beach, B.W. Wessels, S. Gokoglu, 1994, ISBN: 1-55899-234-0

Prior Materials Research Society Symposium Proceedings available by contacting Materials Research Society

PART I

Silicon and Carbon Systems

GAS-PHASE SILICON ATOM DENSITIES IN THE
CHEMICAL VAPOR DEPOSITION OF SILICON FROM SILANE

MICHAEL E. COLTRIN, WILLIAM G. BREILAND, AND PAULINE HO
Sandia National Laboratories, Albuquerque, NM 87175

ABSTRACT

Silicon atom number density profiles have been measured using laser-induced fluorescence during the chemical vapor deposition of silicon from silane. Measurements were obtained in a rotating-disk reactor as a function of silane partial pressure and the amount of hydrogen added to the carrier gas. Absolute number densities were obtained using an atomic absorption technique. Results were compared with calculated density profiles from a model of the coupled fluid flow, gas-phase and surface chemistry for an infinite-radius rotating disk. An analysis of the reaction mechanism showed that the unimolecular decomposition of SiH_2 is not the dominant source of Si atoms. Profile shapes and positions, and all experimental trends are well matched by the calculations. However, the calculated number density is up to 100 times smaller than measured.

INTRODUCTION

The chemical vapor deposition (CVD) of silicon from silane is the simplest system of use in the microelectronics industry. As such, it is a system often chosen for research into the fundamental mechanisms of CVD. Under the usual conditions for low-pressure CVD (LPCVD), homogeneous decomposition of silane is negligible. However, at higher pressures homogeneous pyrolysis and subsequent gas-phase reactions of intermediate species can occur quite readily.

Studying the fundamental chemistry occurring in a CVD system using a commercial reactor is very difficult. The chemical kinetics is strongly coupled to the gas-phase temperature and velocity fields, which can be quite complicated in commonly used commercial CVD reactors. In addition, modeling detailed chemical kinetics coupled to complex fluid flow is still computationally prohibitive. Several years ago, we presented comparisons between a 2-dimensional boundary layer flow model of the silane CVD system [1] and in situ measurements of gas-phase temperatures [2], silane density profiles [2], silicon dimer density profiles [2], and silicon atom density profiles [3] under a wide range of conditions. Generally, good agreement was obtained between model and experiment. However, the idealized 2-D flow in the model was a crude approximation to the actual experimental reactor in which the measurements were performed (which contained various viewing ports). Therefore, when discrepancies occurred between model and experiment it was difficult to be sure that the errors were due to errors in the chemical kinetics portion of the model, or to inaccurate modeling of the fluid flow.

Very simple fluid flow for CVD is found in the rotating disk reactor (RDR). Von Karman showed that the flow above an infinite-radius rotating disk can be reduced to just one mathematical dimension, the distance above the disk. Breiland and Evans [4] presented an analysis and experimental design of a finite-radius RDR which mimics the ideal, 1-D behavior from the von Karman (similarity) solution. Coltrin, Kee, and Evans [5,6] developed a 1-D computer model (SPIN) of the coupled fluid flow, gas-phase and surface chemistry using the rotating-disk similarity solution.

This paper discusses measurement of Si-atom density profiles in a rotating disk reactor and comparisons with calculations using the 1-D model. As has been shown before [4], the fluid flow in this reactor is well-understood and easily modeled; gas-phase temperature profiles match the 1-D solution essentially exactly. The present comparisons of Si-atom profiles provide a well-controlled test of the chemical kinetics portion of the model, with the fluid flow taken as a "given."

EXPERIMENT

The experimental RDR has been described elsewhere [4]. The cylindrical reactor is built with double quartz walls (~ 10 cm i.d.) through which cooling water flows. The rotating disk consists of a 75-mm-diameter stainless steel, resistively heated canister, on which a 51-mm-diameter silicon wafer is placed. The gases (SiH_4, He, or H_2) enter the reactor through a diffuser screen and exit the cell at the bottom. The flow rate is set to match the amount required by the ideal 1-D flow [4] for a desired operating pressure, rotation rate, and disk temperature. Si atoms are detected using laser-induced fluorescence (LIF), essentially as described elsewhere [3]. The excitation source was the frequency doubled and mixed output near 250 nm from a pulsed YAG-pumped dye laser. The fluorescence was collected at 90°, spectrally selected with a 0.5 meter monochromator, and detected with a photomultiplier tube. The Si atom density measurements were done by tuning the laser to the $J=2 \rightarrow 1$ line of the $4s^3 P^0 \rightarrow 3p^2\ ^3P$ transition at 250.7 nm while monitoring the signal at the $J=2 \rightarrow 2$ transition at 251.6 nm. Spatial profiles of Si atom densities were obtained by moving the RDR vertically relative to the laser beam and collection optics. A UV absorption measurement was performed to calibrate the LIF signal, and thus obtain absolute number densities. More detail about the experiment will be given elsewhere [7].

MODEL

The SPIN computer code [6] solves the 1-D equations for the three velocity components, the gas temperature profile, gas-phase mole fractions, surface species site fractions, and deposition rates. The model accounts for homogeneous decomposition of the silane reactant and subsequent reactions of the gas-phase intermediates that are produced. The complete gas-phase silane pyrolysis reaction mechanism used in this study is given in Table 1. The complex temperature and pressure dependence of many of these reactions was parameterized in a Troe form [8,9]. The low-pressure and high-pressure limits of the rate constants are given by

$$k_0 = A_0 T^{\beta_0} \exp(-E_0 / R_c T),$$

$$k_\infty = A_\infty T^{\beta_\infty} \exp(-E_\infty / R_c T).$$

The rate constant as a function of pressure is then

$$k = k_\infty \left(\frac{P_r}{1+P_r}\right) F,$$

$$P_r = \frac{k_0 [M]}{k_\infty}$$

and $[M]$ is the total concentration of species in the mixture. In the Troe form, F is given by

$$\log F = \left[1 + \left[\frac{\log P_r + c}{n - d(\log P_r + c)}\right]^2\right]^{-1} \log F_{cent}.$$

The constants are

$$c = -0.4 - 0.67 \log F_{cent}$$
$$n = 0.75 - 1.27 \log F_{cent}$$
$$d = 0.14$$

and $F_{cent} = (1-a)\exp(-T/T^{***}) + a\exp(-T/T^*) + \exp(-T^{**}/T)$.

The four parameters a, T^{***}, T^*, and T^{**} are specified as the third line of coefficients for reactions G1,2,3,5,6, and 8.

The rate expressions for reactions G1 through G7 represent the best summary of silane, disilane, and trisilane kinetics available from the literature [10-12]. Moffat, Jensen, and Carr have performed a non-linear regression to summarize the available kinetic data available for these systems. The rate constants differ slightly from those reported in [10-12] because they were refit using a heat of formation for SiH_2 of 64.8. The rate constants for these reactions were not varied in any of the calculations reported here. In actuality there are uncertainties of factors of 3 or more in each of these rate constants, which lead to potential errors in our calculated Si atom densities. The isomerization reaction G8 connects the two forms of Si_2H_4. There is no reported rate constant for this reaction. Reaction G9 is the main production / destruction route for Si atoms in the mechanism. Reaction G10 is the main production / destruction route for Si when disilane is used as a starting gas [7], but plays little role in the calculations reported here. If not given explicitly in the table, the reverse rate constants were calculated from the reaction thermochemistry. Thermochemical data from references [5] and [13] were used.

TABLE 1

Reaction		Rate coefficients (see notes)				Notes
G1	$SiH_4(+M) \leftrightarrow SiH_2+H_2(+M)$	3.12E09	1.7	54710		a, b
		5.214E29	-3.545	57550		c
		-0.4984	888.3	209.4	2760	d
G2	$Si_2H_6(+M) \leftrightarrow H_2+H_3SiSiH(+M)$	9.09E9	1.8	54197		a,e
		1.945E44	-7.772	59023		c
		-0.1224	793.3	2400	11.39	d
G3	$Si_2H_6(+M) \leftrightarrow SiH_4+SiH_2(+M)$	1.81E10	1.7	50203		a,e
		5.09E53	-10.37	56034		c
		4.375E-5	438.5	2726	438.2	d
G4	$H_2+H_3SiSiH \leftrightarrow SiH_4+SiH_2$	9.41E13	0	4092.3		a,e
		9.43E10	1.1	5790.3		f
G5	$Si_3H_8(+M) \leftrightarrow SiH_4+H_3SiSiH(+M)$	3.73E12	1	50850		a,g
		4.36E76	-17.26	59303		c
		0.4157	365.3	3102	9.724	d
G6	$Si_3H_8(+M) \leftrightarrow SiH_2+Si_2H_6(+M)$	6.97E12	1	52677		a,g
		1.73E69	-15.07	60491		c
		-3.47E-5	442	2412	128.3	d
G7	$SiH_4+H_3SiSiH \leftrightarrow SiH_2+Si_2H_6$	1.73E14	0.4	8898.7		a,g
		2.65E15	0.1	8473.4		f
G8	$H_3SiSiH(+M) \leftrightarrow H_2SiSiH_2(+M)$	2.54E13	-0.2	5381		a,h
		1.099E33	-5.765	9152		c
		-0.4202	214.5	103	136.3	d
G9	$SiH_4+Si \leftrightarrow H_3SiSiH$	1.0E12	0	5000		a,h
G10	$Si_2H_6+Si \leftrightarrow H_3SiSiH+SiH_2$	1.3E15	0	12600		a,h

Notes:
a Arrhenius coefficients for the high-pressure limit: A_∞ (units depend of reaction order, but are given in terms of moles, sec, and cm^3), β_∞ (unitless), E_∞ (cal/mole)
b Fits to data in Ref. 10
c Arrhenius coefficients for the low-pressure limit: A_0 (units depend of reaction order, but are given in terms of moles, sec, and cm^3), β_0 (unitless), E_0 (cal/mole)
d Troe parameters (see text): a, T^{***}, T^*, and T^{**}
e Fits to data in Ref. 12
f Arrhenius coefficients for the reverse reaction (units as in note a, above)
g Fits to data in Ref. 13
h Estimated, this work

Our earlier modeling studies of silane CVD [1,5] included many more reactions and chemical species. In particular the reaction

$$SiH_2 \leftrightarrow Si + H_2$$

was included, and it was the major production route for Si atoms in the gas phase. A primary result of this paper is that the reaction of SiH_2 cannot be an important route for Si production. It has also been found that species and reactions containing an odd number of hydrogen atoms play little role in the thermal decomposition, and they were thus eliminated from the mechanism. (This, however, would definitely not be the case for plasma-enhanced CVD.) The reaction mechanism has been simplified as much as possible to identify the main pathways for Si atom creation and destruction.

The DAH (dissociative-adsorption/hydrogen desorption) mechanism [14] for heterogeneous reaction of silane at the deposition surface is used in the model. It consists of a two-site dissociative adsorption of silane, followed by first-order elimination of H_2 to the gas [15]. Disilane and trisilane were assumed to react 10 times faster than the silane rate in [14]; SiH_2, Si, H_2SiSiH_2, and H_3SiSiH were assumed to react with the surface with unit probability.

Figure 1 shows the rate of production (and destruction) of Si due to the reverse of reaction G9 as a function of height above the disk. The peak production rate occurs at roughly 1 mm above the disk. The production drops as the reactive precursor species H_3SiSiH is destroyed at the surface. The production rate drops off with increasing distance from the surface as the gas temperature drops. The Si atoms that are produced diffuse both toward and away from the disk. Further from the disk, Si encounters higher concentrations of unreacted silane, and Si is destroyed via reaction G9 (in the forward direction). Thus the same gas-phase reaction is both the dominant production and destruction route for Si in this model.

COMPARISONS BETWEEN MODEL AND EXPERIMENT

Figure 2 compares measured and calculated Si atom density profiles for a baseline set of conditions (200 Torr total pressure, 0.33 Torr silane in He carrier, 450 rpm, 650° C disk temperature). It is significant that the measured Si density profiles have a maximum in the gas phase. This clearly shows that they must be produced via a gas-phase reaction; if they were somehow formed on the surface and then diffused into the gas, the density profile would have its maximum at zero height.

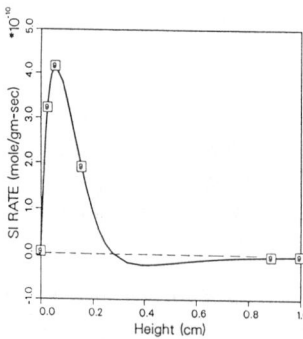

Fig. 1 Production / destruction rate of Si atoms due to reaction G9 as a function of height above the disk.

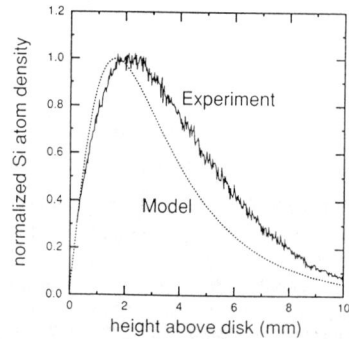

Fig. 2 Measured and calculated Si density profiles.

The shape of the Si density profile calculated by the model agrees well with experiment. The model matches the position of the maximum in the profile, very near the disk (approximately 2 mm). The widths of the profiles also agree well. In the model, the Si density drops at the surface because its production rate goes down (Fig. 1), but also because Si is destroyed by facile reaction with the silicon surface (unit probability). The drop in Si density further from the surface is also explained by the decrease in production rates in Fig. 1.

The absolute number density measured at the peak of the profile is 2×10^8 /cc for the baseline conditions. The Si density calculated by the model is 2.1×10^6 /cc, almost exactly 100 times smaller. A number of factors, discussed later, are sources of error in both the experimental and calculated number densities. However, it appears that the calculated number density cannot be raised much by simply increasing the rate constant for G9 (see below).

Addition of H_2 to the He carrier decreases the measured LIF signal (and thus the density). Figure 3 shows that the maximum Si density measured decreases by about 85% upon addition of 10 Torr hydrogen. The model reproduces the suppression of Si by hydrogen. The experimental and calculated results in Fig. 3 have been normalized at the baseline condition (no H_2) for ease of comparing the hydrogen effects. The model indicates that the addition of H_2 slows the initial silane decomposition by increasing the reverse of G1.

Figure 4 shows that addition of H_2 has virtually no effect on the measured shapes of the Si density profiles. The profiles in Figure 4 have been scaled to the same height, in order to compare the shapes more easily (the scaling of the curves is obtainable from Fig. 3). The Si atom density profiles calculated by the model are shown in Fig. 5. The curves in Fig. 5 have also been scaled to highlight the shapes more clearly. The calculated density profiles using the mechanism presented in Table 1 are also very insensitive to the addition of H_2 to the carrier.

The insensitivity of the Si density profiles with respect to added hydrogen was a key experimental finding. It puts very severe constraints on the possible mechanism for Si production. In fact it is the lack of a "hydrogen effect" that virtually rules out the unimolecular SiH_2 reaction as the primary production route, as discussed next.

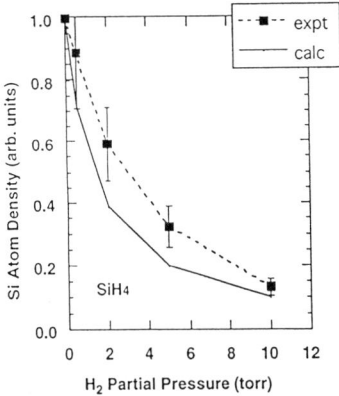

Fig. 3 The effect on the peak Si density of adding H_2 to the He carrier gas.

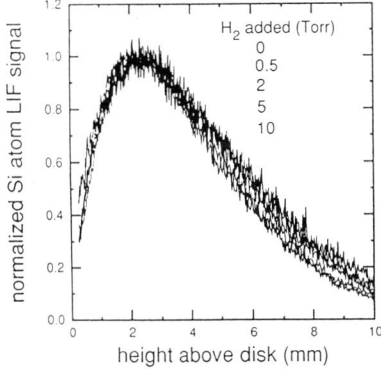

Fig. 4 Shapes of the measured Si density profiles upon addition of H_2. Data has been normalized to the same peak height.

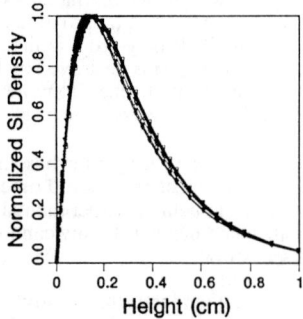
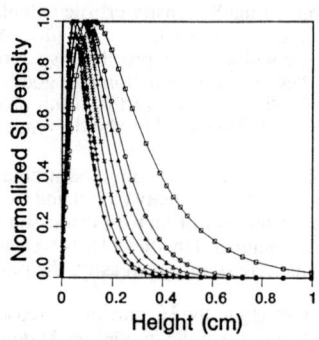

Fig. 5 Shapes of the calculated Si density profiles upon addition of H$_2$. (Curves are normalized. Amounts of added H$_2$ same as in Fig. 4)

Fig. 6 Shapes of the calculated Si density profiles upon addition of H$_2$ when SiH$_2$ reaction G11 produces gas-phase Si.

The effect on the calculated H$_2$-dependence of adding the reaction (which we will call G11)

$$SiH_2 \leftrightarrow Si + H_2$$

to the gas-phase mechanism is shown in Figure 6. A pre-exponential constant of 2×10^{11} and an activation energy of 42600 cal/mol was used for this calculation. (We define this set of reactions and rate constants as Case 2, and the nominal mechanism in Table 1 as Case 1). The peak Si density for the baseline conditions when G11 is included is 1.2×10^8, much closer to the experimental value. However, the calculated Si density profiles narrow significantly as H$_2$ is added to the He carrier. (The broadest profile in Fig. 6 is for the baseline case; the narrowest is for 10 Torr H$_2$ added, with the intermediate curves corresponding to the amounts listed in Fig. 4). The profiles become narrower because, as the amount of hydrogen in the gas increases, the reverse of G11 becomes more important in destroying Si atoms. That is, the Si atoms are produced close to the surface and diffuse away. The higher the hydrogen concentration, the shorter the distance before a Si atom collides with an H$_2$ and is destroyed, thus the narrower profile.

Generally speaking, the rate of the forward direction of G11 is independent of added H$_2$, but as [H$_2$] gets larger, the more quickly the Si atoms that are produced will collide with hydrogen and be destroyed by the back reaction. This will be a fundamental behavior of reaction G11; the behavior depends on the equilibrium constant for G11 (ratio of forward to reverse rate constants) no matter what the value of the forward rate constant k_f itself. The equilibrium constant, in turn, depends on the reaction thermochemistry, which is well established for these species. For a given k_f, the reverse rate constant would have to be decreased by roughly a factor of 100 for the narrowing in Si density profile with additional hydrogen to go away. This factor of 100 would require that ΔG for the reaction be in error by roughly 8500 cal/mol (at 650° C). The estimated error in ΔH is less than 1000 cal/mol. It would require an error of about 9 entropy units for k_r to be in error by a factor of 100, which is also too large.

Due to the marked H$_2$-effect on profile widths, reaction G11 could only be the dominant Si atom production route if there is some other reaction that can compete with the reverse of G11 for Si-atom destruction. Because the H$_2$ concentration is so great at 10 Torr (relative to most other species in the gas), the only plausible collision partner with a concentration large enough to compete with H$_2$ is SiH$_4$, i.e., in reaction G9. To test this possibility, a mechanism including

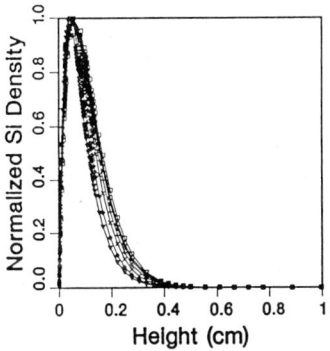

Fig. 7 Shapes of the calculated Si density profiles upon addition of H_2 when SiH_2 reaction G11 produces gas-phase Si and G9 is increased by a factor of 50. (Case 3)

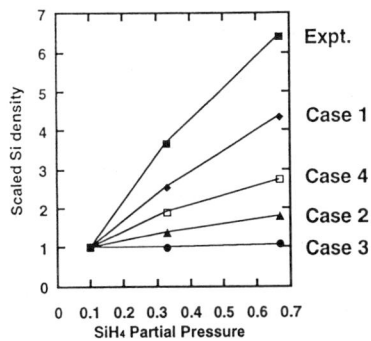

Fig. 8 Dependence of calculated Si density on inlet silane partial pressure. See text for explanation of curves.

Si production by G11 and increasing the rate for G9 by a factor of 50 over that which is reported in Table 1 was used to test the effect of H_2 on the predicted density profile shapes. (We will refer to this combination of reactions and rate constants as Case 3). The results are plotted in Fig. 7.

Figure 7 shows that the calculated Si density profiles shapes become insensitive to added hydrogen when reaction G9 is increased to become the dominant Si-destruction route. However, the profile widths themselves become too narrow, about a factor of three less than experiment. In addition, with these rate constants the calculated Si density (at the profile peaks) becomes virtually independent of inlet SiH_4 partial pressure, as illustrated in Fig. 8.

An almost linear increase in Si-atom density with increasing inlet SiH_4 is seen experimentally, as given in Fig. 8. For the combination of rate constants in Case 3, immediately above, there is essentially no increase in the calculated Si density with increasing silane. For this case, the amount of the precursor SiH_2 increases strongly with silane partial pressure, and so the Si production rate from G11 does, too. However, the rate of destruction of Si also goes up roughly linearly with increasing silane due to reaction G9. The net result is an extremely weak dependence on silane partial pressure. The results calculated for Case 2 (Table 1 mechanism plus reaction G11) also exhibit a very weak dependence of Si density upon inlet silane.

The combination of results just presented essentially rule out reaction G11 as the important producer of Si atoms. Namely, the dramatic narrowing of the Si density profiles with addition of H_2 to the carrier seen in Fig. 6, when experimentally the shapes do not change (Fig. 4). The thermochemistry is known too well to allow the needed decrease of the reverse rate constant for reaction G11 by the necessary factor of 100. When a fast reaction with SiH_4 is allowed to compete with the H_2 for the destruction of Si, the profiles become too narrow (Fig. 7) and the Si density becomes independent of silane partial pressure (Fig. 8).

The dependence on silane partial pressure for the nominal mechanism in Table 1 (labeled Case 1 in Fig. 8) shows a slight sub-linear behavior. Earlier in the paper, it was mentioned that the calculated Si density could not be increased by arbitrarily increasing rate constants in the mechanism. This is illustrated by the curve labeled Case 4 in Fig. 8. This calculation starts with the mechanism just as given in Table 1, but increases the rate of G9 by a factor of 10. With this change, the calculated Si density increases by a factor of 3.5, but at the expense of further deviation from the nearly linear dependence on silane partial pressure seen experimentally.

The absolute number density calculated with the nominal mechanism is roughly 100 times lower than measured experimentally. There are a number of contributing sources of error, the largest of which will be summarized here. A very high activation energy for Si production was observed, about 135 kcal/mol [7]. This high activation energy coupled with an uncertainty in our measured surface temperature of about $15°$ leads to a factor of 3 uncertainty in the Si density. The estimated uncertainty in the UV absorption calibration of the absolute number density is a factor of 2. In the model, we know that the rate constant for the initial silane decomposition is uncertain to within about a factor of 3. In all of the calculations presented here, the rate constants for reactions G1-8 were held fixed. But, for example, if the rate constant for G1 is increased by a factor of 3, the calculated Si density increases by almost exactly the same factor of 3. Each of the other rate constants in the mechanism are also subject to similar uncertainties in their values. With the combination of uncertainties from these sources, the calculated and experimental Si densities are nearly within their joint error ranges. Still, the large discrepancy is not satisfying.

CONCLUSION

The controlled environment of the rotating-disk reactor allowed careful tests of the chemical mechanism for creation of gas-phase Si atoms during CVD from silane. A 10-step reaction mechanism was presented which does a good job in matching experimental density profile shapes, the suppression of Si by hydrogen with the profile shapes unchanged, and the dependence upon silane partial pressure. An important conclusion of this work is that the unimolecular decomposition of SiH_2 is not the primary route to Si-atom formation.

ACKNOWLEDGMENTS

We thank Harry Moffat for providing Chemkin fits to the Troe parameterizations, Robert Kee for collaboration in development of the software used here, Greg Evans for consultations on the rotating disk design, Michael Youngman for technical assistance in the design and construction of the rotating disk, and Fran Rupley for programming assistance. This work was performed at Sandia National Laboratories and supported by the U. S. Department of Energy under contract number DE-AC04-94AL85000.

REFERENCES

1. M. E. Coltrin, R. J. Kee, and J. A. Miller, J. Electrochem. Soc., **131**, 425 (1984).
2. W. G. Breiland, M. E. Coltrin, and P. Ho, J. Appl. Phys. **59**, 3267 (1986).
3. W. G. Breiland, P. Ho, and M. E. Coltrin, J. Appl. Phys. **60**, 1505 (1986).
4. W. G. Breiland and G. Evans, J. Electrochem. Soc. **138**, 1806 (1991).
5. M. E. Coltrin, R. J. Kee, and G. Evans, J. Electrochem. Soc., **136**, 819 (1989).
6. M. E. Coltrin, R. J. Kee, G. H. Evans, E. Meeks, F. M. Rupley, and J. F. Grcar, Sandia National Laboratories Report, SAND91-8003, 1991.
7. P. Ho, M. E. Coltrin, and W. G. Breiland (in preparation).
8. R. G. Gilbert, K. Luther, and J. Troe, Ber. Bunsenges. Phys. Chem., **87**, 161 (1983).
9. W. C. Gardiner and J. Troe, in <u>Combustion Chemistry</u>, edited by W. C. Gardiner (Springer, Berlin, 1984).
10. H. K. Moffat, K. F. Jensen, and R. W. Carr, J. Phys. Chem., **94**, 145 (1991).
11. H. K. Moffat, K. F. Jensen, and R. W. Carr, J. Phys. Chem., **96**, 7695 (1992).
12. H. K. Moffat, K. F. Jensen, and R. W. Carr, J. Phys. Chem., **96**, 7683 (1992).
13. M. E. Coltrin, R. J. Kee, and J. A. Miller, J. Electrochem. Soc., **133**, 1206 (1986).
14. W. G. Houf, J. F. Grcar, and W. G. Breiland, Mat. Sci. and Eng., **B17**, 163 (1993).
15. K. Sinniah, M. G. Sherman, L. B. Lewis, W. H. Weinberg, J. T. Yates, Jr., and K. C. Janda, J. Chem. Phys., **92**, 5700 (1990).

GAS PHASE AND GAS SURFACE KINETICS OF TRANSIENT SILICON HYDRIDE SPECIES

JOSEPH M. JASINSKI
IBM Research Division, Thomas J. Watson Research Center, Yorktown Heights NY 10598

ABSTRACT

This paper presents a summary of the current state of our understanding of the absolute reactivity of transient silicon hydride species, such as SiH, SiH_2 and SiH_3 in the gas phase and at the surface of thin films.

INTRODUCTION

Over the past decade, substantial progress has been made in understanding the absolute reactivity of transient silicon hydride species. The goal of such activities has been to provide a general, transferable set of rate data for use in modelling and understanding various types of silicon CVD processes. The philosophy behind this effort is summarized in a recent review article [1].

Thus far, time resolved gas phase kinetic techniques such as flash photolysis with laser absorption, laser induced fluorescence (LIF), laser magnetic resonance (LMR), and photoionization mass spectrometry have been brought to bear on understanding the absolute reactivity of SiH, SiH_2 and SiH_3 in the gas phase. More recently, LIF, resonance enhanced multiphoton ionization (REMPI) and low energy electron impact ionization mass spectrometry have been used to detect these species as they are consumed at surfaces, thereby allowing the direct determination of β, the overall surface loss coefficient. Individual measurements with appropriate references are presented below.

ABSOLUTE GAS PHASE RATE DATA FOR MONOSILICON HYDRIDE RADICALS

Available absolute rate constants for SiH, SiH_2 and SiH_3 reacting in the gas phase with a range of reaction partners are presented in Table I, Table II and Table III. For comparison the rate constants are given for a specific pressure in a specific buffer gas. All measurements are reported at room temperature unless otherwise noted. Individual references should be consulted for rate constants over broader temperature and pressure ranges. The data are comprehensive through July 1993, and summarize the available kinetic data base for the gas phase reactions of these radicals. It should be noted that in many cases, particularly for reactions of SiH and SiH_2, the rate constants are pressure dependent because the reactions are three body association reactions in the fall off or low pressure regimes. Nothing is known about the temperature dependence of the rate constants for SiH. There is some temperature dependent data for SiH_2. In particular it should be recognized that the rate constant for insertion of SiH_2 into the Si-H bonds of silane has a *negative* temperature dependence (i.e. the reaction becomes slower as temperature increases). The silyl radical SiH_3 has perhaps been the most thoroughly characterized kinetically, with a number of temperature dependent studies. A noteable exception is the reaction of silyl with silane (SiD_4), which was studied only at room temperature, and was too slow to be observed under those conditions.

Table 1 Absolute Rate Constants for SiH Reactions

Reactant	Rate Constant (cm^3 molecule^{-1} s^{-1})	Conditions	Reference
H_2	$\leq (1.2\pm0.2)\times10^{-14}$	5 Torr He	[2]
D_2	$\leq (1.8\pm0.2)\times10^{-14}$	5 Torr He	[2]
NO	$(2.5\pm0.3)\times10^{-10}$	2 Torr Ar	[3]
O_2	$(1.7\pm0.2)\times10^{-10}$	2 Torr Ar	[3]
$C_6H_5SiH_3$	$\sim3\times10^{-10}$	2 Torr Ar	[3]
SiH_4	$(4.3\pm0.3)\times10^{-10}$	5 Torr He	[2]
	$(2.7\pm0.5)\times10^{-10}$	2 Torr He	[2]
	$(2.8\pm0.6)\times10^{-10}$	2 Torr Ar	[3]
	$(3.3\pm0.5)\times10^{-12}$	4-50 mTorr, 500 K	[4]
H_2	$(1.6\pm0.1)\times10^{-11}$	SiH v = 1, 5 Torr He	[2]

Table 2 Absolute Rate Constants for SiH_2 Reactions

Reactant	Rate Constant (cm³ molecule⁻¹ s⁻¹)	Conditions	Reference
H_2	$(1.0\pm0.4)\times10^{-13}$	1.8 Torr He	[5]
	$(2.6\pm0.6)\times10^{-13}$	2 Torr He	[6]
D_2	$(2.6\pm0.7)\times10^{-12}$	5 Torr He	[7]
	$(1.9\pm0.2)\times10^{-12}$	5 Torr SF_6, 268-330K	[8]
SiH_4	$(6.7\pm0.7)\times10^{-11}$	1 Torr He	[6]
	$(1.1\pm0.2)\times10^{-10}$	1 Torr He	[5]
	1.3×10^{-10}	1 Torr He	[9]
	4×10^{-11}	1 Torr He	[10]
	2.5×10^{-10}	1 Torr SF_6	[24]
SiD_4	2.6×10^{-10}	2 Torr He	[26]
CH_3SiH_3	$(3.7\pm0.2)\times10^{-10}$	5 Torr SF_6	[9]
$(CH_3)_2SiH_3$	$(3.3\pm0.3)\times10^{-10}$	5 Torr SF_6	[9]
$(CH_3)_3SiH$	$(2.5\pm0.1)\times10^{-10}$	5 Torr SF_6	[9]
Si_2H_6	$(5.7\pm0.2)\times10^{-10}$	1 Torr He	[5]
	$(1.5\pm0.2)\times10^{-10}$	1 Torr He	[6]
	$(2.8\pm0.3)\times10^{-10}$	5 Torr He	[6]
	$(4.6\pm0.7)\times10^{-10}$	5 Torr Ar	[9]
	1.7×10^{-10}	1 Torr He	[10]
CH_4	$\leq (2.5\pm0.5)\times10^{-14}$	5 Torr He	[11]
C_2H_6	$\leq (1.2\pm0.5)\times10^{-14}$	5 Torr He	[11]
C_2H_4	$(9.7\pm1.2)\times10^{-11}$	1 Torr He	[5]
	$(2.6\pm0.3)\times10^{-11}$	1 Torr He	[11]
	3.9×10^{-11}	1 Torr Ar	[23]
	6.7×10^{-11}	1 Torr SF_6	[23]
C_3H_6 (propylene)	$(1.2\pm0.1)\times10^{-10}$	5 Torr He	[11]
	2.4×10^{-10}	—	[9]

C$_4$H$_6$ (butadiene)	$(1.9\pm0.2)\times10^{-10}$	5 Torr He	[11]
C$_2$H$_2$	$(9.8\pm1.2)\times10^{-11}$	5 Torr He	[11]
HCl	$(6.8\pm1.0)\times10^{-12}$	5 Torr He	[12]
Cl$_2$	$(1.4\pm0.2)\times10^{-10}$	5 Torr He	[12]
NO	$(1.7\pm0.2)\times10^{-11}$	5 Torr He	[12]
O$_2$	$(7.7\pm1.0)\times10^{-12}$	5 Torr He	[12]
CO	$< 10^{-13}$	5 Torr He	[12]
N$_2$	$< 10^{-13}$	5 Torr He	[12]
N$_2$O	$(1.90\pm0.09)\times10^{-12}$	10-100 Torr Ar	[13]

In summary, the monosilicon hydride radicals SiH and SiH$_2$ are extremely reactive with the Si-H bonds of silane. As a result, their gas phase concentrations in most *typical* silane or disilane CVD processes will be low and their ability to diffuse from the bulk gas volume to film growth surfaces will be minimal. The silyl radical, on the other hand is readily reactive only with other radicals or stable molecules with unpaired electrons. This order of relative reactivities readily explains the observed steady state monosilicon hydride radical concentrations in silane plasmas[27], regardless of the relative production rates of each radical by electron induced dissociation of silane. Thus, in plasma or photochemical CVD systems, where all possible monosilicon hydride radicals can be formed, silyl will be the dominant gas phase monosilicon hydride radical at steady state, and will easily reach film growth surfaces. In thermal systems where only silylene, SiH$_2$, is likely to be formed in the initial dissociation of silane, its gas phase concentration will be limited by its subsequent reaction with silane [1].

While the data summarized here provide a consistent qualitative picture, careful inspection of the rate constants, especially in the case of SiH$_2$, reveals the need for further work. For example, while all studies agree that silylene reacts rapidly (in the sense of nearly gas kinetic) with silane, the rate constants are scattered over a factor of two to three in value, while the error bars tend to be $\pm$10-20% for any given study.

Table 3 Absolute Rate Constants for SiH_3 Reactions

Reactant	Rate Constant (cm^3 molecule^{-1} s^{-1})	Conditions	Reference
SiH_3	$\leq (6.1\pm3.5)\times10^{-11}$	9.5 Torr He	[14]
	$(1.5\pm0.6)\times10^{-10}$	0.9 Torr H_2	[15]
	$(7.9\pm2.9)\times10^{-11}$	9.5 Torr He	[16]
	$(1.2\pm0.4)\times10^{-10}$	5 Torr He	[17]
H	$(2.0\pm1.0)\times10^{-11}$	9.5 Torr He	[16]
O_2	$(9.7\pm1.0)\times10^{-12}$	1-27 Torr Ar	[18]
	$(1.3\pm0.4)\times10^{-11}$	1-6 Torr He	[19]
	$(1.3\pm0.3)\times10^{-11}$	1-10 Torr N_2	[20]
	1.2×10^{-11}		[25]
NO	$(2.5\pm0.5)\times10^{-12}$	9.5 Torr He	[14]
	$(8.2\pm0.9)\times10^{-30}$ a	3-11 Torr N_2	[20]
N_2O	$< 5\times10^{-15}$		[19]
NO_2	$(5.1\pm0.9)\times10^{-11}$	3-10 Torr N_2	[20]
	5.05×10^{-11}		[25]
C_2H_4	$\leq (3\pm3)\times10^{-15}$	9.5 Torr He	[14]
C_3H_6 (propylene)	$\leq (1.5\pm0.5)\times10^{-14}$	9.5 Torr He	[14]
C_3H_4 (propyne)	$\leq (1.8\pm0.4)\times10^{-14}$	9.5 Torr He	[14]
NOCl	$(1.3\pm0.3)\times10^{-11}$	5 Torr Ar	[18]
S_2Cl_2	$(2.4\pm0.5)\times10^{-11}$	4-7 Torr He, 326K	[21]
CCl_4	$\leq (5\pm2)\times10^{-14}$	9.5 Torr He	[16]
SiD_4	$\leq (4\pm2)\times10^{-14}$	9.5 Torr He	[16]
Si_2H_6	$\leq (7\pm4)\times10^{-15}$	9.5 Torr He	[16]

HBr	$(1.77\pm0.29)\times10^{-12}$	4.7 Torr He	[22]
HI	$(1.79\pm0.21)\times10^{-11}$	1.8 Torr He	[22]

a Rate constant is linear in total pressure. Units are cm^6 molecule^{-2}s^{-1}

This type of discrepency in kinetic results is not uncommon in early phases of absolute rate constant measurements for new classes of gas phase radical reactions, but is unsatisfactory from the viewpoint of providing a database for CVD modelling. Substantial additional kinetic work will be required to achieve an accuracy comparable to that of databases used in hydrocarbon combustion modelling.

TRANSIENT SPECIES CONTAINING MORE THAN ONE SILICON ATOM

Very little experimental information is available on transient silicon hydride species containing more than one silicon atom. The family of molecules Si_2H_x, x = 1-5 has been extensively studied by *ab initio* theoretical techniques. Ground state structures and the energetics of the ground state potential surfaces have been characterized. This work, along with photoionization data for some of these species is summarized in references [28,29]. To date, only Si_2H_2 in two of its isomeric forms has been detected by high resolution spectroscopy [30,31,32]. The results confirm the *ab initio* predictions, but provide no kinetic data. Jasinski [33] has observed some transient absorption in the near ultraviolet following photolysis of disilane and has argued that this absorption is due to one or more of the disilicon hydride species. It is not clear whether these transients are the same as those reported earlier in a silane flash photolysis experiment [34]. Jasinski [33,35] has also reported gas phase nucleation and growth of particles following excimer laser photolysis of disilane, and has demonstrated the utility of this method for preparing crystalline silicon nanoparticles [35].

SURFACE LOSS COEFFICIENTS FOR MONOSILICON HYDRIDE RADICALS

Techniques have now been demonstrated for direct measurements of the total surface loss coefficient for all three monosilicon hydride radicals. Ho et al. [36] have reported a value of $\beta=0.94$ for the loss of SiH on the surface of a growing amorphous silicon film. This value was determined using a spatially resolved LIF technique. Robertson and Rossi [37,38] have used REMPI in a static bulb experiment to determine a value of $\beta=0.15$ for silylene. Most recently Jasinski [39] has used a discharge flow method to determine a value of $\beta=0.05$ for silyl on a surface containing silicon, hydrogen and chlorine. Krasnoperov et al. [40] have reported a similar experiment, using LMR to detect silyl and report surface loss coefficients as a function of temperature which are in the same range as Jasinski's result. However, the flow dynamics in Krasnoperov's experiment, and hence the determination of β from the experiment, are complicated by thermal gradients which were not fully modelled [41]. While the qualitative ordering of these coefficients appears to reflect the relative gas phase reactivity of the respective radicals towards Si-H bonds, there is as yet no "universal" agreement as to the accuracy or relevance of these values to CVD. The SiH result is the least controversial, while the SiH_3 result is not in good agreement with values derived from PECVD film growth studies [39,42,43].

REFERENCES

1. J. M. Jasinski and S. M. Gates, *Acc. Chem. Res.* **24**, 9 (1991).
2. M. H. Begemann, R. W. Dreyfus and J. M. Jasinski, *Chem. Phys. Lett.* **155**, 351 (1989).
3. M. Nemoto, A. Suzuki, H. Nakamura, K. Shibuya and K. Obi, *Chem. Phys. Lett.* **162**, 467 (1989).
4. J. P. M. Schmitt, P. Gressier, M. Krishnan, G. DeRosny and J. Perrin, *Chem. Phys.* **84**, 281 (1984).
5. G. Inoue and M. Suzuki, *Chem. Phys. Lett.* **122**, 361 (1985).
6. J. M. Jasinski and J. O. Chu, *J. Chem. Phys.* **88**, 1678 (1988).
7. J. M. Jasinski, *J. Phys. Chem.* **90**, 555 (1986).
8. J. E. Baggott, H. M. Frey, K. D. King, P. D. Lightfoot, R. Walsh and I. M. Watts, *J. Phys. Chem.* **92**, 4025 (1988).
9. J. E. Baggott, H. M. Frey, P. D. Lightfoot, R. Walsh and I. M. Watts, *J. Chem. Soc. Faraday Trans.* **86**, 27 (1990).
10. T. R. Dietrich, S. Chiussi, M. Marek, A. Roth and F. J. Comes, *J. Phys. Chem.* **95**, 9302 (1991).
11. J. O. Chu, D. B. Beach and J. M. Jasinski, *J. Phys. Chem.* **91**, 5340 (1987).
12. J. O. Chu, D. B. Beach, R. D. Estes and J. M. Jasinski, *Chem. Phys. Lett.* **143**, 135 (1988).
13. R. Becerra, H. M. Frey, B. P. Mason and R. Walsh, *Chem. Phys. Lett* **185**, 415 (1991).
14. S. K. Loh, D. B. Beach and J. M. Jasinski, *Chem. Phys. Lett.* **169**, 55 (1990).
15. N. Itabashi, K. Kato, N. Nishiwaki, T. Goto, C. Yamada and E. Hirota, *Jpn. J. Appl. Phys.* **28**, L325 (1989).
16. S. K. Loh and J. M. Jasinski, *J. Chem. Phys.* **95**, 4914 (1991).
17. M. Koshi, A. Miyoshi and H. Matsui, *Chem. Phys. Lett.* **184**, 442 (1991).
18. S. A. Chasovnikov and L. N. Krasnoperov, *Khim. Fiz.* **6**, 956 (1987).

19. I. R. Slagle, J. R. Bernhardt and D. Gutman, *Chem. Phys. Lett.* **149**, 180 (1988).
20. K. Sugawara, T. Nakanaga, H. Takeo and C. Matsumura, *Chem. Phys. Lett.* **157**, 309 (1989).
21. L. N. Krasnoperov, E. N. Chesnokov and V. N. Panfilov, *Chem. Phys.* **89**, 297 (1984).
22. J. A. Seetula, Y. Feng, D. Gutman, P. W. Seakins and M. J. Pilling, *J. Phys. Chem.* **95**, 1658 (1991).
23. N. Al-Rubaiey, H. M. Frey, B. P. Mason, C. McMahon and R. Walsh, *Chem. Phys. Lett.* **204**, 301 (1993).
24. R. Becerra, H. M. Frey, B. P. Mason, R. Walsh and M. S. Gordon, *J. Am. Chem. Soc.* **114**, 2751 (1992).
25. R. W. Quandt and J. F. Hershberger, *Chem. Phys. Lett.* **206**, 355 (1993).
26. J. O. Chu and J. M. Jasinski, unpublished.
27. R. Robertson and A. Gallagher, *J. Appl. Phys.* **59**, 3402 (1986).
28. B. Ruscic and J. Berkowitz, *J. Chem. Phys.* **95**, 2416 (1991).
29. L. Curtis, K. Raghavachari, P. W. Deutsch and J. A. Pople, *J. Chem. Phys.* **95**, 2433 (1991).
30. M. Bogey, H. Bolvin, C. Demuynck and J. L. Destombes, *Phys. Rev. Lett.* **66**, 413 (1991).
31. M. Cordonnier, M. Bogey, C. Demuynck and J. L. Destombes, *J. Chem. Phys.* **97**, 7984 (1992).
32. R. S. Grev and H. F. Schaefer, *J. Chem. Phys.*, **97**, 7990 (1992).
33. J. M. Jasinski, *Chem. Phys. Lett.* **183**, 558 (1991).
34. D. Perner and A. Volz, *Z. Naturforsch.* **29a**, 976 (1974).
35. J. M. Jasinski and F. K. LeGoues, *Chem. Mater.* **3**, 989 (1991).
36. P. Ho, W. G. Breiland and R. J. Buss, *J. Chem. Phys.* **91**, 2627 (1989).
37. R. M. Robertson and M. Rossi, *Appl. Phys. Lett.* **54**, 185 (1989).
38. R. M. Robertson and M. Rossi, *J. Chem. Phys.* **91**, 5037 (1989).
39. J. M. Jasinski, *J. Phys. Chem.* **97**, 7385 (1993).
40. L. N. Krasnoperov, V. V. Nosov, A. V. Baklanov and V. N. Panfilov, *Khim. Fiz.* **7**, 528 (1988).
41. L. N. Krasnoperov, personal communication, 1993.
42. A. Matsuda, K. Nomoto, Y. Takeuchi, A. Suzuki, A. Yuuki and J. Perrin, *Surf. Sci.* **227**, 50 (1990).
43. D. A. Doughty, J. R. Doyle, G. H. Lin and A. Gallagher, *J. Appl. Phys.* **67**, 6220 (1990).

Theoretical Prediction of Gas-Phase Nucleation Kinetics of SiO

Michael R. Zachariah and Wing Tsang,
National Institute of Standards and Technology, Gaithersburg, MD 20899

Abstract
This paper describes the application of ab-initio molecular orbital (MO) theories in conjunction with reaction rate theory to obtain thermochemistry and energetics of nucleation processes. The specific example used for the illustration of this approach is the nucleation of SiO. MO computations on the equilibrium structures have shown the polymers up to the tetramer to be planar rings and exothermic to addition of the monomer. Transition state analysis has shown that subsequent addition of the monomer (SiO) most likely proceeds without an energy barrier. Reaction rate theory analysis of the polymerization shows the rate coefficients to be very pressure dependent, with the dimer formation process rate limiting. Oxidation of clusters however showed a substantial barrier which should result in oxygen deficient clusters.

Introduction
One of the prominent issues in the formation of fine particles from the gas phase, particularly as it relates to the synthesis of ceramics and semiconductors (or any very low vapor pressure solid), is how to model such processes. Current models encompass a variety of complex processes, including nucleation, coagulation, coalescence, transport, and sintering [1-4]. Naturally, the quality of the output of these models depends heavily on the quality of the fundamental data inputs to such models. These might include chemical kinetic data for monomer formation rates, nucleation rates, sintering rates, etc. Unfortunately, such data are extremely difficult to obtain.

In this paper, we look at the issues related to the first principles prediction of nucleation rate data. There are several issues and limitations to be addressed. First, it has been recognized for some time that the application of classical nucleation theories [3] for predicting nucleation rates for materials such as ceramics and semiconductors under typical flame and flow reactor conditions are inappropriate. The reason is that the calculated critical cluster size for nucleation is on the order of the size of an atom. Since such calculations employ a bulk quantity (surface tension) to obtain the surface free energy, which is used to obtain a free energy barrier to nucleation via a competition between bulk and surface energetics, it becomes clear that an atomistic approach is needed. These results do suggest that there is no thermodynamic barrier to nucleation and therefore we must look to a kinetics approach.

Calculation Procedure

a. Molecular Orbital Calculation Procedure
All molecular orbital (MO) calculations employed the Gaussian90 [5,6] program for the determination of geometries, energies and vibrational frequencies. Because some of the species of interest have a large number of heavy atoms (> 5) large basis set computations such as the G2 method are not applicable. In order to obtain accuracies sufficient to entertain calculation of kinetic properties we have chosen a procedure which applies a bond additivity correction (BAC) scheme, calibrated against a very high level computation.

Calibration of BAC factors for the O-H and Si-H bonds were computed against the known energies for H_2O and SiH_4 [7]. Because so little is known of an experimental nature of gas phase silicon-oxyhydrides, calibration had to be made against higher level ab-initio calculations. The procedure was to calibrate the Si-O bond against H_2SiO and H_3SiOH using a very high level computation (G2 procedure outlined in reference [8]) and extract corrections needed for the Si-O bond using established bond additivity correction procedures [7]. The correction factor is expressed as $E(Si-O) = 628.3(\exp -2.42r)$; where r is the Si-O bond distance calculated at the HF/6-31G(d) level and E is the energy correction in kcal's calculated at the MP4/6-31G(d,p) level. Spin contamination effects for open shell species have been corrected using the procedure outlined by Schlegel [9] and employed extensively by Melius :E(spin) = E(UMP3)

E(PUMP3).

b. Reaction Rate Theory Approach to Nucleation

One of the first steps to understanding nucleation kinetics is an appreciation of the energetics of small molecule energy transfer, its influence on kinetics of nucleation, and how such processes can be accounted for in the framework of a nucleation model.

When two reactive fragments combine, the newly formed molecule will most likely have excess energy. Unless it is removed via collisions or through other chemical transformations, the excess energy will inevitably lead to the redissociation of the hot species. The consequence of the adduct possessing excess energy, is that at sufficiently low pressures the bimolecular recombination rate constant is dependent on pressure. As the pressure is increased, this dependence becomes weaker until at some limiting value the combination product assumes the standard Boltzman distribution. The pressure at which the rate constant becomes independent of pressure is termed the high-pressure limit of the reaction and is used in the formulation of reaction rate theory. The high pressure limit is unique to a given chemical reaction, bath gas and temperature. The reverse of this process is unimolecular decomposition and the rate constants are related through the equilibrium constant or k(recombination)/k(dissociation)=equilibrium constant. The equilibrium constants are calculated from the microscopic properties determined by the molecular orbital calculations and through standard statistical mechanical formulas.

Based on the assumptions above it is clear that one can apply the already well established procedures employed by unimolecular reaction rate theory (RRKM) to obtain the reverse process of nucleation [10,11].

The individual kinetic steps in a nucleation event can be expressed as follows:

$$A_{n-1} + A = A_n^\#$$
$$A_n^\# + M = A_n + M \quad ; \text{where M is the third body stabilizer.}$$

RRKM formalism provides a general framework for the determination of rate constants for treating nucleation kinetics in the pressure dependent region or equivalently the extent of the departures from the limiting high pressure values. In principle, all the required inputs necessary for RRKM analysis can be obtained via identification of the transition state for a reaction from the MO calculation. However, problems can arise if no defined transition state can be identified by the MO computation. Typically such a condition would be most often encountered if no reaction barrier can be found (no saddle point). If no transition state could be calculated the procedure involved deriving the properties of stable species from MO theory and to use an essentially semi-empirical approach to the application of RRKM for the determination of the rate constants. The empirical contributions involve the selection of high pressure rate expressions for combination and the step-size down parameters derived from experimental studies on organic compounds and other substances in the first row of the periodical table [12]. Since there are no experimental kinetic data for the $(SiO)_x$ systems under consideration, this is a serious approximation. Justification is provided by the observation that the parameters that have been selected are remarkably invariant to molecular type and structure.

Reaction rate coefficients were calculated using the computed thermochemistry, transition state barriers, vibrational frequencies (scaled by 0.89) and rotational constants (obtained from the MO calculations). Due to the high temperatures, energy transfer effects are very important. Thus calculated infinite- pressure rate constants for unimolecular and related processes have been corrected for pressure effects through RRKM calculations under the assumption of weak collisions. This involves the calculation of the specific rate of excited molecules at each particular energy level, based on the solution of the master equation assuming a step- ladder model of the transition probability and a step size down per collision of 400 cm^{-1}. The last number is based on high-temperature studies of the decomposition of organic molecules.

Equilibrium Structures and Energetics of SiO Nucleation

Based on prior experimental and detailed chemical kinetic modeling for a silicon/silica mixed stoichiometry system, it was concluded that SiO should play a significant role in the nucleation process during silane oxidation [13-14]. Two issues remained paramount, however: 1) What would the nuclei look like and what are their energetics; 2) What is a reasonable reaction rate for nucleation. The calculated equilibrium structures for the polymers of SiO are shown to be cyclic and planar as illustrated in Fig. 1.

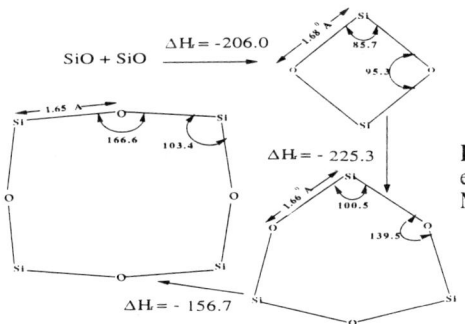

Fig 1. Equilibrium structures and energetics for SiO polymerization. MP4/6-31G(d,p)//HF/6-31G(d)

Straight chain or non-planar ring structures were found to be unstable in agreement with the work of Hastie et. al. who conducted I.R. spectroscopy studies on matrix isolated SiO species concluding that $(SiO)_2$, $(SiO)_3$ and possibly $(SiO)_5$ rings were polymerizing from monomeric SiO [15]. Our higher level computations for the energetics of the process show that insertion of SiO to form cyclic polymers is highly exothermic. The relative stability of the SiO mer's based on the computed exothermicity would suggest that the trimer is the most stable. Si-O bond lengths get progressively shorter with increasing ring size and would indicate increased bond strength. However, the fact that formation of the larger ring results in a decrease in heat release, particularly for the tetramer, suggests that the large Si-O-Si bond angle in the tetramer contributes to ring strain. Computations for a pentamer were not attempted due to the computational requirements.

Nucleation Kinetics

Nucleation kinetics based on the reaction rate theory previously described were computed for the polymerization of SiO. MO computation for the transition state for each subsequent reaction indicated that ring insertion probably proceeds without a potential barrier. The reaction potential energy for dimer formation as a function of monomer separation distance showed that the interaction potential is essentially flat till the monomer-monomer separation is decreased to about 2.8 A at which point the monomer feels a strong attractive force. This is in qualitative agreement with the Hastie et. al. [18] matrix isolation experiments in which the dimer was observed to form at temperatures as low as 5 K.

The rate coefficient for dimer formation as a function of pressure and temperature is shown in Fig 2a. Clearly the rate coefficient itself is pressure dependent in the manner discussed previously regarding reaction rate theory. Note that we are not referring to the nucleation rate itself, which will have pressure dependence simply from a monomer density consideration, but rather the rate constant (clearly not constant here) which has pressure dependence because of energy transfer effects. At the high pressure limit of the reaction (not calculated here) the rate coefficient would be independent of pressure.

Following the rate coefficient's behavior as a function of temperature, we see that even at low temperatures, 400K, the reaction rate is large. This is due to the absence of an energy barrier to dimer formation. As the temperature is increased, for a given pressure, we observe a gradual decrease in the rate constant. This results from a decrease in the ratio of stabilization to decomposition. Behavior for the trimer and tetramer have similar characteristics and are shown in Fig's. 2b,c, although the three body effects are most pronounced for the dimerization due to the

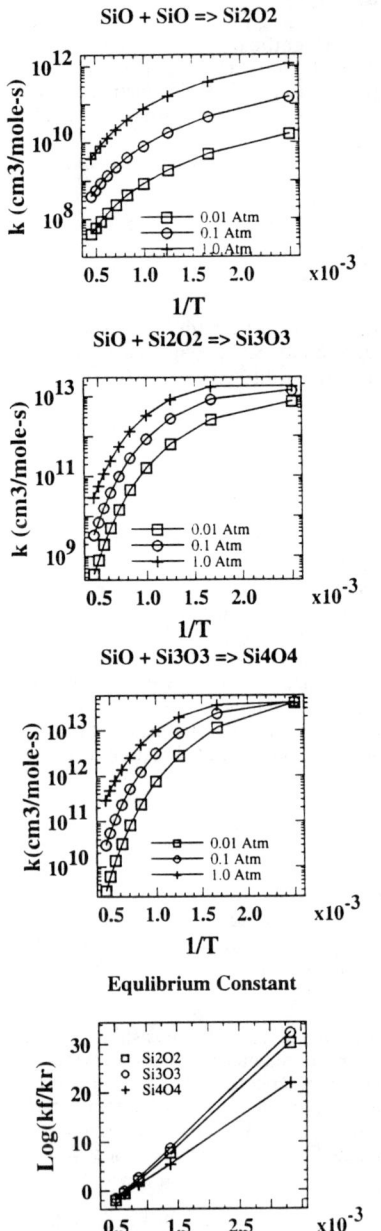

Fig. 2a,b,c. Nucleation rate constants as a function of T and P. d. Equlibrium constant

smaller size of the molecule and the relatively low activation energy for the dissociation. The magnitudes of the nucleation rates are significantly different as are the temperature dependencies. The rate of trimer formation via reaction of monomer with the dimer is two orders of magnitude higher than for dimer formation. This behavior is due again to collision efficiency effects and the energetics of the overall reaction. These results demonstrate that the initial stage of condensation process is dependent not only on the concentration of the reactive species but also directly on the pressure of the bath gas or equivalently a $(p)^{2+n}$ dependence where n is less than 1. With everything else being equal, the increase in molecular size will lead to a gradual reduction of n.

For the tetramer the rate constant at room temperature is at the high pressure limit (i.e. pressure independent). This results from a combination of high third body density and the fact that the tetramer has sufficient number of internal oscillators to dispose of the excess energy more efficiently than the dimer or trimer. As the temperature is increased, the nucleation rate declines again due to third body collision effectiveness and to a relatively small density of states for the tetramer (as compared to the trimer with a much deeper well depth). As such, the pressure dependence of the reaction becomes more pronounced as the temperature increases.

If there are no changes in the general pattern of the chemistry then as one proceeds to higher mers the increasing size of the molecule will lead to values of the rate constant approaching the high pressure limit. Of course this will also be the consequence if an exothermic decomposition reaction should occur after insertion during any step of the overall process. It is clear, however, that the rate coefficients for association are far below the gas kinetic collision rate that has at times been used for modeling purposes. This is in contrast to large cluster growth calculated by molecular dynamics methods which shows unity sticking coefficient resulting from the large number of available oscillators [16].

We know from experience and from the nomenclature of classical nucleation theory that for a nucleating system, increases in temperature result in a lowering of the supersaturation ratio and therefore a lowering in the net nucleation rate. This is clearly evident in the results presented here, albeit from a different perspective. Fig. 2d shows the equlibrium constant for the three reactions. From detailed balancing the cluster fragmentation rate can be obtained, which should only be important at high temperatures.

Cluster Oxidation

During the formation of silicon oxides, as with almost all metal oxides, experimental observation has shown that only partial oxidation takes place during the gas phase synthesis process. In many cases, full oxidation requires post high temperature oxidation to reach stoichometic conversion. In many cases high concentrations of water vapor are either present or required and even under these circumstances oxidation may be incomplete. We investigated cluster oxidation with water vapor by calculating the interaction between H_2O and the dimer and trimer. The initial complex forms through an overlap of the lone pair on oxygen and the empty p orbital on Si (as an out of plane bent structure) at a Si-O bond distance of 2.5 A. One of the hydrogens on H_2O is stabilized by the adjacent O atom in the dimer. The remaining H atom can then migrate through the transition state to the silicon atom as the Si-O bond distance decreases. The analogous reaction with the trimer proceeds in the same manner. Fig. 3 shows the calculated potential energy diagram showing activation barriers 67.3 and 69.0 kJ/mol for the dimer and trimer respectively.

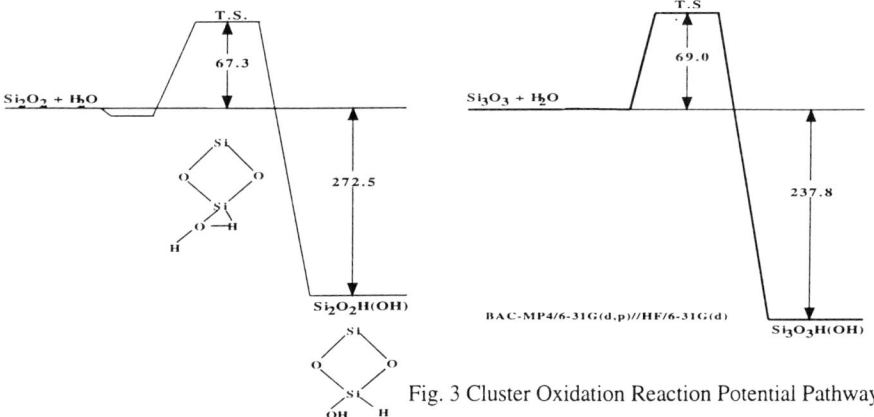

Fig. 3 Cluster Oxidation Reaction Potential Pathway

Qualitatively we can see that the reason for the low oxygen content of metal oxides is that the polymerization of the mono-oxide proceeds without an energy barrier, and whose rate constant is limited only by the collisional stabilization rate. The fits to the computed rate constants for dimer and trimer oxidation are $27.0\ T^{3.15}\ \exp(8256/T)$ and $92.5\ T^{3.02}\ \exp(8832/T)$ respectively. By contrast subsequent oxidation is slow requiring a high temperature for a sustained period.

Calculation of Nucleation and Cluster Oxidation Process

In order to clearly see the effect of the computed rate constants on nucleation a simple mechanism is constructed in which silicon and oxygen are present at time zero in concentrations of 0.05 mole fraction in an argon background at 1000K and 1 atm. For the purposes of the illustration SiO is assumed to be formed by the atomic recombination of Si and O with a rate(moles/cm³-s) = 1.0E09[Si][O] and the final product in the simulation is Si_5O_5 formed by reaction of SiO with Si_4O_4 at a rate = $1.0E12[SiO][Si_4O_4]$. Simulation results are summarised in Fig. 4 and show that the predominant product is the pentamer (Si_5O_5) followed by the dimer and trimer oxide. As the temperature is increased (T>1500K) the oxides of the dimer and trimer become favored over the pentamer.

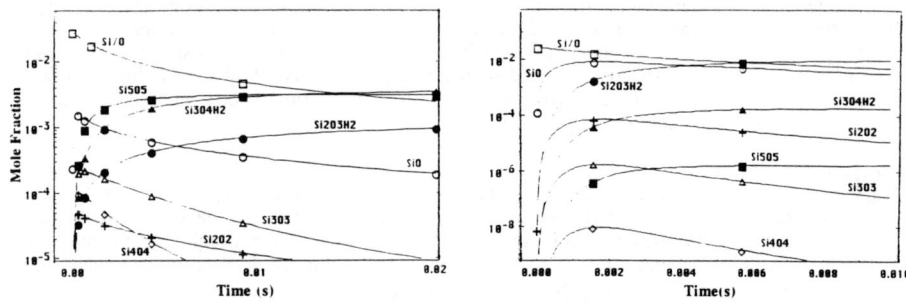

Fig. 3a,b. Species profiles at 1000 and 1500 K

Conclusions
To sum up, a structure exists to describe in detail the initial steps of condensation processes which involve formation of strong covalent bonds. For the specific problem studies here, we have shown the equilibrium structures of SiO mers to be cyclic, planar and relatively stable. The rate coefficients for polymerization are pressure dependent over the temperature regimes of interest in typical processes and proceed without an energy barrier. The primary determining factors in the magnitude of the A factor in the rate coefficient, is the stability of the mer relative to the fragments (a density of states consideration) and the relative number of oscillators in the molecule. Increasing either of these two parameters will increase the rate coefficient for recombination by forcing the activated complex closer to the high pressure limit of the reaction. It is expected that other ceramic oxides would behave in an analogous manner. Cluster oxidation is slow due to the high activation barrier required and qualitatively explains the experimental observation of incomplete cluster oxidation.

References
1. Xiong, Y and , S.E. Pratsinis, *J. Aeros. Sci,* **22** (suppl. 1), s199 (1992)
2. Gelbard, F., Tambour, Y., and Seinfeld, J.H., *J. Colloid Interf. Sci.*, **76**(2), 541 (1980)
3. Zachariah, M.R., and Semerjian, H.G., *AIChE. J.*, **35**, 2003 (1989)
4. Zachariah, M.R. and Dimitriou, P., *Aeros. Sci. Tech.* **13**, 413 (1990)
5. Hehre, J.W and Pople, J.A., "Ab Initio Molecular Orbital Theory, New York, 1986
6. Frisch, M.J., Head-Gordon, M., Trucks, G.W., Foresman, J.B., Schlegel, H.B., Raghavachari, K., Robb, M.A., Binkeley, J.S., Gonzalez, C., DeFrees, D.J., Fox, D.J.,Whiteside, R.A., Seeger, R., Melius, C.F., Baker, L.R., MArtin, L.R., Kahn. L.R., Stewart, J., Topiol, S., Pople, J.A., Gaussian92, Gaussian Inc., Pittsburg, Pa (1992)
7. Melius, C.F., and Binkley, J.S., Twenty First Symposium (International) on Combustion. 1953 (1986)
8. C.L. Darling and Schlegel, H.B., *J.Phys. Chem.* **97**, 8207 (1993)
9. Sosa, C. and Schlegel, H.B., *J. Chem. Phys.* **84**, 4530 (1986)
10. Robinson, P.J. and Holbrook, K.A., "Unimolecular Reactions" Wiley Interscience, New York, (1972)
11. Zachariah, M.R., and Tsang, W., *Aeros. Sci. Tech.* to appear (1993)
12. Tsang, W., *J. Phys. Chem. Ref. Data,* **20**, 609 (1991)
13. Zachariah , M.R., and Tsang, W., "Ab Initio Computation of Thermochemistry and Kinetics in the Oxidation of Gas Phase Silicon Species" in MRS Proceedings on *Chemical Prospectives of Microelectronic Materials* , **282**, 493 (1993)
14. Zachariah, M.R., Ceramic Powder Science III, Ed. G. Messing, **12**, 283 (1990)
15. Hastie, J.W., Hauge, R.H., and Margrave, J.L., *Inorganic Chemica Acta* **3**, 601 (1969)
16. Zachariah, M.R., Carrier, M.J., and Blaisten-Barojas, E., "Molecular Dynamics Simulation of Large Cluster Growth", Published in this issue.

ATOMIC LAYER GROWTH OF SiO$_2$ ON Si(100) USING THE SEQUENTIAL DEPOSITION OF SiCl$_4$ AND H$_2$O

OFER SNEH, MICHAEL L. WISE, LYNNE A. OKADA, ANDREW W. OTT AND STEVEN M. GEORGE
University of Colorado, Department of Chemistry and Biochemistry, Boulder, CO 80309.

ABSTRACT

This study explored the surface chemistry and the promise of the binary reaction scheme:

(A) Si-OH + SiCl$_4$ → Si-Cl + HCl

(B) Si-Cl + H$_2$O → Si-OH + HCl

for controlled SiO$_2$ film deposition. In this binary ABAB... sequence, each surface reaction may be self-terminating and ABAB... repetitive cycles may produce layer-by-layer controlled deposition. Using this approach, the growth of SiO$_2$ thin films on Si(100) with atomic layer control was achieved at 600 K with pressures in the 1 to 50 Torr range. The experiments were performed in a small high pressure cell situated in a UHV chamber. This design couples CVD conditions for film growth with a UHV environment for surface analysis using laser-induced thermal desorption (LITD), temperature-programmed desorption (TPD) and Auger electron spectroscopy (AES). The controlled layer-by-layer deposition of SiO$_2$ on Si(100) was demonstrated and optimized using these techniques. A stoichiometric and chlorine-free SiO$_2$ film was also produced as revealed by TPD and AES analysis. SiO$_2$ growth rates of approximately 1 ML of oxygen per AB cycle were obtained at 600 K. These studies demonstrate the methodology of using the combined UHV/high pressure experimental apparatus for optimizing a binary reaction CVD process.

INTRODUCTION

Many future developments in microelectronics will be closely linked to progress in thin film technology[1,2]. The emergence of the future gigahertz silicon technology requires nanoscale sizes of components on a ULSI chip. This size reduction depends on improving process control, better definition and quality of interfacial regions and developing lower temperature procedures to prevent thermal degradation by diffusion. Central in silicon technology is the growing of silicon oxide - SiO$_2$. It is expected that future technology will require the deposition of gate oxide films as thin as 50 Å. In addition, three dimensional structures will play a dominant role in future ULSI technology. Consequently, a growth method is desired for conformal and atomic layer controlled SiO$_2$ thin films.

One of the promising solutions for the challenges of future silicon technology is the scheme of atomic layer epitaxy (ALE)[3], or atomic layer processing (ALP) for the growth of an amorphous material. This deposition technique involves two separate stages. Each of the individual stages terminates itself after the deposition of a monolayer (ML) or less. The result is thickness control and a precise definition of the interfacial regime. The process is suitable for a cold-wall CVD reactor. Trench and three-dimensional-structures, as well as porous materials, should be evenly covered using this procedure. The highly conformal resulting film on high aspect ratio structures compensates for the time-inefficiency of the ALE process. In addition, a higher capacity for CVD reactors is possible by decreasing wafer spacing without creating uneven deposition profiles[3].

The sequential deposition of two precursors A and B is employed in ALE or ALP. Each precursor should completely react with the surface and deposit a stoichiometric layer. This goal is not straightforward and requires the right selection of the "A" and "B" precursors, as well as adjusting the reaction temperature, pressures and the pumping time between the A and B cycles.

In addition, the reaction time should be long enough that the reactions are never flux-limited. Atomic layer epitaxy has been demonstrated for several systems. In particular, ALE has been successful for depositing III-V and II-VI materials[3], as well as some oxide materials such as Al_2O_3[4-6] TiO_2[7,8] and ITO[3]. Some approaches for silicon ALE have also been suggested[9-11]. The technique has been applied for growing crystalline or amorphous films on a variety of substrates[3-8].

$SiCl_4$ and H_2O have been used to deposit SiO_2 films[12-14]. In one deposition scheme, physisorbed layers of H_2O react with gas phase $SiCl_4$ at temperatures around 300 K[12]. These reactions were not atomic layer controlled according to the strategy of ALE or ALP. We have used the sequential deposition of $SiCl_4$ and H_2O to grow stoichiometric SiO_2 films on Si(100). The experiments were performed on a Si(100) sample which was cleaned and characterized in an ultra high vacuum chamber. The CVD reactions were accomplished in a high-pressure cell inside the UHV chamber. Following the reaction, the cell was evacuated, opened and returned to UHV conditions. This approach allowed the evaluation of the SiO_2 film using Auger electron spectroscopy (AES), laser induced thermal desorption (LITD) and temperature programmed desorption (TPD). Stoichiometric and vitreous SiO_2 films were grown on Si(100) in the temperature range of 400-680 K and pressure range between 10-50 Torr.

EXPERIMENTAL

The experiments were performed in a UHV chamber with a base pressure of $2\text{-}5\times 10^{-10}$ Torr. The chamber was equipped with a LEED/AES optics (Perkin Elmer PHI 15-120) which was used to characterize the clean Si(100) and the deposited films. The chamber was equipped with a sputtering gun for surface cleaning. The low pressure exposures were performed using a glass capillary array doser. A UTI 100C quadrupole mass spectrometer was used in the LITD and TPD experiments.

The surface mounting and preparation has been described elsewhere[15]. Briefly, a ~1 Ω·cm 10x11x0.38 mm Si(100) sample was mounted by means of tantalum/molybdenum clips. A 2000 Å tantalum film deposited on the back of the sample was used to heat the sample to 1200°C. The sample temperature was determined by a Chromel-Alumel thermocouple attached to the crystal. The samples were cleaned in vacuum by means of high temperature annealing cycles. After cleaning, the only surface contamination was carbon with levels ranging from <0.2-1 %. The clean Si(100) surface displayed the expected 2x1 LEED pattern.

A necessary bridge between the UHV techniques and the 10-50 Torr pressures used for CVD deposition is a high pressure chamber inside the UHV chamber. A few schemes have been proposed[16-21] following the pioneering work of Somorjai[16]. The common denominator of all the schemes is minimal surface area exposed to the high pressure to minimize the outgassing following high pressure processing. Our high pressure doser is illustrated in Fig. 1. A combination of a sample holder shaped as a knife-edge sealing-flange and a moving cap is used to seal a ~10 cm^3 volume having about 20 cm^2 of surface area. Once this high pressure cell is sealed, the cell can be pumped by a turbomolecular pump to pressures below 10^{-7} Torr. Typically a pressure of 100 Torr in the high pressure cell will cause a pressure increase in the UHV chamber of less than 4×10^{-10} Torr.

The sample holder is attached to a liquid nitrogen cryostat through a He thermal switch[22]. The thermal switch controls thermal conductivity between 0.05 to 1.25 W/°C by varying the He pressure between 5 to 500 mTorr, respectively. Sample temperatures can be achieved as low as 100 K. In addition, the sample-holder temperature can be quickly raised to 300-350 K before dosing low vapor pressure reactants into the high pressure cell. This temperature rise is accomplished by a combination of evacuating the He thermal switch and rapid resistive heating of the sample holder.

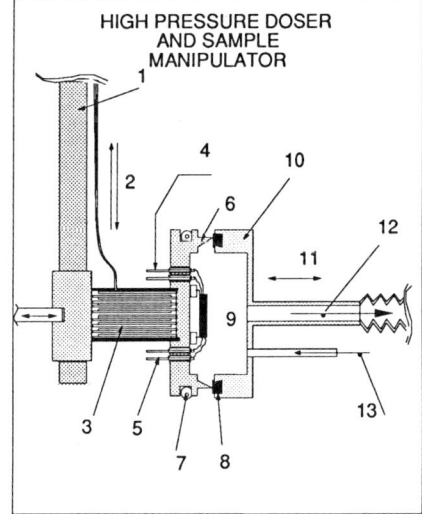

Fig. 1: An illustration of the high pressure dosing apparatus. (1) Liquid nitrogen cryostat; (2) He inlet and Pumping line for the He thermal switch; (3) He thermal switch; (4) Resistive heating electrical feedthroughs; (5) Thermocouple electrical feedthroughs; (6) Knife edge; (7) Sample holder heater; (8) Viton gasket; (9) Si(100) sample; (10) Cap; (11) Translation stage; (12) Pumping line; (13) High pressure dosing line.

Film growth experiments were performed in the high pressure cell. The crystal was cleaned by annealing at 1320 K for a few minutes before each experiment. The sample was then regulated at the desired temperature and the high pressure cell was sealed. High pressures of H_2O (up to 20 Torr) and $SiCl_4$ (1-50 Torr) were introduced into the high pressure cell and allowed to react for a given time. Following each reaction, the precursors were evacuated before exposure to the next reactant.

Chlorine coverage was detected by the laser induced thermal desorption (LITD) of $SiCl_2$ species. The total oxygen coverage was determined from SiO temperature programmed desorption. The integrated SiO mass signals were calibrated by comparison with the SiO signal obtained from a H_2O saturation dose. H_2O exposures on Si(100) are known to saturate at an oxygen coverage of $\theta_O = 0.5$ ML (1 ML $\approx 6.9 \times 10^{14}$ cm^{-2}). The stoichiometry of the film is qualitatively obtained from AES spectra.

RESULTS AND DISCUSSION

Fig. 2 illustrates the experimental procedure for atomic layer controlled growth. The reaction sequence was composed of separate high pressure doses of $SiCl_4$ ("A" reaction) and H_2O ("B" reaction). Initially, the experiments were performed at 400 K. At this temperature, the reaction of H_2O with a chlorine-terminated surface saturates after removing only 65% of the surface chlorine. At temperatures higher than 600 K, both "A" and "B" reactions proceed to completion. The first three "A" and "B" cycles were studied to determine the time required to obtain saturation coverage. The first chlorination stage is performed at 300 K and 1 Torr for 30 sec. to avoid surface etching. The second and third $SiCl_4$ "A" stages saturate after 20 minutes at 50 Torr. In comparison, 10 minutes were necessary for 10 Torr of H_2O to complete the reaction in the first second and third "B" stages.

The results of these experiments are summarized in Fig. 3. In this figure, the chlorine coverage after reaction with $SiCl_4$ is given relative to the first $SiCl_4$ dose. This chlorine is then removed by the H₂O reaction to deposit oxygen on the surface. Evidence that this stage leaves surface hydroxyls is the subsequent reaction with SiCl₄ to yield a chlorine coverage comparable to the first SiCl₄ exposure. Recent FTIR studies on oxidized porous silicon have also observed the growth of SiOH surface species concurrent with the loss of SiCl species versus H₂O exposure at 10 Torr at 600K.

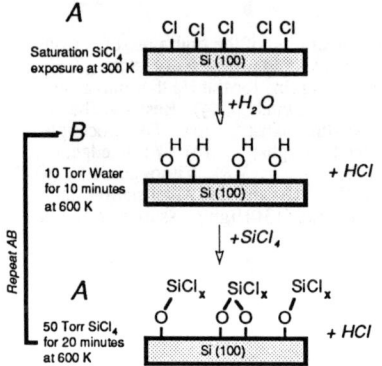

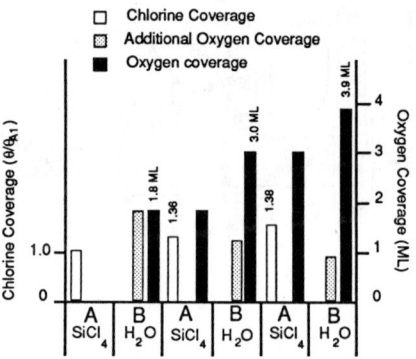

Fig. 2: Atomic layer controlled growth of SiO_2 using $SiCl_4$ and H_2O.

Fig. 3: The first three AB cycles. Bars illustrate the deposited chlorine and oxygen.

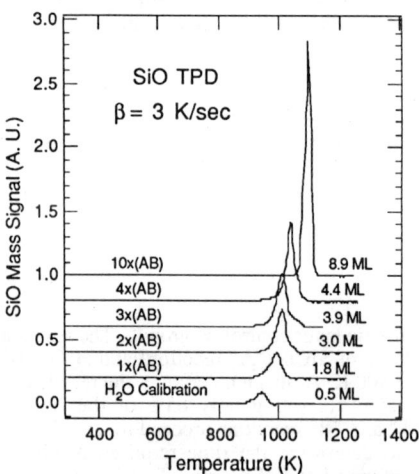

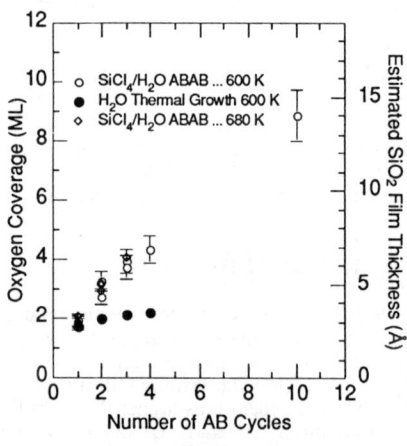

Fig. 4: A series of SiO TPD curves obtained following various AB binary reaction cycles.

Fig. 5: SiO_2 growth as a function of the number of AB cycles. The SiO_2 growth associated with only H_2O exposure for equivalent times is shown for comparison.

The saturation conditions described above were used to deposit SiO_2 films using up to 10 AB cycles. The oxygen coverage was determined from SiO TPD spectra. Some TPD curves are displayed in Fig. 4. Note the shift of the TPD peak to higher temperatures with the increase of the film thickness. This is a typical behavior of SiO desorption from thick SiO_2 films on silicon[23]. The results are summarized in Fig. 5. The surface was also thermally oxidized at 600 K by introducing 10 Torr of H_2O for reaction times equivalent to the times of "B"; 2x"B"; 3x"B" and 4x"B". The comparison in Fig. 5 shows that the first "B" reaction is independent of chlorine on the surface. The thermal oxidation then proceeds very slowly after depositing two oxygen monolayers.

AES spectra taken after each experiment observed the growth of the 63 and 78 eV AES peaks associated with vitreous silica. After the first AB cycle, only the 83 eV AES peak appears. This peak is associated with oxygen on the silicon surface. The 83 eV AES peak then continuously declines and the AES spectrum of the 10x(AB) process resembles the spectrum of a ~15 Å thick thermally grown SiO_2 film[24]. Chlorine incorporation into the film is minor (<1%). Representative AES spectra are displayed in Fig. 6.

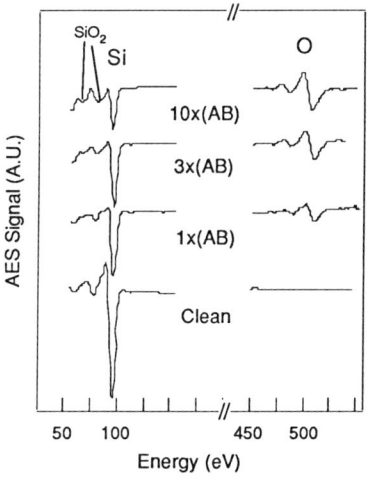 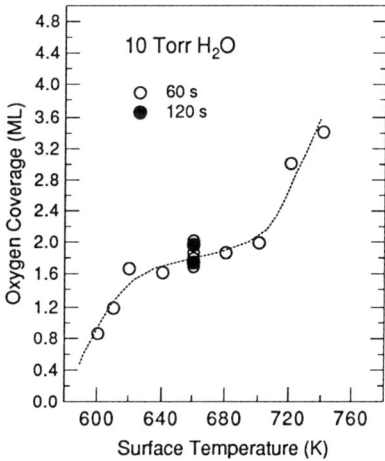

Fig. 6: AES spectra taken after various AB cycles during SiO_2 growth on Si(100).

Fig. 7: Oxygen coverage after 60 s and 120 s H_2O exposures on Si(100) at 10 Torr as a function of temperature.

In order to optimize the reaction conditions to obtain shorter reaction times, the first "B" stage was performed for one minute at 10 Torr in the temperature range of 600-740 K. The results are summarized in Fig. 7. The temperature range below 700 K is suitable for obtaining a self-limiting "B1" reaction on a clean Si(100) surface. At temperatures higher than 700 K, the oxidation proceeds into the silicon bulk in agreement with a previous study[25]. The reaction proceeds to completion in less than one minute in the temperature range of 640-700 K. A similar optimization procedure has determined that the "A2" and "A3" reactions saturate in less than 1 minute at 10 Torr at 660 K. However, saturation of the "B2" and "B3" reactions after less than 1 minute requires a temperature of 680 K.

CONCLUSIONS

Atomic layer controlled growth of SiO_2 has been demonstrated using a binary ABAB... reaction sequence with $SiCl_4$ and H_2O. These studies were facilitated using a high pressure chamber for the ABAB... SiO_2 growth that was internal to the UHV chamber. The SiO_2 film analysis was then performed in the UHV chamber using laser induced thermal desorption, temperature programmed desorption and Auger electron spectroscopy. Controlled SiO_2 atomic layer growth of approximately 1 ML of oxygen per AB cycle was achieved using high pressures of 10-50 Torr for 1-20 minutes at surface temperatures of 600-700 K. These studies illustrate the potential usefulness of the ABAB... binary reaction sequence chemistry for atomic layer controlled deposition of SiO_2 on silicon surfaces.

ACKNOWLEDGMENT: This research was supported by the Office of Naval Research under Contract No. N00014-92-J-1353. The authors thank David Berry for his contribution to the construction of the high pressure doser. M. L. W. acknowledges AT&T Bell Laboratories for a graduate fellowship. O.S. is grateful to the Israeli Academy of Sciences for a Wolfson postdoctoral fellowship. S.M.G. acknowledges the National Science Foundation for a Presidential Young Investigator Award (1988-1993).

REFERENCES

1. T. Venkatesan, Thin Solid Films **216**, 52 (1992).
2. Workshop Explores Commercial and Scientific Opportunities for Future Thin Film Research and Technology, Thin Solid Films **216**, ix (1992).
3. C. H. L. Goodman and M. V. Pessa, J. Appl. Phys. **60**, R65 (1986).
4. J. F. Fan, K. Sugioka and K. Toyoda, Jap. J. Appl. Phys. **6B**, L1139 (1991).
5. C. Soto and T. Tysoe, J. Vac. Sci. Technol. A **9**, 2686 (1991).
6. G. S. Higashi and C. G. Fleming, Appl. Phys. Lett. **55**, 1963 (1989).
7. S. Haukka, E. L. Lakomaa and T. Suntola, Thin Solid Films **225**, 280 (1993).
8. M. Ritala, M. Leskelä, E. Nykänen, P. Soininen and L. Niinistö, Thin Solid Films **225**, 288 (1993).
9. S. M. Gates, D. D. Koleske, J. R. Heath and M. Copel, Appl. Phys. Lett. **62**, 510 (1993).
10. J. A. Yarmoff, D. K. Shuh, T. D. Durbin, C. W. Lo, D. A. Lapiano-Smith, F. R. McFeely and F. J. Himpsel, J. Vac. Sci. Technol. A **10**, 2303 (1992).
11. A. C. Dillon, M. B. Robinson, M. Y. Han and S. M. George, J. Electrochem. Soc. **139**, 537 (1992).
12. D. J. Ehrlich and J. Melngailis, Appl. Phys. Lett. **58**, 2675 (1991).
13. M. Tsapatsis, S. Kim, S. W. Nam and G. R. Gavalas, Ind. Eng. Chem. Res. **30**, 2152 (1991).
14. M. Tsapatsis, and G. R. Gavalas, AIChE J. **38**, 847 (1992).
15. B. G. Kohler, C. H. Mak, D. A. Arthur, P. A. Coon and S. M. George, J. Chem. Phys. **89**, 1709 (1988).
16. G. A. Somorjai, Surf. Sci. **89**, 496 (1979).
17. A. L. Cabrera, N. D. Spencer, E. I. Kozak, P. W. Davies and G. A. Somorjai, Rev. Sci. Instrum. **53**, 1888 (1982).
18. D. W. Blakely, E. I. Kozak, B. A. Sexton and G. A. Somorjai, J. Vac. Sci. Technol. **13**, 1091 (1976).
19. B. E. Koel, B. E. Bent and G. A. Somorjai, Surf. Sci. **146**, 211 (1984).
20. R. A. Campbell and D. W. Goodman, Rev. Sci. Instrum. **63**, 172 (1992).
21. J. M. Campbell and C. T. Campbell, Surf. Sci. **259**, 1 (1991).
22. G. K. White, Experimental Techniques in Low-Temperature Physics, 3rd ed. (Clarendon Press, Oxford 1979), p. 150.
23. Y. Kobayashi and K. Sugii, J. Vac. Sci. Technol. A **10**, 2308 (1992).
24. J. Derrien and M. Commandre, Surf. Sci. **118**, 32 (1982).
25. M. L. Wise, P. Gupta, C. H. Mak, P. A. Coon and S. M. George, in preparation.

INFRARED STUDIES OF OZONE-ORGANOSILICON CHEMISTRIES FOR SiO$_2$ DEPOSITION

J. A. MUCHA and J. WASHINGTON
AT&T Bell Laboratories, Murray Hill, N.J. 07974

ABSTRACT

Gas-phase Fourier transform infrared spectroscopy has been used to monitor reactants and products in the study of O_3 decomposition and the reaction of O_3 with tetramethylsilane and tetraethoxysilane. The results confirm the interpretation that O_3 decomposition is heterogeneous, and a dominant factor in SiO$_2$ deposition using these chemistries. Product analysis shows that the higher reactivity of TEOS with O_3 is probably due to the ability of TEOS to undergo α-hydride elimination of one ethoxy group, per molecule, *and* one oxidative acetic acid elimination, per molecule, during the deposition process. Results also suggest elimination via the Si-center involving a single ethoxy group rather than via a 6-center elimination involving two ethoxy groups.

INTRODUCTION

Previously [1], we showed that SiO$_2$ deposition from ozone (O_3) reactions with tetramethylsilane (TMS) and tetraethoxysilane (TEOS) exhibited the same activation energy (16 kcal/mol). Both systems exhibit [1,2] a saturation (1st- to 0-order kinetics) in deposition rate as O_3 concentration is increased. Recent [3] studies with TEOS exhibited an Arrhenius A-factor about 25 times larger than with TMS, while using less O_3 to reach saturation than TMS (1 O_3-per-TEOS vs. 5 O_3-per-TMS at 330°C). We interpreted these observations by a mechanism in which heterogeneous decomposition of O_3 was rate-limiting, with saturation behavior resulting from surface coverage saturation or complete consumption of one of the principal reagents. Here, we report the application of gas-phase Fourier transform infrared (FTIR) spectroscopy as a probe for reactants and products during deposition of SiO$_2$. The results provide supporting evidence for the interpretations advanced in our previous deposition experiments.

EXPERIMENTAL

The reactor is a modified Plasma Technology, Ltd. parallel-plate unit with a shower-head in the upper electrode to mix and disperse gases. TMS was metered directly into the system, while a bubbler was used to deliver TEOS. For both, He was used as a carrier gas. The resulting gas mixtures were blended with an O_3/O_2 mixture (PCI Ozone ozonizer) near the entrance to the shower-head. Typical reactor feed consisted of about 70% O_2 and 26% He, with a fixed concentration of TMS (0.14 v%) or TEOS (0.09 v%) and variable O_3 concentrations (0.05-3.4 v%). The lower electrode (sample platen) was heated to typical SiO_2 deposition temperatures and the reactor effluent passed through a 15 cm. gas cell for infrared analysis using a Mattson FTIR spectrometer with 1 cm^{-1} resolution. Products were identified and quantified by generating calibration standards for each.

RESULTS and DISCUSSION

All of our previous interpretations were based on observations of only one product (SiO_2) of the reaction of O_3 with the Si-containing precursors TMS and TEOS. The ability to monitor reactants and other products should provide complementary data to test our earlier conclusions and shed light on areas of interpretation that were not definitive. There are several questions that infrared analysis of gas phase might address. By monitoring O_3 loss in a reactor when no Si-source is present, we might expect gain insight into: 1) whether O_3 decomposition or direct reaction with the Si-source initiates SiO_2 deposition; 2) whether initiation occurs homogeneously in the gas-phase or heterogeneously at the growth surface, and 3) the origin of the factor of 25 increase in deposition rate with TEOS.

Estimating the rate of loss of O_3 and its temperature dependence in our experimental apparatus require a mechanistic construct to convert measured concentrations of O_3 to an apparent rate constant for ozone decomposition. The accepted mechanism for O_3 pyrolysis has been given by Benson and Axworthy [4] as:

$$O_3 + M \underset{2}{\overset{1}{\rightleftarrows}} O + O_2 + M \qquad (1,2)$$

$$O + O_3 \overset{3}{\rightarrow} 2 O_2 \qquad (3)$$

where M represents represents a third-body that initiates decomposition. In our experiment (a dilute mixture of O_3 in O_2), M is either O_2 or a surface to distinguish homogeneous and heterogeneous decomposition, respectively. Based on a steady-state analysis of atomic oxygen in the above mechanism, the overall rate of O_3 loss is given by:

$$\frac{d[O_3]}{dt} = \frac{2 k_1 k_3 [O_3]^2 [M]}{k_2 [O_2][M] + k_3 [O_3]} . \qquad (4)$$

For dilute O_3 mixtures, the overall loss becomes second-order in O_3 when

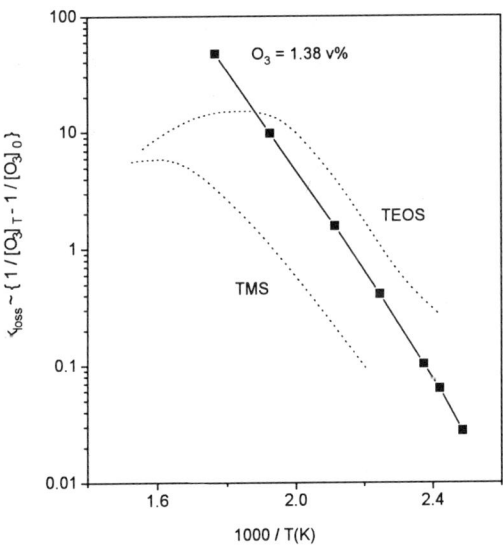

Figure 1. Arrhenius plot of k_{loss} for O_3 decomposition. Dotted lines are scaled, spline fits to SiO_2 deposition rate data for TMS [1] and TEOS [3] reaction with O_3.

$k_3 [O_3] \ll k_2 [M] [O_2]$, and the overall Arrhenius activation energy, E_a, is associated with the ratio $k_1 k_3/k_2$. For purely homogeneous decomposition, the overall E_a has been determined to be 28.8 kcal/mol [4]. When reaction (3) is fast, overall loss of O_3 is first-order in O_3 with a temperature dependence governed solely by k_1. Here, an activation energy of about 24 kcal/mol is expected for homogeneous initiation (reactions 1 & 2, M=O_2).

Figure 1 shows an Arrhenius plot of O_3 loss based on infrared results and the second-order scenario presented above. For comparison, dotted lines showing the temperature dependencies observed for O_3-induced SiO_2 deposition from TMS and TEOS are included. It is apparent that second-order O_3 kinetics gives excellent linearity in the Arrhenius plot of the resulting rate constant for overall O_3 loss, k_{loss}. A weighted [1/(est. error)^2] least-squares fit gives an activation energy of 21.2 ±0.3 kcal/mol, while an unweighted fit gives 18 ±2 kcal/mol. Weighting tends to emphasize the points in the middle of the temperature range more heavily since those at lower T correspond to small differences in large O_3 concentrations while those at higher T correspond to low absolute O_3 concentration. Analyzing the data as first-order kinetics resulted in a noticeable curvature in the Arrhenius plot suggesting that under present experimental conditions $k_2 [M] [O_2] \gg k_3 [O_3]$.

Several points become clear. First, the activation energy for O_3 loss is well below that expected for homogeneous initiation (M = O_2), supporting the interpretation that O_3

decomposition in our system is via a heterogeneous process (M = surface). For surface decomposition, adsorption of ozone is a likely initial step that would reduce the apparent activation energy for reaction (1) by the heat of adsorption. Thus, the observed E_a (21.2 kcal/mol) for O_3 loss here is 7.6 kcal/mol lower than that for homogeneous pyrolysis. Second, these results are also in good agreement with the activation energies (16 kcal/mol) we determined for SiO_2 deposition from TMS and TEOS[1,3]. With TMS or TEOS present, the reaction:

$$O + TMS(TEOS) \rightarrow \rightarrow SiO_2 \quad (5)$$

competes with reaction (3) and a similar steady-state analysis for O atoms gives the overall rate of SiO_2 formation:

$$\frac{d[SiO_2]}{dt} \propto \frac{k_1 k_5 [O_3][TXS][M]}{k_2[O_2][M] + k_3[O_3] + k_5[TXS]} \quad (6)$$

where X = M or EO for TMS and TEOS, respectively. Again, the role of O_3 in the overall kinetics is paramount, although we would expect different *overall* activation energies for SiO_2 deposition from TMS and TEOS, unless the respective $E_a(k_5)$ are accidentally equal or very small. When $k_2[O_2][M]$ is large, SiO_2 deposition is first-order in O_3 and Si-source, with an effective rate constant $\sim k_1 k_5/k_2$. Attributing the suppression in E_a (determined above) to k_1, and applying it (24 - 7.6 = 15.5 kcal/mol) to the deposition scheme (Reactions 1-3, 5) yields good agreement with those measured in our deposition studies [1,3] for this case when O-atom attack on the Si-source has a low barrier. When $k_2[O_2][M]$ is small, agreement with the deposition rate temperature dependence requires reactions (3) and (5) to exhibit nearly equal E_a's for *both* precursors. In either case, the proximity of the activation energies for O_3 decomposition and SiO_2 deposition tends to rule out the possibility that direct reaction of O_3 with the Si-source is initiating deposition.

Using FTIR product analysis, we have gained some insight into the origin of the factor of 25 greater reactivity that TEOS exhibits relative to TMS in O_3-induced SiO_2 deposition. In previous work [1,3], it was suggested that this could arise from the ability of TEOS to undergo spontaneous decomposition by α- and β-hydride elimination [6,7], for example. Thus, once activated by initial attack from an ozone-generated species like atomic oxygen, TEOS could *spontaneously* eliminate C_2-fragments enhancing the probability of SiO_2 formation. At the same time, such eliminations would result in the lower demand for O_3 with TEOS than TMS.

Figure 2 shows portions of the high-resolution FTIR spectrum of an O_3-TEOS reaction mixture with a low O_3/TEOS feed ratio (0.63) and low platen temperature (247°C). Also included are reference spectra of H_2O, acetaldehyde (CH_3CHO), and acetic acid (CH_3COOH). Although H_2O, CO, CO_2, formaldehyde (H_2CO) and formic acid (HCOOH) were common oxidation products in the ozonation of TMS and TEOS, CH_3CHO and CH_3COOH were unique

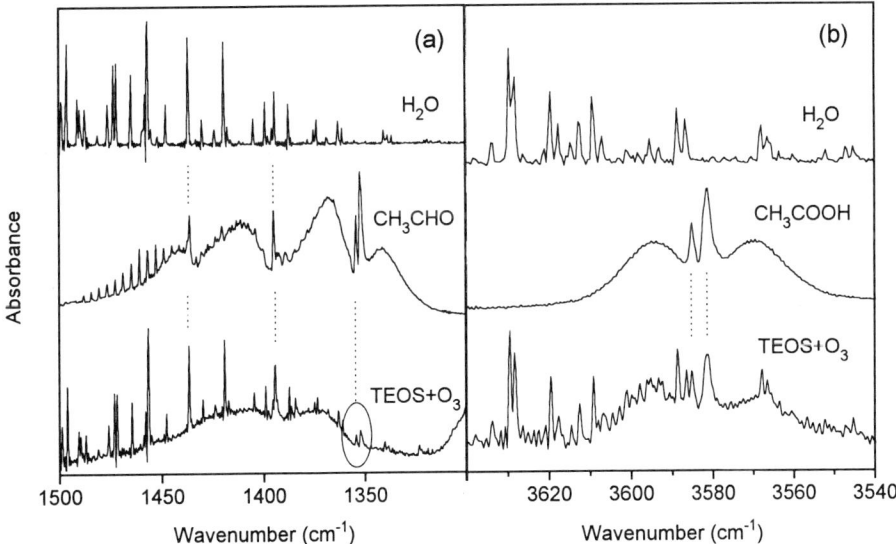

Figure 2. Portions of the high resolution FTIR spectrum of a 247°C, 520 Torr reaction mixture, initially containing 0.08 v% TEOS and 0.05 v% O_3. Spectra of water, acetaldehyde, and acetic acid vapors are overlaid for comparison.

to TEOS and indicative of some degree of spontaneous elimination. By making gaseous standards of these products, we have determined that, on average, for the conditions of Figure 2, each TEOS molecule could be eliminating 1 CH_3CHO *and* 1 CH_3COOH. This is indeed a formidable difference between TEOS and TMS; however, what is more interesting is the absence of other species. We did not detect α-hydride elimination products (C_2H_4 and CH_3CH_2OH) nor C_2H_6, an indicator of 6-center β-hydride elimination. Mechanisms of β-hydride elimination [6] involving 6-center transition states (2 ethoxy groups) require formation of C_2H_6 when acetaldehyde is eliminated. This suggests that acetaldehyde is formed by a 4-center elimination process involving only one ethoxy group and the Si-center. Furthermore, acetic acid has never been reported as an elimination product of alkoxysilanes and yet it appears to be a major contributor to TEOS decomposition under our experimental conditions. We studied the reactions of the possible α,β-hydride elimination products with O_3 in our reactor and ruled out the possibility that acetic acid was formed by their oxidation. By analogy to mechanisms [8] of ozone reactions with hydrocarbons, acetic acid might form from rearrangement of the biradical CH_3HCOO. Such a species could form only from direct oxidative cleavage of an ethoxy group, a process that has never been proposed, let alone detected.

REFERENCES

[1] K. V. Guinn and J. A. Mucha, *Mat. Res. Soc. Symp. Proc.*, **282**, 575 (1992).

[2] K. Fujino, Y. Nishimoto, N. Tokumasu and K. Maeda, *J. Electrochem. Soc.*, **137**, 2883 (1990).

[3] K. V. Guinn and J. A. Mucha, in preparation.

[4] S. W. Benson and A. E. Axworthy, *J. Chem. Phys.*, **26**, 1718 (1957); **42**, 2614 (1965).

[5] For TMS, the rate constant and its T-dependence are comparable to $O + O_3$: L. Ding and P. Marshall, *J. Phys. Chem.*, **97**, 3758 (1993).

[6] S. B. Desu, *J. Am. Ceram. Soc.*, **72**, 1615 (1989).

[7] J. C. S. Chu, J. Breslin, N. S. Wang and M. C. Lin, *Mat. Lett.*, **12**, 179 (1991).

[8] R. Atkinson and W. P. L. Carter, *Chem. Rev.*, **84**, 437 (1984).

DIETHYLDIETHOXYSILANE AS A NEW PRECURSOR FOR SiO_2 GROWTH ON SILICON

MICHAEL L. WISE, OFER SNEH, LYNNE A. OKADA, ANNE C. DILLON, AND STEVEN M. GEORGE
Department of Chemistry and Biochemistry, University of Colorado, Boulder, CO 80309

ABSTRACT

Diethyldiethoxysilane (DEDEOS) is a potential single molecular precursor for SiO_2 growth. Additionally, DEDEOS may be applicable in a deposition scheme that will allow precise atomic layer control of SiO_2 film thicknesses. Deposition of SiO_2 on silicon using DEDEOS was studied in three regimes. In the first regime, the adsorption and decomposition of DEDEOS were studied using temperature–programmed desorption (TPD) and Fourier–transform infrared spectroscopy (FTIR) under UHV conditions. TPD and FTIR results indicated that DEDEOS deposits ethyl and ethoxy species on the surface which decompose by a β–hydride elimination mechanism. In the second regime, step–wise SiO_2 growth using DEDEOS was studied by laser–induced thermal desorption (LITD) and TPD as a function of repetitive growth cycles composed of DEDEOS deposition at 300 K followed by thermal annealing to 820 K. The amount of SiO_2 growth per cycle initially decreased as a function of cycle until reaching a constant value after approximately 10 cycles. Finally, DEDEOS deposition of SiO_2 was studied in a high pressure chamber at surface temperatures between 873 K and 1003 K and at a pressure of 0.5 Torr. The deposition of SiO_2 as a function of exposure was found to display a fast initial growth step followed by a slower step which continued indefinitely. The activation energy for SiO_2 growth by DEDEOS in the slower growth step was found to be 49 ± 6 kcal/mol. This activation energy is very similar to the activation energy of 45 kcal/mol observed for SiO_2 growth by tetraethoxysilane (TEOS).

INTRODUCTION

Many factors affect the choice of the molecular precursors used for the chemical vapor deposition (CVD) of SiO_2 on silicon. Properties of the resultant dielectric layer, such as step coverage and dielectric constant, must be considered as well as process related concerns, such as deposition rate and temperature. Tetraethoxysilane (TEOS) is the most commonly used precursor for the CVD of SiO_2 on silicon substrates. Other molecular precursors have been investigated such as diethylsilane, tetramethylcyclotetrasilane and tetramethylorthosilicate[1]. These alternate precursors have shown both better and worse deposition characteristics than TEOS.

In these experiments, the deposition of SiO_2 on silicon was studied using a new liquid source, diethyldiethoxysilane (DEDEOS). DEDEOS may display very different deposition properties from TEOS because of differences in its size, stoichiometry, reactivity, decomposition products, and sticking coefficient. Most notable is the possibility that DEDEOS may sustain reactive silicon atoms during SiO_2 growth that are not terminated by oxygen. In addition, DEDEOS may allow precise control of SiO_2 deposition by cycling the adsorption of DEDEOS at low temperature with thermal annealing to remove the unwanted reaction products.

SiO_2 deposition on silicon using DEDEOS was studied in three regimes using laser–induced thermal desorption (LITD), temperature–programmed desorption (TPD), Fourier–transform infrared spectroscopy (FTIR), and Auger electron spectroscopy (AES). First, the adsorption and decomposition of DEDEOS were investigated on Si(100)2x1 and porous silicon. Second, step–wise growth of SiO_2 on Si(100)2x1 was studied by cycling DEDEOS exposure with thermal annealing. Finally, SiO_2 deposition was studied at high surface temperatures and high DEDEOS pressures.

EXPERIMENTAL

Three separate vacuum chambers were used in these experiments. The two vacuum chambers and crystal mounting designs used for the LITD, FTIR, TPD and AES experiments are identical to those used previously[2,3]. LITD, TPD and AES experiments were carried out under UHV conditions at base pressures of ~4 x 10^{-10} Torr on flat single-crystal Si(100) samples[3]. The silicon samples were obtained from Virginia Semiconductor and were p-type boron-doped with resistivities of 0.35 to 0.65 Ω cm. Cleaning of the samples between experiments was achieved by annealing to T = 1350 K for several minutes.

Transmission FTIR experiments were conducted at base pressures of 8 x 10^{-8} Torr on Si(100) samples that had been electrochemically etched in a HF/C_2H_5OH solution to increase their surface area[2]. Previous measurements indicate that the surface area is increased by a factor of ~250 using this method[2]. The high pressure thermal growth experiments were conducted in a new UHV chamber. By means of a movable cup that fits over the sample holder, the Si(100) sample could be dosed at high pressure while the remainder of the chamber remained at 10^{-10} Torr. The cup could then be removed and TPD and AES experiments could be performed in UHV conditions.

LITD was used in some of the experiments to determine the amount of DEDEOS adsorbed on the silicon sample after a given exposure. Laser-induced thermal desorption is a technique in which a focused Ruby laser is used to rapidly heat a small spatial area on the silicon sample[3]. Any adsorbates present in this spatial region are thermally desorbed and detected by a quadrupole mass spectrometer with the ionizer located approximately 5 cm in front of the sample. The ion signal for mass 27 (ethylene) was used to detect adsorbed DEDEOS with LITD. This LITD signal was found to be linear with the area under the SiO TPD peak after varying DEDEOS exposures.

Absolute calibration of the DEDEOS coverage on Si(100) was accomplished by comparing the area under the SiO TPD peak for a saturation DEDEOS exposure with the SiO TPD peak area for a saturation H_2O exposure at 300 K. The H_2O exposure saturates on Si(100)2x1 at an oxygen coverage of 0.5 ML[4,5]. DEDEOS was found to saturate at an oxygen coverage of 0.71 ML for adsorption at temperatures between 200 and 350 K.

RESULTS AND DISCUSSION

Silicon Surface Chemistry

TPD and FTIR techniques were used to characterize the adsorption and decomposition of DEDEOS on Si(100)2x1 and porous silicon. Only signals at masses 2, 26, 27, 28, and 44 were detected by TPD. Figure 1 shows the TPD spectra at a heating rate of β = 3.8 K/s for masses 2, 27, and 44 after a saturation DEDEOS exposure at 300 K. The peaks for masses 26-28 at ~750 K are from the same desorption species and scale with the electron impact fragmentation products expected for ethylene, C_2H_4. The mass 2 desorption peak occurs at 790 K and is characteristic of H_2 desorption. Desorption of mass 44 at ~970 K is indicative of SiO desorption.

DEDEOS presumably dissociates upon adsorption on silicon to produce ethyl and ethoxy species. The production of C_2H_4, surface hydrogen, and oxide species upon decomposition of these adsorbates is consistent with a β–hydride elimination mechanism. Ethyl and ethoxy groups deposited on silicon by other related precursors such as $Si(OC_2H_5)_4$ (TEOS)[6,7], C_2H_5OH[7], and $SiH_2(C_2H_5)_2$ (DES)[8] have also been shown to decompose by a β–hydride elimination mechanism at ~700 K.

Additionally, transmission FTIR results on porous silicon are consistent with DEDEOS decomposition by the proposed β–hydride elimination mechanism. After a saturation exposure of DEDEOS at 300 K, C-O, C-C, and C-H are present which are indicative of adsorbed ethyl and ethoxy species. No Si-H is initially present. Annealing the sample in 20 K increments for 1 minute each produced major changes in the spectrum. Figure 2 shows the evolution of the

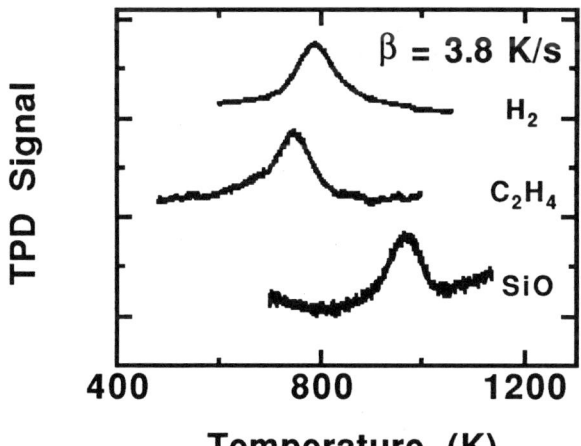

Figure 1: TPD spectra for masses 2, 27, and 44 measured after a saturation exposure of DEDEOS on Si(100)2x1 at 300 K.

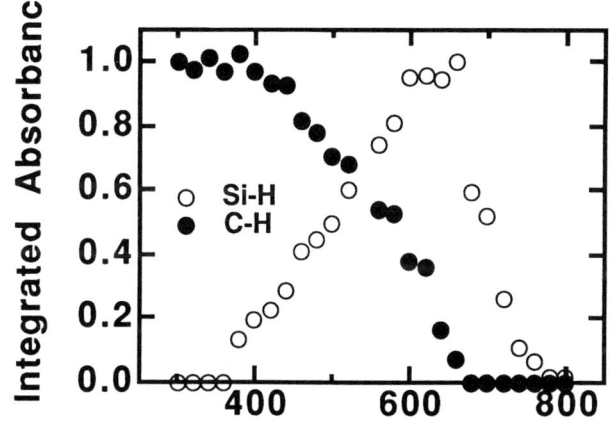

Figure 2: Integrated absorbance for the Si-H stretch (2090 cm^{-1}) and the C-H$_x$ asymmetric and symmetric stretch (~2900 cm^{-1}) as a function of annealing temperature.

integrated Si-H stretch region at 2090 cm^{-1} and C-H$_x$ symmetric and asymmetric stretch region at ~2900 cm^{-1} as a function of annealing temperature. From 400 K to 700 K, the C-H$_x$ species disappear as Si-H grows. The C-O and C-C peaks are difficult to integrate precisely because they are convoluted with other nearby absorption features. However, they qualitatively appear to decrease with the C-H$_x$ species. Additionally, a broad peak at ~1000 cm^{-1} assignable to siloxane bridge species grows as the C-H$_x$ species decompose. No O-H species were detected. At T = 700 K, the Si-H peaks begin to decrease because of H$_2$ desorption.

Step–wise Growth of SiO$_2$ on Silicon

In order to study the deposition of DEDEOS after more than one exposure, repeated cycles of deposition were performed, each consisting of a saturation DEDEOS dose at 300 K followed by annealing the sample to 820 K for 300 s. The annealing temperature of 820 K was chosen because it is high enough to desorb H$_2$ from silicon and to decompose ethyl and ethoxy groups on silicon and silica surfaces. A temperature of 820 K is also low enough to not desorb SiO. The C$_2$H$_4$ LITD signals were used to measure the amount of DEDEOS that adsorbed on each cycle. After a number of cycles, SiO TPD and AES were used to characterize how much oxide had actually been deposited.

Figure 3 shows the amount of DEDEOS that is adsorbing during each cycle for up to 20 cycles. Figure 4 shows the predicted amount of deposited oxygen based on summing the amount of DEDEOS that adsorbed per cycle and multiplying by a factor of 2 because there are two oxygen atoms per DEDEOS molecule. Additionally, Fig. 4 shows the integrated SiO TPD peaks expressed as oxygen coverage whrere one monolayer on Si(100)2x1 is 6.78 x 10^{14} cm^{-2}.

There is very good agreement between the oxygen coverage predicted by the amount of adsorbed DEDEOS per cycle and the oxygen coverage measured by SiO TPD. This indicates that there are no alternative pathways for ethyl or ethoxy loss during adsorption or oxygen loss during annealing. Additionally, Figs. 3 and 4 indicate that the amount of deposited DEDEOS per cycle monotonically decreases until leveling off after 10 cycles at 0.03 ML/cycle. This behavior

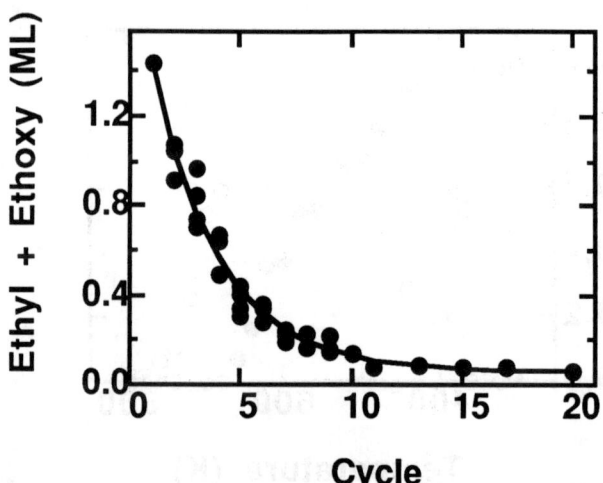

Figure 3: The amount of ethyl and ethoxy species adsorbed as a function of the number of cycles. The adsorbed ethyl and ethoxy species were measured using the C$_2$H$_4$ LITD signals.

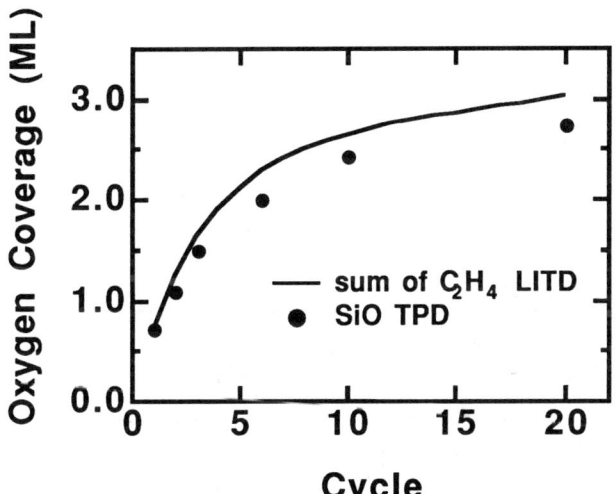

Figure 4: The amount of oxygen deposited as a function of cycle as predicted by adsorbed ethyl and ethoxy species and measured by SiO TPD area.

corresponds to the rapid deposition of ~2.4 ML of oxygen followed by much slower deposition. These observations indicate that DEDEOS may be much more reactive with silicon sites than with SiO_2–like sites.

Thermal Chemistry of SiO_2 Growth

The thermal chemistry of SiO_2 growth was examined using the UHV chamber capable of high pressure dosing. The deposited oxygen coverage was measured after exposing Si(100)2x1 to DEDEOS at 0.5 Torr with surface temperatures between 873 K and 1003 K. Figure 5 shows the results for these experiments. The oxygen coverage is measured in ML relative to Si(100)2x1.

Two regimes are evident from the uptakes displayed in Fig. 5. There is a very rapid initial step followed by a significantly slower step which extends indefinitely. Data points in the rapid uptake regime have only been shown for the data at 913 K. The fast initial uptake step may be a result of the complicated dynamics found in the interfacial region where silicon is being transformed into SiO_2. The oxygen coverage at which the transition from fast to slow steps takes place in the lower temeperature experiments appears to occur at ~4 - 5 ML. The slower growth step represents SiO_2 growth on SiO_2 surfaces.

Figure 6 shows the Arrhenius plot for linear fits to the slow growth rates of SiO_2 on SiO_2 surfaces. The activation energy was found to be 49 ± 6 kcal/mol. This activation energy is very similar to the accepted activation barrier of ~45 kcal/mol for thermal SiO_2 deposition with TEOS[9-11]. Consequently, similar rate-limiting kinetics for deposition can be inferred for these two precursors. The exact mechanisms for SiO_2 deposition by TEOS are still disputed. Mechanisms involving homogeneous and heterogeneous reactions have been proposed[9,10,12]. The similar rate-limiting kinetics for DEDEOS and TEOS suggests that the activation energy may be linked to the β-hydride elimination of ethyl and ethoxy groups to produce C_2H_4.

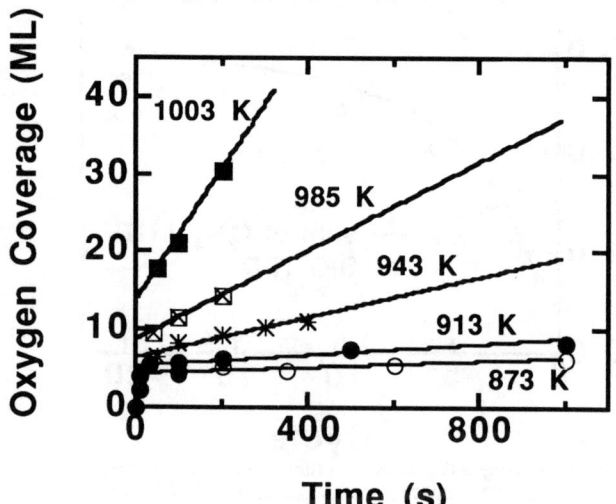

Figure 5: Oxygen coverage as a function of exposure at a DEDEOS pressure of 0.5 Torr for varying surface temperatures.

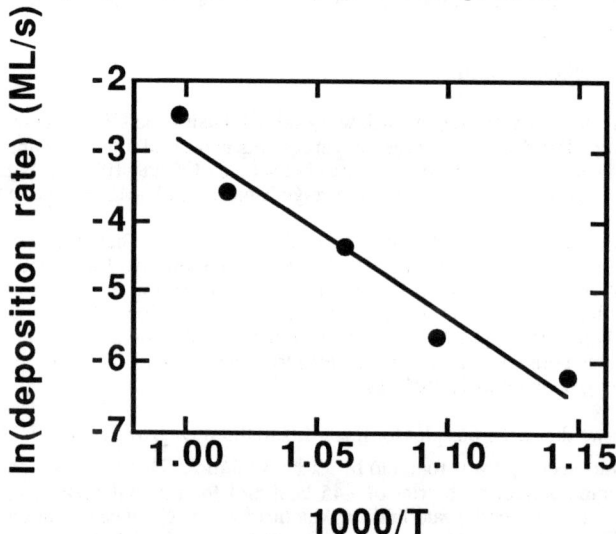

Figure 6: Arrhenius plot for the slow uptake regime of SiO_2 desposition by DEDEOS.

CONCLUSIONS

Diethyldiethoxysilane (DEDEOS) has been shown to be capable of depositing SiO_2 on silicon. Deposition of SiO_2 with DEDEOS was studied in three regimes. For adsorption directly on Si(100)2x1, DEDEOS dissociatively chemisorbs at 300 K to deposit ethyl and ethoxy groups. At higher temperatures, these groups decompose by a β-hydride elimination mechanism resulting in the deposition of silicon and oxygen on the silicon surface. For controlled deposition of ultrathin SiO_2 layers, step–wise SiO_2 deposition on silicon with DEDEOS was achieved by repeated growth cycles composed of DEDEOS adsorption at 300 K and thermal annealing. Finally, for deposition of SiO_2 thin films, SiO_2 deposition was studied at high surface temperatures and DEDEOS pressures. SiO_2 deposition with DEDEOS in this thermal growth regime displayed an activation barrier very similar to the barrier for SiO_2 deposition with TEOS.

ACKNOWLEDGMENTS

This work was supported by the Office of Naval Research under Contract Number N00014-92-J-1353. MLW gratefully acknowledges AT&T Bell Laboratories for a graduate fellowship. OS is grateful to the Israeli Academy of Sciences for a Wolfson Postdoctoral Fellowship. SMG acknowledges the National Science Foundation for a Presidential Young Investigator Award (1988-1993).

REFERENCES

1. P. Singer, Semiconductor Int. **June 1993**, 92 (1993).
2. P. Gupta, V.L. Colvin and S.M. George, Phys. Rev. **37**, 8234 (1988).
3. B.G. Koehler, C.H. Mak, D.A. Arthur, P.A. Coon and S.M. George, J. Chem. Phys. **89**, 1709 (1988).
4. H. Ibach, H. Wagner and D. Bruchmann, Solid State Commun. **42**, 457 (1982).
5. W. Ranke and Y.R. Xing, Surf. Sci. **157**, 339 (1985).
6. J.B. Danner, M.A. Rueter and J.M. Vohs, Langmuir **9**, 455 (1993).
7. J.E. Crowell, L.L. Tedder, H.-C. Cho and F.M. Cascarano, J. Vac. Sci. Technol. **A8**, 1864 (1990).
8. P.A. Coon, M.L. Wise, A.C. Dillon, M.B. Robinson and S.M. George, J. Vac. Sci. Technol. **B10**, 221 (1992).
9. S.B. Desu, J. Am. Ceram. Soc. **72**, 1615 (1989).
10. A.C. Adams and C.D. Capio, J. Electrochem. Soc. **126**, 1042 (1979).
11. F.S. Becker, D. Pawlik, H. Anzinger and A. Spitzer, J. Vac. Sci. Technol. **B5**, 1555 (1987).
12. G.B. Raupp, F.A. Shemansky and T.S. Cale, J. Vac. Sci. Technol. **B 10**, 2422 (1992).

DECOMPOSITION STUDY OF TEOS IN THERMAL CVD

Hideki Takeuchi, Hirohiko Izumi and Atsushi Kawasaki
Nippon Steel Corp., Semiconductor Division, Fuchinobe 5-10-1, Sagamihara, Kanagawa 229, Japan

ABSTRACT

Thermal decomposition of tetraethoxysilane(TEOS) has been investigated in a vertical type LP-CVD reactor. The mass spectrum of reacting gas was found to have a feature corresponding to hexaethoxysiloxane (($H_5C_2O)_3SiOSi(OC_2H_5)_3$). Trapped byproducts were found to include ethanol and ethoxy-based silica. Kinetic analysis of deposition rate profile indicated the existence of two intermediates. Feed gas residence time, as well as substrate temperature, directly affected the film quality.

A vapor phase reaction model was proposed to explain these results. According to this model, TEOS decomposes into ethylene and triethoxysilanol. The latter reacts with TEOS to form hexaethoxysiloxane as observed by mass spectrometry. The triethoxysilanol, hexaethoxysiloxane and TEOS should be actual deposition precursors. Similar reaction continues to form ethoxy-based siloxane polymers which are trapped from the exhaust gas.

INTRODUCTION

Thermal CVD of TEOS(tetraethoxysilane) is widely used for silicon dioxide thin film preparation in integrated circuit fabrication, such as for sidewall spacers of LDD transistors and for inter conduction layer dielectric films. Furthermore we recently proposed a new process concerning all CVD ONO film for inter control and floating gate dielectric for flash memories[1]. This process can improve the reliability of flash memories in the sense that the lowered process temperature can reduce the heat damage to the tunneling oxide and that improved control of the top oxide thickness can improve the leakage current of the total ONO film. Because this process requires severe constraints on electric properties for ultra thin silicon dioxide films of several nanometers thickness, understanding the reaction mechanism is very important to improve the film quality.

On decomposition of TEOS, Desu and Kalidindi assumed that diethoxysilannone (($C_2H_5O)_2Si=O$) might be a vapor phase intermediate[2] and calculated the deposition rate profile utilizing finite element methods[3,4] or bond graph methods[5]. Satake et al.[6] detected ethylene and ethanol by GC/MS measurement and suggested that the vapor phase intermediate might be triethoxysilanol formed by ethylene desorption from ethoxy base of TEOS. Egashira et al.[7] also supported this model in explaining the pressure dependence of step coverage.

In this study we have concluded that TEOS, triethoxysilanol and hexaethoxysiloxane should be actual deposition precursors and that their compositions affect the film quality.

EXPERIMENTAL PROCEDURES

The reactor used for experiments was a vertical batch type for actual 6" wafer fabrication, which means that our experimental results are identical to the actual CVD process in LSI production. By using this reactor we have examined the mass spectrum of the reacting gas, the composition of trapped byproducts, the deposition rate profile and the distribution of film quality.

Mass spectrum of reacting gas.— The reaction gas was directly sampled into the mass spectrometer(UTI;ISS-325) by introducing a quartz tube into the reactor. Raw TEOS was also directly sampled through the bypass line. The electron energy was 70keV and detectable mass range was limited within 300(M/e).

Byproduct analyses.— The atomic composition of byproducts, which were trapped at the outlet pipe from the exhaust gas, was examined. Their composition was analyzed by IR spectroscopy for solid components and by ^{1}H-NMR for liquid components.

Deposition rate profile.— The film thickness was measured in the center of wafers using an optical interferometer. The reactor temperature was maintained isothermal at 675°C. The temperature difference in the bottom region was maintained within 5°C. Reactor pressure at the inlet was maintained at 1.0 torr. TEOS flow rates were varied from 75 to 120 cc/min.

Distribution of film quality.—The film density was calculated from the film thickness and weight change before and after HF etching. The film deposited wafers were cut into 50 cm^2 with a dicing saw. The film stress was evaluated by the distortion of wafers after the film deposition using a laser detection method. The film contraction ratio was calculated from the film thickness change after the heat treatment, whose condition was 900°C in N_2 ambient for 30 min.

IN-SITU MEASUREMENT OF REACTING GAS

The mass spectra of raw TEOS and the reacting gas are shown in Fig. 1(a) and (b). Although they are rather complicated, it is apparent that there exist species that have higher mass than the mass number 208 of TEOS. Specific peaks in the mass spectrum of the reacting gas are summarized in Table 1. They can be well explained as the fragments of several methyl, ethyl or ethoxy bases cracked from hexaethoxysiloxane $((H_5C_2O)_3SiOSi(OC_2H_5)_3; M=342)$. Hexaethoxysiloxane can be regarded as the dimer of raw TEOS. Unfortunately, our measurement was limited within 300 in mass range. Although higher polymers or origommers will produce the same spectrum in the observed mass range, we have concluded that no higher polymers were contained in the reacting gas as deposition precursors from the later kinetic analysis of the deposition rate profile.

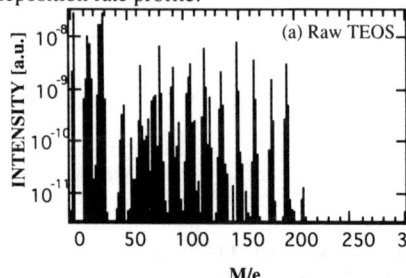 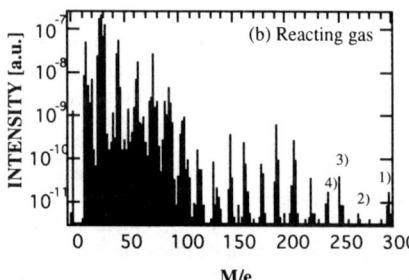

Fig.1 Mass spectrum of (a) raw TEOS and (b) reacting gas.

Table 1 Specific peaks in the reacting gas mass spectrum.

	mass number (difference from 342)	possible structure	bases cracked from hexaethoxysiloxane $(Si_2C_{12}H_{30}O_8:M=342)$
1)	298(44)	$Si_2C_9H_{22}O_8$	CH_3, C_2H_5
2)	253(89)	$Si_2C_5H_{13}O_8$	$4CH_3, C_2H_5$
		$Si_2C_9H_{17}O_7$	CH_3, C_2H_5, C_2H_5O
3)	240(102)	$Si_2C_5H_{12}O_8$	$CH_3, 3C_2H_5$
4)	226(116)	$Si_2C_4H_9O_8$	$4C_2H_5$

BYPRODUCT ANALYSES

The atomic composition of the trapped byproducts was found to be Si(32.9wt%), C(15.2wt%), H(3.59wt%) and O(46.2wt%). The IR spectrum in Fig.2 clearly shows that solid components were comprised of amorphous silica including ethoxy bases. ^{1}H-NMR results, as shown in Fig.3, show that the liquid component was ethanol and did not contained raw TEOS. Ethanol seems to be produced by the decomposition of the ethoxy base of solid components.

Ethoxy-based silica can be regarded as a polymer of TEOS. The existence of dimer and polymer implies a vapor phase polymerizaition process.

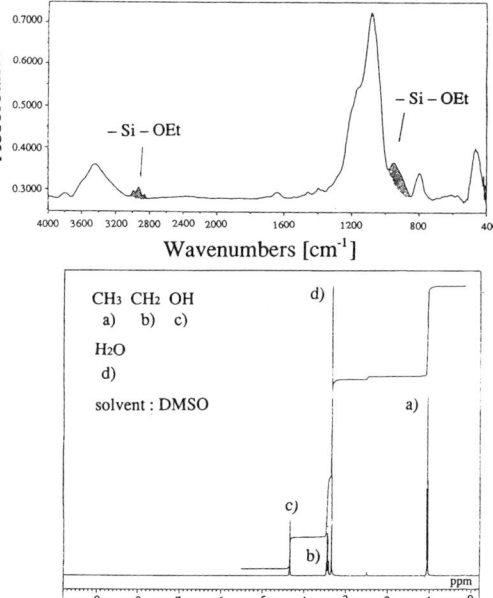

Fig.2 IR spectrum of byproducts.

Fig.3 ^{1}H-NMR spectrum of byproducts.

REACTION MODEL

Figure 4 illustrates the proposed reaction model: TEOS decomposes into silanol and ethylene. Silanol is well known for its high reactivity, and it easily reacts with ethoxy base of TEOS to produce hexaethoxysiloxane(TEOS dimer) as observed in the mass spectrum measurement. The dimer also decomposes into silanol and ethylene. Similar decompositions and condensations also produce polymers.

This model explains not only the detection of ethylene [6], ethanol [6], TEOS dimer and siloxane polymer but also the pressure dependence of step coverage [7] . Step coverage decreases in the low pressure region but increases in the high pressure region. This phenomenon can be explained through the pressure dependence of the concentration of silanol which causes a drop of step coverage for high reactivity and high sticking coefficient. The concentration of silanol is almost proportional to the increase of the total vapor phase reaction amount in the low pressure region, but it is decreased by the consumption for dimer formation in the high pressure region.

Actual deposition precursors can be shown to be TEOS, triethoxysilanol and hexaethoxysiloxane from the following kinetic analysis and distribution of film quality.

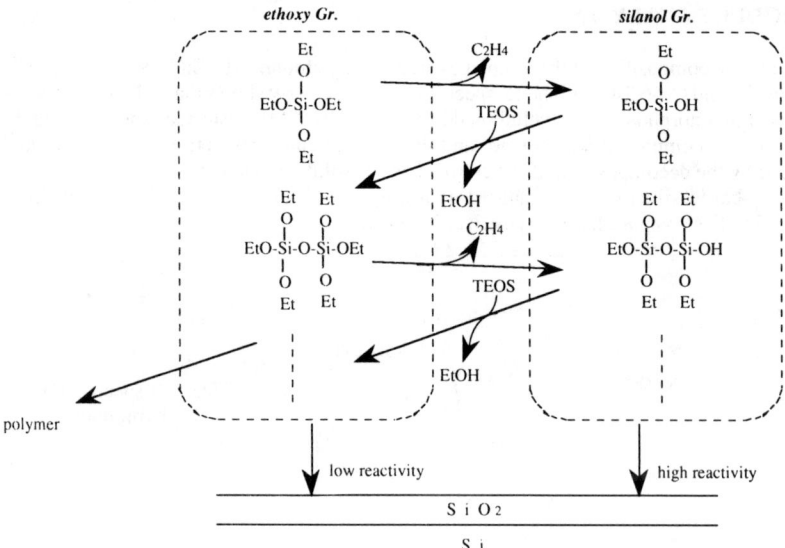

Fig.4 Reaction model.

DEPOSITION RATE PROFILE

Figure 5 shows the deposition rate profile for varying the feed flow rates as a function of feed gas residence time. The rapid increase of deposition rate in the small residence time region is not due to a temperature effect, because the actual temperature difference was within 5°C, whereas there must be more than 50°C difference based on calculation from the activation energy of deposition rate in the case that the deposition precursor is only TEOS. Therefore, the existence of peak in the profile must be due to the time lag for the accumulation of intermediates.

This profile was well matched with a sum of three exponential functions of the feed gas residence time. The solid line in Fig.5 represents the calculated result. The function can be deduced by the proposed reaction model. Equations (1)-(3) represent the possible reactions, where T_1, S_1, and T_2 represent TEOS, triethoxysilanol and hexaethoxysiloxane, respectively.

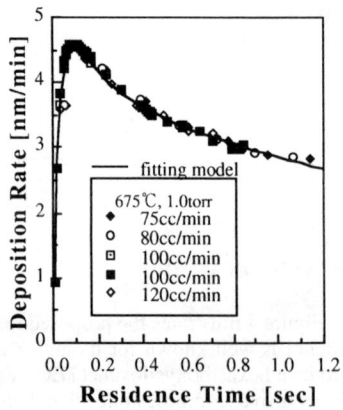

Fig.5 Deposition rate profile and result of calculations.

$$T_1 \xrightarrow{K_1} S_1 + C_2H_4 \quad \cdots\cdots (1)$$

$$T_1 + S_1 \xrightarrow{K_2} T_2 + EtOH \quad \cdots\cdots (2)$$

$$T_1, S_1, T_2 \xrightarrow{J_1, J_2, J_3} SiO_2 \quad \cdots\cdots (3)$$

Equations (4)-(7) represent the simplified one-dimensional kinetic model, where T_1, S_1, and T_2 represent the concentration of TEOS, triethoxysilanol and hexaethoxysiloxane, respectively.

$$\frac{\partial T_1}{\partial \tau} = -K_1 T_1 - K_2 T_1 S_1 - J_1 T_1 \approx -(K_1+J_1)T_1 - K_2' S_1 \quad \cdots\cdots (4)$$

$$\frac{\partial S_1}{\partial \tau} = +K_1 T_1 - K_2 T_1 S_1 - J_2 S_1 \approx +K_1 T_1 -(K_2'+J_1)S_1 \quad \cdots\cdots (5)$$

$$\frac{\partial T_2}{\partial \tau} = +K_2 T_1 S_1 - J_3 T_2 \approx K_2' S_1 - J_3 T_2 \quad \cdots\cdots (6)$$

$$DR(\tau) = J_1 T_1(\tau) + J_2 S_1(\tau) + J_3 T_2(\tau) \quad \cdots\cdots (7)$$

Because the rate of decrease of TEOS is negligible compared with the rate of increase of triethoxysilanol, T_1 can be regarded constant compared with S_1. Therefore, $K_2 T_1 S_1$ can be represented as $K_2' S_1$, which is a function of S_1 only. According to this assumption, equations (4)-(6) can be regarded as linear differential equations composed of three variables, whose solutions consist of three exponential functions. Therefore, the deposition rate profile should be matched with a sum of three exponential functions, as represented in equation (8).

$$DR(\tau) = A_1 \exp(-k_1\tau) + A_2 \exp(-k_2\tau) + A_3 \exp(-k_3\tau) \quad \cdots\cdots (8)$$

Two- or three- dimensional models will also give similar results involving three exponential functions. Similar results are also obtained when one or two deposition precursors are assumed. However, the following experiment showed that all three species are deposition precursors.

DEPENDENCE OF FILM QUALITY ON RESIDENCE TIME

Figure 6 (a),(b) and (c) show the dependence on feed gas residence time of the film density, the film stress, and the film contraction ratio after heat treatment. These results showed that film quality depends not only on temperature but also on the feed gas residence time. The films deposited with long residence time have low density, low stress and high film contraction ratio.

With a single deposition precursor, films formed with a higher deposition rate tend to have low density because of the shorter relaxation time of the surface reaction. However, our experimental result showed that denser films were formed with a higher deposition rate. The observed distribution in film quality must be due to the change of the composition of the deposition precursors as the vapor phase reaction proceeds.

According to the proposed reaction model and the kinetic analysis, the actual deposition precursors are three species: probably TEOS, triethoxysilanol and hexaethoxysiloxane. TEOS and hexaethoxysiloxane have similar low reactivity and triethoxysilanol has high reactivity. Dependence of film density on residence time may be due to the difference of the amount of hexaethoxysiloxanes formed by the vapor phase reaction. Difference between TEOS and hexaethoxysiloxane as a deposition precursor lies in the relaxation time for ethoxy base decomposition which depend on their structure. Dependence on stress and contraction ratio can be explained in the same way as the film density, because dense films have high stress and low contraction ratio.

These results imply that improved control of the vapor phase reaction will result in the improvement of the film quality.

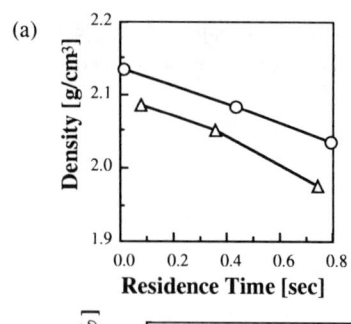
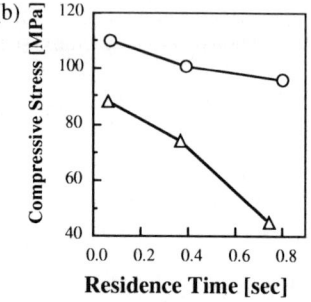
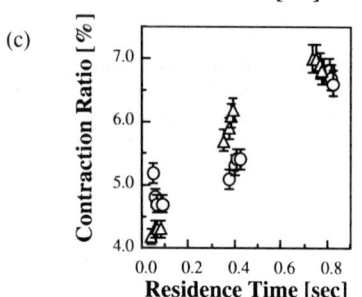

Fig6. The effect of feed gas residence time on
(a) the film density
(b) the film stress
(c) the film contraction ratio with heat treatment.

○, △ keys represent the data for 675°C and 740°C, respectively

CONCLUSION

The results obtained by this work are as follows:
(1) The mass spectrum of the reacting gas shows the features of hexaethoxysiloxane.
(2) The trapped byproducts are found to include ethoxy-based silica which can be regarded as a polymer of TEOS.
(3) The deposition rate profile can be fitted by a sum of three exponential functions, indicating the existence of two intermediate species.
(4) The feed gas residence time, as well as the substrate temperature, directly affects the film density, stress and contraction ratio with heat treatment, implying that the several deposition precursors change their composition as the reaction proceeds.

The experimental results reported here suggest that the actual deposition precursors are probably TEOS, hexaethoxysiloxane and triethoxysilanol, and that vapor phase polymerization is caused by the decomposition of ethoxy base to silanol base and subsequent condensation of silanol base with ethoxy base.

REFERENCES

1. K. Hazama, Y. Tomioka, Y. Sato, S. Iwasa, K. Anzai and T. Wada, Extended Abstracts, 30p-ZV-6(The 40th Spring Meeting,1993); The Japan Society of Applied Physics and Related Societies.
2. S.B. Desu, J. Am. Ceram. Soc., **72**, 1615(1989).
3. S.R. Kalidini and S.B. Desu, J. Electrochem. Soc., **137**,624(1990).
4. S.B. Desu and S.R. Kalidini, Jpn. J. Appl. Phys., **29**,1310(1990).
5. S.R. Kalidini and S.B. Desu, J. Electrochem. Soc., **138**,962(1991).
6. T. Satake, T. Sorita, H. Fujioka, H. Adachi and H. Nakajima, Extended Abstracts, 16p-ZQ-2 (The 53th Autumn Meeting,1992);The Japan Society of Applied Physics.
7. Y. Egashira and H. Komiyama (private communication).

DIRECT MEASUREMENT OF THE REACTIVITY OF NH AND OH ON A SILICON NITRIDE SURFACE

R. J. Buss[*], P. Ho[*], E. R. Fisher[**], and William G. Breiland[*]
[*]Sandia National Laboratories, Albuquerque, New Mexico, 87185-0367
[**]Dept. of Chemistry, Colorado State University, Ft. Collins, CO 80523

ABSTRACT

In order to understand and successfully model the plasma processing used in device fabrication, it is important to determine the role played by plasma-generated radicals. We have used the IRIS technique (Imaging of Radicals Interacting with Surfaces) to obtain the reactivity of $NH(X^3\Sigma^-)$ and $OH(X^2\Pi)$ at a silicon nitride film surface while the film is exposed to a plasma-type environment. The reactivity of NH was found to be zero both during exposure of the surface to an NH_3 plasma and during active deposition of silicon nitride from a SiH_4/NH_3 plasma. No NH surface reaction was detectable for any rotational states of NH and over a surface temperature range of 300-700 K. OH radicals generated in an H_2O plasma were found to have a reactivity of 0.57 on a room temperature oxidized silicon nitride surface. The OH reactivity falls to zero as the temperature of the substrate is raised.

INTRODUCTION

The manufacturing processes used in the fabrication of microelectronic devices, such as chemical vapor deposition and plasma processing, often rely on very complex chemistries. With the continuing need to improve the performance of the processes, there is increased interest in understanding the chemistry so that accurate models can be developed. A detailed model of even a relatively simple plasma deposition such as a-Si:H from SiH_4[1] requires a great body of gas phase reactions, ion reactions, and gas-surface reactions. Chemists have developed a fairly good understanding of gas phase chemistry which allows the modeller to make reasonable guesses for reaction mechanisms and rates. In contrast, much less is known concerning the interaction of molecules at the surfaces of importance in microelectronics. The bulk of the knowledge about surface reactions concerns adsorption of stable molecules on metal surfaces. Of much greater concern to the plasma modeller is the reactivity of free radicals at insulator surfaces.

In 1989, we reported[2] results using a new experimental technique which allows direct measurement of the reactivity of radicals at a surface which is exposed to a plasma-like environment. Imaging of Radicals Interacting with Surfaces, IRIS involves spatial imaging of laser-induced-fluorescence (LIF) from a plasma-generated molecular beam incident on a substrate in a high vacuum. During the experiment, the substrate is bombarded by the full range of plasma species, so it can be argued that the measured reactivity of a radical corresponds closely to radical behavior in the plasma itself.

The IRIS technique has been discussed in detail previously.[2-5] Figure 1 shows the experimental configuration both as a planar view through the apparatus (upper) and as a 3-D view showing the spatial relationships of the components (lower). The method uses spatially resolved LIF to detect simultaneously molecules in a collimated, near-effusive molecular beam and molecules coming from a surface. The selectivity of the LIF is used to monitor one species in a molecular beam made from a rf plasma. Comparison of the LIF signals observed with the surface in and out of the path of the molecular beam yields the fraction of the incident molecules (of the species being detected) that react at the surface.

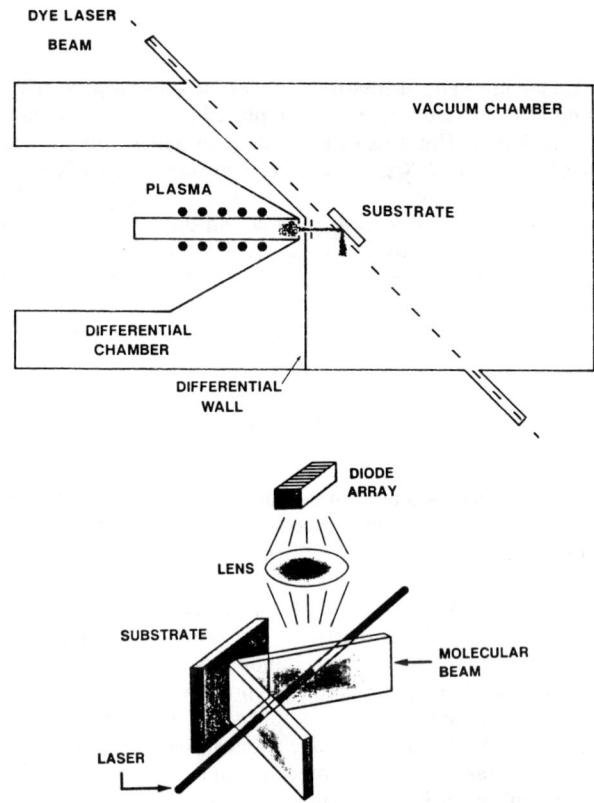

Figure 1. Two schematics of the IRIS experimental arrangement. Above is a top view of the apparatus. Below is a 3-D depiction of the spatial relationships of the components.

We report, here, measurements of the reactivity of $NH(X^3\Sigma^-)$ and $OH(X^2\Pi)$ at a silicon nitride film using IRIS. For the NH studies, either a pure ammonia or an NH_3/SiH_4 plasma is used as the NH source. In the OH reactivity studies, a pure water plasma acts as the OH source. The samples used for these experiments are 1 x 2 cm pieces of silicon with 158 Å of thermally grown nitride. The substrate is placed parallel to the laser beam at a 43.3° angle relative to the molecular beam. For complete experimental details see ref 4 (NH) and ref 5 (OH).

RESULTS AND DISCUSSION

NH

The NH radical is completely unreactive at the silicon nitride surface under all experimental conditions. Under steady state bombardment of the substrate by the NH_3 plasma-beam, 100% (~ >95%) of the NH flux to the surface is observed to rebound without reaction. The IRIS experiment yields the angular distribution of the desorbing radical. For all NH experiments, the desorbing NH has a cosine distribution, consistent with complete energy accomodation with the surface.

IRIS measurements at substrate temperatures ranging from 300-668K result in no increase in reactivity. The angular distribution of desorbing NH remains cosine over the entire temperature range. The LIF of NH is rotational state resolved and experiments with each of the $J = 1, 3$ and 5 states gave identical results (after correction for population changes resulting from thermal accomodation with the surface).

The pure ammonia plasma is neither an etching nor depositing process environment. There is continuous bombardment of the surface by ions and neutrals (eg. H, NH, NH_2, and NH_3) and some transient initial chemistry transforms the surface termination to a steady-state condition. It is plausible that the surface is predominantly terminated by NH_x groups although unpaired electrons and physisorbed molecules are also expected. Although the chemistry of NH at surfaces is unknown, it is energetically possible for NH to insert into an NH bond ($\Delta H \sim -45$ kcal/mol) and very favorable to add to an unpaired electron (eg. NH + SiH_3 → $HNSiH_3$, $\Delta H \sim -83$ kcal/mol). Other possible reaction paths include an NH molecule physisorbing and migrating to a reactive site or remaining adsorbed until a gas phase radical collides with it. We estimate, on the basis of the plasma pressure (30 mTorr) and the distance to the substrate (44 mm), that the collision frequency at a site on the surface is about 150/sec, thus an exceedingly long residence time would be necessary for reaction with incoming molecules. Our observation that essentially all the incident NH desorbs without reaction shows that none of the suggested pathways have significant rates.

When silane is added to the plasma and a film of a-SiN_x:H is being deposited, the steady state surface coverage is expected to include SiH_x groups. Insertion of NH into an X_3SiH bond to give X_3SiNH_2 is energetically very favorable ($\Delta H \sim -110$ kcal/mol). Again we find that NH is unreactive with this surface, indicating that either the SiH_x groups are in very low steady-state concentrations or the insertion reaction has a very low cross section regardless of energetics. The insertion reaction for $NH(X^3\Sigma^-)$ probably has a potential energy barrier because of the triplet electronic configuration.

OH

In contrast to the NH results, OH radicals react to a significant extent on the room temperature substrate. The OH radicals present in the molecular beam generated from a pure H_2O plasma have a reactivity of 0.57 on the silicon nitride surface. The desorbing OH radicals (43% of the incident flux) have a cosine angular distribution suggesting complete thermal accomodation with the surface.

On heating the substrate, the reactivity decreases markedly such that by 475 K, essentially all of the OH desorbs unreacted. The reactivity remains zero at higher temperatures (to 723 K). The temperature dependence experiments are performed for each of six rotational states (0.5 to 5.5). The changes in intensity of scattered signal with rotational state agree well with a model of complete energy accomodation between adsorbed OH and the surface. That is, OH arrives at the surface with a rotational temperature characteristic of the plasma (340 K) and some fraction desorbs at the surface temperature. Figure 2 shows the temperature dependence of the OH reactivity graphed in Arrhenius form.

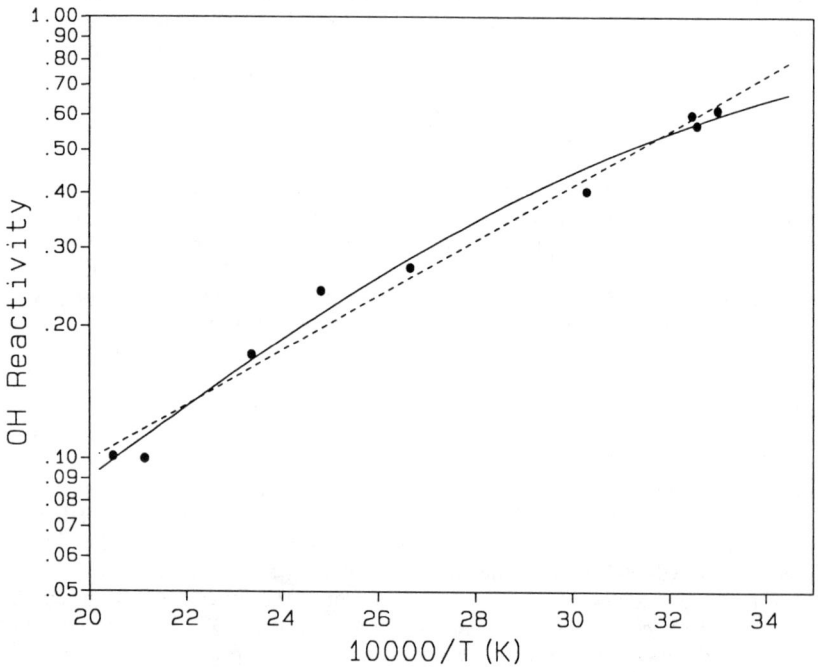

Figure 2. Reactivity of OH graphed vs inverse temperature. The dashed line is a best-fit straight line (E_a = -2.9 kcal/mole). The solid line is a best fit to the data using a model[5] which has a competition between desorption and reaction, both of which are thermally activated.

The pure H_2O plasma, similar to the NH_3 case, neither etches nor deposits a film on the silicon nitride. Auger and XPS measurements of the silicon nitride after exposure to the plasma indicate that the surface has been converted to the more thermodynamically stable oxide[6]. The surface reactions between silicon oxide and water have been the subject of many studies[7]. It is generally believed that, in vacuum at room temperature, a silica surface exposed to water is predominantly terminated by silanol groups. During continuous exposure of the sample to the plasma the surface is apparently converted to an oxide, and, at steady state, the surface presumably has a high surface coverage of silanol groups.

A simple model to explain the temperature dependence of the reactivity of OH is a competition between the desorption of physisorbed OH and its loss through reaction on the substrate[5]. The most probable reaction is the abstraction of hydrogen from a silanol group to give water. The silanol group is later regenerated by reaction with other plasma species (eg. H atoms). The net effect of this model is a temperature-dependent consumption of OH by reaction at the surface, but no net deposition or etching. At room temperature, the adsorbed OH remains on the surface long enough for 57% to find silanol groups and react, whereas, at high temperatures, the residence time becomes shorter and desorption predominates.

In conclusion, we have found that NH is completely unreactive at a silicon nitride surface exposed to NH_3 and SiH_4/NH_3 plasmas. The reactivity of OH on a surface exposed to a water plasma is high, but the reaction does not involve deposition or etching. It is clear that more work is needed to develop general principles governing the reaction of radicals at surfaces.

ACKNOWLEDGMENTS

This work was performed at Sandia National Laboratories and was supported by the U.S. Department of Energy under contract No. DE-AC04-94AL85000, partially from the Office of Basic Energy Sciences, and partially from the Office of Defense Programs through a Technology Transfer Initiative joint program with Texas Instruments. We also thank Pamela Ward for her technical assistance.

REFERENCES

(1) M. J. Kushner, J. Appl. Phys. **63**, 2532 (1988).

(2) P. Ho, W. G. Breiland, and R. J. Buss, J. Chem. Phys. **91**, 2627 (1989).

(3) R. J. Buss, P. Ho, and M. E. Weber, Plasma Chem. Plasma Process. **13**, 61 (1993).

(4) E. R. Fisher, P. Ho, W. G. Breiland, and R. J. Buss, J. Phys. Chem. **96**, 9855 (1992).

(5) E. R. Fisher, P. Ho, W. G. Breiland, and R. J. Buss, J. Phys. Chem. **97**, 10287 (1993).

(6) E. G. Rochow, <u>Chemistry of the Silicones,</u> (John Wiley and Sons: London, 1951).

(7) R. K. Iler, <u>The Chemistry of Silica,</u> (John Wiley and Sons: New York, 1979).

Surface Chemistry of Boron-Doped SiO_2 CVD: Enhanced Uptake of Tetraethyl Orthosilicate by Hydroxyl Groups Bonded to Boron

Michael E. Bartram and Harry K. Moffat
Sandia National Laboratory, Albuquerque, NM 87185-5800

Abstract

Insight into how dopants can enhance deposition rates has been obtained by comparing the reactivities of tetraethyl orthosilicate (TEOS, $Si(OCH_2CH_3)_4$) with silanol and boranol groups on SiO_2. This comparison has direct relevance for boron-doped SiO_2 film growth from TEOS and trimethyl borate (TMB, $B(OCH_3)_3$) sources since boranols and silanols are expected to be present on the surface during the thermal chemical vapor deposition (CVD) process. A silica substrate having coadsorbed deuterated silanols (SiOD) and boranols (BOD) was reacted with TEOS in a cold-wall reactor in the mTorr pressure regime at 1000K. The reactions were followed with Fourier transform infrared spectroscopy. The use of deuterated hydroxyls allowed the consumption of hydroxyls by TEOS chemisorption to be distinguished from the concurrent formation of SiOH and BOH that results from TEOS decomposition at this temperature. It was found that TEOS reacts with BOD at twice the rate observed for SiOD, given equivalent concentrations of BOD and SiOD. This demonstrates that hydroxyl groups bonded to boron increase the rate of TEOS chemisorption. In contrast, additional results show that surface ethoxy groups produced by the chemisorption of TEOS decompose at a slower rate in the presence of TMB decomposition products. Possible dependencies on reactor geometries and other deposition conditions may determine which of these two competing effects will control deposition rates. This has significant implications for microelectronics fabrication since the specific dependencies would be expected to affect process reliability. In addition, this may explain (in part) why the rate enhancement effect is not always observed in boron-doped SiO_2 CVD processes.

Introduction

The addition of trimethyl borate (TMB, $B(OCH_3)_3$) to a thermal tetraethyl orthosilicate (TEOS, $Si(OCH_2CH_3)_4$) chemical vapor deposition (CVD) process produces boron-doped oxide films.[1] Boron-doped films require a lower temperature than undoped SiO_2 films to achieve reflow in the surface planarization step during very-large-scale-integrated (VLSI) device fabrication.[1] Interestingly, the addition of dopant precursors has been observed to enhance the rate for SiO_2 deposition.[2,3] However, this effect is not always observed.[4] Surprisingly, there have not been any studies designed specifically to determine if there is a chemical basis for the rate enhancement effect.

It has been shown that surface boranol groups (BOH) react at a significantly higher rate than silanols when exposed to trimethoxymethylsilane, a TEOS homologue.[5] In view of this reactivity, boranol groups may be anticipated to influence the thermal TEOS deposition chemistry on SiO_2. This is a relevant consideration since silanols (SiOH) and boranols are expected to be present on the surface during boron-doped SiO_2 CVD. For example, silanol groups are known to be consumed and then reformed during the chemisorption and decomposition of TEOS.[6,7] In addition, boranols as well as silanols can be formed from the decomposition of TMB on SiO_2.[8] Boranols can also be produced either by the hydrolysis of TMB chemisorption products[9] (water is produced from TEOS decomposition[10]), or by a more direct reaction between the reaction products of TMB and TEOS chemisorption.[10]

This study is designed to compare the reactivity of TEOS with surface silanol and boranol groups. Reactions were carried out in a cold-wall reactor in the mTorr pressure regime at 1000K using a high surface area silica substrate. The surface contained a mixture of both silanol and boranol groups, formed by the hydrolysis of the products of an initial TMB chemisorption step. The reactions were monitored by moving the substrate into adjoining chambers for Fourier transform infrared spectroscopy (FTIR) and x-ray photoelectron spectroscopy (XPS) analyses.

Experimental

Reactions and surface analyses were carried out in a vertically oriented three-chamber system.[11] The ultra-high vacuum (UHV) surface analysis chamber is located at the top of this system, the reactor is in the middle, and the IR cell is at the bottom. Reactions were carried out in a static mode (flow rate of zero) in the mTorr pressure regime. Changes on the surface were followed by interrupting the reaction, allowing the substrate to cool down, and then moving the substrate into either the UHV chamber for XPS or the IR cell for FTIR. A new background spectrum was acquired before each IR transmission spectrum. The substrate was then moved back into the reactor, and the reaction was resumed. 5mg of high surface area silica, supported on a tungsten screen for resistive heating, was used for the SiO_2 substrates. Cleaning and annealing was accomplished by heating at 900K for 30 minutes in 50 mTorr of O_2.[12] The methods used for preparing the precursors and verifying their purity have been reported.[11]

Results and Discussion

Boranol Formation and the Reactivity of TMB on Silica at 300K

The vibrational spectra associated with the initial step for boranol formation on the silica substrate is shown in fig. 1. Annealing the D_2O treated surface at 1200K prior to TMB exposure forms isolated Si-OD groups (2763 cm^{-1} [7]) and two-membered siloxane (($Si-O)_2$) rings (888 and 908 cm^{-1}).[12] Reaction with 10 mTorr TMB at 300K for 6 sec. results in the formation of methoxy groups (labeled "Ome" in the figure). The vibrational features indicate that methoxy groups are bonded to boron $(B(OCH_3)_n)$ as well as silicon $(Si(OCH_3)_m)$ on the surface.[8] This reaction diminishes the Si-OD concentration by less than 5% while consuming 95% of the $(Si-O)_2$ rings. This suggests that $(Si-O)_2$ rings are ~20 times more reactive than silanols with respect to

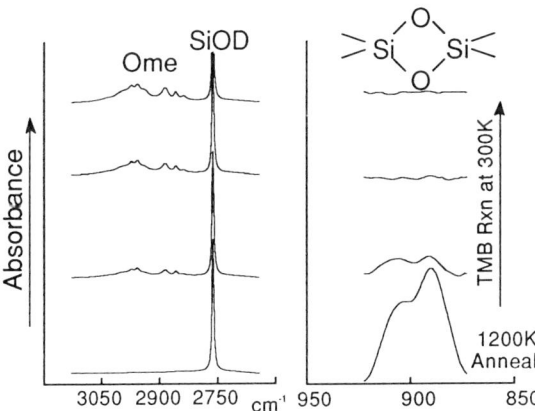

Figure 1. Vibrational spectra for the reaction of TMB at 300K on silica pre annealed at 1200K. Surface methoxy groups are marked as "Ome".

TMB reactions. Further TMB reaction for 10 sec. removes all evidence of the $(Si-O)_2$ rings. Although not shown in the figure, TMB reaction at 100 mTorr for 300 sec. decreases the Si-OD concentration to half of the initial concentration.[9] It has also been shown that subsequent reaction with H_2O at 900K regenerates some of the silanols and produces boranols at the expense of nearly all of the methoxy groups.[9]

This study is designed to compare the relative reactivities of silanols and boranols with TEOS. For this comparison to be valid, it is critical for the relative populations of surface silanols and boranols to be uniform throughout the porous substrate. Achievement of this requirement will be determined by the reactivity of TMB, since TMB is used for the initial boranol preparation step. For example, if the TMB reaction rate with silanols was greater than the TMB diffusion rate, the reactions would be diffusion-rate limited. This would result in undesirable concentration gradients of boranols from the outer to the inner surfaces of the porous substrate. In this regard, the relative reactivities of $(Si-O)_2$ rings and silanols with TMB is particularly significant. Since $(Si-O)_2$ rings are uniformly distributed throughout the porous substrate, the fact that the greater reactivity of $(Si-O)_2$ rings can be observed is clear evidence that TMB reaction is not diffusion-rate limited. This suggests that *TMB exposure will produce a uniform distribution of boron methoxy species throughout the silica substrate* for subsequent hydrolysis to form boranols.

Reactivity of TEOS with Deuterated Silanols and Boranols at 1000 K

The chemisorption of TEOS at 450K followed by heating to progressively higher temperatures has shown that TEOS first consumes hydroxyls and then reforms hydroxyls as a result of the decomposition of ethoxy groups on the surface.[6] In order to measure the rate of hydroxyl consumption without the interference of concurrent hydroxyl formation at 1000K, a surface with deuterated hydroxyls (SiOD and BOD) was used for the TEOS reactions. This surface was prepared by exposing a substrate with SiOH and BOH groups to 500 mTorr of D_2O for 5 minutes at 900K. This was then annealed for 5 minutes in vacuum at 1000K, and then examined with IR and XPS to ensure that the surface was stable on a time-scale comparable to the longest reaction with TEOS at 1000K. Neither the formation of $(Si-O)_2$ rings nor a change in the relative intensities of the hydroxyls, deuterated or otherwise, was observed. As shown by the spectrum at

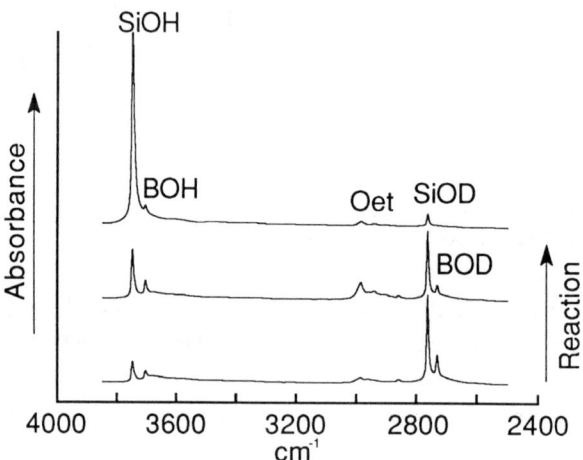

Figure 2. Vibrational spectra representative of the trends observed for fifteen TEOS reaction segments at 1000K. "Oet" denotes surface ethoxy groups.

the bottom of fig. 2, D_2O exposure did not convert all of the SiOH (3748 cm^{-1} [12]) and BOH (3703 cm^{-1} [5]) to SiOD (2763 cm^{-1} [7]) and BOD (2731 cm^{-1} [9]). Additional experiments using extended D_2O exposures suggest this incomplete conversion is due to the initial H/D exchange being reversed by secondary reactions with HOD.[7]

The first reaction segment was carried out at 1000K in 10 mTorr TEOS for 14s. Subsequent reactions were also performed at 1000K using progressively higher pressures and longer reaction times. 200 mTorr and 150s were used for the final reaction segment. The middle spectrum and the top spectrum in fig. 2 are associated with the reaction midpoint and the reaction endpoint, respectively, and are representative of the general trends observed during a total of fifteen reaction segments. As expected, exposure to TEOS clearly reduces the populations of both SiOD and BOD groups while increasing the number of SiOH groups on the surface. Interestingly, there is also an increase in the BOH concentration on the surface, as shown in the middle spectrum. Deuterium-hydrogen exchange reactions could be the cause of the depletion of BOD and the formation of BOH. This could result from reactions with TEOS decomposition products (water or ethanol[10]) formed in the gas-phase or on the surface. However, this possibility is unlikely since both the BOH and BOD concentrations decrease steadily during the initial reaction segments (as made evident by vibrational spectra not shown in the figure). It is more likely that the products from the dissociative chemisorption of TEOS on BOD undergo further reaction and in this way make boranols again available for reaction with TEOS. Although much lower in concentration, the spectrum at the top of the figure shows that BOH groups are still present on the surface when all of the BOD groups have been removed.

At the same time that the BOH concentration reaches a maximum, the ethoxy concentration also reaches a maximum. Evidence of the ethoxy group accumulation can be seen in the middle spectrum in the frequency region labeled with "Oet". Since ethoxy groups maintain a low concentration on a surface which has been exposed to only TEOS at 1000K and not exposed previously to TMB,[7] the relatively high concentration of ethoxy groups seen in the middle spectrum of fig. 2 is evidence that *boron, boranols, or some other product from TMB chemisorption and hydrolysis decrease the rate of ethoxy group decomposition on SiO_2*.

Integrations of the SiOD and BOD vibrational peaks for the initial reaction segments followed by normalization to compensate for their dissimilar initial concentrations allows the SiOD and BOD reactivities with TEOS to be compared directly. These data are plotted in fig. 3. Comparison of the initial slopes shows that the rate constant for TEOS reacting with BOD is twice

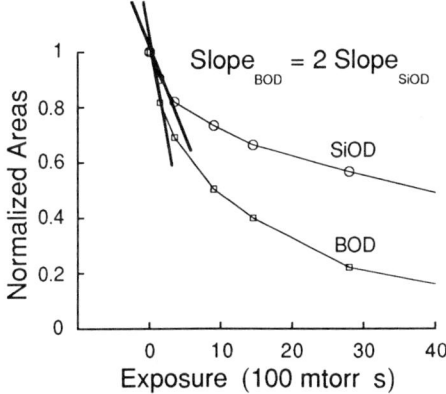

Figure 3.   Normalized integrations of the SiOD and BOD vibrational peaks for the initial TEOS reaction segments at 1000K.

that of SiOD. This result is the first quantitative determination that *boranols increase the rate of TEOS chemisorption on SiO_2 at 1000K.*

Summary

New information about the deposition of doped SiO_2 films from TEOS and TMB sources has been obtained using FTIR. Specifically, the first direct measurements of the relative reaction rates of surface silanols and boranols with TEOS at 1000K have been achieved. To circumvent limitations imposed by the concurrent hydroxyl consumption and reformation which accompany TEOS reaction at 1000K, coadsorbed deuterated silanol and boranol groups were used to study the reaction rates. It was determined that the initial chemisorption of TEOS is *enhanced* by BOD relative to SiOD by at least a factor of two at this temperature. In contrast, the decomposition rate of ethoxy groups (the product of TEOS chemisorption) is *decreased* by the products of TMB chemisorption. Possible dependencies on reactor geometries and other deposition conditions may determine which of these two competing effects will control deposition rates. This has significant implications for microelectronics fabrication since the specific dependencies would be expected to affect process reliability. In addition, this may explain (in part) why the rate enhancement effect is not always observed in boron-doped SiO_2 CVD processes.

Acknowledgements

The technical assistance provided by Larry A. Bruskas is noted with appreciation. This work was performed at Sandia National Laboratories under DOE contract DE-AC04-76DP00789.

References

1   F.S. Becker, D. Pawlik, H. Schafer, and G. Staudigl, J. Vac. Sci. Tech. B **4**, 732 (1986).
2   K. Fujino, Y, Nishimoto, N. Tokumasu, and K. Maeda, J. Electrochem. Soc. **140**, 2922, (1993).
3   D.S. Williams and E.A. Dein, J. Electrochem. Soc. **134**, 657 (1987).
4   P. Smith and C. Apblett, (private communication).
5   M.L. Hair and W. Hertl, J.P. Chem. **77**, 1965 (1973).
6   L.L. Tedder, J.E. Crowell, and M.A. Logan, J. Vac. Sci. Tech. A **9**, 1002 (1991).
7   M.E. Bartram and H.K. Moffat on the reactions of TEOS on silica using deuterated silanols at 300 and 1000K (in preparation).
8   M. Shimizu and H.D. Gesser, Can. J. Chem. **46**, 3517 (1968).
9   M.E. Bartram and H.K. Moffat on the hydrolysis of TMB chemisorption products on silica (in preparation).
10 L.L. Tedder, PhD Thesis, Department of Chemistry, Univiersity of California, San Diego.
11 M.E. Bartram and H.K. Moffat, J. Vac. Sci. Tech. A, (submitted).
12 M.E. Bartram, T.A. Michalske, and J.W. Rogers, Jr., J. Phys. Chem. **95**, 4453 (1991) and references therein.

PREDICTIVE SURFACE KINETIC ANALYSIS: THE CASE OF TiSi$_2$ CVD

M. A. MENDICINO, R. P. SOUTHWELL AND E. G. SEEBAUER
Department of Chemical Engineering, University of Illinois, Urbana, IL 61801

1. INTRODUCTION

Recently, TiSi$_2$ has been the object of considerable study because of its low resistivity among the transition metal silicides and its compatibility with existing ULSI technology [1,2]. Film growth by CVD offers the potential for selective area deposition and high production throughput. However, selective CVD of TiSi$_2$ from gas phase SiH$_4$ and TiCl$_4$ is usually accompanied by a competing reaction which consumes intolerable amounts of the Si substrate [3,4]. Controlling this consumption is crucial in TiSi$_2$ growth; however, no quantitative correlation exists between silicon consumption and growth conditions or film thickness. Additionally, the reaction mechanism for TiSi$_2$ growth is poorly understood, and some disagreement even exists about the reaction stoichiometry [5,6]. The combined CVD/UHV approach we have developed fills many gaps in the current understanding of TiSi$_2$ CVD.

Our studies focus on the selective-area CVD of silicides, particularly TiSi$_2$. Our approach is unique in that it attempts to predict optimal processing conditions from surface reactivity measurements under well-characterized ultrahigh vacuum conditions. These predictions are then tested directly in a CVD chamber, connected to the ultrahigh vacuum system, that operates at normal processing temperatures and pressures. Our work attempts to find ways to lower the deposition temperature and to eliminate substrate consumption. Special effort centers on understanding and controlling the first stages of TiSi$_2$ film growth on the silicon surface. In addition to tackling these problems in the particular case of TiSi$_2$ growth, we are developing a framework of chemical reasoning that will be widely applicable in deposition technology.

2. EXPERIMENTAL

Our apparatus consists of a CVD (Fig. 1) and UHV chamber connected through a load-locked sample transfer arrangement. The UHV chamber is equipped with LEED and AES capabilities for composition and structure analysis before and after growth. The CVD chamber is turbomolecularly pumped and has teflon coated stainless steel walls [7] to avoid reaction with the corrosive TiCl$_4$ ambient. Gas compositions are monitored in real-time by a differentially pumped UTI 100C mass spectrometer which has line of sight to the sample for detection of unstable species. During growth, the sample is suspended from a Cahn 2000 microbalance which measures the deposition rate in real-time. The sample is heated by a 1000W Xe arc lamp with a gold plated ellipsoidal reflector for efficient energy coupling. The sample temperature is monitored by an optical pyrometer detecting 8-14μm light emitted from the sample through a ZnSe window. The pyrometer was calibrated against a thermocouple in separate experiments where the microbalance was not used.

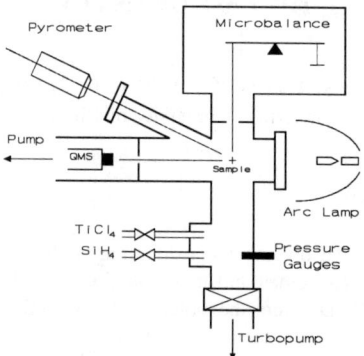

Figure 1   Side view schematic of CVD apparatus. Transfer arm and UHV chamber perpendicular to the plane of the paper (not shown).

3. RESULTS

A. Steady-State Growth on Inert Substrates

Since $TiSi_2$ growth is known to be complicated by Si consumption, polycrystalline tantalum samples were used as substrates to eliminate diffusion effects. Figure 2 shows the effect of source gas ratio and temperature on deposition rate for a fixed $TiCl_4$ pressure of 5×10^{-4} torr. $TiSi_2$ growth displayed both flux-limited behavior at high temperature and reaction limited behavior at low temperature with the flux-limited regime extending to lower temperature as the SiH_4 pressure was increased. At low $SiH_4:TiCl_4$ ratios, the transition from reaction to flux-limited regimes was quite smooth and an activation energy for reaction-limited $TiSi_2$ growth of 50 to 70 kcal/mole was obtained depending on P_{SiH4}. At high $SiH_4:TiCl_4$ ratios, this transition became quite abrupt. The reaction could be quenched from flux-limited growth to no reaction, but only through a substantial hysteresis loop near 700°C. $TiSi_2$ growth could be re-initiated at the flux-limited rate by raising the reaction temperature by roughly 60°C.

The gas phase data for this set of experiments identified the major volatile reaction products as H_2, HCl, $SiCl_2$, and SiH_3Cl. At temperatures below 800°C H_2 and HCl were the primary products. As the reaction temperature was increased, $SiCl_2$ formed for low SiH_4 pressures, and SiH_3Cl formed for high SiH_4 pressures. In all flux-limited areas, a chlorine mass balance was obeyed. As temperature increased, the volatile chlorinated products shifted from HCl to $SiCl_2$ and SiH_3Cl. For reaction temperatures near 700°C and high $SiH_4:TiCl_4$ ratios, the H_2 and HCl data showed the abrupt reaction quenching and hysteresis loops corresponding to the deposition rate data.

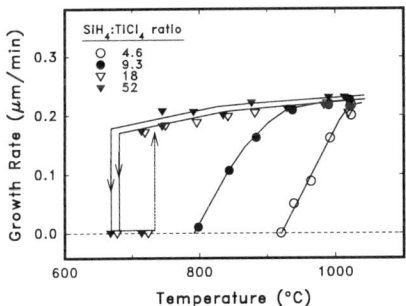

Figure 2 TiSi$_2$ deposition rate on inert substrates.

B. Transient growth on Si Substrates

The initial condition of the Si samples played an important role in film nucleation. All samples for the transient growth experiments were 10 Ω-cm, n-type, phosphorus-doped, Si(100). Some samples were prepared by degreasing in acetone, etching in 48% HF, rinsing in deionized water, then slow heating in vacuum to 950°C to remove any residual oxide. When transferred to UHV, AES of these samples showed only very small amounts of carbon and LEED showed a sharp 2x1 pattern. With such preparation, TiSi$_2$ growth did not begin until the sample was briefly heated to about 900°C in the source gas mixture. Growth could then continue at much lower temperatures. Other samples were simply placed in vacuum and quickly heated to 950°C to remove the native SiO$_2$, however the surface of these samples appeared visually to be slightly hazy, indicative of small etch pits. In these cases, TiSi$_2$ nucleated even at low temperatures with no special treatment. In the end, both preparation procedures yielded high quality TiSi$_2$ films with $\rho_{TiSi2} = 20 \pm 5$ μΩ-cm.

Figure 3 shows a microbalance trace for a transient growth experiment at 825°C using $P_{SiH4} = 5 \times 10^{-4}$ torr and $P_{SiH4} = 5 \times 10^{-3}$ torr. In this experiment, the sample was cleaned by etching in HF and heating slowly in vacuum to 950°C; therefore, growth did not begin spontaneously, but only after the reaction temperature was briefly spiked to 900°C. Figure 3 shows that the sample mass remained nearly unchanged at first, but increased substantially as the film thickened. The gas phase data corresponding to this experiment show very little hydrogenated products at first with large amounts of SiCl$_2$. As the sample gained mass, SiCl$_2$ production decreased and H$_2$ and HCl production increased. SiH$_3$Cl was observed, and did not change appreciably throughout the experiment. TiCl$_4$ conversion was roughly 95% throughout the experiment indicating flux-limited behavior since no other volatile Ti-containing species were formed.

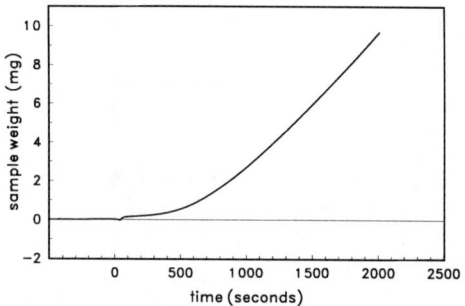

Figure 3 Microbalance trace for a typical transient growth experiment on Si.

Methodology for measuring the instantaneous Si consumption rate has up to now not been available for any deposition system. However, in our experiments this quantity could be easily determined by comparing the $TiSi_2$ deposition rate calculated from gas phase $TiCl_4$ conversion with the microbalance trace. Integrating the Si consumption rate over time gave a quantitative measure of the total amount of Si consumed at any film thickness in a single experiment (Figure 4).

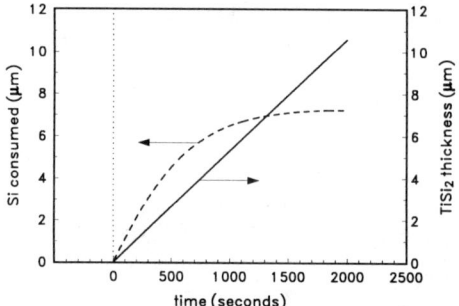

Figure 4 A continuous measure of Si consumed and $TiSi_2$ formed for one transient growth experiment.

It is clear that the relative rates of $TiSi_2$ forming through heterogeneous reaction with SiH_4 and through Si consumption change dramatically as the film thickens. Si is the primary diffusing species [8] in the $TiSi_2$ film. Our transient data suggest that the flux of Si though the $TiSi_2$ film can be written by a Fick's law expression,

$$J_{Si} = D_o \exp(\frac{-E_{diff}}{k_B T}) \frac{\Delta C}{\Delta x}$$

where D_o and E_{diff} are the pre-exponential and activation energy for diffusion; k_B is Boltzmann's constant; T is the substrate temperature, and $\Delta C/\Delta x$ is the Si concentration gradient across the film. From this expression we see that thin films and high temperatures facilitate Si diffusion/consumption.

C. UHV Studies

We have constructed a predictive model for CVD by writing mass balances on all atomic species involved in the two parallel reactions that occur in CVD:

Gaseous: $TiCl_4 + SiH_{4(g)} \rightarrow TiSi_2, H_2, HCl, SiCl_2, SiH_3Cl$

Diffusion: $TiCl_4 + Si_{(s)} \rightarrow TiSi_2, SiCl_2$

If R is an adsorption rate and r is a desorption rate, then the steady state mass balances are:

Ti: $R_{TiCl_4} = r_{TiSi_2}$

Si: $R_{SiH_4} + J_{Si} = 2r_{TiSi_2} + r_{SiH_3Cl} + r_{SiCl_2}$

Cl: $4R_{TiCl_4} = r_{HCl} + r_{SiH_3Cl} + 2r_{SiCl_2}$

H: $4R_{SiH_4} = 2r_{H_2} + r_{HCl} + 3r_{SiH_3Cl}$

Here, J_{Si} is the solid Si diffusion rate, and r_{TiSi_2} is the film growth rate. Detailed rate expressions can be written for these various rates in terms of inputs T, P_{SiH_4} and P_{TiCl_4} and of unknowns θ_{Cl}, θ_H (surface concentrations of Cl and H), R_{SiH_4}, and r_{TiSi_4}.

Using the UHV technique of temperature programmed desorption, we have measured nearly all the other kinetic rate parameters needed to complete the model as follows

Table I. Desorption

Product	Order in H	Order in Cl	$v_o n_o^{m-1}$ (s^{-1})	E_D(kcal/mol)
H_2	2	0	1×10^9	30
HCl	1	1	2×10^{17}	71
$SiCl_2$	0	2	5×10^{14}	78
SiH_3Cl	?	?	?	?

Table II. Adsorption

Reactant	S_0	Cov. Dependence
$TiCl_4$	1	$(1-\theta_{Cl}-\theta_H)$
SiH_4	0.02	$(1-\theta_{Cl}-\theta_H)$

Table III. Diffusion (Fick's law)

Diffusing Species	$D_0\Delta C$ $(cm^{-1}s^{-1})$	E_{diff} (kcal /mol)
Si	5×10^{17}	23

Only the entries with question marks need to be measured before the full-scale predictive model is complete.

This effort will then represent the first-ever construction of a quantitative, predictive rate model based on individual reaction rate laws in a semiconductor-connected reaction process.

4. CONCLUSION

A new experimental approach combining analysis during growth and in ultrahigh vacuum has been used to study the kinetics of $TiSi_2$ deposition. Steady state growth experiments have identified H_2, HCl, $SiCl_2$, and SiH_3Cl as volatile reaction products. Increasing the $SiH_4:TiCl_4$ ratio to 50:1 lowers the flux-limited growth regime near 700°C and gives rise to multiple steady states defining both fast growth and no growth under identical reaction conditions. On Si substrates, nucleation kinetics were shown to correlate with the presence of defects even in the absence of oxide.

REFERENCES

[1] S.P. Murarka, Silicides for VLSI Applications, (New York, Academic), 1983.
[2] S.P. Murarka, *J. Vac. Sci. Technol.*, vol. B4, p. 1325, 1986.
[3] A. Bouteville, J.C. Remy and C. Attuyt, *J. Electrochem. Soc.*, vol. 139, p. 2260, 1992
[4] J. Lee and R. Reif, *J. Electron. Mater.*, vol 20, p. 331, 1991
[5] G.J. Reynolds, C.B. Cooper III and P.J. Gaczi, *J. Appl. Phys.*, vol 65, p. 3212, 1989
[6] V. Ilderem, R. Reif, *J. Electrochem. Soc.*, vol 135, p. 2590, 1988
[7] W. Müller-Markgraf and M.J. Rossi, *Rev. Sci. Instrum.*, vol. 61, p. 1217, 1990.
[8] Y.L Corcoran, A.H. King, N. deLanerolle, and B. 2Kim, *J. Electron Mater.*,vol. 19, p. 1177, 1990.

HYDROGEN PAIRING ON Si(100)-(2x1): A SITE-BLOCKING STUDY

WOLF WIDDRA AND W. HENRY WEINBERG
Department of Chemical Engineering and Center for Quantized Electronic Structures (QUEST), University of California, Santa Barbara, CA 93106

ABSTRACT

The thermal desorption of hydrogen from a Si(100)-(2x1) monohydride surface has been reported previously to follow first-order kinetics. Pairing of hydrogen atoms on a Si-Si dimer prior to desorption is most likely the cause of this unusual behavior. To examine the degree of hydrogen pairing at various surface temperatures, this study examines the blocking of adsorption sites by hydrogen for subsequent acetylene adsorption. Temperature programmed desorption and Auger electron spectroscopy were used to measure the absolute saturation coverage of acetylene for various coverages of preadsorbed atomic deuterium. The observed linear decrease in saturation coverage of acetylene with deuterium coverage, for atomic deuterium adsorption between 550 and 150 K, indicates that deuterium is paired on Si-Si dimers. The observation that pairing occurs even at 150 K suggests that the diffusion of chemisorbed hydrogen is not responsible for pairing.

I. INTRODUCTION

The adsorption and desorption of atomic hydrogen on the Si(100)-(2x1) surface has been the subject of intense research in the past thirty years due to its technologically importance, e.g., for silicon chemical vapor deposition and semiconductor surface passivation [1]. Furthermore, hydrogen on Si(100)-(2x1) is "in principle" one of the simplest systems for an adsorbate chemisorbed on a semiconductor surface.

However, only recently Sinniah, *et al.* discovered that the recombinative desorption of hydrogen from the monohydride phase on the Si(100)-(2x1) surface obeys first-order kinetics in the hydrogen coverage [2]. Further studies by different groups confirmed this result for hydrogen coverages greater than approximately 0.1 [3,4] and showed that desorption is second order on the Si(111)-(7x7) [3] and first order on the Ge(100)-(2x1) surface [5]. Wise, *et al.* proposed that surface dimerization -- present on the (2x1) reconstructed silicon and germanium (100) surfaces and absent on the (7x7) surface -- is responsible for these unusual first-order kinetics [3]. The tendency of the two singly occupied dangling bonds on a Si-Si dimer to form a π-bond lowers the total energy of the Si-Si dimer [6]. Adsorption of a hydrogen atom, on the other hand, breaks this π-bond to form a Si-H bond on one side and a true dangling bond on the other side of the dimer. Therefore, a partially hydrogen-covered surface can lower the total energy by the pairing of hydrogen atoms on dimers, resulting in Si-Si dimers with either no hydrogen or two hydrogen atoms bonded to them. This hydrogen prepairing prior to thermal desorption is proposed to lead to the observed first-order kinetics [3]. Scanning tunneling microscopy (STM) studies by Boland have shown that partial desorption of hydrogen from a saturated Si(100)-(2x1) monohydride surface produces, after cooling to room temperature, only Si-Si dimers which bond either two hydrogen atoms or none [7]. After adsorption of a small hydrogen coverage (θ = 0.08 [8]) at room temperature, STM images show random hydrogen adsorption. Briefly annealing to 630 K which is below the desorption temperature reduces the number of unpaired hydrogen atoms. Based on this work and theoretical studies, it is proposed that hydrogen diffusion prior to desorption establishes the hydrogen pairing responsible for the observed desorption kinetics [7,9].

In this work we present a different approach to examine the degree of hydrogen pairing on Si-Si dimers for various adsorption temperatures and hydrogen coverages between 0.2 and 0.8. The blocking of adsorption sites by hydrogen for subsequent acetylene adsorption is studied quantitatively. Our earlier high-resolution electron energy loss spectroscopy (HREELS) study, supported also by a STM study [10], showed that acetylene is di-σ bonded to the Si-Si dimer on a Si(100)-(2x1) surface [11]. Acetylene adsorbs on the Si(100)-(2x1) surface between 90 and 300 K with an initial probability of adsorption near unity [12,13]. The coverage-dependent decrease of the probability of adsorption is temperature dependent. The saturation coverage has been measured to be approximately 0.8 molecules per nominal Si-Si dimer [12]. The adsorption of acetylene is sensitive to the structure of the Si-Si dimer, and the adsorption of hydrocarbons like acetylene, ethylene and propylene is known to be inhibited if the silicon dangling bonds are saturated by hydrogen [14,15]. Therefore, we use acetylene to titrate the extent of hydrogen pairing on the Si-Si dimers after various exposures to atomic hydrogen. The coverages of adsorbed hydrogen and adsorbed acetylene are determined by temperature programmed desorption (TPD) and Auger electron spectroscopy (AES). HREELS measurements show that the effect of preadsorbed hydrogen is limited to site-blocking only and that no chemical reaction occurs. The measured saturation coverage of acetylene follows a 1-θ behavior in the preadsorbed hydrogen coverage for all examined hydrogen adsorption temperatures. This linear dependence is expected if hydrogen atoms are completely paired on Si-Si dimers. The insensitivity to the adsorption temperature indicates that thermal activated surface diffusion is not responsible for the hydrogen pairing.

II. EXPERIMENTAL DETAILS

The experimental setup and sample preparation has been described previously [11]. Briefly, the experiments were carried out in an ultrahigh vacuum (UHV) chamber (base pressure 5×10^{-11} Torr) equipped with a HREEL spectrometer (LK-2000-14-R), four-grid rear-view LEED optics, an Auger electron spectrometer with a single-pass cylindrical mirror energy analyzer, an ion sputter gun, a pin-hole gas doser, a tungsten filament to produce atomic hydrogen, and a differentially pumped quadrupole mass spectrometer (UTI-100C) with a cryo-shroud. The chamber is pumped by a 1000 l/s turbomolecular pump.

Atomic hydrogen and deuterium were generated by dissociation of the respective molecules with a hot (1800 K) spiral tungsten filament positioned approximately 5 cm from the sample. All reported coverages are relative to the hydrogen coverage for a saturated Si(100)-(2x1) monohydride (6.8×10^{14} atoms/cm^2). For acetylene exposures a 5-μm pin hole doser was used. An estimated acetylene flux of 8×10^{14} molecules/cm^2 was used for the TPD experiments to saturate the surface at 100 K and to avoid significant acetylene adsorption on the sample mounting. Higher fluxes were used to confirm the saturation coverage of acetylene and for AES experiments.

III. RESULTS AND DISCUSSION

Before we describe the experiments in detail, we will emphasize the dependence of the acetylene saturation coverage on the coverage of preadsorbed hydrogen as result of different site-blocking models.

For random adsorption the probability to find an adsorbate at a selected adsorption site is given by the fractional coverage θ. The probability to find no adsorbate at that site is given by 1-θ. Consequently, for random hydrogen adsorption, the probability for a Si-Si dimer to bond 0, 1, or 2 hydrogen atoms is given by $(1-\theta)^2$, $2\theta(1-\theta)$, and θ^2, respectively. In the case of site blocking for subsequent acetylene adsorption, two different cases must be distinguished: (a) Both dangling bonds on a dimer have to be saturated by hydrogen to inhibit subsequent acetylene adsorption. In this case acetylene can adsorb on empty dimers as well as on singly hydrogen-occupied dimers. The acetylene saturation coverage is, in this case, proportional to the added probabilities for both dimer species, $1-\theta^2$. (b) One hydrogen atom bonded to a Si-Si dimer is sufficient to suppress acetylene adsorption. In this case acetylene can adsorb only on empty dimers, and the acetylene coverage is proportional to the probability of finding empty dimers, $(1-\theta)^2$. However, in the case of hydrogen pairing on Si-Si dimers, the probability for a Si-Si dimer to bond 0, 1, or 2 hydrogen atoms is simply given by $1-\theta$, 0, and θ, respectively. In this case the saturation coverage for acetylene is proportional to $1-\theta$. The comparison between these three possibilities shows that the dependence of the acetylene saturation coverage on the coverage of preadsorbed hydrogen can distinguish between hydrogen adsorbed on random sites and hydrogen paired on the Si-Si dimers.

Preadsorbed hydrogen -- in the TPD experiments deuterium is used to distinguish the preadsorbed layer from hydrogen from decomposed acetylene, as described below -- was prepared using two different procedures. The clean Si(100)-(2x1) surface was either exposed to atomic deuterium at surface temperatures of 600, 300, or approximately 150 K, resulting in a surface with a monodeuteride coverage between 0 and 1, or a saturated monodeuteride overlayer was annealed using a temperature ramp of 2 K/s to allow for fractional desorption of the monodeuteride. Subsequently, the surface was exposed to acetylene at 100 K to saturate the available adsorption sites. The acetylene saturation and deuterium coverages were determined by TPD using a heating rate of 2 K/s. TPD spectra were recorded for molecular hydrogen in all three isotopic combinations and for molecular acetylene. Because acetylene largely decomposes upon heating (> 95%)[16], the hydrogen desorption resulting from acetylene decomposition is used for a quantitative determination of the acetylene coverage. After careful calibration for the different quadrupole mass spectrometer sensitivities for H_2, HD, and D_2 using saturated monohydride and monodeuteride overlayers, the deuterium and hydrogen coverages were extracted from the integrated TPD spectra. For selected deuterium coverages AES was used also to determine the acetylene coverage relative to a saturated acetylene overlayer on a clean surface. Precautions have been taken to employ identical AES measurement conditions at low electron fluxes to avoid differences in the AES carbon intensities due to electron induced desorption or decomposition. AES spectra were measured at eight different locations on the surface. A detailed description will be given elsewhere [17].

Figure 1 shows the acetylene saturation coverage after preadsorption of atomic deuterium as a function of the deuterium coverage. Different preparations of the preadsorbed deuterium coverage are indicated by different symbols. Triangles are used to denote preadsorbed deuterium prepared by partial desorption of a saturated monodeuteride overlayer. Adsorption of atomic deuterium at 600, 300, and 150 K is marked by circles, diamonds, and squares, respectively. Open and filled symbols are used to distinguish between TPD and AES results, respectively. The three coverage dependencies for site-blocking that were discussed above are indicated by lines in Fig. 1. All data for the acetylene saturation coverage are, within the experimental error, described by a linear $1-\theta$ dependence on the deuterium coverage, regardless of the deuterium adsorption temperature. The data clearly disagree with both the $1-\theta^2$ and the $(1-\theta)^2$ dependence. The small deviation from linear site blocking for a deuterium coverage of 0.82 adsorbed at 150 K is due to the onset of silicon dideuteride formation, as discussed later. The absolute acetylene saturation coverage of 0.78±0.05 in the absence of hydrogen, as extracted from Fig. 1 and corrected for 5% molecular desorption of acetylene, agrees well with earlier studies [12,16].

Figure 2 shows HREEL spectra of deuterated acetylene adsorbed at 100 K (a) on a clean Si(100)-(2x1) surface and (b) on a Si(100)-(2x1) surface preexposed at 600 K to atomic hydro-

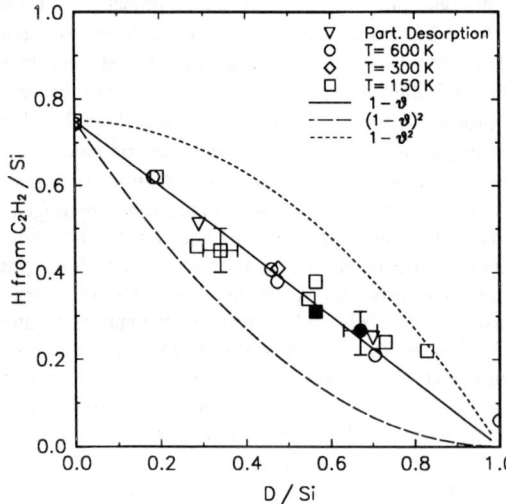

Figure 1: Acetylene saturation coverage as a function of the preadsorbed deuterium coverage for various deuterium adsorption temperatures determined by TPD (open symbols) and AES (filled symbols). Typical error bars are given for both a TPD and an AES result. The expected site-blocking dependence on deuterium coverage is shown for deuterium pairing (solid line) and random deuterium adsorption with two different site-blocking models (dashed and dotted lines).

gen, resulting in a hydrogen coverage of 0.3. The spectrum in Fig. 2(a) is very similar to spectra of a previous acetylene adsorption study on the Si(100)-(2x1) surface [11]. The different vibrational modes have been assigned there. Briefly, the energy loss peaks at 2230, 1040, and 507 cm^{-1} are the stretching, the in-plane bending, and the out-of-plane bending C-D modes of the di-σ bonded acetylene molecule, respectively. The energy loss peak at 1420 cm^{-1} is assigned to the C=C double bond stretching mode which is significantly down-shifted in frequency due to an interaction with the silicon dangling bonds. The loss feature at 670 cm^{-1} is the Si-C stretching mode. In comparison, Fig. 2(b) shows two strong additional energy loss peaks at 2100 and 640 cm^{-1} which are identified as the Si-H stretching and the Si-H bending modes. All other loss features are reduced in intensity due to the reduced acetylene coverage but essentially unchanged. A comparison of the vibrational spectra for acetylene with and without preadsorbed hydrogen shows clearly that the influence of hydrogen for subsequent acetylene adsorption is limited to site blocking only. No chemical reaction occurs between acetylene and hydrogen, as can be seen, for example, by the absence of a C-H and Si-D stretching mode in Fig. 2(b). Furthermore, the Si-H bond seems to be unaffected by the postadsorption of acetylene.

HREEL spectra after different exposures to atomic hydrogen, which are not shown here [17], were used to verify the formation of the monohydride species only. As judged by the appearance of the SiH$_2$ scissoring mode, the spectra indicate that at a low adsorption temperature of 150 K the silicon monohydride phase is populated first with an onset of dihydride formation at a coverage of 0.8, in agreement with earlier studies [18]. At a coverage of 0.8, the relative intensity of the SiH$_2$ scissoring mode is below 0.15 compared to the mode at a higher coverage of 1.3. The formation of dideuteride at 150 K for a coverage greater than 0.8 leads, for a given deuterium coverage of 0.78±0.05 in the absence of hydrogen, as extracted from Fig. 1 and corrected for 5% molecular desorption of acetylene, agrees well with earlier studies [12,16].

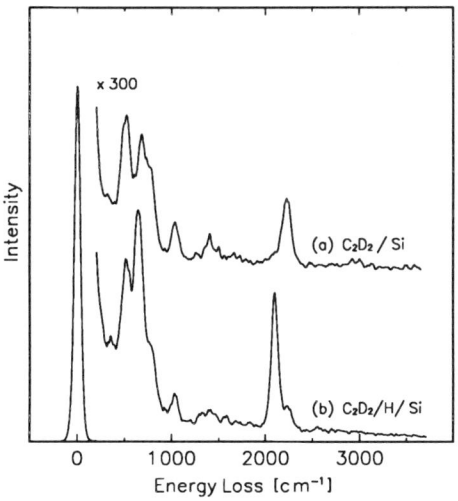

Figure 2: HREEL spectra of adsorbed deuterated acetylene (a) on a clean Si(100)-(2x1) surface and (b) on a Si(100)-(2x1) surface pre-exposed to atomic hydrogen resulting in a hydrogen coverage of 0.3. The spectra are recorded at 100 K in the specular direction with an electron energy of 7.0 eV.

Preliminary results for ethylene adsorption on Si(100)-(2x1), which will be presented elsewhere, indicate a similar hydrogen site-blocking dependence as for acetylene [17]. In this case, the ratio of molecular ethylene desorption to ethylene decomposition on the surface changes with the preadsorbed hydrogen coverage. This complicates a quantitative study and also explains why a previous site-blocking study obtained a different site-blocking result for ethylene adsorption [15].

Theoretical studies have shown that on the Si(100)-(2x1) surface adsorption of two hydrogen atoms on a single dimer is energetically favored over adsorption on different dimers [6,19]. The breaking of the π-bond between the dangling bonds of a clean Si-Si dimer upon hydrogen adsorption determines the energetics. Therefore, preferential hydrogen pairing is expected in thermodynamical equilibrium. For a hydrogen coverage above 0.95 at room temperature which was prepared by hydrogen desorption at 690 K from a saturated monohydride phase, a STM study by Boland has clearly shown complete hydrogen pairing [7]. Furthermore, he demonstrated that at low coverages around 0.08 [8] hydrogen adsorption is random and that subsequent annealing to 630 K promotes the hydrogen pairing. Based on these results, he concluded that hydrogen adsorption is random, and thermal activated diffusion is necessary to generate the thermodynamically favored hydrogen pairing. In our study, the observed linear 1-θ coverage dependence of the site blocking, for hydrogen species prepared by partial desorption of a saturated overlayer or by hydrogen adsorption at 600 K, supports this picture of hydrogen pairing for coverages between 0.2 and 0.8. However, the surprising result that the hydrogen site blocking follows also a 1-θ dependence for atomic hydrogen adsorption at 150 K in the whole observed coverage range indicates hydrogen pairing even in the absence of annealing to higher temperatures. Because of the high activation barriers for hydrogen diffusion on silicon surfaces [20,21,22], the thermodynamical preference for hydrogen pairing cannot account solely for the observed high degree of pairing assuming random hydrogen adsorption.

Further experiments with STM, for example, need to be done to confirm the hydrogen pairing at low temperatures and to reveal the mechanism by which the adsorption process itself can produce the hydrogen pairing at coverages as low as 0.2. Careful examination of the splitting of the

Si-H stretching mode, which is caused by the dipole-dipole interaction between the two Si-H bonds on a dimer [23], using, for example, Fourier-transform infrared spectroscopy, or STM studies at intermediate coverages are well suited to gain a more complete understanding of the hydrogen pairing on dimerized semiconductor surfaces.

IV. CONCLUSIONS

Site blocking of hydrogen for subsequent acetylene adsorption has been employed to examine the degree of hydrogen pairing on the Si-Si dimers on a Si(100)-(2x1) surface. The observed linear decrease in acetylene saturation coverage with hydrogen coverage, for different hydrogen adsorption temperatures between 600 and 150 K, indicates pairing of hydrogen on Si-Si dimers at temperatures as low as 150 K in the coverage range between 0.2 and 0.8. This suggests that this pairing mechanism does not occur by adatom hoping, which might occur at higher temperatures.

ACKNOWLEDGMENTS

This work was financially supported by QUEST, a National Science Foundation Science and Technology Center (grant no. DMR 91-20007), the W. M. Keck Foundation, a MICRO/SBRC grant, and a NATO grant. One of us (WW) thanks the Alexander-von-Humboldt Foundation for a Feodor-Lynen Research Fellowship.

REFERENCES:

[1] See e.g. K. Christmann, Surf. Sci. Rep. **9**, 1 (1988); H. Froitzheim, in *The chemical Physics of Solid Surfaces and Heterogeneous Catalysis*, Vol.5, Eds. D.A. King and D.P. Woodruff (Elsevier, Netherlands, 1988), p.183; J.A. Schaefer, Physica B, **170**, 45 (1991); G.V. Hanson and R.I.G. Uhrberg, Surf. Sci. Rep. **9**, 197 (1988); J.J. Boland, Adv. in Phys., **42**, 129 (1993).
[2] K. Sinniah, M.G. Sherman, L.B. Lewis, W.H. Weinberg, J.T. Yates Jr., and K.C. Janda, Phys. Rev. Lett. **62**, 567 (1989).
[3] M.L. Wise, B.G. Koehler, P. Gupta, P.A. Coon, and S.M. George, Surf. Sci. **258**, 166 (1991).
[4] U. Höfer, L. Li, and T.F. Heinz, Phys. Rev. B **45**, 9485 (1992).
[5] M.P. D'Evelyn, S.M. Cohen, E. Rouchouze, and Y.L.Yang, J. Chem. Phys. (in press).
[6] P. Nachtigall, K.D. Jordan, K.C. Janda, J. Chem. Phys. **95**, 8652 (1991).
[7] J.J. Boland, Phys. Rev. Lett. **67**, 1539 (1991).
[8] We have estimated the coverage by counting the adsorbed hydrogen atoms in the given STM images.
[9] J.J. Boland, J. Vac. Sci. Techn. A **10**, 2458 (1992).
[10] A.J. Mayne, T.R.I. Cataldi, J. Knall, A.R. Avery, T.S. Jones, L. Pinheiro, H.A.O. Hill, G.A.D. Briggs, J.B. Pethica, and W.H. Weinberg, Faraday Discuss. **94**, 199 (1992).
[11] C. Huang, W. Widdra, X.S. Wang, and W.H. Weinberg, J. Vac. Sci. Technol. A **11**, 2250 (1993).
[12] C.C. Cheng, R.M. Wallace, P.A. Taylor, W.J. Choyke, and J.T. Yates, Jr., J. Appl. Phys. **67**, 3693 (1990).
[13] C.C. Cheng, P.A. Taylor, R.M. Wallace, H. Gutleben, L. Clemen, M.L. Colaianni, P.J. Chen, W.H. Weinberg, W.J. Choyke, and J.T. Yates, Jr., Thin Solid Films **225**, 196 (1993).
[14] M.J. Bozack, W.J. Choyke, L. Muehlhoff, and J.T. Yates, Jr., J. Appl. Phys. **60**, 3750 (1986).
[15] L. Clemen, R.M. Wallace, P.A. Taylor, M.J. Dresser, W.J. Choyke, W.H. Weinberg, and J.T. Yates, Jr., Surf. Sci. **268**, 205 (1992).
[16] P.A. Taylor, R.M. Wallace, C.-C. Cheng, M.J. Dresser, W.J. Joyke, W.H. Weinberg, and J.T. Yates, Jr., J. Am. Chem. Soc. **114**, 6754 (1992).
[17] W. Widdra, R. Maboudian, S.I. Yi, G.A.D. Briggs, and W.H. Weinberg (in preparation).
[18] J.A. Schaefer, F. Stucki, J.A. Anderson, G.J. Lapeyre, and W. Göpel, Surf. Sci. **140**, 207 (1984).
[19] M.P. D'Evelyn, Y.L. Yang, and L.F. Sutcu, J. Chem. Phys. **96**, 852, (1992)
[20] G.A. Reider, U. Höfer, and T.F. Heinz, Phys. Rev. Lett. **66**, 1994 (1991).
[21] C.J. Wu and E.A.Carter, Phys. Rev. B **46**, 4651 (1992).
[22] A. Vittadini, A. Selloni, and M. Casarin, Surf. Sci. Lett. **289**, 625 (1993).
[23] Y.J. Chabal, Surf. Sci. **168**, 594 (1986).

Molecular Dynamics Simulation of Large Cluster Growth

Michael R. Zachariah and Michael J. Carrier
National Institute of Standards and Technology, Gaithersburg, MD
and Estela Blaisten-Barojas
George Mason University, Fairfax, VA

Abstract
Classical Molecular dynamics simulation of silicon cluster growth (up to 1000 atoms) have been conducted using the Stillinger-Weber 3-body interaction potential. The cluster binding energy has been fit to an expression that separates the surface and bulk contribution to the energy over a wide temperature and size range. Cluster growth simulations show that large heat release results from new bond formation at gas kinetic rates (i.e. sticking coefficient = unity). Temperature was found to be the primary controlling process parameter in the evolution of cluster morphology from an aggregate to a coalesced cluster below 1000 K, with the impact parameter playing a secondary role.

Introduction
Nanometer processing is receiving considerable interest from a variety of communities, including those from microelectronics and advanced materials. One of the challenges in this area is the processing of very fine particles. This would include their controlled growth, chemical reactivity and transport properties. Considerable attention has been paid to the growth of particles in the range of 100nm and up, however process modelers have implicitly assumed that small particle growth is unimportant. These issues however have reemerged due to the interest in nanometer particle processing[1]. Several outstanding issues are of primary concern in phenomenological models and pedagogical approaches to controlling particle formation[2]. This is true from both the perspective of much of the microelectronics community which hopes to minimize particle growth in the vapor during chemical vapor deposition and the ceramics community which hopes to develop the ability to grow from the vapor, spherical, unagglomerated nanometer scale particles[3,4]. This implies that attention should be addressed toward understanding the nature of the cluster growth kinetics and their morphology.

In this paper we address the dynamics in the formation of silicon particles up to 3 nm from an atomistic view.

Computation Method
The approach used in this work is to apply an atomistic simulation using classical molecular dynamics (MD) methods [5,6]. Computations were conducted using the three body formulation of the silicon potential proposed by Stillinger and Weber (SW) [7]. The three body formulation provides the mechanism by which the directional nature of the bonding can be realistically simulated. While many potentials are available to simulate silicon, the SW potential was chosen because it accurately predicts bulk melting characteristics. Because most cluster formation processes occur at high temperatures, liquid like characteristics should play an important role in any description of cluster growth. Classical MD was conducted by solving the Newtonian equations of motion with a time step of 5.7×10^{-4} ps. All simulations were started by first equilibrating the appropriate size cluster to a specified temperature prior to cluster collision. Each cluster was then given a bulk cluster velocity so that the collision kinetic energy along the line of centers of the clusters corresponds to twice the thermal energy. Both head-on and large impact parameter collisions were included.

Equilibrium Cluster Energetics
Because of the large surface to volume ratio, cluster properties vary with size. The approach to bulk properties as cluster size is increased is in general specific to the property of interest. Fig. 1 shows the effect of cluster size on binding energy per atom at various temperatures. As the cluster size increases the cluster binding energy increases as one would expect for the case where

the ratio of bulk to surface atoms increases. Freund and Bauer [8] have discussed the effect of cluster binding energy with cluster size for metal clusters. Their work has the same qualitative dependence on cluster size, however the effects they observe are approximately twice as great as those we have seen . One possible explanation is that unlike metals which like to form close packed structures, silicon has a lower coordination number (< 4) with highly directional covalent bonding. This would imply that metallic like clusters would have a larger fraction of dangling bonds and are therefore more sensitive to cluster size.

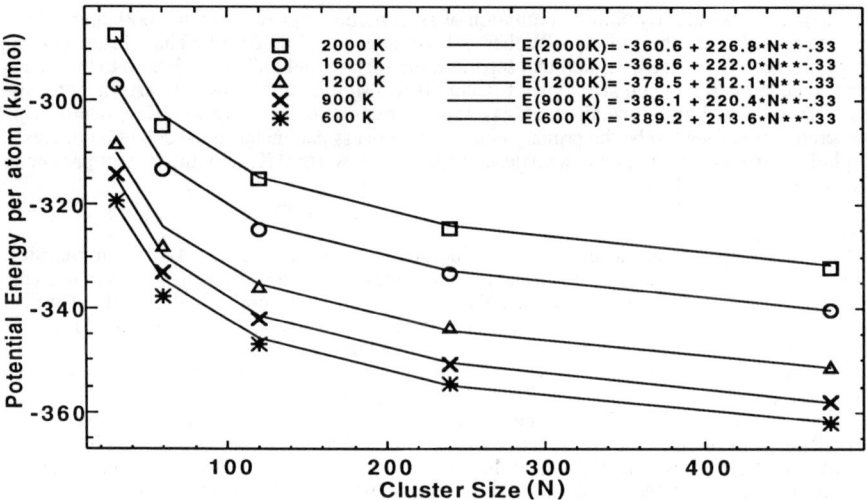

Fig.1 Dependence of cluster binding energy per atom on cluster size (N atoms)

By taking the cluster potential per atom as a function of cluster size we can fit to an expression for the relative importance of bulk (volume) and surface energy (surface area) contribution.

$$\text{Binding Energy} = -E_b N + E_s N^{2/3} \quad (1a)$$
$$\text{Binding Energy/per atom} = -E_b + E_s N^{-1/3} \quad (1b)$$

The fits shown in Fig. 1 are fit to Eq. 1b. By fitting to the temperature dependence one can obtain an expression that encompasses all the data.

$$E_b = 420.0 - T^{0.54}$$
$$E_s = 196.7 - T^{0.44}$$

Vibrational energy transfer processes are very important to the mechanism of cluster stabilization during cluster growth as well as its radiative properties. Fig 2. shows the density of vibrational states as a function of cluster temperature. A comparison with another calculated phonon spectra is also included for comparison purposes [9]. It is clear that the phonon spectra are not significantly altered as a result of the large number of surface atoms. The spectra show the two dominant modes found in bulk silicon (acoustic - 150cm^{-1}; optical- 400 cm^{-1}). As the cluster temperature is increased the phonon density increases, shows broadening and a softening of the modes to lower frequencies.

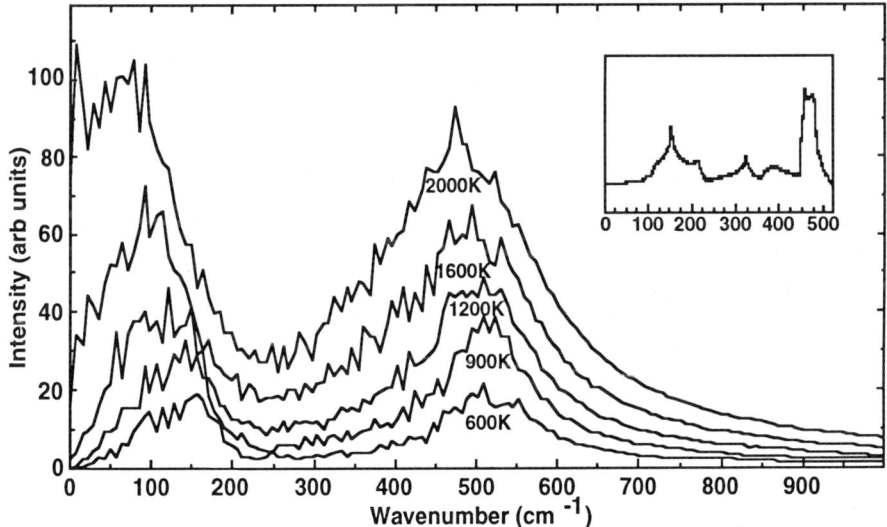

Fig. 2 Phonon density of states for 480 atom cluster as a function of temperature; comparison with spectra for bulk Si calculated with the Bond Charge Model [9].

Cluster Kinetics

Cluster-cluster collision simulations have been conducted on clusters varying in size from 15 atoms up 480 atoms, as a function of temperature and impact parameter. Fig. 3 shows the result of the collision of two 60 atom clusters.

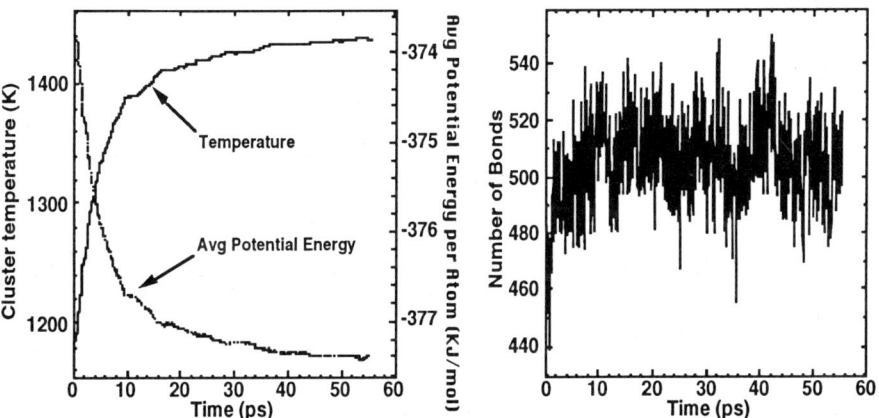

Fig. 3 Dynamics of cluster growth from the collision of two 60 atom clusters at 1200 K

The formation of new chemical bonds during the collision results in a decrease in the internal energy of the resulting cluster with increasing time, as the newly formed cluster finds an increasing stable configuration, by decreasing its surface area and thus the number of dangling bonds. Since this is an adiabatic problem the energy release goes into the thermal motion of each atom within the cluster and is seen as a rise of the cluster temperature. Note that the cluster temperature increases by several hundred degrees, such that following growth, new clusters will be significantly hotter than the surrounding gas. Radiative and conductive cooling will eventually lead to a return of the cluster temperature back to the ambient environment, on a time scale, however, significantly slower than the period of the cluster heating.

For all simulations conducted under thermal energy collision energies, cluster-cluster aggregation kinetics occurred with a sticking coefficient of one. This is an important result in that it provides evidence for the commonly used assumption in phenominological modeling of particle growth, of a gas kinetic growth rate[2,3]. This is in contrast to small cluster nucleation which can show nucleation kinetics that are well below gas kinetic rates, with negative temperature dependence to the rate constants resulting from energy transfer effects [10,11]. Unlike small cluster growth which requires an external collider to dissipate energy, large clusters have sufficient number of internal oscillators to redistribute the excess energy. As a result the smaller clusters experienced the largest temperature rise.

One of the issues of great importance in the formation of nanophase processing of fine particles is the morphology of the resulting clusters during growth. Many, if not most gas phase processes result in chained agglomerates of small particles. Minimization of such structures is a goal to those interested in utilizing the properties of nanoscale particles. The change in morphology is however a very difficult process to study experimentally, although one that we can track through MD in a relatively straight forward manner. Fig. 4 shows the temporal approach to spherical morphology for two colliding (impact parameter =0) 240 atom clusters as a function of cluster temperature.

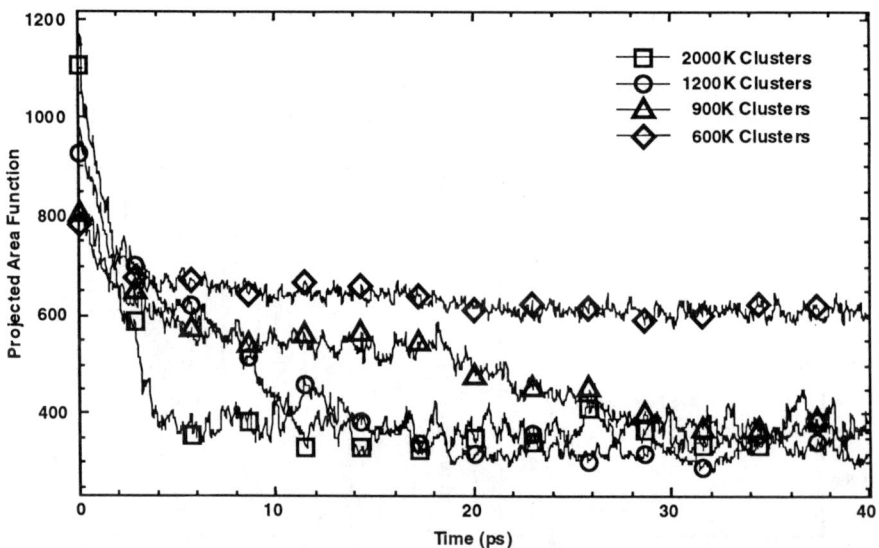

Fig. 4 Temporal behavior of cluster morphology as a function of temperature for two 120 atom clusters.

The projected area coefficient is defined as the area of a circle, obtained from the diameter of the smallest sphere that can be drawn around the two coalescing clusters. The results show that cluster coalescence is very temperature sensitive. Clusters at temperatures above 1000 K show coalescence times that are complete within the simulation period, while showing a substantial decrease in the coalescence rate below 1000 K. Clusters at 600 K shows initial neck growth to be rapid but no evidence for coalescence. Coalescence times at the lower temperatures are very size sensitive, with the larger clusters showing the slowest coalescence rates. At higher temperatures (above 1200 K) cluster coalescence times are independent of size. In general, melted or near melted clusters coalesce spontaneously. Data of this nature could eventually be used to develop scaling relationships for the calculation of sintering rates. An example of the kind of morphology observed, is shown in Fig. 5 for the collision of two 120 atom clusters at temperatures of 600 and 2000 K at 45 ps after the collision.

600 K Cluster
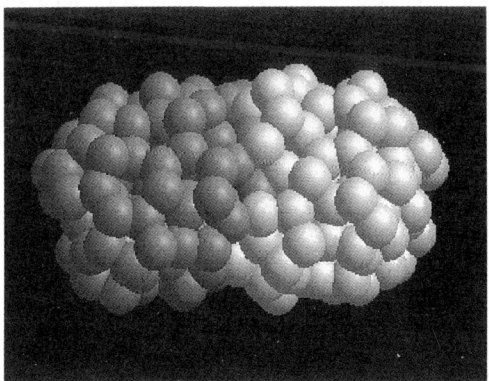

2000 K Cluster
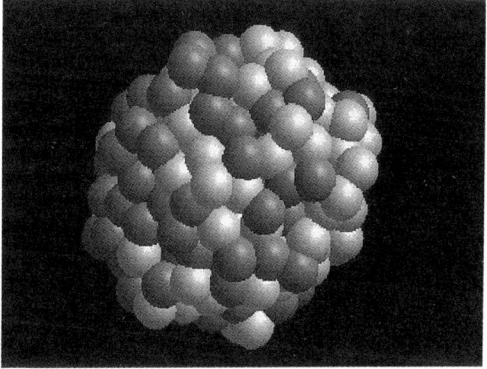

Fig. 5 Cluster morphology during growth at 600 and 2000 K at 45 ps.

Cluster morphology changes require the movement of atoms via internal cluster diffusion. Shown below in Fig. 6 is a measure of the extent of atomic mixing defined as the ratio of the number of nearest neighbors an atom has originating from the other cluster to that of its own cluster. It is clear that the extent of mixing is very temperature sensitive with the coldest clusters showing no evidence for atomic mixing. Even at the highest temperatures studied clusters are still not statistically mixed (a value of 1 would be perfectly mixed). It is clear that particle coalescence is a much faster process than atomic mixing.

Conclusion

Atomistic simulations utilizing classical MD methods have been used to characterize the characteristics of equilibrium and kinetic properties of large silicon clusters undergoing growth via cluster-cluster collisions. The results show that the binding energy of clusters increases with cluster size and decreases with cluster temperature. Cluster kinetics indicate that significant heat release occurs as a result of new bond formation. Cluster morphology effects similar to those observed in nanophase particle processing are also evident. In particular, cluster coalescence is very sensitive to temperature below 1000 K. Atomic mixing is shown to be a slower process than the overall process of cluster coalescence. Future work will be aimed at correlating these observations to scaling laws that can be applied to phenomonological models of particle growth.

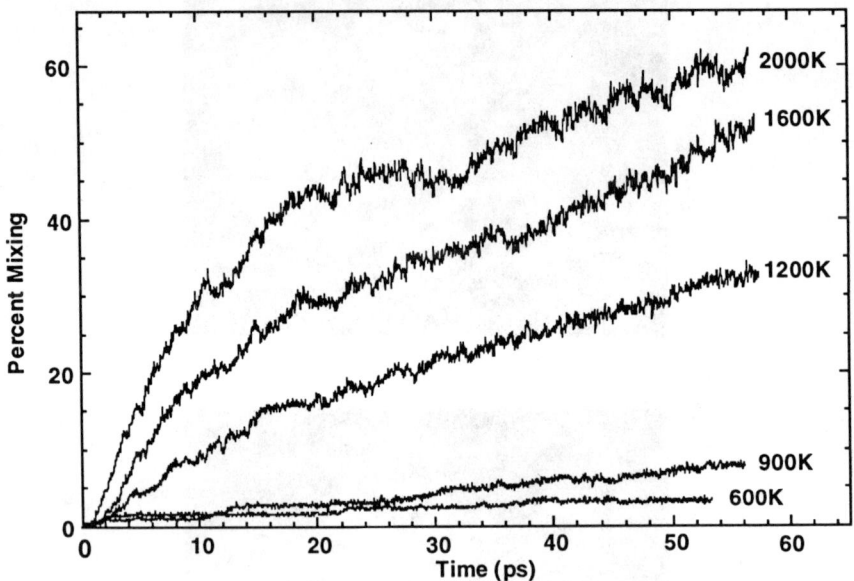

Fig. 6 Atomic mixing as a function of time for various cluster temperatures.

References

1. Xiong, Y and , S.E. Pratsinis, *J. Aeros. Sc*i, **22** (suppl. 1), s199 (1992)
2. Gelbard, F., Tambour, Y., and Seinfeld, J.H., *J. Colloid Interf. Sci.*, **76**(2), 541 (1980)
3. Zachariah, M.R., and Semerjian, H.G., *AIChE. J.*, **35**, 2003 (1989)
4. Zachariah, M.R. and Dimitriou, P., *Aeros. Sci. Tech.* 13, 413 (1990)
5. Blasiten-Barojas, E. and Levesque D., *Phys. Rev. B*, **34**, 3910 (1986)
6. Blasiten-Barojas, E and Zachariah, M.R., *Phys. Rev. B*, **45**, 4403 (1992)
7. Stillinger, F.H. and Weber, T.S., *Phys. Rev. B*, **31**, 5262 (1985)
8. Freund, H.J., and Bauer, S.H., *J. Phys. Chem.* **81**, (1977)
9. Weber, W., *Phys. Rev. B,* **15**, 4789 (1977)
10. Zachariah, M.R., and Tsang, W., *Aeros. Sci. Tech.*, 499 (1993)
11. Zachariah, M.R., and Tsang, W., "Theoretical Prediction of Gas-Phase Nucleation Kinetics of SiO" this volume

ROOM TEMPERATURE FABRICATION OF MICRO-CRYSTALLINE SILICON FILMS FOR TFT'S BY ECR PECVD

Yoo-Chan Jeon, Seok-Woon Lee, and Seung-Ki Joo
Dept. of Metallurgical Eng., Seoul National University, Seoul, 151-742, Korea

ABSTRACT

Microcrystalline silicon films were formed at room temperature without hydrogen dilution by ECR PECVD. Microwave power more than 400 W was necessary to get crystalline films and the crystallinity increased with the power thereafter. Addition of hydrogen and argon enhanced the crystalline phase formation and the deposition rate, the reason of which was found that hydrogen etched silicon films and argon addition drastically increased the etch rate. Annealing of the films showed that microcrystalline silicon films formed by ECR PECVD have a small fraction of amorphous phase. TFT's using silicon nitride and doped/undoped microcrystalline silicon films were fabricatedd with whole processes at room temperature.

INTRODUCTION

Thin film transistors (TFT's) using silicon films as active layers are popular in active matrix LCD's[1-3] and SRAM's.[4-5] In LCD production, in which low temperature process is essential, amorphous silicon TFT's fabricated by PECVD at 250 - 400 ℃ have bee used for pixel driving.[1-3] In recent years the number of pixels in a pannel has been increased to get high resolution display, and it has necesitated the development of new materials to replace amorphous silicon films with low driving capacity and low switching speed. Polycrystalline silicon is thought to be a suitable material for these requirements and the deposition of the crystalline films at a temperature low enough to use inexpensive glass substrates has been studied intensively.[6-8] The most popular method to get crystalline silicon films is crystallization of amorphous films by low temperature furnace annealing or laser annealing. However, these processes need a long process time or show poor uniformity.[9] Therefore the methods to deposit crystalline films at a low temperature without following annealing steps have been tried.[10-13] The results showed that hydrogen radicals enhance the crystallization of silicon films. Since ECR plasma is known to have high electron energy and the high energy electrons can generate hydrogen radicals more readily, ECR PECVD can be thought a desirable method to deposit crystalline silicon films at a low temperature. In this work, microcrystalline silicon films were deposited by ECR PECVD and the effects of the process conditions on the crystallinity of the films were investigated.

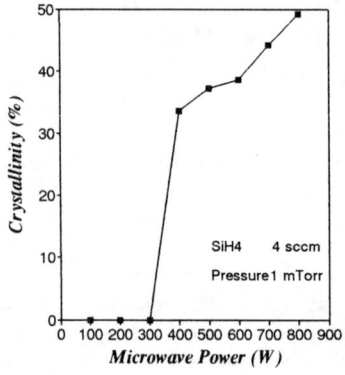

Fig. 1. Effect of microwave power on the crystallinity of silicon films by ECR PECVD.

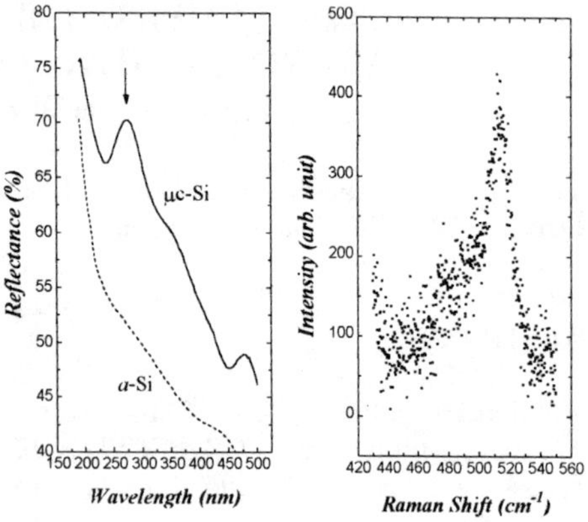

Fig. 2. (a) UV reflectance and (b) Raman spectrum of silicon films by ECR PECVD.

EXPERIMENTS

ECR PECVD system used in this work was shown elsewhere.[14] H_2 and Ar were fed to the top of the plasma chamber and flown through ECR zone to generate high energy species. Pure SiH_4 or 25 %-SiH_4 diluted in Ar was introduced through gas distribution ring above the substrate. Silicon films were deposited on Corning 7059 glasses and the substrates were not heated during deposition. The surface crystallinity of 1000 Å thick films was measured mainly by UV reflectance spectroscopy and confirmed by Raman spectroscopy. The measured reflectance spectra were corrected with a consideration of surface roughness[15] and the peak areas between 238 nm - 318 nm were calculated. The crystalline fractions were determined as the ratio of the peak areas to that of single crystalline silicon wafer.[16] Using room temperature ECR PECVD silicon nitride[14] and undoped/P-doped microcrystalline silicon films TFT's were fabricated with whole processes at room temperature.

RESULTS AND DISCUSSIONS

Fig. 1 shows the effect of microwave power on the crystallinity of silicon films and the major deposition conditions were listed in the inset. In case of microwave power higher than 400 W the deposited silicon films showed crystallinity and the crystalline fraction increased with the power. The UV reflectance spectra of microcrystalline and amorphous silicon films are shown in Fig. 2 (a). The peak around 280 nm, which arise from optical interband transition at X point in Brillouin zone,[16] could be seen for a microcrystalline film while not for an amorphous film. The crystallinity was also confirmed

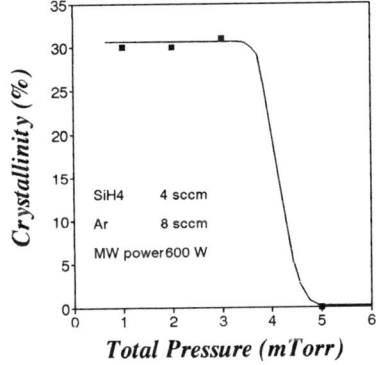
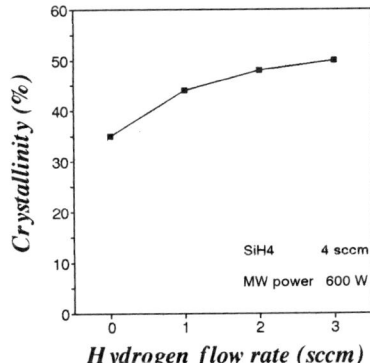

Fig. 3. Effect of total pressure on the crystallinity of silicon films by ECR PECVD.

Fig. 4. Effect of hydrogen addition to silane on the crystallinity of silicon films by ECR PECVD.

by Raman spectrum. In Fig. 2 (b) Raman spectrum shows a peak at 515 cm^{-1} which is shifted from single crystalline position by 5 cm^{-1}. It is noticeable that the silicon films with crystallinity was obtained without heating and hydrogen dilution. The effect of total pressure is shown in Fig. 3. At low pressure up to 3 mTorr the crystallinity was kept constant, but the film deposited at 5 mTorr was amorphous. Considering the above results it can be seen that the high electron energy is responsible for the crystallization because the high microwave power will give high energy to the electrons in the plasma and the mean free path of electron will be longer at lower pressure.

To enhance the crystallinty hydrogen was added to the process gas and the results were shown in Fig. 4. As hydrogen flow rate increased the crystallinity was increased and it means that hydrogen addition helps crystallization of silicon film during deposition. The role of hydrogen in the deposition of crystalline film was investigated with etching experiments using hydrogen gas. As seen in Fig. 5. the etch rate increase with microwave power and it was higher for amorphous silicon than for single crystalline silicon. It means that the deposition and etching occurs simultaneously during silicon film formation and that the higher crystallinity was obtained when the film was deposited in more equilibrated state. Fig. 6 shows argon addition effect. The crystallinity increased monotonically as the argon flow rate increased. Argon is known to have metastable states and it increases the plasma density and dissociation of reactants.[17] The etch rate of silicon by hydrogen was increased about three times by argon addition and the crystallinity improvement by argon addition is thought to be due to enhanced hydrogen etching.

Amorphous and microcrystalline silicon films deposited by ECR PECVD were annealed and the results are listed in Table 1. The films deposited at 300 W (amorphous) shows high crystallinity after furnace annealing while microcrystalline films did not show noticeable increase in crystallinity.

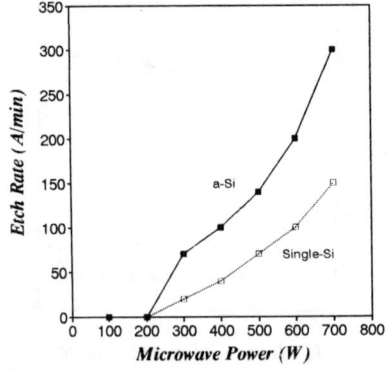
Fig. 5. Variation of silicon etch rate by ECR hydrogen plasma due to microwave power.

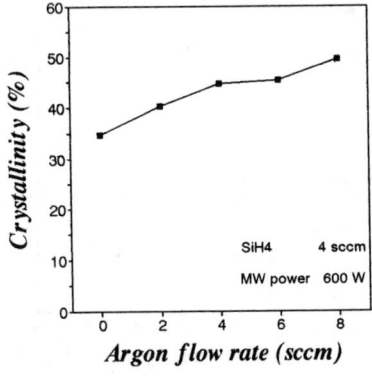
Fig. 6. Effect of argon addition to silane on the crystallinity of silicon films by ECR PECVD.

Amorphous films annealed by RTA shows relatively lower crystallinity than by furnace annealing. However, in the case of microcrystalline films RTAed samples showed higher crystallinity than furnace-annealed. These results inplies that the as-deposited microcrystalline films have only a small amount of amorphous phase because they would exhibited a considerable increase in crystallinity after furnace annealing if they contained a large amount of amorphous phase.

The P-doped silicon films were formed by adding 1 %-PH_3/H_2 and it exhibited conductivity of 0.3 $(\Omega cm)^{-1}$ and Ohmic contact with aluminum electrodes. Using room temperature ECR PECVD doped and undoped microcrystalline silicon films, TFT's were fabricated. Single crystalline silicon wafers were used as substrates, which would serve as gates in transistors. Silicon nitride, undoped and P-doped silicon films were sequentially deposited by ECR PECVD at room temperature and then aluminum was evaporated to

Table 1. Crystallinity change due to annealing. Furnace annealing : 600℃, 30 hrs., RTA : 900℃, 30 sec

structure	amorphous	micro-crystalline
As deposited	0 %	50 %
Furnace annealing	80 %	55 %
RTA	70 %	64 %

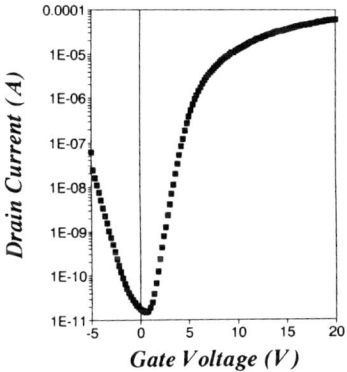

Fig. 7. Transfer characteristic of TFT fabricated by room temperature ECR PECVD.

form electrodes. The transfer characteristic is shown in Fig. 7. The transistor have high ON state current, high I_{ON}/I_{OFF} and Low threshold voltage. It is the first report of room temperature fabrication of TFT's, in which no heat cycle was used, to the best of our knowledge.

CONCLUSION

Microcrystalline silicon films were formed at room temperature without hydrogen dilution by ECR PECVD. Microwave power more than 400 W was necessary to get crystalline films and the crystallinity increased with the power thereafter. It turned out that electron energy in the plasma is an important parameter determining crystallinity. Addition of hydrogen and argon enhanced the crystalline phase formation and the deposition rate, the reason of which was found that hydrogen etched silicon films and argon addition drastically increased the etch rate. Annealing of the films showed that microcrystalline silicon films formed by ECR PECVD have a small fraction of amorphous phase. TFT's using silicon nitride and doped/undoped microcrystalline silicon films were fabricatedd with whole processes at room temperature.

ACKNOWLEDGMENTS

This work has been supported by KOSEF through RETCAM in Seoul National Unversity.

REFERENCES

1. M. J. Tompson et al., IEEE Trans. Electron Dev., **29**, 1643 (1982)
2. K. D. Mackenzie et al., Appl. Phys. A, **31**, 87 (1983)
3. I. Chen, J. Appl. Phys., **50**, 296 (1984)
4. C. T. Liu, K. H. Lee, C. H. D. Yu, J. J. Sung, W. J. Nagy, A. Kornblit, I. Kook, K. R. Olasupo, R. O. Druckemillerm, C. C. Fu, and S. J. Molloy, IEDM 92, 823 (1992)
5. S. Ikeda et al., IEDM 90, 469 (1990)
6. M. Mohri, J. Kakinuma, and T. Tsuruoka, IEDM 92, 673 (1992)
7. K. Shimizu, O. Sukiura, and M. Matsumura, IEDM 92, 669 (1992)
8. M. K. Hatalis, J. Kung, J. Kanicki, and A. Bright, MRS Proc., **182**, 357 (1990)
9. H. Arai, K. Nakazawa, and S. Kohda, Appl. Phys. Lett., **46**, 888 (1986)
10. J. Kanicki, E. Hasan, D. F. Kotecki, T. Takamori, and J. H. Griffith, MRS Proc., **149**, 173 (1989)
11. C. C. Tsai, R. Thompson, C. Donald, F. A. Ponce, G. B. Anderson, and B. Wacker, MRS Proc., **118**, 49 (1988)
12. B. Drevillon, I. Solomon, and M Fang, MRS Proc., **283**, 659, (1993)
13. S. Ishihara, D. He, T. Akasaka, Y. Ariki, M. Nakata, and I. Shimizu, MRS Proc., **283**, 489 (1993)
14. Y. -C. Jeon, H. -Y. Lee, and S. K. Joo, J. Electronic Mat., **21**, 1119 (1992)
15. H. E. Bennett and J. M. Bennett, in Physics of Thin Films, vol. 4, edited by G. Hass and R. E. Thun (Academic Press, London, 1984)
16. G. Habeke and L. Jastrzebski, J. Electrochem. Soc., **137**, 696 (1990)
17. B. Chapman, Glow Discharge Process (John Willy, New York, 1980)

GROWTH OF SILICON FILMS ONTO FLUORORESIN SURFACE BY ArF EXCIMER LASER

T.MIYOKAWA*, M.OKOSHI*, K.TOYODA**,and M.MURAHARA***
*Graduate Student of Tokai University
**The Institute of Physical and Chemical Research (IPCR),2-1 Hirosawa, Wako, Saitama 351-01,JAPAN
***Faculty Engineering of Tokai University, 1117 Kitakaname, Hiratsuka, Kanagawa 259-12, JAPAN

ABSTRACT

Silicon films were deposited on a fluororesin surface. The process was divided into two steps: surface modification process and silicon CVD onto the modified parts. In the modification process, SiH4 and B(CH3)3 mixed gases were used with ArF excimer laser. Fluorine atoms of the surface were pulled out by boron atoms which were photo-dissociated from B(CH3)3 and were replaced with silicon atoms released from SiH4. In the CVD process, SiH4 gas was used with high-density excited ArF excimer laser. Silicon films were deposited onto the nuclei by photodecomposition of SiH4.

Chemical compositions of the modified layers and the deposited parts were inspected by XPS analysis. 1000 Å thickness of the deposited silicon films was confirmed by the surface roughness interference-meter.

INTRODUCTION

Fluororesins such as the polytetrafluoroethylene (PTFE) and the copolymer of tetrafluoroethylene and hexafluoropropylene (FEP) have excellent chemical, electrical and mechanical properties. Because of the chemical stability, the surface is difficult to be made conductive.

If the surface of fluororesin is modified to semiconductor or metallic layers, fluororesin could be employed for a printed circuit for high-frequency,optical use or medical use. Some of the modification of the fluororesin surface to metallic layers reported are carried out by the electroless plating method[1][2].

We have been studying the substitution of functional groups into the fluororesin surface by using the ArF excimer laser[3][4][5]. Based on the photochemical modification method, we tried the deposition of silicon films onto the fluororesin surface at room temperature in order for the surface to be semiconducting. In this method, the fabrication process was simplified because the modification and deposition were performed in the same reaction cell, compared with the electroless plating process.

METHODOLOGY OF SILICON DEPOSITION

In this method, the deposition process of silicon films onto fluororesin surface is divided into two steps: first step is the photochemical modification of the surface; second step, the photo-CVD of silicon films.

Fluororesin is composed of C-F bonds, and its surface shows no affinity for any atoms. Therefore, defluorination of the surface is required to deposit silicon films. Even though C-F bonds of 128 kcal/mol are broken by ArF laser photon of 147 kcal, the fluorine atoms recombine with the carbon atoms since fluorine has the highest electronegativity. Thus, we have employed boron atoms to defluorinate the surface effectively[3]. As the boron atoms have a high bonding energy with the fluorine atoms, the boron atoms easily combine with the fluorine atoms which are broken by ArF laser light. As a result, the surface is effectively defluorinated for the substitution of silicon atoms to become possible. Moreover, the silicon layer of the modified surface functions as a nucleus for silicon film formation.

In the first step, $B(CH_3)_3$ and SiH_4 mixed gases are employed. $B(CH_3)_3$ is photodecomposed by the ArF laser light as follows:

$$B(CH_3)_3 + h\nu \longrightarrow B + 3CH_3$$

The photodissociated boron atoms pull out the fluorine atoms from the fluororesin surface. On the other hand, SiH_4 is dehydrogenated by the methyl radicals which are photodissociated from $B(CH_3)_3$ gas as follows:

$$SiH_4 + CH_3 \longrightarrow SiH_x + CH_4 \qquad (x=0\sim3)$$

The SiH_x radicals are combined with defluorinated carbon atoms of the surface[5].

In the second step, SiH_4 gas is employed to produce silicon atoms. SiH_4 has no optical absorption band at the 193 nm wavelength of the ArF laser; therefore photodecomposition of the gas must be carried out by the high-density excitation of ArF laser beam[6]. In this study, chemical vapor deposition was performed by focussed ArF excimer laser. Silicon films can, thus, be grown only on the irradiated parts at room temperature.

EXPERIMENTAL PROCEDURE

Schematic diagram of the experimental setup is shown in Figure 1. A transparent fluororesin film (FEP) with the thickness of 100 μm and the size of 10X25 mm2 was used. The FEP film was set on the backward surface of the window since the sample has an optical transmittance of 25 % at the 193 nm wavelength of the ArF laser[4]. Thus, the laser fluence was not attenuated by the gas absorption and could be kept constant at the interface between the sample and reactant gases. The reaction chamber was evacuated; in the chamber, SiH_4 (5 % in He) and $B(CH_3)_3$ gases were sealed at the partial pressure ratio of 40:100 Torr. Then, the surface was irradiated the ArF laser beam at the laser fluence of 12.5 mJ/cm2 and the laser shot number of 9000 through

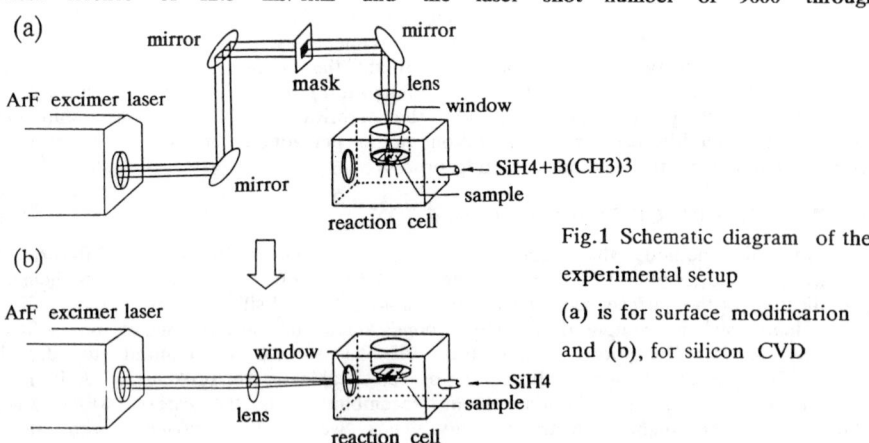

Fig.1 Schematic diagram of the experimental setup
(a) is for surface modificarion and (b), for silicon CVD

a mask-pattern. After modification, the chamber was re-evacuated and filled with SiH4 gas at the pressure of 100 Torr. And the ArF laser beam was irradiated parallel to the surface at the laser fluence of 1 J/cm2 and the laser shot number of 90000 in order to decompose the SiH4.

RESULTS AND DISCUSSION

Chemical composition of the modified and deposited surface

In order to inspect the chemical composition of the photo-modified surface, X-ray photoelectron spectroscopy (XPS) analysis of F 1s, Si 2p and C 1s spectra are carried out, as shown in Figure 2.

(a) exhibits the XPS spectra of the non-irradiated surface. The strong peak of F 1s was measured and was identified as CF2-CF2 bonds of fluororesin structure. The C 1s peak was positioned at 292 eV, which was also due to the CF2-CF2 structure bonds. Naturally, the Si 2p peak was not detected at all. (b) displays the XPS spectra of the modification surface. The F 1s peak remarkably decreased after modification. The C 1s peak was shifted from 292 to 284 eV; the 284 eV peak was identified as Si-C bonds. The Si 2p peak clearly appeared at 100 eV and was also identified as Si-C bonds. Accordingly, the surface was defluorinated and was modified into silicon carbide like.

(c) shows the XPS of the deposited surface after modification. The Si 2p peak was shifted from 100 to 99 eV after the silicon deposition. The 99 eV peak of Si 2p was determined as Si-Si bonds, which means the silicon atoms were deposited on the silicon carbide layers of the photo-modified surface.

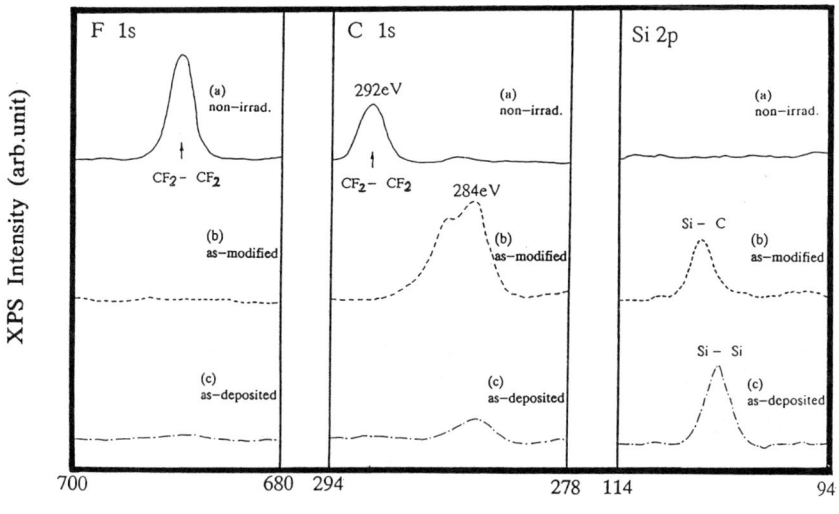

Fig.2 XPS analyses of F 1s, C 1s and Si 2p spectra.

(a) is without irradiation, (b) is as modified and (c), as deposited

Thickness measurement of the silicon deposition.

The deposited surface clearly turned to a metallic brown in contrast with the non-modified surface. Figure 3 indicates the film thickness as a function of the laser shot number. With the laser pulse repetition the film thickness linearly increased. And the 1000 Å of thickness was obtained at the laser shot number of 90000. With the 1000 Å film thickness, the optical bandgap significantly showed 1.8 eV. The 1.8 eV bandgap means the deposition of hydrogenated-silicon films. The deposition rate was 1.1×10^{-2} Å/pulse.

Figure 4 shows the film thickness measured by the surface roughness interference-meter. A selective deposition was confirmed only on the parts irradiated. The line and space was 200 μm.

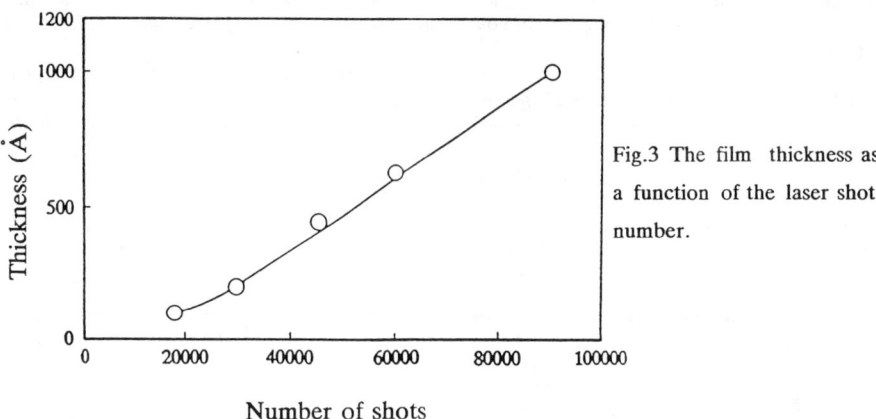

Fig.3 The film thickness as a function of the laser shot number.

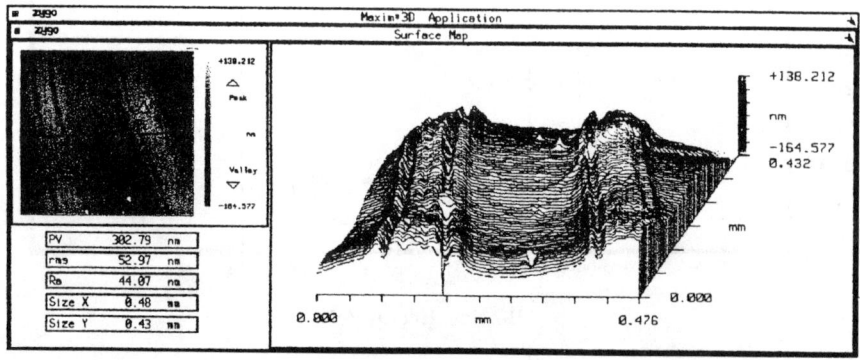

Fig.4 Measurement of the film thickness by the surface roughness interference-meter.

CONCLUSION

Fluororesin surface was photochemically modified to silicon layers, where the silicon films were then deposited only on the nucleus parts by the photo−CVD method. The chemical composition of the nucleus layers was determined as Si−C bond; the Si−C bond was created by the due to chemical bonding of the silicon atoms subustituted for the fluorine atoms and the carbon atoms remained on the surface after defluorination. The silicon were preferentially deposited on the irradiated parts while no silicon atoms, deposited on the non−irradiated parts. The film thickness of 1000 Å was achieved at room temperature, and its optical bandgap was 1.8 eV. The modification and silicon film deposition of the fluororesin surface were successfully demonstrated in the same reaction chamber. To employ fluororesin for developing electronic devices, further study to obtain a high quality of the deposited silicon film is needed.

REFERENCES

1. R. R. Rey, K. M. Chi, M. Hampden−Smith and T. Kodas : Mater. Res. Soc. Fall. Meeting, B3.4, 61 (1991)

2. A. Yabe and H. Ninno : Symposium on generation and application of vacuum ultraviolet (Tokyo), 103 (1993)

3. M. Okoshi and M. Murahara : J. Mater. Res, Vol 7, No.7 , Jul (1992)

4. M. Okoshi and M. Murahara : Mat. Res. Soc. Symp. Proc., Vol.201, 451 (1991)

5. M. Okoshi, T. Miyokawa, H. Kashiura, M. Murahara and K. Toyoda : Laser Science Progress Report of IPCR, 14, 33 (1991)

6. M. Takizawa, H. Saitou, Y. Komatsu, M. Okoshi and M. Murahara, Extend Abstract (The 38th Spring Meeting, 1991) The Japan Society of Applied Physics No.3, 885 (1991)

PYROLYTIC LCVD OF SILICON USING A PULSED VISIBLE LASER - EXPERIMENT AND MODELLING

B. IVANOV, C. POPOV, V. SHANOV AND D. FILIPOV
Technological University of Sofia, Dept. of Semiconductors, 8 Kliment Ohridski St., 1756 Sofia, Bulgaria

ABSTRACT

Maskless deposition of silicon from silane on Si monocrystalline wafer using copper bromide vapor laser (CBVL) is investigated. Morphology and geometric parameters of the stripes obtained are studied and some conclusions for the process mechanism are made. Applying Kirchoff's transform and Green's function analysis non-linear heat diffusion problem for different pulse shapes was solved. The influence of pulse shape on the temperature distribution and its time evolution was studied. Non-isothermal and non-stationary deposition kinetic models using the obtained results were developed.

INTRODUCTION

Laser-induced chemical vapor deposition (LCVD) is an attractive alternative to conventional CVD techniques, whenever direct writing of material patterns is desirable. Several papers concern deposition of Si from SiH_4 induced by cw Ar^+ laser [1-4]. Models of pyrolytic LCVD are developed in case of cw lasers [5,6]. With the exception of works Haba et al.[7,8] in literature there is a lack of information about the use of CVL for LCVD and laser-induced etching. This has directed our research to the activation of pyrolytic deposition of silicon from silane by such a laser and to model the system.

EXPERIMENTAL

The experimental set-up for LCVD was described in our previous work [9]. A focused, 1.2 W CBVL was used to irradiate (111) silicon monocrystalline wafers. The substrates were enclosed within a gas cell containing 1-1000 mbar of electronic grade 5% SiH_4 in Ar and were translated perpendicular to the optical axis with speeds of 40 - 400 μm/s. Typical focal spot diameters were at about 100 μm resulting in intensities of 8 - 48 MW/cm^2 within the focal spot. The width of the stripes obtained were evaluated with a light microscope and their height and profile were measured by Talystep. The morphology of the layers was investigated by scanning electron microscopy (SEM) and Raman spectroscopy.

The profile of the deposited material approximately follows the laser beam power distribution. As shown on Fig.1 the height of the Si stripes deposited from SiH_4 decreases with increase of the scanning speed. This is an expected effect due to the decrease in laser exposure time. Higher laser power results in increase of the stripes height because of the rise of surface temperature.

The width of the lines does not change when increasing the scanning speed. According to Bauerle the width of the deposited stripe does not change substantially when varying the surface temperature at lower activation energies [10]. If we consider the reported activation energy for Si growth from SiH_4 - 25 kcal/mol [11] as relatively low value then we can apply Bauerle's model. It could explain the width behavior when changing the scanning speed.

The morphology of the deposited silicon from silane shows well-defined grains. The grains size depends on the process parameters similar to LCVD written aluminium [9]. Higher laser power results in increase of the silicon grains which is demonstrated on Fig.2 - a) and b). Raman spectra indicate the typical for crystalline silicon shift at 520 cm^{-1} [12].

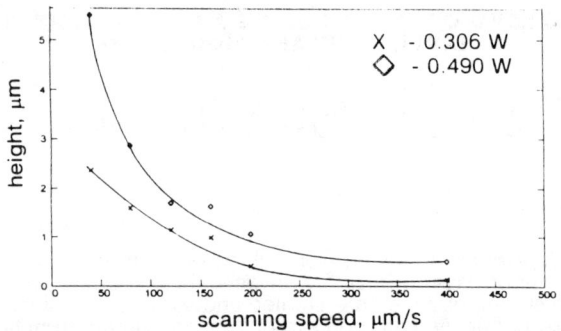

Fig.1 Average stripes height as a function of scanning speed at 1000 mbar 5% SiH_4 in Ar

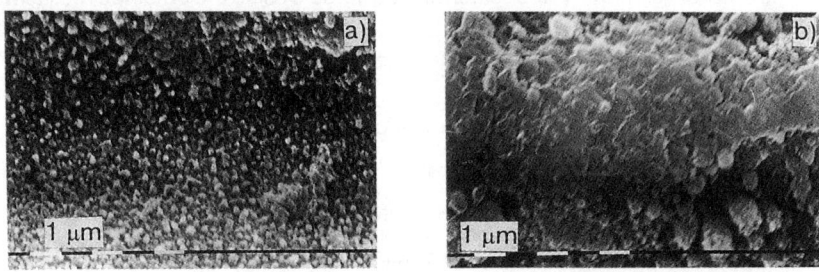

Fig.2 Scanning electron micrograpfs of Si stripes - 40 μm/s scanning speed,1000 mbar 5% SiH_4 in Ar   a) 0.214 W laser power   b) 0.490 W laser power

MODEL

The model is based on the following assumptions: (a) the thermal and optical properties of the substrate and of the growing layer are identical and they do not change during the process; (b) the temperature of the gaseous phase is constant; (c) the nucleation rate is high and it does not limit the total reaction rate; (d) the process is pyrolytic and photochemical effects are not considered. At first the temperature distribution as a result of pulsed laser irradiation is modelled. The obtained results are applied for known models concerning heterogeneous surface decomposition of SiH_4.

Modelling of temperature distribution

The temperature T in the irradiated zone can be described by the following equation

$$\frac{K(T)}{D(T)} \frac{\partial T}{\partial t} = \nabla \{K(T) \nabla T\} + SF(x,y,z,t) \tag{1}$$

where SF is the source function, K(T) the temperature dependent thermal conductivity and D(T) the thermal diffusivity. To solve this equation, Kirchoff's transform can be used which linearizes it by introduction of linearized temperature θ:

$$\theta(T) = \theta(T_0) + \int_{T_0}^{T} \frac{K(T)}{K(T_0)} \tag{2}$$

where $T_o = 300$ K. The boundary conditions are defined by approximation for semi-infinite solid. The consideration of the pulsed character of the irradiation is accomplished by defining of the source function. In case of round spot and Gaussian beam its form will be

$$SF(x,y,z,t) = \frac{P(t)(1-R)}{\pi w_0^2 \lambda} \exp\left[-\frac{(x-V_s t)^2 + y^2}{w_0^2}\right] \exp(-z/\lambda) \qquad (3)$$

where P(t) is the laser power in pulse, R the reflectance, w_o the laser spot radius, V_s the scanning speed and λ the absorption depth.
The power change during a pulse P(t) can be obtained if the average power of the pulsed laser P_a is known

$$P(t) = P_a I(t)/(f I_f) \qquad (4)$$

where I(t) is function, describing the pulse shape, f the laser frequency and I_f is the effective duration of the pulse. After introduction of the non-dimensional variables (5) the linearized equation obtains the form (6)

$$X = \frac{x}{w_o}; Y = \frac{y}{w_o}; Z = \frac{z}{\lambda}; \alpha = \frac{w_o}{\lambda}; D = \frac{K}{\rho C_p}; \tau = \frac{tD}{w_o}; \mu = \frac{V_s w_o}{D} \qquad (5)$$

$$\frac{\partial \theta}{\partial t} = \nabla^2 \theta + \frac{P_a(1-R)}{\pi K(T_0)\lambda f I_f} I(t) \exp\left[-(X-\mu\tau)^2 - Y^2 - \alpha Z\right] \qquad (6)$$

where ρ is the density of silicon and C_p the heat capacity. This equation could be solved using Green's function analysis in assuming that the free term SF is temperature independent. The reflectance is a function of temperature but in order to obtain solution of (6) we suppose that it is constant ($R = R(T_o) = $ const). This assumption is reasonable because for Si the reflectance is weaker function of temperature compared to K. The solution of this boundary-value problem is

$$\theta = \frac{P_a(1-R)}{2\pi K_0 \lambda f I_f} \int_0^t \int_{-\infty}^{+\infty} \int_{-\infty}^{+\infty} \int_0^{\infty} GF(X,Y,Z,\tau,X',Y',Z',\tau')SF(X',Y',Z',\tau')dX'dY'dZ'd\tau' \qquad (7)$$

where the Green's function is defined as

$$GF(X,Y,Z,\tau,X',Y',Z',\tau') =$$
$$= [4\pi(\tau-\tau')]^{-3/2} \exp\left[-\frac{(X-X')^2 + (Y-Y')^2}{4(\tau-\tau')}\right]\left\{\exp\left[-\frac{(Z-Z')^2}{4(\tau-\tau')}\right] + \exp\left[-\frac{(Z+Z')^2}{4(\tau-\tau')}\right]\right\} \qquad (8)$$

The real temperature in the irradiated zone can be obtained if the temperature dependence of K is known. For monocrystalline silicon K(T) could be well approximated by appropriate expression [13]. Applying the reverse Kirchoff's transform we obtain

$$T(\theta) = 99 + (T_0 - 99)\exp\left[C_p D T_0 \theta/299\right] \qquad (9)$$

Since R is very weak temperature function, the solution obtained for $R = R(T_o) = $ const could be used for determination of temperature in case of $R = f(T)$ by simple iteration procedure proposed by [14]. From the solution of the boundary-value problem (7) it can be seen that the temperature distribution strongly depends on the pulse shape so the latter should be precisely defined. The experimentally obtained shape of the laser pulse is defined with mathematical expression and its introduction in (6) allows to obtain the real temperature distribution. Temperature distribution in the time course for different approximations of the pulse shape (rectangular and Gaussian) compared with the temperature distribution for the real pulse is shown on Fig.3. As it can be seen the

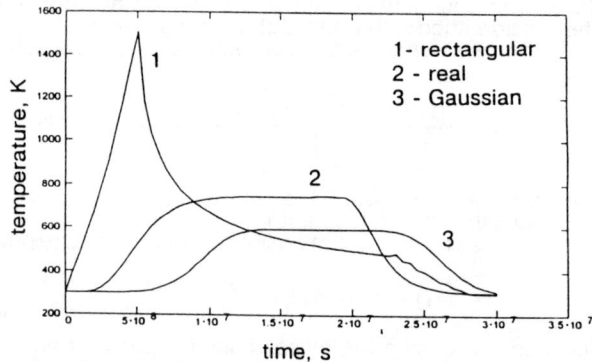

Fig.3 Temperature in the laser spot center for 0.490 W laser power and 100 μm spot diameter as a function of time

pulse shape substantially influence the temperature in the irradiated zone. On the basis of the obtained temperature distribution at the end of the pulse, the distance from the spot center to the point where T = 548 K (the temperature above which the decomposition of SiH_4 on the surface takes place [15]) is determined. It corresponds to the half-width of the deposited stripe. The theoretically calculated and the experimentally measured widths of the stripes are compared on Fig.4. As seen from this figure the model predicts well the experimental widths.

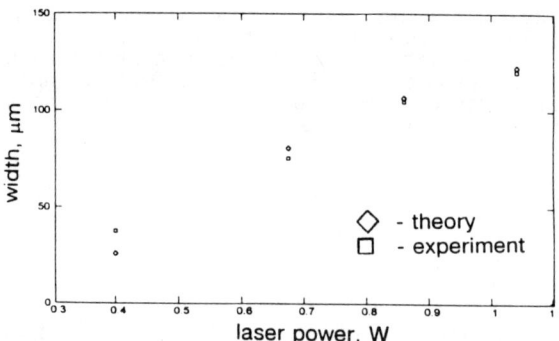

Fig.4 Theoretical and experimental widths of silicon stripes at 1000 mbar 5% SiH_4 in Ar as a function of laser power

Modelling of kinetics

The determination of kinetic parameters during the process is very difficult task that is why we have used cited data from literature. The results of several investigations concerning the thermal decomposition of SiH_4 on heated silicon substrate show that the process has a complex mechanism [16-18] and different pathways. According to the most of models the process could be considered from the point of view of Langmuir - Hinshelwood kinetics, including adsorption, surface chemical reaction and desorption. Our model for laser pyrolysis of SiH_4 using a pulsed laser is based on the following assumptions: (a) process of homogeneous pyrolysis of SiH_4 does not occur; (b) the obtained hydrogen is not adsorbed; (c) LCVD of Si is entirely pyrolytic process; (d) the gas phase diffusion is not taken into account.

According to Buss et al. the heterogeneous decomposition of SiH_4 includes dissociative adsorption, desorption and surface reaction [19]. In this work analytical expression for the reactive sticking coefficient (RSC) as a function of the substrate temperature are presented. In our case the process is non-isothermal and non-stationary and it is necessary RSC distribution along x and y directions and in the time course to be known. The temperature field and its time evolution already obtained are involved in the model of Buss et al. and RSC distributions as a function of coordinates and time are determined. Since RSC is time dependent function, an effective time t_{eff} for the reaction carried out for one pulse with average value of RSC (for x = 0 and y = 0 where RSC is maximum) could be defined.

$$t_{eff} = \int_0^{t_n} RSC(t) \, dt \qquad (10)$$

It is convenient the introduction of another parameter V_m which is constant for definite gas partial pressure:

$$V_m = F M / (\rho N_A) \qquad (11)$$

where F is SiH_4 molecules flux to the surface, M the atomic mass of silicon, ρ the density of monocrystalline silicon and N_A Avogadro's constant. The physical meaning of V_m is the maximum deposition rate assuming that RSC = 1 and its value is 2.71×10^{-3} m/s for 50 mbar partial pressure of SiH_4.

The height of the deposited stripe for one pulse Hp in the center of the laser spot (x = 0, y = 0) will be product of t_{eff} and V_m. The X-Y stepping stage used in the experiment performs step of 2 µm so the number of translations for which the laser beam spot (with diameter $2w_0$) will pass through a geometric point is w_0. It means that for 100 µm laser spot the number of translations will be 50.

The number of pulses at a point N_{ip} can be determined by the following expression

$$N_{ip} = s f / V_s \qquad (12)$$

where s is the step of X-Y stepping stage, V_s the scanning speed and f the laser frequency. Having in mind the character of translation of the X-Y stepping stage and RSC as a function of the coordinate x, along which this translation is carried out, the height of the deposited stripe H can be determined by the expression

$$H = H_p N_{ip} \sum_{i=0}^{50} \int_0^{t_n} RSC(t, X_i) dt \Big/ \int_0^{t_n} RSC(t, X = 0) dt \qquad (13)$$

The obtained results from (14) unfortunately show substantial discrepancy with the experimental data in contrast to the widths modelling. Our opinion is that the kinetics of LCVD of Si from SiH_4 using copper bromide vapor laser cannot be defined with the known kinetic parameters. In literature there is information about photoeffect during the deposition of silicon from silane induced by cw Ar^+ laser, but this effect takes place only at low laser powers [20]. The maximum known value of RSC for decomposition of SiH_4 on Si is 0.2 (for temperatures above 1500 K) [21]. Even if we assume that RSC = 0.2 for the whole range of times during laser irradiation for which the surface temperature is above 548 K - t_r, the theoretical values of the deposit heights in case of flat top beam distribution can be given by

$$H_m = V_m t_r 0.2 w_0 f / V_s \qquad (14)$$

The theoretically calculated and the experimentally obtained values of the stripes height for different incident laser powers, 40 µm/s scanning speed and 50 mbar partial pressure of SiH_4 are presented in Table I.

Table I Theoretical and experimental heights of silicon stripes as a function of laser power

Incident laser power, W	0.490	0.398	0.306	0.214
Theoretical height, µm	1.25	1.19	0.87	absent
Experimental height, µm	5.50	3.13	2.36	2.04

As seen from Table I even for rather high values of RSC the theoretical heights of the stripes are lower in comparison with the experimental data. Besides especially indicative is that no deposition could occur for incident laser power of 0.214 W. In this case the temperature is below 548 K while the actual height is rather high - 2.04 µm. These results show that the kinetics of silicon deposition from SiH_4 induced by CBVL does not follow the known kinetic models for thermal decomposition of silane on silicon. A probable reason for this could be the conciderably lower activation energy of the deposition process enhanced by photoeffect during the laser irradiation. The determination of the process kinetics requires additional labour-consuming investigations because of the non-isothermal and non-stationary character of the process and the probability for appearance of non-linear photoeffects.

CONCLUSIONS

Pyrolytic LCVD of silicon from silane using CBVL is reported for the first time. The influence of the process parameters on the geometry and morphology of the obtained stripes is investigated. Models for surface temperature distribution and RSC are developed. The predicted widths according to the model are close to the experimentally obtained values. Substantial discrepancy is observed between the heights experimentally determined and based on the model which could be due to non-linear photoeffects.

ACKNOWLEDGEMENTS

The authors are grateful to the Ministry of Science and Education for financial support under Project 177-X.

REFERENCES

1. D.J. Ehrlich, R.M. Osgood, Jr., T.F. Deutsch, Appl. Phys. Lett. **39** (12), 957-959 (1981)
2. D. Bauerle, P. Irsigler, G. Leyendecker, H. Noll, D. Wagner, Appl. Phys. Lett. **40** (9), 819-821 (1982)
3. T.D. Binnie, J.I. Wilson, M.J. Colles, J.L. West, J. Appl. Phys. **58** (11), 4446-4448 (1985)
4. S.K. Roy, A.S. Vengurlekar, A.V. Joshi, S. Chandrasekhar, J. Electronic Materials **16** (4), 211-217 (1987)
5. D.C. Skouby and K.F. Jensen, J. Appl. Phys. **63** (1), 198-205 (1988)
6. N. Arnold, R. Kullmer, D. Bauerle, Microelectronic Engineering **20**, 31-54 (1993)
7. B. Haba, B.W. Hussey, A. Gupta, J. Appl. Phys. **69** (5), 2871-2876 (1990)
8. B. Haba, B.W. Hussey, A. Gupta, R.J. Baseman, Mat. Res. Soc. Symp. Proc. **158** (1990) pp. 319-324
9. V. Shanov, B. Ivanov, C. Popov, Thin Solid Films **207**, 71-74 (1992)
10. D. Bauerle, Chemical Processing with Lasers (Springer-Verlag, Heidelberg, 1986), p. 57
11. S. Boughaba and G. Auvert, Appl. Surf. Sci. **54**, 21-29 (1992)
12. I. Herman, F. Magnotta, D.E. Kotecki, J. Vac. Sci. Technol. A **4** (3), 659-664 (1986)
13. Y.I. Nissim, A. Lietoila, R.B. Gold, J.F. Gibbons, J. Appl. Phys. **51** (1), 274-279 (1980)
14. J. Moody and R.H. Hendel, J. Appl. Phys. **53**, 4364 (1982)
15. S.M. Gates, C.M. Greenlief, D.B. Beach, R.R. Kunz, Chem. Phys. Lett. **154** (6), 505-510 (1989)
16. K.F. Jensen and D.B. Graves, J. Electrochem. Soc. **130** (9), 1950-1957 (1983)
17. J.M. Jasinski and S.M. Gates, Acc. Chem. Res. **24** (1), 9-15 (1991)
18. J. Holleman and J.F. Verweij, J. Electrochem. Soc. **140** (7), 2089-2097 (1993)
19. R.J. Buss, P. Ho, W.G. Breiland, M.E. Coltrin, J. Appl. Phys. **63** (8), 2808-2819 (1988)
20. G. Auvert, D. Tonneau, Y. Pauleau, Appl. Phys. Lett. **52** (13), 1062-1064 (1988)
21. M.K. Farnaam and D.R. Olander, Surf. Sci. **145**, 390-406 (1984)

EFFECT OF POST EXPOSED HYDROGEN ON CHEMISORBED ETHYLENE ON Si(100)-(2x1)

WOLF WIDDRA, CHEN HUANG[+], AND W. HENRY WEINBERG
Department of Chemical Engineering and Center for Quantized Electronic Structures (QUEST), University of California, Santa Barbara, CA 93106

ABSTRACT

The effect of post adsorbed atomic hydrogen on the adsorption, desorption, and decomposition of ethylene on Si(100)-(2x1) has been studied using high-resolution electron energy loss spectroscopy (HREELS), temperature programmed desorption (TPD), and low-energy electron diffraction (LEED). Exposures to atomic hydrogen of more than 10^{15} atoms/cm^2 convert the initial (2x1) reconstruction of sp^3-hybridized, di-σ bonded ethylene to a (1x1) structure. Furthermore, after post exposure to atomic hydrogen, the thermal desorption peak of molecular ethylene is shifted up by approximately 100 K and reduced in intensity. HREELS spectra for deuterated ethylene show the formation of a C-H bond after exposure to atomic hydrogen, whereas the C-C bond remains intact. We explain our data by an atomic hydrogen-driven conversion of the di-σ bonded ethylene to a mono-σ bonded surface ethyl. Thermal activation after post exposure to atomic hydrogen leads to decomposition of about 60% of the initial ethylene in contrast to the observed molecular desorption in the absence of hydrogen.

I. INTRODUCTION

The interaction of unsaturated hydrocarbons with silicon surfaces has received attention recently because of its importance for the growth of SiC thin films by chemical vapor deposition (CVD) and chemical beam epitaxy. Some studies have focused on the reaction of precursor molecules like acetylene or ethylene on well-defined silicon surfaces [1,2]. It was shown that acetylene and ethylene are both di-σ bonded to the Si-Si dimers on a Si(100)-(2x1) surface [3-6]. At high temperatures, which are common for CVD, acetylene largely decomposes to SiC liberating molecular hydrogen, whereas ethylene mainly desorbs molecularly [7,8]. Rather small differences in the activation energy for desorption decide whether the molecule will desorb with increasing temperature or undergo thermally activated decomposition. However, the influence of coadsorbates have been neglected. In particular, hydrogen is known to play an important role in many CVD processes, either as a reaction product (e.g., from silane, disilane, and diethylsilane) or as a carrier gas. *Preadsorbed* hydrogen is known to block subsequent hydrocarbon adsorption on silicon surfaces [7,9], but there are few studies concerning the influence of *post exposures* to atomic hydrogen [10]. It was shown previously that low exposures to atomic hydrogen on an ethylene-saturated surface will saturate the silicon dangling bonds which remain after ethylene adsorption [6]. This was used to clarify the structure and bonding of ethylene on Si(100)-(2x1). Here, we report on the chemical reaction between ethylene and atomic hydrogen for higher exposures (> 10^{15} atoms/cm^2). High atomic hydrogen exposures cause formation of an ethyl by breaking one of the Si-C bonds of adsorbed ethylene leading to significant decomposition upon heating in contrast to the mainly molecular desorption in the absence of hydrogen.

II. EXPERIMENTAL DETAILS

The experimental setup and the silicon sample preparation have been described previously [5,11]. Briefly, the ultrahigh vacuum (UHV) chamber with a base pressure of 7 x 10^{-11} Torr contains a HREEL spectrometer (LK-2000-14-R), LEED optics, an Auger electron spectrometer, an ion sputter gun, a pin-hole gas doser, a tungsten filament to produce atomic hydrogen, and a differentially pumped quadrupole mass spectrometer (UTI-100C). The chamber is pumped by a 1000 l/s turbomolecular pump.

Atomic hydrogen and deuterium were generated by dissociation of the respective molecules with a hot (1800 K) spiral tungsten filament positioned approximately 5 cm from the sample.

III. RESULTS AND DISCUSSION

Thermal desorption spectra for a Si(100)-(2x1) surface which was saturated at 100 K with deuterated ethylene and subsequently exposed to various fluences of atomic deuterium at approximately 150 K are shown in Fig.1. The TPD spectrum for ethylene in the absence of hydrogen is shown in Fig. 1(a). The molecular desorption of ethylene occurs around 590 K and shows a slightly asymmetric desorption peak, in agreement with earlier studies [7]. The desorption spectra after exposures of 7×10^{14} and 7×10^{15} D atoms/cm^2 are shown in Fig. 1(b) and (c), respectively. The observed desorption products are molecular D_2 and C_2D_4. The ethylene desorption peak broadens, shifts slightly to higher temperature, and is reduced in intensity after a deuterium post exposure of 7×10^{14} atoms/cm^2. After an exposure of 7×10^{15} atoms/cm^2, the ethylene desorption peak is shifted up by about 100 K to a desorption temperature of 700 K, and

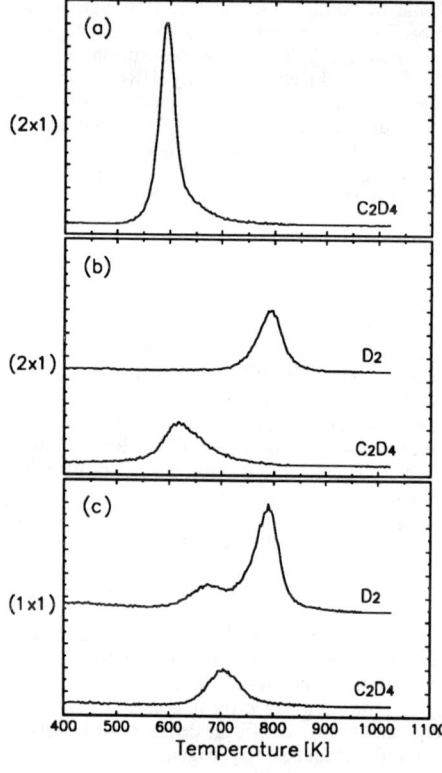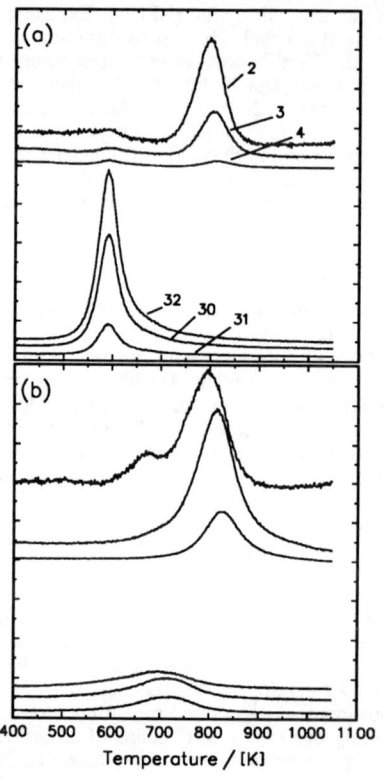

Figure 1: TPD spectra for the C_2D_4-saturated Si(100)-(2x1) surface after post exposure to atomic deuterium: (a) Without exposure to atomic deuterium; (b) after 7×10^{14} atoms/cm^2 deuterium exposure; (c) after 7×10^{15} atoms/cm^2 deuterium exposure. The desorption products are molecular C_2D_4 and molecular deuterium. The heating rate was 2 K/s. The observed LEED patterns for each case are noted on the left side.

Figure 2: TPD spectra for the C_2D_4-saturated Si(100)-(2x1) surface after exposure to atomic hydrogen: (a) 1.4×10^{14} atoms/cm^2 and (b) 1.1×10^{16} atoms/cm^2. Masses 2 (H_2), 3 (HD), 4 (D_2), 32 (C_2D_4), 31 (C_2D_3H), and 30 ($C_2D_2H_2$) are monitored. Masses 30 and 31 have contributions from the cracking pattern of C_2D_4 (60% $C_2D_3^+$) and an isotopic impurity in the C_2D_4 (15% C_2D_3H), respectively.

the area of the desorption peak is reduced to about 40% of ethylene desorption in the absence of hydrogen. Molecular deuterium desorbs around 790 K after an exposure of 7×10^{14} D atoms/cm^2. At the higher exposure an additional desorption peak around 670 K is seen in Fig. 1(c). The deuterium desorption spectra which are shown here are similar to the desorption of deuterium from a clean Si(100) surface after atomic deuterium exposures at low temperatures. The monodeuteride and the dideuteride desorption peaks, originating from deuterium atoms bonded to a Si-Si dimer and from two deuterium atoms bonded to the same silicon atom are clearly seen, as the high- and low-temperature peak, respectively. In addition to the significant upshift of the ethylene desorption temperature and the appearance of a dideuteride desorption peak, the LEED pattern changes from a (2x1) to a (1x1) for exposures to atomic deuterium above 1.5×10^{15} atoms/cm^2.

An isotopic exchange experiment using deuterated ethylene and atomic hydrogen can illuminate the origin of the molecularly desorbing hydrogen and the hydrogen in the desorbing ethylene. Figure 2(a) and (b) show TPD spectra after post exposures to atomic hydrogen of 1.4×10^{14} and 1.1×10^{16} atoms/cm^2. The TPD signal was monitored for masses 2 (H_2), 3 (HD), 4 (D_2), 32 (C_2D_4), 31 (C_2D_3H), and 30 ($C_2D_2H_2$). For the higher exposure, Fig. 2(b), the spectra again show a shift to higher temperatures and a reduction in desorption intensity of the ethylene desorption peaks. For low exposures, the hydrogen desorption shows the typical monohydride peak. The small amount of deuterium, desorbing as HD, originates from some ethylene decomposition which can be seen also on a nominally clean surface [7]. For the higher exposure of 1.1×10^{16} atoms/cm^2, hydrogen desorbs with a monohydride and a dihydride peak. However, the dihydride desorption peak at lower temperatures is only observed for H_2, not for HD or D_2. This indicates that only the isotopic SiH$_2$ dihydride is present on the surface at the onset temperature of thermal desorption. For a proper interpretation of the isotopic exchange data for ethylene (i.e., the origin of masses 31 and 30), the cracking pattern of ethylene must be taken into account ($C_2H_3^+/C_2H_4^+ = 0.6$, as determined independently for our system). Additionally, a C_2D_3H isotopic impurity (15%) in the C_2D_4 gas must be corrected for. The TPD spectra after a low hydrogen exposure, shown in Fig. 2(a), for masses 30 and 31 originate completely from these two effects. In other words, there is no isotopic exchange within the ethylene molecule up to the thermal desorption temperature. This conclusion has been confirmed by HREELS data (*vide infra*) and by TPD results using C_2H_4 and atomic deuterium which are not shown here. The data for an exposure of 1.1×10^{16} atoms/cm^2 in Fig. 2(b), on the other hand, show a strong ethylene peak with mass 31, which indicates isotopic exchange into the ethylene molecule. Clearly, the atomic hydrogen has reacted with the ethylene molecule prior to thermal desorption. After correction for the isotopic impurity and the various cracking patterns, the isotopic mixing in Fig. 2(b) is given by $C_2D_4 : C_2D_3H : C_2D_2H_2 = 1 : 0.45 : 0.32$.

To clarify the bonding on the surface at low temperatures, vibrational spectroscopy has been applied. Figure 3 shows the HREEL spectra of adsorbed C_2H_4 and C_2D_4 on the Si(100)-(2x1) surface at 100 K and saturation coverage. Seven energy loss peaks at 2928, 1416, 1196, 1076, 922, 655, and 465 cm^{-1}, and a shoulder at 760 cm^{-1} are observed in the hydrogenated spectrum, (a). In the spectrum for deuterated ethylene, (b), five energy loss peaks appear at 2185, 1065, 780, 630, and 453 cm^{-1}, with a shoulder around 908 cm^{-1}. Comparison with infrared spectra of organosilicon compounds allows an assignment of the peaks [12], as discussed in detail elsewhere [6]. To summarize, the 2928 cm^{-1} peak in Fig. 3(a) shifts down to approximately 2185 cm^{-1} after deuteration and is a symmetric stretching vibration of the C-H bond. The 1416, 1196, and 922 cm^{-1} peaks in Fig. 3(a) can be assigned as CH$_2$ deformation modes: the CH$_2$ scissoring, the CH$_2$ wagging, and the CH$_2$ rocking vibrations. After deuteration, the frequency of the scissoring mode shifts down and overlaps with the 1065 cm^{-1} peak. A similar down-shift from 1196 to 908 cm^{-1} is also observed for the wagging mode. The downshifted rocking mode is masked by the 780 cm^{-1} peak. The C-C stretching vibration of the sp^3-hybridized, adsorbed ethylene can be seen at 1076 and 1065 cm^{-1} in Fig. 3(a) and (b), respectively. The 655 cm^{-1} peak which is downshifted upon deuteration to 630 cm^{-1} is the Si-C symmetric stretching vibration and the 465 cm^{-1} peak which is downshifted to 453 cm^{-1} is probably the Si-C asymmetric stretching vibration. The peak between 760 and 780 cm^{-1} is due to a small SiC contamination in the near-surface region. However, Auger electron spectroscopy shows only a weak carbon signal for the clean surface: a peak ratio of C(271 eV)/Si(92 eV) below 0.001. The HREELS data after different post exposures to atomic hydrogen are shown for a deuterated ethylene-saturated surface in Fig. 4. The spectrum without

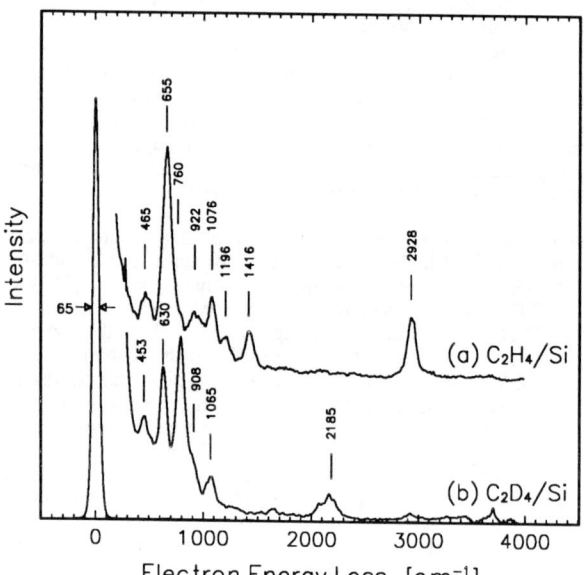

Figure 3: HREEL spectra of adsorbed ethylene on the Si(100)-(2x1) surface at 100 K: (a) C_2H_4/Si(100) and (b) C_2D_4/Si(100). The spectra are magnified by a factor of 300 with respect to the elastic peak. The spectra are measured in the specular direction with an electron energy of 7 eV.

atomic hydrogen has been discussed above and is shown again for comparison. The influence of low exposures ($< 10^{15}$ atoms/cm^2) to atomic hydrogen, as shown in Fig. 4(b) and (c), has been discussed in detail elsewhere [6]. Briefly, for low exposures the ethylene spectrum is only slightly perturbed and hydrogen is bonded to the silicon dangling bonds, as can be seen by the appearance of the Si-H stretching mode. However, with higher atomic hydrogen exposures, as shown in Fig. 4(d) and (e), the Si-H stretching mode at 2100 cm^{-1} increases in intensity and two new loss features are clearly visible: the SiH$_2$ scissoring mode at 980 cm^{-1} and the C-H stretching mode at 2950 cm^{-1}. The latter is increased well above the contribution from the C$_2$D$_3$H isotopic impurity mentioned earlier. Additional HREEL spectra, which are not shown here [13], for high post exposures to atomic deuterium on an ethylene-saturated surface, where the 1000 cm^{-1} region is not masked by C-D deformation modes, show that the C-C single bond is preserved after post adsorption of deuterium.

As the TPD and LEED data clearly indicate, the interaction of post adsorbed hydrogen with an ethylene-saturated Si(100)-(2x1) surface occurs in two distinct regimes. In the case of low exposures ($< 10^{15}$ atoms/cm^2), hydrogen saturates the silicon dangling bonds and there is no chemical reaction between ethylene and atomic hydrogen from the gas phase [6]. For higher exposures, the atomic hydrogen converts the ethylene to an adsorbed ethyl group, which leads to a change of the (2x1) to a (1x1) LEED pattern. The incorporation of atomic hydrogen into the ethyl group and the presence of a C-C single bond are shown by HREELS data (Fig. 4, [13]). We explain the data by an atomic hydrogen-induced breaking of one of the Si-C bonds, leading to formation of an ethyl. This explains the emergence of a (1x1) LEED pattern. The vibrational spectroscopy clearly shows the incorporation of hydrogen into the newly formed ethyl group and the presence of a C-C single bond. The ethyl group can either decompose via α-hydride elimination, or undergo β-hydride elimination, liberating ethylene [14]. In our case, the ethyl group is stabilized by the formation of silicon dihydride on the adjacent silicon site of the former Si-Si dimer. Upon heating, the dihydride must desorb first to reopen the reaction path for β-hy-

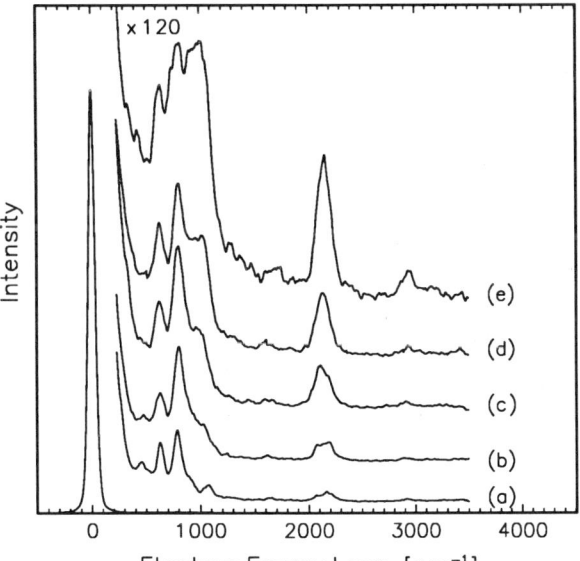

Figure 4: HREEL spectra for the C_2D_4-saturated Si(100)-(2x1) surface after various post exposures to atomic hydrogen: (a) without post exposure; (b) after 2.1×10^{14} H atoms/cm^2; (c) after 5.6×10^{14} H atoms/cm^2; (d) after 1.4×10^{15} H atoms/cm^2; (e) after 3.1×10^{15} H atoms/cm^2.

dride elimination and desorption of ethylene. The expected significant shift in the ethylene desorption temperature above the dihydride desorption temperature has been unambiguously observed. Furthermore, the mechanism predicts the predominance of H_2 dihydride desorption, as is also observed experimentally. At these high temperatures, the conversion between an adsorbed ethyl group and ethylene might be reversible before the molecule finally decomposes to surface carbon and hydrogen or desorbs. This would explain the observed isotopic exchange of more than one hydrogen atom within the ethylene molecule. For the highest exposures to atomic hydrogen, a small probability of a direct reaction between gas-phase atomic hydrogen and the adsorbed hydrocarbon cannot be excluded completely [14].

IV. CONCLUSIONS

We have shown that high exposures to atomic hydrogen ($> 10^{15}$ atoms/cm^2) of an ethylene-saturated Si(100)-(2x1) surface leads to a σ-bonded ethyl, forming a (1x1) surface structure. The ethyl is significantly stabilized by the formation of a silicon dihydride blocking the possibility of hydrocarbon desorption via β-hydride elimination. Therefore, about 60% of the initial ethylene decomposes, leaving carbon on the surface and liberating hydrogen. This is in contrast to the mainly molecular desorption of ethylene in the absence of post adsorbed hydrogen and saturation of silicon dangling bonds for low exposures to atomic hydrogen.

ACKNOWLEDGMENTS

This work was supported by QUEST, a National Science Foundation Science and Technology Center (grant no. DMR 91-20007), the W. M. Keck Foundation, a MICRO/ SBRC grant, and a NATO grant. One of us (WW) thanks the Alexander-von-Humboldt Foundation for a Feodor-Lynen Research Fellowship.

REFERENCES

+ Present address: Battelle-Pacific Northwest Laboratory, P. O. Box 999, MS K2-14, Richland WA 99352.
[1] C.C. Cheng, P.A. Taylor, R.M. Wallace, H. Gutleben, L. Clemen, M.L. Colaianni, P.J. Chen, W.H. Weinberg, W.J. Choyke, and J.T. Yates, Jr., Thin Solid Films **225**, 196 (1993).
[2] P.A. Taylor, M.J. Bozack, W.J. Choyke, and J.T. Yates, Jr., J.Appl.Phys. **65**, 1099 (1989).
[3] J. Yoshinobu, H. Tsuda, M. Onchi, and M. Nishijima, J. Chem. Phys. **87**, 7332 (1987).
[4] M. Nishijima, J. Yoshinobu, H. Tsuda, and M. Onchi, Surf. Sci. **192**, 383 (1987).
[5] C. Huang, W. Widdra, X.-S. Wang, and W.H. Weinberg, J. Vac. Sci. Technol. A **11**, 2250 (1993).
[6] C. Huang, W. Widdra, and W.H. Weinberg, Surf. Sci. Lett. (submitted).
[7] L. Clemen, R.M. Wallace, P.A. Taylor, M.J. Dresser, W.J. Choyke, W.H. Weinberg, and J.T. Yates, Jr., Surf. Sci. **268**, 205 (1992).
[8] P.A. Taylor, R.M. Wallace, C.-C. Cheng, M.J. Dresser, W.J. Choyke, W.H. Weinberg, and J.T. Yates, Jr., J. Am. Chem. Soc. **114**, 6754 (1992).
[9] M.J. Bozack, W.J. Choyke, L. Muehlhoff, and J.T. Yates, Jr., J. Appl. Phys. **60**, 3750 (1986).
[10] M.J. Bozack, P.A. Taylor, W.J. Choyke, and J.T. Yates, Jr., Surf. Sci. **179**, 132 (1987).
[11] X.-S.Wang, et al., J. Vac. Sci. Technol. A (in press).
[12] D.R. Anderson, in *Analysis of Silicones*, A.L. Smith, Ed. (John Wiley, New York, 1978) p.247.
[13] W. Widdra, C. Huang, and W.H. Weinberg (in preparation).
[14] M.A. Rueter and J.M. Vohs, Surf. Sci. **262**, 42 (1992).
[15] C.C. Cheng, S.R. Lucas, H. Gutleben, W.J. Choyke, and J.T. Yates, Jr., Surf. Sci. Lett. **273**, 441 (1992).

THE DECOMPOSITION OF METHYLTRICHLOROSILANE: STUDIES IN A HIGH-TEMPERATURE FLOW REACTOR*

MARK D. ALLENDORF, THOMAS H. OSTERHELD, and CARL F. MELIUS
Combustion Research Facility, Mail Stop 9052, Sandia National Laboratories, Livermore, CA 94551-0969

ABSTRACT

Experimental measurements of the decomposition of methyltrichlorosilane (MTS), a common silicon carbide precursor, in a high-temperature flow reactor are presented. The results indicate that methane and hydrogen chloride are major products of the decomposition. No chlorinated silane products were observed. Hydrogen carrier gas was found to increase the rate of MTS decomposition. The observations suggest a radical-chain mechanism for the decomposition. The implications for silicon carbide chemical vapor deposition are discussed.

INTRODUCTION

Methyltrichlorosilane (MTS) is commonly used in chemical vapor infiltration processes as a precursor to silicon carbide (SiC) [1, 2] Its use is also being explored for the production of thin films for electronics applications [3]. The kinetics of SiC chemical vapor deposition from MTS have been of interest for some time, since computational models are needed to assist in the optimization and scale-up of new synthetic methods. Unfortunately, little is known about the high-temperature reactions of chlorinated organosilanes. In a widely cited paper, Burgess and Lewis measured the MTS pyrolysis rate in hydrogen at atmospheric pressure [4]. Later, Davidson and Dean attempted to measure the unimolecular pyrolysis rates for a series of chlorinated methylsilanes (not including MTS), but found it difficult to achieve non-chain conditions for these systems [5]. Most recently, Niiranen and Gutman measured the $SiCl_3$ + CH_3 recombination rate at 300 K and 2 torr using photoionization mass spectrometry [6]. The limited information relevant to MTS available from these studies indicates that additional experimental data are needed to fully understand the decomposition chemistry of this precursor.

In this paper, we describe measurements of MTS decomposition conducted in a high-temperature flow reactor (HTFR) using a mass spectrometer to monitor the course of reaction. The objectives are: 1) to identify the products of MTS decomposition, 2) to determine the effects of different carrier gases on the decomposition, and 3) to suggest a mechanism for pyrolysis. Following a description of the experimental procedures used, a brief discussion of earlier theoretical predictions is presented to provide useful background for understanding the experimental results.

EXPERIMENTAL METHODS

A schematic view of the HTFR used in these experiments is shown in Figure 1. Reactions occur within a graphite tube with an ID of 5.0 cm and a length of 100 cm. The tube is enclosed in a water-cooled, insulated vacuum chamber. Three independently controlled heating elements surround the tube and can heat the gases flowing within it up to 1500 K. Reactor pressure is controlled by a pressure transducer coupled to a throttle valve in the vacuum line. In a typical experiment, carrier gas (hydrogen or helium) enters the reactor tube at the top and is preheated to the reaction temperature by the first heating element. MTS is then added to the hot carrier gas through a movable, water-cooled injector, allowing its residence time to be varied (from zero to 100 msec in these experiments, based on the average flow velocity). The residence time of the MTS is then varied with respect to a mass spectrometer probe located at the center of

---

*This work was supported by the Advanced Industrial Materials Program of the U.S. Dept. of Energy Office of Industrial Technologies.

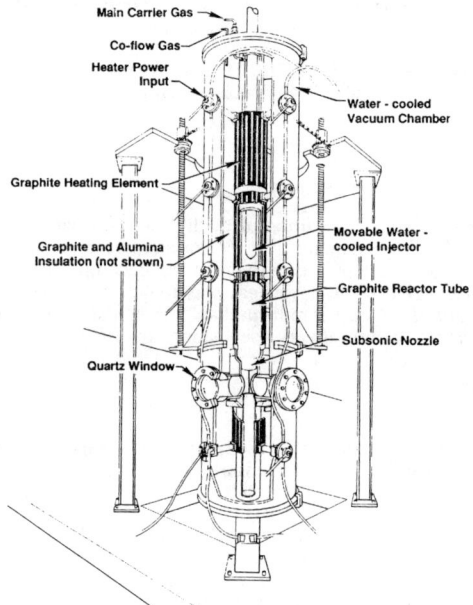

Figure 1: Schematic view of the high-temperature flow reactor.

the interaction region, defined by the intersection of the window ports. The following HTFR conditions were used: total reactor pressure, 25.0 ± 0.2 torr; reactor temperature, 1243 ± 10 K; total flow rate, 5.00 slpm; MTS flow rate, 50 sccm.

MTS and its decomposition products were detected by an Extrel EXM-500 quadrupole mass spectrometer using electron impact ionization. Masses up to 500 amu and species concentrations as low as 5 ppm (based on detection of ^{38}Ar in air) are observable with this instrument. Gases are extracted from the HTFR by a quartz sampling probe with a 475-µm orifice inserted into the center of the flow in the diagnostic region. The pressure inside the probe is maintained at 1.00 torr by a pressure transducer/throttle valve combination. Since the probe pressure is typically a factor of 10 or more lower than the pressure within the HTFR, the rates of chemical reactions (in particular, radical-radical reactions) within the probe are substantially reduced. After extraction, the sampled gases flow past a 200-µm orifice attached to the mass spectrometer chamber. The small amount of the gases leaking through the orifice forms a molecular beam, which is then ionized and detected by the spectrometer.

THEORETICAL PREDICTIONS OF MTS PYROLYSIS

The experiments described here provide an opportunity to test some of the predictions of our earlier theoretical analyses of MTS pyrolysis, in which we describe both the reaction thermochemistry [7] and kinetics [8]. To estimate MTS decomposition rates as a function of temperature and pressure, transition state (RRKM) theory was employed to predict rates for several unimolecular MTS decomposition pathways [8], using transition state structures obtained from ab initio electronic structure calculations [7]. These calculations indicate that the three most important decomposition pathways are:

$$CH_3SiCl_3 + M \rightarrow CH_3 + SiCl_3 + M \quad (1)$$
$$CH_3SiCl_3 + M \rightarrow CH_2SiCl_3 + H + M \quad (2)$$
$$CH_3SiCl_3 + M \rightarrow CH_2=SiCl_2 + HCl + M \quad (3)$$

At the temperatures and pressures typical of SiC CVD (1300 - 1500 K, 10 - 760 torr), the rate of Reaction (1) exceeds that of the other two by at least two orders of magnitude. This is illustrated in Figure 2 for the case of hydrogen carrier gas at 1300 K. Exchanging hydrogen for helium increases the rates of Reactions (2) and (3) relative to Reaction (1), but the rate of Reaction (1) still exceeds that of the other two by at least a factor of 65.

A second prediction of these calculations is that substitution of hydrogen for helium as the carrier gas will decrease the total MTS decomposition rate by about a factor of two, due to the less effective collisional energy transfer expected from hydrogen. A final important finding is that all three reaction channels are sensitive to the total pressure, as illustrated in Figure 2. Decreasing the hydrogen carrier gas pressure from 760 torr to 10 torr results in a factor of 13 decrease in the total MTS decomposition rate.

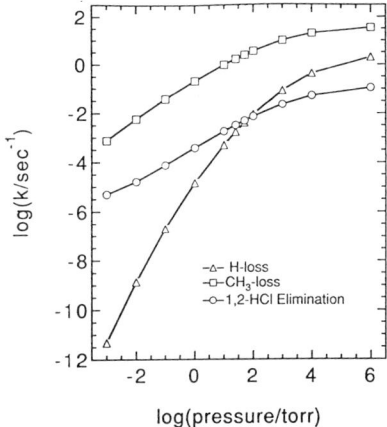

Figure 2: Pressure dependence of the three major MTS unimolecular decomposition pathways.

EXPERIMENTAL RESULTS AND DISCUSSION

Experiments in the HTFR using mass spectrometric detection identified two products of MTS decomposition. The results of these experiments are shown in Figures 3 and 4. Figure 3 shows the mass spectrum obtained for a mixture of 1% MTS in helium with a residence time in the heated zone of 82 msec. Peaks due to background gases were partially removed by subtracting the spectrum obtained in the absence of MTS. Peaks at m/z (mass/charge) ratios of 148, 133, 113, 98, 76, and 63 correspond to the $CH_3SiCl_3^+$, $SiCl_3^+$, $CH_3SiCl_2^+$, $SiCl_2^+$, $CHSiCl^+$, and $SiCl^+$ ions produced by fragmentation of MTS in the spectrometer (smaller peaks 2-6 mass units above each of these correspond to fragments containing the ^{37}Cl isotope). Also apparent in this spectrum are peaks at m/z=36 and 38, corresponding to the two isotopic forms of HCl; this identifies HCl as an MTS decomposition product. Peaks in the range of m/z values between 12 and 30, where hydrocarbon fragments are expected to appear, cannot be readily identified since accurate baseline subtraction is difficult to achieve (often producing negative peaks, for example) due to overlap with much larger background peaks.

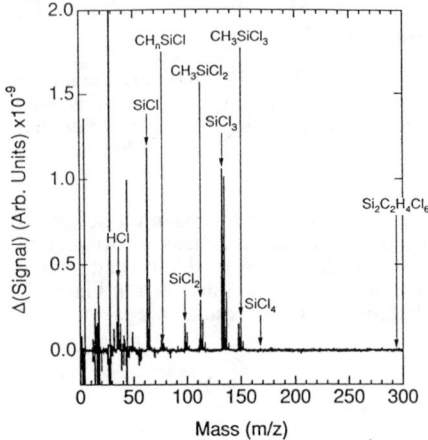
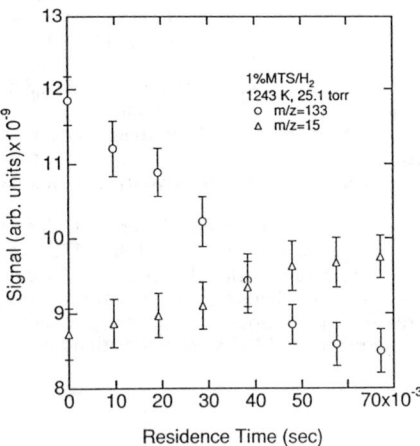

Figure 3: Mass spectrum (background-substracted) of MTS in helium at 25.0 torr and 1243 K, showing the presence of HCl as a product and the absence of chlorosilanes $SiCl_4$ and $Cl_3SiCH_2CH_2SiCl_3$ (labelled $Si_2C_2H_4Cl_6$ in the figure) as products.

Figure 4: Signal-averaged data for m/z=15 (CH_3^+; corresponds to CH_4) and m/z=133 ($SiCl_3^+$; corresponds to MTS) versus reactor residence time, showing that CH_4 is produced as MTS decomposes. Data for each mass averaged over 3.5 minutes.

Notably absent in Figure 3 are peaks associated with chlorosilanes other than MTS, which could form as the result of secondary reactions occuring after Reactions 1-3. No signal was observed at m/z values corresponding to $SiCl_4$ or $HSiCl_3$ (the parent ion of SiH_2Cl_2 cannot be conclusively identified due to its overlap with the $^{28}Si^{37}Cl^{35}Cl$ peak at m/z=100). A compound that would be formed by the reaction of two CH_2SiCl_3 molecules from Reaction (2), $Cl_3SiCH_2CH_2SiCl_3$ (m/z=294), also could not be detected. The absence of signal at these m/z values suggests that silicon-containing radicals formed in the initial stages of MTS decomposition are lost to the reactor or probe walls before reaching the spectrometer. If this is occuring, it may be possible to reduce the rate of wall loss by converting the radical to a more stable species by reacting it with a trapping agent. In the case of MTS decomposition, $SiCl_3$ molecules formed by Reaction (1) are expected to further decompose via the reaction $SiCl_3 \rightarrow SiCl_2 + Cl$. Since silylenes such as $SiCl_2$ are known to react with unsaturated hydrocarbons to form stable alkylsilanes [9], we attempted to trap these molecules by adding C_2H_4 to the carrier gas. The compound that would be formed is $HCl_2SiCH=CH_2$, giving mass peaks at m/z=126, 99, and 91 corresponding to the fragments $HCl_2SiCH=CH_2^+$, $SiCl_2H^+$, and $HClSiCH=CH_2^+$. Efforts to detect these fragments in both helium and hydrogen carrier gas were unsuccessful. This suggests that the $SiCl_2$ molecule, if it forms, is also lost to the walls or that the addition product is not sufficiently stable to be detected.

In addition to HCl, a second product of MTS decomposition was detected by monitoring the signal at m/z=15, which corresponds to the CH_3^+ fragment. As discussed earlier, interference from background gases complicates data collection in the 12-30 amu region. Meaningful data can be obtained, however, by averaging the signal at a particular mass over a period of one to two minutes. Data obtained in this manner are presented in Figure 4, which shows the intensity of the m/z=133 and m/z=15 peaks as a function of reactor residence time. The concentration of MTS decreases as the residence time increases, showing that MTS is decomposing in the reactor. Simultaneously, the m/z=15 peak increases with increasing residence time, indicating that the parent compound is a product of the pyrolysis reaction. The likely source of this signal is methane; plots of the signals at m/z values corresponding to ethane (which could form by recombination of CH_3 radicals) show no such increase. Similar behavior is obtained at m/z=16 (corresponding to CH_4^+), but with reduced precision due to overlap with the very strong O^+ peak. The detection of CH_4 indicates that some of the MTS decomposes via Reaction (1). Quantitative experiments correlating the concentration of MTS and CH_4 are now required to determine if Reaction (1) is the primary decomposition channel.

The effect of exchanging helium for hydrogen carrier gas on the overall MTS decomposition rate was also examined. As discussed above, the unimolecular decomposition Reactions 1-3 are predicted to be a factor of two slower in hydrogen than in helium. Figure 5 compares the amount of MTS decomposition observed in helium versus hydrogen, as indicated by the change in the m/z peaks corresponding to the $SiCl_3^+$ cracking fragment of MTS. In helium, m/z=133 decreases by about 5% for an MTS residence time of 67 msec. In hydrogen, however, the decrease is much larger, about 17% for the same residence time. Thus, addition of hydrogen *increases* the MTS decomposition rate. Since this is the opposite effect of that predicted by collisional energy transfer arguments [8], an additional MTS decomposition mechanism must be operative that is accelerated by the addition of hydrogen.

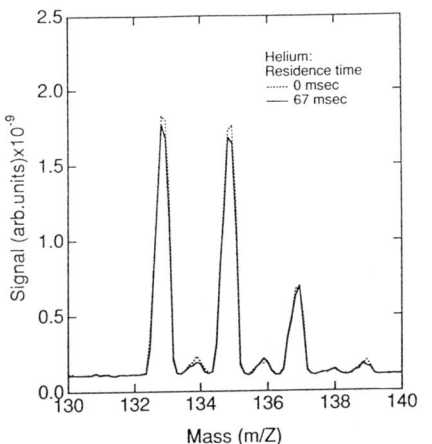
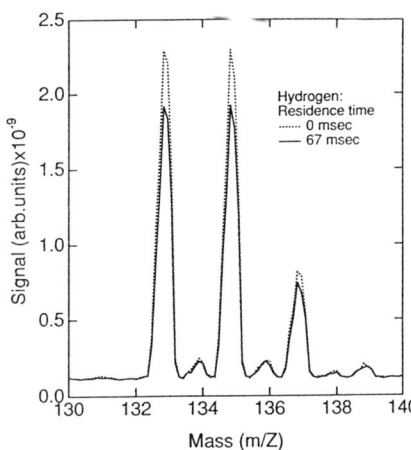

Figure 5: Comparison of mass spectrometer peaks corresponding to $SiCl_3^+$ ions for helium versus hydrogen carrier gas at 1243 K, 25 torr.

The observations described above are consistent with a radical-chain mechanism in which hydrogen atoms and methyl radicals play an important role. Such mechanisms have been proposed for other chlorosilanes [5]. The mechanism is described by the following reactions:

Initiation:	$CH_3SiCl_3 + M \rightarrow CH_3 + SiCl_3 + M$	(1)
Chain Propagation:	$CH_3SiCl_3 + CH_3 \rightarrow CH_4 + CH_2SiCl_3$	(4)
	$CH_3 + H_2 \rightarrow CH_4 + H$	(5)
	$H + CH_3SiCl_3 \rightarrow H_2 + CH_2SiCl_3$	(6)
	$SiCl_3 \rightarrow SiCl_2 + Cl$	(7)
	$Cl + CH_3SiCl_3 \rightarrow HCl + CH_2SiCl_3$	(8)
	$Cl + H_2 \rightarrow HCl + H$	(9)
Chain Termination:	$SiCl_3 + wall \rightarrow SiCl_3(w)$	(10)
	$SiCl_2 + wall \rightarrow SiCl_2(w)$	(11)
	$CH_2SiCl_3 + wall \rightarrow CH_2SiCl_3(w)$	(12)
	$H + H + M \rightarrow H_2 + M$	(13)
	$H + Cl + M \rightarrow HCl + M$	(14)
	$CH_3 + H + M \rightarrow CH_4 + M$	(15)

As predicted by the RRKM calculations, the initation step in this mechanism is the breaking of the Si-C bond. The methyl radical formed in this step can then react with another MTS molecule to produce methane, thereby accounting for the observed production of this species in helium carrier gas. Exchanging hydrogen for helium as the carrier gas accelerates the conversion of the methyl radical to methane by providing a second, presumably faster, pathway (Reaction 5). Since Reaction 5 also produces an H atom, the MTS decomposition rate also increases via Reaction 6.

Formation of HCl occurs through Reactions 7-9. Since the Si-Cl bond strength [7] in $SiCl_3$ is only 68.8 kcal mol^{-1}, thermal energy sufficient to fragment the Si-C bond (whose strength is 96.7 kcal mol^{-1}) is easily sufficient to drive Reaction 7. Other pathways leading to Cl atom formation, such as breaking the Si-Cl bond in MTS, or elimination of HCl via Reaction 3, are expected to be significantly slower than Reaction 7.

The lack of chlorosilane products is accounted for in the mechanism by wall-loss, Reactions 10-12. Although these experiments provide no direct evidence of this, measurements of $SiCl_3$ wall-loss rates at room temperature by Niiranen and Gutman indicate that Reaction 10 at least should be very fast [6]. They obtained a wall-loss rate of 170 s^{-1} for Reaction 10; for comparison, the rate of Reaction 1 is predicted by RRKM calculations to be about 1 s^{-1} in helium at 25 torr and 1243 K.

The mechanism suggested above has significant consequences for SiC CVD. First, the observed activation energy for MTS pyrolysis will be substantially lower than that predicted by RRKM theory, since the chain-propagation reactions (4)-(6) have very low activation energies (8-10 kcal mol^{-1}) compared with the much higher activation energies associated with initiation (>75 kcal mol^{-1} at 25 torr). Thus, MTS decomposition will proceed at reactor temperatures lower than those expected from a purely unimolecular, non-chain process. Second, the rapid loss of silicon-containing species to the walls implied by Reactions (10)-(12) is consistent with the suggestion of previous investigators, namely, that incorporation of silicon during SiC CVD is rapid and the process is limited by the reaction of stable hydrocarbons with the surface [10]. Finally, the identification of methane as the principal carbon-containing product suggests that carbon deposition may be slow, since the reactivity of methane with the silicon carbide surface appears to be low [11].

In summary, the results presented here provide a qualitative picture of MTS decomposition, identifying some of the gas-phase products and suggesting a mechanism for the pyrolysis of MTS at CVD temperatures. Additional data are required to provide quantitative information on the rates of wall reactions, homogeneous decomposition rates, and the effects of temperature and pressure. Experiments to obtain this information are now underway.

REFERENCES

1. J. Schlichting, *Powder Metal. Inter.* **12**, 196 (1980).
2. T. M. Besmann, B. N. Gallois, J. W. Warren, Eds., *Chemical Vapor Deposition of Refractory Metals and Ceramics II* (Mater. Res. Soc. Proc. **250**, Pittsburgh, PA, 1992).
3. C. C. Chiu, S. B. Desu, C. Y. Tsai, *J. Mater. Res.* **8**, 2617 (1993).
4. J. N. Burgess, T. J. Lewis, *Chemistry and Industry* , 76 (1974).
5. I. M. T. Davidson, C. E. Dean, *Organometallics* **6**, 966 (1987).
6. J. T. Niiranen, D. Gutman, *J. Phys. Chem.* **97**, 9392 (1993).
7. M. D. Allendorf, C. F. Melius, *J. Phys. Chem.* **97**, 720 (1993).
8. T. H. Osterheld, M. D. Allendorf, C. F. Melius, submitted to *J. Phys. Chem.*, 1993.
9. I. Safarik, B. P. Ruzsicska, A. Jodhan, O. P. Strausz, *Chem. Phys. Lett.* **113**, 71 (1985).
10. G. S. Fischman, W. T. Petuskey, *J. Am. Ceram. Soc.* **68**, 185 (1985).
11. C. D. Stinespring, J. C. Wormhoudt, *J. Appl. Phys.* **65**, 1733 (1989).

HETEROGENEOUS KINETICS OF THE CHEMICAL VAPOR DEPOSITION OF SILICON CARBIDE FROM METHYLTRICHLOROSILANE

GEORGE D. PAPASOULIOTIS AND STRATIS V. SOTIRCHOS
Department of Chemical Engineering, University of Rochester, Rochester, NY 14627

ABSTRACT

We examine the dynamic behavior of a heterogeneous reaction model for the chemical vapor deposition of carbon, silicon, and silicon carbide from the precursors generated by the thermal decomposition of methyltrichlorosilane (MTS, CH_3SiCl_3). All reactions are treated as reversible in order to account for the strong inhibitory effect of the reaction by-products on the deposition process that was observed in our experiments and in other studies. The equilibrium constants of the adsorption steps of the reactions are treated as model parameters, and those of the other reactions are calculated from the thermodynamic constants of a set of overall deposition reactions. Results are presented on the influence of the various model parameters on the reaction rate, the stoichiometry of the deposit, and the variation of these quantities with the distance in a plug flow deposition reactor.

INTRODUCTION

The good properties and numerous potential applications of silicon carbide have generated considerable interest in the kinetics of its chemical vapor deposition from various precursors, methyltrichlorosilane being among the most frequently used. The results, however, of the various experimental investigations show low reproducibility [1,2] suggesting that the chemistry of the MTS/SiC deposition system is too complex to be described by the simple Arrhenius/power law expression usually employed in the literature. Among the existing models for the deposition of SiC from mixtures of silanes and hydrocarbons, even those that place significant emphasis on the chemistry of the process [3,4] treat the heterogeneous reactions as irreversible. This approach cannot be applied for the deposition of SiC from methyltrichlorosilane since it has experimentally been observed that there is a strong inhibitory effect of the reaction by-products, namely HCl, on the deposition process [5,6].

In this study, we first present some experimental results for the variation of the deposition rate with the distance in a hot-wall CVD reactor and the effect of hydrogen chloride on the deposition process in order to underscore the need for accounting for the reversibility of the surface reactions in the SiC/MTS deposition system. We then describe an expanded version of our heterogeneous reaction mechanism [7] for the deposition of silicon carbide from methyltrichlorosilane and explain how the thermodynamic equilibrium constants of some overall reactions that do not involve surface species can be used to determine conditions that must be satisfied by the equilibrium constants of the surface reactions. Finally, we present some results on the evolution of the reactivity and deposit stoichiometry for some values of the model parameters and compare them qualitatively with the experimental observations.

SURFACE REACTION MECHANISM

The surface mechanism, which was mainly based on arguments outlined in references [7] and [8], is given in Table I. (It should be noted that reaction SR12 was also included in the model presented in [7], where its rate was assumed to be equal to that of SR4.) Surface species are denoted by subscript s, $[S_{Si}]$ and $[S_C]$ represent silicon and carbon sites, respectively, on the deposition surface, and $[S]$ stands for either of these two types of sites. The concentrations of silicon and carbon sites are assumed to be proportional to the concentrations of carbon and silicon atoms in the deposit. The double desorption step of the $SiCl_2$ surface species (reaction SR14) is used to account for the desorption of $SiCl_4$ from silicon carbide surfaces exposed to MTS and hydrogen chloride [9]. To obtain results representative of the experimental arrangement used in our studies, the gas–solid reactions are introduced into the reaction and transport model of a plug flow reactor [7] coupled with the homogeneous chemistry mechanism for the thermal decomposition of MTS [8].

All reactions must be treated as reversible in order to account for the effect of the reaction by-products (HCl) on the deposition process. The heterogeneous kinetic model of Table I encompasses 14 reactions, and as result, 28 reaction rate constants (for the forward and reverse steps of the reactions) must be identified before one begins computations. Since the structure of the silicon carbide deposition surface and those of the adsorbates are not characterized, values for the equilibrium constants of the adsorption steps are not available at present. Some investigators [10–12] have tried to compute the adsorption equilibrium constants for the $Si/H/Cl$ deposition system using first principles, but the predictions of their computations depend strongly [13] on what structures are assumed for the surface species. This has led to significant differences between different studies [10,12].

The reactions of Table I involve 13 non-surface species (C_2H_2, CH_4, C_2H_4, H, H_2, CH_3, $SiCl_2$, $SiCl_3$, $SiCl_4$, HCl, C, SiC, and Si) for which thermodynamic data are available in the literature. It can easily be shown that one can construct at most 9 linearly independent reactions using species from the above set. If the $[S_C]$ and $[S_{Si}]$ are assumed to be energetically equivalent, one can show that the logarithms of the equilibrium constants of the above reactions are linear combinationss of the logarithms of the equilibrium constants of some of the reactions of Table I. Since all of the 9 reactions are linearly independent and their equilibrium constants are known, this yields 9 equations that must be satisfied by the equilibrium constants of the reactions of the heterogeneous mechanism and, on that account, reduces by the same number the number of kinetic constants that must be evaluated.

Table I. Surface Reaction Mechanism for the Deposition of SiC

SR1. $SiCl_2 + [S] \rightarrow [SiCl_2]_s$
SR2. $C_2H_2 + 2[S] \rightarrow 2[CH]_s$
SR3. $C_2H_4 + 2[S] \rightarrow 2[CH_2]_s$
SR4. $[CH]_s + H \rightarrow [CH_2]_s$
SR5. $CH_3 + [S] \rightarrow [CH_3]_s$
SR6. $[CH_3]_s + H \rightarrow [CH_2]_s + H_2$
SR7. $SiCl_3 + [S] \rightarrow [SiCl_3]_s$
SR8. $[SiCl_3]_s + H \rightarrow [SiCl_2]_s + HCl$
SR9. $[CH_2]_s \rightarrow C + [S_C] + H_2$
SR10. $[SiCl_2]_s + H_2 \rightarrow Si + [S_{Si}] + 2HCl$
SR11. $[CH_2]_s + [SiCl_2]_s \rightarrow SiC + [S_{Si}] + [S_C] + 2HCl$
SR12. $[CH]_s + H_2 \rightarrow [CH_2]_s + H$
SR13. $[CH_2]_s + [CH_2]_s \rightarrow C + [S_C] + [S] + CH_4$
SR14. $[SiCl_2]_s + [SiCl_2]_s \rightarrow Si + [S_{Si}] + [S] + SiCl_4$

To determine the kinetic constants, we treated five of the equilibrium constants, specifically, those of the adsorption steps, as model parameters. The kinetic constants of the adsorption steps were estimated as the collision frequency of the corresponding adsorptive species with the surface multiplied by its reactive sticking coefficient. In accord with the results of Buss et al. [14], the reactive sticking coefficients of C_2H_2 and C_2H_4 were assumed to have Arrhenius dependence on temperature with activation energy values 4 and 6 kJ/mol, respectively, while those of the radicals were set equal to unity. The forward rates of reactions SR4, SR6, SR8, and SR12 were assumed to be equal to the collision frequencies of either atomic or molecular hydrogen with the deposition surface multiplied by the fractional surface coverage of the species involved in the reaction. The forward rate constants of reactions SR9, SR11, SR13, and SR14 were set equal to 3.1×10^{-12}, 8.2×10^{-4}, 2.8×10^{-6}, and 3.4×10^{-9} kmole/m^2·s, respectively, and that of reaction SR10 to 8.2×10^{-6} kmole/m^2·s·atm in order to reproduce the order of magnitude of the deposition rates observed in our experiments. (In the above discussion, the various rate expressions are assumed to be expressed in terms of the surface coverages of the surface species and the partial pressures of the gas phase species.) Thermodynamic data for equilibrium constant estimation were obtained from the JANAF [15] tables.

If the surface reactions leading to deposition of Si, C, and SiC are treated as irreversible, one does not have to distinguish between silicon and carbon sites. This approach was taken in some preliminary computations that we presented in [7], where we showed for the first time that a heterogeneous reactor model of the type shown in Table I can explain several of the experimental observations on SiC deposition from MTS. For steady-state operation of the reactor, as it is the case investigated here, the ratio of the concentrations of silicon and carbon sites turns out to be equal to the ratio of the deposition rates of silicon and carbon, $\mathcal{R}$. It follows from the heterogeneous kinetic model that $\mathcal{R}$ is given by

$$\mathcal{R} = \frac{R_{s10} + R_{s11} + R_{s14}}{R_{s9} + R_{s11} + R_{s13}}$$

DISCUSSION OF SOME RESULTS

In order to investigate the variation of the deposition rate with the distance in the reactor, deposition experiments were performed using multiple graphite substrates. The substrates were placed along the axis of a tubular hot-wall deposition reactor in vertical arrangement, and experiments were conducted at atmospheric pressure. Typical results, obtained at 1273 K and with the H_2/MTS ratio in the feed stream equal to 10, are presented in Figure 1. The reactivity initially increases, its increase following the increase of temperature of the reactor. The initial increase is followed by a sharp decrease, occurring within the isothermal zone of the deposition reactor. (Temperature measurements showed that the isothermal zone of the reactor extended between 4 and 16 cm from the top of the heated length of the resistance furnace.) Similar reactivity profiles were observed in experiments conducted at different temperature, flowrate, and reacting mixture composition conditions. Characterization of the deposited films revealed that, apart from the reaction rate, the deposit stoichiometry also varied along the reactor's length. Films deposited in the region of high deposition rates were polycrystalline and contained excess silicon, while those obtained after the sharp decrease in deposition rate occured were largely amorphous and their silicon content was reduced. The variation of the deposition rate and of the stoichiometry of the film with the distance (or equivalently, the residence time) in the reactor may be the main reason for the large discrepancies in the results of different experimental investigations of the deposition of silicon carbide from MTS. As Figure 1 shows, experiments conducted under similar conditions can yield vastly different reactivity results, depending on the point of measurement and the geometry of the experimental apparatus. The analysis of these results may in turn lead to widely different values for the apparent kinetic constant and the order of the overall deposition reaction.

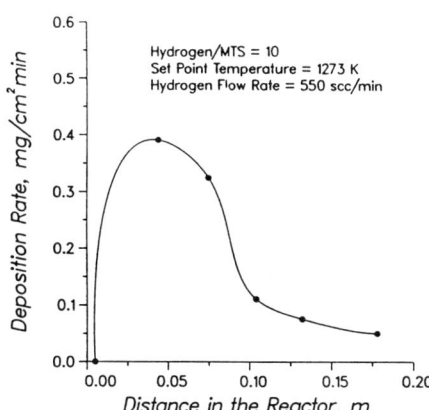

Figure 1. Deposition Rate vs. Distance in the CVD Reactor.

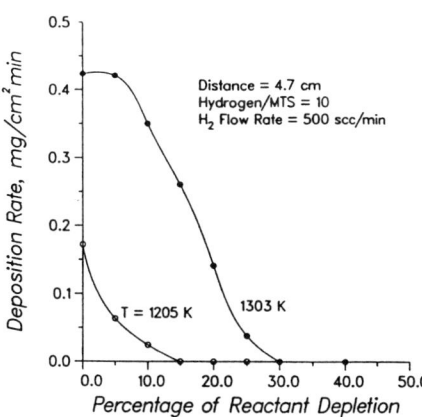

Figure 2. Deposition Rate vs. Reactant Depletion.

The decrease in the deposition rate cannot be attributed to depletion of silicon or carbon bearing species since the feed stream would not have been depleted even if the deposition process occurred at the maximum observed rate throughout the total length of the reactor. Moreover, since it takes place within the isothermal zone of the reactor, it could not be caused by a decrease in the temperature. Experiments were therefore carried out to investigate the effect of the reaction by-products, namely HCl, on the deposition rate. Different levels of reactant depletion were simulated by varying the MTS to HCl ratio in the feed stream on the basis of the generation of three hydrogen chloride molecules for each molecule of MTS consumed. Some of the obtained results are shown in Figure 2 for reaction at 1303 K and 1205 K. It is seen that the deposition rate at 1303 K remains almost invariant for low depletion levels (below 5%), but it subsequently decreases sharply with increasing depletion level. Similar behavior was observed by Besmann et al. [6]. The effect of increasing the HCl concentration on the deposition rate indicates that the drop in deposition rate seen in Figure 1 could be due to the progressive increase of the concentration of hydrogen chloride in the reactor. It should be noted that complete stoppage of all deposition reactions was observed for depletion levels above 30% in the experiments of Figure 2. The corresponding threshold at 1205 K was 15% depletion.

The experimental results of Figures 1 and 2 suggest that reversible surface reactions must be considered in a heterogeneous kinetic mechanism model in order to be able to reproduce the behavior seen there. Typical results predicted by the mathematical model for reversible deposition reactions are shown in Figures 3 and 4, for the evolution of the deposition rate and film stoichiometry, respectively. These results are for the case of a reactor with 1.5 cm diameter, 600 ml/min volumetric flow rate at standard conditions, three pressure levels (0.01, 0.1, and 1 atm), 1300 K temperature, and MTS to H_2 ratio in the feed equal to 0.1. The activity-based equilibrium constants (K_a's) of the adsorption steps were set equal to 320 for $SiCl_2$, 100 for $SiCl_3$ and CH_3, and 1000 for C_2H_2 and C_2H_4.

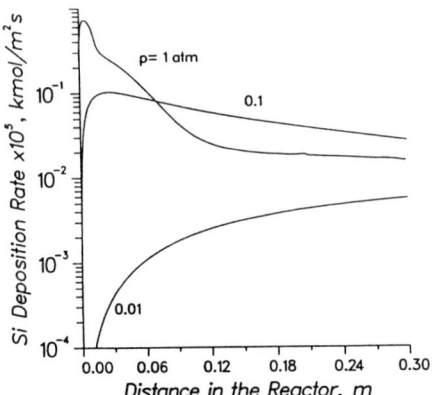
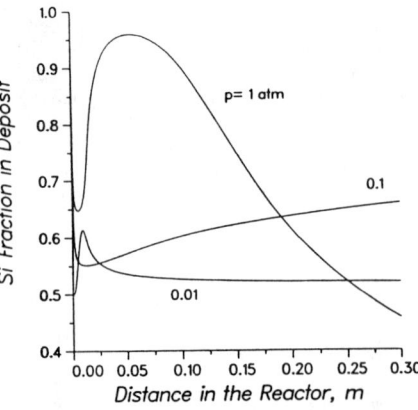

Figure 3. Deposition Rate vs. Distance in the CVD Reactor. Temperature = 1300 K, H_2/MTS = 10, Total Flow Rate = 600 scc/min

Figure 4. Deposit Stoichiometry vs. Distance in the CVD Reactor. Temperature = 1300 K, H_2/MTS = 10, Total Flow Rate = 600 scc/min

Figures 5 and 6 present the variation of the concentrations of some of the carbon and silicon bearing species, respectively, in the gas phase with the distance in the reactor with (chain-dashed lines) and without (solid lines) heterogeneous reactions taking place at 1 atm. The homogeneous chemistry mechanism used in this modelling study was presented in reference [8]. It is seen from the results of these two figures that the occurrence of the heterogeneous reactions has a strong effect on the concentrations of the gas phase species. Comparison of Figures 3, 5, and 6 reveals that the decrease in the Si deposition rate at 1 atm coincides with the decrease in the concentrations of the main deposition precursors (e.g., $SiCl_2$ and $SiCl_3$ in Figure 6 and C_2H_2 and C_2H_4 in Figure 5) and the rise in the concentration of HCl. The rate of reaction SR11 is caused by a reduction in the surface coverage of methylene radicals, which is in turn caused by the decrease of the partial pressures of the carbon deposition precursors in the gas phase. The surface coverage results showed that the surface coverage of $[SiCl_2]_s$, which is the main adsorbate at 1 atm, does not change significantly over the entire length of the reactor. Nevertheless, the rate of reaction SR10 starts decreasing monotonically at the point where the maximum deposition rate occurs because of the increase of the partial pressure of hydrogen chloride in the gas phase. In fact, reaction SR10 proceeds towards silicon etching near the end of the observation window of Figure 3, resulting in a carbon rich deposit (see Figure 4). Therefore, hydrogen chloride not only can affect the reactivity of the system, but also may suppress the deposition of silicon and lead to carbon rich deposits. Ohshita [16] has shown that it is possible to improve the stoichiometry of the deposited film by introducing HCl in the deposition reactor with the reacting mixture.

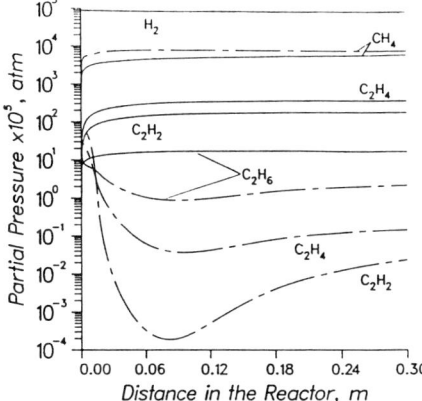

Figure 5. Evolution of the Concentration Profiles of the main Hydrocarbons in the CVD Reactor. Temperature = 1300 K, p = 1 atm, H_2/MTS = 10, Total Flow Rate = 600 scc/min. Solid lines : only gas phase reactions. Chain-Dashed lines : surface reactions included.

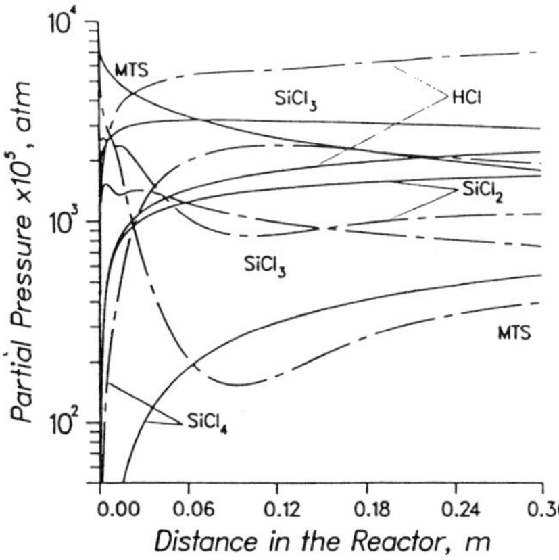

Figure 6. Evolution of the Concentration Profiles of the main Silicon Bearing Compounds in the CVD Reactor. Temperature = 1300 K, p = 1 atm, H_2/MTS = 10, Total Flow Rate = 600 scc/min. Solid lines : only gas phase reactions. Chain-Dashed lines : surface reactions included.

The results of Figures 3 and 5 lead us to conclude that the variation of the carbon content of the deposit follows closely the variation of the partial pressures of the main carbon deposition precursors, C_2H_4 and C_2H_2. It is the effect of the concentrations of C_2H_2 and C_2H_4 on the surface coverages of $[CH_2]_s$ and $[CH]_s$ that is primarily responsible for this behavior. As Figure 4 shows, the deposits obtained for low residence times are silicon rich, while the transition to deposits that are closer to stoichiometry occurs at higher values of residence time, where the high concentration of HCl suppresses silicon formation and the concentration of carbon deposition precursors rises. This result is in qualitative agreement with the behavior seen in our experiments.

We saw in Figures 5 and 6 that the concentration profiles of the major deposition precursors and reaction by-products in conditions of significant Si or SiC deposition (or, equivalently with heterogeneous reactions included in the model) differ significantly from those predicted by the mathematical model without heterogeneous reactions. As one can see in Figures 5 and 6, the presence of surface reactions influences strongly the evolution of partial pressure profiles in the deposition reactor. The decompostion of of methyltrichlosilane is accelerated and the partial pressures of all deposition precursors are lower. On the contrary, the partial pressures of the reaction by-products and other stable compounds increase, by approximately one order of magnitude in the cases of HCl and $SiCl_4$. As Figure 5 shows, the presence of heterogeneous reactions in the system results in a decrease of the partial pressures of acetylene and ethylene by two orders of magnitude. This may very well be the reason for which several gas spectroscopy studies in silicon carbide CVD reactors either did not detect the presence of C_2H_4 and C_2H_2 [2,17] or detected minor quantities of the two compounds [18], whereas methane, silicon tetrachloride and especially hydrogen chloride were found to be the major gas phase species [17].

ACKNOWLEDGEMENT

This work was supported by a grant from the National Science Foundation.

REFERENCES

1. J. Schlichting, Powder Metall. Int., 12, 141 and 196 (1980).
2. T.M. Besmann and M.L. Johnson, in Proc. 3rd Int. Symp. on Ceramic Materials and Components for Engines, (Las Vegas, NE, 1988), p. 443.
3. S. Tanaka and H. Komiyama, J. Am. Ceram. Soc., 73, 3046 (1990).
4. M.D. Allendorf and R.J. Kee, J. Electrochem. Soc., 138, 841 (1991).
5. G. Papasouliotis and S.V. Sotirchos, ACerS Meeting, Cincinatti, OH (1991).
6. T.M. Besmann, B.W. Sheldon, T.M. Moss III, and M.D. Kaster, J. Am. Ceram. Soc., 75, 2899 (1992).
7. S.V. Sotirchos and G.D. Papasouliotis, in Chemical Vapor Deposition of Refractory Metals and Ceramics II, edited by T.M. Besmann, B.M. Gallois, and J.W. Warren, (MRS, Pittsburgh, 1992), p. 35.
8. G.D. Papasouliotis and S.V. Sotirchos, submitted to the J. Electrochem. Soc. (1993).
9. M.D. Allendorf and D.A. Outka, in Chemical Perspectives of Microelectronic Materials III, edited by C.R. Abernathy et al., (MRS, Pittsburgh, 1993), p. 439.
10. J.G.E. Gardeniers and L.J. Giling, J. Cryst. Growth, 115, 542 (1991).
11. J.G.E. Gardeniers, L.J. Giling, F. de Jong, and J.P. van der Eerden, J. Cryst. Growth, 104, 727 (1990).
12. A.A. Chernov and N.S. Papkov, Sov. Phys. Crystallogr., 22, 18 (1977).
13. L.J. Giling, H.H.C. de Moor, W.P.J.H. Jacobs, and A.A. Saaman, J. Cryst. Growth, 78, 303 (1986).
14. R.J. Buss, P.Ho, W.G. Breiland, and M.E. Coltrin, J. Appl. Phys., 63, 2808 (1988).
15. JANAF Thermochemical Tables, edited by M.W. Chase, Jr., et al., Supplement to the J. of Physical and Chemical Ref. Data, 3rd ed. (1985).
16. Y. Ohshita, in Diamond, Silicon Carbide and Related Wide Bandgap Semiconductors, edited by J.T. Glass, R. Messier, and N. Fujimori, (MRS, Pittsburgh, 1990), p. 433.
17. J. Yeheskel, S. Agam, and M.S. Dariel, in Proc. 11th Conf. on Chemical Vapor Deposition, edited by K.E. Spear and G.W. Gullen, (The Electrochem. Soc., Pennington, NJ 1990), p. 696.
18. M.L. Ivanova and A.A. Pletyushkin, Inorganic Materials, 4, 957 (1968).

REACTION KINETICS ON DIAMOND: MEASUREMENT OF H ATOM DESTRUCTION RATES

STEPHEN J. HARRIS AND ANITA M. WEINER Physical Chemistry Dept., General Motors R&D Center, Warren, MI 48090-9055

ABSTRACT

We describe the first measurements of reaction kinetics between diamond and a gas phase species—H atoms—involved in its formation. We develop a remarkably simple method to measure H atom concentrations and use the method to measure γ_d, the destruction probability of H atoms on diamond at 20 torr and 1200 K. We find $\gamma_d = 0.12$. This value is close to that estimated from gas phase alkane rate constants.

INTRODUCTION

In this work we describe a remarkably simple version of the recombination enthalpy technique which we use to obtain H atom concentrations in good agreement with previous measurements. Having validated the analysis for our system, we then use it to make the first direct experimental measurements of reaction kinetics between diamond and a gas phase species involved in its formation. The relationship between gas phase and gas-surface kinetics is discussed.

EXPERIMENTAL

We used a 125 μm diameter Pt/Pt-10%Rh thermocouple as a substrate for diamond growth. It was held about 4 mm above a 250 μm diameter tungsten filament.

To obtain H concentrations, we measured the substrate temperature (thermocouple voltage) in 1% CH_4 in H_2. Experiments to measure the H atom destruction probability began by introducing a thermocouple above a carburized filament at 20 torr. The platinum thermocouple temperature T_p was measured (1100 to 1200 K), a diamond film was then grown overnight, and the temperature of the diamond-coated thermocouple T_d was measured again. The difference between T_p and T_d reflects in part a difference in the H atom recombination rates upon changing from a platinum to a diamond-coated platinum wire. In order to measure the emissivity ϵ_d of the diamond-coated platinum wire relative to the known value for the platinum wire ϵ_p, we heated the thermocouple electrically in vacuum both before and after diamond growth, and we measured the electrical power required to attain a given temperature.

ANALYSIS

Interpretation of recombination enthalpy measurements requires an energy balance analysis which includes radiation heating and cooling of the thermocouple, heat transfer between the gas and thermocouple, heat conduction along the thermocouple wires, and H atom recombination on the thermocouple[1,2,3,4]. Below, we explore the consequences of simplifying the analysis by ignoring all but the largest terms in the energy balance for our system.

In vacuum, radiation heating and cooling control the thermocouple temperature, which rises to approximately 650 K. However, because the radiation terms scale with T^4, that same amount of radiation energy would lead to only a small temperature rise in the range near 1200 K, where our measurements are made. Thus, we ignore radiation heating. When 20 torr of He is added to the system, the thermocouple temperature increases by only an additional 100 K (compared to a rise of around 500 K when H_2 is added, depending on the pressure), testifying to the inefficiency of gas-surface heat transfer under these conditions. This inefficiency is due to the fact that the H atom mean free path is on the order of the wire diameter at 20 torr (Knudsen number ~ 1.) A large temperature discontinuity (up to 1000 K) has been observed at the filament (which has a diameter twice that of the thermocouple). This discontinuity is a demonstration of how poor the heat transfer is at low pressures. Thus, we ignore heat transfer. Finally, an order-of-magnitude estimate of heat conduction along the thermocouple wires shows that this term is small both because the wires are thin and because gradients in the gas temperature and the H atom concentration are small. Thus, we ignore heat conduction.

With these simplifications, the heat balance equation per unit area is

$$0.5 \Delta E \, \gamma \, ([H]v/4) = \epsilon \sigma T^4, \tag{1}$$

in which heating by H atom recombination (the left hand side) is balanced by radiation cooling (the right hand side). In Equation (1) the term in parenthesis is the flux of H atoms to the substrate surface; $[H]$ is the concentration of H atoms at the substrate; γ is the probability for H destruction; ΔE is the heat of recombination of H_2 from two H atoms; v is the velocity of an H atom; σ is the Stefan-Boltzmann constant, and T is the thermocouple temperature. The factor of 0.5 takes into account that two H atoms are required to form an H_2 molecule. Solving for the H atom concentration at the substrate, we have

$$[H] = \left(\frac{8\epsilon\sigma}{\gamma v \Delta E}\right) T^4. \tag{2}$$

In order to interpret the recombination coefficient measurements we evaluate Equation (2) for the platinum and diamond-coated platinum cases and solve for γ_d by ratioing the equations for the two cases. The result is

$$\gamma_d = \frac{[H]_p}{[H]_d} \left(\frac{\epsilon_d}{\epsilon_p}\right) \left(\frac{T_d}{T_p}\right)^4 \gamma_p. \tag{3}$$

The ratio of H concentrations is not 1 because the destruction rates at the different surfaces are different. The effect of the wire on the local H atom concentration can be approximated by treating the wire as a long cylinder in a cylindrically symmetrical field of H atoms and by setting the transport of H atoms to the thermocouple by diffusion equal to the destruction rate at the wire surface. The result is

$$\frac{[H](r)}{[H](R_{tf})} = 1 + \left(\frac{\log(r/R_{tf})}{\log(R_{tf}/R_t) + (4D/\gamma v R_t)}\right) \quad (4)$$

where $H(r)$ is the number density a distance r from the axis of the thermocouple, D is the diffusion coefficient of H, R_{tf} is the distance between the thermocouple and the filament, and R_t is the radius of the thermocouple wire. (At the surface of the thermocouple, $r = R_t$.)

RESULTS AND DISCUSSION

For the concentration experiments the thermocouple temperature was monitored as a function of filament temperature, and the H atom concentration was determined from Equation (2). The results, relative to H_0, the concentration at 2300 K, are shown as large circles in Figure 1. The small circles on the graph are MPI[5], and the triangles are molecular beam[6] data. The solid line is a scaled calculation which assumes thermodynamic equilibrium between H and H_2 at the filament temperature. All of these data show nearly identical slopes, very close to that predicted by the equilibrium calculation[5,6]. Thus, our technique reproduces the relative temperature dependence of other experiments. The dependence of H atom concentration on distance r from the filament is shown in Figure 2. The symbols are our data and the curve is a rough fit to $[H] = c/r$, where r is the filament-thermocouple distance and where c is a constant. The figure shows the empirical result that the H concentration falls off more or less as $1/l$, in agreement with both the H atom and the OH radical laser fluorescence data of Meier et al.[7]. Thus, our technique also reproduces the spatial dependence of other experiments. Finally, Equation (2) can be tested further for the platinum case (with $\gamma_p = 0.1$) by using it to calculate an absolute H concentration. The most direct absolute H atom concentration measurements have been made by Hsu[6] in a molecular beam sampling system. For 0.5% CH_4 in H_2 we find $[H] = 6.5 \times 10^{-10}$ moles/cm^3 from Equation (2), compared with 5×10^{-10} moles/cm^3 at 0.4% CH_4 measured by Hsu.

Equation (3) allows us to estimate the recombination rate on diamond relative to that on platinum—whose recombination coefficient γ_p is about 0.1 above 1000 K[8]—from measurements of (ϵ_d/ϵ_p), $(T_p/T_d)^4$, and $[H]_p/[H]_d$. Our measurements show that $\epsilon_d/\epsilon_p = 1.2 \pm 0.1$ and $(T_p/T_d)^4 = 1.08 \pm 0.05$. Since destruction rates at diamond and platinum surfaces are similar ($T_d \sim T_p$), and since the effect of a platinum wire on the local H atom concentration is seen to be small, we will assume for the moment that $[H]_p/[H]_d = 1$ (see below). Equation (3) then gives $\gamma_d = 0.12$. For $\gamma_p = 0.1$ and $\gamma_d = 0.12$, Equation (4) predicts $[H]_p/[H]_d = 1.06$. Although the equation is only approximate, this result suggests that our assumption above that $[H]_p = [H]_d$ is sufficiently accurate. We estimate an uncertainty of about a factor of 2.

Measurement of γ_d gives us an opportunity to relate rates of reactions taking place on a diamond surface to known rates for alkane reactions. According to our

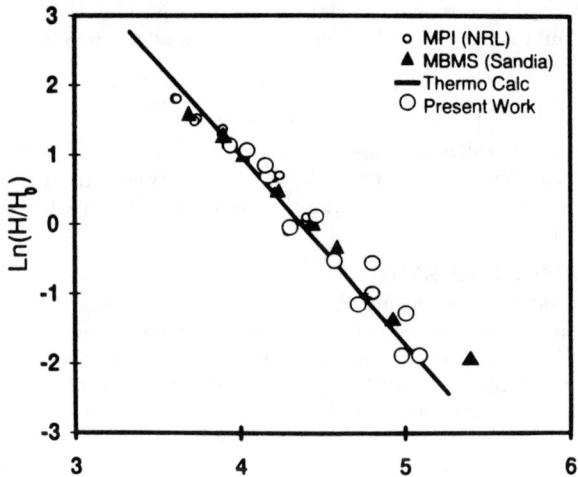

1. H atom concentrations relative to H_0, the concentration at 2300 K, vs filament temperature at 20 torr. Solid curve, thermodynamic equilibrium between H and H_2. Small circles, MPI data from NRL[5]. Open triangles, molecular beam data from Sandia[6]. Large circles, this work.

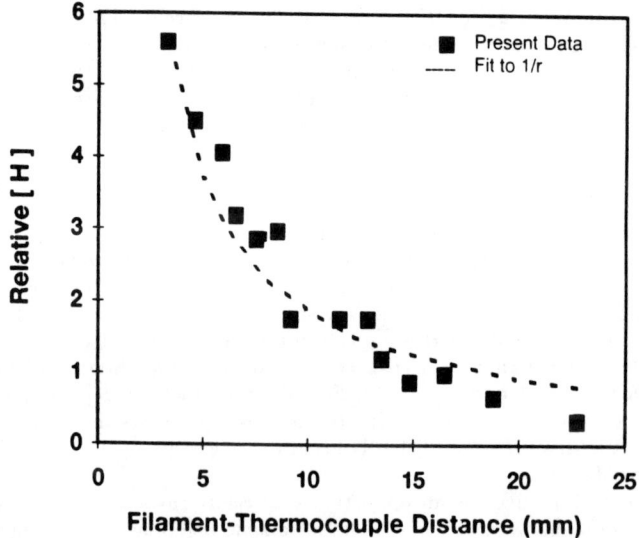

2. Dependence of relative H atom concentration on distance from the filament. There is a 2 mm diamond substrate located 23 mm (solid squares), 15 mm (open squares), or 10 mm (circles) from the filament.

models[9,10], the interaction between H atoms and a diamond surface is dominated by two reactions, abstraction and recombination. If CH represents a hydrogenated diamond surface site and $C*$ the radical site formed when the H is removed, we have

$$CH + H^{gas} = C* + H_2^{gas} \qquad (A)$$

and

$$C* + H^{gas} = CH. \qquad (R)$$

At typical growth temperatures we expect the rates for the reverse of reactions A and R to be relatively slow, so the rate r of H atom destruction per cm^2 of diamond surface can be expressed as

$$r = k_A N_{CH}[H] + k_R N_{C*}[H], \qquad (5)$$

where k_A and k_R are the rate constants per site for abstraction and recombination, respectively, and N_{CH} and N_{C*} are the area densities of hydrogenated and radical sites on the surface, related at steady state by the relationships $N_{CH}/N_{C*} = k_R/k_A$ and $N_{CH} + N_{C*} = N_{tot} \sim 3 \times 10^{-9}$ moles/cm^2. r can also be written in terms of a collision frequency with the surface and a destruction probability,

$$r = \gamma[H]v/4. \qquad (6)$$

Our use of first order kinetics in Equation (6) is supported by the success of Equation (2), in which first order kinetics is assumed. Combining Equations (5) and (6) we obtain

$$\Gamma = \frac{8 k_A k_R}{v(k_A + k_R)} N_{tot}. \qquad (7)$$

(We use the symbol Γ to refer to a destruction probability calculated based on alkane properties. The symbol γ refers to the experimental value for the destruction probability, derived from our measurements and from Equation (3).) For typical alkane reactions we have $k_R \sim 4 \times 10^{13}$ cm^3/mole-s, while

$$k_A = 1.3 \times 10^{14} e^{-7300/RT}. \qquad (8)$$

Evaluating k_A at 1200 K and using Equation (7) gives $\Gamma = 0.25$, double the measured γ_d but within its estimated uncertainty.

SUMMARY AND CONCLUSIONS

From an approximate analysis of the heat flows in our diamond growth apparatus, we found that the reading from a fine platinum thermocouple can be used to obtain absolute H atom concentrations. This technique could be useful in studying CVD growth of other materials such as Si_3N_4 and AlN where hydrogen may be a major species. We used our analysis to measure γ_d, the H atom destruction probability at 1200 K and 20 torr, typical diamond growth conditions. The result is $\gamma_d = 0.12$ with an estimated uncertainty of a factor of 2. For comparison, using alkane rate constants to predict H atom destruction kinetics gives 0.25, while molecular dynamics/Monte Carlo simulations predict 0.12; both results agree with our measurement to within our estimated uncertainty. This result supports our suggestion[9,11] that alkane rate constants can be transferred directly to diamond.

References

[1] K. Tankala and T. DebRoy. Modeling of the role of atomic hydrogen in heat transfer during hot filament assisted deposition of diamond. *Journal of Applied Physics*, 72:712, 1992.

[2] K. Tankala, M. Mecray, T. DebRoy, and W. A. Yarbrough. Hydrogen assisted heat transfer during diamond growth using carbon and tantalum filaments. *Applied Physics Letters*, 60:2068, 1992.

[3] R. Gat and J. C. Angus. Heat transport in hot filament assisted deposition of diamond films. *Journal of Applied Physics*, submitted.

[4] C. Wolden S. Mitra and K. K. Gleason. Heat transfer modelling in hot-filament chemical vapor deposition diamond reactors. *Journal of Applied Physics*, 72:3750, 1992.

[5] F. G. Celii and J. E. Butler. Hydrogen atom detection in the filament-assisted diamond deposition environment. *Applied Physics Letters*, 54:1031, 1989.

[6] W. L. Hsu. Quantitative analyses of the gaseous composition during filament-assisted diamond growth. In *Proceedings of the Electrochemistry Society*, page 217, Electrochemical Society, Pennington, NJ, May 1991.

[7] U. Meier, K. Kohse-Hoinghaus, L. Schafer, and C. Klages. Two-photon excited LIF determination of H atom concentrations near a heated filament in a low pressure H_2 environment. *Applied Optics*, 29:4993, 1990.

[8] B.J. Wood and H. Wise. Kinetics of hydrogen atom recombination on surfaces. *Journal of Physical Chemistry*, 65:1976, 1961.

[9] S. J. Harris. A mechanism for diamond growth from methyl radicals. *Applied Physics Letters*, 56:2298, 1990.

[10] S. J. Harris and D. G. Goodwin. Growth on the reconstructed diamond (100) surface. *Journal of Physical Chemistry*, 97:23, 1993.

[11] D. N. Belton and S. J. Harris. Growth from acetylene on a diamond (110) surface. *Journal of Chemical Physics*, 96:2371, 1992.

A NOVEL MOLECULAR BEAM REACTOR FOR THE STUDY OF DIAMOND SURFACE CHEMISTRY

C. A. Wolden, G. Zau, W. T. Conner, H. H. Sawin, and K. K. Gleason, Massachusetts Institute of Technology, 66-419, Cambridge, MA 02139

ABSTRACT

A novel reactor has been constructed to investigate the fundamental surface kinetics of diamond chemical vapor deposition(CVD). The molecular beam reactor permits independent control of the atomic hydrogen flux, the methyl radical flux and the substrate temperature. The low pressure in the growth chamber ($\approx$ 1 mTorr) minimizes the impact of gas-phase chemistry. The reactive mixture impinging the substrate is sampled through an orifice and analyzed by mass spectroscopy. Differential pumping in the two adjacent chambers quenches the beam, allowing quantitative analysis of radical species such as H and CH_3. In preliminary experiments deposition was achieved onto seeded molybdenum substrates.

INTRODUCTION

Diamond's unique combination of exceptional physical properties (thermal conductivity, hardness, optical transparency, large bandgap) has fueled an intensive research effort during the past decade [1,2]. A number of CVD systems have been employed to produce polycrystalline diamond films, including hot-filament [3], microwave plasma [4], DC arc jet [5], and combustion torches [6]. Upon analyzing the results of a hundred experiments from a variety of CVD systems, Bachmann and coworkers [4] found that the composition of successful mixtures used falls into a narrow range when plotted on an elemental C/H/O phase diagram. In addition, the substrate temperature in all of these systems is typically maintained between 600-900°C. These observations have led to the proposal of a universal growth species, most notably methyl radical [7] and acetylene [8]. Finally, it is understood that atomic hydrogen plays a number of critical roles in diamond deposition among which include stabilization of the growing surface, creating radical sites and preferential etching of graphitic carbon [9].

Despite recent advances, the surface kinetics of diamond growth and defect formation remain unknown. The uncertainty is due in part to the nature of diamond CVD systems, in which critical parameters are highly coupled and the reactors are not easily amenable to quantitative diagnostics. Our reactor is designed to independently control the flux of atomic hydrogen flux, the flux of reactive carbon precursors, and the substrate temperature. By successfully decoupling these variables, we have a unique tool to quantitatively investigate the impact of each parameter on the critical processes of diamond nucleation, growth, and defect formation.

EXPERIMENTAL

The molecular beam reactor is shown schematically in Fig. 1. The system is housed in a six inch stainless steel tube partitioned into three chambers which are differentially pumped. The apparatus used to generate the beams is built onto an eight inch Conflat flange that is mounted on the end of the first chamber. The center beam is the remote microwave discharge source for the production of atomic hydrogen. This source is the unique feature of this reactor, generating a stable, high flux of hydrogen atoms inside the vacuum chamber. The hydrogen atom source is a unique design employing a coaxial waveguide microwave cavity. Microwaves and molecular hydrogen enter on the air-side and generate a microwave plasma in a glass vessel on the vacuum side of the source. High power densities of > 200 W/cm^3 have been achieved, allowing high fluxes of atomic hydrogen to be delivered to the substrate.

A detailed sketch of the coaxial waveguide microwave source is shown in Fig. 2. On the air side the position of a threaded tuning slug (A) is adjusted to produce a resonance cavity. Sliding the microwave coupler (B) allows the cavity to be impedance matched to the microwave source used. Microwave power is also introduced through this coupler. Since, all the tuning elements of the cavity are external to the vacuum, complex vacuum motion feedthroughs are not needed. The center conductor of the waveguide (C) transmits the microwaves from the air side past vacuum seals (D) to a glass vessel (E) at the end of the center conductor. The glass vessel is held by a water-cooled clamp (F) and is designed to minimize gas contact with the walls so that surface treatment of the glass is not required. The microwaves are conducted through the plasma, extending the center conductor to the grounded end of the cavity, completing the microwave circuit. The use of the plasma to form the grounded end of the coaxial cavity makes this atom source extremely compact. Atomic hydrogen is emitted from the nozzle of the glass vessel and transported to the substrate without recombination since the 3 body gas-phase recombination reaction is extremely slow [10]. By varying the design of the glass vessel we can accommodate a wide range of flow rates.

Acetylene and methyl radicals are delivered to the substrate through two 3 mm I.D. quartz tubes which are mounted on either side of the atomic hydrogen source at 38° relative to the substrate normal. Methyl radicals are generated through the thermal decomposition of di-tertiary butyl peroxide (DTBP), $(CH_3)_3COOC(CH_3)_3$. The end of the DTBP quartz tube is

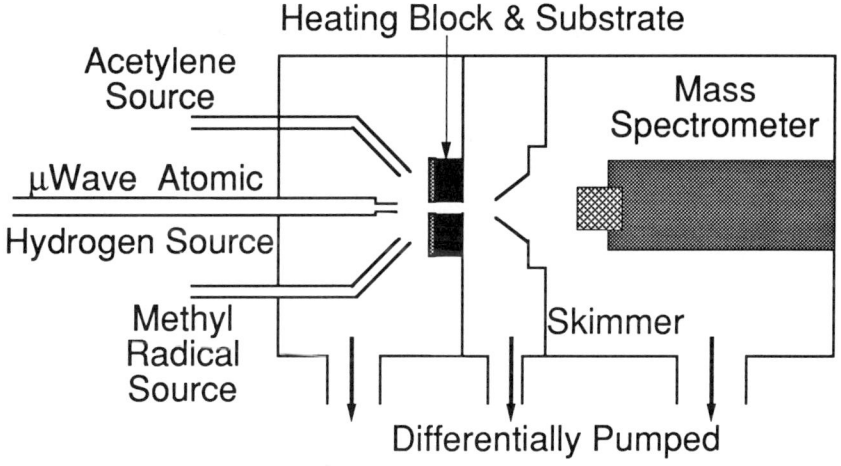

Figure 1: Schematic drawing of molecular beam reactor.

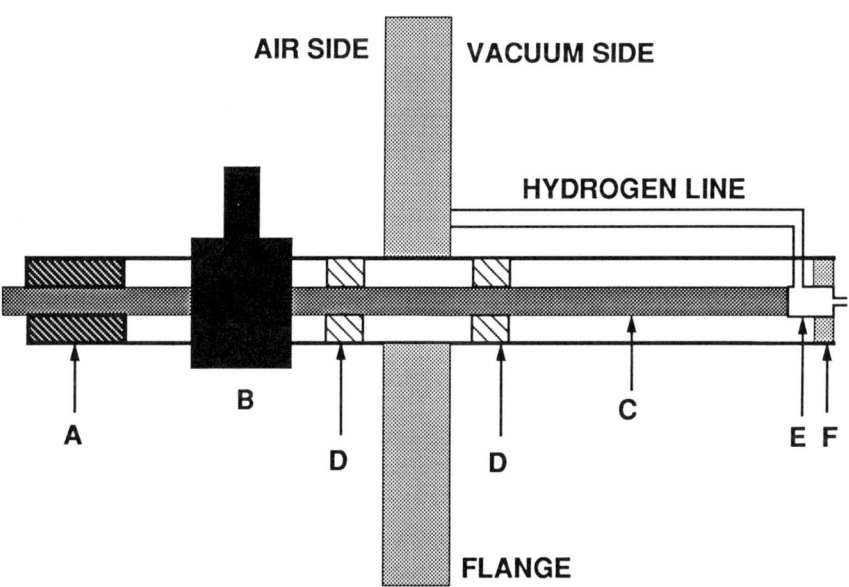

Figure 2: Cross-section of remote microwave discharge source.

125

encircled with heating wire. Acetylene flows through the other tube unheated. All three tubes are mounted approximately 1 cm from the substrate.

Substrates are either molybdenum foils or silicon wafers that are seeded with submicron diamond powder. The substrate is clamped onto a molybdenum block which is resistively heated between 600 and 1000°C, being maintained at the desired temperature by adjusting a transformer. Radiative heating from the heater block necessitates water-cooling of the reactor walls.

Typical flow rates are 8 sccm of the argon/hydrogen through the plasma and 2 sccm of DTBP. Operating pressures are approximately 1 mTorr in the growth chamber, 10^{-5} torr in the second chamber, and 10^{-7} torr in the mass spectrometer chamber. At these pressures the mean free path is longer than the distance from the beams to the substrate, eliminating gas phase chemistry. In addition, molecules withdrawn through the substrate will travel a collisionless path to the mass spectrometer.

RESULTS

The hydrogen dissociation fraction is characterized using molecular beam mass spectroscopy by monitoring the lock-in signal at 1 and 2 amu. When running the plasma with hydrogen alone the 2 amu peak decreased by less than 20%. The dissociation was dramatically improved by diluting hydrogen to 10-30% in argon. Under these conditions the amu 1 signal increased 120X when the plasma was ignited. In addition, the discharge was much brighter by visual inspection. However, at these conditions a significant fraction of the molecules are ionized, such that the amu 2 signal also *increases* when the plasma is ignited. Characterization of the ionized fraction is required to fully quantify the dissociation fraction

When DTBP pyrolyzes, 90% decomposes into two methyl radicals and two acetones. The source appears to be operating satisfactorily. The amu 15 (methyl radical) and amu 43 (acetone) increase sharply as the quartz tube is heated above 300°C and the signals level out at 500°C (Fig. 3). Additional experiments need to be performed to fully quantify the distribution of decomposition products.

Mass spectroscopy confirms the absence of gas-phase chemistry. Monitoring the mass spectra between 12 and 30 amu revealed no observable conversion of methyl radical to methylene or C2 hydrocarbons such as acetylene or ethane when both sources are on. Similarly, the absence of gas phase chemistry is observed when acetylene is the carbon source.

Despite the low absolute pressure, the high dissociation fraction directed nature of the beams should allow respectable deposition rates. By introducing a gas through the directed

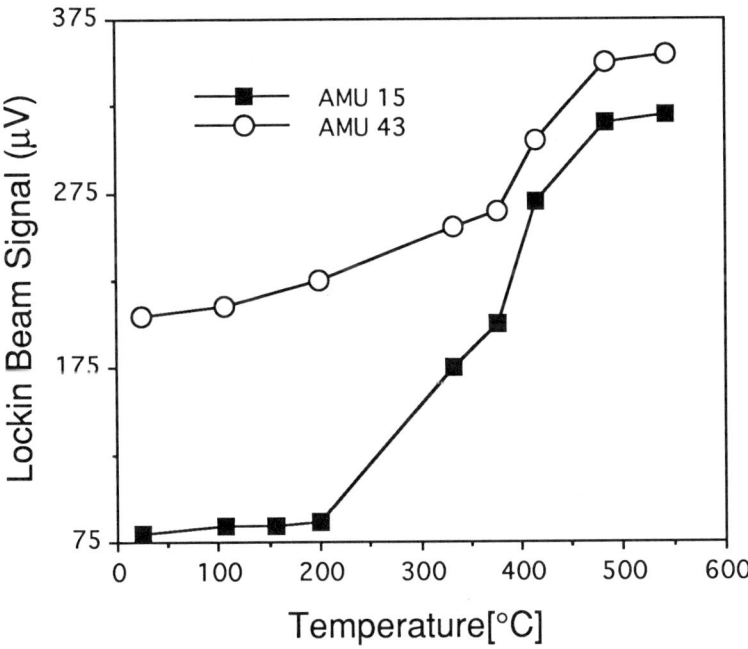

Fig. 3: Lock-in mass spectrometer signal for amu 15, 43 as a function of quartz temperature. 2 sccm DTBP.

beam and comparing the mass spectrometer signal to the case when the gas enters a side port, an enhancement factor between 5 and 15 is observed. Thus, although the base pressure is ≈ 1 mTorr, the effective partial pressures of atomic hydrogen and methyl radical is between 5 and 15 mtorr. These values are similar to fluxes reported for hot-filament systems, which are 40 mTorr and 4 mTorr for atomic hydrogen and methyl radical, respectively [11]. In preliminary experiments crystals between 1 and 20μm were was deposited over a 2 by 1 inch seeded molybdenum foil during 90 minute depositions using the methyl radical source. The deposits have been analyzed by SEM (Fig. 4) and show some faceting and structure, although they are not clearly diamond. The deposits require further characterization by Raman spectroscopy to evaluate the relative sp3 and sp2 fractions.

CONCLUSIONS

A novel reactor has been constructed that will allow diamond surface chemistry to be studied where all the major variables have been decoupled. Low operating pressures eliminate gas-phase chemistry. Deposition has been achieved from a directed beam of highly dissociated atomic hydrogen and a directed flux of methyl radicals.

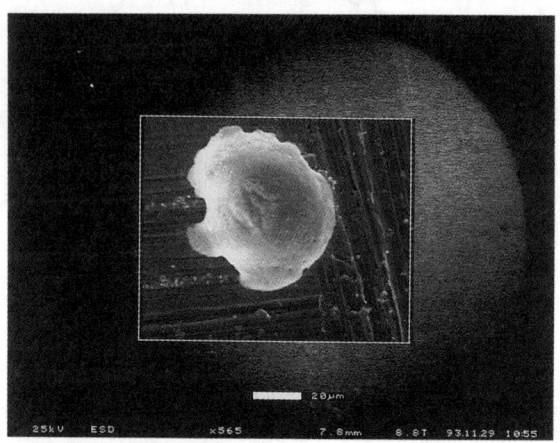

Fig. 4: SEM of material deposited under following conditions; Deposition time = 90 minutes, substrate temperature = 800°C, DTBP flow = 2 sccm, H2 flow =2.5 sccm, Ar flow =6 sccm. Pressure = 0.8 mtorr.

ACKNOWLEDGMENTS

We gratefully acknowledge the National Science Foundation under Grant #CTS 9006705.

REFERENCES
1. J. C. Angus and C. C. Hayman, Science **241**, 913 (1988).
2. F. G. Celii and J. E. Butler, Annu. Rev. Phys. Chem. **42**, 643, (1991).
3. S. Matsumoto, Y. Sato, M. Tsutsumi, and N. Setaka, J. Mat. Sci. **17**, 3106, (1982)
4. P. K. Bachmann, D. Leers and H. Lydtin, Diamond and Related Materials **1**, 1 (1991).
5. K. Suzuki, A. Sawabe, H. Yasuda, T,. Inuzuka, Appl. Phys. Lett. **50**, 728, (1987).
6. Y. Matsui, A. Yukki, M. Sahara and Y. Hirose, Jpn. J. Appl. Phys. **28**, 1718, (1989).
7. S. J. Harris, J. Appl. Phys. Lett. **56**, 2298 (1990).
8. M. Frenklach and K. Spear, J. Mater. Res. **3**, 133 (1988).
9. M. Frenklach, J. Appl. Phys. **65**, 5142 (1989).
10. C. Wolden and K. K. Gleason, Appl. Phys. Lett. **62**, 2329 (1993).
11. W. L. Hsu, Appl. Phys. Lett. **59**, 1427, (1991).

USING MOLECULAR-BEAM MASS SPECTROMETRY TO STUDY THE PECVD OF DIAMONDLIKE CARBON FILMS

I.B. GRAFF, R.A. PUGLIESE, JR.[†], AND P.R. WESTMORELAND
Department of Chemical Engineering, University of Massachusetts, 159 Goessmann Laboratory, Amherst, MA 01003-3110

ABSTRACT

Molecular-beam mass spectrometry has been used to study plasma-enhanced chemical vapor deposition (PECVD) of diamondlike carbon films. A threshold-ionization technique was used to identify and quantify species in the plasma. Mole fractions of H, H_2, CH_4, C_2H_2, C_2H_6 and Ar were measured in an 83.3% CH_4/Ar mixture at a pressure of 0.1 torr and a total flow of 30 sccm. Comparisons were made between mole fractions measured at plasma powers of 25W and 50W. These results were compared to measured concentration profiles and to film growth rates.

INTRODUCTION

Diamondlike carbon films have great potential for industrial uses [1]. These films are produced from ion beams, thermal deposition, and glow discharges [2-4]. Most notably, they have desirable physical properties: low friction coefficient and high density, hardness, and thermal conductivity.
One of the most common thin-film processes is plasma-enhanced chemical vapor deposition (PECVD) because it combines high growth rates with relatively low processing temperatures. Initial, high-energy steps in the deposition take place in the gas phase. It is then important to understand how gas-phase chemistry relates to film characteristics and to operating conditions.
Long used in combustion research [5], molecular-beam mass spectrometry (MBMS) provides a means of studying both stable and reactive species in the gas phase. Molecules and radicals are preserved by supersonically expanding a sample of the gas and then by collimating the free jet into a molecular beam. The technique has been applied in the study of both amorphous silicon and silicon nitride deposition [6-9]. For instance, Smith et al. [9] used the MBMS technique in the study of silicon nitride deposition from a silane-ammonia feed. In that study aminosilane radicals were identified as the depositing species based on good correlation of radical concentrations with film growth rates.

APPARATUS

Films were deposited in the parallel-plate reactor shown in Figure 1. The main reaction chamber was formed from a stainless-steel collar (81 cm diameter and 20 cm high) with 25-mm-thick stainless-steel flanges forming the base and the lid. Access ports were evenly spaced around the walls of the chamber: one 23x30-cm rectangular, clear polycarbonate window; two 15-cm-diameter quartz windows on opposing sides; and a 15-cm-diameter feedthrough that was used for a 3-D positioner. The reactor lid was removable and had several flanges mounted for use with feedthroughs for plasma power, grounding, cooling fluid circulation, pressure monitoring, pumping, and temperature measurement. An additional flange was positioned at the center, 12 inches above the surface of the lid, for supporting the polycarbonate inlet gas tube (13 mm OD). The chamber could be pumped either by a roughing pump (Alcatel 2012 AC, 13 cfm) or by a Roots blower (RGS-HV, size 1024) backed by a rotary-piston pump (Stokes Microvac 412H-10, 300 cfm). Both pumping systems were connected to the reactor through pneumatically actuated valves. Pressure was measured by a capacitance manometer (MKS Baratron Model 127AA-001, 0-10 torr, and MKS Model PDR-C-1C MKS readout) and manually adjusted downstream of the reactor by needle valves that introduced ballast gas.

[†] Present Address: Hewlett Packard, 1020 NE Circle Blvd, Corvallis, OR 97330

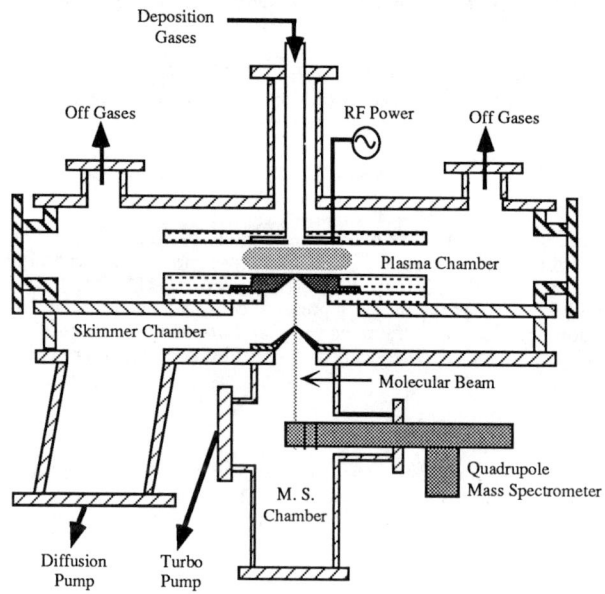

Figure 1. PECVD reactor with molecular-beam system.

The plasma was formed between two 150-mm-diameter aluminum electrodes in the center of the main chamber. The upper electrode was powered by a 13.56 MHz plasma generator (ENI Model ACG-500 with an accompanying ENI Matchwork-5 matching network) and was vertically movable to adjust gap spacing. Inlet gases were metered by mass flow controllers (MKS Model 2259B flow controllers with MKS Model 247-C four-channel readout) and fed to the discharge zone from the inlet flow tube through a 13-mm-diameter holes in the center of the upper electrode and wafer. The grounded, lower electrode had an internal channel for circulation of cooling fluid from a temperature bath (Lauda K-4/R). A 50-mm, flush-mounted disk was at the center of the electrode. This plate could be solid or could have a 0.8-mm orifice in the center, depending on the experiment performed. Both electrodes were electrically insulated by 295-mm-diameter Delrin® rings (19 mm thick). Three Teflon rods (30 mm long by 6 mm OD) could be used to support the wafer against the powered electrode and to space the two electrodes evenly.

A second, lower chamber was formed from a 40-mm-high stainless steel spacing ring mounted on a 72-cm-diameter flange. The flange had two 20-cm-diameter openings: one centered and one offset. A plate supporting a cone-shaped orifice (skimmer cone) rested on the center hole. The offset opening led to an air-cooled baffle and a diffusion pump (Varian M6, 1500 l/s backed by an Alcatel 2012 AC roughing pump). The pipe connecting the opening to the baffle had flanges for ionization and thermocouple gauges (monitored by a Granville-Phillips series 270 controller).

The third and lowest chamber housed the quadrupole mass spectrometer (Balzers QMA 400 quadrupole controlled from a Balzers QMS 420 console). Pumping of the chamber was performed by a turbomolecular pump (Leybold-Heraeus model TMP350, 350 l/s) backed by a roughing pump (Varian SD-300, 13 cfm). A liquid nitrogen shroud (Thermionics SS400/275 with LNC-400 controller) sat at the bottom of this chamber. Optical access was available by removal of a 70-mm flange below the cryotrap, allowing laser alignment of the orifices in the lower electrode and in the skimmer cone. Pressure was monitored by either a thermocouple gauge or an ionization gauge. A five-pin electrical feedthrough was used to control a tuning-fork chopper (Frequency Control Products, Model L2C).

The three differentially pumped chambers formed the basis for a molecular-beam sampling train. Gases were sampled through the 0.8 mm orifice in the lower electrode. This sharp-edged orifice is tapered to limit wall interactions of the highly reactive species. As the gas passed from the main reaction chamber (10^{-1} torr) through the skimmer chamber (10^{-4} torr), the species underwent an isentropic expansion to free-molecular flow. The gas was thereby cooled, preventing collisional deactivation of the reactive species. After traveling for 97 mm, a 2-mm-I.D. skimmer cone collimated a molecular beam, sampling the center of the free jet, in order to remove the outer edges of the spray that were likely to have undergone wall collisions. The collimated beam was modulated at 220 Hz by the tuning-fork chopper, then 183 mm after the skimmer orifice, it entered the cross-beam, electron-impact ionizer of a quadrupole mass spectrometer (10^{-7} torr background). The resulting ions were separated by the quadrupole and were detected by a Cu-Be electron multiplier or Faraday cup. The 220-Hz chopper reference signal and chopped mass-spectrometer signal were sent to a lock-in amplifier (Ithaco Model 395 Narrowband Voltmeter) to filter out the background signal, giving an improved signal-to-noise ratio and allowing detection of weak signals.

Other experimental techniques were also employed to study the plasma and the deposition. Microprobe mass spectrometry employed a tapered quartz probe to produce concentration profiles of stable species in the discharge [10-11]. Both single and symmetric double Langmuir probe methods have been used to determine electron temperature, electron density, and floating point potential (single probe only). An laser interferometry system, consisting of a He-Ne laser (Spectraphysics 155) and a photodiode amplifier circuit (Hamamatsu S1133-12 and TL061CP operational amplifier), monitored the change of refractive index during the deposition. Thickness was determined ex-situ by an ellipsometer (Gaertner Model 110) and by surface profilometers (Dektak-1 and Tencor Alpha-Step 200).

PROCEDURE

Prior to evacuation of the system, a 140-mm silicon wafer with a 13-mm hole in the center was etched for approximately 15 minutes with a 10% HF solution. The wafer was then rested on the three Teflon rods that were spaced evenly along the perimeter of the wafer. The upper electrode was lowered against it, and the hole in the wafer was aligned with the gas flow inlet.

Once system pressures equilibrated, a sampling run could be performed. All instruments were allowed at least a one-hour warm-up period. Mass flow controllers were zeroed, and flows were set to calibration values based on a 53/31/16 mixture of $CH_4/H_2/Ar$ at either 30 or 60 sccm total flow. The lower electrode temperature was regulated by circulating an ethylene glycol-water mixture at 70°C. Pumping of the reaction chamber was switched from the roughing pump to the Roots blower to attain operating pressures of 0.1 or 0.5 torr. Calibration gas flows were started and the gas ballast upstream of the blower was adjusted to set the reactor pressure. Intensity measurements for the calibration gases were determined with the lock-in amplifier.

After the calibration was completed, the flows were set to reaction feed values, a 5:1 mix of CH_4 to Ar at 30 or 60 sccm. The discharge was lit and the RF power for the plasma was set to either 25 or 50 W. Pressure was adjusted as necessary. Intensity measurements were performed as a function of ionization energy using the threshold-ionization method detailed below. After the measurements were completed, the plasma was extinguished and a second calibration was performed as described above.

A threshold-ionization method was used both to identify and to quantify mass spectrometer signals at a given mass number. Ionization efficiency curves for the species of interest were produced by measuring signal intensity as the ionization energy (electron energy) was decreased in small increments to below the ionization potential (*e.g.*, Figure 2). From the ionization efficiency curve, a relative determination of the amount of parent species present could be made. Also, the species could be identified by its ionization potential. The electron-energy scale of the mass spectrometer was not absolute, but it could be calibrated by comparison with literature values of ionization potentials. As the ionization energy was increased above the ionization potential, a linear region could be noted that was indicative of a given species. As the energy rose, additional contributions to or losses from the signal were indicated by a change in slope as the fragmentation

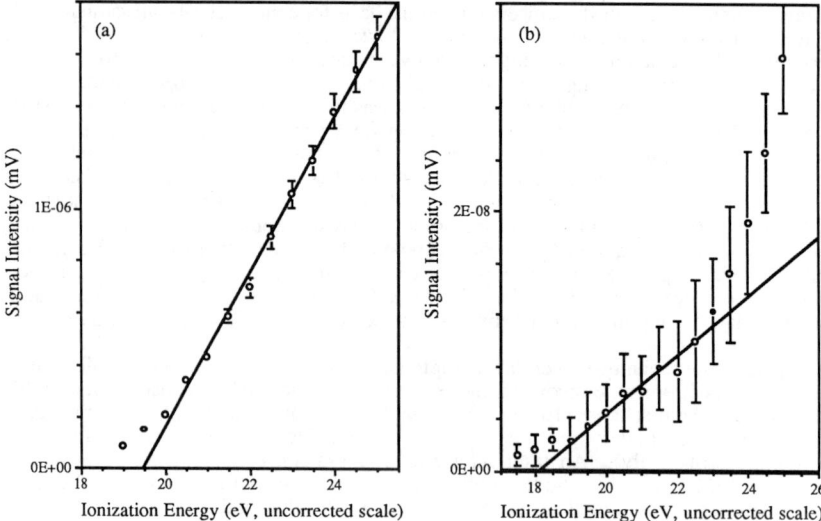

Figure 2. Ionization efficiency curves for (a) mass 16 and (b) mass 1.

potential of the species of interest or the appearance potential of a fragment was reached (*e.g.*, Figure 2). Although broadening was present due to a finite distribution of ionizing-electron energies, this procedure directly identified the parent species entering the ionizer, in contrast to ion fragments from the ionizer.

In the plasma, an ionization efficiency curve was measured for each species. Signal was measured as ionizer energies were varied from below the ionization potential to several electron volts above it (typically to 25 eV).

In terms of quantitative analysis of plasma species concentrations, there were three categories of species: major species which could be calibrated directly, minor species which were uncalibrated, and radicals. Concentrations for the latter two were found from an indirect method.

For the direct calibration, a mixture of hydrogen, methane, and argon was fed to the reactor in proportions representative of the plasma mixture and but at unignited reaction flow conditions. A calibration factor (F_i) was determined from the mole fraction ratios of the species (H_2 or CH_4) relative to argon divided by their intensity ratio at 25 eV. At plasma conditions, intensities of the hydrogen, methane, and argon were determined at 25 eV from the ionization efficiency curve. The following equation related these intensities to the mole fraction ratio of the species to argon.

$$\frac{x_i}{x_{Ar}} = F_i \frac{I_i \text{ at } 25 \text{ eV}}{I_{Ar} \text{ at } 25 \text{ eV}} \qquad (1)$$

Absolute mole fractions were found after the analyses of the radicals and of the other species were performed.

The indirect method [5] made use of the fact that the intensity of a species in the beam was a function of the mole fraction and the ionization cross section of the species. The ratio of the ionization cross section of a species (at a constant value, b, above its ionization potential) to its ionization cross section at 70 eV was assumed to be the same as for the reference species at an equivalent value above its ionization potential and at 70 eV. The mole fraction with respect to the reference species was then found from the following equation:

$$\frac{x_i}{x_{ref}} = \left(\frac{I_i \text{ at } IP+b}{I_{ref} \text{ at } IP+b}\right)\left(\frac{Q_{ref} \text{ at } 70 \text{ eV}}{Q_i \text{ at } 70 \text{ eV}}\right) \qquad (2)$$

To relate the species' mole fraction to argon, the ratio determined above was multiplied by the mole fraction ratio of the appropriate reference species, found from Equation 1. The mole fraction ratios (with respect to argon) of all species present in the plasma were summed. The mole fractions were then found by dividing each mole fraction ratio by this sum.

RESULTS & DISCUSSION

Analysis of the ionization efficiency curves was performed for each species. Curves were also found for mass 28; however, a large nitrogen background signal interfered with any positive determinations of ethylene.

Figure 2a shows an ionization efficiency curve for mass 16. Data were linear in the range from 20 to 25 eV; below 20 eV, tailing of the signal was noted. Regression analysis was performed in the linear region and extrapolated to zero signal (the ionization potential). The x-intercept of 19.47 eV was found and then calibrated to the methane literature value of 12.70 eV [12], a difference of 6.77 eV, typical here for hydrocarbons.

Hydrogen atom was also measured by this method. Figure 2b shows the ionization efficiency curve for mass 1. A linear region was noted between 19 and 23 eV. Tailing was seen below 19 eV, while contributions from fragments were indicated by the change in slope above 24 eV. From the linear region, an ionization potential was found at 18.05 eV(uncalibrated), 4.29 eV above the hydrogen atom literature value of 13.76 eV [12]. This offset compared well with the offset of molecular hydrogen (4.29 eV) from its literature value. The slope and intercept were used later in the analysis to determine concentration. Table I shows the ionization potential measured on the uncalibrated scale and comparisons to the literature values [12].

Table I. Comparison of measured and literature ionization potentials.

Mass	Presumed Species	IP (eV, uncorrected)	IP (eV, literature)	Offset (eV)
1	H•	18.32	13.76	4.29
2	H_2	19.97	15.42	4.55
16	CH_4	19.47	12.70	6.77
26	C_2H_2	17.64	11.41	6.23
30	C_2H_6	18.15	11.52	6.63
40	Ar	21.22	15.76	5.46

It should be noted that the offset values were comparable for similar species. For example, the offset for H-atom was similar to that of H_2, and the offsets for hydrocarbons were in the range of 6.2 to 6.8 eV.

Concentrations of methane, hydrogen and argon were found from direct calibration, while concentrations of hydrogen atom, acetylene and ethane were found from indirect calibrations. Mole fractions were measured at identical flow conditions (0.1 torr and 30 sccm total flow) but for different RF powers (25 W and at 50 W). The mole fractions are shown in Table II, along with comparisons to microprobe measurements within the plasma but close (within 2 mm) to the sampling orifice for the molecular beam.

Table II. Comparison of mole fractions measured by MBMS and microprobe methods.

	25W		50W	
	MBMS	Microprobe	MBMS	Microprobe
H•	0.00482	—	0.00173	—
H_2	0.257	0.270	0.320	0.355
CH_4	0.520	0.561	0.517	0.482
C_2H_2	0.0175	0.018	0.0139	0.018
C_2H_6	0.00978	—	0.00396	—
Ar	0.190	0.151	0.143	0.145

Fairly good agreement was shown among the directly calibrated species between the MBMS and microprobe measurements. The use of the molecular beam increases the sensitivity enough to allow the detection of ethane and of hydrogen atom.

The mole fractions for methane were somewhat lower in the 50 W plasma, indicative of a higher rate of dissociation. Film growth rate was 58.6 Å/min in the 50 W discharge and was 31.2 Å/min at 25 W. A higher rate of hydrogen evolution also is consistent with the increase in power. However it is more surprising that mole fractions of H-atom, acetylene, and ethane are all higher in the 25 W discharge as compared to the mole fractions in the 50 W discharge. The reasons for their differences is under investigation.

CONCLUSIONS & FUTURE PLANS

Molecular-beam mass spectrometry was used to measure concentrations of stable species and H-atom in PECVD of diamondlike carbon. MBMS is shown to be a powerful tool for determining species concentrations in plasma processes.

As the experimental technique on the system is enhanced, many other species will be investigated, including C_2H_4, CH_3, CH_2 and CH. Also, concentrations will be correlated with film properties and deposition rates. Furthermore, MBMS will be employed along with the microprobe technique and Langmuir probing to make determinations of kinetic processes taking place in the plasma.

REFERENCES

1. H. Nakaue, T. Mitani, H. Kurokawa, T. Yonezawa and H. Yoshio, Thin Solid Films, **212**, 240 (1992).
2. S. Aisenberg and R. Chabot, J. Appl. Phys., **42**, 2953 (1971).
3. S. Matsumoto, Y. Sato and M. Kamo, Jap. Jrnl. Appl. Phys., **59**, 3267 (1986).
4. J.C. Angus, P. Koidl and S. Domitz, in Plasma Deposited Thin Films, edited by J. Mort and F. Jansen (CRC Press, Inc., Boca Raton, FL, 1986), pp. 89-127.
5. J.C. Biordi, Prog. Energy Combust Sci., **3**, 151 (1977).
6. S. Bourquard, D. Erni and J.M. Mayor in Proceedings of the 1st International Conference on Plasma Chemistry and Technology, edited by H. Boen (Tecnomic, Lancaster, PA, 1983) p 101.
7. R. Robertson, D. Hills, H. Chatham and A. Gallagher, Appl. Phys. Lett., **43**, 544 (1983).
8. N.P. Johnson, A.P. Webb, and D.J. Fabian in Plasma Processing and Synthesis of Materials edited by J. Szekely and D. Apelian (Mater. Res. Soc. Proc. **30**, 1984), pp. 277-282.
9. D.L. Smith, A.S. Alimonda, C. Chen, S.E. Ready and B. Wacker, J. Electrochem. Soc., **137**, 2, 614 (1990).
10. R.M. Fristrom and A.A. Westenberg, Flame Structure (McGraw-Hill, New York, 1965) pp. 177-203.
11. R.A. Pugliese, Jr., Ph.D. Thesis, University of Massachusetts, 1992.
12. H.M. Rosenstock, K. Draxl, B.W. Steiner and J.T. Herron, J. Phys. Chem. Ref. Data Supplement, **6**, 1 (1977).

ACKNOWLEDGMENTS

We gratefully acknowledge that the material in this paper is based on work supported by the National Science Foundation under grants titled "Research Initiation: Mapping Process Chemistry for Silicon PECVD" (Grant No. CTS-88-10562), "Engineering Research Equipment Grant: Plasma-Enhanced Chemical Vapor Deposition Reactor" (Grant No. CBT-88-05703), and "Presidential Young Investigator Award" (Grant No CTS-90-57406). We thank Charles Musante of the ECE Department for his assistance with the film-thickness measurements.

EFFECT OF OXYGEN ON DIAMOND DEPOSITION IN $CH_4/O_2/H_2$ GAS MIXTURES

H. MATSUYAMA, N. SATO, AND H. KAWAKAMI
Fuji Electric Corporate R&D. Ltd., 2-2-1 Nagasaka Yokosuka 240-01 Japan

ABSTRACT

Diamond growth experiments were carried out by a microwave plasma assisted CVD technique in various gas mixtures of $CH_4(0-100\%)/O_2/H_2$. The phase diagram obtained by this study shows that a diamond growth region exists. With addition of more than 5% O_2 in reactant gases, diamond particles could be included in amorphous or graphitic carbon films even using CH_4/O_2 gas mixtures. Faceted diamond films were obtained if the oxygen gas concentration $[O_2]$ was approximately more than half the methane gas concentration $[CH_4]$ ($[O_2]>[CH_4]/2$). However, no films were grown when $[O_2]$ exceeded half of $[CH_4]$ plus 7% ($[O_2]>[CH_4]/2+7\%$). These results corresponded to the observations by plasma emission spectroscopy. Though oxygen etches carbon films and decomposes methane by forming carbon monoxide, oxygen rarely reacts with hydrogen in a film growth region.

INTRODUCTION

It is an important subject that chemical vapor deposition (CVD) of diamond films increases growth rate and improves quality. Hirose and Terasawa reported that diamond films deposited using oxygen containing organic compounds had better quality and were grown with a faster growth rate compared with films using conventional $CH_4(\leq 1\%)/H_2$ gas mixtures[1]. As a result of their study, it was found that oxygen was useful for CVD diamond growth. Various kinds of reactant gases including oxygen or compounds containing oxygen have been applied to diamond CVD such as CO/H_2 [2,3], C_2H_2/O_2 [4], $CH_4/CO_2/H_2$ [5], $CH_4/H_2/H_2O$ [6] and so on. This data was arranged within a C-H-O phase diagram by Bachmann et al.[7] They described that successful diamond deposition was restricted to a well-defined area near the $CO-H_2$ line where a atomic ratio of C to O in reactant gases was 1. Addition of oxygen was also carried out with low contents of methane in hydrogen gas, so that diamond films with good quality could be quickly deposited[5,8-11]. It is suggested that the role of oxygen is (1) to reduce the concentration of acetylene, which is related to the deposition of non-diamond carbon (amorphous or graphitic carbon), (2) to increase atomic hydrogen, which etches non-diamond carbon selectively, (3) to accelerate the reaction of non-diamond carbon with molecular hydrogen, (4) to act as a selective etchant of non-diamond carbon and so on.

In this paper we study the characteristics of films deposited in $CH_4(0-100\%)/O_2/H_2$ gas mixtures and their discharge properties, especially, in order to investigate the effect of oxygen in a range of high methane content.

EXPERIMENTAL

Films were synthesized by a microwave plasma CVD system using various gas mixtures of $CH_4(0-100\%)$, O_2 and H_2. Si (100) substrates for the film deposition were prepared

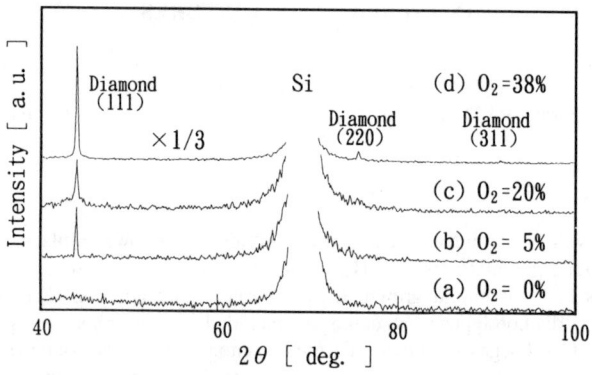

Fig.1. X-ray diffraction patterns of films deposited in CH_4/O_2 gas mixtures.

Fig.2. SEM images of films deposited in $CH_4/O_2/H_2$ gas mixtures.

using ultra sonic abrasion in ethanol mixed with 10 μm diamond powder and they were set in a tube reactor of silica glass. The flow rate of reactant gases and the reaction pressure were maintained at 100 sccm and 30 torr respectively. The substrate temperature rose up to 850 °C using 2.45 GHz microwave plasma discharge of 250 W of power. Plasma emissions were measured using an optical emission spectrometer (OEM) for chemical analysis of the discharge. Deposited films were characterized by X-ray diffractometry (XRD), scanning electron microscopy (SEM) and Raman spectroscopy.

RESULT AND DISCUSSION

Figure 1 shows X-ray diffraction patterns of films deposited in CH_4/O_2 gas mixtures. The film without oxygen has no diffraction peaks of diamond (See in Fig.1(a)), and films deposited in $CH_4(\geq 60\%)/H_2$ gas mixtures also have no peaks. The films in the non-diamond carbon growth region include no diamond particles or diamond particles that are too small. Each film with addition of more than 5% O_2 in reactant gases includes diamond particles, since it obtains a diffraction peak from diamond (111) planes as shown in Fig.1(b) to (d). It is definite that the addition of oxygen is useful to accelerate the diamond formation. Figure 2(a) to (c) show SEM images of films deposited in $CH_4(40\%)/O_2/H_2$ gas mixtures. Their faceted morphology becomes clearer when oxygen gas concentration $[O_2]$ is increased. Oxygen functions to improve the quality of films[5,8-10]. Faceted diamond films are synthesized in a diamond growth region when $[O_2]$ is approximately more than half the methane gas concentration $[CH_4]$ ($[O_2]>[CH_4]/2$). As is shown in Fig.2(d), the film using a CH_4/O_2 gas mixture without hydrogen gas also exhibits faceting, which indicates that the addition of hydrogen gas is not necessary for a diamond formation and that hydrogen generated from methane is enough[4]. Figure 3 shows Raman spectra of faceted diamond films deposited in $CH_4/O_2/H_2$ gas mixtures. It is found that the films shown in Fig.3(b) to (d) are of good quality, since they have sharper diamond peaks at about 1333 cm^{-1} than that in Fig.3(a) for a conventional $CH_4(1\%)/H_2$ gas mixture. Figure 4. shows the growth rate of films deposited with (a) 20% CH_4, (b) 40% CH_4, and (c) 60% CH_4 in $CH_4/O_2/H_2$ gas mixtures as a function of $[O_2]$. Though each growth rate increases with the increase of $[O_2]$, it decreases rapidly in the diamond growth region and vanishes when $[O_2] \geq [CH_4]/2+7\%$. It indicates that oxygen has the function of etching carbon films. The growth rate in the diamond growth region is three times faster at most than the growth rate (0.2 μm/h) for the conventional $CH_4(1\%)/H_2$ gas mixture and it does not increase proportionally to $[CH_4]$.

The above result and one other obtained in our experiment were plotted in a phase diagram and the films can be classified as shown in Fig.5. Here, X- and Y-axis correspond to $[O_2]$ and $[CH_4]$ in reactant gases respectively. The CH_4–O_2 line shows that reactant gases are composed of CH_4/O_2 gas mixtures. The CO–H_2 line shows that the ratio of $[CH_4]$ to $[O_2]$ in reactant gases is 2. The films deposited in a diamond & a-C mixed region have diamond peaks in XRD patterns, though they do not have faceted morphology. Faceted films are obtained in the diamond growth region. No films were deposited in the no growth region. The diamond growth region can be found where $[CH_4]/2<[O_2]<[CH_4]/2+7\%$ in a range of more than 20% CH_4. It suggests that the diamond growth region is independent of hydrogen gas concentration $[H_2]$ and it is roughly equal to the prediction of Bachmann et al.[7]

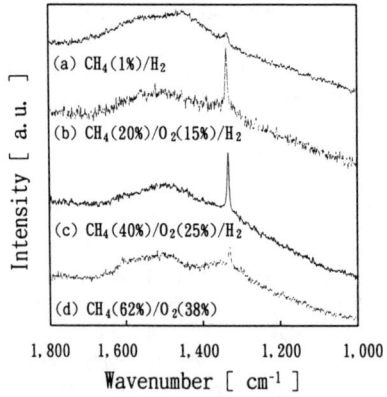

Fig.3. Raman spectra of diamond films deposited in $CH_4/O_2/H_2$ gas mixtures.

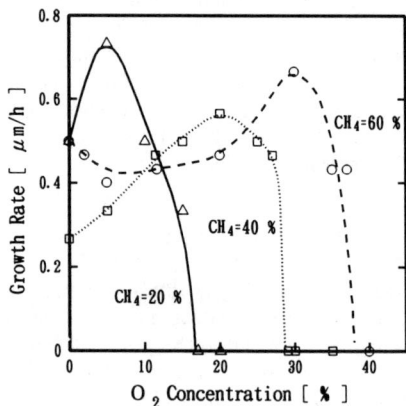

Fig.4. Growth rate of films as a function of oxygen concentration.

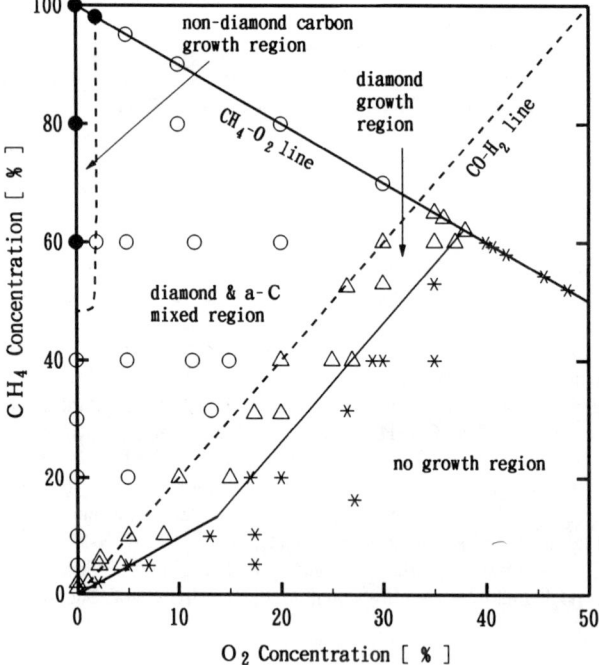

Fig.5. Phase diagram of films deposited in $CH_4/O_2/H_2$ gas mixtures. ✶ : no films were grown, △ : films have diamond morphology, ○ : films have diamond X-ray diffraction pattern.
● : no films have diamond X-ray diffraction pattern.

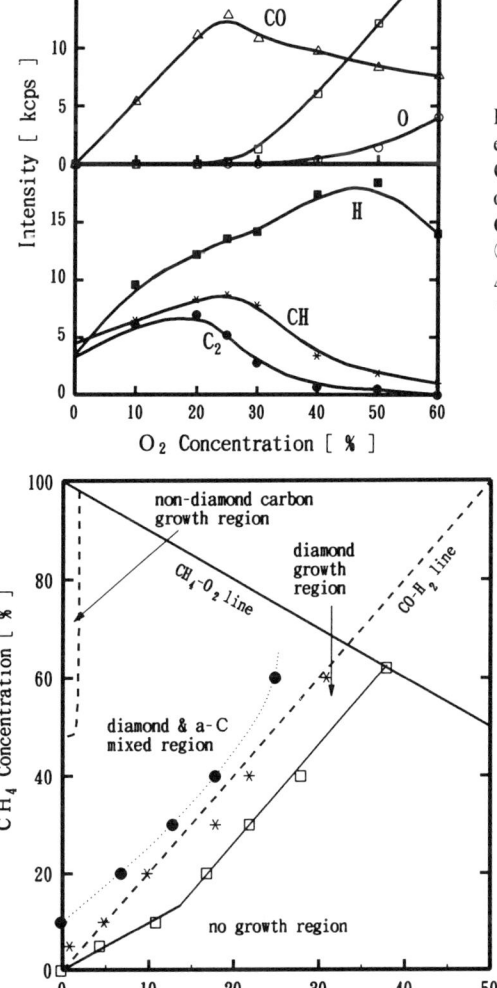

Fig.6. Evolution of the plasma emission intensities of O, OH, CO, C_2, CH, and H radicals as a function of oxygen concentration in $CH_4(40\%)/O_2/H_2$ gas mixtures.
○: O(777nm), □: OH(618nm), △: CO(267nm), ●: C_2(468nm), ∗: CH(431nm), and ■: H(656nm).

Fig.7. Typical gas compositions investigated for the plasma emission, plotted on the phase diagram of Fig.5. Squares correspond to rapid rises of OH(618nm) emission intensity, and asterisks and solid circles to maxima of CH(431nm) and C_2(468nm) as a function of $[O_2]$ for various $[CH_4]$ respectively.

Species in a $CH_4/O_2/H_2$ plasma discharge detected by OEM are CO, C_2, CH, H, O, and OH radicals. Generally speaking, the plasma emission consists of CO, C_2, CH and H lines in a film growth region and it also includes OH and O lines in the no growth region. In regard to the lines for O(777nm), OH(618nm), CO(267nm), C_2(468nm), CH(431nm), and H(656nm), Fig.6 illustrates the plasma emission intensities as a function of $[O_2]$ in $CH_4(40\%)/O_2/H_2$ gas mixtures where $[CH_4]$ is fixed to 40%. The intensities of the CO, CH and C_2 lines go up with the increase of oxygen content and reach maxima near the diamond growth region, which corresponds roughly to the growth rate dependence on oxygen content. It was found that oxygen also has the function of decomposing methane by forming carbon monoxide and accelerating it to generate the CH, C_2 and H radicals. The intensity of the H line also increased[11]. However, the quality change between the non-diamond carbon and the diamond & a-C mixed region could not be explained by our OEM study. Since the intensity of the OH line rose rapidly from the boundary between the diamond growth and the no growth region, oxygen rarely reacted with hydrogen in the film growth region though excess oxygen has the ability to react with hydrogen gas in the no growth region. Additional increase in oxygen allowed O radical emission to appear. The above phenomena are roughly the same as that for different methane contents ($[CH_4]$= 5, 10, 20, 30, and 60%), which means that the plasma conditions in these discharges are equivalent and are controlled by methane and oxygen contents in the film growth region. Some of the specific compositions are plotted on Fig.7. Squares represent the compositions of a rapid rise of the OH radical intensity and are located on the boundary between the diamond growth and the no growth region. Asterisks and solid circles are shown close to the $CO-H_2$ line. These indicate the compositions of maximum emission intensities brought about by the CH and C_2 radicals, respectively, as a function of $[O_2]$ for each $[CH_4]$. This result corresponds to the phase diagram (Fig.5), therefore we believe that the characteristics of the films is determined by the ratio of $[CH_4]$ to $[O_2]$.

CONCLUSION

Diamond growth experiments were carried out by a microwave plasma assisted CVD technique using various gas mixtures of $CH_4(0-100\%)/O_2/H_2$. The effect of oxygen, especially, was investigated in the range of high methane content. The phase diagram obtained by this study shows that a diamond growth region exists. Faceted diamond films were obtained when $[CH_4]/2<[O_2]<[CH_4]/2+7\%$. Oxygen etches carbon films and decomposes methane by forming carbon monoxide. However, oxygen rarely reacts with hydrogen in the film growth region.

REFERENCE

1. Y. Hirose and Y. Terasawa, Jpn. J. Appl. Phys. **25**, L519 (1986).
2. K. Ito, T. Ito and I. Hosoya, Chem. Lett. **4**, 589 (1988).
3. J. Suzuki, H. Kawarada, K. Mar, J. Wei, Y. Yokota and A. Hiraki, Jpn. J. Appl. Phys. **28**, L281 (1989).
4. Y. Matsui, A. Yuuki, M. Sahara and Y. Hirose, Jpn. J. Appl. Phys. **28**, 1718 (1989).
5. A. Inspektor, Y. Liou, T. McKenna and R. Messier, Surf. Coat. Technol. **39/40**, 211 (1989).
6. Y. Saito, K.Sato, H. Tanaka, K. Fujita and S. Matsuda, J. Mater. Sci. **23**, 842 (1988).
7. P.K. Bachmann, D. Leers and H. Lydtin, Diamond Rel. Mater. **1**, 1 (1991).
8. T. Kawato and K. Kondo, Jpn. J. Appl. Phys. **26**, 1429 (1987).
9. C.P. Chang, D.L. Flamm, D.E. Ibbotson and J.A. Mucha, J. Appl. Phys. **63**, 1744 (1988).
10. C.F. Chen, Y.C. Huang, S. Hosomi and I. Yoshida, Mat. Res. Bull. **24**, 87 (1989).
11. J.A. Mucha, D.L. Flamm and D.E. Ibbotson, J. Appl. Phys. **65**, 3448 (1989).

MASS SPECTROMETRIC ANALYSIS OF A HIGH PRESSURE, INDUCTIVELY COUPLED PLASMA DURING DIAMOND FILM GROWTH

PETER G. GREUEL,* HYUN J. YOON,** DOUGLAS W. ERNIE** AND JEFFREY T. ROBERTS*#
*University of Minnesota, Department of Chemistry, 207 Pleasant St. SE, Minneapolis, MN 55455
**University of Minnesota, Department of Electrical Engineering, 200 Union St. SE, Minneapolis, MN 55455
#Author to whom correspondence should be addressed.

ABSTRACT

Determination of the gas phase composition at or near a substrate surface during plasma assisted chemical vapor deposition presents a challenging problem. The species located at a growing surface include highly reactive radicals which are difficult to detect in atmospheric plasma conditions. A system has been designed which consists of an inductively coupled rf plasma reactor linked to a quadrupole mass spectrometer (QMS) via a supersonic convergent-divergent nozzle. Differential pumping in the transient stages allows the plasma chamber to be operated at or near atmospheric pressures, while facilitating the detection of reactant species present in the growth boundary layer with the QMS.

INTRODUCTION

Diamond is a fascinating material with a wealth of unique properties. The economic potential of an efficient diamond fabrication process near atmospheric pressures has led to a great deal of research. Currently, several plasma techniques are used to activate gaseous species for diamond growth, e.g.: hot filament, radio frequency, microwave, arc discharge, oxygen-acetylene flame. Over the past few years several plausible mechanisms have been proposed to explain the growth of diamond in certain plasma assisted chemical vapor deposition (PACVD) processes [1-8]. In most cases, these mechanisms are based upon computer simulations in which the unknown rates of surface reactions are inferred from the rates of analogous reactions in the gas phase. A direct measure of the chemical species present at a diamond growth surface during deposition could help to experimentally verify the mechanism(s) responsible for diamond film growth.

Experimental determination of the gas phase species present in a boundary layer between a plasma and a substrate has proven to be a difficult task. Techniques such as optical emission, laser induced fluorescence and infrared absorption spectroscopy have been used to detect gas phase species present during diamond deposition [9-12]. These techniques are advantageous in that they are non intrusive, but they are applicable only for a limited number of species. Mass spectrometry, on the other hand, enables the detection of a much wider range of species. The primary disadvantage of mass spectrometry is that it is an intrusive diagnostic and therefore requires careful placement of the sampling orifice so as to not greatly disturb the deposition boundary layer.

A number of researchers have employed mass spectrometry in their PACVD systems by introducing a sampling port adjacent to the diamond growth region [13-18]. Typically in these systems, the mean free path of species entering the extraction port is small compared to the distance they travel before detection. This creates a problem in detecting species such as radicals, which can chemically react upon collision. Another problem that develops when using mass spectrometry is that species which are activated by a plasma generally have high thermal energies; this can lead to unimolecular reactions, such as dissociation. Both of these concerns are addressed in the design of the inductively coupled radio frequency system which we discuss herein.

DESIGN

The apparatus is depicted schematically in Figure 1. This system has been designed to detect reactive precursor species present at a substrate surface during rf plasma assisted CVD of diamond. A rf induction plasma was chosen as the activation source to avoid contamination from electrode or filament vaporization. The detection scheme employs a quadrupole mass spectrometer in conjunction with a supersonic sampling orifice to facilitate the detection of both radicals and stable species present at the substrate surface during diamond growth. The system is best described by breaking it down into three sections, each corresponding to a different pressure region within the apparatus, namely the plasma volume, the differential stage and the quadrupole mass spectrometer (QMS) region.

The plasma volume is where activation of diamond growth precursor species occurs. It consists of two concentric cylindrical quartz tubes mounted on a 14" stainless steel support plate with pumping ports. Water flows between the tubes to provide cooling. A 25 kW (Lepel) radio frequency power supply is coupled to the plasma volume with a five turn copper induction coil wrapped helically around the center of the outer quartz tube. Argon is used as the plasma gas and is injected chiefly through the 36 uniformly spaced sheath ports that reside at an angle 45 degrees tangent to the inside quartz cylinder. The pressure maintained in this region can be varied from an upper limit of one atmosphere to a lower limit of a few torr by adjusting the speed of the pump and/or the feed rate of argon. Reactant gases such as hydrogen and methane are generally fed into the system through an injection probe located within the plasma volume. Typical operating conditions of the plasma are shown in Table I. The height of the injection probe end can be adjusted to control precisely where the reactant gases enter the plasma. Gaseous species activated by the plasma impinge on the substrate surface, where diamond growth occurs. Located in the center of the substrate is a small fixed orifice which leads into the intermediate or differential stage of the system.

Table I. Typical discharge operating conditions.

discharge power	8 kW	
hydrogen flow	probe - 3.0 L·min^{-1}	sheath - 0
methane flow	probe - 0.03 L·min^{-1}	sheath - 0
argon flow	probe - 3.0 L·min^{-1}	sheath - 40 L·min^{-1}

The differential stage is enclosed by a water cooled stainless steel disk which has a cylindrical stem protruding from the center. At the top of this stem, a circular stainless steel orifice plate is attached with inner and outer gold O-rings, allowing for water cooling. Located in the center of the orifice plate is a convergent-divergent nozzle. The top of the nozzle is extended through a molybdenum substrate which is attached on top of the orifice plate. The function of the nozzle is

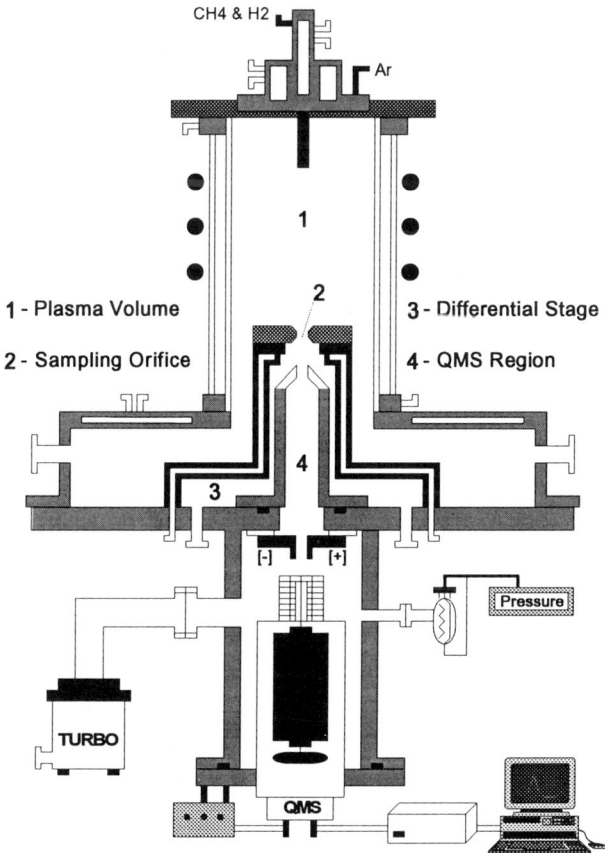

Figure 1. The experimental apparatus.

to promote the supersonic flow of gaseous species present at the nozzle orifice (i.e. the substrate surface) into the differential region while maintaining their chemical composition. The extraction orifice is shown schematically in Figure 2 where subscripts labeled "1" refer to the plasma volume and those labeled "2" indicate the differential region. Equation 1 describes the temperature

$$\frac{T_1}{T_2} = 1 + \frac{\gamma-1}{2}M^2 \qquad (1)$$

change across the supersonic nozzle, where γ is the heat capacity ratio (C_p/C_v) of the gas being expanded and M is the Mach number, which can be estimated using the ratio of the cross sectional areas A and A^* of the nozzle [19]. Equation 1 describes the temperature drop across a supersonic nozzle assuming ideal gas behavior, constant γ and isentropic and continuum flow. From Equation 1, we infer that the temperature drop between the reactor and the differential stage is given by $T_1/T_2 \approx 5$. Molecules which enter the orifice an T = 2500 K would be expelled into the differential stage at 500 K, a drop which is sufficient to prevent unimolecular dissociation. Depending on the pressure in the differential stage, the assumption of continuum flow may break down, in which case the actual temperature drop would be larger. In either case, unimolecular dissociation would be prevented.

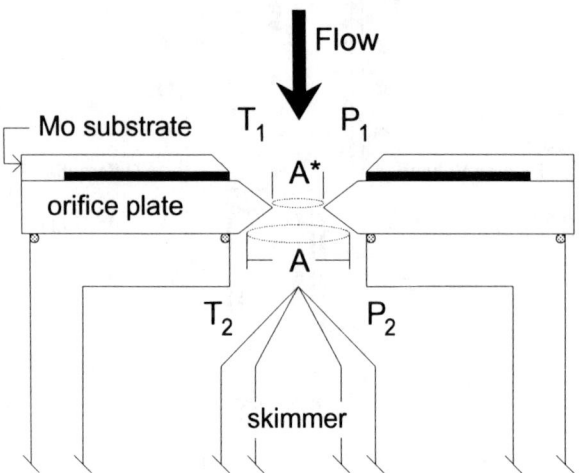

Figure 2. Supersonic flow extraction orifice.

The differential region surrounds a skimmer cone which leads into the high vacuum QMS region. The high vacuum region is composed primarily of a quadrupole mass spectrometer (Balzers QMA 420) positioned below the skimmer cone. A parallel plate ion separator is located between the QMS and the skimmer to keep charged particles from entering the ionizer of the QMS. The region is evacuated with a turbomolecular pump (Balzers TPU-240) which maintains a sufficient vacuum to operate the QMS (minimum $\approx 10^{-5}$ torr) and ensure that a limited number of bimolecular reactions occur.

RESULTS

We present here results concerning the detection of methyl radicals in a 10 kW, 20 torr argon discharge ($H_2/CH_4 \approx 100$). A discriminatory ionization technique is necessary for detection of radical species in a plasma since the parent molecules of the radicals are present in equal or greater concentration. In this instance, detection of methyl radicals is hampered by the presence of methane. Methyl radicals are detected as *m/e* 15; methane also yields a signal at *m/e* 15. By applying a threshold ionization technique, it is possible to eliminate the contribution of methane to *m/e* 15 [20]. The electron impact induced dissociative ionization of methane ($CH_4 \rightarrow CH_3^+$; 14.3 eV) requires an electron energy approximately 4 eV greater than that for the ionization of the methyl radical ($CH_3 \rightarrow CH_3^+$; 9.8 eV). Figure 3 shows experimental verification for the threshold dissociative ionization potential of methane and its fragments. The threshold energies

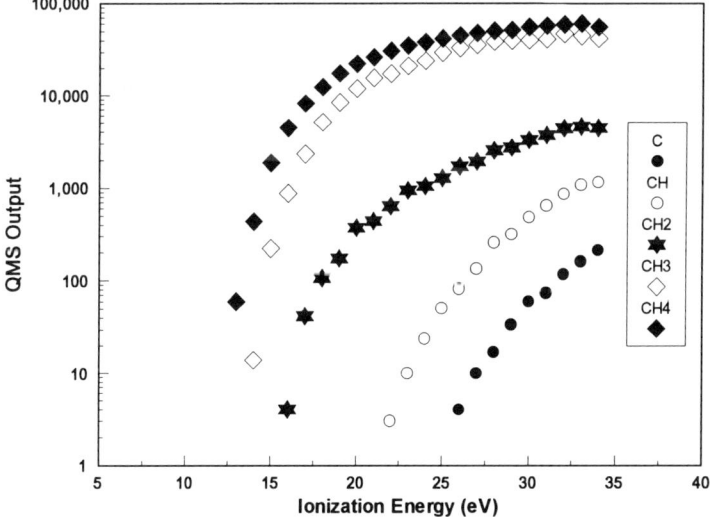

Figure 3. Threshold Ionization of methane.

for detecting CH_x^+, and in particular CH_3^+, from methane are in good agreement with the literature [21].

Figure 4 illustrates a plot of the relative QMS output for CH_3^+ as a function of the electron ionization energy with the plasma discharge on and off. The increased separation of the discharge on and discharge off data as the ionization energy is lowered is attributed to the ionization of CH_3 radicals from the plasma. Therefore, at lower ionization energies, the CH_3^+ signal is proportional to the CH_3 radical density in the plasma.

In the CVD of diamond films, the mechanism(s) responsible for growth are not understood. Several possible routes of diamond growth have been suggested, and debate still continues on the subject. By sampling, chemically freezing and subsequently analyzing the gas phase species present near a substrate during thermal plasma assisted diamond deposition; we hope to offer experimental evidence that will help in the mechanistic understanding of these processes.

Acknowledgment: The authors would like to thank Dr. J. Heberlein and Dr. E. Pfender for their assistance. This work is supported by the National Science Foundation through the Engineering Research Center for Plasma Aided Manufacture.

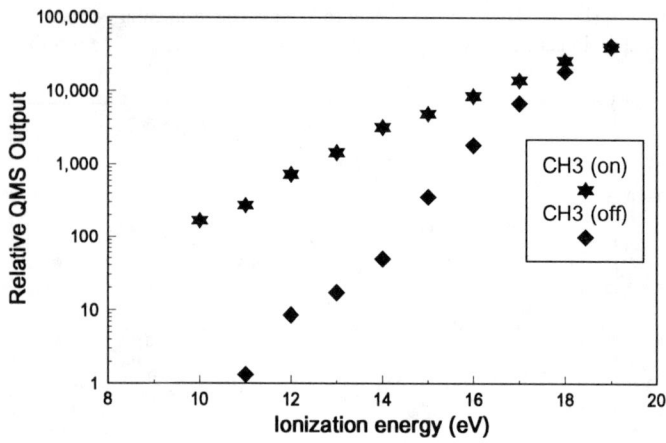

Figure 4. Relative QMS output (m/e 15) vs. ionization energy, with the discharge on and off.

REFERENCES

1. S. L. Girshick, C. Li, B. W. Yu and H. Han, Plasma Chem. Plasma Process. **13**, 181 (1993).
2. M. Frenklach and H. Wang, Phys. Rev. B **43**, 1520 (1991).
3. M. Frenklach and K. E. Spear, J. Mater. Res. **3**, 133 (1988).
4. W. Piekarczyk and W. A. Yarbrough, J. Cryst. Growth **108**, 583 (1991).
5. P. Bou, J. C. Boettner and L. Vandenbulcke, Jpn. J. Appl. Phys. **31**, 1505 (1992).
6. P. Bou, J. C. Boettner and L. Vandenbulcke, Jpn. J. Appl. Phys. **31**, 2931 (1992).
7. S. J. Harris and D. G. Goodwin, J. Phys. Chem. **97**, 23 (1993).
8. D. N. Benton and S. J. Harris, J. Chem. Phys. **96**, 2371 (1992).
9. S. J. Harris, Appl. Phys. Lett. **56**, 2298 (1990).
10. T. Mitomo, T. Ohta, E. Kondoh and K. Ohtsuka, J. Appl. Phys. **70**, 4532 (1991).
11. A. Campargue, M. Chenevier, L. Fayette, B. Marcus, M. Mermoux and A. J. Ross, Appl. Phys. Lett. **62**, 134 (1993).
12. P. W. Pastel and W. J. Varhue, J. Vac. Sci. Technol. A **9**, 1129 (1990).
13. K. R. Stalder and R. L. Sharpless, J. Appl. Phys. **68**, 6187 (1990).
14. W. L. Hsu, Appl. Phys. Lett. **59**, 1427 (1991).
15. S. J. Harris, A. M. Weiner and T. A. Perry, Appl. Phys. Lett. **53**, 1605 (1988).
16. C. E. Johnson, W. A. Weimer and F. M. Cerio, J. Mater. Res. **7**, 1427 (1992).
17. K. Takenchi and T. Yoshida, J. Appl. Phys. **71**, 2636 (1991).
18. L. Okeke and H. Stoeri, Plasma Chem. and Plasma Process. **11**, 489 (1991).
19. James D. Anderson, *Modern Compressible Flow: with Historical Perspective* (McGraw-Hill, New York, 1982).
20. H. Toyoda, H. Kojima and H. Sugai, Appl. Phys. Lett. **54**, 1507 (1989).
21. F.H. Field and J.L. Franklin, *Electronic Impact Phenomena and the Properties of Gaseous Ions* (Academic, New York, 1970).

SELECTIVE GROWTH OF DIAMOND FILMS ON MIRROR-POLISHED SILICON SUBSTRATES

M.Y. MAO, S.S. TAN, X.K. ZHANG AND W.Y. WANG
Shanghai Institute of Metallurgy, Chinese Academy of Sciences, Shanghai 200050, China

ABSTRACT

Polycrystalline diamond thin films have been selectively grown on mirror-polished silicon substrates using bias-enhanced microwave plasma chemical vapour deposition (MPCVD) to increase diamond nucleation density. A slight etching of SiO_2 mask was employed after the nucleation treatment to remove the diamond nuclei on the mask. Perfect diamond patterns with smooth surface (particle size <0.5µm) and sharp boundaries were obtained. The diamond film gears with 400µm in diameter and 5µm in thickness were first fabricated by this technique.

INTRODUCTION

Chemical vapour deposited diamond films have great potential for applications in electronic, optical and micromechanical devices, because of the unique properties diamond has. Film patterning is one of the basic processes to fabricate the devices. Diamond is very high resistant to chemical solutions, therefore, it is very difficult to pattern by chemical etching. Plasma and laser etching are alternate techniques, but both of them require sophisticate and expensive equipment. Selective growth may be the most viable technique to achieve diamond patterning, and many efforts on silicon substrates have been reported [1-4]. However, diamond is very difficult to nucleate on non-diamond substrates, pretreatment with scratching or damaging the substrate surface using diamond or other hard material paste is needed in earlier works. Roughness of substrate surface and relative low diamond nucleation density ($10^8/cm^2$ in common) [5-7] induced by this kind of treatment can only result rough surface and boundary of diamond pattern, which restrict the device with size in micron scale.

Recently, bias-enhanced microwave plasma has been found to be effective at increasing diamond nucleation on unscratched silicon substrate [8-10], and nucleation density up to $10^{11}/cm^2$ has been achieved. But this technology is not available for selective growth, because diamond nucleation on mask materials such as SiO_2 was also enormously enhanced. In this paper we report the selected growth of diamond films on mirror-polished silicon substrate using bias-enhanced MPCVD.

EXPERIMENT

The procedure of selective growth of diamond films is schematically shown in Fig.1.
The substrates used in this study were mirror-polished n- or p-type [111] silicon wafers with resistivity of 5-8 ohm cm. The wafers were first cleaned by convenient RCA process, and thermally oxidized in pyrogenic steam at 1180°C for 120 min to attain a SiO_2 layer with thickness of 1µm. The SiO_2 layer was then lithographically patterned and chemically etched in buffered $HF:NH_4F:H_2O$ (3:6:10) solution.

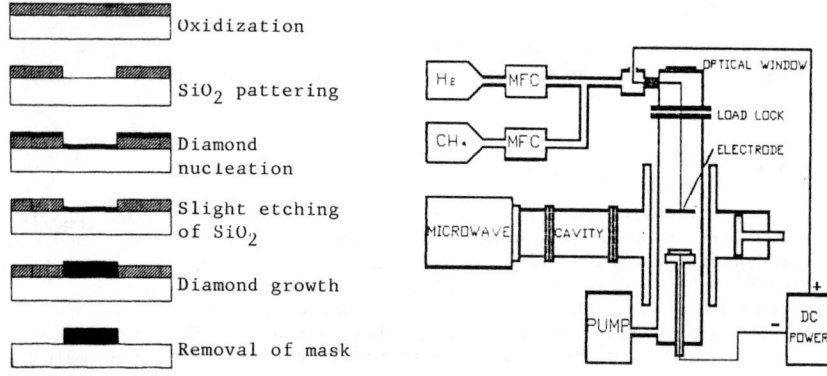

Fig.1 The process of selective growth of diamond films on mirror-polished silicon substrates using bias-enhanced MPCVD.

Fig.2 Schematic diagram of bias-enhanced microwave chemical vapour deposition system.

Then, the SiO_2 patterned silicon wafers were treated for nucleation using bias-enhanced MPCVD system. Fig.2 shows the schematic diagram of the apparatus. The vertically set reactive chamber was made of fused silica, with 46mm in inner diameter. The d.c. bias was applied between grounded molybdenum electrode and graphite substrate holder both located in the reactive chamber 5.0 cm apart from each other, d.c. bias voltage of -300 to 300V was available. The substrate temperature during deposition was detected by radiation pyrometer through the optical window located upper on the reactive chamber. The microwave frequency is 2.45 GHz, and the output power can be set from 500 to 5000 W. The typical parameters for the nucleation treatment were given in Table I.

Table I Experimental conditions for diamond nucleation treatment and normal growth process

Parameters	Nucleation treatment	Normal growth
DC voltage (V)	-130	0
Current (mA)	25	0
Time (Min)	5	180
Pressure (Torr)	22	30
CH_4 flow rate (SCCM)	28	1.0
H_2 flow rate (SCCM)	200	200
Microwave power (W)	1000	600
Substrate temperature (°C)	870	900
Substrate position (cm)*	-2.0	0
Growth rate (μm/h)	...	0.6

*Measured from the centre of cavity, "-" means below the centre.

After nucleation treatment, the wafers were divided into two groups. In the first group, the specimens were etched in the buffered HF solution (40°C) for 20-30 seconds to etch the SiO_2 layer about 50~60 nm in depth, in order to remove the diamond nuclei on the mask induced by the nucleation treatment. In the second group, this etching process was not executed.

Finally, the diamond nucleated substrates were put again into the deposition system, and normal growth of diamond films was carried out. The deposition condition was also listed in Table I.

The obtained products were investigated by scanning electron microscopy (SEM) and Raman sepectroscopy. Fig.3 and Fig.4 correspond to the specimens of the first group, and Fig.5 to the second group.

RESULTS AND DISCUSSION

Fig.3(a) is a SEM photograph of a selectively deposited polycrystalline diamond films with thickness of 1.8μm after 3h growth. The gap width of the two patterns is 1.8μm. Diamond nucleation density are measured to be $10^{10} cm^{-2}$ on mirror-polished Si and $10^{4} cm^{-2}$ on SiO_2 mask. The selectivity is high, and sharp boundaries are seen. The diamond films grown on unscratched silicon are flat with mean particle size about 0.4μm, and the film roughness is estimated to be 0.3μm(peak to peak). Fig.3(b) gives the Raman spectrum of the film. The appearance of the characteristic crystalline diamond peak at $1332 cm^{-1}$ clearly shows diamond phase in the film. However the peak is not well shaped due to the small size of crystals, this result is in agreement with previous measurements by Jiang et al [9].

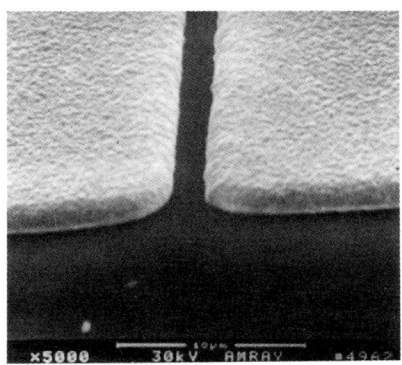
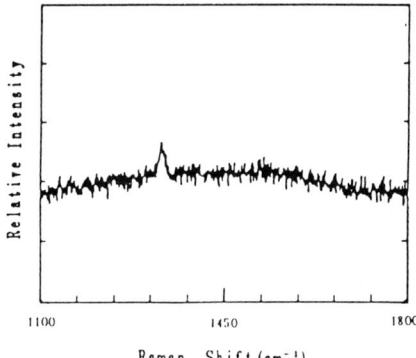

Fig.3 SEM photographs of selectively deposited polycrystalline diamond films with thickness of 1.8μm on mirror-polished silicon substrate. (a) patterns with gap width 1.8μm; (b) the Raman spectrum of the film.

Fig.4(a) and (b) show the SEM photographs of diamond gears fabricated by the technique described above. The diamond gear is 400μm in diameter, and 5μm in thickness. The surface of gear is smooth, and the aspect ratio is high. This result implicates the technology developed in this work is hopeful for the fabrication of diamond microelectronic mechanical systems.

Fig.5 shows the contrast of diamond growth on unscratched silicon and SiO_2. The sample illustrated here is grown for 3h, without etching in buffered HF solution after nucleation treatment. Above $10^7/cm^2$ diamond

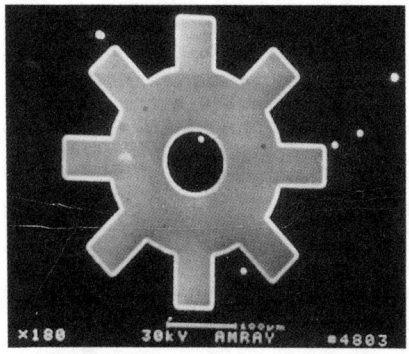 

Fig.4 SEM photographs of diamond gear with 400μm in diameter and 5μm in thickness, fabricated by selective growth method using bias-enhanced MPCVD on mirror-polished silicon substrate. (a) top view; (b) side view.

Fig.5 SEM photograph of selectively deposited diamond film on unscratched Si in which the nucleated substrate had not been etched in buffered HF solution before normal growth. The right half of the photograph corresponds to the part shielded by SiO_2 layer.

particles have grown on the SiO_2 mask (See the right part of the photograph). This figure is in three magnitude order larger than that on untreated SiO_2 ($10^4 cm^{-2}$). Therefore, the high nucleation density on the mask is reasonably induced by the treatment in negative biased plasma. Furthermore, the results of Fig.3 and Fig.4 suggest that the nuclei on SiO_2 mask can be removed effectively by the process of slight etching of SiO_2 just after nucleation treatment.

Bias-enhanced diamond nucleation on SiO_2 has not been reported by previous work. The mechanism has not been throughly understood. We suggest the enhanced defects on the SiO_2 surface acted as diamond nuclei sites induced by the bias treatment may pay an important role in nucleation. Some authors [10-11] proposed the following reaction existed when SiO_2 exposured under C/H plasma:

$$SiO_2 + 3C \rightarrow SiC + 2CO$$

Therefore, when negative d.c. bias is applied, an increase in the flux and kinetic energy of $C_xH_y^+$ species to the surface may fasten this reaction, and a reasonable enhancement of surface defects is obtained.

The phenomenon of enhanced diamond nucleation on SiO_2 is interesting, it may make it possible to serve SiO_2 as sacrificial layer in diamond microstructure fabrication if higher nucleation density can be achieved by setting up the optimum treatment condition. Further study is being conducted.

CONCLUSIONS

A technique for selective growth of polycrystalline diamond thin films on mirror-polished silicon substrate using bias-enhanced MPCVD has been developed. Bias-enhanced diamond nucleation on SiO_2 was observed, and a shallow etching of SiO_2 after nucleation treatment is found effective to remove these nuclei on the mask. Patterned diamond films, including diamond microgear, with smooth surface and sharp boundaries, have been obtained.

REFERENCES

1. K. Hirabayashi and Y. Tanguchi, Appl. Phys. Lett. 53, 1815 (1989).
2. J.L. Davidson, C. Ellis and R. Ramesham, J. Electron. Mater. 18, 711 (1989).
3. Jingsheng Ma, Hiroshi Kawarada, Takao Yonehara, Jun-Ichi Suzuki, Jin Wei, Yoshihiro Yokota and Akio Hiraki, Appl. Phys. Lett. 55, 1071 (1989).
4. R. Ramesham and T. Roppel, J. Mater. Res. 7, 1144 (1991).
5. M.P. Everson and M.A. Tamor, J. Vac. Sci. Technol. B, 9, 1570 (1991).
6. Y. Avigal, Diamond Relat. Mater. 1, 216 (1992).
7. Ewa J. Bienk and Svend S. Eskildsen, Diamond Relat. Mater. 2, 432 (1993).
8. S. Yugo, T. Kanai, T. Kimura and T. Muto, Appl. Phys. Lett. 58, 1036 (1991).
9. X. Jiang, R. Six and C.P. Klages, Diamond Relat. Mater. 2, 407 (1991).
10. B.R. Stoner, G.H. Ma, S.D. Wolter and J.T. Glass, Phys. Rev. B, 45, 11067 (1992).
11. D.J. Pickrell, W. Zhu, A.R. Badzian, R.E. Newnham and R. Messier, J. Mater. Res. 6, 1264 (1991).

PART II

Compound Semiconductors

REAL TIME CONTROL OF III-V SEMICONDUCTOR SURFACES DURING MOVPE GROWTH BY REFLECTANCE ANISOTROPY SPECTROSCOPY

K. PLOSKA*, W. RICHTER, F. REINHARDT, J. JÖNSSON, J. RUMBERG, M. ZORN

Technische Universität Berlin, Institut für Festkörperphysik, Sekr. PN 6-1, Hardenbergstr. 36, 10623 Berlin,
* GOS e.V. Rudower Chaussee 6, 12489 Berlin, Germany

ABSTRACT

Reflectance anisotropy spectroscopy (RAS) is presented as real time analytical tool for metalorganic vapourphase epitaxy (MOVPE) of III-V-semiconductors. This optical method derives its surface sensitivity from the anisotropy of surface structures. It is shown that it is possible to monitor with RAS the oxide desorption from the substrate and that the substrate surface conditions thereafter, still in the pregrowth stage, can be correlated with certain reconstructions of the (001)-surfaces of InP and GaAs. The latter is possible through simultaneous RAS and RHEED measurements during MBE (molecular beam epitaxy) or MOMBE (metalorganic molecular beam epitaxy). Characteristic spectral features are also observed for other binary or ternary III-V-semiconductors. Time resolved measurements during growth give monolayer resolution for the growth rate in the case of GaAs. In the study of heterointerface growth exchange reactions between As and P together with their corresponding reaction time constants can be monitored and conclusions for the epitaxial growth procedure can be drawn.

INTRODUCTION

Many observations exist in the literature about the anisotropic optical response of solids with cubic symmetry. For semiconductors one of the first observations was made already in 1966 in the course of reflectance measurements on Si(110) [1]. However, it took nearly twenty years until it was first recognized that these anisotropies can be exploited for surface analysis [2]. For the GaAs(001) surface the equivalence of reflectance anisotropy spectra, taken as difference between the reflectance in [$\bar{1}$10] and [110] directions, under gasphase and under UHV conditions was shown [3]. It was concluded that surfaces in both cases exhibit identical surface reconstructions. Grazing incidence X-ray scattering has proven by now that this is the case [4] and that reflectance anisotropy spectroscopy (RAS) indeed can be utilized to determine the surface reconstructions of semiconductors. Moreover, during MOVPE (metal organic vapour phase epitaxy) growth of GaAs at standard pressures oscillations equivalent to those in the RHEED intensity for MBE were observed recently in the RAS signal [5]. They showed the potential of this spectroscopic method for analysis and monitoring of MOVPE growth. First observations on a number of other semiconductor surfaces than GaAs(001) have by now been performed and characteristic RAS spectra have been obtained under MOVPE conditions in each case [6,7]. Moreover, even at room temperature, spectra typical for the state of substrate oxidation are observed and it seems, therefore, that RAS is capable of monitoring and analyzing most of the stages of MOVPE growth. In this paper we will illustrate this potential of RAS with examples mostly taken from the by now best understood cases of GaAs and InP growth. We will follow the standard steps in MOVPE growth and the results section is correspondingly organized starting with the substrate deoxidation which is followed by a discussion of pregrowth conditions. Finally homoepitaxial as well as heteroepitaxial growth will be discussed. MBE or MOMBE results will be presented in order to underline the equivalence of pregrowth surface reconstructions.

EXPERIMENTAL

The experiments were performed in a low-pressure MOVPE horizontal quartz reactor equipped with a hydrogen purged "strain-free" quartz window on the top and a small hole in the inner (liner) tube. The RAS equipment is mounted vertically above the reactor and has principally the same design as published by Aspnes et al. [8] (Fig.1).

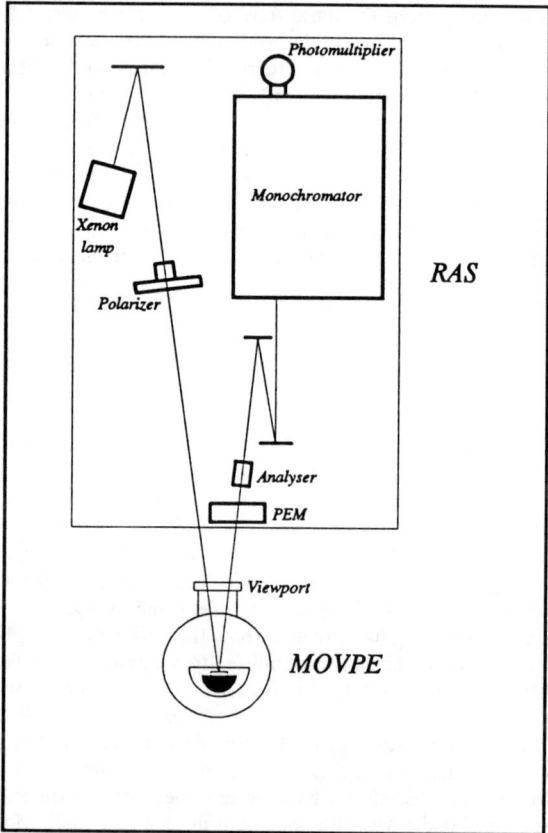

FIG. 1 Schematic drawing of the RAS-MOVPE experiment showing the components of the RAS spectrometer and the quartz reactor

The anisotropy in reflectance was measured along the optical eigenaxes [$\bar{1}10$] and [110] of the (001) surface and the real part of the complex reflectance difference

$$\Delta r/r = 2(r_{\bar{1}10} - r_{110})/(r_{\bar{1}10} + r_{110}) \tag{1}$$

is quoted in the RAS spectra shown throughout this paper.

For growth the precursor trimethylgallium (TMGa), pure arsine (AsH$_3$), trimethylindium (TMIn) and pure phosphine (PH$_3$) were used together with hydrogen as carrier gas. Substrates were exactly oriented GaAs(001) and InP(001) wafers (ACROTEC, MCP, SUMITOMO). Standard values for total pressure and flow rate were 100 mbar and 3 l/min. The ratio of group V to group III precursors and the growth temperature were kept in the range of typical growth conditions. Additional surface studies reported were performed within a MBE (VG 80 with solid As and Ga) and a MOMBE (VG 80 with precracked PH$_3$ and TMIn) system. Both, MBE- and MOMBE-system were equipped with a RHEED utility.

RESULTS AND DISCUSSION

InP substrate deoxidation

In preliminary experiments we found that all III-V semiconductor substrates exhibit an anisotropic response even measured in air at room temperature. The spectral response, however, depends very strongly on the individual history and treatment of the sample. By following the development of these spectral features starting from a freshly grown layer it was found that the strongest structures, appearing in the range from 4 to 6 eV, are due to oxidation of the layer surface. This is especially true for so-called epi-ready wafers. The detailed spectral shape, however, seems to depend on the specific procedure the oxide was generated with. This opens up the possibility to characterize the substrates before loading them into the reactor and to qualify the preparation procedure (or the "epi-ready" wafer). Moreover, the heating process and the pregrowth stabilisation can be monitored in the MOVPE reactor until the oxide is removed.

As an example the spectral changes during heating of an InP wafer (epi-ready from ACROTEC) is shown in Fig.2 (for GaAs deoxidation see [5]). At 300 K the oxidized substrates display a characteristic anisotropic structure with a maximum around 5.5 - 6.1 eV and a minimum at 4.7 eV. The amplitude of these features depends strongly on the preparation of the samples and is much less pronounced for example for a sample etched freshly with 12 H$_2$SO$_4$: H$_2$O$_2$: 4H$_2$O. Moreover, the spectra of epi-ready wafers change too with increasing time after the package has been opened. Thus, the heights of these spectral features seem to be correlated with the oxide thickness and/or the composition of the oxides.

With increasing temperature the oxide related structures in the spectra of Fig.2 disappear and a new spectral shape develops. At 725 K a spectrum is obtained which is typical for MOVPE grown InP measured at this temperature within the reactor. We define this temperature as the oxide desorption temperature. In a realistic pregrowth process of course the temperature does not increase slowly stepwise but continuously at a much faster rate. The deoxidation can then be monitored through the time development of the RAS signal at a fixed wavelength the choice of which is only governed by the desire to obtain a large signal change indicating the deoxidation. Fig. 3 shows such a time resolved measurement at fixed wavelength of 2.6 eV. The two peaks at 240 s and 370 s are related to the change of spectral features at 2.6 eV. At 755 K, however, there is a sharp rise of the signal caused by the large peak at 3 eV of the oxide-free InP surface. Comparing to Fig.2 the oxide desorption temperature seems to be higher in this case. The reasons for that might be either the faster heating rate or the fact that the epi-ready substrate in this case was already exposed some time to air.

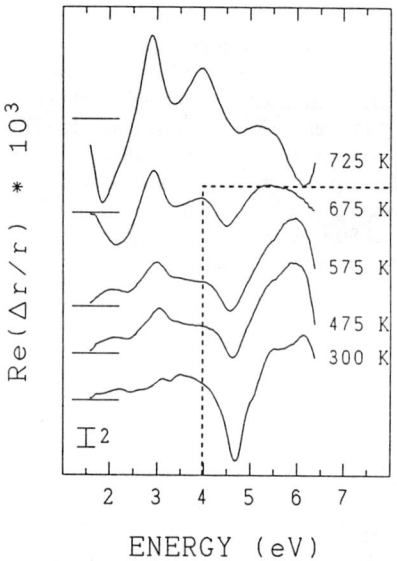

FIG. 2. RAS spectra of an oxidized InP(001) surface under hydrogen (10^4 Pa) and PH_3 (190 Pa) flow obtained during stepwise heating up to the growth temperature. The features enclosed by the dashed lines are characteristic for the oxide related anisotropy.

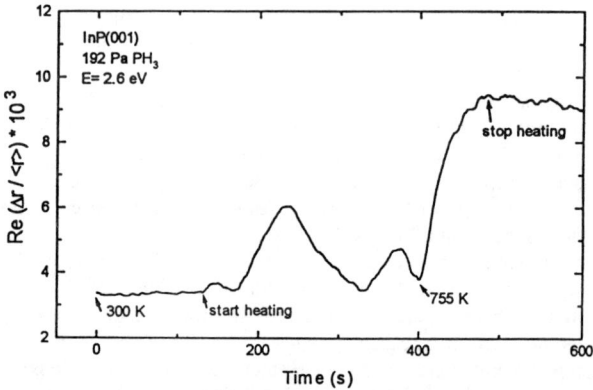

FIG. 3. RAS transients at 2.6 eV taken during the heating phase of an oxidized InP(001) wafer at 10^4 Pa hydrogen and 172 Pa phosphine.

GaAs surfaces under different pregrowth conditions

By now it has been well established that under typical MOVPE conditions, before growth is started, the (001)-surface of GaAs is reconstructed identical to the surface in the UHV environment of MBE at equivalent As-fluxes and temperatures [3,4]. The RAS spectra which sample the dielectric response of the surface dimers are identical if measured on the same sample or very similar on different samples. This is shown in Fig.4 where firstly a GaAs buffer layer was grown in order to improve the crystalline quality of the material and then cooled down to room temperature under As_4 or AsH_3 flux, respectively. Afterwards the RAS spectra shown where taken with increasing temperature monitoring the subsequent As desorption from the surface. In the MBE experiment RHEED data were taken simultaneously with the RAS spectra. In the MOVPE case up to 675 K the disordered (4x4) reconstruction, formed by additional As atoms on the c(4x4) structure, is stable. The latter is characterized by a double layer of As at the surface [9]. The spectra of both surface structures differ essentially by the spectral shape and width of the peak around 4 eV.

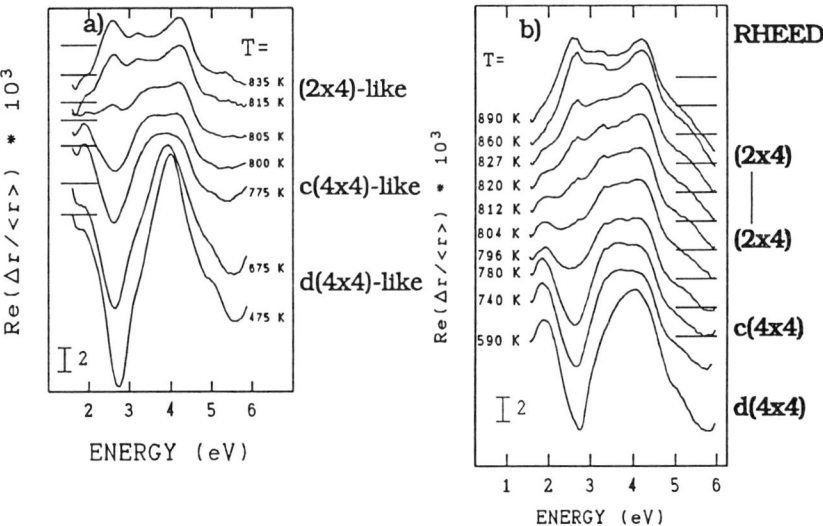

FIG. 4. RAS spectra of GaAs(001) surfaces in (a) a MOVPE reactor under hydrogen atmosphere (10^4 Pa) and in (b) a MBE growth chamber.

The c(4x4) reconstruction appears in the temperature range from 775-800 K. The stability range of these different surfaces depends of course not only on the temperature but also on the unintentional flux of As from the walls or the susceptor to the GaAs surface. At elevated temperatures the (2x4) reconstruction is present, indicated by the change of sign in the RAS signal at 2.6 eV. This sign reversal of the signal is correlated with the directional change of the As dimers in the surface layer. The RAS signal at this photon energy as well as the peak at 4.2 eV have been shown to be related to electronic transitions into empty dimer states from either lone pair states (2.6 eV) or filled dimer states (4.2 eV) [10].

The difference occurring between the two sets of RAS data in Fig. 4 at the lowest temperatures is very probably caused by the incomplete decomposition of AsH3 and the correspondingly different adsorbates in MBE and MOVPE. For even higher temperatures than those in Fig.4 the surfaces start to detoriate. This is demonstrated in Fig. 5 where the GaAs surface is exposed to 1000 K without arsine stabilization only in hydrogen flow. Starting from the (2x4) reconstructed surface As continues to desorb from the surface and large changes in the measured spectra occur. They are due to elastically scattered light which must originate from an anisotropic surface roughening. The spectra should allow in principle also to draw more quantitative conclusions about the surface morphology by performing appropiate simulations with an effective medium approximation [11].

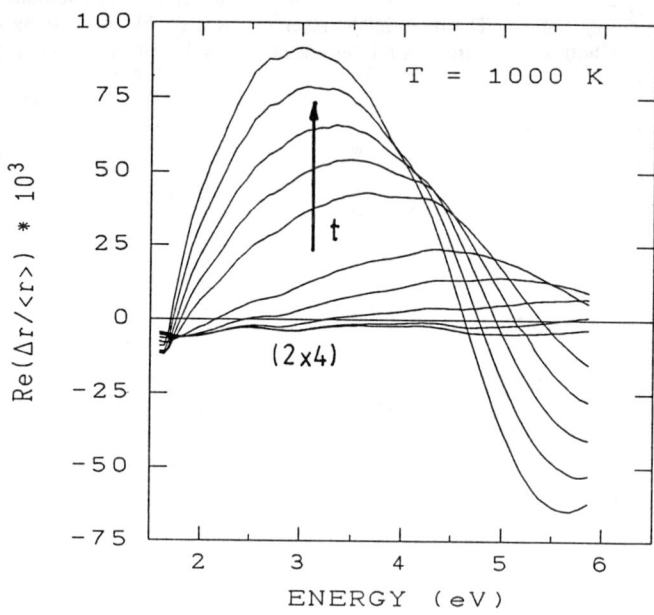

FIG. 5. RAS spectra of (001)GaAs in MOVPE under hydrogen flow without arsine at 10^4 Pa. Spectra were taken every 5 minutes.

InP surfaces under different pregrowth conditions

InP surfaces have not been previously studied by RAS under MBE conditions. In order to correlate the spectral shapes with certain surface reconstructions, we performed therefore a parallel study under MOVPE and MOMBE condititions. The growth chamber for the latter was equipped with a RHEED facility. In order to prepare a high quality InP surface firstly an InP buffer layer was grown at 875 K and the sample was then cooled down under PH_3 or P_2 flux to room temperature. The surface was heated again and RAS spectra were taken at certain temperatures (Fig. 6). Looking first at the spectra obtained in the MOVPE reactor under hydrogen flow it is striking that the magnitude of the spectral features is much larger compared to those from GaAs surfaces. Up to 605 K the spectra are dominated by the features already known from Fig. 2. There is one minimum at about 2 eV, and maxima at 3 and 4 eV. At 755 K the peaks at 2 and 3 eV becomes smaller. Above 775 K there is a "three-buckle structure" and a minimum at 1.8 eV. Comparing these spectra with spectra taken in a MOMBE system from InP surfaces we observe similar features (Fig. 6b). RHEED showed at 640 K a (2x4) reconstruction and at 715 K a (4x2). From the similarity of the spectral structures and especially from their spectral positions (even in spite of the limited temperature and spectral range in Fig. 6b) we conclude that in the MOVPE environment the surfaces around 755 K are In-rich with a (4x2)-like reconstruction but at lower temperatures the more P-rich (2x4) reconstruction is present. The larger strength of the peak at 3eV in the MOVPE spectra can be explained by a higher P-supply in the MOVPE environment. This is verified also by the increase of this peak with increasing partial pressure of PH_3.

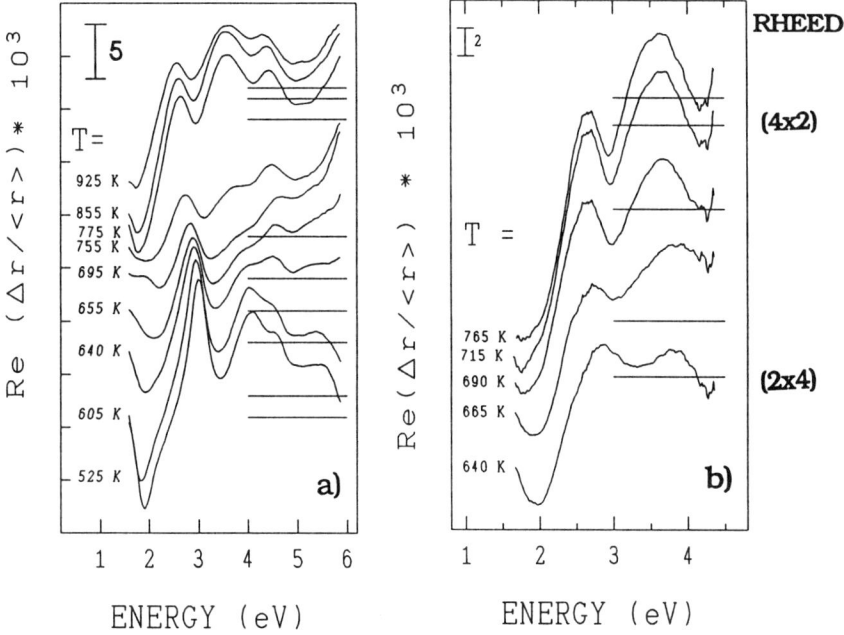

FIG. 6. RAS spectra of InP(001) surfaces under (a) hydrogen atmosphere (100 mbar) in a MOVPE reactor and (b) in a MOMBE growth chamber. The data in (b) were made through a standard viewport and have therefore only a limited spectral range. This made also necessary the subtraction of a larger anisotropic background originating from the birefringent window.

Other III-V materials before growth

Fig. 7 shows the spectral RAS response for a number of different III-V (001)surfaces within the InGaAsP material system. Strong differences exists between all of these spectra. They can be used therefore for an *in-situ* monitoring of the material under growth. From detailed measurements of the InGaAs and the GaAsP material systems it has become clear by now that RAS can be used to probe the surface concentration of group III as well as group V elements. In order to achieve a high accuracy of composition control, operation conditions should preferably be near a phase transition between surface reconstructions. This is for example the case for InGaAs on GaAs where the RAS spectra gradually transform from that of a (4x4) reconstructed GaAs surface to that of a (1x3) reconstructed InAs layer, and also for GaAsP on GaAs where the spectrum corresponding to the (4x4) reconstructed GaAs surface changes into a spectrum correlating with the (2x4)-like P-terminated structure of InP(001).

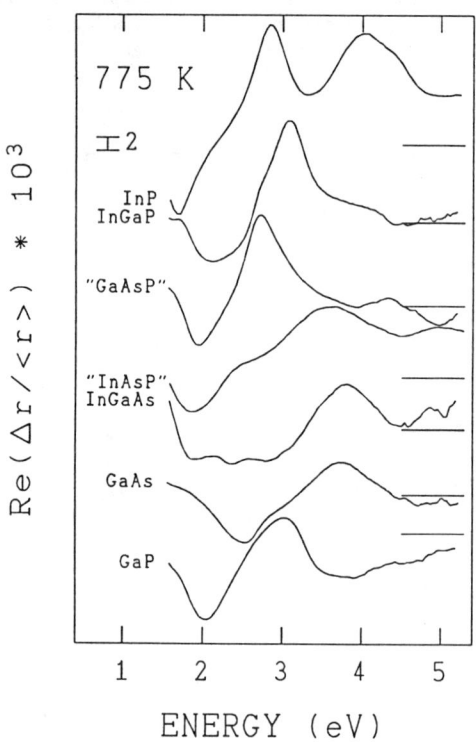

FIG. 7. RAS spectra under 10^4 Pa hydrogen flow for several III-V materials: GaP was grown on a (001) GaP substrate, InGaAs is grown lattice-matched on (001)InP, InGaP is lattice-matched on (001)GaAs, the spectra "GaAsP" and "InAsP" were obtained by exposing GaAs- and InP-surfaces to phosphine and arsine, respectively.

Growth studies

The status of growth of course is the most important question to be answered by a surface diagnostic tool sensitive in a MOVPE-environment. In the following we discuss the homoepitaxial growth on GaAs(001) in more detail.

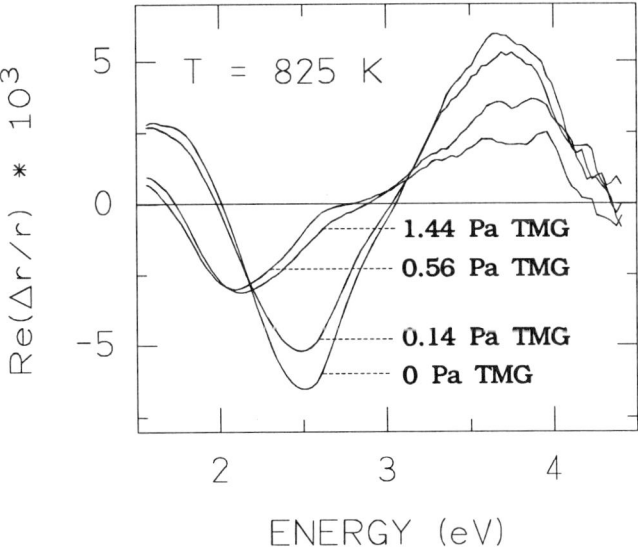

FIG. 8. RAS spectra before (0 Pa TMGa) and during growth of GaAs(001) at different TMGa partial pressures as indicated. Total pressure: 100 mbar, AsH$_3$: 32 Pa.

Fig. 8 shows stationary time averaged RAS spectra after growth has been initiated by adding different amounts of TMGa into the reactor. The spectrum with 0 Pa TMGA gives the already shown c(4x4) as the starting surface. The spectra before and during growth are different. The magnitude of the change depends on the partial pressure of trimethylgallium and the temperature. The largest difference occurs at this temperature for the highest partial pressure (1.44 Pa) of TMGa. A similar behaviour is displayed at higher temperatures. At lower temperatures, however, the situation is different and the most strongest deviations from the As-stabilized c(4x4) are obtained at relative small partial pressures [12]. In general these spectral changes indicate that as a consequence of additional Ga species on the surface a certain number of As-dimers directed along [110] is either destroyed or switches to the [$\bar{1}$10] direction appropriate for the (2x4)-like surface.

For time resolved studies the photon energy was chosen according to the spectral position where the largest difference occured in the stationary spectra before and during growth. The results for a temperature of 785 K and different TMGa partial pressures are shown in Fig. 9a. As expected from the difference in stationary spectra, a strong rise in the RAS signal is seen when TMGa is switched on. This is followed by a number of well-resolved oscillations. Their period decreases with increasing partial pressure of TMGa. This indicates already strongly that the oscillation period corresponds to the growth of 1 ML of GaAs. *Ex-situ* layer thickness determinations indeed have verified this interpretation [5].

Varying the total pressure with constant TMGa and AsH$_3$ partial pressures at a temperature of 775 K in the range of 20-200 hPa changes the oscillation period too, because of the different hydrodynamic conditions (Fig. 9b). The growth rates derived from these oscillations as a function of the different parameters correspond exactly to the results obtained a decade ago in MOVPE studies by performing *ex-situ* growth rate determinations [13]. A summary of growth rates derived from the oscillation periods is given in Fig. 10a (total pressure dependence) and Fig. 10b (temperature dependence). The latter shows the well known Arrhenius plot clearly indicating the temperature ranges for mass transport limited growth at higher temperatures and the kinetically limited regime at lower temperatures. The practical advantage of exploiting the oscillations for such data in comparison to *ex-situ* determinations of course is that the data are obtained on one sample only and within the short time just needed for observation and changing the epitaxial parameters.

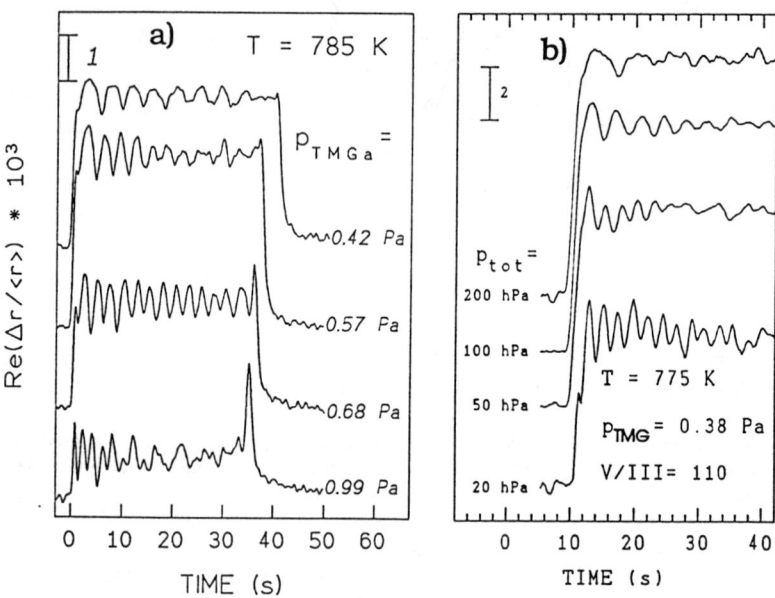

FIG. 9. RAS signals at a fixed photon energy (2.6 eV) as a function of time from a GaAs (001) surface (a) when starting growth by adding TMGa at different partial pressures between $t=0$ and $t=35$-40s (total pressure 100 hPa) and (b) for different total pressures at constant TMGa-partial pressure. Arsine partial pressure and temperature were held constant. Carrier gas was nitrogen.

These results should help also in interpreting the oscillations in terms of surface structures. A microscopic model has to take into account their dependence on the different growth parameters, including gasphase processes as well as their dependence on different vicinal surfaces.

Heterostructure growth

Although RAS measurements have shown already a number of impressive results for homoepitaxy, growth of heterostructures constitutes an field where even more diagnostic tools are needed. In order to follow the growth of a heterointerface again time resolved studies are necessary which, with the presently available equipment, can performed only at a fixed photon energy.

Fig. 11 shows the time resolved RAS signal at 2.5 eV when a GaAs surface is exposed to phosphine. It is seen that the response consists of two parts with significantly different time constants. An initial fast rise (at t = 60 s in Fig. 11) with estimated time constants in the range of 100 to 200ms is followed by a very much slower response (from t = 60 s to t = 200 s). The latter is also accompanied by a strong decrease in the total reflectance and an increase of elastically scattered light.

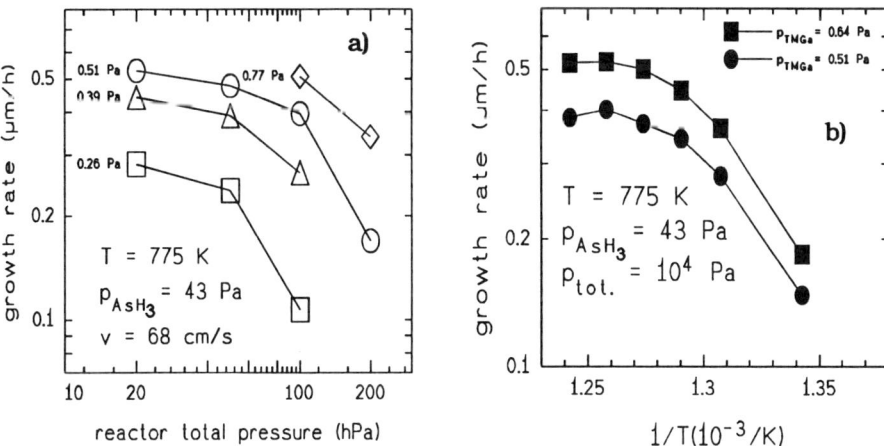

FIG. 10. Growth rates derived from time resolved "monolayer" oscillations in the RAS spectra for different partial pressures of TMG: (a) as a function total pressure, (b) as a function of temperature.

A detailed analysis of these features made us conclude that the initial reaction originates from an exchange of the (4x4) reconstructed As double layer to a P-layer while the slower time constant corresponds to a surface roughening [7]. The reaction rate for the exchange, k, is given by $k=10^{9.7} * p[PH_3]^{1/3} * \exp(1.64eV/RT)$. For standard growth conditions (T=600 °C, p[PH$_3$]=50Pa) the time constant for the As by P exchange is comparable to the experimental time resolution. The activation energy of 1.64 eV should be compared with the desorption energy of As from GaAs, which has been reported to be 1.84eV under MOVPE conditions [14]. The fact that the energy for exchange is lower than the energy for desorption suggests that the presence of phosphine enhances the desorption of As. This was also confirmed by comparing the RAS signal for the exchange reaction with the signal for desorption of As. The time constant associated with the formation of the single layer P-terminated structure, by exchange of As is a factor of 3 to 5 shorter than the time constant for As desorption in pure hydrogen. According to the data presented here, one should consequently use a very short PH$_3$-only step in gas switching sequences over As-terminated surfaces to avoid surface roughening. The step should, however, still be included in order to allow the As-terminated c(4x4) reconstructed surface to reconstruct into the P-terminated single layer structure since this surface state is much more stable in the phosphine environment.

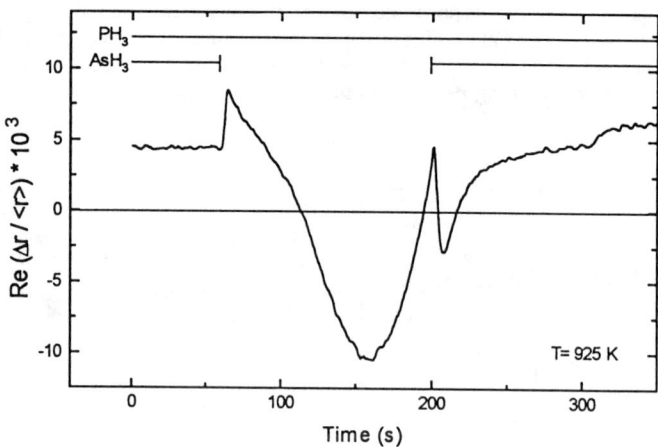

FIG. 11. Time resolved RAS response when a GaAs(001)-(4x4) is exposed to 40 Pa PH_3 at 925 K. Strong anisotropies develop which indicate that the surface is not stabilized anymore but becomes rough and anisotropic.

SUMMARY

We have shown that the Reflectance Anisotropy Spectroscopy technique gives information on all stages of the MOVPE growth. It allows to obtain information on substrate quality, pregrowth conditions, growth rate, interface structure and growth morphology. These quantities are clearly of technological relevance, even when there is at presence a lack of understanding in the microscopic nature of the RAS signals. Information about the corresponding surface structures may gained by performing parallel studies in UHV based growth chambers which allow for electron diffraction experiments and other electron based techniques. A deeper understanding of the growth mechanism may be further obtained through detailed studies of the growth oscillations under various epitaxial conditions. They should reveal at least some of the basic factors controlling growth. Nevertheless, microscopic theoretical studies of the dielectric response of the surfaces in question are urgently needed in order to allow for a more quantitative interpretation of the RAS spectra.

ACKNOWLEDGEMENT

The support given to us and necessary to obtain the MBE (D. Woolf, D. Westwood, R. H. Williams - UWC Cardiff) and the MOMBE data (H. Künzel, M.Schellhase, HHI - Berlin) is very gratefully acknowledged. We would like also to thank I. Kamiya for interesting discussions concerning the InP surface problem. We acknowledge in particular P. Kurpas for his permanent advice in MOVPE growth procedures. The work was partly supported by the ESPRIT Basic

Research Action 6878 "EASI" and the DAAD ARC program. One of us (J. J.) acknowledges support received through a grant from the Alexander von Humboldt Stiftung.

REFERENCES

[1] M. Cardona, F. H. Pollak, K. L. Shaklee, J. Phys. Soc. Jpn. Suppl. **21**, 89 (1966)
[2] D. E. Aspnes, J. Vac. Sci. Technol. **B 3**, 1498 (1985)
[3] I. Kamiya, D. E. Aspnes, H. Tanaka, L. T. Florez, J. P. Harbison, R. Bhat, Phys. Rev. Letters **68 (5)**, 627 (1992)
[4] D. W. Kisker, P. H. Fuoss, K. L. Tokuda, G. Renaud, S. Brennan, J. L. Kahn, Appl. Phys. Lett. **56**, 2025 (1990)
[5] F. Reinhardt, W. Richter, A. B. Müller, D. Gutsche, P. Kurpas, K. Ploska, K. C. Rose, M Zorn, J. Vac. Sci. Technol. **B 11(4)**, 1427 (1992)
[6] K. Ploska, F. Reinhardt, M. Zorn, J. Jönsson, K. C. Rose, W. Richter and P. Kurpas, Proc. ICFSI 4, 1993, Jülich, Germany
[7] J. Jönsson, F. Reinhardt, M. Zorn, K. Ploska, W. Richter, J. Rumberg, Appl. Phys. Lett., submitted
[8] D. E. Aspnes, J. P. Harbison, A. A. Studna and L. T. Florez, J. Vac. Sci. Technol. **A6**, 1327 (1988)
[9] D. K. Biegelsen, R. D. Bringans, J. E. Northrup, L.-E. Swartz, Phys. Rev. B **41(9)**, 5701 (1990)
[10] Y.-C. Chang, S. F. Ren, D. E. Aspnes, J. Vac. Sci. Technol. **A 10(4)**, 1857 (1992)
[11] S. M. Scholz, A. B. Müller, W. Richter, D.R.T. Zahn, D. I. Westwood, D. A. Woolf, R. H. Williams, J. Vac. Sci. Technol. **B 10**, 1717 (1992)
[12] F. Reinhardt, K. Ploska, J. Jönsson, M. Zorn, J. Rumberg, P. Kurpas, W. Richter, J. Vac. Sci. Technol. **B** (to be published)
[13] H. Heinecke, E. Veuhoff, N. Pütz, M. Heyen, P. Balk, J. Electron. Mater. **13**, 815 (1984)
[14] N. Kobayashi, Y. Kobayashi, Jpn. J. Appl. Phys. **30** (1991) 1699.

GAS-PHASE AND SURFACE DECOMPOSITION OF *TRIS*-DIMETHYLAMINO ARSENIC

MING XI, SATERIA SALIM, KLAVS F. JENSEN AND DAVID A. BOHLING*
Department of Chemical Engineering, Massachusetts Institute of Technology, Cambridge, MA 02139, *Air Products and Chemicals, Allentown, PA 18195

ABSTRACT

Gas phase and surface decomposition reactions of a novel arsenic precursor *tris*-(dimethylamino) arsenic (DMAAs) have been studied. Optical fiber-based Fourier transform infrared spectroscopy was used to monitor, *in situ*, the gas-phase pyrolysis of DMAAs. Homolysis of the arsenic-nitrogen bond with formation of dimethylamine radicals was identified as the key gas-phase reaction pathway. Formation of methylmethyleneimine from reactions with the decomposition products was directly observed. Surface decomposition of DMAAs on GaAs(100) was investigated by low energy electron diffraction and temperature-programmed reaction under ultra-high vacuum conditions. DMAAs adsorbed onto GaAs(100)-(4×6) was found to decompose with 100% efficiency and two surface reaction pathways were identified. The first reaction channel was homolysis of the arsenic-nitrogen bond with formation of dimethylamine radicals, whereas the second pathway involved a β-hydrogen transfer.

INTRODUCTION

There has been considerable interest in the development of liquid organometallic arsenic reagents as replacements for the highly toxic, gaseous source arsine. Alternative sources also have potential for reduced growth temperatures and lower V/III ratios in epitaxy of III-V semiconductors. A wide range of organoarsenic compounds have been explored including trialkylarsenics [1,2], alkylarsenic hydrides, such as tertiarybutylarsine (t-BuAsH$_2$) [3, 4], and phenylarsine [5]. Carbon incorporation into the film grown from precursor molecules or from pyrolysis fragments has been a major concern for organometallic vapor phase epitaxy (OMVPE). Current understanding of the underlying chemistry suggests that some hydrogen must be bonded to the As to avoid carbon contamination. In fact, t-BuAsH$_2$ has been reported to be successful in producing GaAs films with similar properties to those achieved with arsine [3-5]. In the case of metalorganic molecular beam epitaxy (MOMBE) or chemical beam epitaxy (CBE), the desirable precursors must not only lead to low level carbon incorporation, but must also decompose at low temperatures to avoid having to precrack the organometallic reagent [6].

Tris-(dimethylamino) arsenic (DMAAs) has been synthesized as a promising alternative arsenic source for both OMVPE or MOMBE. This compound is distinct from previous arsenic precursors by having arsenic bonded directly to nitrogen and thus no carbon-arsenic bond. It also has the required moderate vapor pressure and long term stability to be of practical use in OMVPE. Abernathy *et al.* [7] have shown in MOMBE studies that GaAs films, with no detectable carbon, may be grown at relatively low temperatures (~450°C) using trimethylgallium and DMAAs. The use of DMAAs in OMVPE has also been explored with encouraging results [8] and the compound has also been utilized recently in atomic layer epitaxy (ALE) of GaAs and AlAs [9].

There have been several recent studies of the decomposition mechanisms of DMAAs on GaAs(100). Salim *et al.* used *in situ* mass spectroscopy at MOMBE conditions to show that the thermal decomposition of DMAAs was completed at 450°C, the major products being dimethylamine, hydrogen, aziridine or methylmethyleneimine. Two decomposition pathways of DMAAs on GaAs(100) were suggested: simple scission of the arsenic-nitrogen bond followed by reaction of the dimethylamine radical with hydrogen to form dimethylamine; and hydrogen elimination reaction to form aziridine or N-methylmethyleneimine and surface hydrogen. Fujii *et al.*, on the other hand, have proposed that a monomethyl amine species (NCH_3) is the key decomposition intermediate in the dissociative adsorption of DMAAs onto GaAs based on transient mass spectrometry measurements during MOMBE or ALE [9, 11]. To gain more insight into reactions occurring during OMVPE or MOMBE processes, we present more detailed studies of DMAAs decomposition reaction in the gas phase and on GaAs(100).

EXPERIMENTAL

Details of the fiber optics-based Fourier transform infrared system for *in situ* monitoring of the OMVPE gas-phase reaction have been described previously [12]. Briefly, a 1:20 mixture of DMAAs and carrier gas flowed through a 12.7 cm long, restively heated, quartz sample tube with CaF_2 windows at both ends. The temperature was measured by a K-type thermocouple inserted into a well on the wall of the tube. Residence time analysis revealed that the cell could be considered well mixed. The cell pressure was measured by a baratron mounted in the outlet and maintained at 600 Torr. The flow rate was kept constant at 60 sccm. Chalcogenide optical fibers with frequency window between 1000 to 3300 cm^{-1} were used to transmit the IR light. All the IR spectra were taken under steady state flow conditions with a resolution of 4 cm^{-1}.

The surface spectroscopy studies were performed in an ultra-high vacuum chamber equipped with an ion sputtering gun, rear-view LEED/Auger optics and a quadrupole mass spectrometer (QMS). With liquid nitrogen cooling, typical chamber working pressure was ~ 5×10^{-10} Torr. A 1.5 cm×1 cm rectangular piece of GaAs (100) ± 0.5° wafer (Si doped, at 2-4×10^{17} cm^{-3}) served as the sample. A 300 nm of Ta film was sputtered onto the backside of the wafer for resistive heating. The surface temperature was measured by a K-type thermocouple clamped to the back of the crystal by a Ta strip, and the thermocouple reading was calibrated by the intrinsic GaAs congruent evaporation temperature [13]. In the studies presented here, the temperature-programmed reaction (TPR) experiments were performed with the crystal line-of-sight to a shielded QMS, but the shield had slots cut open to the main chamber. The heating rates in the TPR experiments were 2 K/s, and up to 10 masses could be monitored simultaneously.

DMAAs was dosed onto the GaAs(100) surface by back-filling the chamber via a leak valve. Exposures were reported in Langmuirs (1 L = 1×10^{-6} Torr·s) and uncorrected for differing ion gauge sensitivity. Atomic deuterium exposures were generated by dissociating D_2 on a hot tungsten filament placed ~5 cm in front of the surface. Since the degree of dissociation is not known, exposures of molecular deuterium are reported as a relative scale. The DMAAs reagent was donated by Air Products and Chemicals. Three kinds of carrier gases were used in the studies: H_2 (99.9995%, Matheson), D_2 (99.5 atom %, Matheson), and He (99.999%, Matheson).

RESULTS AND DISCUSSIONS

Gas-Phase Pyrolysis of DMAAs

Figure 1 shows FTIR spectra of DMAAs in hydrogen under steady state flow at 50 °C and 450°C. The peak frequencies in the spectrum taken at 50°C were assigned on the basis of equivalent bands reported for *tris*-(dimethylamino) phosphorus [14]; specifically, 1464 cm^{-1} (CH_3 bending), 1252 cm^{-1} (asymmetric streching of C-N-C bonds), 1156, 1195 and 1058 cm^{-1} (deformation of CH_3). The As-N-C bending frequency was not observed since it is lower than the cut-off frequency of the optic fiber (1000 cm^{-1}). The spectrum taken at 450°C is dramatically different from the one at 50°C. Intensities of modes related to the parent molecule, in particular the peak at 1195 cm^{-1}, are reduced, while new peaks related to decomposition products have appeared. Methylmethyleneimine ($H_2C=N-CH_3$) is observed as one of the major reaction products based on the apearance of a new peak at 1660 cm^{-1} (C=N streching), a broad shoulder around 1475 cm^{-1} (CH_3 bending) and a sharp peak at 1025 cm^{-1} (CH_3 deformation). The deformation of methane is observed as a sharp peak at 1306 cm^{-1}. The peak at 1155 cm^{-1} remains unchanged indicating that dimethylamine is also a product [15]. Vibrations associated with aziridine, the the thermodynamically more stable isomer of methylmethyleneimine, were not observed, nor was there any evidence of the formation of trimethylamine.

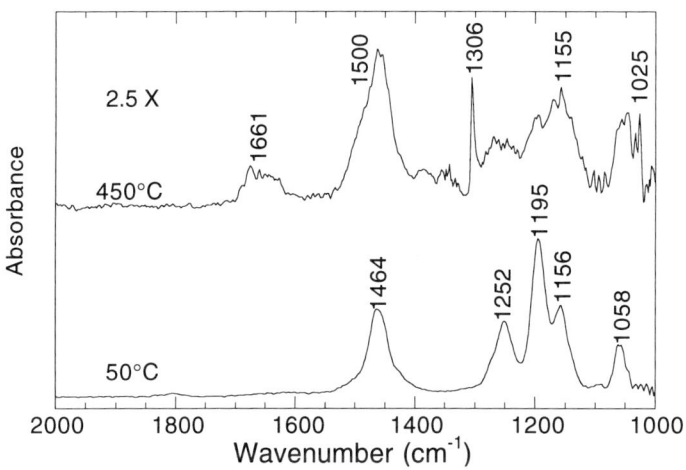

Figure 1. FTIR spectra of DMAAs in H_2 at 50°C and 450°C.

Changes in the normalized intensities of the characteristic modes of DMAAs and its pyrolysis products are plotted in Figure 2 as functions of temperature. The gas-phase decomposition of DMAAs is initiated at ~350°C and completed at ~450°C. The products, dimethylamine and methylmethyleneimine, appear as soon as DMAAs starts to dissociate. The decrease in the concentration of the two products above 500°C is presumably caused by the decomposition of dimethylamine radicals to form methyl radicals and methyleneimine [17]. The latter product has a characteristic vibration at 1640 cm^{-1}. The decomposition of DMAAs is slightly enhanced by the presence of H_2 or D_2 relative to the case of He carrier gas. This behavior suggests that H• and D• radicals, formed in reactions of dimethylamine radicals with H_2 and D_2, accelerate the decomposition of the parent molecule. Based on these observations, we propose the following pyrolysis mechanism where scission of the arsenic-nitrogen bond is the first step:

$$As(N(CH_3)_2)_3 \rightarrow As(N(CH_3)_2)_2 + \bullet N(CH_3)_2$$
$$As(N(CH_3)_2)_2 \rightarrow AsN(CH_3)_2 + \bullet N(CH_3)_2$$
$$AsN(CH_3)_2 \rightarrow As + \bullet N(CH_3)_2$$

Subsequent steps include reactions of dimethylamine and H• radicals with $As(N(CH_3)_2)_n$ (n = 1-3), as well as disproportionation of dimethylamine radicals:

$$\bullet N(CH_3)_2 + As(N(CH_3)_2)_n \rightarrow As(N(CH_3)_2)_{n-1} + HN(CH_3)_2 + CH_3N=CH_2$$
$$\bullet N(CH_3)_2 + H_2 \rightarrow HN(CH_3)_2 + H\bullet$$
$$H\bullet + As(N(CH_3)_2)_n \rightarrow As(N(CH_3)_2)_{n-1} + HN(CH_3)_2$$
$$\bullet N(CH_3)_2 + \bullet N(CH_3)_2 \rightarrow HN(CH_3)_2 + CH_3N=CH_2$$
$$\bullet N(CH_3)_2 \rightarrow CH_2=NH + \bullet CH_3$$
$$\bullet CH_3 + \bullet N(CH_3)_2 \rightarrow CH_4 + CH_3N=CH_2$$

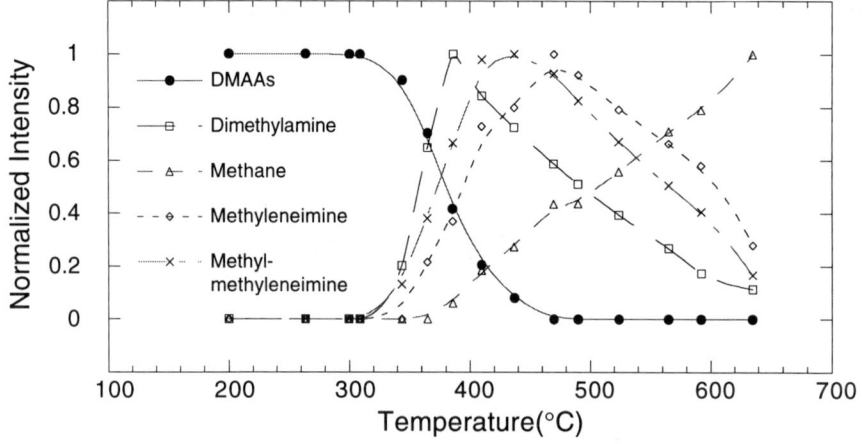

Figure 2. Gas-phase decomposition products of DMAAs in H_2

<u>Surface Reactions of DMAAs on GaAs(100)</u>

Complete pyrolysis of DMAAs and deposition of arsenic on GaAs(100) surfaces have been followed by the LEED pattern change. A well developed (4×6) LEED pattern was repeatedly obtained after sputtering and annealing the crystal at 480°C. The Ga-rich C(8×2) surfaces were formed by annealing the (4×6) surface at 550°C. As-rich C(2×8) surfaces were produced from either of the Ga-rich surfaces by dosing DMAAs while annealing the surface at 400°C. Specifically, a sharp C(2×8) LEED pattern was observed after dosing the (4×6) surface twice at 100°C with 10 L of DMAAs, then flash heating to 480°C. These observations are consistent with the results of Hamaoka *et al.* [11], who systematically studied the conversion of C(8×2) surfaces into C(2×8) reconstructions with DMAAs as an arsenic source under different temperatures and pressures. Large exposures (> 2000 L) of DMAAs to the C(2×8) surface at 350°C made the surface more As-rich with a blurred "(1×1) like" LEED pattern.

Deposition of arsenic from DMAAs was also evidenced by temperature-programmed desorption (TPD) spectra of As_2^+. Figure 3 shows the TPD spectra following m/e=150 (As_2^+) as a function of DMAAs exposure at 350°C. The appearance of the As_2^+ signal around 600°C from GaAs (100)-(4×6) without DMAAs dosing is caused by congruent evaporation from the Ga-rich surface. Deposition of arsenic from the precursor is revealed in Figure 3 by the reduced desorption temperature of As_2 with the exposure of DMAAs. In fact, the decrease in the starting desorption temperature was evident after as little as 5 L of DMAAs exposure (data not shown). In a study of arsenic deposition on GaAs(100) from arsine, Banse and Creighton observed the development of the low temperature desorption state of As_2^+ at comparable arsine exposures [19].

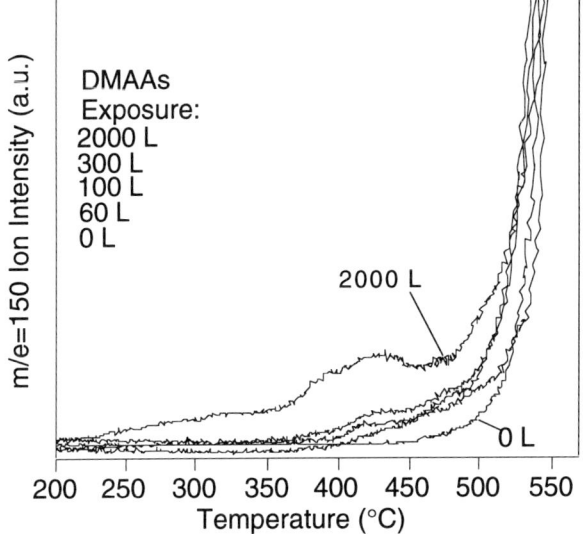

Figure 3. TPR data for As_2 desorption for different DMAAs exposures.

To understand the decomposition of DMAAs on GaAs(100), TPR studies on the desorption products were performed. Results are presented in Figure 4. The products ions corresponding to m/e =42, 43, 44, 45 were followed after a 8 L dose of DMAAs onto the (4×6) surface at 100°C. No other species with higher masses were detected desorbing from the surface during the temperature ramping from 100°C to 500°C. Ions with m/e=15 and 29 were observed, but showed no features additional to those of m/e =42 and 44. After each TPR ramp, the surface was annealed at 480°C to ensure the recovery of (4×6) LEED pattern. No build up of decomposition products of DMAAs was found on the surface to a degree that would affect the reproducibility of the LEED patterns or the TPR spectra. It was thus concluded that all the dimethylamine ligands desorb from the surface in the temperature ramp. In correlation with the results from the gas-phase study, possible product candidates are dimethylamine (MW=45), dimethylamine radical (MW=44) and methylmethyleneimine (MW=43).

The most significant feature in Figure 4 is the difference in desorption profiles between m/e=42, 43 and those of m/e=44, 45. Since m/e=44, 45 can result only from a dimethylamine molecule or a dimethylamine radical, whereas m/e=42, 43 can derive either from fragmentation of dimethylamine/dimethylamine radicals or from methylmethyleneimine, the absence of the desorption peak ~280°C in the TPR spectra of m/e=44, 45 indicates the presence of two reaction pathways for DMAAs decomposition on GaAs(100)-(4×6). Methyl-methyleneimine is formed, presumably, through β-hydride elimination, at ~280°C.

To elucidate the nature of the reaction pathway at 160°C, we performed a D atom co-adsorption experiment. Previous studies have established that a submonolayer of D atoms adsorbed on GaAs(100) recombines and desorbs from the surface above 220°C [20]. After co-adsorption of submonolayer of D atom and DMAAs onto GaAs(100)-(4×6) at 100°C, the peak positions and intensities of m/e=44, 45 remained approximately identical to those from after the same DMAAs exposure, but without D atoms co-adsorption. Only the peak at m/e=46, which had contributions solely from the ^{13}C isotope when there was no D atom exposure, showed a small, but definite increase. As a reaction involving hydrogen addition from the surface would be affected by the presence of D atom on the surface, this results strongly suggests that the product evolving at ~160°C is a dimethylamine radical, and the signal of m/e=45 and the small increase of signal at m/e=46 were due to abstraction of hydrogen or deuterium atoms adsorbed on the inside wall of the mass spectrometer shield. Similar abstraction reactions by methyl radicals are well documented, and their interference in mass spectrometric detection has been described previously [20, 21].

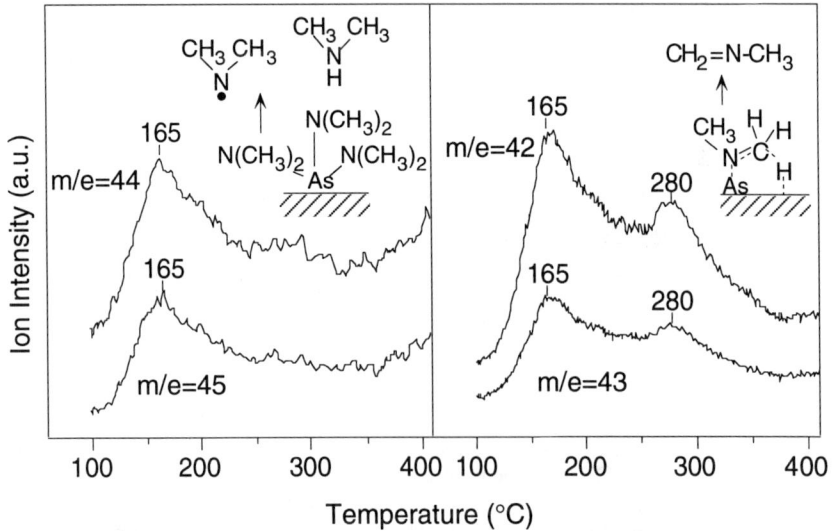

Figure 4. TPR data for product desorption. 8L exposures of DMAAs at 100°C

CONCLUSION

Decomposition of DMAAs in the gas phase and on the GaAs(100) surface has been studied. Gas-phase pyrolysis of DMAAs was found to start at ~350°C, and the decomposition was complete at ~450°C. Sequential scission of the arsenic-nitrogen bond and formation of dimethylamine radicals was identified as the main decomposition pathway. Further reactions of the dimethylamine radicals lead to the formation of methylmethyleneimine. No trimethylamine was detected. Deposition of arsenic from monolayers of DMAAs adsorbed onto GaAs(100) at 100°C was evidenced by both LEED pattern changes and As$_2$ temperature programmed desorption spectra. TPR results indicate that DMAAs decomposes on GaAs(100)-(4×6), transferring dimethylamine ligands to the surface. These ligands subsequently desorb from the surface by two pathways: ejection of dimethylamine radicals at ~160°C and formation of methylmethyleneimine at ~280°C.

ACKNOWLEDGMENTS

This work was supported by the National Science Foundation (DMR-9023162 and CTS-9212984)

REFERENCES

1. A. Braners, O. Kayser, R. Kall, H. Heinecke, P. Balk and H. Hoffman, *J. Crystal Growth* **93**, 7 (1988).
2. D. M. Speckman and J. P. Wendt, *Appl. Phys. Lett.* **50**, 676 (1987).
3. G. Haacke, S. P. Wakins and H. Burkhard, *Appl. Phys. Lett.* **54**, 2029 (1989); S. P. Watkins and G. Haacke, *Appl. Phys. Lett.* **59**, 2263, 1991.
4. T. Kikkawa, H. Tanaka and J. Komeno, *J. Appl. Phys.* **67**, 7576 (1990).
5. R. M. Lum, J. K. Klingert, D. W. Kisker, D. M. Tennant, M. D. Morris, D. L. Malm, J. Kovalchick and L. A. Heimbrook, *J. Electron. Mater.* **17**, 101 (1988).
6. A. C. Jones, *J. Crystal Growth* **129**, 728 (1993).
7. C. R. Abernathy, P. W. Wisk, D. A. Bohling and C. T. Muhr, *Appl. Phys. Lett.* **60**, 2421 (1992); *J. Crystal Growth* **142**, 64 (1992).
8. M. H. Zimmer, R. Hovel, W. Brysch, A. Brauers and P. Balk, *J. Crystal Growth* **107**, 348 (1991).
9. K. Fujii, I. Suemnne and M. Yamanishi, *Appl. Phys. Lett.* **60**, 1498 (1992); 61, 2577 (1992); 62, 1420 (1993).
10. S. Salim, J. P. Liu, K. F. Jensen and D. A. Bohling, *J. Crystal Growth* **124**, 16 (1992).
11. K. Hamaoka, I. Suemune, K. Fujii, T. Kishimoto and M. Yamnishi, Japan *J. Appl. Phys.* **30**, L579, (1991).
12. S. Salim, C.K. Lim, K.F. Jensen, and R. Driver, *Proc. Int. Soc. Optical Eng.* **2069**, xxx (1994).
13. C. T. Foxon, J. A. Harvly and B. A. Joyce, *J. Phys. Chem. Solids* **34**, 1693 (1973).
14. R. R. Shagidullum, A. V. Chemova, V. S. Vvnogradava and F. S Mukhametov eds. *Atlas of IR Spectra of Organophosphorous Compound*, Nauka Publishers, Kluwer Academic Publishers, 1990.
15. B. Schrader, Ed. *Raman/Infared Atlas of Organic Compound*, VCH, 1989.
16. I. Stolkin, T.-K. Ha and Hs. H. Gunthard, *Chem. Phys.* **21**, 327 (1977).
17. J. Seetula, K. Kalliorinne, and J. Koskikallio, *J. Photochem. Photobio.*, A **43**, 31 (1988)
18. B. A. Banse and J. R. Creighton, *Appl. Phys. Lett.* **60**, 856 (1992).
19. J. R. Creighton, *J. Vac. Sci. & Technol.* **A8**, 3984 (1990).
20. J. R. Creighton, *Surf. Sci.* **234**, 287 (1990).
21. J.-L. Lin and B. E. Bent, *J. Am. Chem. Soc.* **115**, 2849 (1993).

KINETICS OF LOW-PRESSURE MOCVD OF GaAs FROM TRIETHYL-GALLIUM AND ARSINE

N.K. INGLE*, C. THEODOROPOULOS*, T.J. MOUNTZIARIS*, R.M. WEXLER** AND F.T.J. SMITH**#
* Department of Chemical Engineering and Center for Electronic and Electro-optic Materials, State University of New York, Buffalo, NY 14260.
** Eastman Kodak Company, Corporate Research Laboratories, Rochester, NY 14650.

ABSTRACT

The growth of undoped GaAs films from $Ga(C_2H_5)_3$ (TEG) and AsH_3 by Metalorganic Chemical Vapor Deposition (MOCVD) has been studied in a low-pressure reactor operating at 3 Torr. This precursor combination is known to produce GaAs films with very low carbon incorporation when compared to films grown from $Ga(CH_3)_3$ (TMG) and AsH_3. A kinetic model of the growth process has been developed that includes both gas-phase and surface reactions based on reported decomposition mechanisms of TEG and AsH_3. The kinetic model was coupled to a transport model describing flow, heat and mass transfer. Finite element simulations were performed to determine the rate constants of the growth reactions that provide the best fit between predicted and observed growth rates. Under typical operating conditions the surface reactions were found to dominate the growth process and a reduced surface kinetic model was identified by sensitivity analysis. The proposed reaction-transport models can successfully predict observed growth rates of GaAs films and they can be used for identifying optimal operating conditions in MOCVD reactors.

INTRODUCTION

Metalorganic Chemical Vapor Deposition (MOCVD) is the most versatile technique for growing thin films of GaAs and related III-V semiconductors for advanced electronic, microwave and optoelectronic devices. The low-pressure (LP) MOCVD of GaAs has very attractive features that can produce interface abruptness during heterostructure growth and selective growth suitable for planarization of optoelectronic devices [1]. The growth of GaAs films from TEG and AsH_3 has been found to result in a much lower carbon incorporation when compared to the most commonly used precursor pair of TMG and AsH_3 [2,3]. A reactor pressure of about 3 to 5 Torr has been reported to yield GaAs films from TEG and AsH_3 with minimal concentrations of both ionized donors [4], typically silicon impurities from the TEG source [5], and acceptors, typically carbon impurities from the TEG decomposition byproducts [4]. Understanding the underlying kinetics is essential for optimal design and operation of MOCVD reactors and for large-scale commercial development of the process [6].

The homogeneous thermal decomposition of TEG can proceed by both homolytic fission of ethyl radicals, i.e $Ga(C_2H_5)_3 \rightarrow Ga(C_2H_5)_2 + C_2H_5$, and β-elimination, i.e. $Ga(C_2H_5)_3 \rightarrow GaH(C_2H_5)_2 + C_2H_4$ [7-9]. Mass spectroscopy studies indicate that the second pathway is probably the dominant one at typical MOCVD conditions [8,9]. The lower levels of carbon incorporation when using TEG instead of TMG can be attributed to a stronger Ga-C bond in TMG, which favors the formation of carbenes ($Ga-CH_2$) through abstraction reactions [10]. On the other hand, a weaker Ga-C bond in TEG can be cleaved more easily to yield saturated hydrocarbons, when attacked by radicals. From observations of TEG decomposition on GaAs surfaces [11-14] it appears that TEG will adsorb dissociatively on GaAs surfaces at MOCVD conditions. Species desorbing from the surface included C_2H_4, C_2H_5 and Ga sub-alkyls [11-14]. Arrhenius parameters for several surface reaction steps of TEG decomposition on GaAs surfaces have been estimated [13,14] and they can be used in the development of a surface kinetic model of the MOCVD process.

Present Address of F.T.J. Smith: LORAL Infrared Imaging Systems, Lexington, MA 02137.

The thermal decomposition of AsH_3 is believed to be mostly heterogeneous, but the actual mechanism is still not well understood and may also include some homogeneous steps [15]. It has been postulated that under MOCVD conditions AsH* is the dominant adsorbed group-V precursor resulting from dissociative adsorption of AsH_3 on GaAs surfaces [10].

The objective of this work is the development and testing of a kinetic model for LP-MOCVD of GaAs from TEG and AsH_3 using mechanistic information from decomposition studies of the two precursors reported in the literature and growth rate data from an experimental reactor.

EXPERIMENTAL

An MOCVD reactor was used for the growth studies in which the reactant gases flow horizontally toward a heated substrate placed on a SiC-coated graphite susceptor. The susceptor is positioned in the middle of a 4-inch-diameter horizontal channel at a 10° angle to the vertical. This arrangement results in a stagnation point flow against a slightly tilted surface. The substrates were 2-inch-diameter wafers of semi-insulating GaAs with (100) surfaces misoriented 2° in the direction of the (110) plane. Under typical operating conditions the total pressure was 3 Torr, the TEG flow rate 0.61 sccm, the arsine flow rate 16.75 sccm (i.e. V/III = 27.5) and the flow rate of the hydrogen carrier gas 150 sccm. The average horizontal velocity of the gases at the inlet of the reactor was 8.7cm/s. The susceptor temperature was varied from 450 to 750°C. The thickness uniformity of the GaAs epilayers was measured by the groove and stain technique at two radii, 1cm and 2cm from the center of the wafers.

KINETIC MODEL

The kinetic model of the growth process includes both gas-phase and surface reactions, which are listed in Table 1. There is experimental evidence [8,9] that β-elimination (G1) will be the major decomposition pathway of TEG in the gas phase at the operating conditions used in this work. This reaction was assumed to proceed with the rate of a fission reaction, postulated to be the rate limiting step with estimated Arrhenius rate constants $k_o = 10^{15.7}$ s^{-1} and E = 46.6 kcal/mole in [7], although this value of k_o is rather high for β-elimination. Rate constants for reactions G2-G6 were obtained from the combustion literature [16]. Since the pressure is low, the mixture dilute, and the residence time in the reaction zone small, the secondary bimolecular reactions (G4-G6) were found to be negligible. The gas-phase decomposition of $GaH(C_2H_5)_2$ produced by G1 was neglected, because it is also a secondary reaction. It should be mentioned that the C_2H_5· radicals participating in gas-phase reactions are produced by surface reactions.

The surface mechanism includes dissociative chemisorption of the precursors, surface decomposition and recombination reactions, and growth reactions. The chemisorption steps (S1, S2, S4-S9) were assumed to occur with collisional rates obtained from statistical mechanics, with a coverage-dependent sticking coefficient equal to 1 at zero coverage for stable Ga species and all the unsaturated species. The activation barrier for chemisorption of arsine (S6) was taken to be 5 kcal/mole [17]. The concentration of C_2H_5* was found to be much larger than that of H* and additional surface reactions involving the latter were neglected. Arrhenius rate parameters for reactions S3, -S4, S10 and S11 were obtained from [13], for -S7 from [17] and for -S8 and -S9 from [14]. The rate parameters for -S4 were assumed to be $k_o = 10^8$ s^{-1} and E = 27.4 kcal/mole following observations reported in [13]. The coverage dependence of E of -S4 at high coverage [13] has not been included in this study. Finally, the rate constants of the two growth reactions (S12, S13) were taken to be the same and their values were fitted to obtain growth rates near the observed ones. The best fit was obtained for $k_o = 2 \times 10^{17}$ cm^2 $mole^{-1}$ s^{-1} and E = 25 kcal/mole.

TRANSPORT MODEL

The transport model is based on the fundamental equations describing momentum, heat and mass balances in an ideal compressible gas with temperature dependent properties [6,10,17,18].

Table 1. Kinetic Model of the LP-MOCVD of GaAs from Triethyl-Gallium and Arsine.

	GAS-PHASE REACTIONS		
(G1)	$Ga(C_2H_5)_3$	$\rightarrow$	$GaH(C_2H_5)_2 + C_2H_4$
(G2)	$C_2H_5\cdot + H_2$	$\rightarrow$	$C_2H_6 + H\cdot$
(G3)	$C_2H_5\cdot + M$	$\rightarrow$	$C_2H_4 + H\cdot + M$ (M = carrier gas)
(G4)	$C_2H_5\cdot + H\cdot$	$\rightarrow$	C_2H_6
(G5)	$C_2H_5\cdot + C_2H_5\cdot$	$\rightarrow$	C_4H_{10}
(G6)	$H\cdot + H\cdot + M$	$\rightarrow$	$H_2 + M$
	SURFACE REACTIONS		
(S1)	$Ga(C_2H_5)_3 + S_A$	$\rightarrow$	$[Ga(C_2H_5)_2]_A{}^* + C_2H_5\cdot$
(S2)	$GaH(C_2H_5)_2 + S_A$	$\rightarrow$	$Ga(C_2H_5)_A{}^* + C_2H_6$
(S3)	$[Ga(C_2H_5)_2]_A{}^*$	$\rightarrow$	$Ga(C_2H_5)_A{}^* + C_2H_5\cdot$
(S4)	$Ga(C_2H_5)_2 + S_A$	$\leftrightarrow$	$[Ga(C_2H_5)_2]_A{}^*$
(S5)	$Ga(C_2H_5) + S_A$	$\leftrightarrow$	$Ga(C_2H_5)_A{}^*$
(S6)	$AsH_3 + S_G$	$\rightarrow$	$(AsH)_G{}^* + H_2$
(S7)	$AsH + S_G$	$\leftrightarrow$	$(AsH)_G{}^*$
(S8)	$C_2H_5\cdot + S_A$	$\leftrightarrow$	$(C_2H_5\cdot)_A{}^*$
(S9)	$C_2H_5\cdot + S_G$	$\leftrightarrow$	$(C_2H_5\cdot)_G{}^*$
(S10)	$(C_2H_5\cdot)_A{}^*$	$\rightarrow$	$C_2H_4 + H_A{}^*$ (1)
(S11)	$(C_2H_5\cdot)_G{}^*$	$\rightarrow$	$C_2H_4 + H_G{}^*$ (1)
(S12)	$Ga(C_2H_5)_A{}^* + (AsH)_G{}^*$	$\rightarrow$	$GaAs + C_2H_6 + S_A + S_G$
(S13)	$[Ga(C_2H_5)_2]_A{}^* + (AsH)_G{}^*$	$\rightarrow$	$GaAs + C_2H_6 + C_2H_5\cdot + S_A + S_G$

S_A : Site corresponding to an As surface atom S_G: Site corresponding to a Ga surface atom
(1) : The H* coverage was assumed to be negligible   *: Indicates adsorbed species

Such models are becoming powerful tools for realistic simulations of the growth of both Si [18] and compound semiconductor films [6,10,17] by CVD. Since the mixture of reactants in the carrier gas is dilute, the heat of reaction can be neglected. Then, the flow and heat transfer equations can be solved separately from the mass transfer and kinetics. In addition to Fickian diffusion, thermal diffusion of reactants in the carrier gas has been included in the model. The unknown diffusion and thermal diffusion coefficients of the gaseous species have been estimated by using group contribution methods [17]. The resulting systems of nonlinear partial differential and algebraic equations were solved in two dimensions by the Galerkin Finite Element Method. The mass transfer and kinetics simulations presented here are for a susceptor perpendicular to the direction of flow, i.e. for stagnation point flow. In such a case, a similarity transformation leads to a system of ordinary differential and algebraic equations, which can be used for developing and testing kinetic models, because it can be solved more efficiently than the full two-dimensional model. A complete description of reaction-transport models developed in this study and comparisons between predictions and experimental observations will appear in a forthcoming publication [19].

RESULTS AND DISCUSSION

Several experiments were conducted using the LP-MOCVD reactor at the operating conditions described above and both growth rate and thickness uniformity data were obtained for the GaAs films. The experimental results were used for developing the surface kinetic model of the process and for fitting its parameters by comparing predicted and observed growth rates.

The pathlines of H_2 predicted by the detailed two-dimensional flow and heat transfer model of the actual reactor are shown in Figure 1. In these simulations it was assumed that the reactor walls are uniformly cooled to 300K and that the susceptor is isothermal at 873K. The flow pattern is almost an ideal stagnation flow, slightly distorted because of the small tilt of the susceptor. There are no buoyancy driven recirculations, because the low operating pressure leads to a low gas density making the buoyancy term in the momentum equation to be negligible. Small

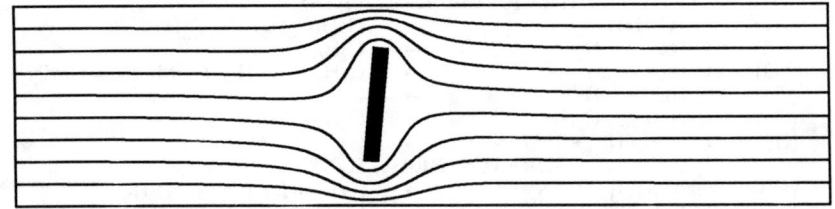

Figure 1. Predicted pathlines of hydrogen in the LP-MOCVD reactor for $T_s = 873K$, $v_o = 8.7$ cm/s and $P = 3$ Torr.

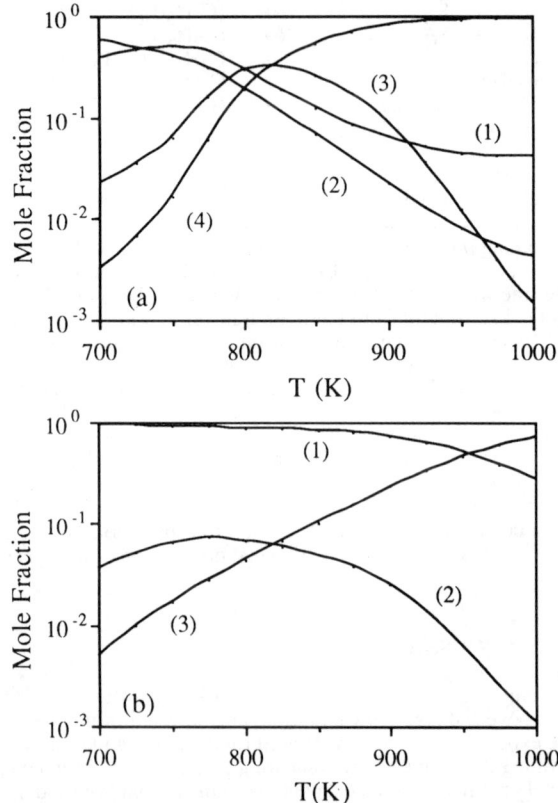

Figure 2. Predicted mole fractions of adsorbed species at different susceptor temperatures.
(a) Species adsorbed on As sites: (1) $Ga(C_2H_5)_A^*$, (2) $[Ga(C_2H_5)_2]_A^*$, (3) $(C_2H_5\cdot)_A^*$, (4) S_A.
(b) Species adsorbed on Ga sites: (1) $(AsH)_G^*$, (2) $(C_2H_5\cdot)_G^*$, (3) S_G.

vortices (not shown) can be formed behind the susceptor at high inlet velocities due to flow separation.

To develop and test the kinetic model, flow, heat and mass transfer simulations were performed by using the finite element code describing GaAs deposition in a stagnation point flow reactor, with a susceptor perpendicular to the direction of flow. The computational expense for running these simulations was much smaller compared to the full two-dimensional finite element simulations of the actual reactor. Furthermore, since the small tilt of the actual susceptor is not expected to have a significant effect on the observed average growth rates, the growth rate data obtained from the experimental reactor can be used for fitting purposes. The tilt has a small effect on the thickness uniformity of the films [19]. A sensitivity analysis was also carried out to study the effects of the values of the kinetic parameters on the growth rate.

The gas-phase reactions were found not to contribute significantly to the growth rate and can be omitted from the mechanism, when only growth rate predictions are sought, but not when accurate predictions of the species distribution in the gas phase and on the surface are desirable. The mole fractions of adsorbed species, predicted by the full model at different susceptor temperatures, are shown in Figure 2. The coverage was normalized for each individual type of site (A or G), i.e. the sum of the mole fractions of adsorbed species and free sites is unity for each type of site. By comparing the coverage of the two Ga surface precursors it is obvious that $Ga(C_2H_5)_A*$ is the dominant one. Furthermore, the concentration of $(C_2H_5 \cdot)*$ is significant only on As sites at temperatures between 750K and 950K. Adsorbed ethyl radicals can compete for sites with the adsorbed Ga precursors, which are the limiting growth species since V/III is larger than one. This will affect the growth rate, if the number of available free sites is limited, as is the case for susceptor temperatures between 750K and 850K (Figure 2a). This effect should be smaller at temperatures below 750K, due to the abundance of the adsorbed Ga species, and negligible at temperatures above 850K, because the concentration of the free sites is high. The only important species adsorbed on Ga sites is $(AsH)_G*$, whose coverage drops off above 900K due to fast growth reactions and desorption.

From the above observations and from the sensitivity analysis results, a reduced surface kinetic model is proposed, which can accurately predict the observed growth rates. The reduced surface model has only five reactions. It starts with the reaction: $Ga(C_2H_5)_3 + S_A \rightarrow Ga(C_2H_5)_A* + 2 C_2H_5 \cdot$ and it also includes reactions S5,S6,S7 and S12 from the mechanism shown in Table 1.

The predicted growth rates of GaAs using the two kinetic models and the stagnation point flow geometry are compared to experimentally observed growth rates in Figure 3. The experimental growth rates reach a plateau between 830K and 950K (i.e, for 1000/T between 1.2 and 1.05) and decrease at lower and higher susceptor temperatures. This behavior is typical for MOCVD processes and is due to kinetic limitations from slow growth reactions at low susceptor temperatures and to increased desorption rates of surface precursors at high temperatures. At intermediate temperatures the growth is limited by the transport of Ga precursors to the surface. The growth rates predicted by the reduced surface model match very well the ones predicted by the full model. At low temperatures the reduced model predicts slightly higher growth rates than the full model, because of the elimination of adsorbed ethyl radicals from the reduced model.

CONCLUSIONS

A kinetic model describing the growth of GaAs films by LP-MOCVD using TEGa and AsH_3 has been developed. GaAs films have been grown in an experimental LP-MOCVD reactor, at substrate temperatures between 450°C and 750°C, to obtain data on film growth rates and thickness uniformities. Large-scale simulations of transport phenomena coupled to chemical reactions underlying the growth process were performed and the unknown kinetic parameters of the surface growth reactions were fitted to predict observed growth rates. A sensitivity analysis of the effects of individual reaction rates on the growth rate of the films was performed. The gas-phase reactions were found to be insignificant in determining the growth rate of the films. A reduced surface model based only on surface reactions is proposed. This model can accurately predict the observed growth rates.

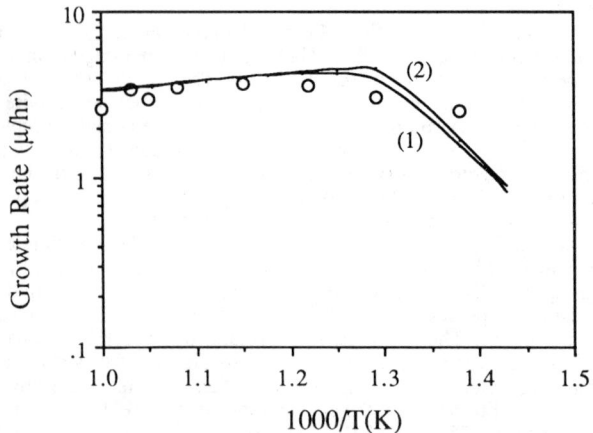

Figure 3. Comparison between predicted (solid lines) and observed (o) growth rates of GaAs. Predictions using: (1) The full kinetic model. (2) The reduced surface kinetic model only.

ACKNOWLEDGMENTS

Financial support of this work by Eastman-Kodak Company and by the Center for Electronic and Electro-optic Materials of SUNY-Buffalo is gratefully acknowledged. Supercomputing resources were provided by the Pittsburgh Supercomputing Center. We also thank Dr. V.M. Donnelly of AT&T Bell Laboratories for helpful discussions on the kinetics.

REFERENCES

1. F.T.J. Smith, *Prog. Solid St. Chem.* **19**, 111 (1989); F.T.J. Smith, *MRS Symp. Proc.* **204**, 117 (1991).
2. N. Kobayashi and T. Makimoto, *Jpn. J. Appl. Phys.* **24**(10), L824 (1985).
3. T.F. Kuech, *Mat. Sci. Rep.* **2**, 1 (1987); and references within.
4. K. Kimura, S. Takagishi, S. Horiguchi, K. Kamon, M. Mihara and M. Ishii, *Jpn. J. Appl. Phys.* **25**(9), 1393 (1986).
5. T.F. Kuech and R. Potemski, *Appl. Phys. Lett.* **47**(8), 821 (1985).
6. K.F. Jensen, *J. Crystal Growth* **98**, 148 (1989).
7. M.C. Paputa and S.J.W. Price, *Can. J. Chem.* **57**, 3178 (1979).
8. M. Yoshida, H. Watanabe and F. Uesugi, *J. Electrochem. Soc.* **132**, 677 (1985).
9. P.W. Lee, T.R. Omstead, D.R. McKenna and K.F. Jensen, *J. Crystal Growth*, **85**, 165 (1987).
10. T.J. Mountziaris and K.F. Jensen, *J. Electrochem. Soc.* **138**, 2426 (1991).
11. J.A. McCaulley, V.R. McCrary and V.M. Donnelly, *J. Phys. Chem.* **93**, 1014 (1989).
12. J.S. Foord, N.K. Singh, A.T.S. Wee, C.L. French and E.T. Fitzgerald, *MRS Symp. Proc.* **204**, 3 (1991).
13. V.M. Donnelly, J.A. McCaulley and R.J. Shul, *MRS Symp. Proc.* **204**, 15 (1991); V.M. Donnelly and A. Robertson, *Surf. Sci.* **293**, 93 (1993).
14. M.L. Yu, U. Memmert, N.I. Buchan and T.F. Kuech, *MRS Symp. Proc.* **204**, 37 (1991).
15. R. Lückerath, P. Tommack, A. Hertling, H.J. Koss, P. Balk, K.F. Jensen and W. Richter, *J. Crystal Growth* **93**, 151 (1988).
16. W. Tsang and R.F. Hampson, *J. Phys. Chem. Ref. Data* **15**(3), 1087 (1986).
17. T.J. Mountziaris, S. Kalyanasundaram and N.K. Ingle, *J. Crystal Growth*, **131**, 283 (1993).
18. M.E. Coltrin, R.J. Kee and J.A. Miller, *J. Electrochem. Soc.* **133**, 1206 (1986).
19. N.K. Ingle, C. Theodoropoulos, T.J. Mountziaris, R.M. Wexler and F.T.J. Smith, manuscript in preparation.

THE ROLE OF UNPRECRACKED HYDRIDE SPECIES ADSORBED ON THE GaAs(100) IN THE GROWTH OF GaAs BY ULTRAHIGH VACUUM CHEMICAL VAPOR DEPOSITION USING TRIMETHYLGALLIUM AND TRIETHYLGALLIUM

SEONG-JU PARK, JEONG-RAE RO, JAE-KI SIM, JEONG SOOK HA, AND EL-HANG LEE

Electronics and Telecommunications Research Institute, P.O. Box 8, Daeduk Science Town, Daejeon City, 305-606, Republic of Korea.

ABSTRACTS

We have grown GaAs epilayers by ultrahigh vacuum chemical vapor deposition(UHVCVD) using adsorbed hydrides and metalorganic compounds via a surface decomposition process. This result indicates that unprecracked arsine(AsH_3) can be used in chemical beam epitaxy(CBE) and that a new hydride source with a low decomposition temperature, monoethylarsine(MEAs) can replace the precracked AsH_3 source in CBE. The impurity concentrations in GaAs grown with trimethylgallium(TMG) and triethylgallium(TEG) were found to be very sensitve to growth temperature. It was also found that the uptake of carbon impurity is significantly reduced when TMG is replaced with TEG. The carbon concentrations in epilayers grown using TMG and TEG with unprecracked AsH_3 and MEAs were reduced by 2-3 orders of magnitude compared to those by CBE process employing TMG and arsenics from precracked hydrides. We have also found that the hydrogen atoms play an important role in the reduction of carbon content in GaAs epilayer. Intermediates like dihydrides from MEAs decomposed on the surface are considered to supply hydrogen atoms and hydrides during growth, which may remove other carbon containing species.

INTRODUCTION

It is generally accepted that it is not possible to grow GaAs using unprecracked AsH_3 at very low pressure because cracking of AsH_3 into arsenic at the substrate surface alone may be negligible and thus not sufficient for achieving growth below 570°C. To be used in the CBE process, therefore, AsH_3 must first be precracked at high temperatures prior to growth[1-3]. In an attempt to understand the effect of a solid surface on AsH_3 the thermal chemistry of AsH_3 on GaAs(100) has been studied, but not extensively. Surface catalyzed decomposition of AsH_3 is believed to play a dominant role during arsenic decomposition in GaAs MOCVD and ALE[4]. Thus the present study examines the role of the most important precursor, AsH_3 adsorbed on GaAs, in the growth of GaAs.

It is also known that AsH_3 plays a role in reducing carbon incorporation during GaAs epitaxial growth, especially when TMG is used as the gallium source. The actual arsenic species arriving at the growth surface as a result of precracking process in CBE may not be suitable for use with gallium-containing organometallic compounds, since they offer no chemistry to reduce the carbon contamination anticipated from the decomposition of carbon-containing organometallic compounds at the growth surface. During growth by CBE little or no AsH_3 impinges on the surface and carbon concentrations as high $10^{19-21}cm^{-3}$ can be produced[2,5]. Therefore, we have also investigated the effect of carbon incorporation in GaAs epilayer grown by UHVCVD using unprecracked AsH$_3$.

Due to the increasing concern for safety, there is a growing need to replace extremely toxic AsH$_3$ with less toxic arsenic compounds. Organoarsines have been the focus of most of these studies[6,7], because of their more desirable physical properties and their potential for high purity growth. Moreover, it was observed that GaAs growth is possible at fluxes comparable to that of cracked AsH$_3$ in CBE using unprecracked AsH$_3$ source[8]. This fact also induced us to explore the possibility of using MEAs which has a lower decomposition temperature in the gas phase than AsH$_3$.

In this study, we have investigated the characteristics of GaAs growth using unprecracked AsH$_3$ and MEAs adsorbed on the surface together with TEG and TMG. The role of hydrogen atoms dissociated from hydrides in reducing carbon incorporation in GaAs epilayers was also studied. To study growth characteristics in detail, GaAs growth rates were measured as a function of growth temperature. Impurity type and its concentration were also characterized by Hall measurement. This study showed that GaAs epilayers can be successfully grown by unprecracked AsH$_3$ and MEAs coupled with TMG and TEG under an ultrahigh vacuum condition where the chemical species which participate in the growth reaction are produced only at the growth surface. The epilayer surfaces were also examined by an atomic force microscopy(AFM) in air to inspect the surface morphologies caused by different source gases. From the results of these UHV-growth experiments, it was suggested that the unprecracked AsH$_3$ or MEAs serve as a successful arsenic reagent for use in CBE. Our results indicated also that the hydrogen atoms and partially-cracked hydrides from the catalytically-decomposed AsH$_3$ or MEAs on the GaAs surface play an important role in the significant reduction of carbon impurities in epilayers.

EXPERIMENT

Growth experiment was carried out in a CBE system which consists of an ultra-high vacuum chamber and a gas handling system. The growth chamber is pumped by a turbo-molecular pump. The chamber which is not equipped with a liquid nitrogen shroud, was maintained at a base pressure of less than 3×10^{-10} Torr. The gases were introduced through pressure controlled gas inlet tubes without the use of a carrier gas. In our study, we used TMG and TEG as gallium sources together with unprecracked AsH$_3$ and MEAs. To avoid the thermal decomposition of source gases by radiation from the hot susceptor during the growth, we kept the temperature of gas tubes below 118°C by locating the tubings 8 inches away from sample holder. In this way, we could investigate the role of unprecraced hydrides adsorbed on GaAs in the growth characteristics by excluding the possible gas phase processes. Substrate was Cr-doped GaAs(100) with 2° off towards (011). Before growth, the substrate was annealed at 910 K for 15 min at a hydride pressure of 5×10^{-4} Torr in order to remove the oxide layer from the surface. The gas pressure in the chamber during growth was in a range of 10^{-5} - 10^{-4} Torr depending upon the V/III ratio. Layer thickness was measured by scanning electron microscopy(SEM) on stained and etched cleavage planes. The samples were inspected by optical microscope and SEM for the evaluation of their morphology. Epilayer samples were transferred from UHVCVD system into an AutoProbe scanning force microscope(Park Scientific Instruments, Sunnyvale, CA 94089) through air. The surface morphologies of epilayers were then examined by SFM in an error mode using a force of about 10 nN under ambient condition. Low temperature PL and Hall measurements at room temperature were also made for optical and electrical characterization of grown epilayers.

RESULTS AND DISCUSSION

Figure 1(a) and 1(b) show atomic force microscope(AFM) images of epilayers grown using TMG with AsH$_3$ and TEG with MEAs, respectively. The surface morphology in fig. 1(b) is smoother than that in fig. 1(a). Furthermore, both fig. 1(a) and 1(b) show island-like growth patterns on the surfaces. To measure the heights of islands, the height scale for images were also obtained and shown in the fig. 1. The typical heights of grown features are 8 Å for fig 1(b) and about 4 Å for fig. 1(a), respectively. These heights which correspond to 2-3 monolayers of GaAs indicate that the 2-dimensional islands were grown on these two surfaces. The surface roughness shown in AFM images suggests that the use of TEG and MEAs produces smoother surface than that of TMG and AsH$_3$ even at lower growth temperature.

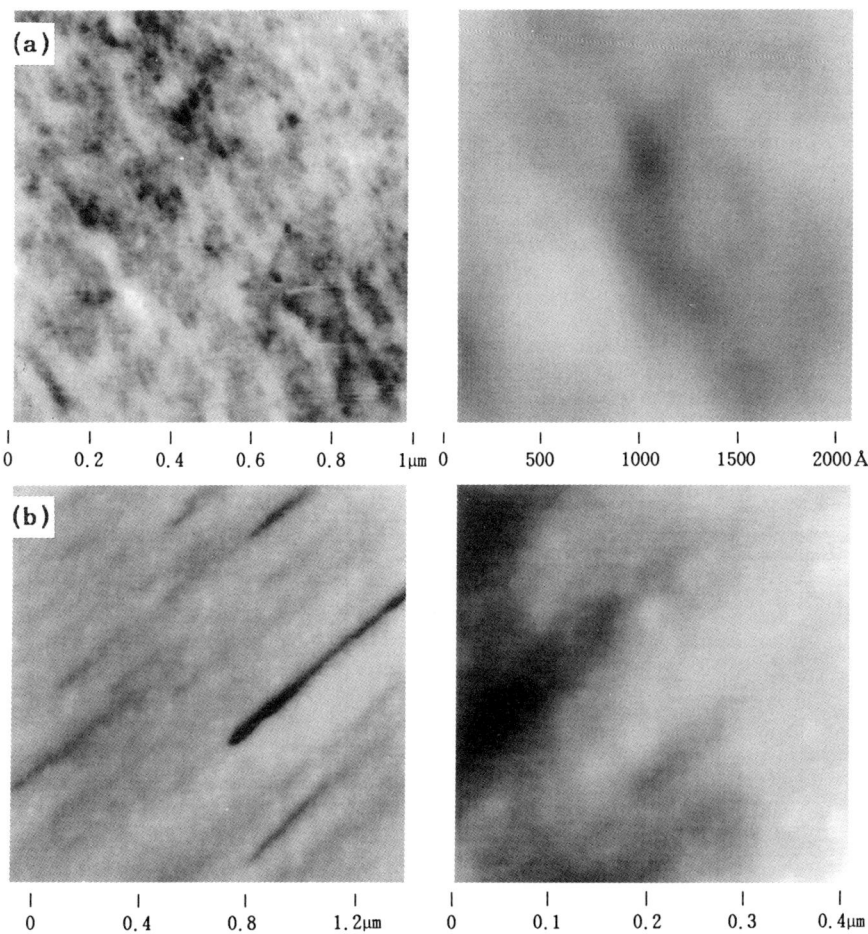

Figure 1. AFM images of GaAs epilayers grown by UHVCVD using (a) TMG and AsH$_3$ at 630°C and (b) TEG and MEAs at 600°C. The scanning areas are (a) 1μm x 1μm and 2000Å x 2000Å and (b) 1.4μm x 1.4μm and 0.4μm x 0.4μm.

Figure 2 represents the temperature-dependent growth rates measured by SEM using a cleave-and-stain method. The growth rate for TMG as shown in fig. 2 is divided into three distinct regions. Below 600°C, the growth rate is believed to be limited by the decomposition process of TMG on the surface. Between 600 and 630°C, the temperature is high enough to completely dissociate all the TMG resulting in a constant growth rate with increasing temperature. Above 630°C, the gradual decrease in growth rate is considered to be caused by the desorption of gallium atoms or partially cracked TMG. The growth rate for TEG, however, has two humps as shown in fig. 2. Our observation is similar to an anomalous temperature dependence of growth rate of GaAs grown by CBE using TMG and arsine[9]. This anomalous behavior of growth rates for TEG may also be explained by a growth model which assumes that TEG is decomposed into two forms of reaction intermediates depending on the growth temperature resulting in the two humps in the growth rate.

The carrier concentration and the corresponding Hall mobility were obtained from van der Pauw measurements at room temperature. The thicknesses of the grown layer were in the range of 2-3μm. The hole concentrations of TMG-grown samples are in the range of 10^{16}-10^{19} cm^{-3} as shown in fig. 3. The hole concentrations in TMG samples decrease as substrate temperature increases below 660°C, which suggests that carbon-containing species drastically desorb from the growth surface with increasing growth temperature. Above 660°C, the hole concentration increases again mainly due to the incorporation of carbon atoms in arsenic sites on the growth surface at high temperature. The different temperature dependence of carbon incorporation, which results in U-shaped curve, has been also reported for AlGaAs[10] suggesting that the carbon incorporation is dominated by the different mechanisms in the different temperature regimes. The carbon impurity concentrations measured in our samples are significantly lower than those reported in other studies[11,12] where CBE was employed for the growth of GaAs using arsenics obtained from precracked arsine. This result may be explained by an observation that high temperature exposure to AsH$_3$ consumes the CH$_2$ adsorbate, apparently by hydrogenating them back to CH$_3$ groups that then desorb[13,14]. It is often thought that AsH$_3$ facilitates the removal of adsorbed CH$_3$ via CH$_4$ formation[15,16], thus leading to a reduction in carbon doping, even though attempts to directly monitor CH$_4$ production via a surface reductive elimination process have failed[17,18].

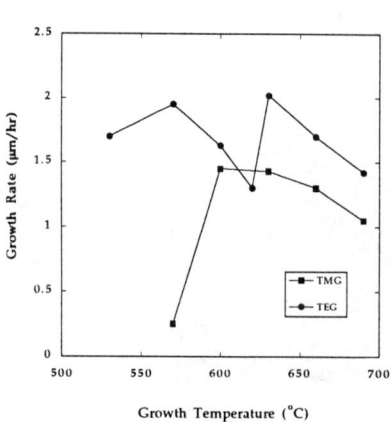
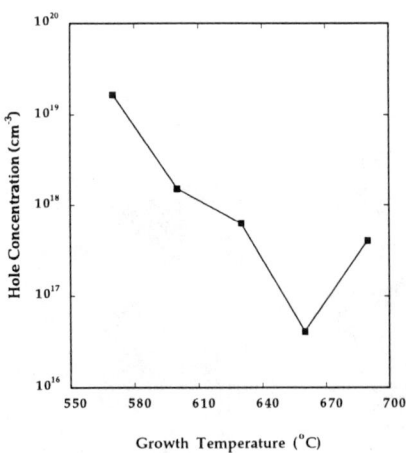

Figure 2. Effect of growth temperature on the GaAs growth rates using unprecracked AsH$_3$ with TMG(V/III=10, P=1.2x10^{-4} Torr) and TEG(V/III=20, P=1.2x10^{-4} Torr).

Figure 3. Dependence of hole concentration in the TMG-grown GaAs film on the substrate temperature(V/III=10, P=1.2x10^{-4} Torr).

Figure 4 represents the temperature dependent growth rates of GaAs grown using MEAs with TMG and TEG. The growth rates for TMG in fig. 4 are divided into three distinct regions which were also observed in the case of growth with unprecracked AsH_3 and TMG[19,20]. The growth rate for TEG, however, increases gradually with increasing substrate temperature presumably due to the low decomposition temperature of TEG as illustrated in fig. 4. Whereas the growth rate for TMG is very small at a low growth temperature of 550 °C, a growth rate of about 1 μm/h can be obtained at the same temperature when TEG is used as gallium source. All materials obtained by growth with MEAs

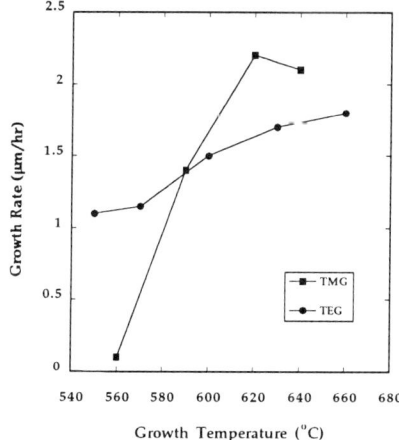

Figure 4. Effect of growth temperature on the GaAs growth rates using unprecracked MEAs with TMG(V/III=10, P=5.5x10^{-5} Torr) and TEG(V/III=20, P=5.7x10^{-4} Torr).

and TMG exhibited p-type background doping of 10^{15}-10^{17}cm^{-3} with 300K mobility of 230-330 cm^2/V·sec. The hole concentrations in TMG-grown samples decrease as substrate temperature increases, which suggests that carbon containing species may drastically desorb from the growth surface with increasing growth temperature. Epilayers grown with TEG were mostly semi-insulating with p-type background doping of less than 10^{14}cm^{-3}. At the growth temperature of 600 °C and V/III ratio of 5, n-type background doping of 4.9x10^{15} cm^{-3} with 300 K mobility of 4290 cm^2/V·sec was observed in TEG-grown epilayer. These values indicate that the samples are moderately compensated.

CONCLUSIONS

We have shown that AsH_3 and MEAs without precracking can be used in the growth of GaAs by UHVCVD using TMG and TEG via a surface decomposition process. Three distinct temperature-dependent regions of growth rates by TMG were observed. The growth rates by TEG, however, produced an anomalous dependence on the growth temperature showing two humps, which are indicative of two hydride intermediates. Growth of GaAs epilayers was also successful at fluxes compatible to that of precracked arsine together with TMG and TEG. This points to the fact that a sufficient amount of arsenics is produced by the thermal and catalytic decomposition of hydrides adsorbed on the GaAs surface. It was also found that the uptake of carbon impurity is significantly reduced when TMG is replaced with TEG for both AsH_3 and MEAs. Our results indicate also that the significant drop in hole concentration results from the efficient removal of carbon-containing species by surface hydrogens and partially decomposed hydrides from the adsorbed AsH_3. Intermediates like AsH_2 from MEAs are also considered to supply hydrogen atoms during growth, which may remove carbon containing species as CH_3 or CH_4.

ACKNOWLEDGMENTS

This work has been supported by Korea Telecom and by the Ministry of Communications, Korea.

REFERENCES

1. E. Veuhoff, W. Pletschen, P. Balk, and H. Lüth, J. Cryst. Growth **55**, 30(1981).
2. N. Pütz, E. Veuhoff, H. Heineke, M. Heyen, H. Lüth, and P. Balk, J. Vac. Sci. Technol. **B3**, 671(1985).
3. N. Pütz, H. Heineke, M. Heyen, P. Balk, M. Weyers, and H. Lüth, J. Cryst. Growth **74**, 292(1986).
4. F. T. J. Smith, Prog. Solid State Chem. **55**, 111(1989).
5. C. R. Abernathy, S. J. Pearton, R. Caruso, F. Ren, and J. Kovalchik, Appl. Phys. Lett. **55**, 1750(1989).
6. D.M. Speckman and J.P. Wendt, J. Crystal Growth **105**, 275(1990); D.M. Speckman and J.P. Wendt, Appl. Phys. Lett. **56**, 1134(1990).
7. J. Musolf, M. Weyers, P. Balk, M. Zimmer, and H. Hofmann, J. Cryst. Growth **105**, 271 (1990).
8. S. J. Park, J. R. Ro, J. K. Sim, and E. H. Lee, Mat. Res. Soc. Symp. Proc. **281**, 37(1993).
9. T. Isu, M. Hata and A. Watanabe, J. Cryst. Growth **105**, 209(1990).
10. B.J. Lee, Y.M. Houng, and J.N. Miller, J. Cryst. Growth **105**, 168(1990); J.L. Benchimol, X.Q. Zhang, Y. Gao, G. Le Roux, H. Thibierge, and F. Alexandre, J. Cryst. Growth **120**, 189(1992).
11. C.R. Abernathy, S.J. Pearton, F. Ren, W.S. Hobson, T.R. Fullowan, A. Katz, A.S. Jordan, and J. Kovalchik, J. Cryst. Growth **105**, 375(1990).
12. M. Konagai, T. Yamada, T. Akatsuka, S. Nozaki, R. Miyake, K. Saito, T. Fukamachi, E. Tokumitsu, and K. Takahashi, J. Cryst. Growth **105**, 359(1990).
13. J. R. Creighton, B. A. Bansenauer, T. Huett, and J. M. White, J. Vac. Sci. Technol. **A11**, 876(1993).
14. X. -Y. Zhu, M. Wolf, and J. M. White, J. Chem. Phys. **97**, 605(1992).
15. D. H. Reep and S. K. Ghandhi, J. Electrochem. Soc. **130**, 675(1983).
16. C. A. Larsen, S. H. Li, N. I. Buchan, G. B. Stringfellow, and D. W. Brown, J. Cryst. Growth **102**, 126(1990).
17. U. Memmert and M L. Yu, Appl. Phys. Lett. **56**, 1883(1990).
18. D. W. Squire, C. S. Dulcey, and M. C. Lin, Mat. Res. Soc. Symp. Proc. **101**, 301(1988).
19. A. Robertson, Jr., T. H. Chiu, W. T. Tsang, and J. E. Cunningham, J. Appl. Phys. **64**, 877(1988).
20. B. W. Liang, T. P. Chin, and C. W. Tu, J. Appl. Phys. **67**, 4393(1990).

CONTROLLED IMPURITY INTRODUCTION IN CVD : CHEMICAL, ELECTRICAL, AND MORPHOLOGICAL INFLUENCES

T. F. KUECH[†], J. M. REDWING[†], J.-W. HUANG[†], AND S. NAYAK[††]
University of Wisconsin, [†]Department of Chemical Engineering, [††]Materials Science Program, 1415 Johnson Drive, Madison, WI 53706

ABSTRACT

We present here an overview of recent studies of the influence of oxygen doping on the electrical and structural properties of semiconductors grown through the metal organic vapor phase epitaxy (MOVPE) technique. In particular, we have measured the impact of oxygen introduction on several of the principal aspects of the growth process: incorporation, activation, and influence on the growing surface structure. The gas phase chemistry and dopant incorporation was investigated for two different precursors: dimethyl aluminum methoxide and diethyl aluminum ethoxide. The simple change in the structure of the oxygen source leads to significant changes in the oxygen incorporation behavior. Complementary studies of the gas phase decomposition of these oxygen sources have indicated that the decomposition mechanism is substantially different for these two sources leading to the change in incorporation behavior. The impact of the selective incorporation of oxygen at heterointerfaces has been studied here through the growth of superlattice structures. Glancing angle X-ray diffraction and atomic force microscopy measurements have shown that the incorporation of oxygen at the GaAs-to-$Al_xGa_{1-x}As$ interface leads to modest increases in the average roughness of the heterointerface with more significant changes in the interfacial structure. The structure of this interfacial roughness was also studied through measurements of the diffuse X-ray scattering about a Bragg peak. Measurements of the component of the roughness which is correlated between the superlattice layers show significant changes with oxygen addition.

INTRODUCTION

The metal-organic vapor phase epitaxy (MOVPE) process is a versatile growth technique that offers high quality epitaxial compound semiconductors and the ability to produce atomically abrupt interfaces. The fabrication of a wide variety of electronic and photonic devices has been achieved with this technique. Incorporation of impurities in these device structures is a critical issue since device performance depends strongly on clever and controlled manipulation of doping. There have been a great number of growth-based studies of the controlled incorporation of impurities during MOVPE. These studies have generally focused on the incorporation of shallow impurities which provide electrical conductivity in the grown layers. The detailed chemistry of dopant incorporation and the impact of these dopants on the structure of the growing surface have not received as much attention despite its importance. MOVPE growth is a consequence of complex and coupled chemical, thermal, hydrodynamic, and mass transport effects. The doping process in the MOVPE environment is similarly complicated. A simplified picture of the doping process, however, can be described as involving five different steps : (1) mass transport of dopant precursors to the growth front, (2) any gas phase reactions, (3) adsorption of dopant precursors on the growing surface, (4) surface decomposition and product removal, and (5) re-evaporation of dopant species. Gas phase and surface reactions determine the chemical nature of adsorbates on the growth front and can therefore affect their subsequent incorporation behavior. The incorporated impurities, generating the shallow or deep levels, then provide for the electrical and optical properties of materials.

We have studied this doping process in the MOVPE growth of GaAs and $Al_xGa_{1-x}As$ as model systems. In particular, we have studied oxygen incorporation and its impact on materials growth and properties. The MOVPE process is used to grow high purity $Al_xGa_{1-x}As$ and high quality $Al_xGa_{1-x}As$/GaAs interfaces. The most common impurities in $Al_xGa_{1-x}As$ have been found to be carbon and oxygen [1]. Carbon arises from the alkyls used in the growth and its incorporation can be controlled by the specific growth conditions [1, 2]. Oxygen can be introduced into the growth ambient through a variety of mechanisms due to the reactive nature of both the Al-bearing alkyl and the $Al_xGa_{1-x}As$ growth surface. Alkoxide contaminants in the metal organic precursors (such as trimethyl aluminum, TMAl, and trimethyl gallium, TMGa), as well as trace levels of H_2O and O_2 in the process gases, have been cited as the source of oxygen incorporation in $Al_xGa_{1-x}As$[2]. In the latter case, the oxygen incorporation mechanism has been speculated to involve an O_2-TMAl gas phase reaction to form volatile dimethyl aluminum methoxide (DMALO, $(CH_3)_2AlOCH_3$) [3]. The initial studies of MOVPE $Al_xGa_{1-x}As$ often reported materials which were highly compensated and possessing a low photoluminescence efficiency [4, 5], possibly due to the unintentional oxygen incorporation. Oxygen has been known to form a variety of deep levels in both GaAs and $Al_xGa_{1-x}As$. We have previously demonstrated that oxygen can, however, be controllably incorporated into GaAs through the independent introduction of DMALO [6]. The trace amounts of Al on the growth surface and in the gas phase provide a driving force for oxygen incorporation. Oxygen concentration in excess of 10^{18} cm^{-3} was readily achievable, and oxygen-related deep levels were generated, compensating shallow Si impurities. These oxygen-doped epitaxial layers had specular surface morphology and high crystal quality, indicating that there is little macroscopic disruption of the growth process by oxygen incorporation.

We have extended the controlled oxygen incorporation study through the alteration of carbon-bearing ligands in oxygen precursors. DEALO, $(C_2H_5)_2AlOC_2H_5$, is used as a new oxygen-doping precursor [7]. DEALO is a commercially available source which is a liquid at room temperature (melting point is about 2.5°C) with suitable vapor pressure (about 0.063 Torr at 25°C). It can therefore be used in a conventional bubbler arrangement, leading to reproducible oxygen doping results. In contrast, DMALO is a solid at room temperature with a melting point of 35°C and a relatively high vapor pressure of 7.4 Torr at 76°C [8]. Through the use of these precursors, we can add oxygen in a controlled fashion and systematically probe its influence.

Many of the effects of impurity addition are strongly influenced by surface phenomena. In general, surface processes occurring during MOVPE growth are generally not well understood due to a lack of *in-situ* growth monitoring techniques compatible with the MOVPE environment. Recent developments, including reflection difference spectroscopy [9] and grazing angle X-ray diffraction [10], have contributed substantially to our initial understanding of these processes. Epitaxy is generally considered to occur in a step-flow growth mode at the high substrate temperatures used during MOVPE [11]. Very little is known, however, about the influence of surface step density or step character on this process. In this study, we have used controlled impurity incorporation as a probe of MOVPE growth morphology. Impurities are known to preferentially incorporate on certain types of surface sites during growth. For example, the incorporation of an impurity at the edge of a surface step may disrupt the step-flow growth process resulting in the development of the film morphology. These present studies, based on X-ray scattering and atomic force microscopy (AFM) measurements, should be considered as providing one of the first systematic determinations of the impact of such impurity incorporation on the surface structure during growth.

These chemical and physical aspects of the doping process are presented in this paper through the study of controlled oxygen incorporation into MOVPE GaAs using aluminum-oxygen bonding based precursors: dimethyl aluminum methoxide (DMALO) and diethyl aluminum ethoxide (DEALO). Growth studies, combined with material characterization, were used to determine the chemistry and incorporation behavior. The change in surface morphology due to impurity adsorption was investigated through the selective adsorption of oxygen at heterointerfaces.

EXPERIMENTAL PROCEDURES

Samples for these studies were grown in a conventional horizontal low pressure (78 Torr) reactor [7]. TMGa, TMAl and AsH_3 were used with H_2 as the carrier gas. The growth temperature was varied over the range 600-800°C and the V/III ratio (AsH_3/TMGa) from 10 to 120. The growth rate was held constant at 0.05µm/min., corresponding to a TMGa mole fraction of about 1.8×10^{-4} in the reactor. DMALO and DEALO were used as oxygen precursors. All oxygen-doped layers were co-doped with Si using disilane (Si_2H_6) or C using carbon tetrachloride (CCl_4) for n- or p-type doping to allow for standard electrical characterizations. Secondary ion mass spectroscopy (SIMS) was used to determine the physical concentration of the impurities in the epitaxial layer. Superlattice (SL) structures consisting of 40 periods of 70Å GaAs/70Å $Al_{0.3}Ga_{0.7}As$ on a 0.6 µm $Al_{0.3}Ga_{0.7}As$ buffer layer were grown by MOVPE at 650°C. The layers were grown on semi-insulating (100) GaAs wafers miscut approximately 0, 2 and >4° toward the <110> direction. Oxygen incorporation at the growth front was carried out by the addition of DEAlO (1×10^{-5} mole fraction) to the inlet gas stream during the 30 sec. purge periods between the growth of successive superlattice layers. Oxygen was typically added only after the growth of the GaAs superlattice layers (GaAs-to-$Al_xGa_{1-x}As$ interface). Identical SL structures without oxygen doping were grown for comparison. A Si-doped GaAs/ $Al_{0.3}Ga_{0.7}As$ multi-layer sample was also grown with DEAlO addition between layers to verify the presence of oxygen at the heterointerface using capacitance profiling.

Gas phase decomposition measurements were carried out in a flow-tube reactor described elsewhere.[12] This mass spectroscopy based system allows for the study the pyrolysis products under controlled thermal and residence time conditions. The analysis of the effluent from the reactor is used to postulate gas phase reaction pathways.

Small angle X-ray diffraction was used to determine the total average or RMS interfacial roughness, correlated roughness and lateral correlation length of roughness of the superlattice structures. The theory of X-ray diffraction from multilayers and details of the kinematic model used in this study have been previously discussed [13,14,15]. X-ray reflectivity from superlattices at small angles arises from constructive interference of individual layers of the superlattice (SL). The X-ray reflectivity typically consists of a diffuse background with intense Bragg peaks. The structure or roughness of the interface can be characterized by the RMS roughness and, in the case of multilayers, the degree of correlation in the roughness found at sequential interfaces. The reduction in the intensity of Bragg peaks in a θ-2θ scan is predominantly due to the RMS interfacial roughness. The RMS interfacial roughness is determined here by comparing the measured peak intensity with that predicted for a perfect SL mirror assuming that the roughness follows a Gaussian distribution in the direction perpendicular to the growth front. Correlated roughness results from morphology that is copied from layer to layer. The diffuse background in an X-ray spectrum contains information about this correlated roughness and correlation length. The intensity distribution of the diffuse scattering can be mapped by performing a transverse scan around a specific Bragg peak. The correlated roughness and correlation length can then be determined by comparing the experimental data with theoretical predictions.

These X-ray (Cu-K_a) measurements were carried out using a conventional Nicolet X-ray diffractometer. The rotation of the detector and sample can be controlled to an accuracy better than 0.01º and 0.005º, respectively. The aperture of the entrance slit limits the divergence of the incident beam to 0.03º in the diffraction plane. Two types of scans were measured: θ-2θ scans, which map out reflected intensity in reciprocal space perpendicular to the interface and rocking curves on transverse scans, which measure the reflected intensity parallel to the interface. In the latter case, the detector slit integrates the scattered intensity perpendicular to the diffraction plane. All scans were made at a grazing incident angle ($0º \leq θ \leq 3º$). The kinematic model outlined in ref. 4 was used to determine, from the experimental X-ray data information, the total RMS roughness, correlated roughness and lateral correlation length. The interfacial roughness values determined by the X-ray measurements were compared to RMS surface roughness measured by atomic force microscopy (AFM). The AFM images were obtained on a Digital Instruments Nanoscope III using constant height mode.

RESULTS

Oxygen Incorporation Behavior

Both the DMALO and DEALO-doped GaAs had a specular growth morphology at the oxygen concentrations investigated in this study. The oxygen doping efficiency, however, depends on the oxygen precursors. DEALO is found to be substantially less efficient in incorporating oxygen [13]. Typical concentrations of Al and O, obtained from SIMS measurements, are given as a function of DEALO mole fraction in fig. 1. These DEALO-doped samples used in fig. 1 were all grown at 600ºC with a V/III ratio of 40. Equivalent oxygen concentrations in the DEALO-based samples require a DEALO mole fraction approximately 60 times greater than the equivalent amount of DMALO. Even with the higher DEALO mole fraction, the Al concentration in the film is nearly two orders of magnitude smaller than that found in the DMALO-doped GaAs. While DEALO doping leads to a linear dependence of the Al content on oxygen precursor mole fraction, the oxygen incorporation exhibits a nonlinear power law with an order of about 4. Oxygen concentration in fig. 2 is usually lower than that of Al, but exceeds that of Al at high DEALO mole fraction. These observations imply that multiple-oxygen-bearing species are involved in the incorporation process.

The most surprising difference between the DMALO and DEALO-based GaAs:O was the dependence of oxygen concentration on the V/III ratio [13]. Higher V/III ratios cause decreasing oxygen concentration in both the nominally undoped $Al_xGa_{1-x}As$ [16] and DMALO cases. The GaAs:O samples grown using DEALO, however, indicate that the use of higher V/III ratios greatly enhances the incorporation of oxygen as well as aluminum by more than an order of magnitude over the investigated range. When using DMALO, in sharp contrast, the oxygen doping decreases and the aluminum content stays constant as the V/III ratio increases.

The effect of Si doping, using Si_2H_6, on the oxygen incorporation was studied for the case of DEALO-based doping [13]. The Si doping efficiency was not altered by the presence of the DEALO, and the Si concentration was proportional to the Si_2H_6 mole fraction in the reactor as shown in fig. 1. The Si_2H_6 mole fraction did affect the degree of compensation of shallow donors as described below, indicating that Si participates in the formation of deep levels.

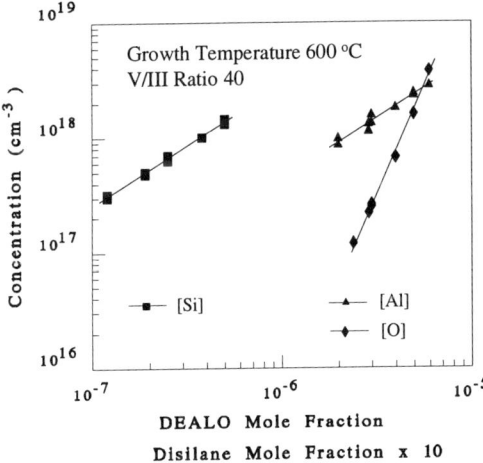

Figure 1: The concentrations of Al and O from DEALO, and Si from Si_2H_6 in GaAs as a function of DEALO and Si_2H_6 mole fractions [13].

Electrical Behavior

Free carriers from both n- and p-type shallow impurities were found to be compensated by the incorporation of oxygen or oxygen-based defect complexes from both DMALO [12] and DEALO [13]. A reduction in the free carrier concentration is observed in fig. 2, and is directly correlated with the oxygen concentration. Figure 2 shows the SIMS O profile obtained from a sample where the DEALO mole fraction was changed in a stepwise fashion. The corresponding profile of the electron concentration obtained from capacitance-voltage measurements is also shown in fig. 2. The free carrier concentration drops as the oxygen concentration is increased. The degree of compensation is shown in fig. 3 as a function of DEALO/Si_2H_6 mole fraction ratio as well as the absolute mole fraction of Si_2H_6. The compensation exhibits a power-law dependence on DEALO/Si_2H_6, approximately given as:

$$\frac{\Delta n}{n} \propto \left[\frac{DEALO}{Si_2H_6}\right]^x, x \approx 2.$$

We have studied the role of Si-based species in compensation by growing two more samples at different Si_2H_6 concentrations as illustrated in fig. 3. Three parallel lines (all with a power of ~2) were obtained, corresponding to these three separate Si_2H_6 mole fractions. Higher compensation levels were correlated with a higher Si_2H_6 mole fraction at the same value of the DEALO/Si_2H_6 ratio, indicating that Si-O related species participate in compensation. This additional compensation appears to be proportional to the Si_2H_6 mole fraction, or to the Si concentration in the sample. If there was no additional compensation associated with the Si dopant, curves obtained from these three samples should coincide. A second-order power-law dependence of Δn on DEALO mole fraction can be established for a given Si_2H_6 mole fraction. This should be compared with the fourth-order power-law behavior of the total oxygen concentration obtained from SIMS (fig. 2). These combined results indicate that not all the incorporated oxygen is electrically active. Given the strong bonding energy between Si and O, it is only natural to postulate that Si-O related species should form and be active in free carrier trapping and responsible for part of the observed compensation. The responsible oxygen-related deep levels have been investigated using deep level transient spectroscopy (DLTS)[12, 13,17].

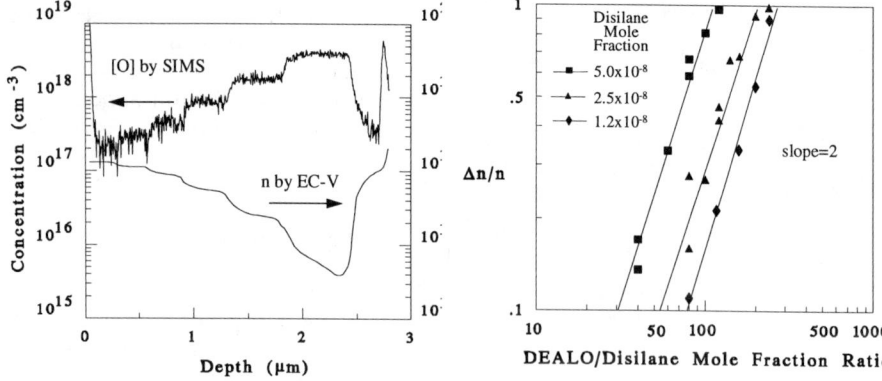

Figure 2: The profiles of SIMS oxygen and C-V carrier concentrations of a GaAs multilayer sample grown at 600°C and a V/III ratio of 40 [14]. Si_2H_6 mole fraction is kept constant at 5×10^{-8} throughout with the DEALO mole fraction changing in each layer.

Figure 3. The degree of shallow dopant compensation as a function of normalized DEALO mole fraction at different Si_2H_6 mole fractions [14] indicates that both Al and Si can form O-based deep levels.

The addition of DEAlO to the inlet gas stream during the purge time between layer growth results in oxygen incorporation and carrier compensation at the heterointerface as shown in fig. 4. Oxygen incorporation, which creates deep level traps in the material, results in a drop in the carrier concentration of the Si-doped multilayer structure at the $Al_{0.3}Ga_{0.7}As$ /GaAs interface. Electrical compensation at the layer interface increases with a corresponding increase in DEALO mole fraction.

DEALO Decomposition Study: Gas Phase Chemistry

DEALO was independently introduced into the flow tube reactor system for the decomposition study. The major pyrolysis product was found to be ethylene (C_2H_4) as registered by the peak enhancements at mass number 28, 27, and 26 [18]. DEALO therefore decomposes and releases ethylene at temperature as low as 350°C. The corresponding alkyl to DEALO is $(C_2H_5)_3Al$ or TEAl. TEAl is known to undergo a uni-molecular β-hydride elimination reaction where an Al-C_2H_5 bond is replaced by an Al-H bond, and an alkene, C_2H_4, is released [7]. This β-elimination reaction results in the gas phase formation of $(C_2H_5)_{3-x}AlH_x$ with an end point perhaps being alane, AlH_3. Alane is quite unstable at typical growth temperatures. This gas phase compound would decompose on the warm walls of the reactor, upstream from the growth area, removing reactants from the gas stream and resulting in the low Al incorporation efficiency at the growth front. The gas phase decomposition of DEALO could be very similar to TEAl, i.e.

$$(C_2H_5)_2 AlOC_2H_5 \rightarrow (C_2H_5)HAlOC_2H_5 + C_2H_4$$
$$(C_2H_5)HAlOC_2H_5 \rightarrow H_2AlOC_2H_5 + C_2H_4$$

with the alkoxide decomposing well outside the growth area.

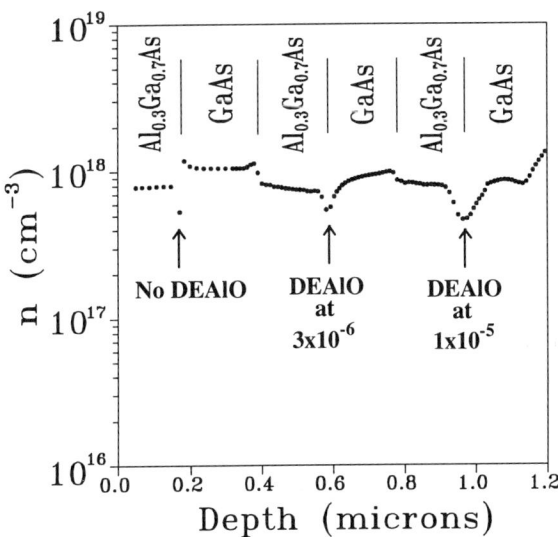

Figure 4: Carrier concentration as a function of film depth for Si doped $Al_{0.3}Ga_{0.7}As$/GaAs layers grown with oxygen incorporation from DEAlO at the interface. The mole fraction of DEAlO was 1×10^{-5}, 3×10^{-6} and zero. Electrical compensation due to oxygen-related deep levels results in a drop in the carrier concentration at the interface. The compensation increases with increasing DEAlO mole fraction. Growth temperature was 650°C.

X-ray Analysis of Superlattice Structures: Effect of Oxygen

The effect of oxygen addition on interfacial roughness was studied using small angle X-ray scattering. Figure 5 is a typical plot of the measured and calculated X-ray profile for an $Al_{0.3}Ga_{0.7}As$/GaAs superlattice grown without oxygen addition. The RMS roughness is determined by comparing the measured and calculated integrated specular intensity of the 3rd, 5th and 7th order peaks in the X-ray spectrum. The 1st order peak is neglected since it is not described well by the kinematic theory used here. The even order peaks are neglected due to their lower intensity. The probe depth of the sample is 1 μm at θ=2°. The lower and upper limit of roughness is determined by fitting the integrated intensity of the 7th and 3rd order peak, respectively. An RMS roughness of 5.0 Å was calculated for the sample in fig.5.

The effect of oxygen addition on interfacial roughness is given in Table I. The addition of DEAlO to the inlet gas during the purge period after growth of a GaAs layer leads to oxygen incorporation at the interface, as electrically determined by capacitance measurements. The presence of oxygen at this heterointerface can, at times, increase the interfacial RMS roughness as described in Table I. The RMS roughness of the oxygen doped superlattice shown in fig. 5 however was determined to be 5.0 Å, identical to the undoped sample. Oxygen addition does significantly alter the correlated roughness and correlation length, as shown in Table I. A smaller correlation length was measured for the oxygen doped sample, indicating that impurity addition is altering the growth process in a systematic fashion. Such an effect would arise if oxygen were to preferentially incorporate at surface step edges during growth.

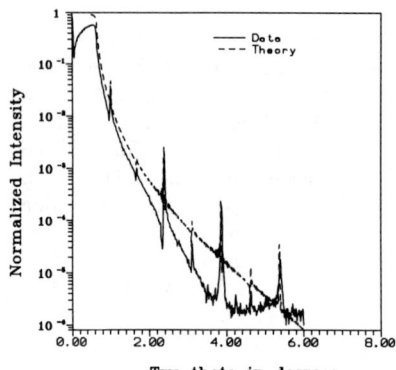

Figure 5: A fit of the theoretical model (dashed line) to experimental X-ray data obtained from a small angle θ-2θ scan (solid line) yields information on the RMS roughness, the superlattice period and individual layer thickness.

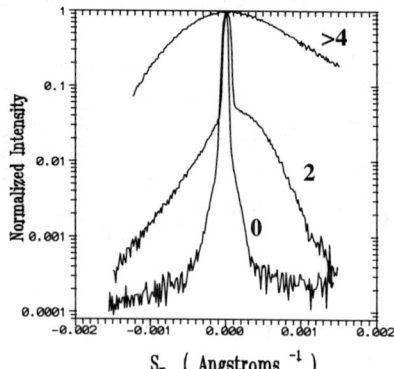

Figure 6: X-ray rocking curves (transverse scan) of 3rd order Bragg peaks of $Al_{0.3}Ga_{0.7}As$/GaAs superlattices grown on (100) GaAs wafers miscut 0, 2 and >4° toward <110> with oxygen doping at GaAs to $Al_xGa_{1-x}As$ interface. The peak broadness increases with increasing miscut angle corresponding to a decrease in roughness correlation length.

Table I: The total RMS roughness (σ_{tot}) and correlated roughness (σ_{corr}) of $Al_{0.3}Ga_{0.7}As$/GaAs superlattices determined by X-ray diffraction. The thickness of each layer and RMS roughness are obtained by fitting θ-2θ scans. The correlated roughness and correlation length (ξ) is determined by fitting the diffuse component of the intensity of the 3rd order rocking curve.

Periods $Al_{0.3}Ga_{0.7}As$/GaAs	Oxygen doping at the interface	Total RMS Roughness σ_{tot} (Å)	Correlated Roughness σ_{corr}(Å)	Correlation Length ξ (Å)
60/64Å	No doping	5.0±0.5Å	1.2Å	2500Å
57/61Å	GaAs-to-$Al_xGa_{1-x}As$ interface	5.0±0.5Å	*	<1000Å

* Could not be determined with present model.

Table II: The effect of oxygen doping on the roughness of $Al_{0.3}Ga_{0.7}As/GaAs$ superlattices grown on (100) GaAs substrates miscut at various angles toward the <110> direction. RMS surface roughness values obtained from Atomic Force Microscopy measurements of the oxygen doped samples are also included for comparison.

miscut angle	No doping			Oxygen on GaAs surfaces			AFM
	σ_{tot} (Å)	σ_{corr} (Å)	ξ (Å)	σ_{tot} (Å)	σ_{corr} (Å)	ξ (Å)	††$\sigma_{surface}$ (Å)
0°	4.8±0.5	1.2	2000	3.8±0.5	<1000	1500	11.7
2°	5.0±0.5	1.2	1500	†5.0	<100	<1000	N.A.
>4°	†10.0	*	<100	†7.0	*	<100	30.0

* Could not be determined with present model.
† RMS roughness is the lower limit of roughness.
†† averaged over a 5x5 μm area

    The change in correlation length with oxygen addition was further investigated through the growth of oxygen doped superlattice structures on (100) GaAs wafers with increasing miscuts of 0, 2 and >4° toward the <110> direction. The surface step density increases with increasing miscut angle and thus provides a means to explore the effect of step density on oxygen incorporation and interfacial roughness. The results are outlined in Table II. Increasing miscut does not significantly alter the RMS roughness or correlation length in superlattice samples grown without oxygen addition on 0° or 2° miscut substrates. The RMS roughness increases and correlation length decreases with increasing miscut angle, however, for superlattices grown with oxygen addition. This behavior is clearly depicted in X-ray rocking curves of the oxygen doped superlattices shown in Figure 6. The X-ray spectra were obtained from transverse scans of the 3rd order Bragg peak. The increase in X-ray peak broadness with increasing miscut angle corresponds to a decrease in the correlation length of roughness.

    AFM images of the surface of the oxygen doped $Al_{0.3}Ga_{0.7}As/GaAs$ superlattices also show an increasing roughness and decreasing correlation length with increasing substrate miscut as listed in Table II. The surface roughness is a combination of the interfacial roughness of the multilayers and the surface oxide layer. The AFM measurements do not spatially average over as broad an area as the X-ray measurements and therefore a difference in the quantitative values of the roughness is expected. The RMS surface roughness measured by AFM is a factor of 3-4 times greater than the interfacial roughness determined by X-ray, but the trends with miscut angle are similar.

DISCUSSION

    The complete description of a doping process requires that the surface adsorbed species and the chemical kinetic processes be determined. The subsequent changes in interfacial structure can be understood only after the general features of the chemistry are determined. In the present case, the incorporation behaviors of Al and O into MOVPE-grown GaAs using DMALO and DEALO are complicated functions of the precursor chemistry and growth conditions. We have proposed a mechanism for the observed oxygen doping behaviors using DMALO [14]. The decomposition pathway of DMALO most likely involves bond cleavage of the Al-C bond as the bond strength of Al-O (121.3 kcal/mole) [18] is much higher than that of Al-CH_3 (65 kcal/mole) [19]. Whatever the specific steps involved in the initial decomposition of the precursor, the last step probably occurs on or near the growth surface. A proposed reaction for the methyl-based growth chemistry, for example, involves an atomic hydrogen from an adsorbed arsine hydride, AsH_x, reacting with surface-adsorbed CH_3 and leading to the removal of carbon through the formation of CH_4 [3]. A similar reaction between the surface adsorbed O and an AsH_x species can be postulated. The rational for this type of reaction is the thermodynamic driving force for a reaction between arsine and aluminum monoxide :

$$AlO_{(g)} + AsH_{3(g)} \leftrightarrow AlAs_{(s)} + H_2O_{(g)} + 1/2\ H_{2(g)}$$

The free energy for this reaction, ΔG_r, is highly favorable (-92 kcal/mole at 900 °K [20]) for the reduction of surface oxides, mainly due to the high ΔG_f associated with the formation of H_2O. This reaction may occur in the gas phase or on the growth front, as a temperature activated process. The observed activation energy, ~0.9-1.5 eV, when compared with the ΔG_r derived above, indicates that a surface-mediated reaction is most probable. Additionally, there are aluminum suboxides which can form at high temperatures. These species, if formed, would also provide a temperature activated pathway for the removal of oxygen from the surface. The aluminum concentration, however, does remain constant with changes in growth temperature and V/III ratio, supporting the view that oxygen is primarily removed through the reduction of oxygen bearing species at the growth front.

Many of the above arguments concerning the trends in oxygen incorporation with changing growth temperatures and V/III ratios should apply to the case of DEALO-based doping. There are, however, several key differences in the oxygen incorporation between DMALO and DEALO which must be addressed : 1) the low efficiency of the DEALO source, and 2) the increase in both oxygen and aluminum concentration with the V/III ratio using DEALO. These discrepancies can be rationalized by a tentative explanation which draws an analogy between the alkoxide and the alkyl of Al [14], TEAl. In this case, the AsH_3 serves to preserve the O and Al-containing species within the reactor allowing for its transport to the growth front.

The controlled addition of oxygen impurities during the MOVPE growth of $Al_{0.3}Ga_{0.7}As$ /GaAs superlattices was used to study the evolution and correlation of interfacial morphology. The samples were characterized by glancing angle X-ray diffraction and AFM. Oxygen incorporation on GaAs surfaces of the superlattice can change the total RMS roughness, but more significantly decreases the length scale of correlated roughness. Increasing surface step density, resulting from increased wafer miscut angle, results in the increased oxygen incorporation and total interfacial roughness with a decreased lateral correlation length. These changes in the magnitude and character of the interfacial roughness, when combined with electrical measurements, indicate that both more oxygen is incorporated as the substrate miscut increases and the interface becomes rougher. Preferential incorporation of oxygen at surface steps which disrupts the step-flow growth process is used to explain this behavior. Such observations may serve to indicate that minimal impurity incorporation can be achieved with perfectly 'on-axis' substrates. This work also demonstrates the new techniques described above as useful tools by which the interaction of impurities with the growing surface can be studied and eventually understood.

CONCLUSIONS

A full understanding of the dopant incorporation and activation process requires a wide variety of complementary studies. The gas phase and surface chemistry of the doping source will determine the nature of the adsorbed species. These species may be preferentially adsorbed at steps and other sites on the growing surface. These impurities can disrupt the normal pattern of growth at the surface leading to the formation of morphological structures or roughness at subsequently formed interfaces. We have studied aspects of this process through the intentional doping of oxygen during growth and at heterointerfaces. The efficiency of oxygen incorporation is very dependent on the chemical nature of the oxygen-containing doping source. Once incorporated into the crystal, either in bulk or at interfaces, oxygen can cause the roughening of surfaces as well as the introduction of electrically active defect sites.

ACKNOWLEDGMENTS

The authors would like to acknowledge and thank the Army Research Offiice (DAAL03-92-G-0227), Naval Research Laboratory (NC0014-92-K-2004), and the National Science Foundation (DMR-9201558 and the Materials Research Group on Chemical Vapor Deposition) for support of various aspects of this work.

REFERENCES

1. T. F. Kuech, D. J. Wolford, E. Veuhoff, V. Deline, P. M. Mooney, R. Potemski, and J. Bradley, J. Appl. Phys. 62, 632 (1987).
2. T. F. Kuech, E. Veuhoff, T. S. Kuan, V. Deline, and R. Potemski, J. Crystal Growth 77, 257 (1986).
3. H. Terao and H. Sunakawa, J. Crystal Growth 68, 157 (1984).
4. G. B. Stringfellow and G. Hom, Appl. Phys. Lett. 34, 794 (1979).
5. R. H. Wallis, M-A di Forte Poisson, M. Bonnet, G. Beuchet and J-P Duchemin, Inst. Phys. Conf. Ser. 56, 73 (1981).
6. M. S. Goorsky, T. F. Kuech, F. Cardone, P. M. Mooney, G. J. Scilla, and R. M. Potemski, Appl. Phys. Lett. 58, 1979 (1991).
7. J. W. Huang, D. F. Gaines, T. F. Kuech, R. M. Potemski, and F. Cardone, presented at the 1993 EMC, Santa Barbara, CA, 1993.
8. N. Davidson and H. C. Brown, J. Am. Chem. Soc. 64, 316 (1942).
9. D.E. Aspnes, R. Bhat, E. Colas, L.T. Florez, S. Gregory, J.P. Harbison, I. Kamiya, W.E. Quinn, S.A. Scwartz, H. Tanaka and M. Wassermeier, Mat. Res. Soc. Proc. 222, 63 (1991).
10. P.H. Fuoss, D.W. Kisker, G. Renaud, K.L. Tokuda, S. Brennan and J.L. Kahn, Phys. Rev. Lett. 63, 2389 (1989).

11. G.H. Gilmer, J. Crystal Growth 49, 465 (1980).
12. J. M. Redwing, T.F. Kuech, D. Saulys, and D.F. Gaines, to be published in J. Crystal Growth.
13. D.E. Savage, J. Kleiner, N. Schimke, Y.H. Phang, T. Jankowski, J. Jacobs, R. Kariotis and M.G. Lagally, J. Appl. Phys. 69, 1411 (1991).
14. D.E. Savage, N. Schonke, Y.H. Phang and M.G. Lagally, J. Appl. Phys. 71, 3283 (1992).
15. Y.H. Phang, D.E. Savage, T.F. Kuech, M.G. Lagally, J.S. Park and K.L. Wang, Appl. Phys. Lett. 60, 2988 (1992).
16. T. F. Kuech, R. Potemski, F. Cardone, and G. Scilla, J. Electron. Mater. 21, 341 (1992).
17. J. W. Huang and T. F. Kuech, presented at 1993 MRS Fall Meeting, Boston, MA, 1993 (to be published).
18. CRC Handbook of Chemistry and Physics, 65th Edition (CRC Press, Inc. 1984).
19. A.C. Jones, J. Crystal Growth 129, 728 (1993).
20. O. Knacke, O. Kubaschewski, and K. Hesselmann, eds., Thermochemical Properties of Inorganic Substances, 2nd ed(Springer-Verlag Berlin, Heidelberg, Germany, 1991).

STUDY OF SILICON INCORPORATION IN GaAs MOVPE LAYERS GROWN WITH TERTIARYBUTYLARSINE

J.M. Redwing*, T.F. Kuech*, H. Simka** and K.F. Jensen**
*University of Wisconsin, Dept. of Chemical Engineering, Madison, WI 53706
**Massachusetts Institute of Technology, Dept. of Chemical Engineering, Cambridge, MA 02139

ABSTRACT

MOVPE growth experiments have been used to evaluate the role of gas phase and surface chemistry in the silicon doping of GaAs films grown using SiH_4 and Si_2H_6 dopant sources and tertiarybutylarsine (t-$C_4H_9AsH_2$ or TBAs) as the Group V source. The use of TBAs with SiH_4 results in a greater Si doping efficiency and weaker dependence on growth temperature than is typically observed with AsH_3. Gas phase pyrolysis studies combined with reactor residence time experiments indicate that heterogeneous chemistry is responsible for this enhanced Si doping process. TBAs has much less of an effect on Si incorporation from Si_2H_6, a doping process controlled by homogeneous Si_2H_6 chemistry. Increased Si doping from Si_2H_6 with TBAs can be attributed to contributions from the enhanced doping efficiency of SiH_4, a product of Si_2H_6 decomposition. The influence of TBAs on SiH_4 doping chemistry was found to improve the doping uniformity of GaAs films grown in our reactor, from +/-12% for films grown with AsH_3 to +/-5% for TBAs-based material.

INTRODUCTION

Safety concerns surrounding the use of AsH_3 in the metal organic vapor phase epitaxy (MOVPE) process have led to the development of low vapor pressure alkylarsine sources, such as tertiarybutylarsine (t-$C_4H_9AsH_2$ or TBAs). High quality GaAs layers have been grown with TBAs [1] and the use of this source in MOVPE is rapidly increasing. A change from AsH_3 to TBAs does not alter the GaAs growth rate [2], but it does significantly impact the level of impurity incorporation in the films, both from residual carbon [3] and intentional dopant sources such as CCl_4 [4], SiH_4 [5] and H_2S [6]. An understanding of this effect is needed for the development of reliable doping techniques for use with TBAs.

Silane, SiH_4 and disilane, Si_2H_6 are common sources used for n-type doping of GaAs grown from trimethylgallium (TMGa) and arsine, AsH_3. Characteristic features of Si doping from SiH_4 are a low doping efficiency and a strong dependence on substrate temperature and crystallographic orientation [7]. These results combined with numerical modeling studies [8] indicate that heterogeneous SiH_4 chemistry controls the doping process. Si_2H_6, which has a much lower pyrolysis temperature than SiH_4, decomposes in the gas phase at typical MOVPE growth conditions forming SiH_2 a highly reactive intermediate [9]. Si doping from Si_2H_6 is limited by gas phase SiH_2 formation and its subsequent reactions [8]. A high Si doping efficiency and temperature independent incorporation are commonly observed with Si_2H_6 [7].

The use of TBAs in place of AsH_3 has been found to increase the doping efficiency of SiH_4 by approximately one order of magnitude and Si_2H_6 by a factor of 3 [5,10,11]. Kikkawa [10] observed differences in the dependence of Si incorporation on growth temperature, V/III ratio and flow rate between the two Group V sources. In the case of SiH_4, the concentration of Si in TBAs-based GaAs films was found to be independent of growth temperature, in contrast to the strong temperature dependence typically observed with AsH_3. Kikkawa [10] attributed the change in Si doping behavior to gas phase chemistry arising from the low pyrolysis temperature of

TBAs [12]. A model was proposed wherein AsH and AsH_2 produced from TBAs decomposition react with SiH_4 in the gas phase to form monosilylarsine, SiH_3AsH_2. SiH_3AsH_2 then functions as the primary precursor for Si incorporation in the film. Gas phase co-decomposition studies of TBAs and SiH_4, however, produced no evidence of gas phase interaction or homogeneous SiH_3AsH_2 formation [13]. These conflicting results combined with the growing importance of TBAs in MOVPE warrant a more thorough examination of this doping process.

In this study, MOVPE growth experiments were used to evaluate the importance of gas phase and surface chemistry in the Si doping of TBAs-based GaAs films. This work was carried out in a low pressure, horizontal MOVPE reactor with a simple geometry in order to facilitate future numerical modeling studies of this doping process.

EXPERIMENTAL METHODS

Silicon doped GaAs films were grown in a horizontal MOVPE reactor operating at 78 Torr using TMGa and TBAs or AsH_3. The inlet TMGa mole fraction was 1.13×10^{-4}, resulting in a GaAs growth rate of ~0.05 µm/min. Palladium purified H_2 was used as the carrier gas. Dilute gas mixtures of SiH_4 (15.7 ppm) and Si_2H_6 (10.4 ppm) in H_2 were used as dopant sources. The effect of growth conditions on Si doping was investigated by growing multilayered structures in which either the substrate temperature, total flow rate or V/III ratio were varied, keeping the other parameters constant. The carrier concentration of the films was measured using electrochemical capacitance profiling which has a measurement accuracy of ~5%. Doping uniformity was evaluated by measuring the variation of dopant concentration of films grown on 2" diameter GaAs wafers in our MOVPE reactor.

RESULTS

The effect of growth temperature on Si doping was investigated, comparing the results obtained for films grown under identical conditions using TMGa and either AsH_3 or TBAs. Figures 1(a) and (b) show the results obtained using SiH_4 and Si_2H_6, respectively. Si doping from SiH_4 is strongly dependent on growth temperature with AsH_3 and the Arrhenius type behavior yields an activation energy for doping of 1.4 eV, similar to previously reported values [7]. An order of magnitude increase in Si doping efficiency was obtained with TBAs and the doping exhibits a much weaker dependence on substrate temperature becoming independent of temperature in the range from 700 to 800°C.

The data obtained for Si doping from Si_2H_6 with AsH_3, shown in Figure 1(b), exhibits the characteristic high doping efficiency and weak temperature dependence that is typically reported for Si_2H_6 under these conditions [7]. The use of TBAs has much less of an impact on doping efficiency than was observed for SiH_4 and is significant only at lower temperatures (~650°C).

Changes in total flow rate effect the residence time of reactant gases in the heated region of the reactor. Doping processes that are limited by homogeneous chemistry are more sensitive to gas residence time than those controlled by heterogeneous reactions. A series of Si doping runs was completed in which the total flow rate was varied from 2.7 to 6.45 slm by increasing the flow of H_2 carrier gas keeping the mass flow of TMGa, AsH_3 or TBAs, and SiH_4 or Si_2H_6 constant. The carrier concentration is plotted versus the estimated residence time of gases in the heated region of our reactor in Figure 2. Si doping from SiH_4 is independent of gas residence time with both AsH_3 and TBAs. Si_2H_6, on the other hand, exhibits a strong dependence on residence time with AsH_3 as previously reported [8] and this dependence is slightly reduced when TBAs is used.

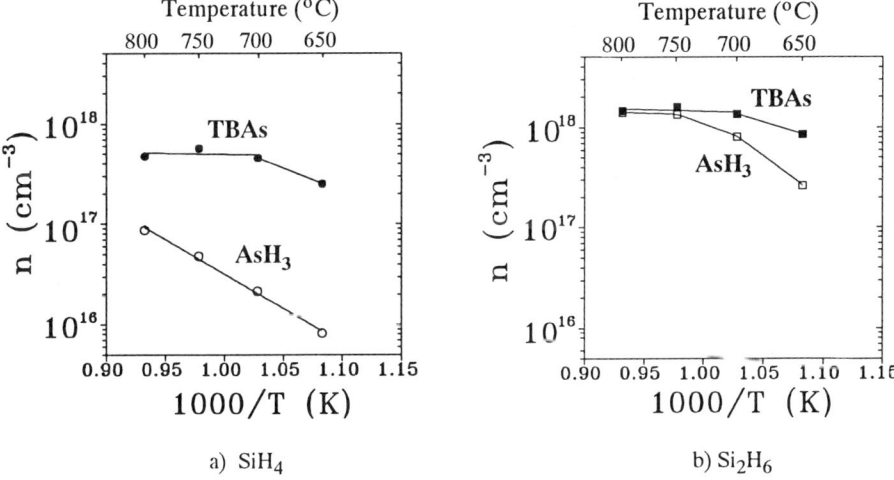

a) SiH$_4$ b) Si$_2$H$_6$

Figure 1: Carrier concentration as a function of growth temperature for GaAs films grown using TMGa, AsH$_3$ or TBAs and either a) SiH$_4$ or b) Si$_2$H$_6$ as the dopant source. The TMGa mole fraction was 1.13×10^{-4}, V/III = 30 and the dopant/TMGa ratio was 1.56×10^{-4}.

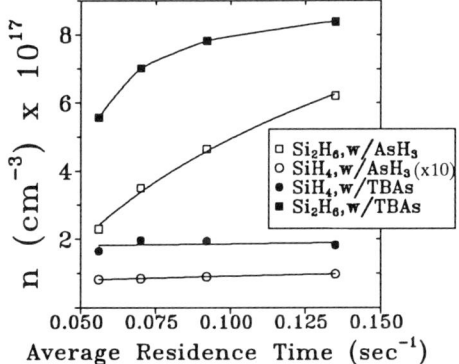

Figure 2: The effect of gas residence time on Si doping from SiH$_4$ and Si$_2$H$_6$ sources. The mass flow of TMGa, AsH$_3$ or TBAs, and SiH$_4$ or Si$_2$H$_6$ were held constant and the flow rate of the H$_2$ carrier gas was varied to change the total flow rate from 2.7 to 6.45 slm. The residence time is estimated as the average time that gas spends in the heated region above the susceptor in our MOVPE reactor. The temperature was 650°C, V/III=30 and the dopant/TMGa ratio was 9×10^{-5}.

A series of runs was completed examining the effect of V/III ratio on Si doping, comparing the two Group V sources. The mass flow rate of TBAs or AsH$_3$ was varied keeping the mass flow of TMGa and SiH$_4$ and all other growth parameters constant. A decrease in carrier concentration, from 9.4 to 6.5×10^{16} cm^{-3}, was observed as the AsH$_3$/TMGa ratio was raised from 30 to 100, similar to reported trends [14]. The same effect was obtained with TBAs, with the carrier concentration decreasing from 2.6 to 1.9×10^{17} cm-3 as TBAs/TMGa is increased from 10 to 100.

DISCUSSION

Silane and TBAs

The use of TBAs in the GaAs growth process resulted in a Si doping efficiency from SiH_4 about one order of magnitude greater than obtained with AsH_3-based growth, similar to previous reports [5,10,11]. Si doping is less sensitive to growth temperature with TBAs and is independent of temperature from 700 to 800°C in our reactor. Temperature independent doping was also reported by Kikkawa [10] using TBAs and SiH_4 for GaAs films grown at atmospheric pressure. Chichibu [11] observed temperature dependent doping with TBAs from 570 to 640°C at a reactor pressure of 152 Torr. This temperature range corresponds to the region in which we also observed a weak dependence on temperature. These results indicate that high pressures and high temperatures lead to temperature independent doping with SiH_4 and TBAs.

A similar effect of pressure and temperature on Si doping is typically observed with Si_2H_6 and AsH_3, as shown in Fig. 1(b). As a result, it was originally proposed that the same dopant mechanism that is operative with Si_2H_6 and AsH_3 is also responsible for the enhanced Si doping with SiH_4 and TBAs. Si_2H_6 decomposes in the gas phase forming SiH_2 which readily reacts with AsH_3 to form SiH_3AsH_2 [13]. Numerical modeling [8] has confirmed that SiH_3AsH_2 formation is an important pathway to Si incorporation with Si_2H_6. TBAs decomposes in the gas phase at low temperatures (~400°C) forming As subhydrides such as AsH and AsH_2 [15,16]. Kikkawa [10] proposed that AsH and AsH_2 react with SiH_4 in the gas phase forming SiH_3AsH_2, which controls Si doping leading to an increased doping efficiency with SiH_4 and TBAs. In a previous study [13], we examined the co-decomposition of a mixture of 3%TBAs and 3%SiH_4 in H_2 carrier gas in a flow tube reactor system using mass spectrometry but found no evidence of gas phase interaction or SiH_3AsH_2 formation. This work indicated that gas phase chemistry was not significant in this doping process and that surface chemistry instead was the controlling factor. The results of our gas phase residence time study, shown in Figure 2, provide further support for this hypothesis. Doping processes controlled by surface chemistry are insensitive to the amount of time the reactant gas spends in the heated region above the susceptor during growth. Si doping from SiH_4 with AsH_3, a process known to be limited by heterogeneous chemistry, is independent of gas residence time. Similar behavior is observed with SiH_4 and TBAs indicating that surface chemistry is controlling Si incorporation in this doping process.

The strong dependence of Si doping on growth temperature using SiH_4 in AsH_3-based growth reduces the uniformity of doping that can be achieved in conventional MOVPE reactor systems. Veuhoff [7] estimated that a 1° change in substrate temperature at 650°C leads to a 2% variation in carrier concentration with SiH_4. Stringent tolerances on the threshold voltage of charge control devices require doping uniformity better than +/-2% [17]. The use of SiH_4 as a source for n-type doping of TBAs-based GaAs results in dopant incorporation that is relatively independent of growth temperature, V/III ratio and gas residence time. We have found that this insensitivity to growth conditions results in improvements in Si doping uniformity in our MOVPE reactor. Figure 3 is a plot of the carrier concentration of Si doped GaAs layers grown on 2" diameter GaAs wafers at 700°C and V/III=30. The carrier concentration was measured approximately every 10 mm from the front of the wafer moving in the direction of gas flow. The variation of doping from SiH_4 is +/-12% with AsH_3 which can be attributed to temperature non-uniformity across the wafer during growth. The use of TBAs results in an improved doping uniformity of +/-5% which is within the 5% measurement accuracy of the capacitance profiling technique that was used to measure the carrier concentration.

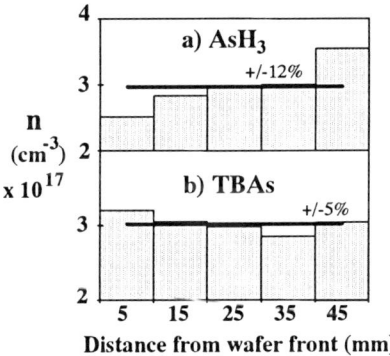

Figure 3: The uniformity of the carrier concentration of Si doped GaAs layers grown on 2" diameter GaAs wafers in our MOVPE reactor using a) AsH$_3$ and b) TBAs as the Group V source. The SiH$_4$ mole fraction was adjusted in order to obtain n~3x10^{17}cm^{-3} with AsH$_3$ and TBAs. The growth temperature was 700°C and V/III=30. The carrier concentration was measured using electrochemical capacitance profiling which has a measurement accuracy of ~5%.

Disilane and TBAs

The Si doping efficiency of Si$_2$H$_6$ is not as affected by the use of TBAs as is SiH$_4$. Increased doping occurs only at lower temperatures ($\leq$700°C). Kikkawa [10] reports a factor of ~3 increase in Si doping from Si$_2$H$_6$ with TBAs for films grown at 650°C, which is similar to the enhancement we observe at this temperature. Si$_2$H$_6$ decomposes in the gas phase at typical growth temperatures forming SiH$_4$ and SiH$_2$. The highly reactive SiH$_2$ species is the dominant Si precursor in this doping process and the contribution to doping from SiH$_4$ formed in the gas phase is negligible in AsH$_3$-based growth [8]. The enhanced doping efficiency of SiH$_4$ with TBAs, however, would result in significant contributions to doping from SiH$_4$ produced by Si$_2$H$_6$ decomposition. The gas phase concentration of SiH$_4$ can be estimated assuming that Si$_2$H$_6$ is fully decomposed, a good assumption at these growth conditions [8]. The Si doping resulting from this mole fraction of SiH$_4$ can be estimated assuming the doping efficiency obtained with TBAs. Adding this contribution to the amount of Si incorporated from Si$_2$H$_6$ in AsH$_3$-based growth results in an increase in Si doping comparable to that obtained from Si$_2$H$_6$ with TBAs shown in Figure 1(b). The enhanced doping efficiency of Si$_2$H$_6$ with TBAs can therefore be attributed to increased doping from SiH$_4$. This suggests that TBAs is not directly interacting with Si$_2$H$_6$ in the gas phase and that SiH$_2$ is still the major contributor to Si doping with TBAs. The results of the gas phase residence study, shown in Figure 2, support this conclusion. The amount of Si incorporated from Si$_2$H$_6$ with AsH$_3$ increases by a factor of ~3 as the residence time is increased from 0.056 to 0.135 sec. Si doping from Si$_2$H$_6$ with TBAs is also sensitive to the gas residence time increasing by a factor of ~1.5 over the same range. This dependence on residence time indicates that gas phase chemistry is controlling Si doping with Si$_2$H$_6$ and TBAs.

CONCLUSIONS

We have investigated the effect of MOVPE growth conditions on Si doping of GaAs layers grown using SiH$_4$ and Si$_2$H$_6$ as dopant precursors and TBAs as the Group V source. The results were compared to doping behavior typically obtained with AsH$_3$-based growth. A higher SiH$_4$ doping efficiency and weaker temperature dependence was obtained with TBAs. Residence time studies support previous gas phase decomposition results which indicated that heterogeneous SiH$_4$ chemistry is responsible for the enhanced Si doping observed with TBAs. The insensitivity of SiH_4 doping to temperature, gas residence time and V/III ratio leads to improvements in doping uniformity in the TBAs-based growth process. A small increase in Si doping from Si_2H_6 with TBAs occurs at lower growth temperatures and this difference diminishes with increasing growth temperature. The enhanced doping efficiency of Si_2H_6 with TBAs can be attributed to the increased doping efficiency of SiH_4, a product of Si_2H_6 gas phase decomposition. It has been found that TBAs, which is fully decomposed in the gas phase at the growth conditions of our study, has a significant impact only on doping processes that are limited by surface chemistry.

ACKNOWLEDGMENTS

The authors would like to acknowledge the financial support of the National Science Foundation through grant DMR-9106633 and funding of a Materials Research Group on Chemical Vapor Deposition (DMR-9121074).

REFERENCES

1. G. Haacke, S.P. Watkins, H. Burkhard, Appl. Phys. Lett. **56**, 478 (1990)
2. R.M. Lum, J.K. Klingert and M.G. Lamont, Appl. Phys. Lett. **50**, 284 (1987)
3. S.P. Watkins and G. Haacke, Appl. Phys. Lett. **59**, 2263 (1991)
4. W.S. Hobson, S.J. Pearton, D.M. Kozuch and M. Stavola, Appl. Phys. Lett. **60**, 3259 (1992)
5. H.B. Serreze, J.A. Baumann, L. Bunz, R. Schachter and R.D. Esman, Appl. Phys. Lett. **55**, 2532 (1989)
6. R.M. Lum, J.K. Klingert and F.A. Stevie, J. Appl. Phys. **67**, 6507 (1990)
7. E. Veuhoff, T.F. Kuech and B.S. Meyerson, J. Electrochem. Soc., **132**, 1958 (1985)
8. H.K. Moffat, T.F. Kuech, K.F. Jensen and P.J. Wang, J. Crystal Growth **93**, 594 (1988)
9. B.S. Meyerson, B.A. Scott and R. Tsui, Chemtronics **1**, 151 (1986)
10. T. Kikkawa, H. Tanaka and J. Komeno, J. Electron. Mater. **21**, 305 (1992)
11. S. Chichibu, A. Iwai, S. Matsumoto and H. Higuchi, Appl. Phys. Lett. **60**, 489 (1992)
12. R.M. Lum and J.K. Klingert, J. Appl. Phys. **66**, 3820 (1989)
13. J.M. Redwing, T.F. Kuech, D. Saulys and D.F. Gaines, presented at the 1993 EMC Conference, Santa Barbara, CA, 1993 (to be published in J. Crystal Growth)
14. S.J. Bass, J. Crystal Growth **47**, 613 (1979)
15. P.W. Lee, T.R. Omstead, D.R. McKenna and K.F. Jensen, J. Crystal Growth **93**, 134 (1988)
16. C.A. Larsen, N.I. Buchan, S.H. Li and G.B. Stringfellow, J. Crystal Growth **94**, 663 (1989)
17. P.M. Solomon, Proc. IEEE **70**, 489 (1982)

PREPARATION AND EVALUATION OF ERBIUM *TRIS*(AMIDE) SOURCE COMPOUNDS FOR ERBIUM DOPING OF SEMICONDUCTING MATERIALS

WILLIAM S. REES, JR.*, UWE W. LAY* AND ANTON C. GREENWALD**
*Department of Chemistry and Materials Research and Technology Center
The Florida State University, Tallahassee, FL 32306-3006
**Spire Corporation, One Patriots Park, Bedford, MA 01730-2396

ABSTRACT

Rare earth elements are important as dopants in the preparation of semiconductors, particularly for temperature-independent optical devices. The compound $Er\{N[Si(CH_3)_3]_2\}_3$ has been demonstrated to be a suitable source compound for erbium doping of semiconducting materials. Further investigations of this and related compounds have been carried out. New compounds have been characterized by ^1H NMR, FTIR, TGA and elemental analysis techniques. Vapor pressures and decomposition profiles have been examined for these compounds.

BACKGROUND

Erbium doped III-V semiconductors attract increasing interest as optoelectronic devices [1]. The sharp and temperature-independent intra-4f shell emission ($^4I_{13/2}$ ---> $^4I_{15/2}$) at 1.54 µm (0.801 eV) nearly matches the lowest loss transmission window of silica fibers. The most widely used source compounds for the preparation of erbium-doped semiconductors by OMVPE (organometallic vapor phase epitaxy) to date have been *tris*(cyclopentadienyl)erbium compounds $Er(Cp^R)_3$ ($Cp^R = C_5H_5$ [2, 3], $C_5H_4CH_3$ [3, 4], $C_5H_4(i-C_3H_7)$ [4]) and $Er(tmhd)_3$ [5] (tmhd = 2,2,6,6-tetramethyl-3,5-heptanedionate). $Er(Cp)_3$ suffers from a very high incorporation of carbon in the deposited films. If AsH_3 (unsafe!) is used as an arsenic source, a reduction of the carbon incorporation emanating from erbium precursors in the deposited films seems possible [6]. If, however, the preferred As source compound $t-C_4H_9AsH_2$ is used, large amounts of carbon are found in the deposited films, presumably arising from incomplete removal of elemental carbon originating in Cp⁻ ligands present in the dopant source compound. Therefore, attention was turned to compounds which do not contain any direct metal-carbon interactions. Recently, we have reported the successful use of a new class of compounds, namely erbium *tris*[*bis*(trialkylsilyl)amides] as source compounds for the doping of semiconducting GaAs by OMVPE [7].

RESULTS AND DISCUSSION

Although *bis*(trimethylsilyl)amides of some of the f-series elements have been reported previously [8], no prior report, to our knowledge, described compounds containing the erbium amide linkage. Transition metal amides have first been described by Bürger *et al.* [9] and are synthesized by the metathesis reaction between a lithium or sodium amide and the appropriate metal halide. We previously have employed zinc *bis*(amides) with great success in the preparation of p-type ZnSe by OMVPE [10] and, therefore turned our attention to the erbium *tris*(amide) Er{N[Si(CH$_3$)$_3$]$_2$}$_3$ (**Figure 1**) as a possible source compound for erbium doping of semiconducting materials. The synthesis of this compound was accomplished by the metathesis reaction of ErCl$_3$ with LiN[Si(CH$_3$)$_3$]$_2$ and purification was carried out following the procedure given in **Table I**.

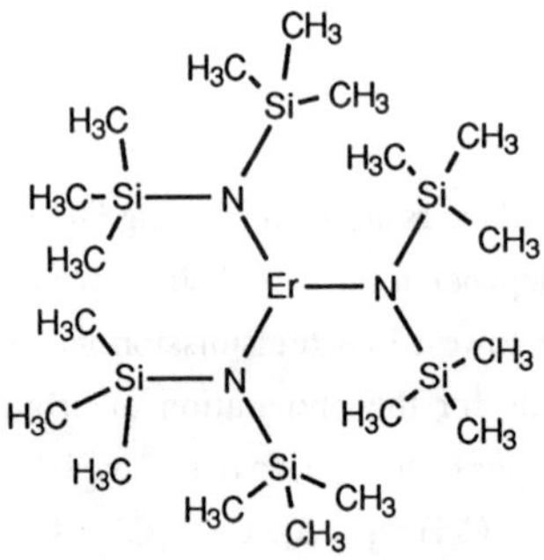

Figure 1: Chemical structure of Er{N[Si(CH$_3$)$_3$]$_2$}$_3$.

Table I: Purification of erbium *tris*(amides).

step	procedure	purification
1	extraction with ether/hexanes	removal of LiCl and unreacted $ErCl_3$
2	recrystallization from ether/hexanes	removal of unreacted $LiNR_2$
3	sublimation at 5×10^{-2} Torr	removal of unreacted $LiNR_2$
4	sublimation at 5×10^{-4} Torr	isolation of volatile material

The erbium *tris*(amide) $Er\{N[Si(CH_3)_3]_2\}_3$ decomposes cleanly with little trace of silicon, nitrogen, or carbon contamination in contrast to $Er(Cp)_3$, which leaves large amounts of carbon in the resulting film [7]. In order to obtain erbium *tris*(amide) precursors with improved characteristics for use in OMVPE, especially compounds with increased volatility, and/or lower melting point, a series of other erbium *tris*(amide) precursors was investigated. The motivation driving the selection of potentially liquid precursors is the occurrence of more reproducible and controllable vapor pressure as a consequence of bubbling versus percolating in the case of ambient condition solid precursors. The starting secondary amines for the preparation of these new erbium *tris*(amides) are closely related to hexamethyldisilazane (**1a**), and are depicted in **Figure 2**. Syntheses of the new erbium *tris*(amides) were carried out using the standard procedure shown in the equations (1) and (2). Purification to a level sufficient for OMVPE was carried out in a four step process, depicted in **Table I**.

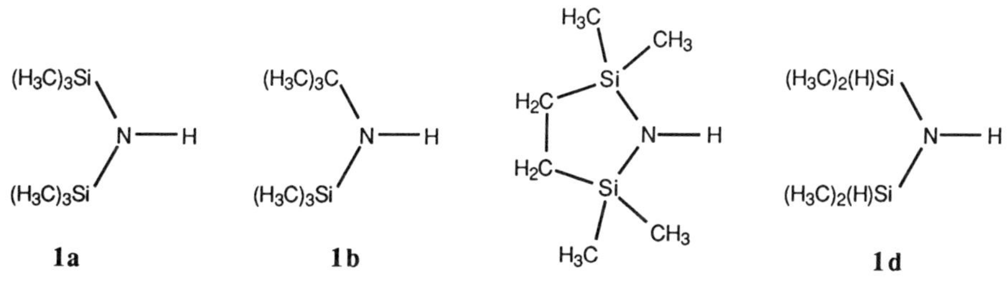

Figure 2: Amines used for the preparation of erbium *tris*(amides).

$$(R)(R')NH + {}^nBuLi \xrightarrow[\text{THF, 2 h}]{-78°C - 25°C} (R)(R')NLi + {}^nBuH \qquad (1)$$
$$\mathbf{1} \qquad\qquad\qquad\qquad\qquad \mathbf{2}$$

$$3\ (R)(R')NLi + ErCl_3 \xrightarrow[\text{THF, 24 h}]{0°C - 25°C} Er[N(R)(R')]_3 + 3\ LiCl \qquad (2)$$
$$\mathbf{2} \qquad\qquad\qquad\qquad\qquad \mathbf{3}$$

3a: R = R' = Si(CH$_3$)$_3$, **3b**: R = C(CH$_3$)$_3$, R' = Si(CH$_3$)$_3$,
3c: N(R)(R') = NSi(CH$_3$)$_2$CH$_2$CH$_2$Si(CH$_3$)$_2$, **3d**: R = R' = Si(H)(CH$_3$)$_2$

One of the most important factors for successful OMVPE is the reproducible and high volatility of the source compounds. To ensure the suitability of all prepared erbium *tris*(amides) for OMVPE processes, sublimation of all compounds and a study of their decomposition by TGA was initiated. Further characterization of the compounds was obtained from FT-IR, and ^{1}H-NMR. The erbium *tris*(amide) **3b** decomposed on attempts to sublime it and an orange-brown involatile material, with a melting point above 320°C, was obtained. Compound **3c** did not sublime at temperatures up to 190°C/10^{-3} Torr and has a melting point of 240°C. Compound **3d** sublimes similarly to **3a** at 130 -140°C and 5 x 10^{-4} Torr. Their melting points also are close (**3a**: 162°C, **3d**: 160°C) and both **3a** and **3d** decompose at temperatures above 250°C, as indicated by TGA (**Figure 3**) and visually observed in the melting point capillary.

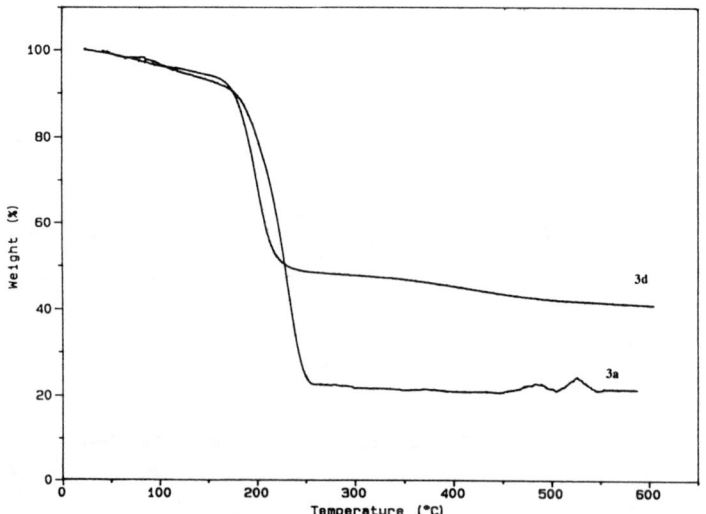

Figure 3: TGA-plot of Er{N[Si(CH$_3$)$_3$]$_2$}$_3$ (**3a**) and Er{N[Si(H)(CH$_3$)$_2$]$_2$}$_3$ (**3d**)
(1 atm., N$_{2(g)}$ flow, 10°C/min).

SUMMARY

From the studies carried out to date, the erbium *tris*(amides) Er{N[Si(CH$_3$)$_3$]$_2$}$_3$ and Er{N[Si(H)(CH$_3$)$_2$]$_2$}$_3$ possess characteristics making them suitable candidates for potential application in OMVPE processes directed at the growth of thin films of erbium-doped semiconducting materials. Previously the successful employment of Er{N[Si(CH$_3$)$_3$]$_2$}$_3$ in the growth of erbium-doped GaAs has been reported by us. Future work is directed at extensions of this new mode of doping.

ACKNOWLEDGMENTS

We gratefully acknowledge the Air Force Office of Scientific Research for partial financial support of this project under the umbrella of the SDIO Small Business Innovative Research Program. U.W.L. thanks the Deutsche Forschungsgemeinschaft (DFG) for a Postdoctoral Fellowship.

REFERENCES

1. Rare Earth Doped Semiconductors, edited by G.S. Pomrenke, P.B. Klein, D.W. Langer (Mater. Res. Soc. Proc. **301**, Pittsburgh, PA, 1993).
2. K. K. Uwai, H. Nakagome, K. Takahei, Appl. Phys. Lett. **51**, 1010 (1987).
3. K. K. Uwai, H. Nakagome, K. Takahei, J. Crystal Growth **93**, 583 (1988).
4. J. Weber, M. Moser, A. Stapor, F. Scholz, G. Bohnert, A. Hangleiter, A. Hammel, D. Wiedmann, J. Weidlein, J. Crystal Growth **104**, 815 (1990).
5. A.J. Neuhalfen, B.W. Wessels, Appl. Phys. Lett. **59**, 2317 (1991).
6. F. Scholz, J. Weber, K. Pressel, A. Dörnen, in Rare Earth Doped Semiconductors, edited by G.S. Pomrenke, P.B. Klein, D.W. Langer (Mater. Res. Soc. Proc. **301**, Pittsburgh, PA, 1993) pp. 3-13.
7. A.C. Greenwald, W.S. Rees, Jr., U.W. Lay, in Rare Earth Doped Semiconductors, edited by G.S. Pomrenke, P.B. Klein, D.W. Langer (Mater. Res. Soc. Proc. **301**, Pittsburgh, PA, 1993) pp. 21-26.

8. D.C. Bradley, J.S. Ghotra, F.A. Hart, J. Chem. Soc., Dalton 1973, 1021.
9. H. Bürger, U. Wannagat, Monatsh. Chem. **94**, 1007 (1963).
10. (a) W.S. Rees, Jr., D.M. Green, W. Hesse, T.J. Anderson, B. Pathangey, in <u>Chemical Perspectives of Microelectronic Materials</u>, edited by C.R. Abernathy, C.W. Bates, Jr., D.A. Bohling, W.S. Hobson (Mater. Res. Soc. Proc. **282**, Pittsburgh, PA, 1993) pp. 63-67. (b) W.S. Rees, Jr., D.M. Green, T.J. Anderson, E. Bretschneider, B. Pathangey, J. Kim, J. Electronic Materials **21**, 361 (1993). (c) W.S. Rees, Jr., T.J. Anderson, D.M. Green, E. Bretschneider; in <u>Wide Band-Gap Semiconductors</u>, edited by T.D. Moustakas, J.I. Pankove, Y. Hamakawa, (Mater. Res. Soc. Proc. **242**, Pittsburgh, PA, 1992) pp. 281-286. (d) W.S. Rees, Jr., D.M. Green, W. Hesse; Polyhedron **11**, 1697 (1992). (e) W.S. Rees, Jr., O. Just, presentation at the MRS fall meeting Boston 1993, W2.9.

TETRAALKYLDIARSINES AS POTENTIAL PRECURSORS FOR ELECTRONIC MATERIALS: SYNTHESIS AND CHARACTERIZATION OF VARIOUS *ISO*-PROPYL ARSENIC COMPOUNDS

LAWRENCE F. BROUGH, LIU GANG, MATTHEW A. LIPKOVICH, THOMAS J. COLACOT, VIRGIL L. GOEDKEN,‡ AND WILLIAM S. REES, JR.*
Department of Chemistry and Materials Research and Technology Center, The Florida State University, Tallahassee, Florida 32306-3006

ABSTRACT

Tetrakis(*iso*-propyl)diarsine was synthesized by the reaction of $(i\text{-Pr})_2\text{AsLi}$ with $(i\text{-Pr})_2\text{AsI}$. The lithium salt of the secondary arsine was produced following deprotonation of $(i\text{-Pr})_2\text{AsH}$, obtained by reduction of $(i\text{-Pr})_2\text{AsI}$, which was prepared by the thermolysis of $(i\text{-Pr})_3\text{AsI}_2$. The X-ray crystal structure of $[(i\text{-Pr})_3\text{AsI}][I]$ has been determined on the product of the reaction of $(i\text{-Pr})_3\text{As}$ and I_2. Compounds of the general form $E=As(i\text{-Pr})_3$ (E = O, S, Se) have been prepared.

INTRODUCTION

Several recent publications have discussed the potential of utilizing *tetrakis*(alkyl)diarsines as *in situ* sources of arsenic for employment in the preparation of arsenic-containing electronic materials.[1] Such routes to, for example, GaAs presently rely on toxic and highly volatile precursors (*e.g.*, AsH_3). The motivation for this new approach is depicted in equation 1. Although the molecularity of the arsenic-containing product is uncertain, such routes have seen some moderate success. The precise stoichiometry of the reaction product may be 1/2 (As_4) or 2 As; however, the confirmation of this synthetic procedure for *in situ* generation of elemental arsenic-containing species relies less on the accretion coefficient, and more on the lack of any carbon-containing species in the product. Such contamination by the disadvantageous decomposition of pendant organic groups highly is detrimental to the final electronic properties of such materials.

‡Deceased 22 December 1992
*Address all correspondence to this author at: School of Chemistry and Biochemistry and School of Materials Science and Engineering, Georgia Institute of Technology, Atlanta, Georgia 30332.

As one component of a program directed at an examination of the generality of this approach,[1c] several *iso*-propyl arsenic compounds have been prepared. The synthesis and characterization of some of these organometallic compounds is reported in this contribution.

$$3 \ R_2As\text{-}AsR_2 \rightleftharpoons 4 \ AsR_3 + As_2 \qquad (1)$$

RESULTS AND DISCUSSION

Credit is given to Cadet for the initial report of a compound containing a metal-carbon bond.[2] His 1760 paper describes the simplest tetraalkyldiarsine, $(H_3C)_2As\text{-}As(CH_3)_2$, given the trivial name of cacodyl. A variety of methods have been developed over the years for the preparation of diarsines.[3] In order to address the above-described motivation for exploration of tetraalkyldiarsines as potential precursors in the OMVPE growth of high-quality semiconducting materials, *tetrakis*(*iso*-propyl)diarsine was prepared (Scheme I).

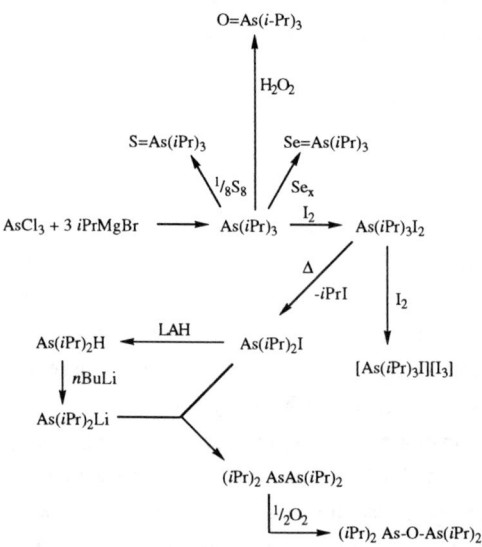

Scheme I

Crystal structures of [(i-Pr)$_3$AsI][I] and [(i-Pr)$_3$AsI] [I$_3$]

The general format for the oxidative addition of a dihalogen molecule to a tertiary organoarsine (equation 2) can proceed according to two alternate pathways.

$$R_3As + X_2 \xrightarrow{\text{Route A}} R_3AsX_2$$
$$\xrightarrow{\text{Route B}} [R_3AsX][X] \quad (2)$$

In Route A, the ~T_d R$_3$As is transformed into the molecular R$_3$AsX$_2$ with a trigonal bipyramidal geometry. In Route B, the resultant product is an [arsonium][halide] ionic moiety. Factors influencing the path of choice include the electronic and spatial considerations of the alkyl groups represented in the tertiary arsine, as well as the $\Delta\chi$ present between the halogen and arsenic. For example, both [Me$_3$PX][X] and Me$_3$SbX$_2$ have been characterized.[4] In all likelyhood, the difluorides are covalently bound. Previous infrared spectroscopic work has indicated that R$_3$AsI$_2$ compounds (R = Me, Et) are ionic in the solid state.[5] In the present investigation, [(i-Pr)$_3$AsI][I] was determined, by X-ray diffraction, to be the conformation adopted by the reaction product of (i-Pr)$_3$As and I$_2$. If the stoichiometry was not carefully controlled (equation 3), a different product was obtained. There have been reports of comparable reactions observed for Ph$_3$As.[6]

$$[(i\text{-Pr})_3\text{AsI}][I] + I_2 \longrightarrow [(i\text{-Pr})_3\text{AsI}][I_3] \quad (3)$$

A single crystal structure determination also was conducted on [(i-Pr)$_3$AsI][I$_3$]. It was, however, suspended prior to complete refinement. Once the identical atomic linkages were ascertained to be present in the organometallic cation, the structural refinement was not pursued further. There was little new structural information revealed in the anion replacement species.

Pertinent crystal and refinement data along with selected interatomic distances and angles will be published elsewhere. An ORTEP representation of the structure is presented (Figure 1), including both covalent and ionic interactions between As and I atoms.

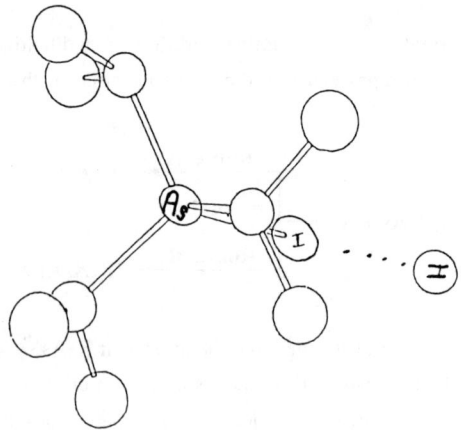

Figure 1 ORTEP representation of [(*i*-Pr)₃AsI][I]
Open circles represent Carbon atoms. Hydrogen
atoms have been omitted for clarity of presentation.

Preparation of E=As(*i*-Pr)₃ (E = O, S, Se)

The preparation of O=As(*i*-Pr)₃ was effected under mild conditions (equation 4).

$$\text{As}(i\text{-Pr})_3 + \tfrac{1}{2} \text{H}_2\text{O}_2 \longrightarrow \text{O=As}(i\text{-Pr})_3 \quad (4)$$

Either a solvent-mediated or a direct reaction can be utilized in the synthesis of S=As(*i*-Pr)₃ (equation 5 - 6).

$$\tfrac{1}{8} \text{S}_8 + \text{As}(i\text{-Pr})_3 \xrightarrow[\text{N}_{2(g)},\ 18\ \text{h}]{\text{THF, r.t.}} \text{S=As}(i\text{-Pr})_3 \quad (5)$$

$$\tfrac{1}{8} \text{S}_8 + \text{As}(i\text{-Pr})_3 \xrightarrow[100^\circ\text{C}]{\text{neat}} \text{S=As}(i\text{-Pr})_3 \quad (6)$$

The most efficient route for the preparation of Se=As(*i*-Pr)₃ was determined to be the reaction of As(*i*-Pr)₃ with elemental selenium (equation 7).

$$\text{As}(i\text{-Pr})_3 + \tfrac{1}{x}\text{Se}_x \xrightarrow[\text{10 min}]{150°C} \text{Se}=\text{As}(i\text{-Pr})_3 \qquad (7)$$

ACKNOWLEDGEMENTS

We acknowledge support of this project by the Office of Naval Research Chemistry Division. L.F.B. was the recipient of a Florida State University Summer Faculty Professional Enhancement Program stipend during the summer of 1990.

REFERENCES

1. a) C. W. Porter and P. Borgstrom, J. Am. Chem. Soc. **41**, 2048 (1919). b) Wide Gap II-VI Semiconductors, W. Kuhn, H. P. Wagner, H. Stanzi, K. Wolf, K. Worle, S. Lankes, J. Betz, M. Woraz, D. Lichtenberger, H. Leiderer, W. Gebhard and R. Triboulet, R. L. Aulombard and J. B. Mullin, eds., (Adam Hilger, Bristol, UK), 1991, p. A105. c) V. L. Goedken, L. F. Brough and W. S. Rees, Jr., J. Organomet. Chem. **449**, 125 (1993).
2. C. Elschenbroich and A. Salzer, Organometallics: A Concise Introduction, (VCH, New York), 1992.
3. G. O. Doak and L. D. Freedman, Organometallic Compounds of Arsenic, Antimony, and Bismuth, (Wiley-Interscience, New York), 1970.
4. a) J. Goubeau and R. Baumgärtner, Z. Electrochem. **64**, 598 (1960). b) E. F. Wells, Z. Kristallogr. **99**, 367 (1938).
5. a) M. H. O'Brien, G. O. Doak, and G. G. Long, Inorg. Chim. Acta **1**, 34 (1967). b) M. H. O'Brien, Ph.D. Thesis, North Carolina State University, 1968.
6. a) K. R. Bhaskar, S. N. Bhat, S. Singh, and C. N. R. Rao, J. Inorg. Nucl. Chem. **28**, 1915 (1966). b) S. N. Bhat and C. N. R. Rao, J. Amer. Chem. Soc. **88**, 3216 (1966).

NEW ZINC-bis(DIALKYLAMIDES) POTENTIALLY USABLE AS SITE-SELECTIVE DOPANTS FOR p-TYPE ZnSe.

WILLIAM S. REES, JR. AND OLIVER JUST.
MS: B-164, Department of Chemistry and Materials Research and Technology Center, The Florida State University, Tallahassee, Florida, 32306-3006.

ABSTRACT

In earlier work, the effective utilization of $Zn\{N[Si(CH_3)_3]_2\}_2$ as a site-selective dopant for the production of p-type ZnSe by OMVPE was demonstrated. Several new zinc-bis(dialkylamides) of the general form $(R)(R')NZnN(R'')(R''')$ now have been prepared. They have been characterized by ^1H- and ^{13}C-NMR, GC/MS and elemental analysis. Vapor pressures and gas phase decomposition profiles have been examined. Correlations of vapor pressure and structure are discussed for this series of compounds. A mechanism for the site-selectivity observed in the incorporation of nitrogen is proposed.

INTRODUCTION

Among the various potential candidates of semiconducting materials for utilization in optoelectronic and electroluminescent devices operating in the blue region of the visible electromagnetic spectrum, p-type ZnSe recently has emerged as a promising leading contender for near-term practical applications.[1-6] Therefore a variety of efforts have been undertaken to gain access to suitable, readily-prepared and purified precursors for the production of this desired composition. Recently, we demonstrated that compounds of the general form $(R)(R')NZnN(R'')(R''')$ retain a zinc-nitrogen bond during their decomposition to produce a thin film of N-doped ZnSe by OMVPE (organometallic vapor phase epitaxy).[7] As a consequence of these promising results, an investigation into the scope of the process was initiated.[6] In previous reports[8,9] it was shown that acceptable results could be achieved either with the homoleptic symmetrical $Zn\{N[Si(CH_3)_3]_2\}_2$ or the homoleptic unsymmetrical $Zn\{N[Si(CH_3)_3][C(CH_3)_3]\}_2$ regarding their suitability as potential dopant sources for the preparation of high quality p-type ZnSe thin films by OMVPE. This success is based on the selective incorporation of nitrogen atoms on the native selenium lattice site, which is enforced by the lack of an appreciable antisite defect density for zinc atoms in the native ZnSe lattice. Thus, in the instances when the Zn-N bond of the precursor dopant molecule is retained intact throughout the deposition process, the nitrogen atom has an enforced residence on a lattice site adjacent to a native zinc location, *i.e.*, in an electronically active position on a native selenium atom lattice site. The proposed mechanism for the gas phase decomposition profile of the heteroleptic $Zn\{[N(Si(CH_3)_3)_2][N(Si(CH_3)_3)(C(CH_3)_3]\}$[10] provided for the existence of the desired penultimate structural fragment "RNZn". Subsequent to undergoing a β-hydride elimination by the appropriate alkyl group, observation of co-product olefin and "HNZn" significantly contributed to the overall understanding of the nitrogen incorporation process occurring during the epitaxial growth of p-type ZnSe. As a consequence of the limited volatility of these prior precursors, a priority was given to extend these studies to other diverse substituted zinc-*bis*(dialkylamides) and an examination of their employment as possible dopants by OMVPE.

EXPERIMENTAL

An effective access to compounds depicted by the general formula $Zn[N(R)(R')]_2$ is provided by a synthetic route[10] comprising the salt elimination reaction between one equivalent of zinc dichloride and two equivalents of alkali metal dialkylamide under inert gas (eq. 1) yielding the target molecules as stable compounds. The preparation of the alkali metal dialkylamide salt was accomplished by treating the corresponding primary amine with trimethylchlorosilane (eq. 2) and deprotonation of the generated trimethylsilylalkylamine with butyllithium or potassium hydride (eq. 3a or 3b):

$$ZnCl_2 + 2\, LiNRR' \longrightarrow Zn[N(R)(R')]_2 + 2\, LiCl \qquad (1)$$

$$2\, R'NH_2 + ClSi(CH_3)_3 \longrightarrow HN(R')[Si(CH_3)_3] + R'NH_2 \cdot HCl \qquad (2)$$

$$HN(R)(R') + BuLi \longrightarrow LiN(R)(R') + BuH \qquad (3a)$$

$$HN(R)(R') + KH \longrightarrow KN(R)(R') + H_2 \qquad (3b)$$

R = $Si(CH_3)_3$; R' = Et, *i*-Pr, *n*-Pr, *t*-Bu, *c*-Hex, Ph.

After demonstrating the usefulness of the homoleptic symmetrical Zn{N[Si(CH$_3$)$_3$]$_2$}$_2$ as a dopant for deposition of p-type ZnSe films, attention was focused toward performing experiments in order to obtain a zinc-*bis*(dialkylamide) containing the closely related tetramethyldisilaazacyclopentane-ligand as a minor modification of the *bis*(trimethylsilylamido)-ligand (eq. 4). The anticipation in seeking this compound was to increase the stability of the intrinsically labile zinc-nitrogen bond under normal conditions by potentially protecting it from nucleophilic oxygen attack. Additionally, the retention of this desired interaction throughout the thermal decomposition is ensured by primary elimination of the fragment [Me$_2$SiCH$_2$CH$_2$SiMe$_2^+$].

$$2\ \begin{array}{c}\text{Si}\\ \text{N-Li}\\ \text{Si}\end{array}\ +\ ZnCl_2\ \xrightarrow{-2\ LiCl}\ \begin{array}{c}\text{Si}\quad\text{Si}\\ \text{N-Zn-N}\\ \text{Si}\quad\text{Si}\end{array}\quad (I) \quad (4)$$

Based on encouraging results obtained during the investigation with *bis*(triorgano-silylsubstituted)amide derivatives of zinc(II), an effort was instigated to examine the potential for extension of a comparable synthetic and purification scheme to encompass *bis*(diorgano-silylsubstituted)amide analogs. As an initial entry into this new class of compounds, the homoleptic derivative of *bis*(dimethylsilyl)amide was selected. Examples of this new class of compounds were isolated by application of an analogous synthetic route (eq. 5) to that depicted in eq. 1. It was postulated that replacement of one of the methyl groups on silicon by a hydrogen atom might have a substantial impact on the properties of this next generation of zinc compounds, particularly with respects to enhancing their vapor pressures.

$$ZnCl_2\ +\ 2\ LiN(R)(R')\ \xrightarrow{}\ Zn[N(R)(R')]_2\ +\ 2\ LiCl \qquad (5)$$

R = Si(H)(CH$_3$)$_2$; R' = *i*-Pr, *t*-Bu, Si(H)(CH$_3$)$_2$.

Additionally, capitalizing on a previously developed and described method[9] it was possible to obtain a new heteroleptic zinc-*bis*(alkylsilylamide); (R = Si(CH$_3$)$_3$, R' = *i*-Pr), (eq. 6a and 6b):

$$ZnCl_2\ +\ LiNR_2\ \xrightarrow{}\ ClZnNR_2\ +\ LiCl \qquad (6a)$$

$$ClZnNR_2\ +\ LiN(R)(R')\ \xrightarrow{}\ Zn[N(R)(R')]_2\ +\ LiCl \qquad (6b)$$

RESULTS AND DISCUSSION

The isolation and purification process for these new generations of zinc-*bis*(dialkylamide) compounds has revealed several characteristic and unconventional properties. Following eq. 1 the isolated compounds are colorless and viscous distillable liquids (R = Et, *i*-Pr, *c*-Hex, Ph) as well as ambient condition colorless solids (R = *t*-Bu, *n*-Pr). The highest vapor pressure to date is displayed by the *i*-propyl representative (b.p. = 56°C / 0.05 Torr). This property diminishes dramatically with increasing size of the ligands: *n*-Pr > *t*-Bu > *c*-Hex > Ph, but remarkably the smallest member of this group of compounds, represented by the ethyl derivative, exhibits in a discrepancy with the expected trend an extremely decreased volatility. This result indicates that the corresponding zinc compound is most likely not a monomer, which is consistent with ^{1}H-NMR studies. An additional feature is that the heaviest substituted zinc amides (R = *c*-Hex, Ph) are colorless and viscous oils, while zinc amides carrying medium size ligands such as *t*-Bu and *n*-Pr are solids. The analogous synthesis of compounds containing the dimethylsilyl group (eq. 5) revealed that the isolated *t*-butyl product possesses (in comparison with the corresponding trimethylsilyl compound) an uncomparably enhanced vapor pressure. On the contrary the volatility of the *i*-propyl as well as the dimethylsilyl representative (both are highly viscous oils), is decreased greatly.

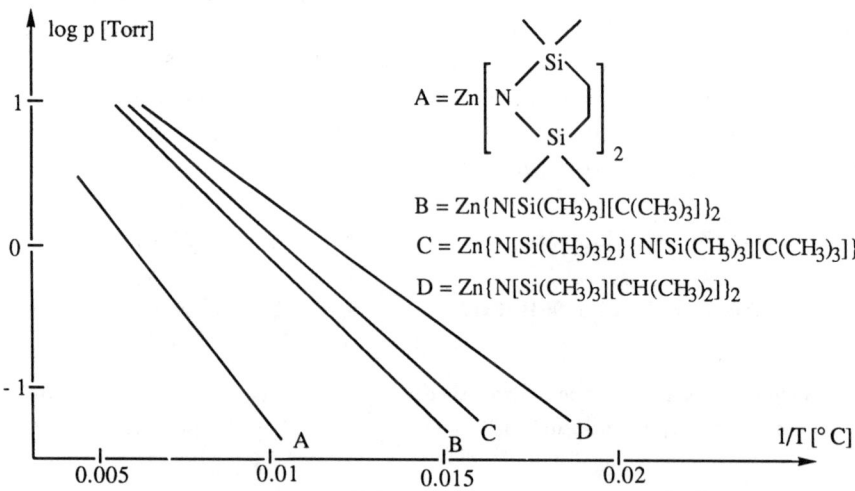

Fig. 1: Vapor pressure profiles determined for a series of various zinc-*bis*(amides).

By the established method (eq. 6a and 6b) separated new heteroleptic i-propyl species is distinguished by its extreme volatility which only slightly varies from that found for the analogous homoleptic product. The symmetrical zinc-bis(dialkylsilylamide) (I) generated by the successfull execution of the reaction outlined in eq. 4 turned out to be a colorless crystalline solid (m.p. = 47°C) and under normal conditions for a limited time the most environmentally stable zinc compound prepared in this study. However, compared with $Zn\{N[Si(CH_3)_3]_2\}_2$, it displays a negligible depressed vapor pressure.

The investigated mechanisms of processes occurring in the gas phase during thermal decomposition of three alternate types of zinc-bis(amide) examples show in two cases the presence of the thermodynamically stable byproduct NR_3 as a driving force for formation of "HNZn" and elimination of "ZnN$_2$", respectively. Since experiments conducted on the heteroleptic unsymmetrical zinc amide[9] have revealed the existence of site-selective deposition of p-type ZnSe in conjunction with essential structural fragments NR_3, RNZn and "HNZn" our investigations on the produced compounds concentrated on searching for these important fragments. The detailed studies of the gas phase decomposition profile of the heteroleptic $Zn\{N[Si(CH_3)_3]_2\}\{N[Si(CH_3)_3][C(CH_3)_3]\}$ have confirmed in the preliminary investigations[9] reported pattern containing N(TMS)$_3$ and ^{t}BuNZn and subsequent to β-hydride elimination, the presence of "HNZn". On the contrary, those signals are absent in the mass spectrum of the homoleptic unsymmetrical $Zn\{N[Si(CH_3)_3][C(H)(CH_3)_2]\}_2$, which is consistent with the requirement for a bimolecular mechanistic pathway. The most characteristic observed fragments which have been split off the parent peak can be interpreted as: [i-Pr$_2$N$^+$] and [i-PrNSiMe$_3^+$].

Due to the well-known stability of the fragment [Me$_2$SiCH$_2$CH$_2$SiMe$_2^+$], the elimination of "ZnN$_2$" has been observed in the decomposition pattern of the homoleptic symmetrical zinc amide depicted in eq. 4. We note, that in other work, erbium[$tris$(trimethylsilyl)amide] has been employed successfully in the OMVPE growth of Er-containing electronic materials.[11,12]

The experiments which were carried out to date on the available zinc-bis(amides) allow the arrangement into three categories regarding their potential suitability for OMVPE:
Precursors which should have utility in the epitaxial growth of p-type ZnSe:
$Zn\{N[Si(CH_3)_3][C(H)(CH_3)_2]\}_2$; $Zn\{[N(Si(CH_3)_3)_2][N(Si(CH_3)_3)(C(H)(CH_3)_2]\}$; $Zn\{N[Si(H)(CH_3)_2][C(CH_3)_3]\}_2$; $Zn\{NSi(CH_3)_2CH_2CH_2Si(CH_3)_3\}_2$ (I).
Products which probably are well suited to achieve this goal:
$Zn\{N[Si(CH_3)_3][n-Pr]\}_2$; $Zn\{N[Si(CH_3)_3][C(CH_3)_3]\}_2$.
Compounds with chemical and physical properties which most likely will not serve this purpose as well as the examples enumerated above:
$Zn\{N[Si(CH_3)_3][CH_3CH_2]\}_2$; $Zn\{N[Si(CH_3)_3][c-Hex]\}_2$; $Zn\{N[Si(CH_3)_3][Ph]\}_2$; $Zn\{N[SiH(CH_3)_2][CH(CH_3)_2]\}_2$; $Zn\{N[SiH(CH_3)_2]_2\}_2$).

ACKNOWLEDGMENTS

We gratefully acknowledge the Office of Naval Research for financial support of this work and the Deutsche Forschungsgemeinschaft (DFG) for granting a post-doctoral research fellowship to Oliver Just.

REFERENCES

1. K. Nakanishi, I. Suemune, Y. Fujii, Y. Kuroda, M. Yamanishi, Appl. Phys. Lett. **59**, 1401 (1991).
2. C.A. Zmudzinski, Y. Guan, P.S. Zory, IEEE Photon. Technol. Lett. **2**, 94 (1990).
3. M.A. Haase, J. Qiv, J.M. DePuydt, H. Cheng, Appl. Phys. Lett. **59**, 1272 (1991).
4. J. Qiv, J.M. DePuydt, H. Cheng, M.A. Haase, Appl. Phys. Lett. **59**, 1992 (1991).
5. M.A. Haase, H. Cheng, D.K. Misemer, T.A. Strand, J.M. DePuydt, Appl. Phys. Lett. **59**, 3619 (1991).
6. H. Jeon, J. Ding, W. Patterson, A.V. Nurmikiko, W. Xie, D.C. Grilla, M. Kobayashi, R.L. Gunshor, Appl. Phys. Lett. **59**, 3619 (1991).
7. W.S. Rees, Jr., T.J. Anderson, D.M. Green, E. Bretschneider, in Wide Band-Gap Semiconductors, edited by T.M. Moustakas, J.I. Pankove, Y. Hamakawa (Mater. Res. Soc. Proc. **242**, Pittsburgh, PA, 1992) pp. 281 - 286.
8. W.S. Rees, Jr., D.M. Green, T.J. Anderson, E. Bretschneider, B. Pathangey, J. Kim, J. Electronic Materials **21**, 361 (1992).
9. W.S. Rees, Jr., D.M. Green, W. Hesse, T.J. Anderson, B. Pathangey in Chemical Perspectives of Microelectronic Materials, edited by C. R. Abernathy et al. (Mat. Res. Soc. Proc. **282**, Pittsburgh, PA, 1992) pp. 63 - 67.
10. W.S. Rees, Jr., D.M. Green and W. Hesse, Polyhedron, **11**, 1667 (1992).
11. A.C. Greenwald, W.S. Rees, Jr., U.W. Lay, in Rare Earth Doped Semiconductors, edited by G.S. Pomrenke, P.B. Klein, D.W. Langer (Mater. Res. Soc. Proc. **301**, Pittsburgh, PA, 1993) pp. 21-26.
12. W.S. Rees, Jr., U.W. Lay, A.C. Greenwald, presentation at the MRS Fall Meeting, Boston, MA, 1993, Abstr. # W2.7.

GROWTH OF InAs AND $(InAs)_1(GaAs)_1$ SUPERLATTICE QUANTUM WELL STRUCTURES ON GaAs BY ATOMIC LAYER EPITAXY USING TRIMETHYLINDIUM-DIMETHYLETHYLAMINE ADDUCT

NOBUYUKI OHTSUKA AND OSAMU UEDA
FUJITSU LABORATORIES LTD., 10-1 Morinosato-Wakamiya, Atsugi 243-01, Japan

ABSTRACT

Atomic layer epitaxy (ALE) of InAs has been developed using trimethylindium-dimethylethylamine adduct (TMIDMEA) as a novel In source. Distinct self-limiting growth of InAs was successfully carried out over a wide temperature range from 350°C to 500°C because of the high thermal stability of TMIDMEA. The possible growth temperature range for ALE-InAs was extended by using TMIDMEA. These results lead us to conclude that the use of TMIDMEA enables us to grow InAs/GaAs heterostructures at a single growth temperature. Using this technique, $(InAs)_1(GaAs)_1$ short period superlattice (12 periods) quantum-well structures were grown on a GaAs(100) substrate at 460°C. A photoluminescence peak at 1.3 μm was observed in these structures at room temperature.

INTRODUCTION

Much attention has been paid to the growth of $(InAs)_m(GaAs)_n$ short-period strained-layer superlattice structures on GaAs. These systems are among the most attractive systems from the view point of application to long-wavelength optical fiber communication devices [1,2].
Atomic layer epitaxy (ALE) is currently the most promising technique for fabricating this system because it can control growth at an atomic level with a self-limiting mechanism. We have previously developed an ALE technique called pulsed jet epitaxy, or PJE, which can grow self-limited III-V compounds under a wide range of conditions [3-6]. But a severe problem arises in fabricating InAs/GaAs superlattices using ALE. It is the difference between the growth temperatures for ALE among the binary semiconductors. We have grown GaAs by PJE using trimethylgallium (TMGa) from 450 to 550°C. We have also grown InAs using trimethylindium (TMIn) from 300 to 400°C. However, there is no overlap in ALE growth temperature for the two materials. It is very important for the fabrication of InAs/GaAs superlattices that each material can be grown in the same temperature region with a self-limiting mechanism [7]. One of the factors determining the growth temperature range for ALE has been the decomposition temperature of source gases for group III elements. From the view point of the crystal quality, an expansion of the temperature range for InAs ALE was required [8]. Therefore, we expected that the thermal stability of TMIn which is a conventional In source might be improved by an addition of some adduct. By analogy with the thermal stability results for AlH_3, we chose $N(CH_3)_2C_2H_5$ as the adduct [9]. Subsequently, we developed a trimethylindium-dimethylethylamine adduct (TMIDMEA) as a new In source for ALE.
In this paper, we describe the growth of InAs on a InAs (100) substrate by ALE using TMIDMEA. This technique provides a significant expansion of temperature range for ALE-InAs. Using this method, we also demonstrate the fabrication of a $GaAs/(InAs)_1(GaAs)_1$[12 periods]/GaAs single quantum well structure on a GaAs (100) substrate. This sample shows a long wavelength (1.3 μm) photoluminescence (PL) peak at room temperature.

EXPERIMENTAL

We used a low-pressure metalorganic vapor phase epitaxy (MOVPE) apparatus with a chimney reactor. This system was designed for PJE, in which source gases are supplied in a fast, pulsed stream. For the gas handling system, we used a fast switching manifold with a pressure-balanced vent-and-run configuration. An RF coil was used to heat the graphite susceptor. We used TMIDMEA (for InAs ALE), TMGa (for GaAs ALE), triethylgallium (TEGa) (for GaAs MOVPE) and arsine (AsH_3) as source materials. Substrates were exactly (100)-oriented InAs and GaAs. We grew InAs under 2000 Pa from 300 to 540°C. The source gases for group III and AsH_3 were alternately supplied to the reactor with H_2 carrier gas, separated by H_2 purging pulses. For ALE, a sequence of "$H_2 \rightarrow$ group III $\rightarrow H_2 \rightarrow AsH_3$" is defined as one ALE cycle. TMIDMEA, TMGa, and TEGa were kept at 19.2°C, 3.0°C and 17.0°C respectively.

The growth thickness of InAs was measured by a surface profiler (DEKTAK 3030) after removing the SiO_2 mask. The optical properties of $(InAs)_1(GaAs)_1$/GaAs single quantum well structures grown on GaAs were evaluated by PL. The PL measurements were performed at room temperature using an Ar^+ laser (λ = 488 nm) as excitation source, 0.75 m SPEX spectrometer and a liquid nitrogen cooled Ge detector.

RESULTS AND DISCUSSION

TMIDMEA is a solid at room temperature with a melting point of 45°C. At 80°C, its vapor pressure is about 1200 Pa. The thermal stability of TMIDMEA might be higher than that of TMIn which is a conventional In source. An additional benefit of TMIDMEA in MOVPE or chemical beam epitaxy (CBE) is its reactivity in air, which is nonpyrophoric, thus enabling safe handling.

We investigated the ALE growth of InAs using TMIDMEA and AsH_3 on InAs substrates. Figure 1 shows the relationship between the group-III pulse duration and the growth thickness per cycle at 460°C. When we use TMIn, the thickness per cycle increases monotonously with the TMIn pulse duration at 460°C. This may be due to an increase in the thermal decomposition rate of TMIn in the boundary layer. However, when we use TMIDMEA, a complete self-limiting growth can be achieved: the growth thickness per cycle saturated at one monolayer of InAs. These results indicates that TMIDMEA can be used to grow InAs by ALE at high temperatures.

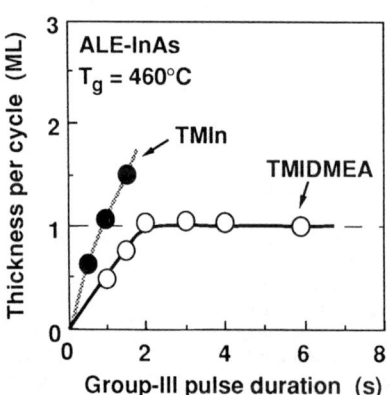

Fig. 1 Growth rate dependence of InAs (100) as a function of group III source pulse duration at 460°C.

It is very important to control the purging time after supplying AsH_3 accurately for InAs ALE using TMIn [10]. We studied the dependence of the InAs growth rate on H_2 purging time in ALE using TMIDMEA. Figure 2 shows the relationship between the H_2 purging time and the growth thickness per cycle at 440°C. In the case of using TMIDMEA, the same results as for

TMIn were observed. Thus, a decrease in the growth rate was observed only at the purging time after supplying AsH$_3$. This might be due to the desorption of As atoms from the growing surface. These results indicate that the desorption of As atoms from the growing surface of InAs at the high temperature range is independent of In source gases. Consequently, we used very short H$_2$ purging time (0.1 s) after supplying AsH$_3$ to avoid As desorption in this temperature range.

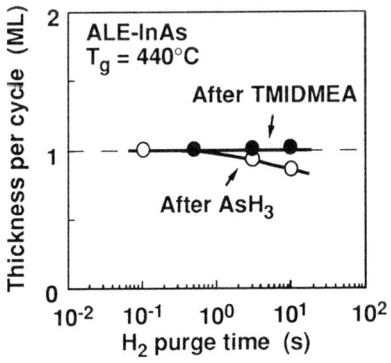

Fig. 2 Growth rate dependence of InAs as a function of H$_2$ purge time at 440°C.

The temperature dependence of growth rate is shown Fig. 3. For InAs ALE using TMIDMEA, a self-limiting mechanism in the growth rate was observed over a temperature range from 350 to 500°C. The upper limit of this temperature region is more than 50°C higher than that using the conventional TMIn In source. Thus, the possible growth temperature range for ALE-InAs was extended by using TMIDMEA. Thus, we can grow InAs and GaAs in the same temperature region with a self-limiting mechanism. These results lead us to conclude that the use of TMIDMEA can enable us to grow InAs/GaAs heterostructures at a single growth temperature.

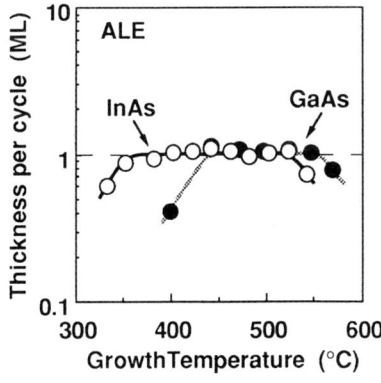

Fig. 3 Growth rate dependence on temperature for InAs grown by ALE using TMIDMEA.

Using this technique, we have successfully fabricated a GaAs/(InAs)$_1$(GaAs)$_1$/GaAs single quantum well structure grown on a GaAs substrate. This structure was grown by the combined use of ALE and conventional MOVPE on a semi-insulating GaAs (100) just substrate. Figure. 4 schematically shows the structure grown in this study. Starting from the substrate, the structure consists of a 500 nm thick undoped GaAs buffer layer, 12 periods (InAs)$_1$(GaAs)$_1$ layer (7 nm) and a 30 nm thick undoped GaAs cap layer. The buffer layer and cap layer were grown by conventional MOVPE at 460°C using TEGa and the (InAs)$_1$(GaAs)$_1$ layer was grown by ALE at 460°C.

The room-temperature PL spectrum of such a single quantum-well structure grown on a GaAs(100) substrate at 460°C by ALE is shown Fig. 5. A PL peak at 1.343 μm with a full width at half maximum (FWHM) of 28.20 meV was observed in this structure. The FWHM of this study was sharper than that previously reported using similar single quantum well structures [11]. The PL intensity using TMIDMEA was stronger than that using TMIn. These might be due to structural differences in the GaAs/(InAs)$_1$(GaAs)$_1$/GaAs single quantum wells. A precise evaluation of the microstructure in the GaAs/(InAs)$_1$(GaAs)$_1$/GaAs single quantum well region is very difficult, but this point does need further clarification.

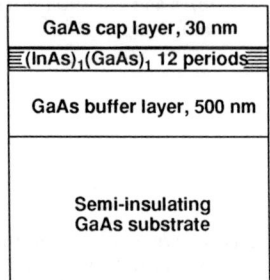

Fig. 4 Schematic of single quantum well structure used this work.

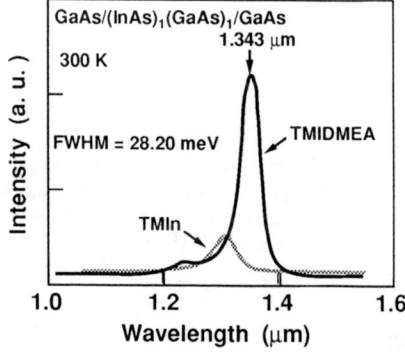

Fig. 5 The room-temperature PL spectrum for (InAs)$_1$(GaAs)$_1$ short period superlattice (12 periods) quantum-well structure grown on a GaAs(100) substrate at 460°C.

SUMMARY

We have developed TMIDMEA as a new In source for ALE-InAs and studied the growth of InAs by ALE using TMIDMEA at temperatures from 300 to 540°C. The thermal stability of TMIDMEA is higher than that of TMIn which is a conventional In source. The possible growth temperature range for ALE-InAs was extended by using TMIDMEA. These results lead us to conclude that the use of TMIDMEA enables us to grow InAs/GaAs heterostructures at the same growth temperature. Using this technique, $(InAs)_1(GaAs)_1$ short period superlattice (12 periods) quantum-well structures were grown on a GaAs(100) substrate at 460°C. A PL peak at 1.3 μm was observed in these structures at room temperature.

ACKNOWLEDGMENTS

We thank K. Nakajima for his continuing encouragement, T. Ashino for help with the PL measurements, and H. Hosoi for their technical assistance.

REFERENCES

1. E. J. Roan and K. Y. Cheng, Appl. Phys. Lett. 59, 2688 (1991).
2. N. K. Dutta, N. Chand, J. Lopata and R. Wetzel, Appl. Phys. Lett. 60, 924 (1992).
3. M. Ozeki, K. Mochizuki, N. Ohtsuka, and K. Kodama, Appl. Phys. Lett. 53, 1509 (1988).
4. M. Ozeki, K. Mochizuki, N. Ohtsuka, and K. Kodama, Thin Solid Films 174, 63 (1989).
5. Y. Sakuma, K. Kodama and M. Ozeki, Japan. J. Appl. Phys. Lett. 27, L2189 (1988).
6. M. Ozeki, N. Ohtsuka, Y. Sakuma and K. Kodama, J. Cryst. Growth 107, 102 (1991).
7. K. Mori and S. Usui, Appl. Phys. Lett. 60, 1717 (1992).
8. D. I. Westwood, D. A. Woolf and R. H. Williams, J. Cryst. Growth 98, 782 (1989).
9. S. M. Olsthoorm, F. A. J. M. Driessen and L. J. Giling, Appl. Phys. Lett. 60, 82 (1992).
10. Y. Sakuma, M. Ozeki and K. Nakajima, J. Cryst. Growth 130, 147 (1993).
11. E. J. Roan, K. Y. Cheng, P. J. Pearah, X. Liu, K. C. Hsieh and S. G. Bishop, Inst. Phys. Conf. Ser. 120, 577 (1991).

NEW ORGANOMETALLIC SELENIUM REAGENTS FOR LOW TEMPERATURE OMCVD OF ZnSe

M. DANEK[*a], J-S. HUH[*b], K. F. JENSEN[*c], D. C. GORDON[**], and W. P. KOSAR[**]
[*]Massachusetts Institute of Technology, Departments of Chemistry [a], Materials Science [b], and Chemical Engineering [c], Cambridge, Massachusetts 02139
[**]Advanced Technology Materials, Danbury, Connecticut 06810

ABSTRACT

ZnSe epitaxial films have been grown on (100) GaAs by reduced pressure organometallic chemical vapor deposition (OMCVD) from *tertiary*-butyl(allyl)selenium (tBASe) and dimethylzinc triethylamine adduct (DMZnNEt$_3$) at temperatures of 325-450°C. Good surface morphology, film crystallinity and interface quality have been found with scanning electron microscopy (SEM), double crystal X-ray diffraction (DCD) and Rutherford back scattering spectroscopy (RBS). Secondary ion mass spectrometry (SIMS) shows negligible carbon concentration (below 5x10^{17} atoms/cm^3). Low temperature photoluminescence (PL) exhibits a strong near band-edge emission with a dominant donor-bound peak. Gas-phase pyrolysis of tBASe has been probed at reduced pressure in a molecular beam mass spectrometric system in hydrogen and deuterium carrier gases. The precursor decomposes above 200°C by ß-hydrogen elimination and by homolysis of the Se-C bonds. High isobutene *vs.* isobutane ratios (50-100) indicate a predominance of ß-hydrogen elimination over homolysis at temperatures below 400°C. Diallylselenium is present in the gas-phase in low concentrations at temperatures of 200-350°C. Diallylselenium, methylallylselenium and dimethyl-selenium have been observed as minor by-products during pyrolysis of co-dosed tBASe and DMZnNEt$_3$. The effect of the retro-ene decomposition pathway of allylselenium reagents on carbon incorporation into ZnSe films is further probed by growth experiments with *in situ* generated 2-methylpropaneselenal.

INTRODUCTION

Epitaxial films of ZnSe are promising materials for short wavelength optolectronic devices, such as green and blue light emitting diodes and lasers. Although ZnSe films of excellent optoelectronic quality can be grown by organometallic chemical vapor deposition (OMCVD) from hydrogen selenide and dimethylzinc (DMZn) or its triethylamine adduct (DMZnNEt$_3$), a parasitic gas-phase reaction between the precursors results in a poor surface morphology and thickness uniformity [1]. In addition, the extreme toxicity of H$_2$Se, in combination with a high vapor pressure, imposes serious safety problems. A large variety of organoselenium compound has been tested as replacement for H$_2$Se [2]. However, the high temperatures required for the growth with organoselenium precursors, typically above 450°C, are detrimental to the optoelectronic properties of the ZnSe films. Particular attention has therefore been paid to organoselenium reagents with thermally labile alkyl groups, which would facilitate low temperature growth. Methylallylselenium (MASe) and diallylselenium (DASe) allow growth of epitaxial ZnSe at 400-450°C [3,4]. Unfortunately, the films are heavily contaminated with carbon. A gas-phase pyrolysis study under OMCVD growth conditions has shown that a retro-ene decomposition pathway of the allyl precursors, resulting in formation of reactive selenoaldehydes, is likely the source of the carbon incorporation [5].

We report on a new allyl-based selenium precursor, *tertiary*-butyl(allyl)selenium (tBASe), which has been designed to decompose at low temperatures by homolysis of the Se-C bonds or ß-hydrogen elimination. Since the t-butyl substituent does not facilitate the retro-ene rearrangement, low carbon levels in the films would be expected. Alkyl redistribution under growth conditions, however, complicates the pyrolysis mechanism by the formation of DASe and MASe. Since the presence of these compounds in the gas-phase might be critical to the film quality, we have carried out extensive OMCVD growth experiments to assess the feasibility of tBASe for preparation of high optoelectronic quality ZnSe films. The material characteristics are correlated with results of gas-phase pyrolysis conducted under the growth conditions. In order to further explore the effect of the selenoaldehydes on the composition and properties of ZnSe films, we have included bicyclo-[2.2.1]-2-selena-3-(2-methyl)ethylhept-5-ene (BCSe) into our study. This compound yields 2-methylpropaneselenal under pyrolysis conditions, a species closely related to the intermediate selenoaldehydes formed from DASe and MASe.

EXPERIMENTAL

ZnSe films were grown in a vertical downflow OMCVD reactor equipped with a laser interferometer for *in-situ* growth rate monitoring [6]. Semi-insulating (100) GaAs substrates misoriented 2° towards <110> were prepared according to standard procedures [7]. The native oxide was removed at 600°C in hydrogen flow containing 0.7 Torr partial pressure of H_2Se to passivate the GaAs surface. The films were grown at reactor pressure of 300 Torr. The delivery rates of the selenium organometallics (Advanced Technology Materials) were varied between 10-120 µmol/min, while the delivery rate of $DMZnNEt_3$ (Epichem) was maintained at 20 µmol/min in all experiments. The total flow rate of the hydrogen carrier gas was 1000 standard cm^3/min (sccm).

Surface morphology of the deposited films was examined by scanning electron microscopy. X-ray rocking curves were measured by a double crystal diffractometer (BEDE Scientific, Inc., Model 300) with $CuK_{\alpha 1}$ radiation from a rotating anode X-ray generator (Rigaku, Model RU-200). Rutherford back scattering spectrometry with 2 MeV helium ions was employed to further probe the structural quality of the films. Secondary ion mass spectrometer measurements of carbon concentrations were carried out on a Cameca IMS-4f spectrometer using a cesium primary ion beam with an incident energy of 10keV and ion current 33 nA rastered over an area of 100x100 µm. Undoped ZnSe films implanted with ^{12}C were used as internal standards for determination of carbon concentration. Photoluminescence spectra were measured on a 0.85m Spex 1404 spectrometer using a He-Cd laser ($\lambda = 325$ nm) as the excitation source.

Gas-phase pyrolysis was carried out at 30 Torr in a molecular beam mass spectrometric system (MBMS) described previously [8]. The delivery rates of the organometallics were maintained at 20 µmol/min using hydrogen or deuterium carrier gas with a total flow of 25 sccm. The relative concentrations of the reaction products were determined from intensities at the characteristic *m/z* values. The contribution of the fragmentation patterns of the source compounds was subtracted from the mass spectra and the data were corrected for sampling flux at various susceptor temperatures. Ionization cross sections for isobutane and isobutene were determined in a separate experiment with mixtures of known composition.

RESULTS AND DISCUSSION

Growth Characteristics
The temperature dependence of the growth rate with tBASe is shown in Figure 1. The apparent activation energy of kinetic limited growth is approximately 21 kcal/mol. The transition to the mass transport limited growth is at ~400°C. This value is significantly reduced in comparison to the transition temperatures for DASe (440°C) and MASe (480°C) growth systems under identical experimental conditions [3,4,6].

The films grown from tBASe exhibit a pronounced variation of surface morphology with growth temperature (Figure 2), but little change with VI/II ratio. Films deposited at 325°C are specular and almost featureless even at a high magnification. At 350°C, the films possess a smooth surface with a step-like texture. Above 400°C, the films appear dull and the surface is composed of hexagonal features of nearly uniform size.

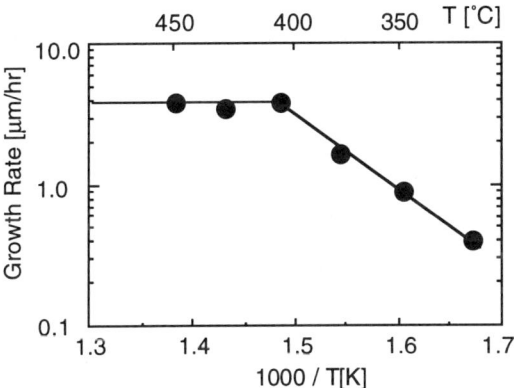

Figure 1. Arrhenius plot for the growth rate of ZnSe from tBASe at reactor pressure 300 Torr and VI/II = 2.

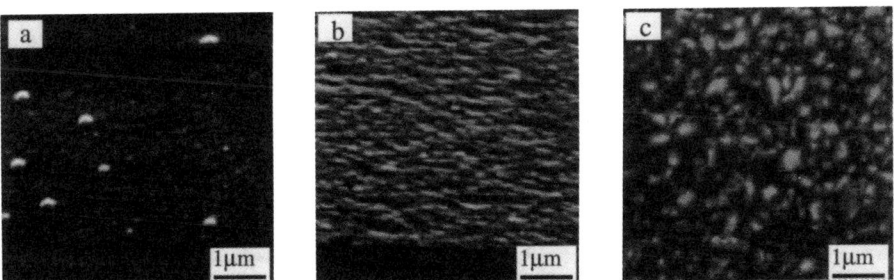

Figure 2. SEM micrographs of the surface morphology of ZnSe films grown at 325° (a), 350°C (b) and 400°C (c) (VI/II = 2).

X-ray diffraction has confirmed that all the films grown at temperatures of 325-450°C and VI/II ratio of 1-6 are epitaxial. Narrow DCD rocking curves for the (004) diffraction with FWHM ranging from 220 arc/sec to 400 arc/sec indicate that tBASe allows growth of films with a low density of structural defects. Rutherford back scattering spectra in channeling configuration along <001> further demonstrate an excellent crystallinity of the films and good quality of the epitaxial interface (Figure 3). An increase of the minimum background yield, $\chi_{min} = Y_{channeled}/Y_{random}$, from 7% to 11% can be observed with the increase of the film thickness from 0.2 µm to 1.2 µm. This result may be interpreted in terms of relaxation of the ZnSe/GaAs misfit trough dislocation formation [9].

Photoluminescence spectra (PL) acquired at 10 K exhibit a strong near band-edge emission and a very weak emission from self activated centers around 2.3 eV (Fig. 4). The near band-edge PL consists of doublets of free exciton (E_x^{lh} and E_x^{hh}) at 2.8004 eV and 2.8026 eV, and donor bond exciton (I_2^{lh} and I_2^{hh}) at 2.7954 eV and 2.7970 eV. The near band-edge PL spectra are similar to those previously reported for ZnSe grown from DMZnNEt$_3$ and H$_2$Se [10].

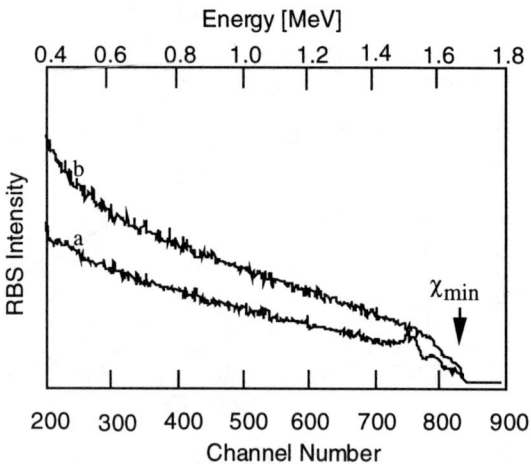

Figure 3. RBS spectra measured in the channeling configuration perpendicular to the ZnSe/(100) GaAs interface for film thickness of 0.2 μm (a) and 2.0 μm.(b).

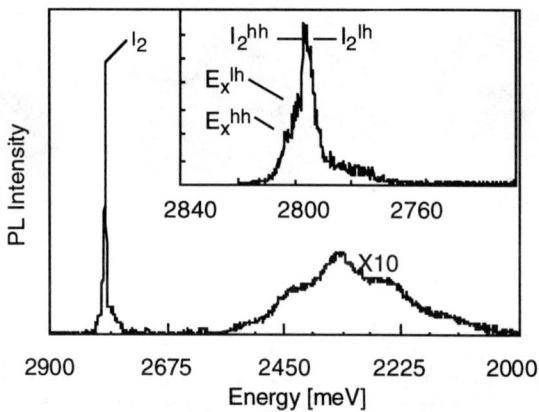

Figure 4. Low temperature (10 K) photoluminescence spectra of ZnSe grown at 350°C and VI/II = 2. The inset shows the near band-edge emission.

Secondary ion mass spectrometry (SIMS) analysis has shown a carbon concentration below the detection level of 5×10^{17} atoms/cm^3 for all the films grown at VI/II ratios of 1-6. It is worthy to point out that MASe and DASe yield heavily contaminated films with carbon level up to 10^{21} atoms/cm^3 under similar growth conditions [3,4,6].

Gas-Phase Decomposition of tBASe

Pyrolysis of tBASe was probed in the MBMS system in hydrogen at 30 Torr. The precursor decomposes above 200°C with apparent activation energy of ~20 kcal/mol. The decomposition is complete by 350°C (Figure 5). The major reaction products observed in the gas-phase are isobutane (m/z=43), isobutene (m/z=56), propene (m/z=42), and 1,5-hexadiene (m/z=67). The intensities of propene and isobutene increase sharply with the onset of the pyrolysis, but drop above 300°C. Isobutane and 1,5-hexadiene formation becomes favorable above 350°C. Traces of DASe can be detected in the 200-350°C range, with a peak concentration at 240°C.

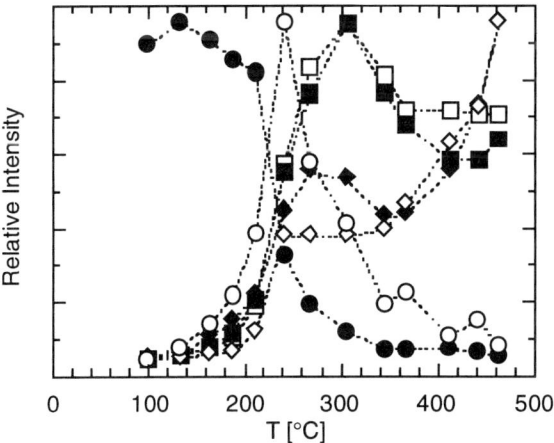

Figure 5. Pyrolysis of tBASe in H_2 at 30 Torr. Relative concentrations of species in the gas-phase as a function of susceptor temperatures. The following symbols are used : ● parent (m/z=178), ◻ isobutene (m/z=56), ◆ isobutane (m/z=43), ◊ 1,5-hexadiene (m/z=67), ■ propene (m/z=42), ○ DASe (m/z=162)].

In order to determine whether the hydrogen carrier gas participates in the gas-phase reactions, pyrolysis in deuterium has been carried out under the same experimental conditions. The presence of hydrogen radicals in the gas-phase, or activation of dihydrogen on the selenium-rich surface of the graphite susceptor, should result in isotope exchange. However, neither formation of deuterated hydrocarbons nor isotope exchange leading to HD have been observed at pyrolysis temperatures up to 450°C. This result demonstrates that the carrier gas is not directly involved in the pyrolysis mechanism of tBASe under the reduced pressure conditions.

The proposed mechanism for pyrolysis of tBASe under reduced pressure is shown in Scheme I below. The primary steps of the tBASe decomposition can be ß-hydrogen elimination and homolysis of the Se-C bonds. The ß-hydrogen elimination pathway (**A**) is facilitated by the t-butyl group. The expected products are isobutene and allylselenol. At elevated temperatures selenols may undergo fission of the Se-C bond or bimolecular hydrogen abstraction in analogy with pyrolysis of thiols [11]. Therefore, allylselenol is likely a reactive intermediate that cannot easily be observed. The homolysis pathway (**B**) should be accessible at relatively low temperatures since t-butyl and allyl groups may yield highly stabilized free radicals. However, a significantly lower decomposition temperature of tBASe, compared to that for DASe [5], indicates that the homolysis of the Se-C(allyl) bond is probably not the initial pyrolysis step below 400°C. The lability of the t-butyl-Se moiety is consistent with the reduction of the OMCVD growth temperatures of ZnSe with tBASe and di(t-butyl)selenium [12] compared to those for DASe and MASe [3,4,6].

The radical species formed either by decomposition of intermediate allylselenol, or by the homolysis of the Se-C bonds may give rise to a complex gas-phase radical mechanism. Radical recombination accounts for the formation of DASe and 1,5-hexadiene. Other expected products of radical recombination, e.g. 4,4-dimethylpentene, di(t-butyl)selenium and dialkyldiselenides, have not been observed. Disproportionation of t-butyl radicals should result in an equimolar ratio of isobutene and isobutane. The presence of a hydrogen donor, such as allylselenol or hydrogen selenide, would shift this ratio in favor of isobutane. The large isobutene *vs.* isobutane ratios (50-100) observed below 400°C indicate that ß-hydrogen elimination rather than a radical mechanism is dominant in the pyrolysis mechanism at low temperatures.

Scheme I.

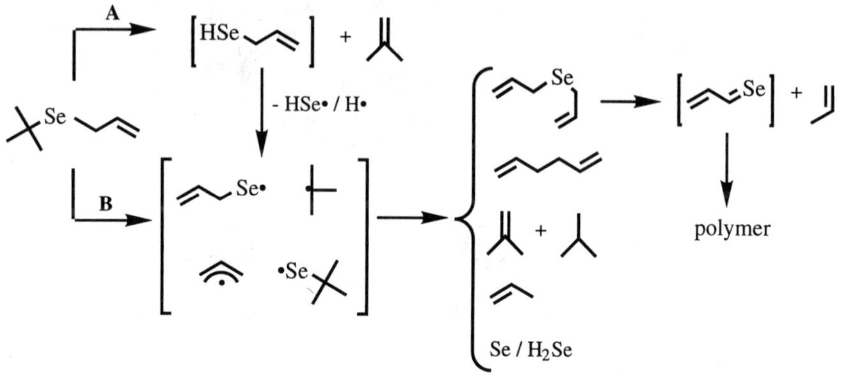

The decay of the DASe signal above 250°C is consistent with the thermal stability of the compound [5]. DASe decomposes by a combination of homolysis of the Se-C bond or by retro-ene rearrangement. The radical pathway yields 1,5-hexadiene and elemental selenium. The rearrangement results in the formation of propene and propeneselenal. The unstable selenoaldehyde will likely polymerize under the pyrolysis conditions. Since the latter decomposition pathway may result in carbon contamination of the films, the concentration of DASe in the gas-phase during the growth will be critical for the film quality.

The presence of DASe in the gas-phase pyrolysis products indicates that the OMCVD chemistry of tBASe may involve formation of other dialkylselenium compounds *via* radical recombination reactions. Indeed, pyrolysis of an equimolar mixture of tBASe and DMZn under the reduced pressure conditions yielded DASe, MASe and dimethylselenium (DMSe) as minor by-products [13]. The growth experiments with MASe and DASe have shown that the carbon level in the ZnSe films can be significantly reduced at low VI/II ratios [6]. Thus, the marginal conversion of tBASe to the allylselenium compounds under the pyrolysis conditions correlates well with the low carbon content in the ZnSe films grown from tBASe, even at VI/II=6. In order to maintain a low carbon level, the films should be grown at a low partial pressure of tBASe to supress the formation of the allylselenium by-products. In addition, elevated carbon levels can be avoided by OMCVD growth at temperatures below 400 °C. At these temperatures, the contribution of MASe and DASe to the growth should be marginal [3,4,6].

Further insight into the mechanism of carbon incorporation during the growth of ZnSe from allyl-based reagent has been gained from pyrolysis and growth experiments with BCSe, as a model compound. The pyrolysis of BCSe under reduced pressure conditions has confirmed a retro-Diels-Alder decomposition pathway, resulting in formation of 2-methylpropaneselenal and cyclopentadiene [13]. The *in situ* generated selenoaldehyde was the actual selenium precursor for ZnSe. Therefore, a high carbon level in the films was expected based on the results reported for MASe and DASe. Surprisingly, the films grown from BCSe contain carbon below 5×10^{17} atoms/cm^3 and show excellent crystallinity and photoluminescence properties [6]. This result indicates that the molecular structure and reactivity of the indermediate selenoaldehydes generated along the retro-ene pathway play an important role in the mechanism of the carbon incorporation.

CONCLUSION

The growth experiments have demonstrated that tBASe is a successfully designed organometallic precursor for low temperature OMCVD growth of high quality ZnSe films. Epitaxial ZnSe can be prepared with acceptable growth rates from tBASe and DMZnNEt$_3$ at temperatures as low as 350°C. The films are free of carbon contamination. The gas-phase pyrolysis study has shown that the reagent is likely to decompose by a combination of ß-hydrogen elimination and homolysis of the Se-C bonds. DASe is formed as a minor intermediate during pyrolysis of tBASe. Pyrolysis of an equimolar mixture of tBASe and DMZn results in formation of DASe, MASe and DMSe. The conversion of tBASe to the allylselenium compounds is sufficiently low so as not to interfere with growth of good quality ZnSe films, even at VI/II=6.

ACKNOWLEDGMENTS

We would like thank to Dr. John Turner, Hewlett Packard, for the SIMS analysis; Dr. A. C. Jones, Epichem, for the Zn source; and the NSF for financial support.

REFERENCES

[1] K. P. Giapis, D. C. Lu, K. F. Jensen, *Appl. Phys. Lett.* **54**, 353 (1989).
[2] A. C. Jones, *J. Cryst. Growth* **129**, 728 (1993).
[3] K. P.Giapis, K. F. Jensen, J. E. Potts, S. J. Pachuta, *J. Electron. Mater.* **19**, 453 (1990).
[4] S. Patnaik, K. F. Jensen, K. P. Giapis, *J. Cryst. Growth* **107**, 390 (1991).
[5] S. Patnaik, K-L. Ho, K. F. Jensen, D. C. Gordon, R. U. Kirss, D. W. Brown, *Chem. Mater.* **5**, 305 (1993).
[6] J-S. Huh, PhD thesis, MIT, 1993.
[7] W. Stutius, *J. Electron. Mater.* **10**, 95 (1981).
[8] P. W. Lee, T. R. Omstead, D. R. McKenna, K. F. Jensen, *J. Cryst. Growth* **85**, 165 (1987).
[9] H. Mitsuhashi, I. Mitsuishi, H. Kukimoto, *J. Cryst. Growth* **77**, 219 (1986).
[10] J,S. Huh, S. Patnaik, and K.F. Jensen, *J. Electron. Mater.* **22**, 509 (1993).
[11] S. W Benson, *Chem Rev.* **78**, 23 (1978); *J. Chem. Soc. Faraday Trans. 2* **83**, 791 (1987).
[12] W. Kuhn, A. Naumov, H. Stanzl, S. Bauer, K. Wolf, H. P. Wagner, G. Gebhardt, U. W. Pohl, A. Krost, W. Richter, U. Dumichen, K. H. Thiele, *J. Cryst. Growth* **123**, 605 (1992).
[13] M. Danek, J-S. Huh, K. F. Jensen, D. C. Gordon, W. P. Kosar, in preparation for publication in *Chem. Mater.*

THE SURFACE CHEMISTRY OF CdTe MOCVD

*WEN-SHRYANG LIU AND **GREGORY B. RAUPP
*Science and Engineering of Materials Program
**Department of Chemical, Bio and Materials Engineering
Center for Solid State Electronics Research
Arizona State University, Tempe, Arizona 85287-6006

ABSTRACT

Temperature programmed desorption (TPD) studies in ultra high vacuum revealed that diethyltellurium (DETe) and dimethylcadmium (DMCd) adsorb weakly on clean Si(100) and desorb upon heating without decomposing. These precursors adsorb both weakly and strongly on CdTe(111)A, with DMCd exhibiting the stronger interaction with the surface than DETe. Dimethylcadmium partially decomposes to produce Cd adatoms; a large fraction of the excess Cd atoms desorb upon heating. In contrast, DETe desorbs without decomposing, suggesting that the rate limiting step in CdTe MOCVD on CdTe(111)A is surface decomposition of the tellurium alkyl.

INTRODUCTION

Metalorganic chemical vapor deposition (MOCVD) is a potentially attractive method for producing epitaxial CdTe or HgCdTe films since this process can in principle be scaled to large substrates and offers the potential for reasonable throughputs [1,2]. The process is more complex than competing processes such as molecular beam epitaxy (MBE) since deposition rate, film thickness uniformity and quality are complex functions of the gas phase transport and homogeneous reactions coupled with the heterogeneous reaction kinetics. Neither the precise nature of the reactions nor the relative importance of the homogeneous *vs.* heterogeneous reaction routes has not been established. Several groups have proposed that gas-phase adduct formation between the Cd and Te precursors or their decomposition fragments is an essential step in the film growth process [3,4]. Hicks has recently tested two reaction schemes using literature deposition rate data and has concluded that heterogeneous reaction steps control the initial decomposition of the precursors [5]. Gas phase reactions between hydrocarbon radicals formed through surface-adsorbed metal alkyl decomposition reactions were thought to be important in determining the gas byproduct distribution and the deposition rate through readsorption on the surface.

In this study we investigate the interaction of diethyltellurium (DETe) and dimethylcadmium (DMCd) with Si(100) and CdTe(111) surfaces using UHV thermal desorption techniques and Auger electron spectroscopy. These experiments were performed under conditions for which homogeneous reactions and byproduct readsorption are negligible, so that the results can be interpreted purely in terms of heterogeneous chemical reactions.

EXPERIMENTAL

Experiments were performed in the ultra high vacuum (UHV) chamber described elsewhere [6,7]. Si(100) substrates were chemically etched to remove surface contaminants and to form a well-defined thin oxide layer prior to introduction into the UHV chamber. The oxide layer was removed *in situ* by heating the sample in UHV to 1173 K for several minutes. Analysis of the pretreated surface with retarding field Auger electron spectroscopy (AES) showed that this pretreatment could reproducibly yield an oxygen-free surface with a trace (less than 3 atomic %) of carbon contamination. Single crystals of polished p-type (16 Ω-cm) CdTe(111) purchased from II-VI Incorporated were cleaned by immersion in E-solution [8] at room temperature for 30 s and then in boiling dithionate solution for 60-180 s prior to introduction into the UHV chamber. The sample was exposed to ultrapure hydrogen at 723 K for 30 minutes and then annealed in vacuum at 623 K for 30 minutes. Auger analyses of the pretreated surface revealed that these treatments produced a nearly stoichiometric CdTe surface with residual O and S impurities.

In a typical TPD experiment the substrate was cooled to a temperature below 180 K and then dosed with a controlled amount of DETe (Morton-Thiokol, 99.995%) or DMCd (Morton-Thiokol, 99.995%). A stainless steel syringe connected to a pressure-controlled 1 liter gas ballast reservoir was used to provide directed dosing of the source gas. Following gas exposure, samples were heated at *ca.* 15 K/s (Si) or 10 K/s (CdTe) using a tungsten filament placed behind the sample as a radiative heat source. Sample temperature was measured with a fine wire chromel-alumel thermocouple spot welded to a small tantalum spring fixed to the edge of the Si or CdTe crystal. Line-of-sight desorption flux spectra were collected using a microcomputer-controlled multiplexed VG Spectralab 1-300 amu mass spectrometer.

RESULTS

DETe and DMCd Interaction with Si(100)

The low temperature range of DETe TPD spectra from Si(100) following DETe exposure at 160 K are shown in Figure 1. The most abundant ion in the DETe cracking pattern, the parent $(C_2H_5)_2Te^+$, was tracked in these experiments. The desorption flux spectra for the lower mass ions were identical in shape and position to those for the parent ion. A single low temperature, asymmetric desorption peak at 177 K at low initial DETe coverage shifts to lower temperature with increasing coverage. Assuming first order desorption kinetics typical of molecular adsorption and desorption, and a first order pre-exponential factor of 10^{13} s^{-1}, we estimate a low coverage desorption activation energy of 43 kJ/mol. This relatively low value is consistent with a physisorption mechanism; the downshift in peak temperature with increasing initial coverage is consistent with the presence of repulsive interactions in the DETe adlayer.

Figure 2 shows the DMCd desorption flux spectra from Si(100) for three different initial DMCd coverages. In these experiments the $(CH_3)Cd^+$ ion (129 amu) was tracked since it's signal is more intense than that for the parent ion. The single desorption peak exhibits an asymmetric peak shape suggesting molecular adsorption/desorption. Unlike DETe, the

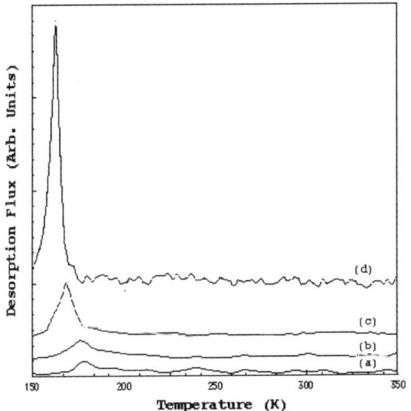

Fig. 1 Thermal desorption of DETe from Si(100) following (a) 4300, (b) 6500, (c) 9100 and (d) 15600 L DETe exposure at 160 K. (1 L = 1 Langmuir = 10^{-6} Torr-s)

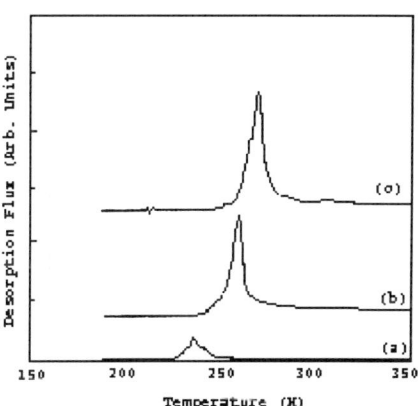

Fig. 2 Thermal desorption of DMCd from Si(100) following (a) 3900, (b) 9100 and (c) 11700 L DMCd at 180 K.

desorption peak maxima shift to a higher temperature with increasing initial coverage, suggesting the presence of attractive interactions between adsorbed DMCd molecules. The estimated desorption activation energy is 57 kJ/mol at low coverage, indicating a stronger interaction between DMCd and Si than between DETe and Si.

DETe and DMCd Interaction with CdTe(111)

The TPD spectra of DETe from the CdTe(111)A surface shown in Figure 3 contain two distinct desorption peak maxima - a narrow, intense peak at approximately 260 K and a broad, less intense peak at about 410 K. The asymmetric shape and nearly invariant peak temperature of the low temperature state are characteristic of first order kinetics, suggesting desorption of a weakly-held chemisorbed molecule. The broad width of the high temperature peak suggests that a distribution of strong molecular adsorption sites for DETe exist on the CdTe surface. Assuming first order kinetics and a 10^{13} s^{-1} pre-exponential factor, estimated desorption activation energies are 66 and 107 kJ/mol for the low and high temperature states, respectively. Figure 4 shows retarding field AES spectra of the surface collected during different stages of a single TPD experiment. The top curve was collected prior to DETe exposure, the middle curve after dosing at low temperature but prior to flashing, and the bottom curve after flashing to 700 K. The 1:0.9 Cd:Te ratio on the original surface decreased to *ca.* 1:3 after DETe exposure, reflecting the presence of adsorbed DETe on the CdTe surface. Following rapid heating to 700K the ratio decreased to 1:0.83. Within the experimental uncertainty caused by low Auger signal:noise, we conclude that the Cd:Te ratio returned to it's original value; *i.e.*, the data are consistent with desorption of DETe without deposition of Te. However, because of the overlap of the Cd and C peaks near 270 eV, it is unclear whether or not C was deposited on the surface.

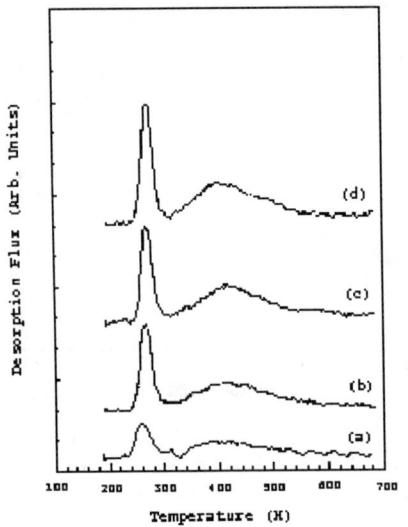
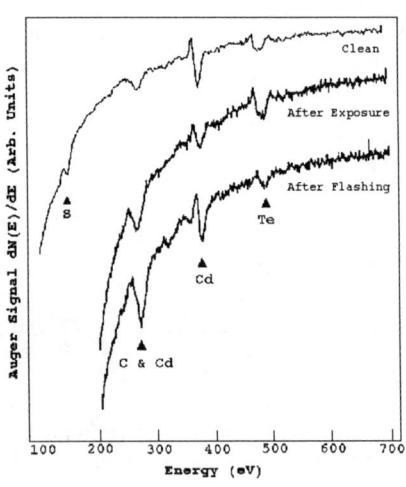

Fig. 3 Thermal desorption of DETe from CdTe(111) following (a) 10400, (b) 14300, (c) 18200 and (d) 23400 L exposure at 180 K.

Fig. 4 AES spectra of CdTe(111) at various stages of a DETe TPD experiment.

Thermal desorption spectra of DMCd from CdTe(111)A summarized in Figure 5 show three distinct desorption features as follows: (i) a low temperature, sharp desorption peak at about 320 K which shifts to higher temperature with increased DMCd exposure, and (ii) a broad peak at about 460 K which overlaps with (iii) a second broad high temperature peak at ~ 540 K. The relative invariance or slight upshift in peak temperature with increasing initial coverage are consistent with molecular adsorption/desorption. Both the high temperature and low temperature states desorb at higher temperatures than the corresponding DETe states on CdTe, revealing a stronger interaction between DMCd and CdTe than between DETe and the CdTe surface. Estimated desorption activation energies are 80, 110 and 135 kJ/mol for the three observed states. Figure 6 shows AES spectra of the CdTe(111) surface recorded following DMCd exposure at 160 K (top curve) and following flashing to 660 K (bottom curve). Following DMCd dosing the Cd:Te ratio increased to 5:1 from the near stoichiometric initial value prior to exposure, reflecting the adsorption of DMCd. Heating to 660 K reduced the ratio to 2.6:1, suggesting that a fraction of the originally adsorbed DMCd decomposed to deposit Cd. This conclusion is supported by comparison of the apparent cracking ratio of the various ion fragments detected by the mass spectrometer during desorption of DMCd from Si(100) and from CdTe(111) as summarized in Table I. The ratio of the mono-methylcadmium to dimethylcadmium ion is essentially identical for the two surfaces, but the signal corresponding to Cd ions is significantly higher for experiments performed with the CdTe surface. On the basis of these measurements in conjunction with the AES characterization of the surface, we conclude that DMCd readily decomposes on CdTe to produce a Cd-rich surface, and that a fraction of these excess Cd atoms desorb into the gas phase upon heating.

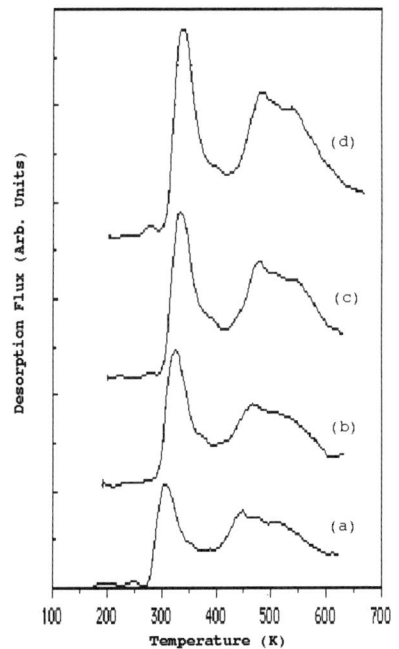

Fig. 5 Thermal desorption of DMCd from CdTe(111) following (a) 10400, (b) 14300, (c) 18200 and (d) 22000 L DMCd exposure at 180 K.

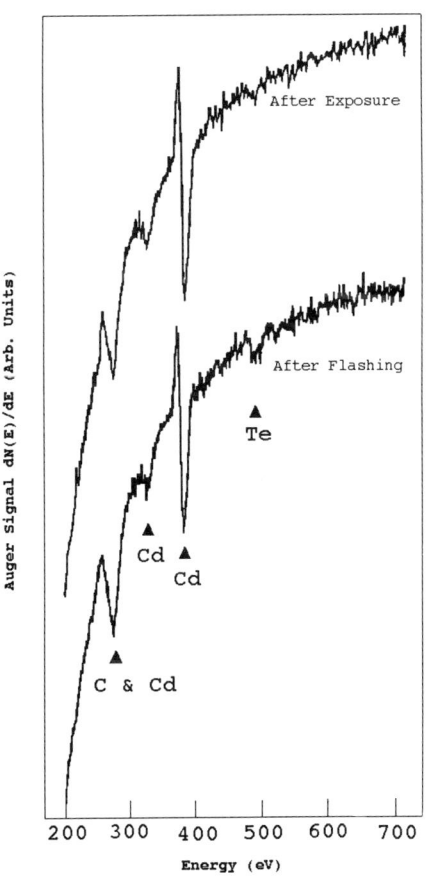

Fig. 6 AES spectra of CdTe(111) at different stages of a DMCd TPD experiment.

TABLE I. Comparison of the Apparent Ion Intensity Patterns of DMCd Desorbed from Si and CdTe

Mass:charge ratio (amu)	Ion Fragment	DMCd / Si(100)	DMCd / CdTe(111)A
114	Cd^+	4.10	9.56
129	$(CH_3)\text{-}Cd^+$	2.50	2.44
144	$(CH_3)_2\text{-}Cd^+$	1.00	1.00

DISCUSSION AND CONCLUSIONS

High metal alkyl exposures (thousands of Langmuirs) were required to achieve a significant extent of adsorption on the CdTe surface, and hence the sticking factors for the source gases on CdTe(111) are low. Although the sticking coefficients could not be quantified, this qualitative finding is consistent with the low measured initial sticking coefficients for dimethylcadmium and dimethyltellerium on GaAs(100) of 7×10^{-3} and 3×10^{-5}, respectively [9]. Liu et al. [10] extracted a sticking factor for DMCd and DETe of 1.5×10^{-4} from deposition rate data on CdTe(100). In spite of it's low sticking probability, DMCd readily decomposes on CdTe with little or no activation barrier to produce Cd atoms. In contrast, DETe does not decompose under the conditions of our experiments. During CdTe MOCVD, it is likely that excess Cd atoms produced through heterogeneous decomposition of DMCd desorb into the gas phase, where they may participate in homogeneous reactions above the substrate surface. It appears that the decomposition of DETe is not catalyzed by the Cd-rich surface, at least in the absence of adsorbed hydrogen [11]. Snyder et al. [12] have observed a near quenching of the CdTe deposition rate on the CdTe(111)A surface in an impinging jet reactor when the carrier gas was switched from hydrogen to helium. We conclude that, at least on this surface, the rate limiting step for MOCVD film growth is heterogeneous decomposition of adsorbed tellurium alkyl.

ACKNOWLEDGMENTS

We gratefully acknowledge the financial support of the National Aeronautics and Space Administration, Office of Space Science and Applications (NASA NAGW-1654).

REFERENCES

[1] T. F. Keuch, *Mater. Sci. Rpts.* 2(1), 1 (1987).
[2] D. W. Hess, K. F. Jensen and T. J. Anderson, *Rev. Chem. Eng.* 3, 97 (1985).
[3] J. B. Mullin, S. J. C. Irvine and D. J. Ashen, *J. Cryst. Growth* 55, 92 (1981).
[4] P. I. Kuznetsov, L. A. Zhuravlev, I. N. Odin, V. V. Shemet and A. V. Novoselova, *Inorg. Mater. USSR* 18, 779 (1982).
[5] R. F. Hicks, *Proc. IEEE* 80(10), 1625 (1992).
[6] G. B. Raupp and G. T. Hindman, in: <u>Tungsten and Other Refractory Metals for VLSI VI</u>, R. S. Blewer and C. M. McConica (MRS, Pittsburgh, PA, 1989) p. 231.
[7] G. T. Hindman, Ph. D. Dissertation, Arizona State University, 1989.
[8] M. Inoue, I. Teramoto and S. Takayanagi, *J. Appl. Phys.* 33, 2578 (1962).
[9] C. D. Stinespring and A. Freedman, *Chem. Phys. Lett.* 143(6), 584 (1988).
[10] B. Liu, A. H. McDaniel and R. F. Hicks, *J. Cryst. Growth* 112, 192 (1991).
[11] R. F. Hicks, National Meeting of the AICHE, St. Louis, MO, November 1993.
[12] D. W. Snyder, P. J. Sides and E. I. Ko, *J. Electrochem. Soc.* 139(7), L66 (1992).

NANOSECOND AND FEMTOSECOND UV LASER ABLATION OF CdTe (100): TIME-OF-FLIGHT AND ANGULAR DISTRIBUTIONS

P.D. Brewer[*], M. Späth, and M. Stuke
Max-Planck Institut für biophysikalische Chemie, Abteilung Laserphysik,
Am Fassberg, D-3400 Göttingen, Germany.

ABSTRACT

Angularly resolved time-of-flight (TOF) measurements have been used to probe the velocity and angular distributions of Cd atoms and Te_2 molecules ejected from CdTe (100) substrates under irradiation by 248 nm nanosecond and sub-picosecond laser pulses. These experiments employ a dye laser TOF mass spectrometer with resonance enhanced multiphoton ionization for sensitive, high resolution detection of the desorbed products. The velocity distributions are well described by Maxwell-Boltzmann distributions for low fluence nanosecond (<60 mJ/cm^2) and sub-picosecond (<3.3 mJ/cm^2) pulses. Angular flux distributions for nanosecond irradiation are observed to be highly forward peaked about the surface normal, whereas, for sub-picosecond irradiation the distribution approaches $\cos^3\theta$.

INTRODUCTION

Pulsed-laser ablation has become a successful technique for surface preparation and deposition [1-4], however little is known about the detailed mechanisms and pathways operative in the ablation process. Ultrashort UV laser pulses offer several advantages (e.g. lower fluence ablation thresholds) compared to normal UV excimer laser irradiation [5] and ablation of UV transparent samples [2], including the possibility of time resolved measurements using optical autocorrelation techniques [6]. The pulsed-laser ablation of compound semiconductors using above band-gap radiation has been the subject of numerous studies [1,7,8]. However, a comprehensive description of the phenomena has been hindered due to complications arising from the effects of near surface collisions on the desorption dynamics [7,9], and the influence of structurally and compositionally modified surfaces on the ablation process [1]. It is well established that in order to derive detailed information about the surface-particle interactions from ejected products, multiple gas phase collisions must be eliminated [11-13]. As a rule of thumb for gas densities comparable to ablation of 0.5 monolayer in 10 nsec, a limited number of gas phase collisions are known to occur and obscure the nascent velocity and angular distributions [11-13]. Multiple gas phase collisions have lead to inconsistencies among experimental results in which the ejected products exhibit angularly varying translation temperatures [1] and vastly dissimilar translational temperatures for different mass species [1,4]. The ablation process is generally further complicated by fluence-dependent changes in the surface composition and structure [1,7-9]. The effect of this damage on the ejected particles may be manifested by non-stoichiometric and unstable yields, and complicated flux characteristics due to bulk diffusion processes.

In an attempt to identify a material system for use in developing a comprehensive model of the pulsed laser ablation of compound semiconductors, we have systematically studied the KrF excimer laser ablation of CdTe (100) [1]. It has been demonstrated, using nanosecond pulses over a limited fluence range, that stoichiometric and atomically ordered surfaces can be maintained after multiple pulse ablation. One important finding of this work is that the composition and structure of CdTe (100) can be reversibly controlled by excimer laser irradiation [3]. In effect damaged surfaces (i.e. non-stoichiometric surface composition and amorphous surface layers) can be removed and the surface restored to an ordered and stoichiometric condition by low fluence laser irradiation. These results are in sharp contrast to other material systems such as CdS, InP(100) and GaP(111) in which irreversible damage to the

---

*Permanent Address: Hughes Research Laboratories, Malibu, California, USA.

surface occurs under all conditions above the ablation threshold [7,9,17].

In this paper we investigate the effects of laser pulse duration on the time-of-flight distributions and angular distributions of the ejected products from CdTe (100) using above band-gap radiation (248 nm). These experiments employ a dye laser time-of-flight system with resonance enhanced multiphoton ionization for sensitive, high resolution measurements of the ablated products. We will show for nanosecond and femtosecond irradiation of the CdTe (100) surface that the time-of-flight distributions of the ejected products are well described by Maxwell-Boltzmann distributions and that a clear deviation from Knudsen laws are found for the angular flux distributions. These results support the position that although thermal effects contribute to the surface processes, non-thermal desorption pathways dominate much of the ablation phenomena.

EXPERIMENTAL

A schematic of the experimental apparatus is shown in Figure 1. The UHV system consists of a main chamber (10^{-7} mbar), into which the substrates are introduced and a differentially pumped linear time-of-flight (TOF) mass spectrometer [15]. A nanosecond KrF excimer laser (16 nsec, 248 nm) and an ultrafast UV laser (600 fsec, 248 nm) [16] are used for ablating the samples. The ionizing laser beam, generated from an excimer laser pumped dye laser (Lambda Physik FL3002), was apertured and focused above the sample surface. The ablation products were detected by resonance enhanced ionization at 228.8 nm for Cd atoms and 419 nm for Te_2 dimers. Ions generated at the focus are extracted through a grid by a weak electric field (-400 V/cm) and then accelerated (2200V/cm) before entering a field-free drift section serving for mass discrimination. Ions are detected by a tandem channel plate and the resulting signal is preamplified and transferred to a Tektronix digital signal analyzer (DSA 602). The TOF mass spectrometer is also highly sensitive to the emission of ions from the ablated surface since they enter the extraction region of the ion optics. Typically 25-100 shot averaging of the mass spectrum was required for suitable signal-to-noise levels. Data was ultimately transferred to a Macintosh computer for analysis.

High-resistivity single-crystal CdTe (100) samples were used in this study. All samples were

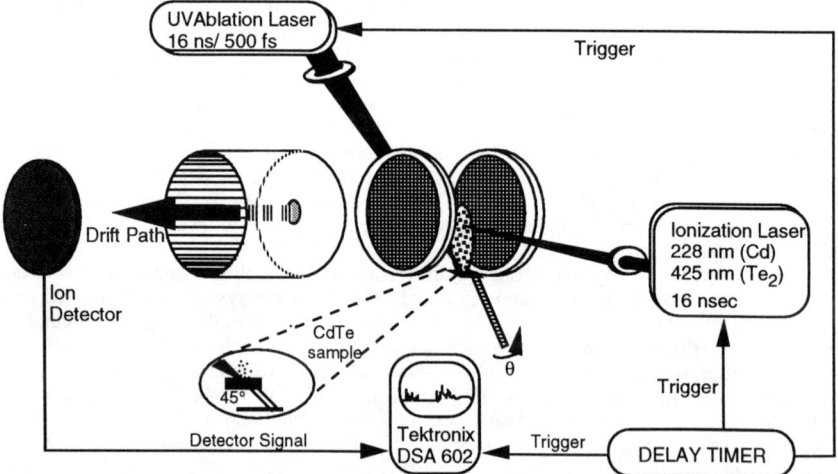

Figure 1. Experimental setup for the angular flux distribution measurements of ejected products from the UV laser ablation of CdTe (100).

degreased and chemically etched with a bromine/methanol solution (1/16%) to remove contaminants and native oxides just prior to introduction into the vacuum system. The CdTe substrates are bonded to a sample holder which is positioned 0.5-3.5 cm below the axis of the linear TOF mass spectrometer under the ion optics. The time-of-flight distributions of the particles ejected along the surface normal are measured by systematically varying the delay times between the ablation and ionization lasers with the ablation laser aligned at 90° to the sample surface. Angular resolved measurements employ a different geometry in which the ablation beam is positioned collinear to the rotational axis of the sample holder and at a compound angle of 45° with respect to the surface normal (see Figure 1). The angular distributions are measured by rotating the sample holder.

RESULTS AND DISCUSSION

The ablation yield of CdTe (100) per pulse for nanosecond and sub-picosecond laser irradiation as a function of laser power is shown in Figure 2. The yields were determined by measuring the depth of material removal after multiple pulse irradiation by stylus profilometry and averaging over the total number of shots. TOF mass spectrometry measurements of the ejected product flux indicate that the removal rate per pulse is constant for multiple pulse exposure at a constant fluence for both nanosecond (<40 mJ/cm^2) and sub-picosecond (<3.3 mJ/cm^2) lasers. It has been found using Auger electron spectroscopy (AES) and reflection high energy electron diffraction (RHEED) that the surface structure [14] and composition [1] of CdTe (100) is not altered by multiple-pulse 248 nm nanosecond irradiation for fluences <40 mJ/cm^2.

The effects of collisions on the velocity and angular flux distributions of ejected particles generated thermally using nanosecond pulses have been observed experimentally [7] and treated theoretically [11-13]. For most ablation experiments the yield of ejected particles is so high that multiple gas phase collisions occur in the near surface region. These collisions result in ablated fluxes exhibiting angular dependent translation temperatures and modified angular distributions. For CdTe (100) exposed to nanosecond irradiation under low fluence (15-40 mJ/cm^2) conditions, the yield is sufficiently low (0.06 to 0.5 monolayers per 10 nsec) that collisions should not influence the velocity or angular distributions. For ablation using sub-picosecond irradiation the collisionless regime is dependent on the desorption time and therefore has not yet been established.

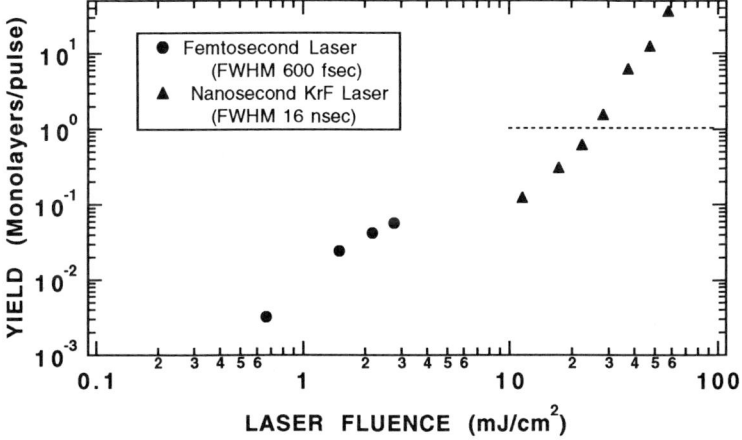

Fig. 2 Yield of removed material as a function of laser fluence for CdTe (100). The dashed line indicates the limit below which post-desorption collisions can be neglected for nanosecond pulse regime.

A dye laser TOF mass spectrometer was used to identify the desorption products and to measure the time-of-flight distributions of the species ejected from the CdTe (100) surface. The only products detected when the CdTe (100) surface is irradiated by low fluence nanosecond (<60 mJ/cm^2) or sub-picosecond (<3.3 mJ/cm^2) pulses are neutral Cd atoms and Te$_2$ molecules. A threshold is observed for ejecting of ionized species when the CdTe surface is irradiated at higher fluences.

Typical TOF spectra are shown in Figure 3 for neutral Cd atoms and Te$_2$ molecules ejected from the CdTe(100) surface by nanosecond (21 mJ/cm^2) and sub-picosecond (3.3 mJ/cm^2) irradiation. The signal intensity from the laser ionization is plotted as a function of delay time between the 248 nm ablation laser pulse and the dye laser pulse. The experimental data are fit by the solid curve using a Maxwell-Boltzmann distribution expressed in terms of the ablated flux dn/dt as a function of delay time **t**:

$$dn/dt = Ct^{-4}\exp(-ml^2/2kT_{trans}t^2) \qquad (1)$$

where **C** is a constant, **m** is the mass of the ejected species, **l** is the flight distance, **k** is Boltzmann's constant and T_{trans} is the most probable translational temperature. Over the fluence range of 15-40 mJ/cm^2 for nanosecond pulses and 0.8-3.5 mJ/cm^2 for sub-picosecond pulses the time-of-flight distributions of neutral Cd atoms and Te$_2$ molecules are well described by a single component Maxwell-Boltzmann distribution with similar translational temperatures for both species at a given fluence. The deviation of the slower velocity component of Cd atoms from a single component Maxwell-Boltzmann distribution is attributed to the existence of loosely

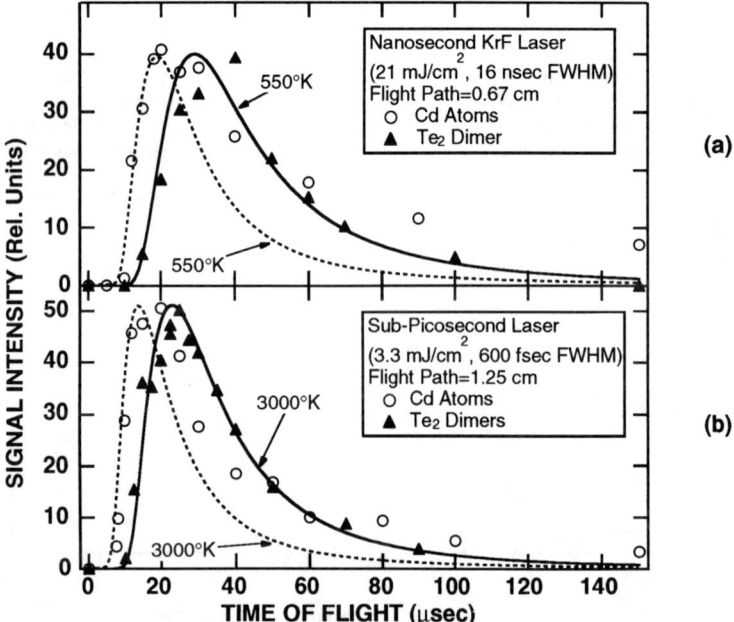

Fig. 3 Time-of-flight distributions of Cd atoms and Te$_2$ molecules ejected from CdTe (100) surface by nanosecond (Fig. 3a) and sub-picosecond (Fig. 3b) laser irradiation. Solid and dashed lines are fits of Maxwell-Boltzmann distributions to the experimental data.

bonded Cd on the surface, formed and desorbed during the transient heating. The time-of-flight distributions observed for the nanosecond irradiation are the nascent distributions since, as stated earlier, the presence of multiple gas phase collisions can be ruled out. Several issues relating to the sub-picosecond ablation are currently being investigated in order to establish the desorption time scale and to determine the conditions for eliminating multiple gas phase collisions.

Representative angular distributions for Cd atoms and Te_2 molecules ejected from the CdTe (100) surface by 248 nm nanosecond and sub-picosecond (3 mJ/cm^2) irradiation are shown in Figure 4. For nanosecond (16 mJ/cm^2) ablation ~0.12 monolayers are removed during the ~20 nsec pulse, conditions under which post-desorption collisions do not occur. The ejected species are extremely forward peaked toward the surface normal. The extreme deviation from the Knudsen law (i.e. $\cos^n(\theta)$, n=1) signifies that the nascent distributions are not formed via a thermal evaporative mechanism. Other examples of highly forward peaked angular distributions have been reported whose origins have been ascribed to a photoejection or photochemical mechanism [14,15]. Since the angular distributions of ejected particles under collisionless conditions are determined by the final interatomic forces occurring during desorption, the highly forward peaked distribution indicates that the local forces acting on the ejected particles have symmetry about the <100> direction. One plausible mechanism has been proposed [19] which involves a sudden momentum excitation of the solid by the laser pulse resulting in a linear collision cascade which causes the highly forward peaked flux distribution. Another proposed mechanism is the desorption over a potential barrier, also resulting in a strong forward orientation of the ablated atoms or molecules [21].

Angular flux distributions for Cd atoms and Te_2 molecules ejected from the CdTe (100)

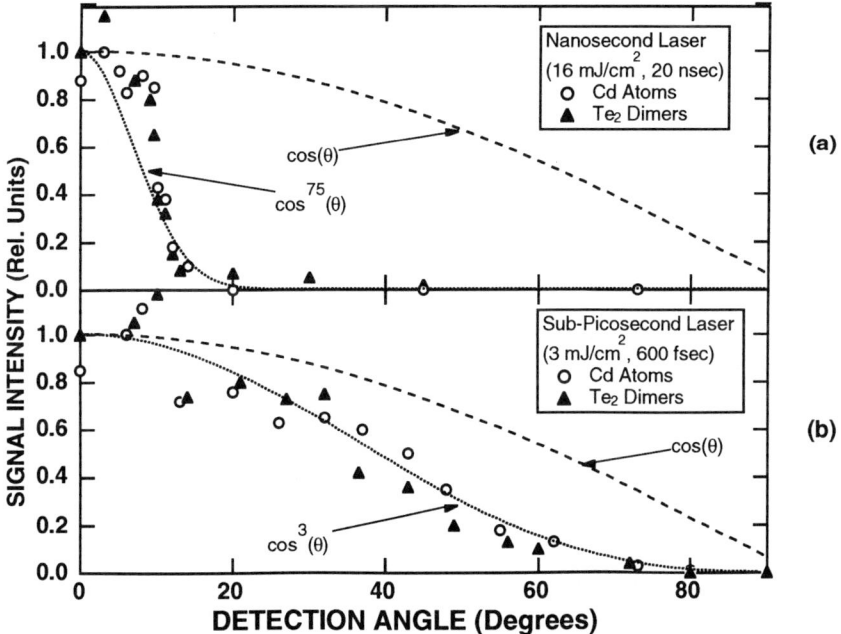

Fig. 4 Angular flux distributions of Cd atoms and Te_2 molecules ejected from CdTe (100) surface by nanosecond (Fig. 4a) and sub-picosecond (Fig. 4b) laser irradiation.

surface by sub-picosecond 248 nm pulses are also shown in Figure 4. The ejected species are not as forward peaked as those observed for the nanosecond irradiation. At a fluence of 3 mJ/cm^2 the distribution approaches a $\cos^3\theta$ form. This could be due to excitation pathways accessible with sub-picosecond laser irradiation, such as two-photon processes or a plasma formation, which may not be possible with nanosecond laser excitation. Further work is underway to more fully understand the nature of these angular flux distributions.

SUMMARY

We have investigated the velocity and angular flux distributions of Cd atoms and Te$_2$ molecules ejected from CdTe (100) surfaces by 248 nm nanosecond (FWHM 20 nsec) and sub-picosecond (FWHM 600 fsec) laser irradiation. CdTe (100) surfaces have been shown to remain stoichiometric and atomically ordered during multiple pulse ablation with nanosecond pulses under low fluence conditions (<40 mJ/cm^2). Yields for nanosecond laser ablation under those conditions are sufficiently low to ensure that multiple gas phase collisions can be neglected. Time-of-flight distributions for the ejected products are well described by Maxwell-Boltzmann distributions for both low fluence nanosecond and sub-picosecond (<4 mJ/cm^2) irradiation. A deviation from a Maxwell-Boltzmann distribution is observed for slower Cd atoms Nascent angular flux distributions from nanosecond ablated products are highly forward peaked about the surface normal under conditions where the ejected particles do not suffer any post-desorption collisions. Angular flux distributions of products ejected from sub-picosecond irradiation follow a $\cos^3\theta$ form.

REFERENCES

1. P.D. Brewer, J.J. Zinck, and G.L. Olson, Appl. Phys. Lett. **57**, 2526 (1990). P.D. Brewer, J.J. Zinck, and G.L. Olson, Mat. Res. Soc. Symp. Proc. **201**, 543 (1991).
2. S. Küper and M. Stuke, Appl. Phys. Lett. **54**, 4 (1989).
3. F. Bachmann, Chemtronics **4**, 149 (1989).
4. J.R. Lankard Sr., G. Wolbold, Appl. Phys. **A54**, 355 (1992).
5. S. Küper and M. Stuke, Appl. Phys. **B44**, 199 (1987).
6. S. Preuss, M. Späth, Y. Zhang, and M. Stuke, Appl. Phys. Lett. **62**, 3049 (1993).
7. A. Namiki, T.Kawai, and K. Ichige, Surf. Sci. **166**, 129 (1986). A. Namiki, H. Fukano, T.Kawai, Y. Yasuda and T. Nakamura, J. Phys. Soc. Jpn. **54**, 3162 (1985).
8. T. Nakayama, H. Ichikawa, and N. Itoh, Surf. Sci. **123**, L693 (1982). N. Nakayama, Surf. Sci. **133**, 101 (1983).
9. W. Sesselmann, E.E. Marinero, and T.J. Chuang, Appl. Phys. **A41**, 209 (1986).
10. V.M. Donnelly, V. McCrary, and D. Brasen. Mat. Res. Soc. Symp. Proc. **75**, 567 (1987).
11. J. Cowin, D.J. Auerbach, C. Becker, and L. Wharton, Surf. Sci. **78**, 545 (1978).
12. R. Kelly, J. Chem. Phys. **92**, 5047 (1990). R. Kelly and R.W. Dreyfus, Nucl. Instru. Meth. Phys. Res. **B33**, 341 (1988).
13. I. NoorBatcha, and R. Lucchese, J. Chem. Phys. **86**, 5816 (1987). I. NoorBatcha, R. Lucchese, and Y. Zeiri, J. Chem. Phys. **89**, 5251 (1989).
14. J.J. Zinck and P.D. Brewer to be published.
15. M. Stuke, Appl. Phys. Lett. **45**, 1175 (1984).
16. S. Szatamari and F.R. Schäfer, Appl. Phys. **B46**, 305 (1988).
17. B. Stritzker, A. Pospiezczyk, and J.A. Tagle, Phys. Rev. Lett. **47**, 356 (1981).
18. T. Yamamoto, K. Hattori, Y. Nakai, and N. Itoh, Rad. Eff. Def. Sol. **109**, 213 (1989). Y. Nakai, K. Hattori, and N. Itoh, Appl. Phys. Lett. **56**, 1980 (1990).
19. C.-C. Cho, J.C. Polanyi, and C.D. Stanners, J. Phys. Chem. **92**, 6859 (1988).
20. B.J. Garrison and R. Srinivasan, J. Appl. Phys. **57**, 2909 (1985).
21. K.D. Rendulic, Appl. Phys. **A47**, 55 (1988).

PICOSECOND OPTICAL SECOND HARMONIC STUDIES OF ADSORBATE REDUCTION KINETICS ON CADMIUM SULFIDE

T. W. SCOTT[a], J. MARTORELL[b] AND Y. J. CHANG[c]
[a] Department of Chemistry, New York University, New York, NY.
[b] Department of Physics, University of Barcelona, Barcelona Spain.
[c] Department of Chemistry, Brookhaven National Laboratory, Upton, NY.

ABSTRACT

Time resolved surface second harmonic generation has been used to probe the photoreduction kinetics of malachite green adsorbed onto single crystal cadmium sulfide. A detailed analysis is presented of how the adsorbates and the non-centrosymmetric substrate contribute separately to the total second harmonic signal. Conditions under which the adsorbates can be cleanly detected are described. To complement kinetic measurements of adsorbate reduction, the time evolution of conduction band carriers was determined using sum frequency up conversion of the recombination luminescence. In addition, the formation and decay of surface trapped carriers was monitored using near infrared transient absorption. Comparing the time scale for photoreduction with the relaxation kinetics of mobile and trapped charge carriers indicates that short lived mobile carriers rather than longer lived surface trapped carriers dominate interfacial charge transfer in this system.

Second Harmonic Generation By Surface Adsorbates is a comparatively new tool for the study of surface and interfacial phenomena. It is a technique with intrinsic surface sensitivity since SHG is forbidden in the bulk of a medium with inversion symmetry but allowed at the surface or interface. Even when the solid has a non-centrosymmetric structure, recent results have shown that information from processes occurring at the surface may be extracted using a proper orientation of the crystalline structure or appropriate polarization of the incident and reflected fields [1]. The second harmonic light generated by adsorbates on a non-centrosymmetric substrate has contributions from both the adsorbate polarization $P_A(2\omega)$ and the substrate polarization $P_S(2\omega)$. The reflected second harmonic field $E_R^{\parallel}(2\omega)$ polarized parallel to the plane of incidence is determined by the boundary conditions on the electric $E_X(2\omega)$ and magnetic $H_Y(2\omega)$ field components [2,3]

$$\frac{E_R^{\parallel}}{4\pi} = i\left(\frac{2\omega}{c}\right)\frac{\sin\theta_T P_{AZ} - \cos\theta_T P_{AX}}{\sqrt{\varepsilon_T}\cos\theta_R + \sqrt{\varepsilon_R}\cos\theta_T} + \frac{\left[\sqrt{\varepsilon_T} - \sqrt{\varepsilon_S}\cos(\theta_T - \theta_S)\right]P_{SX} + \sqrt{\varepsilon_S}\sin(\theta_T - \theta_S)P_{SZ}}{(\varepsilon_T - \varepsilon_S)\left[\sqrt{\varepsilon_T}\cos\theta_R + \sqrt{\varepsilon_R}\cos\theta_T\right]} \quad (1)$$

where the dielectric constant of the substrate is denoted $\varepsilon_S(\omega)$ at the fundamental frequency and $\varepsilon_T(2\omega)$ at the second harmonic frequency. The dielectric constant for the reflected second harmonic field is $\varepsilon_R(2\omega)$. The optical geometry used for experiments on CdS is shown in Fig.1. Equation 1 was used to interpret measurements of the second harmonic intensity for a film of malachite green deposited onto CdS. The polarization of the incident beam, the angle of incidence and the orientation of the crystalline substrate were examined. Figure 2 shows how the reflected second harmonic intensity depends on the angle of incidence for the optimum substrate orientation. When the angle of incidence is $\approx 40°$ the adsorbates are readily detected for any polarization of the fundamental beam. The reflected intensity when the polarization of the fundamental beam is rotated through 90⁰ is shown in Fig. 3 for two different crystal orientations. It can be seen that the substrate contribution is reduced and the adsorbates are readily detected if the c-axis of the crystal is oriented parallel to the surface normal.

For analyzing the data shown in Figs. 2 and 3 the polarization of the adsorbates is assumed to be independent of crystal orientation and described by

$$P_{AX} = 2\chi^{(2)}_{XZX}\left(f_P^T\right)^2 \cos^2\varphi \sin\theta_s \cos\theta_s E_I^2 \quad (2a)$$

$$P_{AZ} = \left\{\left(f_P^T\right)^2 \cos^2\varphi\left[\chi^{(2)}_{ZZZ}\sin^2\theta_s + \chi^{(2)}_{ZXX}\cos^2\theta_s\right]\right.$$
$$\left. + \left(f_S^T\right)^2 \chi^{(2)}_{XZX}\sin^2\varphi\right\}E_I^2 \quad (2b)$$

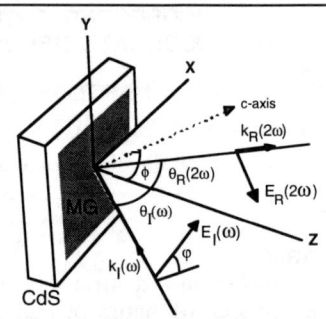

FIGURE 1 - SKETCH OF THE OPTICAL GEOMETRY USED FOR SURFACE SECOND HARMONIC GENERATION. THE ANGLE OF INCIDENCE FOR THE FUNDAMENTAL BEAM IS DENOTED $\theta_I(\omega)$ AND THE REFLECTION ANGLE OF THE SECOND HARMONIC BEAM IS DENOTED BY $\theta_R(2\omega)$. THE ELECTRIC FIELD VECTOR FOR A LINEARLY POLARIZED INCIDENT BEAM MAKES AN ANGLE φ WITH RESPECT TO THE PLANE OF INCIDENCE.

where the angle (φ) gives the tilt of the in coming fundamental electric field vector with respect to the plane of incidence. The substrate polarization depends on the crystal orientation. The cadmium sulfide crystals used in this work have the Wurzite structure and belong to the space group P63mc. There are six non-vanishing tensor elements [4] which take on three distinct values, d_{15}, $d_{31}(-)$ and $d_{33}(-)$. To make use of the crystalline symmetry the components of the bulk nonlinear polarization (P_{SX}, P_{SZ}) which can contribute to a p-polarized output are described in terms of the polarization components $(P_{SX'}, P_{SZ'})$ along the principal axes (X', Y', Z') of the crystal

$$P_{SX} = P_{SX'}\cos\phi + P_{SZ'}\sin\phi \quad \text{and} \quad P_{SZ} = P_{SZ'}\cos\phi - P_{SX'}\sin\phi \quad (3)$$

Equations 3 assume that the c-axis of the crystal lies in the plane of incidence and makes an angle ϕ with respect to the z-axis of the coordinate system shown in Fig. 1. The crystal polarization components are

$$P_{SX'} = 2\left(f_P^T\right)^2 d_{15}\cos^2\varphi \sin(\theta_s + \phi)\cos(\theta_s + \phi) E_I^2 \quad (4a)$$

$$P_{SZ'} = \left\{\left(f_P^T\right)^2 \cos^2\varphi\left[d_{31}\cos^2(\theta_s + \phi) + d_{33}\sin^2(\theta_s + \phi)\right]\right.$$
$$\left. + \left(f_S^T\right)^2 d_{31}\sin^2\varphi\right\}E_I^2 \quad (4b)$$

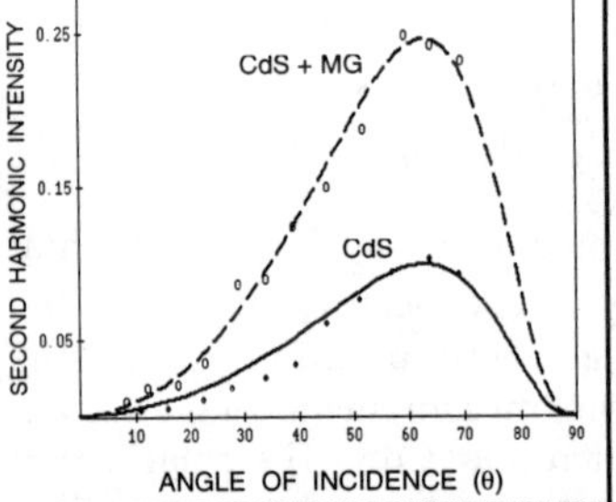

FIGURE 2 - ANGLE OF INCIDENCE DEPENDENCE FOR THE P-POLARIZED REFLECTED SECOND HARMONIC INTENSITY FROM A CdS CRYSTAL WITH AND WITHOUT MG. THE C-AXIS IS NORMAL TO THE SURFACE. ONLY P-POLARIZED SECOND HARMONIC LIGHT WAS MEASURED.

The Fresnel factors f_P^T and f_S^T refer to transmission of a p-polarized and s-polarized incident beam, respectively

$$f_S^T = \frac{2\sqrt{\varepsilon_I}\cos\theta_I}{\sqrt{\varepsilon_I}\cos\theta_I + \sqrt{\varepsilon_s}\cos\theta_s} \quad (5a)$$

$$f_P^T = \frac{2\sqrt{\varepsilon_I}\cos\theta_I}{\sqrt{\varepsilon_S}\cos\theta_I + \sqrt{\varepsilon_I}\cos\theta_S} \quad (5b)$$

All the data points shown in Figs 2 and 3 were fit simultaneously using Eqns. 1 through 5.

Kinetic Measurements Of Adsorbate Photoreduction are of interest because electrons and holes generated in a bulk semiconductor by the absorption of light have the ability to reduce or oxidize adsorbates at the semiconductor surface. The efficiency of charge transfer depends on a number of competing processes taking place on different time scales and in different spatial regions of the sample. The surface second harmonic technique is capable of isolating interfacial reactions in the presence of simultaneous bulk processes.

Figure 4 shows kinetic measurements made by exciting single crystal CdS with a 35 ps, 354.7 nm light pulse and probing the 532 nm SHG signal at a series of time delays. In the absence of malachite green no transients were observed in the SHG signal. The crystal was rastered to a fresh position after each excitation pulse. The transient reduction in signal is attributed to electron capture

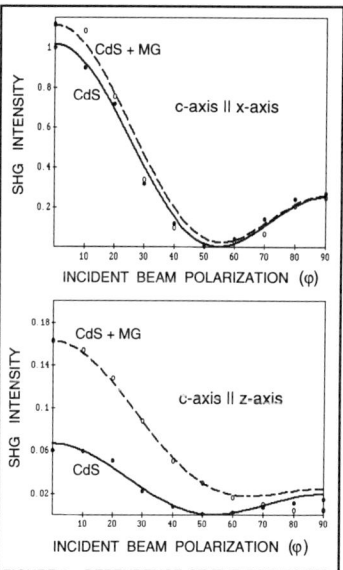

FIGURE 3 - DEPENDENCE OF THE REFLECTED SECOND HARMONIC INTENSITY ON THE POLARIZATION ANGLE OF THE FUNDAMENTAL LASER BEAM FOR CADMIUM SULFIDE WITH AND WITHOUT ADSORBED MALACHITE GREEN.

by malachite green which has been shown previously to shift the first absorption band to higher energies. It can be seen in Fig. 4 that photoreduction is rapid and there is no back transfer on the time scale of a few nanoseconds. Since only 20-40% of the adsorbates in the irradiated area are reduced and the number of photons per unit area is 100 times greater than the number of adsorbates, the reduction yield per photon absorbed is of the order $\approx 3\times 10^{-3}$.

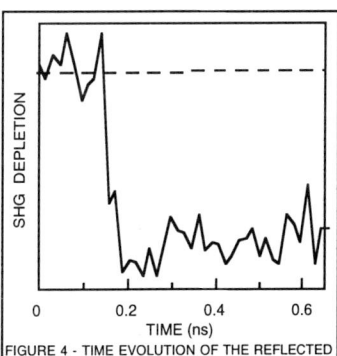

FIGURE 4 - TIME EVOLUTION OF THE REFLECTED SURFACE SECOND HARMONIC INTENSITY WHEN CdS WITH A MALACHITE GREEN OVERLAYER IS IRRADIATED AT 354.7 NM. THE ENERGY DENSITY FOR EXCITATION AND PROBE BEAMS WAS 2 x 10^{-4} J/cm^2 and 5 x 10^{-4} J/cm^2, RESPECTIVELY.

An alternate explanation for the decrease in total second harmonic intensity by loss of adsorbates through thermal desorption was found to be unlikely. Excitation of CdS with a 10 ns pulse at 345.7 nm induced the same change in reflected SHG as did excitation by a 35 ps pulse. Since higher peak temperatures are expected for the short pulse duration thermal desorption appears not to be significant. The temperature rise at the semiconductor surface was estimated from the time dependent heat equation [5]. Assuming that heat radiation into air is slow compared to heat transport into the bulk semiconductor, the time dependent surface temperature $\Delta T_{surf}(t)$ is given by

$$\Delta T_{surf}(t) = \frac{\alpha(1-R)(h\nu - E_g)}{\pi^{3/2}\omega^2 \rho c \tau} \int_{-\infty}^{t} dt' \; \exp\left[-\frac{t'^2}{\tau^2} + \alpha^2 \kappa (t-t')\right] erfc\left[\alpha\sqrt{\kappa(t-t')}\right] \qquad (7)$$

where $\kappa = 0.57$ cm^2/sec is the thermal diffusivity, $\rho c = 0.125$ cal/cm^{3}deg is the specific heat, $R = 0.6$ is the surface reflectivity and $2\tau\sqrt{\ln(2)}$ is the full width at half height of a Gaussian temporal heat source. On the time scale of several tens of picoseconds it is assumed that the electron and lattice temperatures are the same [6] and that the heat capacity is dominated by the lattice. Figure 5 shows plots of the time dependent surface temperature for the two laser pulse widths used.

To complement SHG measurements of the adsorbates, time resolved absorption and emission was performed to detect the photogenerated carriers in the substrate. A transient absorption measurement at 1064 nm is shown in Fig. 6. It is interesting that a substantial absorption remains

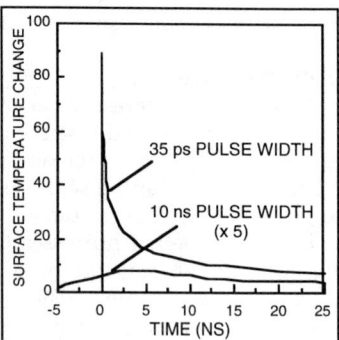

FIIGURE 5 - SIMULATED SURFACE TEMPERATURE RISE FOR CdS FOLLOWING 354.7 NM EXCITATION USING 200 µJ/pulse WITH EITHER A 35 ps OR 10 ns PULSE DURATION.

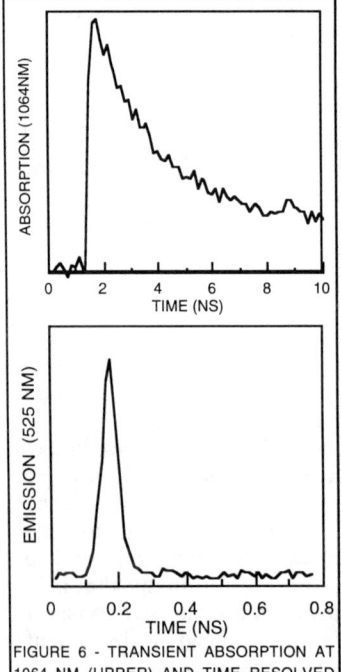

FIGURE 6 - TRANSIENT ABSORPTION AT 1064 NM (UPPER) AND TIME RESOLVED EMISSION AT 525 NM (LOWER) FOLLOWING 354.7 NM EXCITATION OF A CDS CRYSTAL.

well after the adsorbate reduction is complete. We interpret this long lived component as trapped carriers which are not effective oxidizing agents for oxazine dyes. Figure 6 also shows the time resolved band gap emission at 525nm. This measurement detects only the excess carriers in the conduction band. It is apparent from Fig. 6 that the conduction band carriers are short lived. It appears that only the conduction band carriers give rise to adsorbate reduction. The trapping of these carriers also gives rise to the long lived absorption seen in Fig. 6.

Figure 7 shows additional kinetic measurements, but for methylene blue photoreduction on CdS powder. The measurement was performed by multiple light scattering and shows a superposition of transient absorption by the semiconductor and transient bleaching by methylene blue. The difference between these two decay curves is attributed to methylene blue reduction. It is similar to the SHG transient in Fig. 4 and supports the interpretation that only short lived conduction band carriers are effective oxidizing agents. Performing transient absorption measurements at a series of wavelengths confirms that more than one species contributes to the signal. Figure 8 shows absorption transients measured at 630nm and 1544nm. The longer wavelength measurement, which is expected to favor the detection of conduction band carriers, has a rapid decay similar to the emission and reduction kinetics.

Modeling Adsorbate Reduction Using The Smoluchowski Equation with reasonable physical parameters can be used to analyze the competitive decay pathways available to the conduction band carriers. The model includes carrier transport, bulk trapping and charge pair decay at the interface. The time evolution of excess carriers n(z,t) generated at a distance z from the surface is described [7,8] by

$$\frac{1}{D}\frac{\partial n(z,t)}{\partial t} = \frac{\partial^2 n(z,t)}{\partial z^2} + \frac{1}{k_B T}\frac{\partial}{\partial z}\left[\frac{dV}{dz}n(z,t)\right] - \frac{n(z,t)}{\tau_B} \quad (7)$$

where D is the ambipolar diffusion constant and τ_B is the bulk carrier lifetime. The boundary condition for the solution of Eqn. 7 is

$$D\frac{\partial n(z,t)}{\partial z}\bigg|_{z=0} = k \cdot n(0,t) \quad (8)$$

where k is the total reaction velocity. It includes a component k_S associated with reduction of adsorbates. Under strong illumination conditions $dV/dz \approx 0$ allowing Eqn. 7 to be solved by making the substitution $n(z,t) = \rho(z,t)e^{-t/\tau_b}$. This transforms Eqns. 7 and 8 into a problem with a known Green function solution [9]. The initial distribution g(z') of mobile carriers generated by the excitation pulse is attenuated according to Beer's Law $g(z') = \alpha e^{-\alpha z'}$. The survival probability of mobile carriers in the bulk $N_m(t)$ and the probability that an adsorbate will be reduced at the interface $N_S(t)$ have the analytical solutions

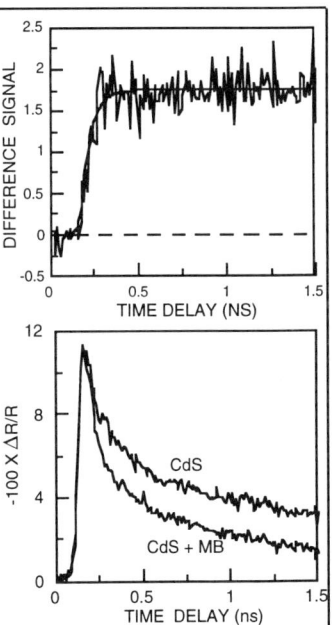

FIGURE 7 - LOWER PANEL: TRANSIENT LIGHT SCATTERING SIGNALS AT 630 NM FOLLOWING 354.7 NM EXCITATION FOR CdS POWDER WITH AND WITHOUT METHYLENE BLUE. UPPER PANEL: METHYLENE BLUE TRANSIENT BLEACHING SIGNAL.

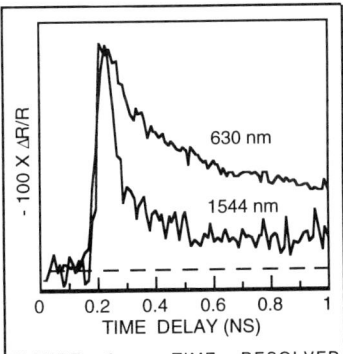

FIGURE 8 - TIME RESOLVED ABSORPTION AT 1544 NM AND 630 NM FOLLOWING 354.7 NM EXCITATION OF POWDERED CdS CRYSTALS.

$$N_m(t) = \frac{e^{-t/\tau_B}}{k-\alpha D}\left\{ke^{+\alpha^2 Dt}erfc[\alpha\sqrt{Dt}]\right.$$
$$\left. - \alpha D e^{+k^2 t/D}erfc[k\sqrt{t/D}]\right\} \quad (9)$$

$$N_S(t) = \frac{\alpha k_S \tau_B}{(1-k/\alpha D)(1-\alpha^2 D\tau_B)}\{1$$
$$-e^{-t/\tau_B}e^{+\alpha^2 Dt}erfc(\alpha\sqrt{Dt}) - \alpha\sqrt{D\tau_B}\cdot erf(\sqrt{t/\tau_B})\}$$
$$+ \frac{\alpha k_S \tau_B}{(1-\alpha D/k)(1-k^2\tau_B/D)}\{1$$
$$-e^{-t/\tau_B}e^{+k^2 t/D}erfc(k\sqrt{t/D}) - k\sqrt{\tau_B/D}\cdot erf(\sqrt{t/\tau_B})\} \quad (10)$$

255

Equation 10 was used to model the data in the upper part of Fig. 7 and is shown as a solid line in that figure. Figure 10 shows plots of both Eqns. 9 and 10 using the parameters $D=1.2 \text{cm}^2/\text{sec}$, $k_S=2\times10^3 \text{cm/sec}$, $k=10^6 \text{cm/sec}$, $\tau_B=10\text{ns}$ and $\alpha=1.3\times10^5 \text{cm}^{-1}$. The bulk carrier lifetime τ_B was measured using two-photon excitation which generates carriers uniformly through out the 2mm crystal. The decay of these carriers is shown in Fig. 9 and should be much less sensitive to surface effects than carriers generated by one-photon excitation. The photoreduction yield is given by the long time limit of Eqn. 10

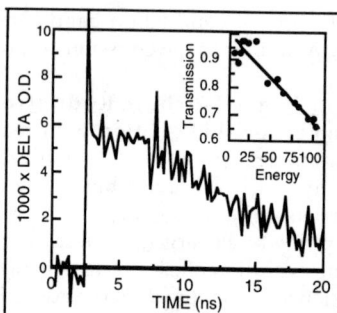

FIGURE 9 - TIME RESOLVED ABSORPTION AT 1064 NM FOLLOWING TWO-PHOTON EXCITATION OF CDS AT 630NM. INSET SHOWS NONLINEAR DEPENDENCE OF TRANSMISSION ON ENERGY OF EXCITATION LIGHT PULSE.

$$N_S(t \to \infty) = \frac{\alpha k_S \tau_B}{\left(1+\alpha\sqrt{D\tau_B}\right)\left(1+k\sqrt{\tau_B/D}\right)} \quad (11)$$

The parameters listed above give $N_S(t \to \infty) \approx 10^{-2}$. This low reduction yield is consistent with experiment.

CONCLUSIONS

Time resolved surface second harmonic generation has been used to characterize the photoreduction of malachite green on single crystal cadmium sulfide. The description of SHG in terms of boundary conditions on the electric and magnetic fields is shown to work well for adsorbates on a non-centrosymmetric substrate. Comparing the time scale for photoreduction with the relaxation times of mobile and trapped charge carriers indicates that short lived mobile carriers rather than longer lived surface trapped carriers dominate interfacial charge transfer in this system.

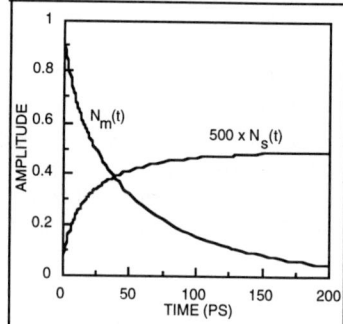

FIGURE 10 - MOBILE CARRIER RELAXATION AND ADSORBATE REDUCTION KINETICS ACCORDING TO EQNS. 9 AND 10 OF THE TEXT.

REFERENCES

1. T. Stehlin, M. Feller, P. Guyot-Sionnest and Y. R. Shen, Opt. Lett. **13**, 389 (1988).
2. N. Bloembergen and P. S. Pershan, Phys. Rev. **2**, 606 (1962).
3. Y. R. Shen, Ann. Rev. Phys. Chem. **40**, 327 (1989).
4. C. K. N. Patel, Phys. Rev. Lett. **16**, 613 (1966).
5. J. H. Bechtel, J. Appl. Phys. **46**, 1586 (1975).
6. F. Budde, T. F. Heinz, M. M. Loy, J. A. Misewich, F. de Rougemont and H. Zacharias, Phys. Rev. Lett. **66**, 3024 (1990).
7. M. Evenor, S. Gottesfeld, Z. Harzion, D. Huppert and S. W. Feldberg, J. Phys. Chem. **88**, 6213 (1984).
8. J. W. Orton and P. Blood, The Electrical Characterization of Semiconductors: Measurement of Minority Carrier Properties (Academic Press, 1990), p. 28.
9. H.S. Carslaw and J. C. Jaeger, Conduction of Heat in Solids (Oxford University Press, 1990), p. 358.

SELECTIVE DEPOSITION OF ZnS THIN FILMS BY KrF EXCIMER LASER ON PATTERNED Zn SEEDS

K.YAMANE* and M.MURAHARA**
*Graduate Student of Tokai University
**Faculty Engineering of Tokai University, 1117 Kitakaname, Hiratsuka, Kanagawa, 259-12, Japan

ABSTRACT

The patterned Zn nucleation and the ZnS growth onto the Zn seeds on a thermal oxidized silicon substrate was demonstrated at room temperature with the excimer laser chemical vapor deposition method.
The formation of ZnS films was realized by the method based on the two-step process consisting of the nucleation and the subsequent ZnS growth. In the nucleation, a gaseous dimethylzinc was sealed in a reaction chamber and was then evacuated immediately. Then, the substrate surface which was uniformly adsorbed by dimethylzinc molecules was exposed with a single shot irradiation of a patterned KrF laser; Zn seeds were created only on the irradiated parts by a photodecomposition. And the subsequent growth of ZnS was performed by the parallel or perpendicular irradiation methods. As a result, in the perpendicular irradiation method, the high selectivity and crystallinity of the film were performed by irradiating the whole substrate surface with very low fluence of the KrF laser such as 3 mJ/cm2.

INTRODUCTION

Recently, several reports on a thin film formation using a nucleation technique have been made, and metal deposition such as aluminum has been accomplished [1-3]. However, there is few reports such as a film formation of compound semiconductor. It is considered that appling the nucleation technique to a film formation of compound semiconductor makes it possible for its film formation to be selective, be improved in quality and be achieved at low temperature. We have been studying a film formation of ZnS using an excimer laser chemical vapor deposition (CVD) method in order to create a high quality material for a luminous device [4].
ZnSe and ZnS have a wide bandgap of 2.7 and 3.7eV, respectively, and are expected as materials for luminous devices with short-wavelength. ZnSe is extensively studied for a blue luminous laser [5], whereas ZnS is few reported due to the difficulties in obtaining a high quality crystal and controlling a conductive type although it has a possibility of luminous region from blue to ultraviolet wavelength. Therefore, we tried the growth of patterned ZnS films onto Zn seeds on thermal oxidized silicon substrate using nucleation technique at room temperature. As a result, the selective growth of ZnS thin films was successfully performed by the surface irradiation of very weak laser fluence after forming Zn seeds by a single shot laser irradiation.

PRINCIPLE OF NUCLEATION AND FILM GROWTH

The formation of ZnS films was realized by the method based on the two-step process consisting of the nucleation and the subsequent ZnS growth. And dimethylzinc (DMZ) and H2S gases were employed for a source of Zn and S supply, respectively.

Figure 1 illustrates the principle models of this work. First, the nucleation of the Zn seeds. As shown in Figure 1 (a), a gaseous DMZ is sealed in a reaction chamber and is then evacuated immediately. DMZ molecules were, thus, uniformly adsorbed on a substrate surface. Then, the surface is irradiated with an excimer laser light through a reticle-pattern; Zn seeds are created only on the irradiated parts by a photodecomposition. Next, the subsequent growth of ZnS. The ZnS growth is performed by the parallel or perpendicular irradiation, as shown Figure 1 (b) or (c). Namely, these irradiations leading to deposition occur either in the vapor-phase or on the substrate surface. These film formations are considered to be carried out as described below. In case of the parallel irradiation, DMZ and H2S gases are photodecomposed by an excimer laser light in vapor-phase; then, the radicals which were produced are deposited on the nucleation parts. On the other hand, in case of the perpendicular irradiation, DMZ and H2S gases which were adsorbed on the Zn seeds are photodecomposed by an excimer laser; then, the radicals are grown on the seeds. Therefore, a weak laser fluence is required in order for only the nucleated parts to react with.

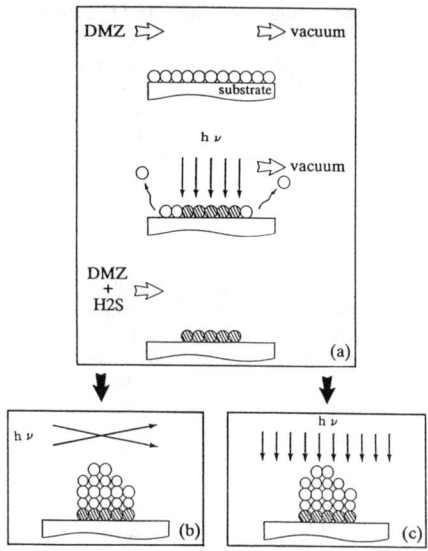

Fig.1 Principle models of the formation of Zn seeds and growthing of the ZnS thin films by the excimer laser CVD method.
a) Zn nucleation. b) film growth of the parallel irradiation. c) film growth of the perpendicular irradiation.

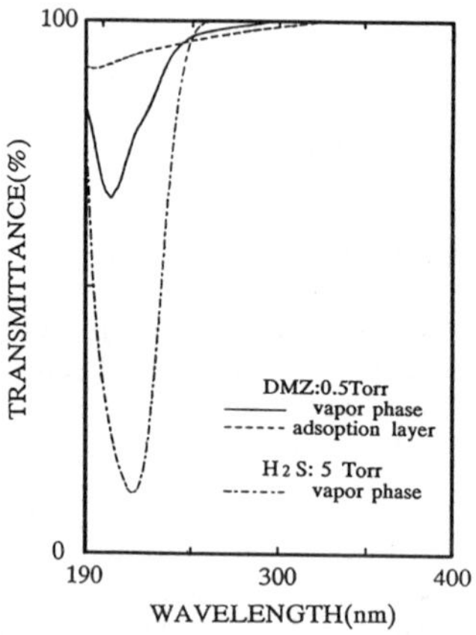

Fig.2 The UV absorption spectra of DMZ and H S gases.

Figure 2 displays the UV absorption spectra of DMZ and H2S gases. When a gaseous DMZ was sealed in the chamber at the pressure of 0.5 Torr, the absorption was shown with the solid line and the peak was placed at 205nm wavelength. Even when the DMZ gas was evacuated, the absorption spectrum was observed and indicated with the broken line; which means the DMZ molecules will be easily adsorbed on a substrate. And the UV absorption spectrum of the adsorbates was independent of the DMZ gas pressure. From these results, experiments can be demonstrated using either an ArF or KrF laser as the source of UV light supply. However, when irradiating with an ArF laser, the photodecomposition of DMZ gas takes place to deposit the Zn metal on the quartz incident window because of the strong absorption at the 193 nm wavelength. Therefore, the Zn deposition attenuates the laser light remarkably, and the film growth is ineffectively carried out. On the other hand, DMZ and H2S gases have only about 5 % of absorption at the 249 nm wavelength of a KrF laser. The adsorbates of DMZ also have same absorption as in the vapor-phase at the 249 nm. Therefore, with a KrF laser, the nucleation and the film growth can be performed effectively. The perpendicular irradiation of a KrF laser sufficiently reaches to the substrate surface, while the parallel irradiation requires a high density excitation.

EXPERIMENTAL METHOD AND RESULTS

NUCLEATION PROCESS

Figure 3 shows the experimental setup for nucleation process. Thermal oxidized silicon was used as a substrate. This substrate was put in a reaction chamber. As soon as gaseous DMZ was enclosed in the evacuated chamber, the gas was exhausted from the chamber in order for the adsorbed molecules to be selectively which was remained. Then, a patterned KrF laser of 100mJ/cm2 fluence, scaled down to one-half, was perpendicularly irradiated onto the adsorbates. As a result, the patterned Zn seeds were created by a photodecomposition of the adsorbates.

To find the optimum condition of Zn nucleation the DMZ adsorbed layers which were irradiated with the KrF laser were inspected by the X-ray photoelectron spectroscopy (XPS). Figure 4 shows the dependence of the laser shot number on the deposition amount of Zn atoms. The KrF laser was irradiated at 100 mJ/cm2 fluence and the range of 1 to 200 shots. As the shot number inceased, the deposition amount of Zn went up. But the deposition amount decreased at 40 shots and more. The decrease of deposition amount seemed to be caused by the desorption of Zn occurred with the laser irradiating. With the fluence of 100mJ/cm2, therefore, the shot number of 40 and below is suitable for the nucleation. And, even with a single shot irradiation, the adsorbates were photodecomposed to create Zn seeds.

A high-resolution of pattern projection can be realized by the nucleation with a single shot irradiation. Therefore, the nucleation in the experiments stated below were conducted with a single shot irradiation.

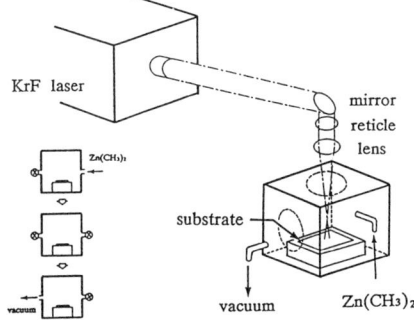

Fig.3 The experimental setup for nucleation process.

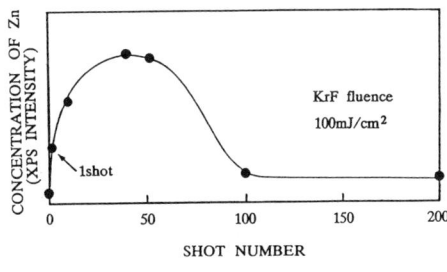

Fig.4 The condition of Zn nucleation. The dependence of the laser shot number and the deposition amount of Zn atoms.

GROWTH OF ZnS FILMS

Figure 5 and 6 illustrate the experimental setup for the growth by the parallel and perpendicular irradiations. After the nucleation, DMZ and H2S gases were sealed in the chamber at the partial pressure ratio of 1:10. The total pressure was 11 Torr. In the case of parallel irradiation, the KrF laser was focussed upon the area 1 cm above the substrate surface for a high density excitation; the laser was irradiated at 1000 mJ/cm2 fluence and 6000 shots. On the other hand, in the perpendicular irradiation, the KrF laser of 3 mJ/cm2 fluence and 6000 shots was irradiated perpendicularly to the whole substrate surface for growing a film. In order to compare the above, the ZnS film was formed on the non-nucleated substrate by the identical method.

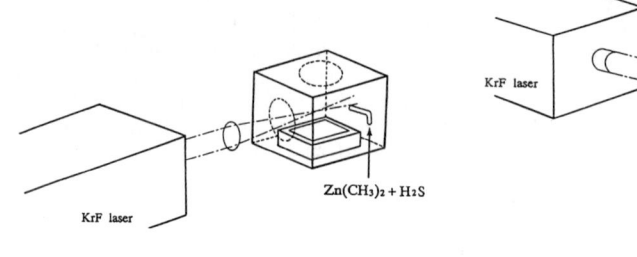

Fig.5 The experimental setup for film growth process by parallel irradiation.

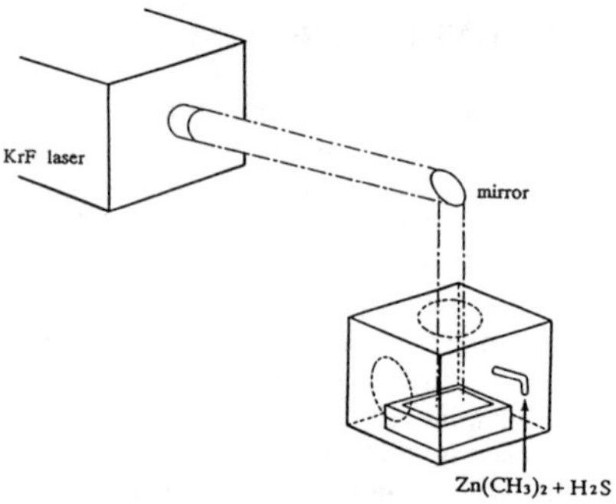

Fig.6 The experimental setup for film growth process by perpendicular irradiation.

Figure 7 indicates the optical microscope measurement of the patterned ZnS thin film formed by the parallel irradiation and the stylus surface profile as well. This displays the ZnS film preferentially deposited on the nucleated parts. However, its selectivity was not fine. The film thickness was about 800 Å. The patterned thin film which was formed by the perpendicular irradiation was measured by the interference roughness meter, as shown in Figure 8. It was observable that the ZnS thin film was grown only on the nucleated parts; the film thickness was about 400 Å.

On the contrary, as for the formation of thin film on the non-nucleated substrate, it was observed that nothing was deposited in the perpendicular irradiation; however, the whole substrate surface was covered with white powder in the parallel irradiation. This powder was similarly observed on the non-nucleated parts of the nucleated substrate which was performed by the parallel irradiation.

The nucleated and non-nucleated parts of the film, which was formed by the parallel and perpendicular irradiation, were analyzed by the XPS.

Figure 9 exhibits the results of the analysis of the depth direction composition. As to the nucleated parts which were formed by both parallel and perpendicular irradiation methods, the chemical composition of Zn-S was clearly detected from the results as follows; Zn LMM peak was shifted about 3 eV from ordinary Zn LMM peak of 261.2 eV and S 2p/3 peak was also shifted about 2 eV from ordinary S 2p/3 peak of 164.1 eV by the depth direction with Ar+ ion etching; which means the deposition was effectively carried out on the nucleated parts. On the other hand, the XPS peaks of the non-nucleated parts in film which was formed by the parallel irradiation were very weak; the weak peaks resulted from the powder of ZnS compound. It is evident, consequently, the ZnS growth was preferentially performed on the nucleated parts.

And then the crystallinity of the film which was formed by the perpendicular irradiation was inspected with reflection high-energy electron diffraction (RHEED) system, which observed halo pattern. This pattern means amorphous structure.

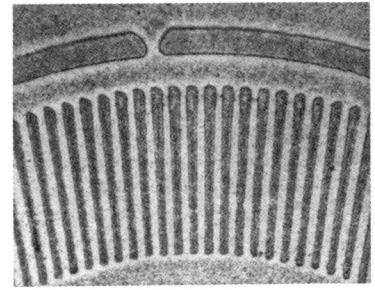 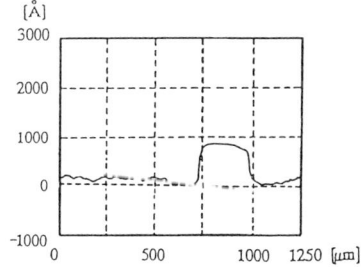

Fig.7 The microscope photograph and the result of film thickness measurement of ZnS thin film obtained by the parallel irradiation method.

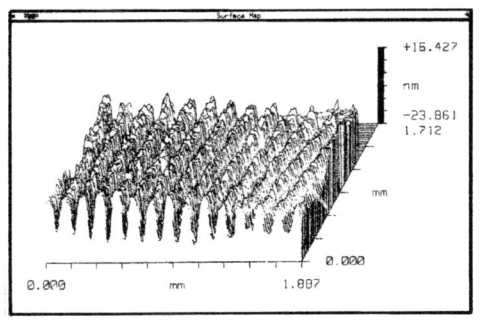

Fig.8 Interference roughness meter image of a patterned ZnS film obtained by the perpindicular irradiation method.

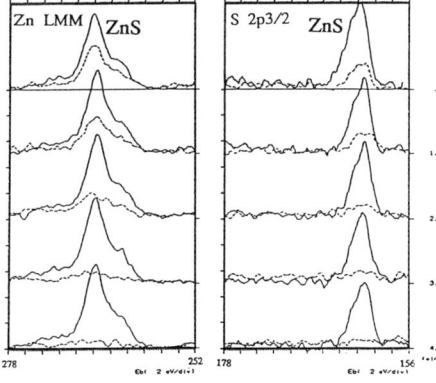

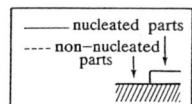

Fig.9 XPS spectra of the nucleated and non-nucleated parts in film which was performed by the parallel irradiation method.

DISCUSSION

It was observed that the parallel irradiation created the surface morphology worse than the perpendicular irradiation performed. Following were considered as its causes. In the parallel irradiation, the radicals which were produced by a photodecomposition inconsistently behaved on the way to reach the surface. Therefore, some of the radicals combined each other in the vapor–phase to produce ZnS powder, as seen in the XPS observation. The powder contained in the films and deposited on the non–nucleated parts, therefore, made the crystallinity and selectivity worse.

On the contrary, the parallel irradiation method induced no damages to the grown films because of no direct irradiation to the substrate. Because of no activation energies supplied to the substrate by the laser irradiation either, the surface migration was not occured, and no effective film formation was performed. An other problem was that not only Zn–CH3 bond but also C–H bond were photodissociated by the high density excitation of the KrF laser; therefore, impurities of carbon atoms were contained in the formed films. It is considered that the parallel irradiation method is not suitable to achieve fine ZnS films. In the perpendicular irradiation, ZnS was grown on the nucleated substrate, while nothing was deposited to the non–nucleated substrate. Therefore, it is considered that Zn seeds play an important role in the perpendicular irradiation. In general, the formation of thin film is carried out by substrate heating in order to activate surface[5]. Active layer can also be created without heating by the nucleation. Therefore, in the case of perpendicular irradiation, only DMZ and H2S molecules adsorbed on the Zn seeds were possibly dissociated with the assist of an excimer laser; then, the film growth was selectively occured. Namely, to achieve a high selectivity requires a weak laser fluence. In addition, in the perpendicular irradiation even if irradiating with the laser perpendicular to the whole substrate, the thin films were not damaged because of its very low fluence such as 3 mJ/cm2. And the surface migration took place; the ZnS was effectively grown on the Zn nucleation parts. Moreover, the low decomposition rate of the gases over the substrate surface resulted in the growth of the fine film.

CONCLUSION

The selective growth of ZnS thin film via Zn nucleation was successfully performed with the excimer laser CVD method. Firstly, the optimum condition for Zn nucleation was found to create Zn seeds. Secondly, ZnS thin films were formed on the Zn seeds by both parallel and perpendicular irradiation methods. As a result, it was confirmed that the growth of ZnS thin film was selectivity performed on the Zn seeds by both methods. Next, the quality of the films created by both methods was compared. The results showed that the films were clearly different in quality. In the parallel irradiation, because of the powder which was produced by the vapor excitation, the selectivity and quality of the film were not fine. On the other hand, in the perpendicular irradiation, the high selectivity and crystallinity of the film were performed by irradiating the substrate surface directly.

Even though a low temperature has been expected in case of using a metal–organic (MO) gas, impurities of carbon atoms were contained into a crystal because the hydrocarbon in MO gas became difficult to desorb at low temperature[6]. While, by the perpendicular irradiation of KrF laser at very low fluence, desorption of carbon atoms could be avoided, so that the film contained no impurities. We successfully demonstrated the growth of the high quality film by using the nucleation technique at room temperature.

In the future, we will study further on a photodecomposition process of the gases by the vacuum−ultra−violet spectroscopy and to clarify the reaction mechanism for film formation.

REFERENCES

1. T. Kawai, and M. Hanabusa, Jpn. Appl. Phys., 32, pp4690−4693 (1993)

2. Nongfan Zhu, Ted Cacouris, Rob Scarmozzino, and Richard M. Osgood, Jr., J. Vac. Sci. Technol. B10, 3, pp1167−1176 (1992)

3. G. S. Higasi, Applied surface Science, 43, pp6−10 (1989)

4. K. Yamane, and M. Murahara, The Institute of Electrical Engineers of Jpn. Symp. Proc., OQD−93−50 (1993)

5. K. Ohkawa, and T. Mituro, A monthly publication of The Jpn. Sci. of Appl. Phys, 63, 3, pp113−119 (1993)

6. S. Iwai, H. Ishiki, and T. Meguro, Extended Abstracts (The 53rd Autumn Meeting), Jpn. Sci. of Appl. Phys, 16aZE/I9, pp177 (1992)

VAPOR DEPOSITED Ag_2S FILMS AS HIGH RESOLUTION PHOTOREGISTERING MATERIALS IN THE INFRARED REGION

J. ENEVA[*], S. KITOVA[*], A. PANOV[*] AND H. HAEFKE[**]
[*]Central Laboratory of Photoprocesses, Bulgarian Academy of Sciences, BG-1113 Sofia, Bulgaria
[**]Institute of Physics, University of Basel, CH-4056 Basel, Switzerland

ABSTRACT

Ag_2S as a narrow band gap semiconductor is appropriate for photoimaging in the infrared (IR) region. Co-evaporation of Ag and S from two separate sources was used for preparing of thin Ag_2S films with different Ag/S ratio. Gelatine subbed glass plates were used as substrates. The structure of the films obtained was examined by transmission electron microscopy and electron diffraction. The effects of chemical composition, film thickness and processing conditions on the photographic parameters were studied.

It is shown that after appropriate processing thin Ag_2S films with stoichiometric composition can be successfully used as high resolution (1600 lines/mm) photographic materials in the IR region.

INTRODUCTION

In the last years there has been a growing interest in silver chalcogenide thin films because of their optical and electrical properties. Applications of Ag_2S in devices like photoconducting cells[1], solar-selective coatings[2], photovoltaic cells[3], infrared detectors[4], etc., have been reported. Moreover, the absorption of silver sulfide in a wide spectral region makes it very attractive as a photoimaging system. Kinoshita has observed that Ag_2S layers as well as bilayer systems of Ag_2S/Ag_2Se, Ag_2S/Ag_2Te change reversibly their optical transmission on irradiation with He-Ne laser ($\lambda = 633$ nm)[5].

It has also been found that a developable latent image can be created in Ag_2S crystals, treated in a solution of noble metal ions before or during illumination[6-8]. This process called *sensitized photolysis of semiconductors* has been employed for developing of emulsions, photosensitive in the IR region[6,7]. The Ag_2S microcrystals have been grown in gelatin suspensions by chemical precipitation. The resolution of Ag_2S emulsions has been determined to be 30-40 lines/mm[8].

Thin-film technology has the potentials to improve the quality of the Ag_2S photographic materials. Evaporation is one of the most important methods for preparing thin films. As known, vapor deposited thin films have some advantages - uniform thickness, high purity, optical homogeneity, etc. Since Ag_2S decomposes before a sufficiently high vapor pressure is reached, its deposition by conventional resistance heating is excluded.

In this study, thin films of Ag_2S were deposited by co-evaporation of silver and sulfur from two separate sources. The structure, optical and photographic characteristics of the Ag_2S films have been investigated to point out their potential as IR-sensitive photographic materials.

EXPERIMENTAL

The Ag_2S thin films used were obtained by simultaneous thermal evaporation of silver and sulfur[9] on gelatin subbed glass substrates[10] in a vacuum of about 1×10^{-4} Pa. Silver was evaporated from a graphite effusion cell situated in an indirectly resistance-heated quartz tube. The silver deposition rate was calibrated through the cell temperature, measured with a Pt/Rh - Pt thermocouple. Sulfur was sublimated from a tubular quartz crucible heated indirectly with a helical tungsten wire. The crucible temperature was also controlled with a thermocouple. The Ag deposition rate was preset to the desired value. After it became constant, the sulfur crucible was heated and the S sublimation rate regulated to a determined value. In order to avoid sulfur deficiency in the evaporated Ag_2S films, all films were grown with a stoichiometric sulfur excess. The films were left in the vacuum chamber until the excess S was reevaporated from the substrate due to its high vapor pressure. The Ag_2S films were deposited at rates $0.15 - 0.2$ nm s^{-1}, defined by the Ag evaporation rate. The film thickness was varied from 2 to 100 nm. The co-evaporation technique described above also allows to obtain well-defined films with different Ag/S ratios.

The structure of the films was examined under high resolution transmission electron microscope (TEM) and electron diffraction (ED). The absorbance of the Ag_2S films was evaluated on the basis of their reflectivity and transmittance, measured with Perkin Elmer spectrophotometer (VIS - IR).

For studying the photographic characteristics, the Ag_2S films were exposed with 1000 W halogen lamp through red filter ($\lambda \geq 650$ nm) and sensitometric step tablet. For determining the resolution of the films contact printing of a chromium test mask (line width ranging from 0.3 to 100 μm) was applied.

The exposed films were treated successively in: 10-40%(wt) etching solution of $AgNO_3$, gold latensification bath[11], stabilized Fe^{2+}/Fe^{3+} physical developer[12] and 2-5%(wt) fixing solution of thiourea containing 0.1n HCl. The developed image was inspected with an optical transmission microscope. The optical density of the image was measured with a Macbeth densitometer.

RESULTS AND DISCUSSION

<u>Structure and optical properties</u>

Under the above-mentioned experimental conditions the structure of the Ag_2S films depends on the thickness, but not on the deposition rate. In the initial stage the Ag_2S films grow in the form of three-dimensional islands. With increasing amount of Ag_2S the formation of larger islands and their coalescence was observed. Owing to liquid-like coalescence, the Ag_2S islands disappear and bare substrate areas are formed in which secondary nucleation occurs. The occurrence of grain boundaries in some islands was also observed. At nominal film thickness larger than 4 nm continuous films with polycrystalline structure were obtained.

Fig. 1 shows an electron micrograph of a 5 nm thick Ag_2S film grown on a gelatin subbed substrate. The film possesses a fine-grained structure which is retained by continued growth of grains during the deposition. The ED pattern inset in Fig. 1 reflects the polycrystalline structure and stoichiometric composition of the Ag_2S films.

The spectral dependence of the absorbance of Ag_2S films is given in Fig. 2. As seen, the evaporated thin Ag_2S films absorb relatively well in the red and IR region. Their absorbance in the visible and deep UV region is higher, but of no practical importance because of the existence of well developed silver halide photographic materials.

Fig. 1 Electron micrograph (a) and electron diffraction pattern (b) of a 5 nm thick Ag_2S film vapor deposited onto a gelatin subbed glass substrate.

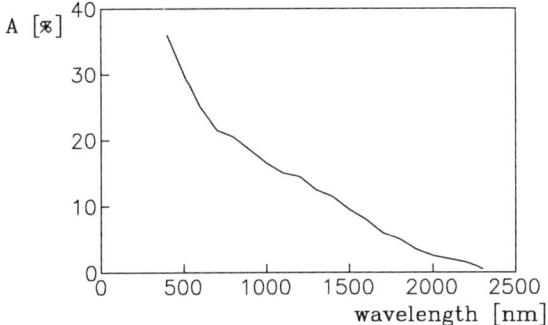

Fig. 2 Dependence of the absorbance (A) on the wavelength for 10 nm thick Ag_2S film.

Photographic response

Ag_2S films with stoichiometric composition and thickness of 2-100 nm were exposed with light energy density within the range of 1×10^{-1} - 1×10^2 J/cm^2. It is found that the direct treatment of the exposed Ag_2S films in the physical developer, acting practically on the film surface, does not lead to a developed image.

It should be noted that a similar effect of undevelopable surface image has been observed in unsensitized, vacuum deposited silver bromide layers[13,14]. In this case it was found that latent image centers formed on the surface were destroyed by the generated photoholes, which are migrating to the surface. As a result, latent image centers are formed predominantly under the surface and in the volume of the silver bromide microcrystals. Their development, however, becomes possible either by etching of the silver halide surface or by using internal developers, which dissolve partially or even completely the silver bromide layers[13,15].

By analogy with silver halide layers, attempts are made to etch the surface of Ag_2S films. Solutions of cyanide or thiocarbamide are used as complexing agents for silver ions, and concentrated solutions of $AgNO_3$, $Hg(CH_3COO)_2$, $Hg(NO_3)_2$ - for sulfide ions. We found that the most appropriate etching agent is a concentrated solution of $AgNO_3$, while cyanide and thiocarbamide solutions dissolve partially the latent image centers. The time of etching as well as the concentration of $AgNO_3$ depend on the thickness of the Ag_2S films[16].

After etching, a treatment of Ag_2S films in a monovalent gold latensification bath and in a stabilized physical developer leads to the growth of the revealed latent image centers to a visible silver image[16]. Typical dependence of the optical density (D) of the image obtained on the exposure energy (H) for a Ag_2S film with stoichiometric composition is given in Fig. 3. The maximal optical density depends on the developing time which is limited, however, by the appearance of "fog". Most probably this "fog" is due to the heating of the films during exposure since the unexposed films treated under the same processing conditions have no "fog".

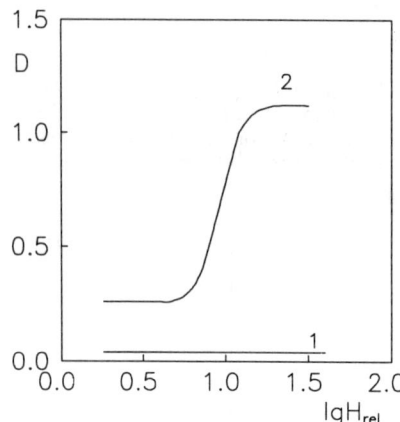

Fig. 3 Characteristic curves (D/lgH) of physically developed Ag_2S films (d = 10 nm) without etching before developing (1) and with etching in 20% solution of $AgNO_3$ (2).

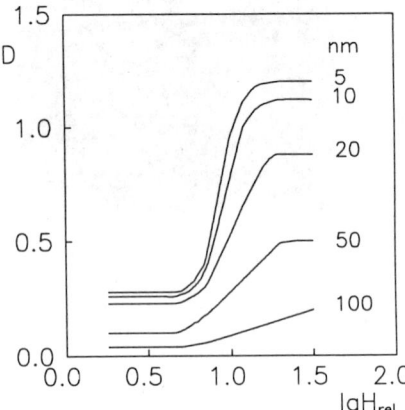

Fig. 4 Characteristic curves of etched and physically developed Ag_2S films with indicated film thickness.

The results obtained are of interest both in theoretical and practical aspects. It is usually assumed[6,7] that developable latent image centers cannot be formed in unsensitized Ag_2S emulsion materials because of the rapid recombination between the photogenerated charge carriers. Only *sensitized photolysis* is possible in semiconductors such as Ag_2S by making use of ions of noble metals adsorbed on their surface[6,7]. Our results, however, show that the excess amount of Ag in the volume leads neither to sensitized photolysis nor to fog in physically developed Ag_2S films.

The image developed in thin, vapor deposited Ag_2S films after etching of their surface is an indication that on illumination the process of photolytic decomposition prevails over the electron-hole recombination even in the absence of sensitizers. Our results, however, should not be regarded as contradicting to the work of the authors cited above. The results only indicate that stable latent image centers are formed predominantly in the volume of Ag_2S microcrystals. As known, due to the very low solubility product of Ag_2S only physical developers acting practically on the surface can be used. Therefore the latent image centers or excess amount of Ag in the volume are not able to contact with them and cannot catalyze the development.

The so called *sensitized photolysis* makes possible the formation of surface latent image centers, developing directly in the physical developer. This effect was also observed in evaporated Ag_2S films treated in $AgNO_3$ solution before or during exposure. The image developed however has considerably coarser grained structure than that obtained after etching.

The experiments performed show a substantial decrease of the photographic response with increasing of the Ag_2S film thickness (Fig. 4). The result is unexpected because of the higher light absorption per unit area for the thicker films. It should be noted that the analogous drop of photographic response is also observed in the films with an amount of sulfur over the stoichiometric ratio. Most probably the accumulation of photolytic or surplus sulfur is the reason for decreasing of the photographic response. One of the most acceptable explanations of the results described is that the accumulation of sulfur (photolytic or overstoichiometric) in the Ag_2S volume increases the possibility of both the recombination process between photogenerated charge carriers and the destroying of the photolytic Ag specks. This assumption is further supported by the observation that the polymer overlayer with acceptors of sulfur leads to increasing the photographic response.

Resolution

Fig. 5 shows optical micrographs of a copy of the chromium test mask obtained with 10 nm thick, stoichiometric, underexposed Ag_2S films by contact printing and processing. The smallest line of 0.3 μm of the Cr test mask is resolved, but the quality of the reproduced details is not very high. Although the film is underexposed, black dots are observed in the white regions, probably due to the heating of the film. Because of the underexposure, the recorded details have coarse grained structure. It is seen from the figure that the structure of the edges additionally exposed with light reflected from the substrate is better.

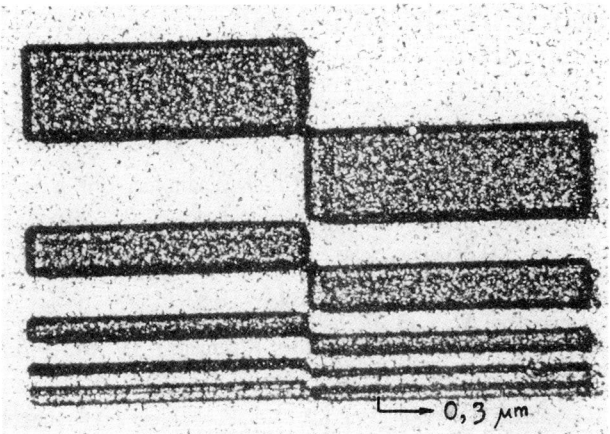

Fig. 5 Optical micrographs (transmitted light) of a copy of Cr test mask obtained by underexposure, etching, physical developing and fixing.

The obtained copy gives only a rough evaluation of the reproduction ability of Ag_2S films in the micron and submicron region. The further optimization of their preparation, exposure and processing conditions is necessary. Preliminary experiments show that the quality of the recorded details may be improved by an overlayer consisting of gelatin and sulfur acceptors as salts of H_2SO_3, HNO_2, H_2PO_3. It may further be expected that cooling of the films during exposure and applying heat filter will substantially decrease the "fog", thus allowing the optimal exposure of 5-10 J/cm^2, leading to finer grained structure of the details to be used.

It should be stressed that resolving of the smallest line of 0.3 μm by contact printing corresponds to resolution of 1600 lines/mm. The resolution of photographic materials for the IR region, known so far, is 30-40 lines/mm at a sensitivity of $1 \times 10^{-3} J/cm^2$. The vapor deposited Ag_2S films have 800-1000 times lower content of Ag and about 4 orders of magnitude lower sensitivity but the high resolution makes them appropriate for spectroscopy, semiconductor laser technique, temperature relief registration. It is also shown that the Ag_2S films with stoichiometric composition and excess amount of silver can be successfully used for the development of a non-erasable laser recording medium, type "Drexon"[17].

CONCLUSION

The results show that the method of co-evaporation used in this work allows deposition of stoichiometric Ag_2S films with reproducible properties. Further well-defined films with different Ag/S ratio can also be obtained. The structure of the films depends on their thickness, but not on the deposition rate. The experiments show that films with thickness ≥ 4 nm are continuous with polycrystalline structure and can be used as IR-sensitive photographic materials. Upon illumination photolytic decomposition takes place in thin vapor deposited Ag_2S films. As a result latent image centers or photolytic Ag specks are formed predominantly in the volume of Ag_2S microcrystals. It is shown that their growth in a stabilized physical developer become possible after appropriate etching of the exposed Ag_2S films, followed by treating in a latensification bath.

It is found that the Ag_2S films exhibit high resolution of 1600 lines/mm at a sensitivity of 5-20 J/cm^2. The quality of the recorded micron and submicron details depends strongly on the film thickness, Ag/S ratio and processing conditions. It should be noted that the results described are obtained with very thin (5-20 nm) Ag_2S films. For thicker films, the accumulation of photolytic sulfur creates serious problems.

REFERENCES

1. M.J. Mangalam, K.N. Rao, N. Rangarajan and C.V.Syryanarayana, Brit. J. Appl. Phys. **2**, 1643 (1969).
2. A.K. Abass, Solar Energy mater. **17**, 375 (1988).
3. S.S. Dhumureand and C.D.Lokhande, Solar Energy mater. Solar Cells, **28**, 159 (1992); **29**, 183 (1993).
4. T.S. Hsu, H. Buhay and N.P. Murarka, SPIE, **259**, 39 (1980).
5. A. Kinoshita, Japan J. Appl. Phys. **13** 1027 (1974).
6. G. Parizki and S. Rivkin, Zh. Nauch. Prikl. Fotogr. Kinematogr. **15**, 184 (1970).
7. A. Sablin-Yavorski, V. Vigant and V. Fridkin, Zh. Nauch. Prikl. Fotogr. Kinematogr. **18**, 389 (1973).
8. A.C.Heimann, B.I. Shapiro, V.P. Donatova, K.V. Vendrowski and A.I.Haritonova, SU Authorship No. 593 568 (25 March 1979).
9. H. Haefke, A. Panov and V. Dimov, Thin Solid Films, **188**, 133 (1990).
10. S. Kitova, H. Alexandrova and J. Malinowski, J. Imaging Technol. **15**, 126 (1989).
11. T. James, W. Vanselow and R. Quirk, RSA Journ. **14**, 349 (1948).
12. J.P. Archambault, N.H. Nushua and W.S. Holmes, U.S. Patent No 3 672 899 (26 May 1972)
13. J. Eneva and J. Malinowski, J. Photogr. Sci. **22**, 273 (1974).
14. H. Haefke, M. Krohnt, J. Eneva, J. Photogr. Sci. **40**, 70 (1992).
15. J. Eneva, J. Photogr. Sci. **29**, 181 (1981).
16. H. Haefke, J. Eneva, S. Kitova and A. Panov, BD Patent DD No. 286 885 (7 February 1991).
17. H. Haefke, J. Eneva, S. Kitova and A. Panov, BD Patent DD No. 276 938 (14 March 1990).

PART III

Metallization

PROBING SELECTIVE DEPOSITION OF ALUMINUM USING IN SITU INFRARED SPECTROSCOPY

WAYNE L. GLADFELTER, MICHAEL G. SIMMONDS, LARRY A. ZAZZERA, and JOHN F. EVANS
Department of Chemistry, University of Minnesota, Minneapolis, Minnesota 55455

ABSTRACT

Dimethylethylamine alane (DMEAA) has been used to deposit thin films of aluminum selectively on gold in the presence of silicon oxide. This paper presents studies of the effect of temperature, pressure, and substrate pattern on the selectivity. A specially designed reactor allowed us to probe the structure of the species on the silica surface using infrared spectroscopy. At room temperature two absorptions in the Al-H stretching region were assigned to two species; weakly bound molecules of the intact precursor and a strongly bound dihydride formed from the reaction of DMEAA with surface bound (and H-bonded) hydroxyls (from H_2O or silanol groups). At higher temperatures the CH vibrations of the amine disappeared, and the Al-H stretch shifted to higher energy. A weak absorption at 2250 cm^{-1} attributable to a Si-H also appeared. The impact of these observations on the loss of selectivity is discussed.

INTRODUCTION

Pure aluminum films can be deposited from amine complexes of alane (AlH_3) at high deposition rates.[1] One of the most challenging goals in the area of CVD is to control the chemistry of a deposition in such a way that the process becomes selective and film growth occurs only on one surface in the presence of others. Incorporation of selective deposition into device production may eliminate some of the lithographic steps. The current study presents an appraisal of selective deposition of aluminum using dimethylethylamine alane (DMEAA).[2]

Throughout the study, selectivity was defined in terms of the quantifiable parameter, $^{gs}S_{ns}$, a description of which has already been provided in the literature[3] and which is defined in equation 1.

$$^{gs}S_{ns} = (\theta_{gs} - \theta_{ns})/(\theta_{gs} + \theta_{ns}) \qquad (1)$$

θ_{gs} and θ_{ns} are the fractional surface coverages of Al on the growth and non-growth surfaces, respectively, and were determined in the present study by scanning electron microscopy (SEM). To understand the reason the growth was slower on insulating surfaces such as SiO_2 we studied the nature of the interaction of this surface with DMEAA using infrared spectroscopy.[4]

EXPERIMENTAL

The wafers used for this study were fine line test patterns (6x6 mm) containing varying widths of alternating Au and SiO_2 strips adjacent to one another. These were cleaned using the following sereies of treatments; a 4:1 H_2SO_4(98%):H_2O_2(30%) aqueous solution, HF (0.5%), deionized water (18.2 MΩ-cm), and exposure to UV irradiation in air. The sample was then immediately transferred into the reactor via a load-lock.

The details pertaining to the ultra-high vacuum (UHV) compatible stainless steel system used to deposit Al films, as well as to the use of DMEAA, have already been reported.[5] The tube furnace described in the previous study was replaced by a six way cross having a resistively heated filament at the center. The wafers were positioned over the filament and supported on a stainless steel/Mo holder which was heated to 100-200°C. Under typical deposition conditions, a thermocouple was permanently attached to the heating system and another was in contact with the substrate during the time allowed for equilibration of the system.

Most of the depositions were peformed with a H_2 flow rate of 100 sccm at a partial pressure of 3.0 torr. During the flow of DMEAA, the total system pressure was 3.3-3.4 torr. Experiments were also performed at 2.5×10^{-4} and 2.5×10^{-5} torr using a turbomolecular pump to pass only DMEAA through the system. Under these conditions, Al films were grown on the Au/SiO_2 test patterns at a temperature of 100°C for 3.5 hours.

Analysis of the coverage of the wafer surfaces was performed with a SEM, and selectivity was calculated according to equation 1. Auger electron spectroscopy (AES) and X-ray photoelectron spectroscopy (XPS) established the chemical composition of the surfaces.

In Situ Infrared Analysis

The design and construction of the attenuated total reflection (ATR) cell used here has been reported elsewhere.[6] The cell was modified for operation under rough vacuum by using flexible stainless steel tubing with VCO™ connectors for the introduction and removal of the various processing agents. The deposition pressure was measured with an Inficon™ capacitance manometer. Infrared spectra were recorded with a Bruker model IFS-113V spectrometer equipped with a liquid nitrogen cooled MCT detector. The spectral range from 4000-1500 cm^{-1} was investigated at a resolution of 2 cm^{-1} by signal averaging 60, 120 or 240 scans acquired during one, two or four minute intervals, respectively. The beam, normally incident on the input bevel of the crystal, was internally reflected a total of 25 times from both sides. However, due to the geometric constraints of the cell, only 8 of these reflections sampled the area of the Si (100) crystal exposed to the DMEAA. Absorbance spectra were generated by taking the log of the ratio of a single beam spectrum collected when only N_2 gas was flowing over the substrate to a single beam spectrum recorded during or after DMEAA exposure.

Substrate Cleaning Procedure for the IR Studies

The Si(100) substrates were first cleaned ex situ by a $H_2SO_4:H_2O_2$ (4:1)-water rinse-$HF:H_2O$ (1:100)-water rinse procedure, followed by a UV/ozone treatment. The ex situ cleaning step was designed to remove both organic and metal impurities from the silicon surface.[7] The final UV/ozone cleaning was performed in situ by exposing the silicon surfaces to deep UV radiation (184.9 nm) in an oxygen ambient for at least 10 minutes. This cleaning procedure yielded a thin (1 nm) hydrated layer of silicon oxide on the Si(100) surface.

Deposition Procedure

Following the final in situ UV/ozone clean, the IR cell was purged with N_2 at a flow rate of 100 sccm. A 90 minute purge time was required to allow the internal reflection element (IRE) to stabilize at the desired temperature, thus minimizing fluctuations in the IR absorbance due to lattice vibrations. The Si(100) surface was exposed to DMEAA vapor at substrate temperatures of 25°C, 100°C and 145°C by redirecting the N_2 flow through a glass bubbler containing 10-20 ml of the liquid precursor. Depositions were concluded by again allowing the N_2 to bypass the bubbler and to purge the IR cell. This procedure was carried out at both atmospheric pressure and 8 torr.

The experiments conducted at 8 torr were performed by pumping on the system (using a mechanical pump fitted with a liquid nitrogen trap) that initially had a N_2 flow of 100 sccm at 1 atm. The capacitance manometer was connected on the upstream side of the cell via 20 inches of 1/4 inch diameter stainless steel tubing so that the actual pressure inside the cell was slightly lower than the nominal value of 8 torr. Because of the pumping arrangement, the flow rate of N_2 increased substantially. The slow flow of N_2 through the precursor vessel at atmospheric pressure allowed us to estimate the partial pressure of DMEAA in the cell at close to the room temperature equilibrium vapor pressure (1.5 torr) of DMEAA. At the higher flow rate, the partial pressure of DMEAA was estimated to be a few tenths of a torr similar to its known partial pressure under in the actual CVD process.[5]

RESULTS

Selective Deposition on Patterned Wafers

Depicted in Figure 1 are SEM images of Au and SiO_2 strips present on a fine line test pattern following 4 minutes exposure to DMEAA in a flow of H_2, and at substrate temperatures of 100,

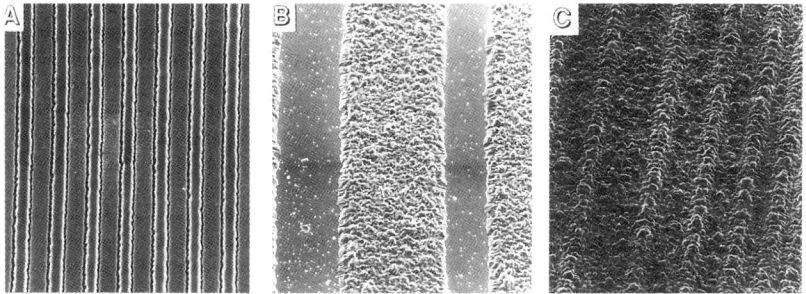

Figure 1. Effect of increasing the Au/SiO_2 wafer temperature. Wafer A (T = 100°C, $^{Au}S_{SiO_2}$ > 0.99, Magnification = 2000x). Wafer B (T = 160°C, $^{Au}S_{SiO_2}$ = 0.65, Mag. = 700x). Wafer C (T = 200°C, $^{Au}S_{SiO_2}$ = 0, Mag. = 1000x).

160, and 200°C. In all cases the Au regions were fully covered. At 100°C the surface of the SiO_2 strips appeared featureless, even when examined at high magnification (150 K), however, at 160°C, Al particles appeared on SiO_2 while a rough continuous Al film grew on Au. The selectivity of this deposition was 0.65. No selectivity was observed for films deposited at 200°C ($^{Au}S_{SiO2} = 0$).

Examination of a wafer processed at a temperature of 160°C in cross-section established that Al had encroached over the Au/SiO$_2$ boundaries and grew onto the edges of the SiO$_2$ strips. At longer DMEAA exposure times for wafers at 100°C, selectivity was eventually diminished as evidenced by the presence of Al particles on SiO$_2$. It was interesting that in the Au/SiO$_2$ boundary regions for this wafer, not only had Al encroached onto the SiO$_2$ surfaces, but regions depleted in Al particles were present along the edges of the SiO$_2$ strips. This was a common observation for those films which were processed at 100°C which led to enhanced values of $^{Au}S_{SiO2}$ on the narrower SiO$_2$ strips.

Aluminum films grown at 10^{-4} torr and 10^{-5} torr at 100°C also selectively deposited onto Au (grain size > 1000 nm). From a film thickness of 2000 nm, the growth rate on Au at 10^{-4} torr was 10 nm/min and the selectivity of the deposition was 0.98 after 3.5 hours.

In Situ Infrared Spectroscopy

Figure 2 displays a typical infrared spectrum of the native-oxide-covered Si(100) surface during exposure to DMEAA (P_{DMEAA}~1 torr, P_t=1 atm. at 25°C). Three general features are shown which were also observed at reduced pressure and elevated substrate temperatures. The positive absorptions at ca 1800 cm^{-1} and 2900 cm^{-1} indicated net increases in the amount of Al-H and C-H species, respectively, on the oxide surface during DMEAA exposure. The broad negative value at 3300 cm^{-1}, assigned[8] to OH stretching. represents the net consumption or removal of adsorbed water. At higher temperatures this negative peak shifts towards higher energy consistent with consumption of H-bonded silanols.

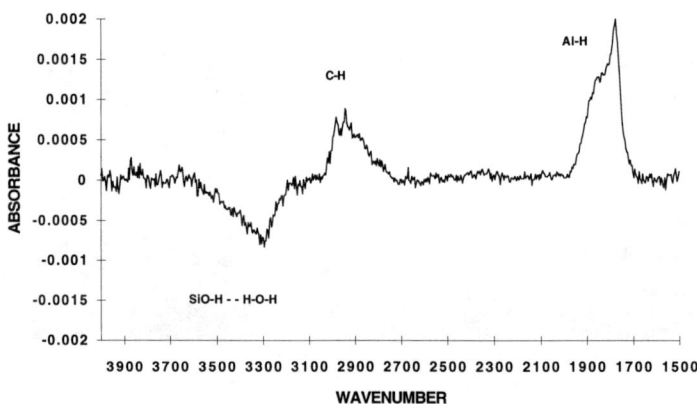

Figure 2. Infrared spectrum of the oxide covered Si(100) surface during exposure to DMEAA vapor (approx. 1 Torr in a 100 sccm flow of N$_2$ at ambient pressure and 25°C).

The in situ infrared spectral measurements allowed an examination of the kinetics of the DMEAA-substrate reaction. This established that the Al-H and C-H intensities increased concurrently, when the DMEAA flow was initiated, reached a plateau after 30 minutes of DMEAA exposure, and decreased concurrently when the DMEAA flow was stopped.

It was of particular interest to monitor the Al-H signal as a function of adsorption time in order to observe the possible interactions between the DMEAA precursor and the oxidized surface. Expanded views of this region after exposure to DMEAA vapor for 1, 4, and 60 minutes are shown in Figure 3. The Al-H band is comprised of two distinct contributions indicating that at least two different chemical environments were present. During the early stages of adsorption, the precursor was characterized by a relatively broad band centered about 1850 cm^{-1}. As the reaction time elapsed, an additional peak centered at 1780 cm^{-1} evolved.

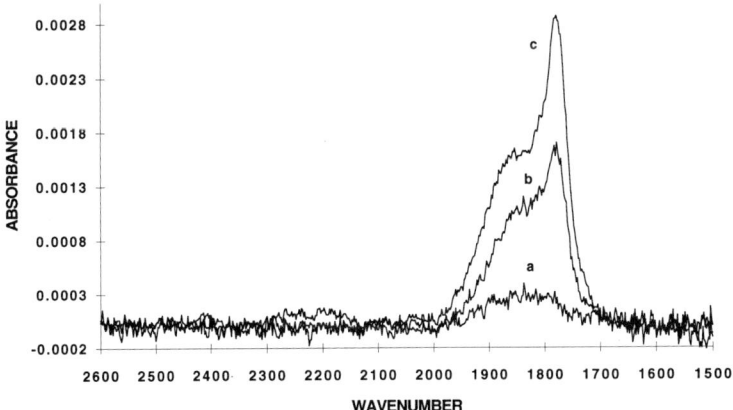

Figure 3. Infrared spectra of the Al-H stretching region during DMEAA adsorption onto the silicon oxide surface after (a) 1 minute, (b) 4 minutes and (c) 60 minutes of DMEAA exposure. The exposure conditions and temperature were the same as in Fig. 1.

At the reduced total pressure (nominally 8 torr), DMEAA adsorption onto native Si oxide substrates yielded spectra in the Al-H stretching region which differed somewhat from those obtained in atmospheric pressure experiments. Under these conditions the intensity of the Al-H stretch at 1780 cm^{-1} was less pronounced with respect to the higher wavenumber Al-H contribution, and the intensity did not increase with longer exposure times.

Because the deposition of Al films commenced when metal substrates were heated to temperatures as low as 100°C in the presence of DMEAA, the adsorption of DMEAA on the oxide surface was also characterized at this temperature (typical for a selective deposition), as well as at 145°C (a temperature where selectivity begins to degrade). Figure 4 depicts the Al-H stretching region during adsorption of DMEAA vapor onto the native oxide of Si at substrate temperatures of 25°C, 100°C and 145°C. The cell was at ambient pressure and the elapsed time of DMEAA adsorption was 60 minutes in each case. At 100 and 145°C the negative peak in the O-H stretching region shifted to between 3500 and 3600 cm^{-1} indicative of the consumption of hydrogen-bonded silanol (Si-OH) groups. Figure 4 shows that the intensity of the Al-H band at 1780 cm^{-1} decreased upon heating. Heating the substrate above 25°C also shifted the relatively broad Al-H band at 1850 cm^{-1} to higher energy. Specifically, at 100°C the band maximum occurred at 1885 cm^{-1} and reached a steady state value after only 10 minutes. The integrated C-H intensity decreased by a factor of 2-3 times. At 145°C the Al-H band occurred near 1915 cm^{-1} and reached intensity values 2-3 times greater than those found on the cooler substrates. In the case of the 145°C experiment, the Al-H signal continued to increase throughout the time of the experiment. In the spectra obtained at 145°C (Figure 5), the appearance of a weak absorption at

2250 cm^{-1} became noticeable. This band was assigned to a Si-H stretch which resulted from partial reduction of the native silicon oxide by reaction with DMEAA.

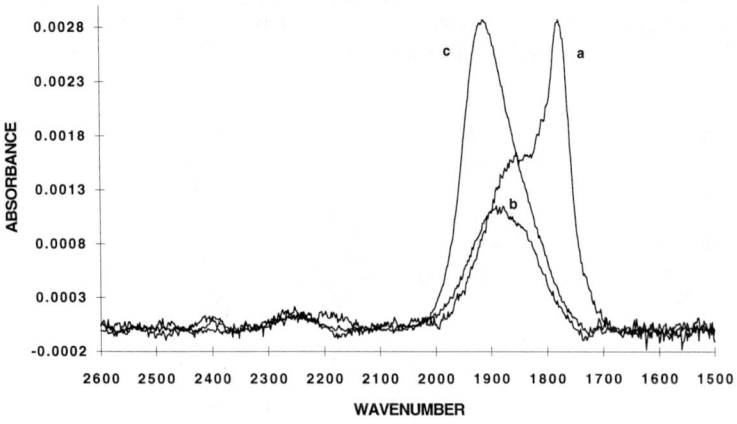

Figure 4. Infrared spectra of the Al-H stretching region during DMEAA adsorption onto the silicon oxide surface at substrate temperatures of (a) 25°C, (b) 100°C and (c) 145°C.

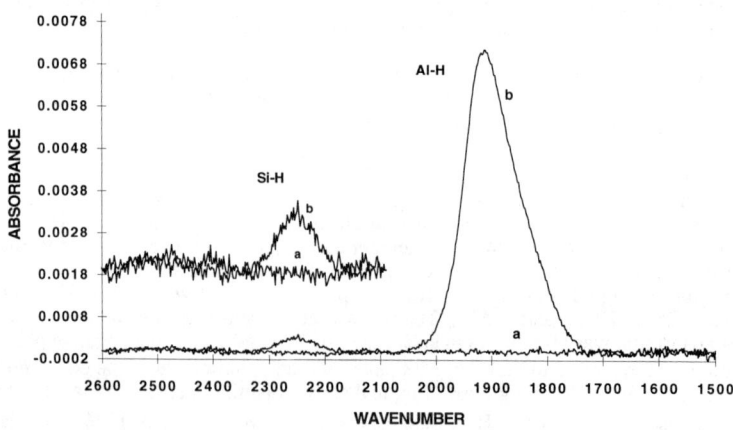

Figure 5. Infrared spectra of the Si-H and Al-H stretching regions (a) before and (b) during DMEAA adsorption onto the silicon oxide surface maintained at 145°C.

DISCUSSION

The present study demonstrates that DMEAA deposited Al readily on a metallic Au surface but not on SiO_2. At a substrate temperature of 100°C, and after a 4 minute DMEAA exposure time, Al was deposited with nearly perfect selectivity ($^{Au}S_{SiO_2} > 0.99$). At longer times, selectivity was eventually diminished. The effect of increasing the wafer temperature was more dramatic. Above 140°C, selectivity dropped off rapidly and by 180°C the process was unselective ($^{Au}S_{SiO_2} = 0$). Decreasing the partial pressure of DMEAA in the system by several orders of magnitude and removing the H_2 carrier gas had no significant effect.

Deposition on Metallic Surfaces

Although the mechanism of DMEAA decomposition on Au has not been investigated, the decomposition mechanisms of trimethylamine alane and triethylamine alane on single crystal surfaces of Al have been investigated under UHV conditions.[9,10] While the rate determining step was not the same, the elimination of hydrogen from the metallic surface was the ultimate source of electrons for the reduction of Al. This may help to explain the facile growth of Al from amine-alane adducts on Au and other metallic surfaces. When the precursor adsorbs to metals, hydrogen atoms are dispersed over the surfaces and H_2 loss is able to occur under relatively mild conditions. Hydrogen desorption takes place on clean Au surfaces at around 200K.[11] Similar mechanisms are not likely on insulating surfaces such as SiO_2.

DMEAA Interactions with Silicon Oxide Surfaces

The results of the current study indicated that chemistry occurred when DMEAA was exposed to the surface of silicon oxide at all of the temperature and pressure conditions examined. This was evidenced by the appearance of positive and negative absorptions in the surface IR spectra. A typical spectrum (Fig. 2) for an ambient cell pressure and substrate temperature shows the presence of three main peaks. The positive band at 2900 cm^{-1} was characteristic of C-H stretching and was interpreted as being indicative of the presence of molecularly adsorbed DMEAA or adsorbed amine that was no longer complexed to aluminum.

The positive signal occurring at around 1800 cm^{-1} (Fig. 3) resulted from a convolution of absorption bands of at least two different species. The sharper band appearing at lower energy has a peak maximum occurring at 1780 cm^{-1} which was similar to the stretching frequency for Al-H in the gas phase molecule of DMEAA (1790 cm^{-1}).[5] We assigned this signal to the presence of the intact precursor complex which was weakly adsorbed to the silicon oxide surface. The band at 1850 cm^{-1} was considerably broader and had a similar energy to related alkoxide model compounds.[12] For example, the IR spectrum of [i-$C_4H_9OAlH_2$]$_2$ was reported to have an Al-H stretching band centered at 1842 cm^{-1}. We assigned the band at 1850 cm^{-1} to an Al-H stretch in an adsorbed dihydride species of the type $H_2AlOSiO_3$ formed from the reaction of DMEAA with H2O (at room temperature) and SiOH groups (elevated temperatures).

The breadth in the Al-H peak suggested inhomogeneity in the chemical environment around Al. For example, it may represent the presence of both $H_2AlOSiO_3$ and $HAl(OSiO_3)_2$ species (mono and dihydride) and mixing of different donors in the fourth coordination site. In support of this, the alkoxide compound (t-$C_4H_9O)_2AlH$ (a monohydride with v_{Al-H} (Nujol) at 1859 cm^{-1}) was reported to exhibit an Al-H stretching signal shifted to higher wavenumbers by 17 cm^{-1} relative to (t-$C_4H_9O)AlH_2$.[13] The amorphous nature of the oxide surface also may have contributed to the width of this band. The negative absorption of the O-H stretching band after exposure of the silicon oxide surface to DMEAA resulted from a reaction between DMEAA and residual adsorbed water molecules and, at higher temperatures, surface SiOH groups to form H_2 and strong Al-O bonds.

A recent XPS study of the adsorption behaviour of TMAA in a UHV environment concluded that a weakly bound layer of TMAA is present at room temperature on surfaces of SiO_2.[14] It was suggested that these species are coordinated to the surface via a lone pair of electrons from the oxygen atom in a Si-O-Si suface structure. Because of the broad nature of the Al-H bands in our IR spectra, we cannot rule out the possibility that these types of species were also present. We anticipate that the five-coordinate species suggested by those authors would have an absorption band centered at significantly lower wavenumber than observed in our study. This argument is based upon the reported range of 1711-1725 cm^{-1} for the Al-H stretch in the gas phase IR spectrum of $[(CH_3)_3N]_2AlH_3$.[15]

The evidence presented above suggested that the precursor was present in both a molecularly adsorbed state and a partially reacted state on hydroxylated silicon oxide surfaces during exposure to DMEAA. Figure 6 illustrates the proposed initial interaction between the SiO_2 surface and the precursor. The band at 2250 cm^{-1}, which was observed most readily at elevated temperatures, was assigned to Si-H stretching in species of the type O_3Si-H. Extensively oxidized silicon hydrides formed during the air oxidation of hydrogenated a-Si films[16] exhibit a peak at 2260 cm^{-1}. Similar bands have been observed when hydride terminated silicon surfaces were exposed to UV/ozone in the same cell. Outlined in Fig. 6 is a possible reaction which accounts for the presence of the Si-H species. The reaction involves cleavage of bridging siloxane bonds on the silicon surface and concomitant formation of Si-H and Al-O bonds. This helps to account for the greater number of oxygen atoms thought to be bound to Al at the higher temperatures. A related reaction involving methyl groups has been reported for trimethylaluminum adsorbing onto heated surfaces of silicon oxide.[16] It is noteworthy that the temperature at which selectivity is lost corresponds to the temperatures where Si-H groups are observed in the infrared spectrum. Whether or not a cause-and-effect relationship exists between these observations is unknown.

SUMMARY

The selective deposition of aluminun on gold in the presence of SiO_2 has been demonstrated, and the effect of parameters such as time, temperature, and pressure have been studied. In situ infrared spectroscopy established that the aluminum precursor, DMEAA, when passed over the surface of silicon oxide, was found to adsorb weakly as an intact molecule and more strongly after reacting with surface-bound hydroxyls in the form of water or silanols. When the temperature of the SiO_2 surface was increased, an additional band in the infrared spectrum, attributed to the presence of Si-H bonds, indicated that the substrate itself was partially reduced. This behavior differs greatly from the facile deposition of metallic Al that occurs with this precursor on metal surfaces and which leads to the selective chemical vapor deposition of aluminum.

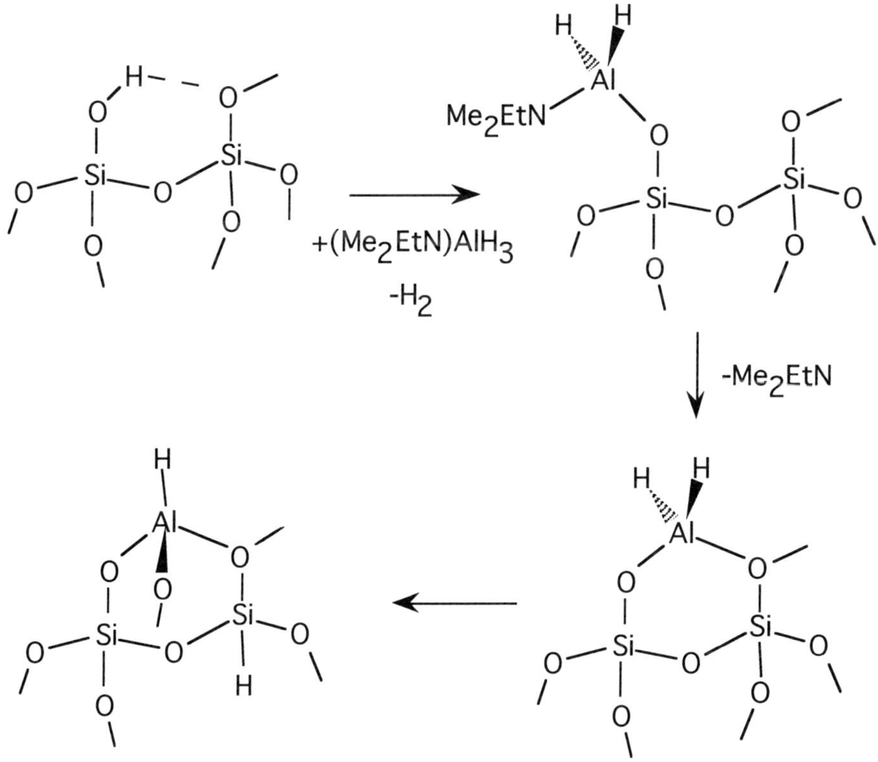

Figure 6. Pictorial representation of the proposed chemical reactions between DMEAA and hydroxylated SiO_2.

ACKNOWLEDGEMENTS

This work was supported by the Center for Interfacial Engineering, a National Science Foundation Engineering Research Center.

REFERENCES

[1] M. G. Simmonds and W. L. Gladfelter, in *Chemical Aspects of Chemical Vapor Deposition for Metallization*, edited by T. T. Kodas and M. J. Hampden-Smith (VCH Publishers, New York, 1993) pp. in press.

[2] M. G. Simmonds, I. Taupin and W. L. Gladfelter, Chem. Mater. submitted for publication (1993).

[3] W. L. Gladfelter, Chem. Mater. **5**, 1372 (1993).
[4] M. G. Simmonds, L. A. Zazzera, J. F. Evans and W. L. Gladfelter, Chem. Mater. submitted for publication (1993).
[5] M. G. Simmonds, E. C. Phillips, J.-W. Hwang and W. L. Gladfelter, Chemtronics **5**, 155 (1991).
[6] L. A. Zazzera and J. F. Evans, J. Vac. Sci. Technol. **A11**, 934 (1993).
[7] L. A. Zazzera and J. F. Moulding, J. Electrochem. Soc. **136**, 484 (1989).
[8] Iler, in *The Chemistry of Silica*, edited by (John Wiley and Sons, New York, 1979) pp. 622.
[9] L. H. Dubois, B. R. Zegarski, C.-T. Kao and R. G. Nuzzo, Surf. Sci. **236**, 77 (1990).
[10] L. H. Dubois, B. R. Zegarski, M. E. Gross and R. G. Nuzzo, Surf. Sci. **244**, 89 (1991).
[11] E. Lisowski, L. Stobinski and R. Dus, Surf. Sci. **188**, L735 (1987).
[12] S. Cucinella, A. Mazzei and W. Marconi, Inorg. Chim. Acta, Rev. **4**, 51 (1970).
[13] v. H. Nöth and H. Suchy, Zeitschrift für Anorg. Allg. Chem. **358**, 44 (1968).
[14] F. M. Elms, R. N. Lamb, P. J. Pigram, M. G. Gardiner, B. J. Wood and C. L. Raston, J. Chem. Soc., Chem. Commun. 1423 (1992).
[15] C. W. Heitsch and R. N. Kniseley, Spectrochimi. Acta **19**, 1385 (1963).
[16] M. E. Bartram, T. A. Michalske and J. W. Rogers Jr., J. Phys. Chem. **95**, 4453 (1991).

INFLUENCE OF FLOW DYNAMICS ON THE MORPHOLOGY OF CVD ALUMINUM THIN FILMS

DONNA M. SPECKMAN, DENISE L. LEUNG, and JERRY P. WENDT
The Aerospace Corporation, Electronics Technology Center, P.O. Box 92957, Los Angeles, CA 90009

ABSTRACT

Aluminum thin films were deposited by chemical vapor deposition on SiO_2 substrates using trimethylamine alane (TMAA) in a low pressure CVD reactor system. A high TMAA flow rate during deposition, combined with an initial burst of added argon during the nucleation of the substrate surface resulted in the growth of aluminum thin films with excellent purity and surface morphologies. Film resistivities averaged 3.4 $\mu\Omega$-cm, and the average surface peak-to-valley height for each film was found to be <4% of the film thickness. The surfaces of films with thicknesses of ≤ 1 μm were extremely smooth and reflective. In contrast, the use of a high alane flow rate in the absence of any added argon resulted in the growth of films with extremely textured surface morphologies. Furthermore, films grown using an argon carrier gas, but with a slow alane flow rate, exhibited both textured surface morphologies and whisker growth.

INTRODUCTION

Inadequate aluminum step coverage over increasingly harsh contact window and via hole topographies continues to be a problem for advanced silicon integrated circuits. Increasingly, the conventional, line-of-site sputtering techniques that are currently employed to deposit these materials are not meeting the step coverage demands for state-of-the-art circuits. Chemical vapor deposition (CVD) is a technique that can deposit thin films conformally over a variety of topographies, and the use of aluminum CVD for interconnect applications has received ongoing attention. Trimethylamine alane ($\{CH_3\}_3N)AlH_3$; TMAA), a stable, white solid with an appreciable vapor pressure (~2 Torr at 25°C), has shown promise for the deposition of high purity aluminum thin films [1]. However, despite relatively high film purities, aluminum films grown using TMAA typically suffer from a highly textured surface morphology which is unsuitable for VLSI photolithographic processing. Earlier work performed in this laboratory demonstrated these highly textured films, and in addition, showed significant aluminum whisker growth on SiO_2 substrates [2]. We wish to report here the results from a series of aluminum deposition experiments we have carried out using TMAA in a new, low pressure CVD reactor. The influence of precursor and carrier gas flow rates on aluminum thin film quality and morphology will be described.

EXPERIMENTAL

Two variations of a cold wall, low pressure organometallic chemical vapor deposition (OMCVD) system were built for our aluminum deposition experiments. The first variation, used for high alane flow rate studies, is shown in Figure 1a. A stainless steel deposition chamber was attached via a gate valve at the base of the chamber to both a mechanical roughing pump and a turbomolecular pump, and the TMAA bubbler was attached directly to an inlet port at the top of the chamber. The substrate was situated on a resistively heated aluminum stage that was mounted perpendicular to the precursor gas stream. An argon line was attached to the chamber for backfill purposes, but no carrier gas line was connected to the TMAA bubbler. This configuration allowed for a high alane flow rate through the deposition chamber by allowing the bubbler to be open to the roughing pump during deposition.

The second CVD reactor variation, shown in Figure 1b, was a modification of the first design. An argon carrier gas line and mass flow controller were added to the inlet side of the

TMAA bubbler, to allow for carrier gas transport of the alane to the deposition chamber. A pressure transducer/flow valve feedback loop was placed between the TMAA bubbler outlet and the chamber inlet, in order to regulate the TMAA bubbler pressure. By maintaining the bubbler at a constant pressure, a constant TMAA/carrier gas mole fraction could be maintained. This second configuration allowed for a study of the effects of low alane flow rate combined with a carrier gas transport.

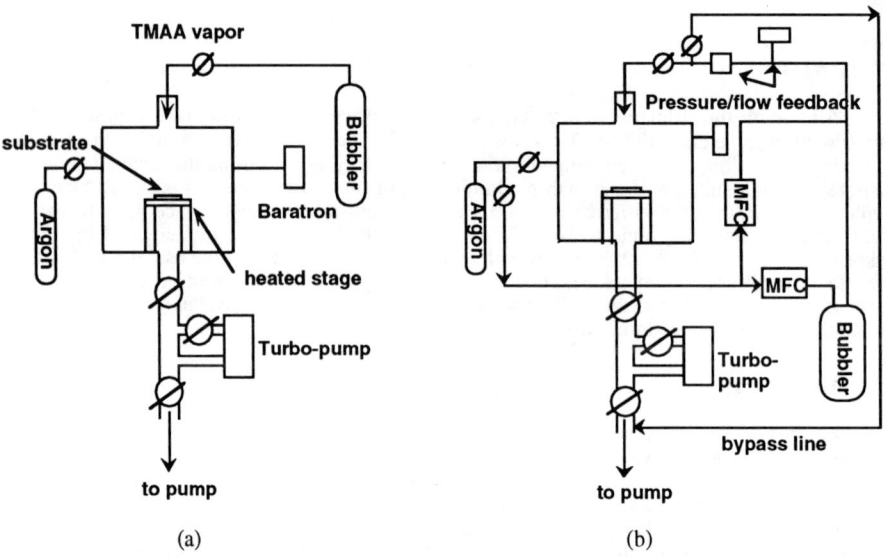

Figure 1. Deposition reactor designs: (a) high alane flow rate configuration, and (b) carrier gas transport, low alane flow rate configuration.

For all of our experiments, the substrates used were 2" diameter, 1000 Å SiO_2 on Si(100). The TMAA bubbler, which was maintained at 25°C, was also filled with one atmosphere of argon whenever a deposition experiment was not in progress, in order to minimize premature trimethylamine dissociation and subsequent decomposition of the TMAA precursor. All of the films deposited from TMAA for these studies were measured for sheet resistivity using a Veeco four-point probe. Film thicknesses and surface roughness measurements were determined by patterning and etching the aluminum films, and profiling the resultant aluminum lines with a Sloan DekTak stylus. The morphology of the films was studied using a Hitachi scanning electron microscope (SEM).

RESULTS AND DISCUSSION

High Alane Flow System

For our initial experiments, we used the reactor design shown in Figure 1a to examine the influence of a high alane flow rate on aluminum thin film growth. We maintained a maximum TMAA flow rate during deposition by fully opening the TMAA bubbler outlet valve to the deposition chamber as it was actively evacuated by the system roughing pump. Carrier gas was not flowed through the TMAA bubbler during growth; however, the argon stored in the bubbler between experiments was present in the bubbler at the beginning of each new deposition experiment. Prior to deposition, the reaction chamber and all lines leading to the TMAA bubbler were first evacuated to approximately 5×10^{-6} Torr using the turbomolecular pump. The substrate was then heated to 275°C under vacuum, and the turbomolecular pump was closed off, leaving only the roughing pump active. The TMAA bubbler (containing one atmosphere of argon) was then opened to the actively pumped deposition chamber. The chamber pressure initially rose to ~10 Torr, and then dropped to 0.30 Torr after 30 seconds (as the initial burst of argon from the bubbler was evacuated). At the end of growth, the bubbler was closed and the heater was turned off. The chamber and bubbler were then backfilled with argon to a pressure of one atmosphere. Growth times ranged from 30 seconds to 15 minutes.

In terms of reproducibility, purity, and surface morphology, these films were substantially superior to those previously grown in this and other laboratories [1,2]. Film thicknesses ranged from 0.8 to 8.0 μm, depending on deposition time, and all of the films exhibited very high purities. Resistivities averaged 3.4 μΩ-cm, which compares very favorably to the bulk aluminum value of 2.7 μΩ-cm. The films all had relatively smooth surfaces, and were highly oriented ($I_{(111)}/_{(200)}$ = 70-700). Moreover, these aluminum films exhibited *no whisker growth*. The average surface peak-to-valley step height for each film was found to be <4% of the film thickness. For thin films (0.8-1 μm) this low percent surface roughness corresponds to actual surface feature heights of only ~400 Å, and these films are extremely smooth and reflective (Figure 2). Since most interconnect films are ≤1 μm, the films grown using this deposition process should be suitable for interconnect processing.

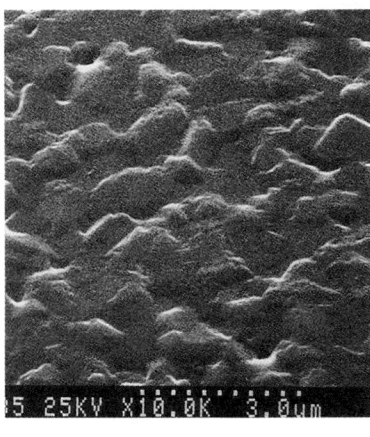 

(a)   (b)

Figure 2. Morphology of a 0.8 μm thick aluminum film deposited using the high alane flow rate system: (a) scanning electron micrograph, and (b) surface reflectivity across a 2" wafer.

The very smooth morphologies obtained for our thin films were somewhat surprising. Most of the literature reports of aluminum CVD on silicon or SiO_2/Si substrates indicate that very rough, textured aluminum film morphologies are typically obtained unless the substrates have been subjected to a $TiCl_4$ or plasma pre-treatment [1,3]. Pre-treatment is believed to improve the surface nucleation of aluminum on these substrates. However, despite our lack of substrate surface pre-treatment, we did not observe a similar rough surface texture for our films. In order to identify the factor(s) responsible for this difference, we compared our deposition process with those of other groups. One feature unique to our system was the presence of argon in the TMAA bubbler at the beginning of growth. In order to assess the importance of this argon, two experiments were carried out. For the first experiment, we modified our typical deposition procedure such that the TMAA bubbler was evacuated prior to deposition. Thus, no argon was present in the bubbler at the beginning of growth. For the second experiment, the alane bubbler was filled as usual with one atmosphere of argon prior to growth. Thus, argon was present in the bubbler at the beginning of aluminum deposition. Growth time was 2 minutes. The aluminum films deposited from these two experiments were then examined by SEM.

The SEM micrographs of these two aluminum films are shown in Figure 3, and the differences in surface morphology are dramatic. Evacuating the bubbler of argon prior to aluminum film growth was found to severely degrade the surface quality of the resultant films. The aluminum film deposited using the evacuated TMAA bubbler exhibits an extremely rough surface with many voids, and the nominal surface roughness as measured by Dektak profilometry is approximately 50% that of the average film thickness. In contrast, the aluminum film deposited using the argon filled bubbler exhibited a very dense, smooth surface morphology, with a surface roughness of only 4% relative to film thickness. The burst of argon during the initial stages of growth therefore appears to aid the surface nucleation of aluminum on SiO_2 in our system, and contributes to the production of smooth, dense aluminum films.

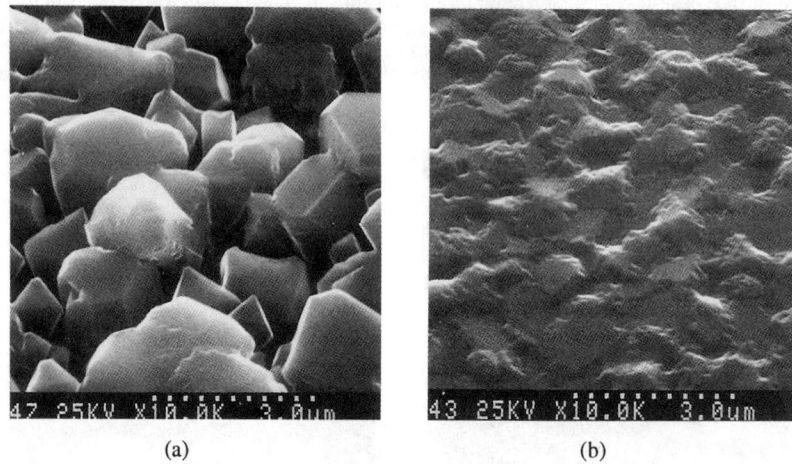

Figure 3. Scanning electron micrographs of aluminum thin films grown (a) with no argon in the TMAA bubbler, and (b) with 1 atm of argon in the TMAA bubbler

This phenomenon of improved film morphology with added argon was studied further to help elucidate the mechanism of improvement. The existing reactor was modified such that argon could be added to the alane gas flow during a typical deposition. The added argon was introduced through the inlet side of the TMAA bubbler, and the amount of added argon was controlled using a mass flow controller. Incorporating the new argon line into the system resulted in re-positioning the TMAA bubbler much further from the chamber than in the original system. A series of growth experiments using varying amounts of added argon was then carried out and the resultant thin film morphologies were analyzed by SEM. With the exception of the added argon, standard deposition conditions were used: one atmosphere of argon was initially present in the bubbler, the bubbler was open to the roughing pump during deposition, and the substrate temperature was kept at 275°C. Argon flows of 0, 50, 100, and 150 sccm were allowed to flow through the bubbler during growth.

Repositioning the bubbler farther from the chamber resulted in poorer nucleation and slower growth rate, apparently due to loss of alane on the gas transport line walls. With 0 sccm added argon flow (but with argon initially in the bubbler), the substrate surface following a 3 minute deposition was covered with individual, discontinuous, polyhedron-shaped aluminum nuclei. As argon flow was introduced however, the deposition rate increased, and the aluminum nuclei began to coalesce into a single smooth continuous film. At lower added argon flow rates, voids were still apparent between the coalesced nuclei, but at 150 sccm of added argon, the films were smooth, dense, and continuous. An increased deposition rate due to the added argon flow was not entirely responsible for the improved morphology, since longer deposition times at low argon flow rates did not produce the same smooth films grown at high added argon flows. Argon flow thus appears necessary for improved film morphology.

The initial nucleation of a substrate surface during CVD is crucial to the morphology of the resultant thin film. The presence of an argon flow during the initial stages of aluminum deposition, when combined with a high TMAA flow rate, appears to lead to improved surface nucleation. Several possible mechanisms may be responsible for this improved nucleation. It may be that the argon aids in the rapid delivery of alane to the substrate surface, allowing for a quick, dense nucleation of the substrate surface. Argon might also provide a greater thermal gradient at the substrate surface, enhancing any gas phase decomposition reactions. The flow of argon past the substrate also may help disperse the precursor more evenly over the substrate during nucleation. In addition to these possibilities, other factors may be at work as well. We plan to continue efforts to understand this phenomenon.

Low Alane Flow System

In order to further examine the impact of precursor flow on aluminum thin film morphology, we also carried out a set of experiments in which the influence of a low alane flow on aluminum thin film growth was studied. For these experiments, we used the reactor design shown in Figure 1b. By maintaining the TMAA bubbler at 400 Torr and using an argon carrier gas to transport the TMAA vapor, the flow rate of alane to the deposition chamber was greatly reduced relative to our previous experiments. The TMAA flow was also equilibrated prior to being introduced into the deposition chamber, by flowing the carrier gas through the bubbler and then out via a chamber bypass valve. The substrate temperature for all of these growth experiments was 275°C. Depositions were carried out using a variety of argon carrier gas flow rates and growth durations, and the morphologies of the resultant films were monitored by SEM.

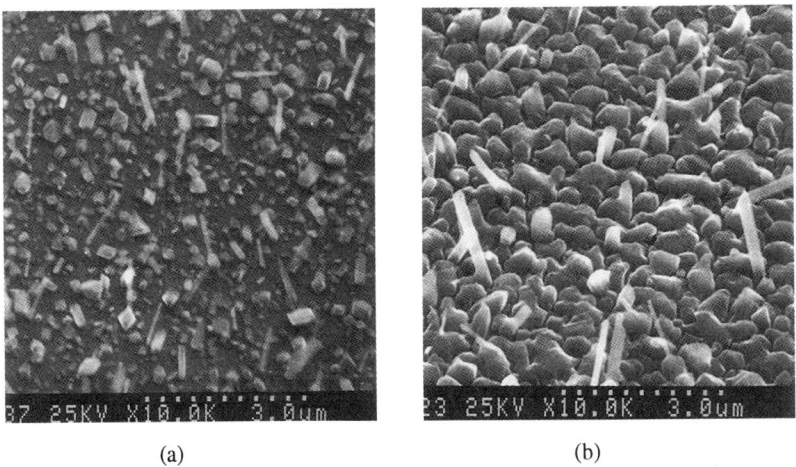

(a)　　　　　　　　　　　　(b)

Figure 4. Scanning electron micrographs of aluminum films deposited using the low alane flow rate system, with growth times of (a) 15 minutes, and (b) 60 minutes.

The results from these deposition studies were found to differ dramatically from those of our high alane flow rate experiments. As anticipated, the aluminum deposition rate was significantly lower than our previous experiments, due to the reduced alane delivery rate of this system. However, more significant were the differences in aluminum nuclei shape and in thin film morphology. Figure 4 shows SEM micrographs of two aluminum films grown using the low alane flow apparatus and an argon carrier gas flow of 100 sccm. The film in Figure 4a was deposited for 15 minutes, and that in 4b was grown for 60 minutes. The reduced aluminum growth rate is observed in Figure 4a, where even after 15 minutes, nucleation of the SiO_2 surface is sparse. Surprisingly, rod-shaped nuclei appear on the substrate surface, which were not observed in our high alane flow system. Longer growth time (Figure 4b) produced a continuous film, but with the formation of whiskers from the rod-shaped nuclei. These films also exhibited small grain sizes and a highly textured surface. Whiskers were observed on all films grown using the low alane flow deposition system. The low alane flow rate, combined with the argon carrier flow, somehow enhances the formation of the rod-shaped nuclei. We do not yet know if argon is essential to the formation of these rod-shaped crystals, or if the rod-shaped nuclei are primarily the products of gas phase or substrate-mediated reactions [4]. These issues are currently under investigation.

SUMMARY

Excellent quality aluminum thin films were reproducibly grown in a low pressure reactor, using a high flow rate of TMAA combined with an argon gas flow. The quality and morphology of the films grown with this deposition system were exceptionally good. Resistivities averaged 3.4 $\mu\Omega$-cm (compared to bulk aluminum at 2.7 $\mu\Omega$-cm), and the average surface peak-to-valley height for each film was found to be <4% of the film thickness. The surfaces of films with thicknesses of ≤ 1 µm were extremely smooth and reflective. These film characteristics should be adequate for most aluminum interconnects for advanced integrated circuits. The presence of an argon flow during at least the initial stages of growth was found to be essential to the production of smooth surface morphologies, provided a high flow rate of alane into the growth chamber is maintained. Absence of an initial argon flow resulted in the deposition of highly textured aluminum films. However, none of the films grown using the high alane flow rate system exhibited any whisker growth.

Slowing the flow rate of alane into the deposition chamber by maintaining the TMAA bubbler at 400 Torr and using argon to transport the alane vapor resulted in slower aluminum film growth, and also promoted the growth of aluminum whiskers. Whiskers were produced from rod-shaped nuclei that formed during deposition. These whiskers are extremely undesirable since they wreak havoc with integrated circuit processing techniques. The factors controlling the formation of the rod-shaped nuclei and subsequent whisker growth are currently not well understood, and are under further investigation.

Acknowledgements

We wish to thank the R. Robertson and M. Isaac for their assistance with wafer processing and Dektak measurements, and the Aerospace Sponsored Research Program for financial support.

References

1. M.E. Gross, K.P. Cheung, C.G. Fleming, J. Kovalchick, and L.A. Heimbrook, J. Vac. Sci. Technol. A **9**, 57 (1991); D.B. Beach, S.E. Blum, and F.K. LeGoues, J. Vac. Sci. Technol. A **7**, 3117 (1989); W.L. Gladfelter, D.C. Boyd, and K.F. Jensen, Chemistry of Materials **1**, 339 (1989).

2. D.M. Speckman, Mat. Res. Soc. Symp. Proc. **282**, 305 (1993).

3. A. Weber, U. Bringmann, K. Schiffmann, and C.-P. Klages, Mat. Res. Soc. Symp. Proc. **282**, 311 (1993).

4. M.G. Simmonds, W.L. Gladfelter, H.Li, and P.H. McMurry, Mat. Res. Soc. Symp. Proc. **282**, 317 (1993).

FIRST PRINCIPLES STUDY OF ALUMINUM DEPOSITION ON HYDROGEN-TERMINATED Si(100) SURFACE

CARLOS SOSA
Cray Research, Inc., 655 E lone Oak Dr., Eagan, MN 55121

ABSTRACT

The deposition of Aluminum on Si(100) surface has been investigated using density functional methods. This has been accomplished by adoption of a Si_9H_{16} cluster to model the H terminated Si(100) 1X1 surface and Si_9H_{15} cluster to model the surface with an unpaired electron. The predicted NLSD dissociation energy for the $Si_9H_{16} \rightarrow Si_9H_{15}$ + H is 86.3 ± 2.0 Kcal/Mol. This is in agreement with previous theoretical calculations on similar systems.

Introduction

Selective growth of metals on silicon surfaces is an important process in deep-submicron very large scale integrated circuits (VLSI) [1-3]. In particular, Al is used for interconnecting materials. The chemistry of silicon surfaces is of obvious importance to chemical vapor deposition techniques. Clearly, the specific values of the bond energies for the different chemical species involved in the reaction are crucial for interpreting experimental results.

Recently, Tsubouchi and Masu [3] have proposed a mechanism for selective growth of Al on Si surfaces. They have investigated the Al growth on Si by means of chemical vapor deposition (CVD). This has been carried out using dimethylaluminum hydride (DMAH) and H_2 [1-3].

In their work they have assumed that DMAH interacts mainly with the Si(100) 1X1:2H surface[1-3] (See Figure I).

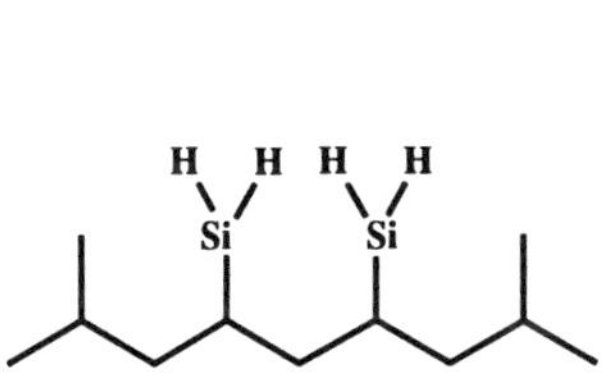

Figure I. Si(100) 1X1

First, the DMAH is adsorbed on the H terminated Si(100) surface. Then, the CH_3 group of the $RAl-CH_3$ interacts with the neighboring hydrogen to form CH_4 and Al gets deposited on the Si(100) surface. Tsubouchi and Masu [3] have pointed out that the bonding energy for Al-H is higher than the bonding energy of $Al-CH_3$, thus, hydrogen remains bonded (Al-H) after Al is being deposited on the Si(100) surface. The additional H that is required to obtain a H terminated surface is provided by the H_2 molecule. A more detailed discussion of the mechanism can be found in refs. [1-3].

In this study, Density Functional Theory (DFT) has been used to calculate the strength of the Si(100) 1X1 Si-H, R_2Al-CH_3, and R_2Al-H bonds. This information is important for the understanding of DMAH/H_2 interaction with the Si(100) surface. Similar to previous studies [4-6], we have adopted a Si_9H_{16} cluster model of the Si(100) 1X1 surface.

Method

The DFT calculations were carried out using the DGauss [7] program. This program uses a Gaussian orbital basis set to solve the Kohn-Sham equations [8, 9]. Gaussian type functions are also used to expand exchange-correlation potential [10, 11]. Geometries are optimized in the local spin density approximation using Dirac exchange functional and the Vosko, Wilk, and Nusair (LSD) correlation energy functional [12].

Single point calculations including nonlocal corrections (NLSD) are carried out at the LSD optimized structures using the Becke [12] exchange and Perdew [14] correlation functional. For the nonlocal corrections (Becke-Perdew model), the gradient corrected exchange-correlation terms were added perturbatively to the local spin density energy (NLSD). Cartesian Gaussian-type basis sets are used to represent orbital and auxiliary fitting functions. Calculations have been performed with double-zeta-split-valence plus polarization (DZVP) basis sets [15].

The geometries of the two clusters were optimized as described in refs. [4-6]. In this section we only briefly summarize all the constraints that were used. All the Si atoms for layers four and three were fully frozen in the bulk-like positions (See ref. 5). All of the coordinates for the hydrogen atoms used to terminate dangling bonds were also fully fixed. The four Si atoms in the second layer were allowed to move in the x and z directions. The position of the surface Si atoms was chosen to be 3.84 Å [16]. The associated hydrogen atoms were fully optimized. This set of constraints were chosen to reproduce bulk-like interactions.

Results and Discussion

The optimized geometries for CH_3, CH_3AlH, CH_3AlCH_3, $(CH_3)_2AlH$, and Si_9H_x (x= 15 and 16) are predicted using density functional methods. Selected optimized geometrical parameters for these systems are summarized in Table I

Table I. Selected Optimized parameters for $(CH_3)_2$Al-H, CH_3Al-H, and $(CH_3)_2$Al.

System	Parameter		
	r(Al-H)	r(C-Al)	θ(C-Al-X)
$(CH_3)_2$Al-H	1.615	1.951	121.1
CH_3Al-H	1.632	1.968	119.9
$(CH_3)_2$Al		1.965	118.0

The optimized LSD Al-H bond lengths for $(CH_3)_2$Al-H and CH_3Al-H are 1.615 and 1.632 Å, respectively. The effect of dissociating one methyl group from $(CH_3)_2$Al-H ($(CH_3)_2$Al-H → CH_3Al-H + CH_3) is to elongate the Al-H bond distance by about 0.02 Å when compared to CH_3Al-H. Similar effects may be observed for the C-Al bond lengths. On the other hand, the predicted C-Al-X angle for $(CH_3)_2$Al-H is 121.1 degrees. The calculated angles for CH_3Al-H and CH_3AlCH_3 are only 2 to 3 degrees smaller than the calculated angle for $(CH_3)_2$Al-H. In general, the DMAH geometry is only slightly distorted when a methyl group or hydrogen atom is removed to give the corresponding radicals.

Figure II summarizes the Si-H bond lengths for the Si_9H_x clusters. Two types of Si-H bond lengths are predicted for the Si_9H_{16} cluster. The first type corresponds to the Si-H lateral bonds (Si-H_l, See Figure II). The second type corresponds to the Si-H internal bonds (Si-H_i). These are predicted to be shorter by 0.016 Å. The shortening of the Si-H_i bond is principally due to the H_i---H_i interaction. This may be seen in Figure IIb, in this case, there is no H_i---H_i interaction, Si-H_l and Si-H_i bond lengths are predicted to be almost identical.

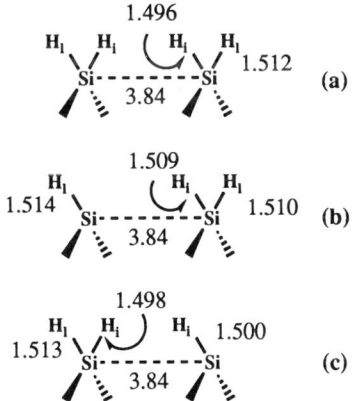

Figure II. Optimized Parameters for the Si_9H_x clusters.

The LSD and NLSD relative energies for the different chemical reactions reported in this study are summarized in Table II. The energies predicted with LSD and NLSD methods for the Si-H_l bonds are ($Si_9H_{16} \rightarrow Si_9H_{15}$ + H): 93.98 and 87.92 Kcal/Mol, respectively. The calculated relative energies for the Si-H_i bonds are: 91.64 and 84.63 Kcal/Mol, respectively. The calculations performed with the Si_9H_{16} cluster are giving energies of Si-H bonds in agreement with results obtained using an Si_9H_{14} cluster [4-6] ($Si_9H_{14} \rightarrow Si_9H_{13}$ + H).

Table II. Energies of Reaction in Kcal/Mol.

Reaction	LSD	NLSD
$(CH_3)_2Al-H \rightarrow (CH_3)_2Al + H$	91.78	86.19
$(CH_3)_2Al-H \rightarrow CH_3AlH + CH_3$	96.09	80.12
$(CH_3)_2Al-H + 1/2 H_2 \rightarrow Al + 2CH_4$	42.76	30.80
$2[(CH_3)_2Al-H] \rightarrow 2Al + 2C_2H_6 + H_2$	119.16	97.59
$Si_9H_{16} \rightarrow Si_9H_{15} + H$ (l)	93.98	87.92
$Si_9H_{16} \rightarrow Si_9H_{15} + H$ (i)	91.64	84.63

Recently, it has been proposed that a CH_3 group of the DMAH reacts with the H terminated Si(100) surface[3]. This is the preferred interaction since the Al-H dissociation energy is larger than Al-CH_3. LSD calculations (See Table II) predict a dissociation energy for Al-CH$_3$ of 96.09 Kcal/Mol. This value is almost 5 Kcal/Mol larger than the Al-H dissociation energy. This is opposite to what is reported experimentally. On the other hand, Al-H dissociation energy computed with the NLSD approximation is larger by about 6 Kcal/Mol when compared to Al-CH$_3$. This is in agreement with experimental results.

Also, our results indicate that Aluminum deposition without H$_2$ is energetically less favorable than DMAH/H$_2$. Relative energies for the (CH$_3$)$_2$Al-H/H$_2$ and (CH$_3$)$_2$Al-H reactions are summarized in Table II. The reaction of (CH$_3$)$_2$Al-H + 1/2H$_2$ is energetically more favorable than 2[(CH$_3$)$_2$Al-H] by more than a factor of 3.

In summary, our theoretical calculations indicate that the RAl-CH$_3$ bond breaking is thermodynamically more stable than RAl-H breaking. However, the difference is only of the order of 5 Kcal/Mol. The value of the energy of the Si-H bond is predicted in good agreement with previous theoretical results [4-6]. The interaction energies between DMAH and Si surfaces is important for the understanding of the selective growth of Al on Si(100). This is currently the subject of another publication.

Acknowledgements

The authors wish to thank P. Nachtigall and Prof. K. D. Jordan for valuable discussions and the Corporate Computing and Networking Division at Cray Research is kindly acknowledged for providing computational resources to carry out this work.

References

1. K. Tsubouchi and K. Masu, J. Vac. Sci. Technol. **A10**, 856(1992).
2. K. Tsubouchi, K. Masu, K. Sasaki, and N. Mikoshiba, *Technical Digest of 1991 IEEE International Electron Devices Meeting*, Washington, DC 1991 (IEEE, NY, 1992) pp. 269.
3. K. Tsubouchi and K. Masu, *Surface Chemical Cleaning and Passivation for Semiconductor Processing*, San Francisco 1993, Materials Research Society Symposium Proceeding 315 (Materials Research Society, San Francisco, CA, 1993) pp. 59.
4. P. Nachtigall, K. D. Jordan, and K. C. Janda, J. Chem. Phys. **95**, 8652(1991).
5. C. Sosa, C. Lee, P. Nachtigall, and K. D. Jordan, *Surface Chemical Cleaning and Passivation for Semiconductor Processing*, San Francisco 1993, Materials Research Society Symposium Proceeding 315 (Materials Research Society, San Francisco, CA, 1993) pp. 273.
6. P. Nachtigall, K. D. Jordan, and C. Sosa, J. Phys. Chem. **97**, 11666(1993)
7. (a) J. Andzelm and E. Wimmer, J. Chem. Phys. **96**, 1280(1992); (b) J. Andzelm, in *Density Functional Methods in Chemistry*, edited by J. Labanowski and J. Andzelm (Springer-Verlag, NY, 1991) pp. 155.
8. P. Hohenberg and W. Kohn, Phys. Rev. **B136**, 864 (1964).
9. W. Kohn and L. J. Sham, Phys. Rev. **A140**, 1133 (1965).
10. H. Sambe, and R.H. Felton, J. Chem. Phys. **B62**, 1122 (1975).
11. B.I. Dunlap, J.W.D. Connolly, and J.R. Sabin, J. Chem. Phys. **71**, 3396 (1979).
12. S. H. Vosko, L. Wilk, and M. Nusair, Can. J. Phys. **58**, 1200(1980).
13. (a) A. D. Becke, Phys. Rev. **A38**, 3098(1988); (b) A. D. Becke, J. Chem. Phys. **88**, 2547(1988).
14. J. P. Perdew, Phys. Rev. **B33**, 8822(1986).
15. N. Godbout, D. R. Salahub, J. Andzelm, and E. Wimmer, Can. J. Chem. **70**, 560(1992).
16. B. W. Holland, C. D. Duke, and A. Paton, Surface Sci. **140**, 1269(1984).

PROPERTIES OF A NEW LIQUID ORGANO GOLD COMPOUND FOR MOCVD

Hiroto Uchida, Norimichi Saitou, Masamitu Satou,
Masayuki Tebakari and Katsumi Ogi,
Mitsubishi Materials Corporation, Central Research Institute,
1-297 Kitabukuro-cho, Omiya, Saitama 330, Japan

ABSTRACT

Dimethyl(1,1,1,5,5,5-hexafluoroaminopenten-2-onato)Au(III), DMAu(hfap) was found to be very promising compound for CVD Au wiring. It is liquid at r.t. and has high vapor pressure (0.5 Torr at r.t.) and high thermal stability (dec.t. 171°C, $\Delta E=25.4$ Kcal/mol) and can be used as a precursor for MOCVD. CVD gold films were grown on Si(100) from DMAu(hfap) in a quartz vacuum chamber, using thermal activation by an electric heater or photo activation through quartz windows by excimer lasers(XeCl 308nm). A growth rate of 10 nm/min was achieved by thermal activation at 250 °C, (evaporation temperature of 30 °C, H2 flow rate at 100 sccm and chamber pressure of 30 Torr). Resistivities ranged between 2.8 and 5 $\mu\Omega$-cm, depending on the film morphology. The growth rate of laser CVD gold films had the tendency of increasing with decreasing the substrate temperature. This phenomenon indicates that photolysis of adsorbed species on the substrate plays an important role in film growth by laser CVD.

INTRODUCTION

Gold films are a potentially important metallization material in electronic and photonic devices due to their high conductivity, superior electromigration resistance and chemical stability. The CVD technique has advantages to sputtering or evaporation of conformal growth and better coverage on sharp steps. For CVD use, organogold complexes must be stable but at the same time volatile enough so that a sufficient vapor pressure can be obtained at moderate heating and high vacuum in order to give a high rate of plating. Recently, various kinds of gold complexes have been studied extensively. The most widely studied class of organogold compounds are dimethyl(2,4-pentanedionato)gold and its fluorinated analogues[1-4]. The high vapor pressure and room temperature liquid nature of DMAu(hfac) make it particularly suitable for CVD, but the thermal stability of DMAu(hfac) was not sufficient to achieve a constant supply to the CVD apparatus at elevated temperature. The thermal stability of some dimethylgold(β-diketone)complexes, dimethylgold(β-thioketone)complexes and dimethylgold(β-iminoketone)complexes were studied by I.K.Igumenov et al.by DTA analysis[5]. The thermal stability of dimethylgold complexes were increased by the kind of donor atoms of the chelate ring in the order of (O,S)< (O,O) < (O,NH). We have developed a new dimethylgold complex of hexafluorinatediminoketone(Hhfap). It is a liquid at r.t. and has a high vapor pressure (0.5 Torr at r.t.) and a thermal stability superior to DMAu(hfac).

EXPERIMENT

The thermal properties of the dimethylgold complexes 1.DMAu(acac) 2.DMAu(hfac) 3.DMAu(ap) and 4.DMAu(hfap) were examined by thermal gravimetric analysis(TGA) and differential thermal analysis(DTA). TGA data were obtained by the RIGAKU(TG-8110) instrument using an open Al pan which was purged with N2 (30ml/min) at a heating rate of 10°C/min DTA data were obtained on the same instrument using a sealed Al pan at the following conditions. The heating rates were selected at five points in the range of 2.5 - 20°C/min; N2 flow rate was 30ml/min; and the amount of gold complexes were 3 mg. The activation energies for the thermal decomposition reactions of the dimethylgold complexes were obtained by the Kissinger method[6].

The gold metal films were then deposited from these CVD precursors by using a horizontal quartz CVD reactor that has a quartz windows for laser CVD. For excimer laser irradiation experiments, a LAMBDA PHYSIK (LPX-315i) excimer laser apparatus was used. The reactor pressure was varied from 2.0 to 100 torr. The growth temperature was varied from 50 to 350°C. The precursors were loaded in a glass cylinder, and the cylinder pressure was adjusted by a variable conductance valve. The cylinder temperatures of DMAu(hfap) and DMAu(hfac) were held at 35°C and 24°C, respectively. Ultrahigh-purity argon and hydrogen were used as carrier gases. The flow rates in the cylinders were H_2=100sccm or Ar=10sccm. (100)silicon wafers were used as substrates.

AES, SIMS and GD-MS analysis and conductivity measurements were carried out on the deposited gold films. AES were recorded by using a Perkin-Elmer(SAM670) instrument. SIMS were recorded by using a HITACH(IMA-3) instrument. GD-MS were recorded by using a MICROTRACE(VG9000) instrument. Electro conductivities of the film were recorded by a Mitsubishi Petroleum Chemical Co. (LORESTA) instrument.

RESULTS

Fig.1 shows TGA curves for the dimethylgold complexes. The thermal stabilities and vaporization temperatures of the complexes were compared. The amount of decomposition residue indicates the thermal stabilities. According to the amount of decomposition residue, the stabilities of dimethylgold complexes increase in the order of DMAu(acac) < DMAu(ap). The vaporization temperature of dimethylgold complexes decrease in the order of DMAu(ap) > DMAu(ac) > DMAu(hfap) > DMAu(hfac). Table 1 shows the the activation energies for thermal decomposition of these dimethylgold complexes determined by the Kissinger method[6].

Table 1 Thermal analysis data of dimethyl goldcomplexes

complexes	m.p. °C	dec.temp. °C	activation energy kcal/mol
DMAu(acac)	76	154	25.7
DMAu(ap)	64	196	22.4
DMAu(hfac)	liq.	164	21.7
DMAu(hfap)	liq.	171	25.4

Fig.2 shows the vapor pressure data of dimethylgold complexes. The vapor pressures of DMAu(hfap) and DMAu(hfac) are ten times higher than those of DMAu(acac) and DMAu(ap).

Thermal MOCVD Results

The growth rates of the gold films were determined as a function of the partial pressures of dimethylgold complexes and the substrate temperature. Fig.3 shows the growth rates of DMAu(hfap) and DMAu(hfac) as a function of reciprocal growth temperature at reactor pressures of 30 torr. Comparing the film growth rates, that of DMAu(hfap) is five times slower than that of DMAu(hfac). The difference in the growth rates may be explained by the difference in the activation energies for thermal decomposition, which are 25.4 and 21.7 kcal/mol, respectively (Table1). But the observed activation energies (Ea) for the film growth from DMAu(hfap) and DMAu(hfac) shows the opposite tendency, which are 6.6 kcal/mol and 9.8 kcal/mol in the reaction-limited region, respectively. This phenomenon may indicate that the decomposition rate of the decomposition intermediates of DMAu(hfap) and DMAu(hfac) on the substrate are increasing in the order of DMAu(hfac) < DMAu(hfap). Fig.4 shows the growth rate as a function of reciprocal growth temperature at reactor pressures of 2.0, 30, and 100 torr. The activation energies (Ea) of film growth in the reaction-limited region were 10.3, 6.6, 4.7 kcal/mol, respectively. The activation energy decreases as the reactor pressure increases. A similar phenomenon was reported in the thermal CVD of DMAu(hfac) by M.Hosino et al., and the mechanism was proposed that the gas phase decomposition rate increases as the reactor pressure increases[4].

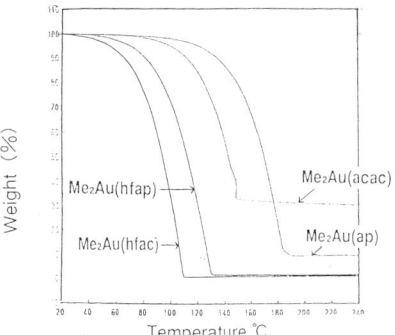

Fig.1 TGA curves of dimethylgoldcomplexes

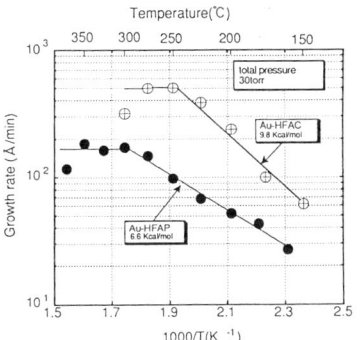

Fig.3 Growth rate of DMAu(hfap) and DMAu(hfac) as a function of temperature

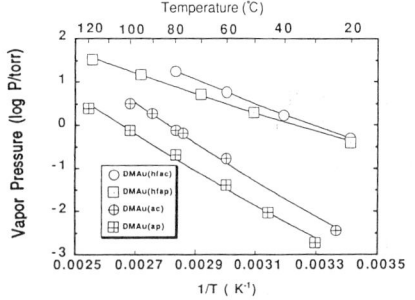

Fig. 2 Clausius-Clapeyron plots for dimethylgoldcomplexes

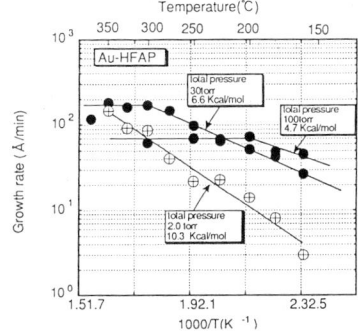

Fig. 4 Growth rate of DMAu(hfap) as a function of temperature

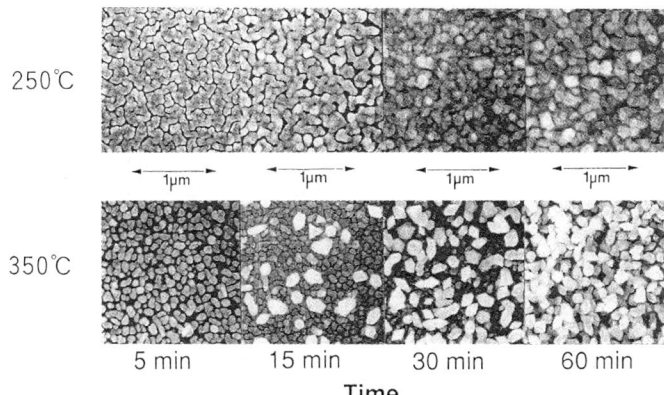

Fig.5 Morphology of gold films deposited on Si(100) by thermal CVD of DMAu(hfap) at 250°C and 300°C.
total pressure 2.0 torr, DMAu(hfap) 0.9 torr, carrier gas Ar

Fig.5 shows the change of the surface morphology of the gold films grown by thermal CVD of DMAu(hfap) at 250°C and 350°C on Si(100) as a function of time(Ar was used as carrier gas). Fig.6 shows the weights of the gold films as a function of growth time under the same conditions. When the growth temperature was at 250°C, the growth rate of the films at the first 5min was very slow. After this nucleation period, the growth rate increased rapidly, and islands were fused together. After the gaps between furrows were filled(after 30min) there was almost no weight change. When the growth temperature was at 350°C, at the first five minutes the growth rate of the films was slow. After this nucleation period, the growth rates increased rapidly for the next 10min and then slowed down from 15min to 30min as the gaps between grains were filled. A second nucleation was started concurrently in this stage and the growth rate then increased again.

Fig.7 shows resistivities of films as a function of substrate temperature. The reactor pressure was 30torr and the carrier gas were H_2(100sccm). The minimum resistivities are obtained in the temperature range from 190°C to 230°C, and the resistivity ranges from $2.8\mu\Omega$-cm to $5.0\mu\Omega$-cm. Above 250°C the resistivity rapidly increases. Resistivities above 300°C were very large. This temperature range is in the diffusion controlled region, and the morphology of the films shows very low density similar to the sample at 350°C(60min) shown in Fig.5.

Fig.8 shows a depth profile of the gold film measured by Auger electron spectroscopy (AES). The film was grown at 2torr at 250°C in the reaction limited region. The film thickness was 1200Å. At the film surface some amount of adsorbed carbon was observed, but the incorporation levels of carbon, oxygen and fluorine were below the detection limit.

Laser MOCVD Results

The growth rate of gold films by laser CVD of DMAu(hfac) and DMAu(hfap) at various substrate temperatures were compared. Fig.9 shows the growth rate of DMAu(hfac) as a function of reciprocal growth temperature at a reactor pressure of 30 torr with or without XeCl(308nm) irradiation, and Fig.10 shows the growth rate of DMAu(hfap) as a function of reciprocal growth temperature under the same conditions. The power and frequency of XeCl(308nm) were 150mJ/p-cm2 and 20Hz, respectively. In the low temperature regime below 150°C, both of the film growth rates by laser CVD of DMAu(hfac) and DMAu(hfap) increase as the substrate temperature decreases. This phenomenon indicates that the photolysis of adsorbed species on the substrates plays an important role at low temperature. In the high temperature regime above 150°C, the film growth rates by laser CVD of both DMAu(hfac) and DMAu(hfap) were improved, but comparing the rate of improvements between DMAu(hfac) and DMAu(hfap), the growth rates of gold films by laser CVD of DMAu(hfap) are ten times bigger than those of thermal CVD, but the growth rate of gold films by laser CVD of DMAu(hfac) are only two or three times bigger than those of thermal CVD. This phenomenon indicates that the growth rate of thermal CVD of DMAu(hfap) is slower than that of DMAu(hfac), but laser irradiation accelerates the film growth from DMAu(hfap) more effectively than that from DMAu(hfac).

Table 2 shows the atomic purity of the deposited gold films from different dimethyl gold complexes as determined by SIMS, GD-MS and LSIMS. SIMS and GD-MS analysis indicates, that all the impurity elements, (Na, K, Ca, Fe, Ag, Cu) in the gold films were less than 0.1ppm. To compare the Cu concentration in Au films from different dimethyl gold complexes, laser ionized SIMS(LSIMS) analysis was carried out. The Cu concentration in Au films grown from DMAu(hfap) was found to be the smallest.

Table 2 Purities of the LCVD gold films by SIMS, GD-MS and LSIMS (ppm)

compounds	SIMS				GD-MS		LSIMS
	Na	K	Ca	Fe	Ag	Cu	Cu
DMAu(acac)	< 0.1	< 0.1	< 0.1	< 0.1	< 0.1	< 0.1	0.1
DMAu(ap)	< 0.1	< 0.1	< 0.1	< 0.1	< 0.1	< 0.1	0.23
DMAu(hfap)	< 0.1	< 0.1	< 0.1	< 0.1	< 0.1	< 0.1	0.02

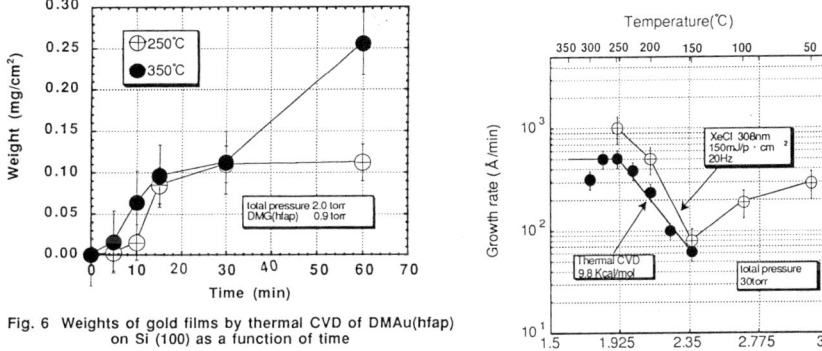

Fig. 6 Weights of gold films by thermal CVD of DMAu(hfap) on Si (100) as a function of time

Fig. 9 Growth rate of gold film by thermal and laser CVD of DMAu(hfac) as a function of temperature

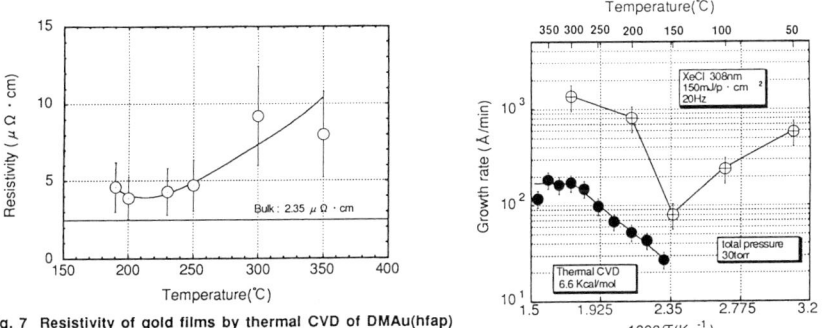

Fig. 7 Resistivity of gold films by thermal CVD of DMAu(hfap) as a function of growth temperature

Fig. 10 Growth rate of gold film by thermal and laser CVD of DMAu(hfap) as a function of temperature

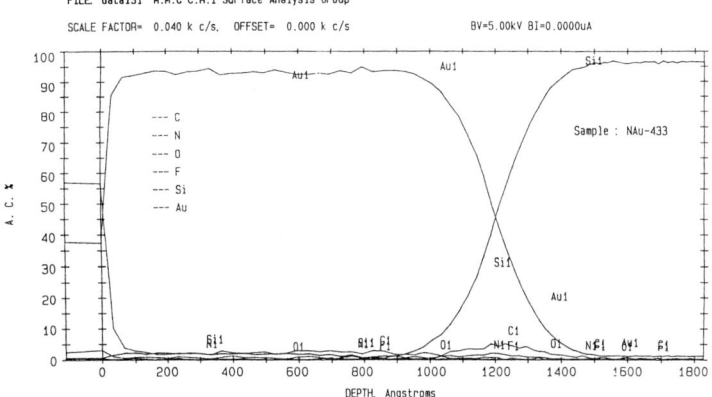

Fig.8 AES spectrum of CVD gold films as a function of the sputtering time.

DISCUSSION

Comparing the thermal CVD films from DMAu(hfap) and DMAu(hfac), faceted dendrites were formed from DMG(hfac) at a total pressure of 30torr, (growth temp.225°C) as is mentioned in the literature[3,4], but at the same film growth condition only round crystal formation was observed from DMG(hfap). This results indicates that the surface migration length of surface adsorbed Au molecules from DMAu(hfap) is shorter than that from DMAu(hfac). This is explained by the fact that the activation energy of film growth from DMAu(hfap)(6.6Kcal/mol) is smaller than that from DMAu(hfac)(9.8Kcal/mol) and this is reflecting that the lifetime of adsorbed Au molecules from DMAu(hfap) that can migrate on the surface is shorter than that from DMAu(hfac).

The changes of the surface morphology (Fig.5) and the weight (Fig.6) of the gold films grown by thermal CVD of DMAu(hfap) at 250°C and 350°C on Si(100) as a function of time indicates that the film growth proceeds stepwise. When the growth temperature was at 250°C, in the reaction-limited regime, only the surface reactions were occurring: surface nucleation at first and then islands were fused together. After the gaps between furrows were filled there was almost no film growth. When the growth temperature was at 350°C, in the diffusion-limited regime, gas phase nucleation was occurring and on the high density layer formed at the first period, a low density powdery layer was piled.

In laser CVD of DMAu(hfac) and DMAu(hfap) in the high temperature regime above 150°C, the film growth rates were improved for the compounds, and the films growth rates showed a temperature dependency. This fact explains why XeCl irradiation promotes the formation of film-forming intermediate species effectively.

CONCLUSION

DMAu(hfap) is liquid at r.t. and has high vapor pressure (0.5 Torr at r.t.) and high thermal stability (dec.t. 171°C, ΔE=25.4Kcal/mol). CVD gold films deposited thermally from DMAu(hfap) at 225°C-250°C were quite pure. Auger analysis indicates that incorporated carbon, nitrogen and fluorine in the films are less than 1.0 at% and the deposited films have a low resistivity ranging from 2.8 $\mu\Omega$-cm to 5.0 $\mu\Omega$-cm.

The Growth rate of thermal CVD of DMAu(hfap) is slower than that of DMAu(hfac) because of its thermal stability, but laser irradiation accelerates the film growth from DMAu(hfap) more effectively than that from DMAu(hfac). The atomic purities of LCVD gold film from DMAu(hfap) is 6N.

ACKNOWLEDGEMENTS

This work was conducted in the program; "Advanced Chemical Processing Technology", consigned to the Advanced Chemical Processing Technology Research Association from the New Energy and Industrial Science and Technology Frontier Program enforced by the Agency of Industrial Science and Technology, the Ministry of International Trade and Industry.

REFERENCES

1) C.E. Larson, T.H. Baum and R.L. Jackson, J. Electrochem. Soc. **134**, 226 (1987).
2) T.H. Baum and C.R. Jones, Appl. Phys. Lett. **47**, 538 (1985).
3) K. Holloway, S.P. Zuhoski, S. Reynolds and C. Matuszewski,
   Mat. Res. Soc. Symp. Proc. **204**, 409-414 (1991).
4) M. Hoshino, K. Kasai and J. Komeno, Jpn. J. Appl. Phys. **31**, 4403-4406 (1992).
5) P.P. Semyannikov, G.I. Zharkova, V.M. Grankin, N.M. Tyukalevskaya,
   I.K. Igumenov, Metalloorg. Khim. **1(5)**, 1105-12 (1988).
6) H.E. Kissinger, J. Res. Natul. Bul. Std. **57**, 217 (1956).

DEPOSITION OF TUNGSTEN FILMS BY PULSED EXCIMER LASER ABLATION TECHNIQUE

A.M. DHOTE AND S.B. OGALE
Center for Advanced studies in Materials Science and Solid State Physics, Department of Physics, University of Poona, Pune - 411 007, India

ABSTRACT

Low temperature deposition of W on Semiconductor substrates is vital for microelectronics technology. Laser ablation technique using KrF (248 nm) excimer laser has been employed to deposit W films from $W(CO)_6$. The substrates used are Si(100) and SiO_2. The influence of substrate temperature on the film growth rate was investigated in a broad temperature range (20 - 500 °C), keeping the laser fluence fixed at 0.4 Jcm^{-2}. The substrate temperature is found to have a strong influence on the resistivity of the deposited films. Film resistivities within a factor of 3 of the value for pure bulk W have been observed in the substrate temperature range of 300 - 500 °C. X-ray diffraction data were also obtained. These revealed that the crystal structure of the film deposited in this temperature range corresponds specifically to the α-phase. Optical emissions from the plasma generated during the pulsed excimer laser ablation of $W(CO)_6$ are also examined by an optical Multichannel Analyser (OMA).

INTRODUCTION

The refractory metal film deposition is of interest for the fabrication of low resistance contacts and interconnects for semiconductor devices. In order to develop faster and smaller Si large scale integration devices, tungsten is an attractive alternative to poly - Si or Al for gate and interconnection materials. In addition, tungsten can be used as a diffusion barrier.

Various tungsten deposition techniques such as PVD (sputtering, electron beam evaporation etc.) and CVD (TCVD [1-4], PECVD [5] as well as LCVD [6-11]) have been proposed and investigated. Conventional (TCVD) deposition of W on SiO_2 requires temperature in excess of 700°C and at lower temperature deposition is difficult with poor adhesion of the deposited films. We have used pulsed KrF excimer laser ablation (PLD) technique to deposit adherent low resistivity W films on SiO_2 at low temperature (<500°C). Also, the use of PLD technique has made higher deposition rates (300×10^7 nm/min.) possible at low temperatures as compared to thermal and LCVD techniques. VLSI microelectronics, X-ray absorbing tungsten lithographic mask upon polymer membranes etc. demand for low temperature deposition process.

EXPERIMENTAL

The experimental apparatus consisted of a stainless steel chamber, evacuation system, and KrF excimer laser (LPX 200, 248 nm, 20 ns). The rotating tungstenhexacarbonyl target in the form cylindrical pellet and substrate were kept 5 cm apart facing each other. The impact of the laser pulse produces a dense pulsed vapor cloud concentrated around the normal to the target surface. The vapor is condensed on the target to form a thin film. Depositions are carried out at laser fluence of 0.4 Jcm^{-2} in the substrate temperature range of 20-400°C for Si and 200-500°C for SiO_2/Si. Electrical resistivity measurements were carried out using a four point probe method. Low angle X-ray diffraction observation of the film was performed using a CuKα ray to determine phase content. In addition, the study of optical emission spectra of the laser plasma produced during the deposition process was carried out using Optical Multichannel Analyser (OMA) to diagnose plasma state which is important for the deposition process.

RESULTS AND DISCUSSION

The deposition rate as a function of temperature was investigated from 20°C to 400°C for Si and from 200°C to 500°C for SiO_2. The activation energy of the W film formation for the thermal reaction is 16 Kcal/mole [12,13]. The activation energy reported for LCVD by Deutsch and Rathman [5] and Shintani [14] are 9.7 Kcal/mole and 8.2 Kcal/mole, respectively. Arrhenius plot of W deposition rates on Si using KrF excimer laser ablation technique is shown in Fig.1. The activation energy of W film formation by PLD technique is calculated to be ~ 1.3 Kcal/mole for Si which is much less than that in LCVD, thereby increasing the deposition rate on Si and allowing significant deposition on SiO_2 at temperatures in the 200-500°C range.

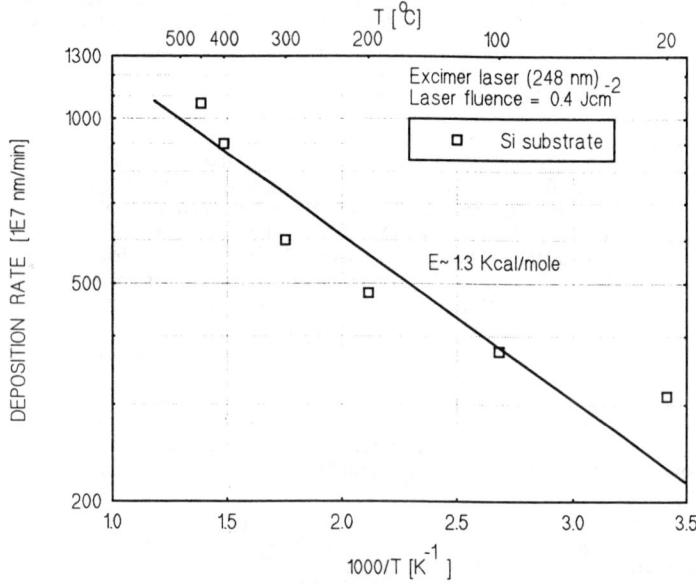

Fig. 1 : Temperature dependence (Arrhenius plot) of the deposition rate for pulsed laser deposited W film on Si(100) at laser fluence of 0.4 Jcm^{-2}.

Temperature variation of resistivity of W film on both Si and SiO_2 by laser ablation deposition technique is depicted in Fig.2. The resistivity of W film both on Si and SiO_2 decreased with increasing deposition temperature. The film resistivity value of 17 μohm-cm,

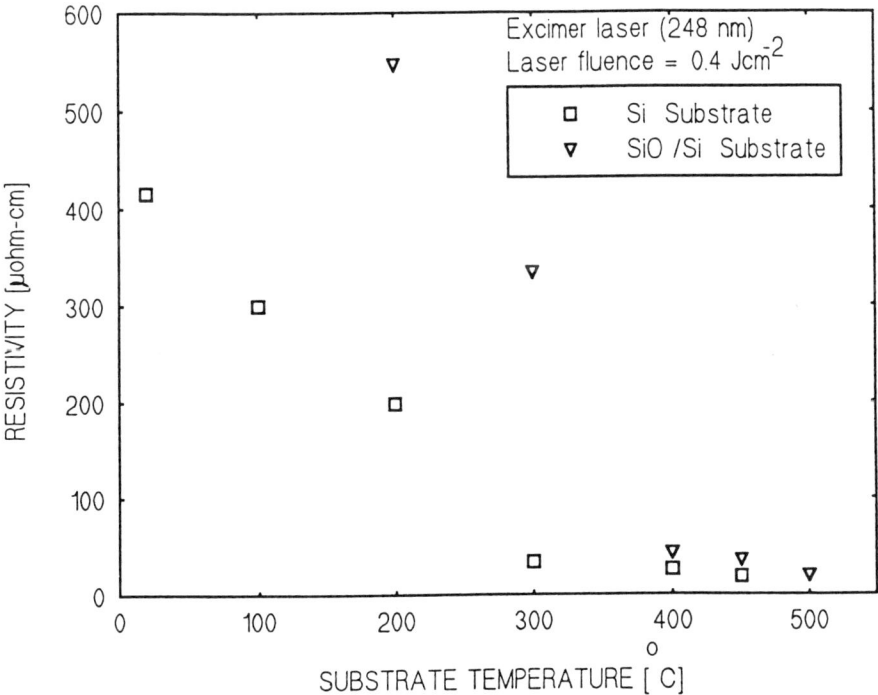

Fig. 2 : Temperature dependence of resistivity for pulsed laser deposited W films on Si(100) and SiO$_2$/Si(100) at laser fluence of 0.4 Jcm^{-2}.

as low as 3 times the value of that of bulk (5.3 μohm-cm) was observed above 300°C and 400°C for Si and SiO$_2$ respectively and the corresponding film exibited a good adhesion. The resistivity values have a strong correlation with the phase content of the samples.

Fig. 3 shows the result of X-ray diffraction measurements on W film deposited by PLD technique on Si substrate at three different substrate temperatures. At lower temperatures below 300°C, the incorporation of the high resistivity β-W phase is found partly in the W film. The amount of incorporation of β-W phase depends on the substrate temperature. Above 300°C , the deposited W film is uniquely of the α phase. This change in phase contribution is responsible for the steep decrease in resistivity with deposition temperature as shown in Fig.2.

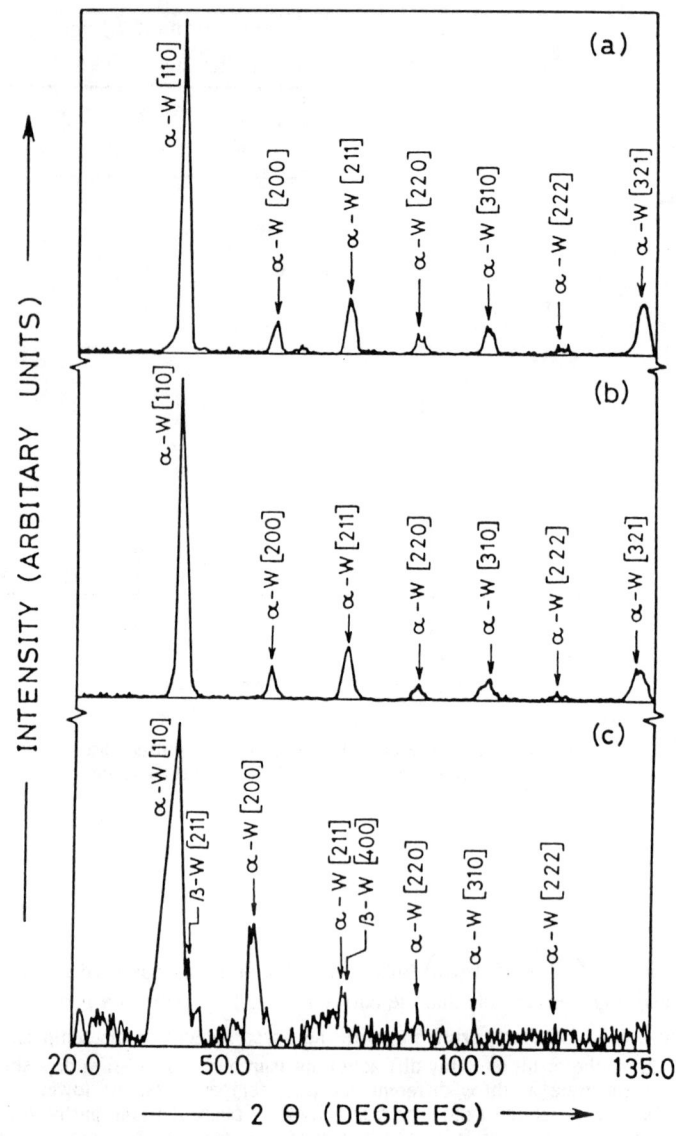

Fig. 3 : X-ray diffraction pattern for pulsed laser deposited (PLD) W films on Si(100) substrate at (a) 450°C, (b) 300°C, (c) 200°C

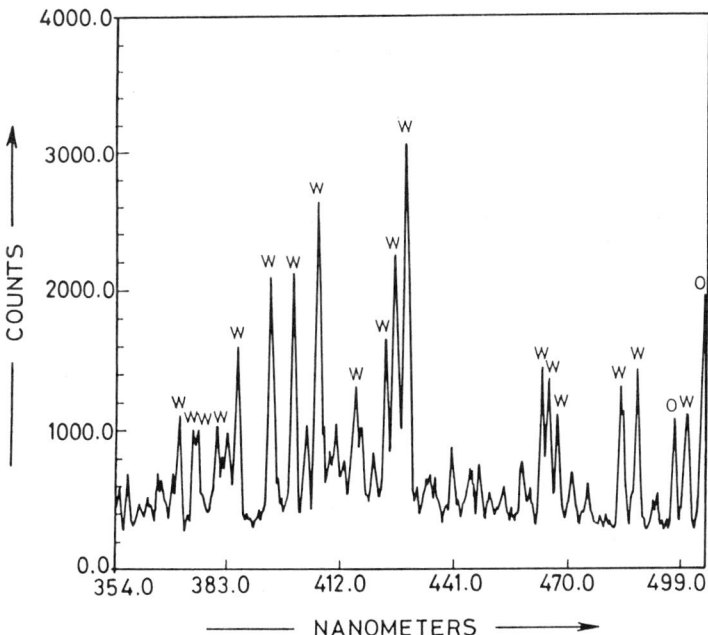

Fig. 4 : Optical emission spectrum from plume generated by KrF excimer laser ablation of $W(CO)_6$ target (laser fluence = 0.4 Jcm^{-2}).

Fig. 4 shows the optical emission spectrum produced during KrF excimer laser ablation of tungstenhexacarbonyl at laser fluence of 0.4 Jcm^{-2} in vacuum. A thorough examination of all possible emission lines across the wavelength range allowed by the optical multichannel analyser showed strong emission lines mostly attributable to W and O lines in minority. An analysis of optical emission produced by laser ablation of $W(CO)_6$ targets showed that 248 nm radiation produced mostly neutral atomic species. In the laser ablation process, these already dissociated neutral W atomic species demand no thermodynamic energy contribution from the substrate surface to dissociate $W(CO)_6$ into elemental W as in the conventional TCVD case, resulting into a decrease in the activation energy of thin film formation thus leading to an enhancement in the growth rate.

CONCLUSION

Pulsed excimer laser ablation technique can be used to form adherent low resistivity W films on both Si and SiO_2 substrate. Much lower activation energy of ~ 1.3 Kcal/mole for Si calculated for W film formation by PLD technique makes high deposition rates possible on Si and allows deposition on SiO_2 at low temperature range of 200-500°C. The unique α-W phase contribution is found above 300°C for Si giving film resistivities within 3 times that of the bulk value. In the laser ablation process, already dissociated neutral W atomic species demand no thermodynamic energy contribution from the substrate surface to dissociate $W(CO)_6$ into elemental W, and cause an enhancement in the growth rate.

The authors wish to acknowledge the support of National Laser Program (NLP) and Department of Electronics (DOE), India.

REFERENCES

1. J.O. Carlsson, Crit. Rev. Solid State Mater. Sci **16**, 161 (1990).
2. H. Korner, Thin Solid Films **175**, 55 (1989).
3. D.L. Hitchman, A.D. Jobson and L.F. Tz. Kwakman, Appl. Surf. Sci. **38**, 312 (1989).
4. Y.T. Kim, S.K. Min, J.S. Hong, and C.K. Kim, Appl. Phys. Lett. **58**, 837 (1991).
5. T.F. Deutsch and D.D. Rathman, Appl. Phys. Lett. **45**, 623 (1984).
6. H. Matsuhashi, S. Nishikawa, and S. Ohno, Mater. Res. Soc. Symp. Proc. **129**, 63 (1989).
7. R.L. Krans, A. Bernsten, and W.C. Sinke, Mater. Res. Soc. Symp. Proc. **181**, 597 (1990).
8. H. Matsuhashi, S. Nishikawa, and S. Ohno, Jpn. J. Appl. Phys. **27**, L2161 (1988).
9. A.J.P. Van Maaren, R.L. Krans, E. de Haas, and W.C. Sinke, Appl. Surf. Sci. **38**, 386 (1989).
10. A.J.P. Van Maaren, R.L. Krans, E. de Haas, and W.C. Sinke, J. Vac. Sci. Technol. **B9**, 89 (1991).
11. Peter Mogyorosi and Jan-Otto Carlsson, J. Vac. Sci. Technol. **A10**(5), 3131 (1992).
12. E.K. Broadbent and C.L. Ramiller, J. Electrochem. Soc. **131**, 1427 (1984).
13. W.A. Bryant, J. Electrochem. Soc. **125**, 1534 (1978).
14. Akira Shintani, J. Appl. Phys. **61**(6), L365 (1987).

PART IV

Dielectrics and Transitional Layers

GASDYNAMICS AND CHEMISTRY IN THE PULSED LASER DEPOSITION OF OXIDE DIELECTRIC THIN FILMS

JOHN W. HASTIE, DAVID W. BONNELL, ALBERT J. PAUL, AND PETER K. SCHENCK
National Institute of Standards and Technology, Gaithersburg, MD 20899

ABSTRACT

In the context of "chemistry and its effects on film quality," we and a number of other research groups have developed spectroscopic and modeling approaches to better define the pulsed laser deposition process. An overview of these approaches is given here, using the results of recent work performed in our laboratory on the oxide dielectric systems of $BaTiO_3$ and $PbZr_{0.53}Ti_{0.47}O_3$ (PZT).

INTRODUCTION

Efforts to optimize control of thin film deposition processes require fundamental understanding of chemistry and gasdynamics. A recent study by the National Research Council [1], for example, deals with the incomplete state of knowledge and future work needed to design expert process control systems. The pulsed laser deposition (PLD) technique appears to be especially amenable to the development of *in situ* process control, particularly for chemically complex systems such as metal oxides [2]. With PLD, there are essentially three regions where chemistry can influence the properties of a deposited film: at the target surface, in the plume/plasma region between the target and substrate, and at the substrate surface.

At the target, surface microstructure and the distribution of chemically distinct components can strongly influence plume properties (particularly species angular distributions and energies), with consequent effects on the resulting film quality, as demonstrated by us and others elsewhere [3, 4]. The target surface morphology can also change dramatically as successive laser shots erode material, and special target movement schemes, target repolishing, or other specialized procedures are often employed to minimize such changes during deposition.

In the gas phase, the plume/plasma can undergo complex chemical and gasdynamic processes (near both target and substrate surfaces) that vary in space and time and determine the final species identities, kinetic energies, concentrations, mobilities, and sticking coefficients at the substrate surface. Interaction of translationally hot plume species or electrons with inert or active background gases provides an additional opportunity for charge transfer processes, excitation, and even neutral molecule formation to occur.

At the substrate surface, the energetics and order of events as each gas pulse arrives at the surface can be crucial to the growth rate and film morphology. Both the previous morphology of the substrate surface and the interaction of successive gas pulses with the growing film govern the growth behavior. Gas rebounding from the substrate surface can interact with the incoming plume gas, and any ambient background gas, to alter species energetics and excite or ionize some fraction of the near-substrate gas. Chemical events at the deposit surface (and also the target) are not readily amenable to *in situ*, real-time characterization. Hence, the experimental focus has generally been on the gas phase with extension to the deposit surface being made through *ex situ* surface characterization and gasdynamic models.

In the present work, we have emphasized the development and application of optical emission and molecular beam mass spectrometric *in situ* techniques, coupled with models describing the plume dynamics. Case studies include the technologically important dielectric thin film systems of BaTiO$_3$ and PZT, deposited using relatively high laser fluences, in the range of 2-10 J/cm^2, and 20 ns pulse times.

EXPERIMENTAL

In situ Probes for PLD

A schematic representation of the PLD process and the location of various *in situ*, real-time monitors we employ is given in Fig. 1. Note that the laser plume is depicted as having three regions of characteristic behavior, following Kelley and Dreyfus [5]. These include, a stagnation zone or Knudsen layer (typically located within 10^{-3} cm of the target), a region of adiabatic gasdynamic-dominated expansion, and a collisionless free-flight region. The coupling of a multi-chamber, differentially pumped molecular beam mass spectrometer (MS) with the laser-target interaction process has been described in detail elsewhere [6]. Two types of optical emission detection systems, indicated in Fig. 1 as OMA (optical multichannel analyzer), have been employed, a linear photodiode array (PDA)-based OMA, and a CCD (charge coupled detector array) imaging/OMA. Both systems are described below.

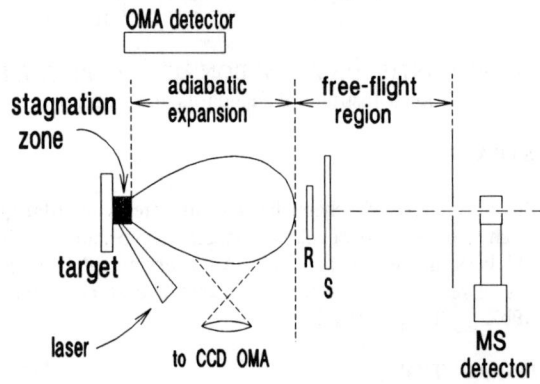

Figure 1. Schematic of laser plume and various *in situ* probes. **R** is a deposition monitor and **S** the substrate.

TABLE 1. Optical Emission Probes

Probe (λ = 185 - 1100 nm)	Dynamic Range	Spatial Scale
Conventional spectroscopy	<8 bits(single λ)	Point source
PDA OMA: • intensified linear photodiode array • point to line UV/Vis. fiberoptic • XYZ image point mount	256:1 (8 bits)	Spectral dispersion, 512 λ elements
CCD imaging/OMA: • charge coupled detector with image intensifier • time resolution 5 ns • stationary (imaging) mount line imaging, or UV/Vis. fiberoptic	256,144:1 (18 bits)	2D (camera mode) 512 x 512 pxls white light (~350-900 nm) *or* narrow pass filters 1D (spectral mode) 512 pxls x 512 λ elements

Optical Emission Probes

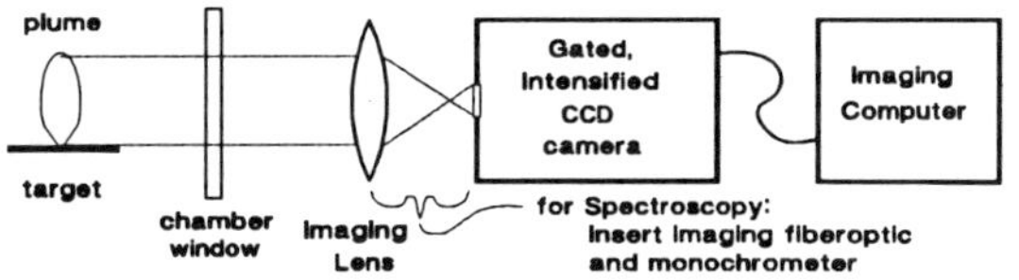

The technology available for probing spectrally emissive systems has improved dramatically in recent years with the advent first of linear photodiode array OMA's and, then, CCD-based OMA's and cameras. We have

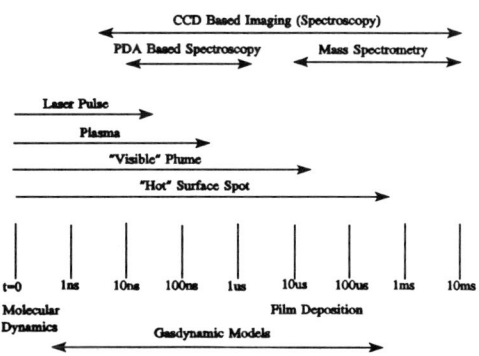

Figure 2. Schematic of the CCD imaging/OMA optical arrangement.

used both types, utilizing a fast gated image intensifier to provide time resolution below 10 ns, and within one gate time of the end of the laser pulse. A fused silica lens is used to image (1:1) the plume onto a 2 mm diameter UV fiber optic cable connected to the spectrograph of either detector. The lens and fiber optic are mounted to permit computer-controlled rastering over the plume profile. The CCD also allows direct narrow- or broad-band imaging, typically using a high resolution 55 mm or 105 mm focal length macro-optimized lens.

A comparison of the typical dynamic range and spatial scale capabilities of these OMA systems with those of conventional spectroscopy is given in Table 1. Details of the application of the PDA OMA system have been described by us elsewhere [7]. Fig. 2 shows, schematically, the essential features of the newer CCD imaging/OMA system.

Figure 3. Time scales for *in situ* monitoring of PLD.

The CCD system has a time-scale capability, noted in Fig. 3, that conveniently includes and bridges those of the earlier available plume probing techniques of PDA OMA detection and molecular beam mass spectrometry. The time frames associated with various stages of the PLD process, and those covered by various molecular and gasdynamic models are also indicated, for comparison, in Fig. 3.

There is considerable interest (eg., see [8]) in the possible application of OMA devices to the *in situ* monitoring, and intelligent feedback control, of industrial processes. It is thus pertinent to note that a practical correlation can be found between OMA emission intensities and actual deposition rates, as shown in Fig. 4. Indeed, this type of excitation and detection approach now forms the basis of a new analytical technique — "laser induced plasma spectroscopy" [9]. Application of these emission probes to several dielectric thin film deposition systems is demonstrated below.

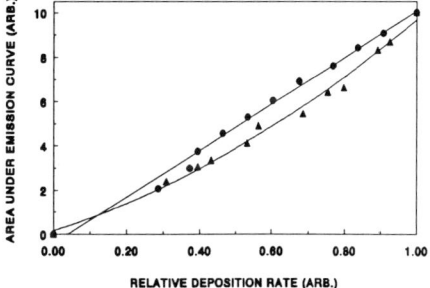

Figure 4. Integrated spectral emission intensities *vs.* measured deposition rates for Fe_3O_4 (▲, ∫ 300-600 nm, 20 mm above target) and Ag (●, ∫ 200-500 nm, 10 mm above target).

307

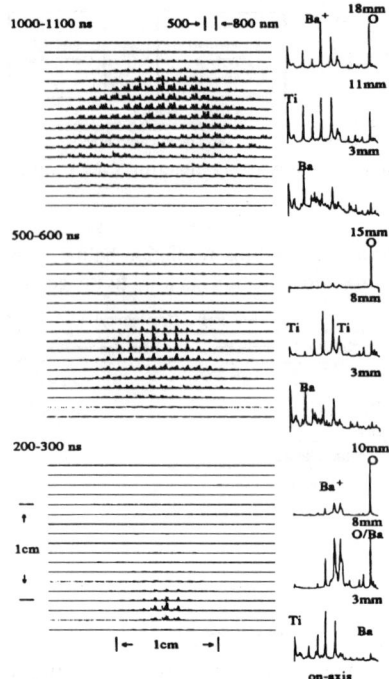

Figure 5. Emission spectra maps for a BaTiO$_3$ plume, taken at three different times (200, 500, & 1000 ns) following the laser pulse. The plume axis is vertical; laser focus is located at bottom center of maps.

BaTiO$_3$ Plume Emission

Results of earlier work [10] on BaTiO$_3$ plume emission spectra, obtained using the PDA OMA system, are reproduced here in Fig. 5. Major emission lines for each of the elements, Ba, Ti and O are conveniently visible in the spectral region 500-800 nm. The overall pattern produced by the individual spectral scans, obtained at different spatial and temporal plume locations, reflects the actual plume geometry. Inspection of the magnified spectral scans at the right hand side of Fig. 5 reveals the tendency for lighter (Ti, O) and heavier (Ba) species to segregate in space. This observation will be recalled again below in connection with the predictions of plume gasdynamic models. It should be noted that at least part of the downstream O-atom emission in Fig. 5 originates from energetic electrons colliding with the added O$_2$ background gas.

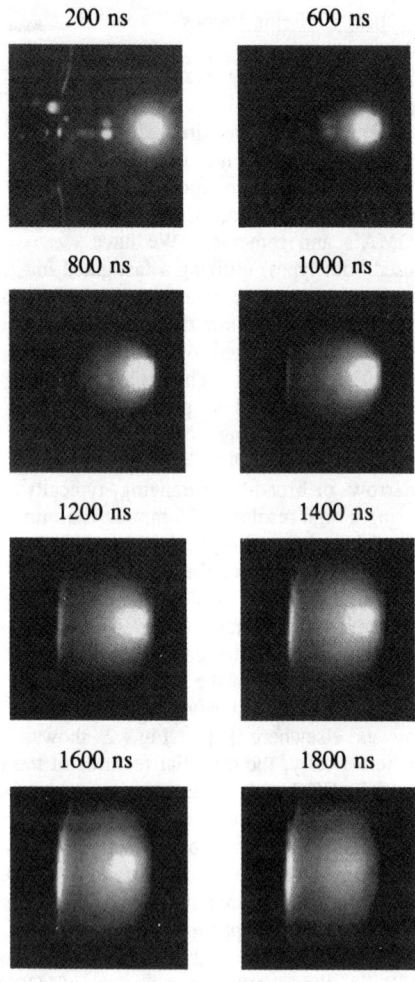

Figure 6. CCD images of BaTiO$_3$ plumes, taken at indicated times after laser pulse, with a 50 ns gate. Target and substrate are separated by 2.5 cm. O$_2$ background pressure < 0.1 Pa. Each image is self-scaled.

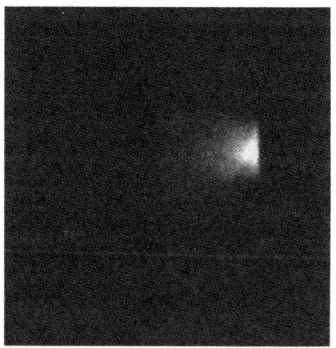

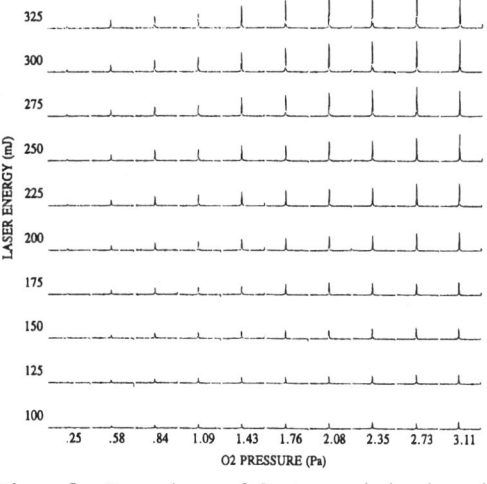

Figure 8. Dependence of O atom emission intensity on laser energy and added (background) O_2 pressure, sampled at a centerline position 1.8 cm from the target.

Figure 7. CCD imaging/OMA frames of $BaTiO_3$ emission, taken with narrow pass filters for Ti (top, 410 nm) and Ba (bottom, 620 nm) at 1200 ns delay (100 ns exposure) after the laser pulse; background pressure < 0.1 Pa.

Detailed PDA OMA monitoring of the formation and decay of PLD plumes, such as that represented in Fig. 5, requires separate experiments for each spectrum. The 2D array capability of the CCD imaging/OMA system allows these plume observations to be performed with far fewer experiments and with greater dynamic range, as shown in Fig. 6. Plume transport from the target (right side) to the substrate (left side) is readily visualized in this sequence of frames. Of particular note is the reappearance of emission when the plume interacts with the substrate surface (see also [11]). The identity of the responsible emitting species in this region has not yet been determined. Another example of the utility of the CCD imaging/OMA system is given in Fig. 7. Here, imaging with narrow band pass filters allows us to further demonstrate the tendency for spatial segregation of species to occur in the plume. In this case, the filters isolate prominent Ba^+ emissions (620 ± 5 nm), and Ti, Ti^+ emissions (410 ± 5 nm). The downstream halo, evident in the top frame of Fig. 7, shows clearly that the Ti emission is distributed well ahead of the Ba emission, displayed in the bottom frame.

Research is in progress to determine the key process intermediates and optimum parameters for deposition of high dielectric constant $BaTiO_3$ thin films. Previous work on YBaCuO complex oxide systems has indicated the need for additional O_2 (added as background gas) to maintain stoichiometric deposition [12]. Evidence has also been presented correlating the presence of CuO (rather than Cu) in the plume with formation of a high T_c film [13, 14], suggesting the possible general importance of transported molecular species in complex-stoichiometry films. For $BaTiO_3$, relationships between film morphology, crystal structure, dielectric behavior, substrate temperature and support material, and process variables such as O_2 pressure, laser fluence, and wavelength have been investigated and will be reported on elsewhere [15]. Here, we present results relating key plume species and deposition rates with process parameters. In Fig. 8 the dependence of the O atom 778 nm emission feature on both laser energy and added O_2 pressure is indicated. For oxides, O_2 has a relatively low sticking probability and in order to maintain oxygen stoichiometry it is considered desirable to enhance the flux of O atoms to the substrate. This condition is favored by high laser energy and high O_2 pressure conditions, as shown in Fig. 8. However, these conditions do not favor high deposition rates, as Fig. 9 shows, and some compromise must be made in the selection of process parameters.

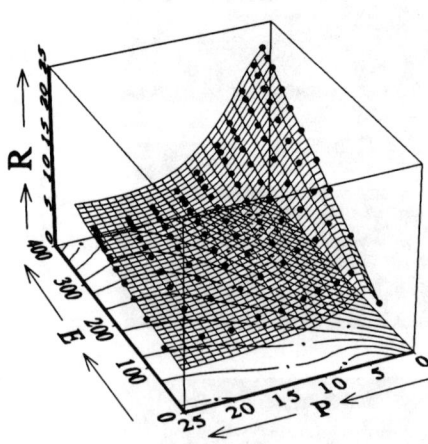

Figure 9. Process diagram of $BaTiO_3$ deposition rate (R, Å/s) as a function of added O_2 background pressure (P, Pa) and laser energy (E, mJ).

PZT Plume Species Identities and Velocities

Control of PZT thin film stoichiometry is more difficult than for $BaTiO_3$ owing to the relatively high volatility, and likely low accommodation coefficient, of Pb atoms. Earlier work by us, and others, suggests that PbO rather than Pb is a more effective species for film retention of Pb [4]. We have not been able to identify PbO emission spectroscopically but this does not preclude the presence of PbO in a non- or weakly-emitting state. Molecular beam mass spectrometry is not dependent on the presence of excited molecular species and this technique has been used to monitor PbO and all other species present in PZT plumes. The MS technique also provides time-of-arrival (TOA) information which can be transformed into species-specific velocity distributions [7, 10]. Fig. 10 shows such distributions for some of the major species observed. Note that PbO was not observed with the shorter wavelength laser source. This behavior parallels that found by others for CuO in YBaCuO plumes [13]. The 248 nm case also favors much faster velocity distributions, typical of a non-thermal ablation process. These distributions have been fit to full-range Maxwell velocity distribution curves from which beam velocity and temperature information may be derived, as shown by the example in Fig. 11, and discussed in the next section.

PLUME MODELS

Velocity and Temperature Models

The molecular beam of plume species extracted by differential pumping for mass spectrometric (MS) analysis represents an end result of the overall plume formation and decay process. In the collisionally-dominated plume, species kinetic energies may be modified by at least three dynamic processes prior to the attainment of a free-flight condition and mass spectrometric detection: (1) desorption, vaporization, and/or ablation, (2) plume formation, and (3) adiabatic expansion. In order to relate molecular beam mass spectrometric results to the earlier stages of desorption (particularly in the present high-yield case), dynamic models of plume formation and adiabatic expansion need to be invoked. These models fall into roughly three classes: gasdynamic, numerically integrated flow simulations, and detailed molecular-dynamic simulations.

Kelly and Dreyfus have developed gasdynamic models for both low- and high-yield surface desorption cases [5, 16, 17]. The chronology of events in their models

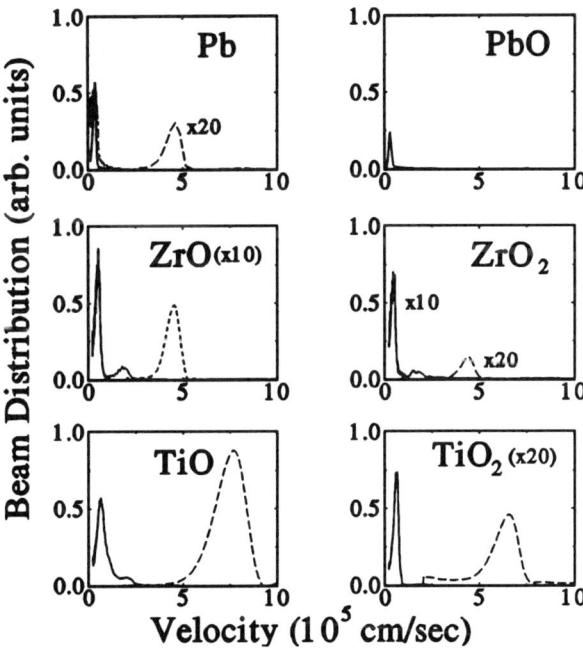

Figure 10. Beam velocity distributions (I'(v) as defined in [10]) of plume species for the PZT system. Solid curves result from 1064 nm Nd/YAG laser excitation; dashed curves are for the 248 nm excimer case.

begins with 1) interaction of the laser with the target surface, leading to species ejection and hydrodynamic-controlled flow, 2) formation of a "Knudsen layer" (KL) stagnation zone in the close vicinity of the surface, and 3) transition from stagnation to a region of "unsteady" or non-steady-state adiabatic expansion, followed by 4) an abrupt transition to free molecular flow, the so-called "frozen" expansion assumption. These events are depicted in Fig. 1.

By placing a differential orifice downstream from the sudden freeze point, a representative molecular beam can be extracted for mass analysis [6]. However, the presence of the KL effectively masks the relationship between this beam and target surface processes. Thus, the effect of gasdynamic processes in each region on the gas velocity profile must be accounted for in order to relate the MS observations to the events at the target surface, or at a substrate located within the plume region.

For a collision-dominated flow expanding into a molecular "beam" in the direction of the MS (Z axis), the corrected (for number density detector case) beam velocity distribution function, $P_b(v_Z)$, should take the form of a full-range Maxwellian with a center-of-mass flow:

$$P_b(v_Z) \propto v_Z^3 \exp[-(m/2kT_c)(v_Z - u_c)^2] . \tag{1}$$

The species velocities perpendicular to the surface have an imposed center-of-mass flow away from the target due to collisions above the target that result in scattering of species above the surface; v_Z are the velocity magnitudes observed at the MS detector; u_c is the overall group or center-of-mass flow velocity of the gas in the free-flight region, T_c the local (terminal) gas temperature, m the species mass, and k the Boltzmann constant. The MS data, I(t) vs. t, are converted to beam velocity distributions, $I'(v) \propto P_b(v_Z)$ vs. v_Z [see refs. 7, 10]. Leuchtner, et. al. [4] present an alternative, equivalent treatment, using I(t) directly, recognizing that the transformation from I(t) for a

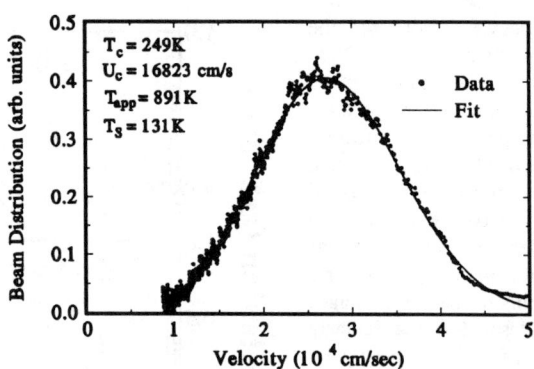

Figure 11. Maxwell distribution, $P_b(v_Z)$, fit to experimental, $I'(v)$, for PbO from PZT (conditions as for Fig. 10). Derived values defined in text; T_{app} an "apparent" surface temperature defined in [16].

number density detector, giving I(v), yields a result proportional to a cavity model velocity distribution. Ref. [7] discusses the two distributions in more detail. Values of T_c and u_c are obtained by non-linear least-squares fitting of Eqn. (1) to the MS-determined velocity distributions.

The free-expansion process begins at the downstream KL boundary where the flow velocity and speed of sound are equal; i.e., $M_K=1.0$. The Mach number for flow in a gas is defined as, $M = u/a$, the ratio of the gas flow velocity, u, to the local velocity of sound, $a=(\gamma kT/m)^{1/2}$, where $\gamma=(j+5)/(j+3)$ is the heat capacity ratio, and j is an average number of internal degrees of freedom of the gas. Values of γ typically range from 1.667 ($j \approx 0$), for an ideal ground state monatomic gas, to as low as about 1.2 ($j > 6$) for highly excited plasmas containing polyatomic species. The adiabatic expansion accelerates the gas to Mach numbers higher than unity, with a terminal Mach number, M_c, being achieved at the abrupt cessation of collisions as the gas enters free molecular flow at the end of the collisionally dominated adiabatic expansion. Also, values of

$$M_c = u_c/(\gamma kT_c/m)^{1/2} \qquad (2)$$

can be determined from the experimentally derived μ_c and T_c data. The Mach number values can then be used to calculate the effective gas temperature at the beginning of the free expansion, the outer edge of the KL.

Formation of the KL will occur during or at the end of the laser pulse, depending upon the density of material ejected from the surface. If the KL forms at some time $t \leq \tau$, where τ is the laser pulse width, a transition to adiabatic expansion will occur before the end of the laser pulse, leading to gas acceleration and M values greater than one. If the KL forms at the end of the laser pulse ($t = \tau$), the plume undergoes an abrupt transition into free-flight without undergoing an adiabatic expansion and M is unity. In low yield cases, typical of many etching processes, a KL may not form, and collisional interaction will yield M values less than unity.

Kelly [16, 17] grafted a solution to unsteady adiabatic expansion, obtained for a one-dimensional gun, to the outer, expansion boundary of the KL, thereby obtaining the following expression relating T_c to the temperature, T_K, at the Knudsen layer:

$$T_c/T_K = [(\gamma+1)/(2+(\gamma-1)M_c)]^2 \quad . \tag{3}$$

Kelly and Dreyfus [5, 17] also derived the following expression relating T_K to the effective surface temperature, T_S:

$$(T_K/T_S)^{1/2} = -(\pi\gamma/2)^{1/2}/2(j+1) + [1 + (\pi\gamma/2)/4(j+1)^2]^{1/2} \quad . \tag{4}$$

It follows from Eqns. (2) - (4) that T_S can be derived from the experimentally derived values of μ_c and T_c. Kelley [16] also gives the following expression, valid for thermal release conditions, relating T_S to average velocity $(\hat{v}^2)$ and geometric factors (θ is the angle between the direction normal to the surface and the observed beam direction; in this application, $\theta = 0°$:

$$T_S = (m\hat{v}^2/2k)/(\eta T_c/T_S) \text{, where} \tag{5}$$

$$\hat{v}^2 = 2kT_c/\pi m \text{, and} \tag{6}$$

$$\eta = (\gamma M^2 \cos^2\theta/8) [1 + (1 + 16/\gamma M_c^2 \cos^2\theta)^{1/2})]^2 \quad . \tag{7}$$

Equations (3) — (5) may also be combined to give an independent value for T_S. The internal consistency between these two approaches has been tested, using our MS data for PZT ablation, and the T_S values differed by as much as a factor of two, even though we see reasonable evidence of local thermal equilibrium among species from TOA vs $m^{1/2}$ plots for the 13 observed species over PZT.

Varying the laser wavelength also had a very strong effect on values of T_S obtained by this analysis approach. Calculated T_S values were either very high (10,000-20,000 K for 248 nm excimer irradiation), or very low (< 100 K for 1064 nm Nd/YAG ablation).

Kelley and Dreyfus [5] reported reasonable agreement between this model approach and their observations for cases where the yield was low enough that relatively small Mach numbers were obtained. One explanation for our observed discrepancies, is that the Kelley and Dreyfus formalism was primarily derived for processes where M is close to unity, whereas for our PZT case Mach numbers of the order of 3-5 were observed. These high Mach numbers are supported by observation of strong forward peaking, generally much higher than $\cos^4\theta$. Experimentally, high T_S values may be attributed to non-thermal interaction of the high energy excimer photons with the plume, or with direct laser/plume interactions altering the dynamics. The Nd/YAG data suggest extreme adiabatic cooling, consistent with the derived high Mach numbers.

Although the gasdynamic model has a desirable simplicity, it is very sensitive to parameters difficult to determine directly, such as true values for γ. The degree of ionization within the KL may strongly influence the effective value of j and hence γ. For a monatomic gas, $T_K/T_S = 0.669$, which may not be a very realistic assumption, even for atomistically simple targets. There are also several simplifying assumptions used, valid near M=1, but possibly poor at higher yields. One such assumption is that the KL boundary is stationary. Another assumption is that γ is constant throughout the collision-dominated regions. We have presented downstream evidence for species segregation and, at high yield, the KL itself may persist long enough to invalidate model assumptions. It is also likely, that for certain conditions, the laser sputtering process may evolve directly to an unsteady adiabatic expansion without the presence of a Knudsen layer, as discussed more recently by Kelly and Braren [18]. Formation of clusters and recondensation at the target are also factors that complicate development of more reliable gasdynamic models [19]. The computation of M is also highly dependent on the nature of the expansion. To resolve these questions, more detailed modeling of the expansion process is needed. Stepwise numerical integration of a suitable flow model is one method that has the potential for providing that detail.

Plume Spatial Evolution Model

Singh *et. al.* have developed a flow model to simulate the formation of a high-temperature high pressure gaseous plume in the vicinity of the target surface [20, 21]. A finite difference method is used to simulate gasdynamic isothermal plume expansion that occurs during the laser pulse width τ and, at the end of the laser pulse, abruptly evolves into an adiabatic expansion. The equations of motion and the continuity equation are used to transform the expressions for the time-dependent density, pressure, and velocity into force equations that govern the dynamics of the plume's outer edges (X(t),Y(t),Z(t)) during isothermal expansion:

$$X(t)\left[\left(\frac{1}{t}\right)\frac{dX}{dt} + \frac{d^2X}{dt^2}\right] = Y(t)\left[\left(\frac{1}{t}\right)\frac{dY}{dt} + \frac{d^2Y}{dt^2}\right]$$
$$= Z(t)\left[\left(\frac{1}{t}\right)\frac{dZ}{dt} + \frac{d^2Z}{dt^2}\right] = \frac{kT_o}{m}, \quad t \leq \tau, \tag{8}$$

where T_o is the isothermal plume temperature. Inclusion of the equation of energy and the adiabatic equation of state leads to gasdynamic equations for the adiabatic expansion of the laser plume:

$$X(t)\left[\frac{d^2X}{dt^2}\right] = Y(t)\left[\frac{d^2Y}{dt^2}\right] = Z(t)\left[\frac{d^2Z}{dt^2}\right]$$
$$= \left[\frac{kT_o}{m}\right]\left[\frac{X_o Y_o Z_o}{X(t)Y(t)Z(t)}\right]^{\gamma-1}, \quad t > \tau. \tag{9}$$

Values X_o, Y_o, and Z_o are components of the plasma edge at the boundary of the adiabatic expansion. One appealing feature of this model is that it predicts that the plume's extents are largest in the directions where the spot size is least--in keeping with the observation that the major axis of the elliptical deposit pattern, arising from an elliptical laser footprint, is orthogonal to the footprint major axis. Strong forward peaking in the Z direction is a natural consequence of the narrowness of the isothermal vapor zone, analogous to the KL discussed earlier.

Solutions to these equations may be determined by computer simulation using the Runge Kutta iterative integration for second order differential equations. For the isothermal region (Eqn. 8), the initial boundary conditions are given, under our conditions, by the spot size of the excimer laser which was 0.04 cm and 0.01 cm for X(0) and Y(0), respectively. The initial dimension in the Z direction was determined from the initial velocity of gas determined by the surface temperature, i.e.,

$$<v> = dX(0)/dt = dY(0)/dt = dZ(0)/dt$$
$$= (3kT_S/m)^{1/2}, \text{ and } Z(0) = <v>t, \tag{10}$$

where t is the incremental step size of 10^{-14} s. The step size was changed to 10^{-10} s after 1 ns had elapsed. Based on values obtained experimentally in our earlier work, we assumed typical values for the plasma temperature of 20,000 K, the surface temperature of 4700 K, and γ of 1.67 (monatomic gas). An accurate knowledge of these temperatures is not critical to the present discussion. Results indicate that the displacement of the plume edge becomes linear and the acceleration goes to zero well before typical deposition distances are reached (≈ 3 cm). Attainment of maximum velocities are predicted at post laser pulse times of about 100 ns. Typical results, given in Figs. 12 and 13, may be compared with the experimental observations of Figs. 5 and 7, where Figs. 5, 7, and 13 show a similar spatial segregation of light and heavy species, with the Ti and O atom distribution peaks being located well down-stream of the Ba atoms. These observations are also supported by the MS velocity distributions (for the PZT system) where the O atom distribution peaks far earlier than other neutral species. The predicted plume front locations in Fig. 12 are also consistent with the observed locations given in Fig. 6, although we do not necessarily expect an exact coincidence between the location of a plume emission front and that predicted for non-emitting species. Dye *et. al.*, however, cite evidence that a reasonable coincidence between these two classes of species can occur [11].

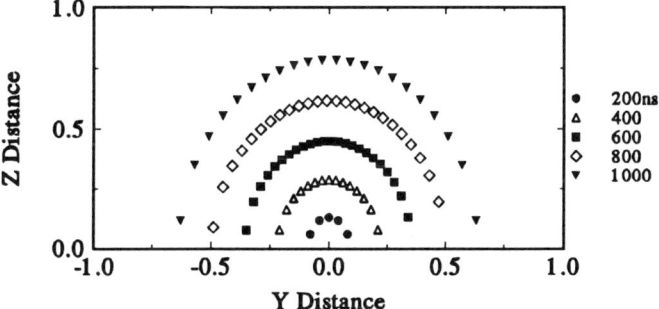

Figure 12. Model prediction of plume front location for TiO_2 species (PZT and $BaTiO_3$) at different delay times following the laser pulse. Distances are in cm and the plume originates at Y = 0, Z = 0.

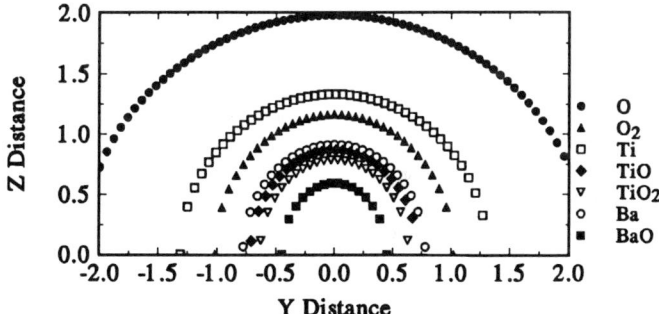

Figure 13. Model prediction of plume front location at 1 μs delay time for various species in the $BaTiO_3$ system. Distances are in cm and the plume originates at Y = 0, Z = 0.

SUMMARY AND CONCLUSIONS

Several representative examples have been considered to demonstrate the application of *in situ* spectroscopic techniques and transport models to the PLD of thin films. For the plumes, a reasonably definitive picture of the chemical species present and their kinetic energies has been developed. Indirect indications of the effects of plume chemistry on film quality can be derived from such studies. For instance, plume species such as O, PbO, CuO [14], are considered beneficial to the formation of stoichiometric film compositions. In other cases, where the oxide species is more volatile (and less condensable) than the metal, such as for YO vs. Y, the metal atom is believed to be more effective in deposition [22]. *In situ* surface measurement techniques and gas-solid kinetic models need to be developed to relate plume chemistry more directly to substrate film properties; *e.g.* see [23, 24].

REFERENCES

1. *Plasma Processing of Materials*, (Nat. Acad. Press, Washington, DC, 1991).
2. *MRS Bulletin*, 27, "Pulsed Laser Deposition," (Special Edition), February, (1991).
3. A.J. Paul, D.W. Bonnell, J.W. Hastie, P.K. Schenck, R.D. Shull, and J.J. Ritter, *Mat. Res. Soc. Symp. Proc.* 236, 455 (1992).
4. R.E. Leuchtner, J.S. Horwitz, and D.B. Chrisey, *Mat. Res. Soc. Symp. Proc.* 285, 87 (1993).
5. R. Kelly and R.W. Dreyfus, *Nucl. Instr. and Meth. Phys. Res.* B32, 341, (1988).
6. J.W. Hastie, D.W. Bonnell, and P.K. Schenck, *High Temp.-High Press*, 20, 73 (1988).
7. P.K. Schenck, D.W. Bonnell, and J.W. Hastie, *High Temp. Sci.*, 27, 483 (1990).
8. I. Nir, J. Winniczec, M.P. Splichal, H.M. Anderson, and A. Stanton, *Photonics Spectra*, August, 1993, p.87.
9. R. Nordstom, *Lasers and Optronics*, Dec. p. 23 (1993).
10. J.W. Hastie, D.W. Bonnell, A.J. Paul, and P.K. Schenck, *Mat. Res. Soc. Symp. Proc.* 285, 39, (1993) (Note in Eqn. (3) of [10], v^2 factor should be v^3).
11. R. C. Dye, R. Brainard, S. R. Foltyn, R. E. Muenchausen, X. d. Wu, and N. S. Noger, *Mat. Res. Soc. Symp. Proc.* 285, 15 (1993).
12. A. Inam, M.S. Hegde, X.D. Wu, T. Venkatesan, P. England, and P.F. Micell, *Appl. Phys. Lett.* 53, 908 (1988).
13. C. E. Otis and R.W. Dreyfus, *Phys. Rev. Lett.* 67, 2102 (1991).
14. O. Eryu, K. Yamaska, and K. Masuda, *Mat. Res. Soc. Symp. Proc.* 285, 57 (1993).
15. P.K. Schenck, M. D. Vauden, and Cheol-Seong Hwang, *Crystal Structure and Crystal Texture in Laser-Deposited Thin Films of $BaTiO_3$*, to be published (1993).
16. R. Kelly, *SPIE Photochemistry in Thin Films*, 258 (1989).
17. R. Kelly, *J. Chem. Phys.* 92 (8), 5047 (1990).
18. R. Kelly and B. Braren, *Appl. Phys.* B53, 160 (1991).
19. R. Kelly, *Springer Series in Surface Sciences* 19, 135 (1990).
20. R.K. Singh and J. Narayan, *Phys. Rev. B.*, 41, 8843 (1990).
21. R.K. Singh, O.W. Holland, and J. Narayan, *J. Appl. Phys.* 68, 233 (1990).
22. S. Yukya, J. Tsujino, N. Tatsumi, and Y. Shishara, *J. Mater. Res.* 8, 709 (1993).
23. C.H. Chen, R.C. Phillips, and P.W. Morrison, *Mat. Res. Soc. Symp. Proc.*, 285, 45 (1993).
24. M.R. Predtechensky, A.V. Bulgarov, A.P. Mayorov, and A.V. Roshchin, *Appl. Supercond.*, 1, 1995 (1993).

ADSORPTION AND REACTION OF TiCl4 ON W(100)

WEI CHEN AND JEFFREY T. ROBERTS*
University of Minnesota, Department of Chemistry
Minneapolis, MN 55455

ABSTRACT

The adsorption and reaction of titanium tetrachloride ($TiCl_4$) on W(100) was investigated using temperature programmed desorption mass spectrometry (TPRS), x-ray photoelectron spectroscopy (XPS), Auger electron spectroscopy (AES), and low energy electron diffraction (LEED). $TiCl_4$ adsorbs molecularly on W(100) at 100 K. Desorption from the molecularly bound state occurs near 220 K. Competing with desorption is dissociation to adsorbed $TiCl_3$, which reacts to form gaseous $TiCl_4$ near 450 K. $TiCl_3$ also decomposes into atomically adsorbed Ti and Cl on the surface upon heating to 700 K.

INTRODUCTION

The chemical vapor deposition (CVD) of titanium carbide has been the subject of much recent interest because of the potential applications of TiC thin films in hard coatings. TiC thin films are typically fabricated using titanium tetrachloride as the titanium source and propane as the carbon source [1, 2]. Although the steady-state kinetics for growth of a TiC thin film have been recently determined [2], the *initial* growth steps, about which there remain several unresolved issues, have not been well studied. Of particular interest to us is the fact that TiC growth is readily initiated on tungsten, but not at all on platinum [3]. Studies of adsorption and reaction of $TiCl_4$ on tungsten and platinum surfaces under a well controlled ultrahigh vacuum environment could provide insight into the basis for this selectivity. In this contribution, we discuss recent results concerning adsorption and reaction of titanium tetrachloride ($TiCl_4$) on the W(100) surface.

* Author to whom correspondence should be addressed.

EXPERIMENTAL

Experiments were conducted in an ultrahigh vacuum chamber (base pressure $\approx 10^{-8}$ Pa) equipped for x-ray photoelectron spectroscopy (XPS), Auger electron spectroscopy (AES), temperature programmed desorption mass spectrometry (TPRS), and low energy electron diffraction (LEED). $TiCl_4$ (Aldrich, 99.9%), which is air sensitive, was transferred into a sealed bottle under an argon atmosphere. $TiCl_4$ was degassed via several freeze-pump-thaw cycles before use every day. The W(100) single crystal was purchased from Metal Crystals Limited (Cambridge, England) and cleaned according to established methods [4]. Briefly, this involved heating by electron bombardment under oxygen (P $\approx 10^{-6}$ Pa) at 1400 K for 5 minutes. The surface was then flashed to 2300 K for 30 seconds thus removing oxygen as a volatile tungsten oxide. Cleaning cycles were repeated until no trace of contamination could be detected using AES.

RESULTS

1) Temperature Programmed Reaction Mass Spectrometry (TPRS)

In these experiments, $TiCl_4$ was adsorbed on W(100) at 100 K, and the gas phase products detected mass spectrometrically as the surface was heated to 700 K. $TiCl_4$ was the sole desorption product detected. A representative temperature programmed desorption spectrum is shown in Fig. 1. Masses monitored were Ti^+ (m/e 48), $TiCl_3^+$ (m/e 153) and $TiCl_4^+$ (m/e 188), the three principle cracking fragments of $TiCl_4$. Three desorption features are observed with peak maxima at 175, 220, and 450 K. The low temperature state does not saturate with increasing $TiCl_4$ exposure and is assigned to sublimation of the multilayer. The states at 220 K and 450 K are attributed to desorption of molecularly adsorbed $TiCl_4$ and to disproportionation or recombination of $TiCl_3$, respectively. Justification for these assignments will be given below.

$TiCl_4$ desorption kinetics were investigated using the method of heating rate variation [5]. Based upon this analysis, we conclude that the process at 220 K, assigned as molecular desorption, is well described by a first order rate expression. The activation energy for desorption from this state is 54 kJ·mole^{-1} and the pre-exponential factor is 3×10^{12} s^{-1}. The reaction order for the high temperature desorption state(s) could not be determined using the heating rate variation method. However, the peak maximum temperature decreases with increasing $TiCl_4$ coverage, suggesting that the rate expression is greater than first order in coverage of adsorbed $TiCl_4$. Also, careful analysis of the spectra suggests that $TiCl_4$ evolution near 450 K may result from a superposition of two or more states.

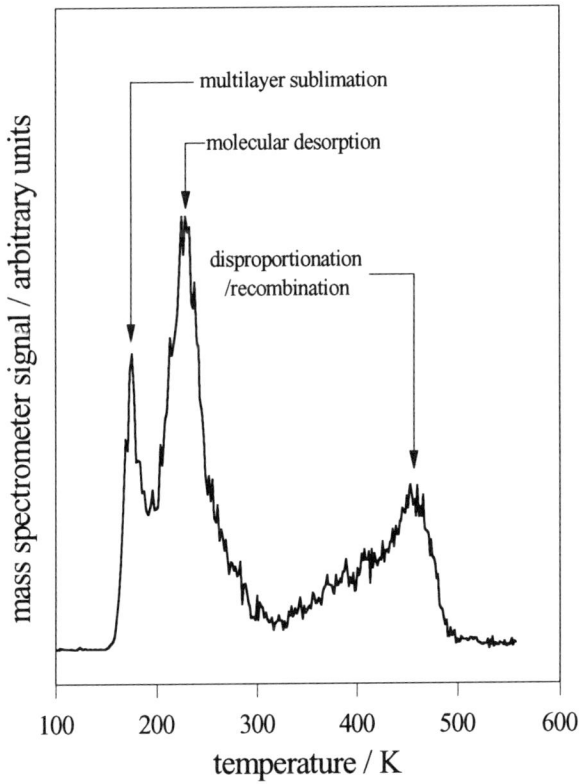

Figure 1. Temperature programmed desorption of a TiCl$_4$ multilayer on W(100). TiCl$_4$, the only gas product, was detected as m/e 153, TiCl$_3^+$. The heating rate was 10 K·s^{-1}.

2) X-ray Photoelectron Spectroscopy (XPS)

XPS spectra of the Ti(2p) region were recorded after annealing the surface to 180, 330, and 700 K (Fig. 2). The Ti(2p$_{3/2}$) binding energy of the TiCl$_4$ multilayer is 459.0 eV (Fig. 2a). After the surface is annealed to 180 K, the binding energy of Ti(2p$_{3/2}$) remains at 459.0 eV (Fig. 2b), and the state is therefore assigned to titanium in the +4 oxidation state. Further annealing to 330 K, well past the molecular desorption feature, results in a significant shift down in binding energy, to 458.2 eV (Fig. 2c), a value which is consistent with titanium in the +3 oxidation state. By 700 K, all remaining Ti has a 2p$_{3/2}$ binding energy of 455.8 eV (Fig. 2d), consistent with atomically adsorbed titanium [6]. The total amount of residual Ti and Cl on the surface after thermal desorption of a TiCl$_4$ multilayer is 0.5 ML.

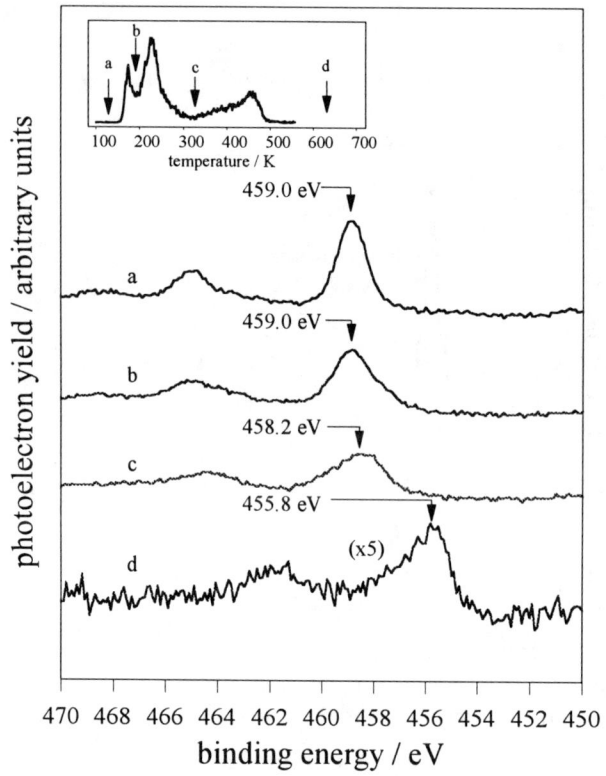

Figure 2. XPS spectra of Ti(2p) at different annealing temperatures: (a) 100 K, (b) 180 K, (c) 330 K, and (d) 700 K. The inset shows the temperature programmed reaction of a $TiCl_4$ multilayer.

3) AES and LEED

AES results are consistent with the x-ray photoelectron spectra. Ti and Cl features are evident in spectra recorded after reaction of $TiCl_4$. The Ti/Cl Auger peak ratio is unchanged as a function of annealing below 1000 K. The Cl signal vanishes at 1500 K, and Ti disappears at 1900 K. Upon adsorption of a monolayer of $TiCl_4$, no new LEED patterns are observed. Furthermore, the same (1×1) pattern seen on the clean W(100) surface is observed after heating the $TiCl_4$ precovered surface to 700 K.

DISCUSSION

We propose the following reaction pathway for $TiCl_4$ on W(100):

$$TiCl_{4(g)} \underset{220\ K}{\overset{100\ K}{\rightleftarrows}} TiCl_{4(a)} \qquad (1)$$

$$TiCl_{4(a)} \xrightarrow{220\ K} TiCl_{3(a)} + Cl_{(a)} \qquad (2)$$

$$TiCl_{3(a)} + Cl_{(a)} \xrightarrow{450\ K} TiCl_{4(g)} \qquad (3a)$$

or

$$2TiCl_{3(a)} \xrightarrow{450\ K} TiCl_{4(g)} + Ti_{(a)} + 2Cl_{(a)} \qquad (3b)$$

$TiCl_4$ adsorbs molecularly on W(100) at 100 K. Upon heating to 200-400 K, some $TiCl_4$ desorbs, and some dissociates to form adsorbed $TiCl_3$. $TiCl_3$ reacts near 450 K, either via recombination with $Cl_{(a)}$ (Step 3a) or via a disproportionation process (Step 3b), thereby generating the second $TiCl_4$ state in the thermal desorption spectra. This pathway, although by no means proved, is amply supported by the data.

The first step of the reaction sequence involves the desorption of molecularly adsorbed $TiCl_4$ near 220 K. This is consistent with the fact that the $TiCl_4$ desorption rate is first order in the $TiCl_4$ coverage *and* with the observation that below 200 K, titanium is in the +4 oxidation state. The higher temperature state near 450 K cannot, however, be a simple molecular desorption process. Between 180 K and 330 K, there is an abrupt shift in the $Ti(2p_{3/2})$ binding energies, from 459.0 to 458.2 eV. This direction of the shift is consistent with a reduction in oxidation state, and the magnitude is consistent with a change of one in oxidation state. For instance, the $2p_{3/2}$ binding energy difference between solid $TiCl_4$ and $TiCl_3$ is 0.9 eV [7]. Moreover, the Ti(II) and Ti(I) oxidation states are exceedingly rare [8]. We therefore conclude that the $TiCl_4$ derived intermediate between 220 K and 400 K is in the +3 oxidation state; the intermediate is most reasonably assigned to $TiCl_{3(a)}$.

Adsorbed $TiCl_3$ reacts to form gaseous $TiCl_4$ near 450 K. There are two possible pathways for $TiCl_4$ formation from $TiCl_{3(a)}$: recombination of $TiCl_{3(a)}$ and $Cl_{(a)}$ (Step 3a), and disproportionation of two $TiCl_3$ units to $TiCl_{4(g)}$, $Ti_{(a)}$ and $Cl_{(a)}$ (Step 3b). Although we cannot distinguish between these two possibilities, we consider the former to be unlikely. Atomic chlorine forms a very strong bond to tungsten: chlorine can be removed from tungsten only as a volatile tungsten chloride by heating above 1500 K. Moreover, the direct recombination pathway does not rationalize the fact that some total decomposition to $Ti_{(a)}$ and $Cl_{(a)}$ also occurs. If $TiCl_4$ is formed via recombination, then there must be a fourth step in above pathway which competes with recombination, i.e.:

$$TiCl_{3(a)} \longrightarrow Ti_{(a)} + 3Cl_{(a)} \qquad (4)$$

Given the W-Cl bond strength, we believe that disproportionation is a more likely step. Consistent with disproportionation is the fact that the amount of adsorbed Ti drops by approximately a factor of two between 330 K and 700 K (see Fig. 2), as required by the stoichiometry imposed by Step (3b).

In summary, we have studied the reaction of $TiCl_4$ on W(100). $TiCl_4$ decomposes on W(100) via a $TiCl_3$ intermediate. $TiCl_3$ likely decomposes by way of a self-disproportionation reaction, which also liberates gaseous $TiCl_4$. Decomposition of $TiCl_4$ is extensive, however, with the amount of residual Ti and Cl ≈ 0.1 and 0.4 monolayer, respectively.

ACKNOWLEDGMENT

This work was supported by the University of Minnesota, by the Dreyfus Foundation through a New Faculty Award, and by the National Science Foundation through the University of Minnesota Center for Interfacial Engineering.

REFERENCES

1. J. E. Sundgren and H. T. G. Hentzell, J. Vac. Sci. Technol. A4, 2259 (1986).
2. E. A. Haupfear and L. D. Schmidt, in press.
3. E. A. Haupfear (private communication).
4. K. A. Pearlstine and C. M. Friend, J. Phys. Chem. 90, 4341 (1986).
5. P. A. Redhead, Vacuum, 12, 203 (1962).
6. M. C. Asensio, M. Kerkar, D.P. Woodruff, A. V. de Carvalho, A. Fernández, A. R. González-Elipe, M. Fernández-Garcia and J. C. Conesa, Surf. Sci., 273, 31 (1992).
7. C. Mousty-Desbuquoit, J. Riga and J.J. Verbist, Inorg. Chem. 26, 1212 (1987).
8. F. A. Cotton and G. Wilkinson, Advanced Inorganic Chemistry, 5th ed., (John Wiley and Sons, New York, 1988).

ECR PLASMA ENHANCED MOCVD OF TITANIUM NITRIDE

M.E. Gross[†], A. Weber[††], R. Nikulski[††], C.-P. Klages[††], R.M. Charatan[†], W.L. Brown[†], E. Dons[†] and D.J. Eaglesham[†]
[†]AT&T Bell Laboratories, 600 Mountain Ave., Murray Hill, NJ 07974;
[††]Fraunhofer-Institut fur Schicht- und Oberflachentechnik, Hamburg, F.R.G.

ABSTRACT

The ability to deposit high purity TiN by a CVD process that can provide improved conformality over current sputter deposition processes is critical for multilevel Si ULSI device applications. Low temperature deposition of high quality TiN films has been achieved by an ECR plasma enhanced CVD process using the metal-organic precursor, tetrakis(dimethylamido) titanium, $Ti(N(CH_3)_2)_4$ (TDMAT). TDMAT is introduced into the downstream region of either a N_2 or NH_3 ECR plasma to deposit highly conducting films at deposition temperatures of as low as 100°C. The electrical resistivity of the TiN films decreases from 100-150μΩ-cm at a deposition temperature of 100°C to 45μΩ-cm at 600°C. The choice of plasma gas, N_2 or NH_3, affects the composition as well as the crystallographic texture of the films, which exhibit preferred (200) and (111) orientations, respectively. Pretreatment of the Si substrate with N_2 or NH_3 plasmas also affects the film morphologies, as demonstrated by switching the plasma gas between pretreatment and deposition.

INTRODUCTION

The interest in the properties of titanium nitride (TiN) films for Si ultra large scale integrated (ULSI) circuit fabrication is generating much activity aimed at developing improved deposition processes. TiN is currently deposited by reactive sputtering of Ti with N_2.[1,2] This line-of-sight physical vapor deposition process is approaching its limit in being able to provide adequate sidewall and bottom coverages in the increasingly high aspect ratio 0.25μm windows and vias. Thus, the interest in chemical vapor deposition (CVD) processes is being driven by the potential for achieving conformal deposition profiles. TiN is a refractory, metallic material with good compatibility to the metals (Al, W, Cu) and dielectrics (SiO_2) found in the typical integrated circuit (IC) metallization structure. It is used as a diffusion barrier between Si and Al or Cu, a protective layer against aggressive reagents such as WF_6, an adhesion layer, and as a current shunt and strengthening layer for Al interconnects.

The first CVD TiN processes were based on the thermal reaction of $TiCl_4$ with N_2 and H_2 or NH_3.[3-6] Depositions can be carried out at lower temperatures using the more reactive NH_3 but it is still necessary to maintain temperatures of at least 650°C to achieve low levels of residual Cl. Cl is a concern in IC metallization because of its potentially corrosive effect on adjacent Al metallization. The >650°C deposition temperatures are tolerable at the contact level; however, once Al has been deposited, further processing is restricted to <450°C. In the interest of eliminating the Cl problem and lowering the deposition temperatures even further, Gordon, et al. developed a metal organic CVD (MOCVD) process based on tetrakis(dimethylamido) titanium, $Ti(N(CH_3)_2)_4$ (TDMAT).[7] Despite the presence of four Ti-N bonds in TDMAT, it is necessary to add NH_3 as both a reducing agent and N source to achieve carbon-free deposits.

Recently, Weber, et al., and Intemann, et al., reported lowering deposition temperatures even further by using plasma enhanced MOCVD (PE-MOCVD) with TDMAT.[8,9] In these experiments, the TDMAT is introduced in the downstream region of either a microwave electron cyclotron resonance (ECR) or rf plasma reactor. High purity, crystalline TiN could be deposited at temperatures of as low as 100°C.[10] Resistivities of 45 to 150μΩ-cm for depositions at 600 to 100°C, respectively, rival those achieved with reactive sputtering.[10,11] In a previous publication on the ECR PE-MOCVD of TiN, we noted a striking difference in the crystallographic texture of the TiN films depending upon whether N_2 or NH_3 was used as the plasma gas. In this paper we report our recent results on the effects of the plasma gas and pretreatment of the Si and SiO_2 surfaces on the structure and properties of the TiN films.

EXPERIMENTAL

Depositions were carried out in cold wall reaction chamber (base pressure: 7×10^{-7} Torr) outfitted with an ASTeX HPM/M ECR plasma source and a resistively heated substrate holder placed 4 inches below the ECR source flange. The details of the experimental apparatus have been described elsewhere.[8] All depositions were carried out at a microwave power of 400 W. In each run, films were simultaneously deposited Si(100), Si(111) (cleaned in 1:100 HF:H_2O) and 0.5μm thermal SiO_2 on Si(100) substrates. TDMAT (Solvay, Germany) was introduced through a gas dispersion ring downstream of the plasma cavity at a rate of 2.5 sccm, controlled by the temperature of the TDMAT container. The chamber pressure was maintained at 0.8 mTorr and the flow rate of the N_2 or NH_3 plasma gas was 15 sccm. Depositions were carried out at 100 to 600°C, as measured on the susceptor, but in this study we focus on films deposited at 200°C, a temperature that is compatible with both inorganic and potential polymeric interlevel dielectric materials. In the downstream configuration that we use, with no applied bias to the substrate, we do not expect significant heating from the plasma.

The film compositions were determined by channeled Rutherford backscattering (RBS) spectrometry for Ti, N, O, and C, and by forward recoil scattering (FRS) spectrometry for H. The details of these measurements have been published elsewhere.[11] Crystallographic texture, structure, and morphology of the films were examined by X-ray diffractometry (XRD), transmission electron microscopy (TEM), and atomic force microscopy (AFM).

RESULTS AND DISCUSSION

It is quite remarkable that high purity, crystalline films of refractory TiN can be deposited at temperatures of as low as 100°C by plasma activation of N_2 or NH_3. For the purpose of this study of factors affecting film morphologies, all depositions were carried out at 200°C. Typical compositions for N_2 and NH_3 films deposited at 200°C are listed in Table I. Note that the NH_3 films have a higher defect density and are less pure than the N_2 films.

Plasma gas	Ti	N	H	C	O
NH_3	1.00	0.74	0.42	0.01	0.20
N_2	1.00	1.02	0.16	<0.01	<0.01

Table I. Representative composition of TiN films deposited at 200°C as determined by RBS (Ti, N, O, C) and FRS.

Previous results showed that there is little variation in electrical resistivity and Ti:N ratio between 200 and 500°C, although the H content decreases with increasing temperature.[10] Early in this work we noted differences in the crystallographic texture of the films depending on the plasma gas that was used. Films deposited with a N_2 plasma exhibit predominant (200) or random texture, whereas films deposited with an NH_3 plasma exhibit predominant (111) texture. To a lesser extent, the texture shows some dependence on whether the substrate is crystalline (Si(111) or Si(100)) or amorphous (SiO_2).

In a typical experiment, the plasma is ignited and stabilized before introduction of the TDMAT. Therefore, the substrate is briefly exposed to the downstream plasma before deposition is initiated. To see whether this pre-deposition exposure has any effect on the subsequent film deposition, we exposed the substrates to the plasma gas alone for 10 minutes before starting the TDMAT flow. We also looked at the effect of pretreating with one plasma gas (*i.e.*, N_2) then switching to the other (*i.e.*, NH_3) for deposition. No significant differences were observed between Si(111) and Si(100) substrates in any of the depositions.

The XRD results for films deposited at 200°C with NH_3 on Si(111) and SiO_2 are shown in Figure 1. These reveal the strong preference for (111) orientation regardless of pretreatment with either NH_3 (Figs. 1a,b) or N_2 (Figs. 1c,d). Pretreatment of the Si(111) surface with N_2 leads to some loss of the preferred orientation and an increase of the very weak (200) reflection observed with the NH_3 pretreatment (Figs. 1c,a, respectively). Broadening of the (111) reflection indicates a decrease in grain size that may result from a higher density of nucleation sites on the starting surface, In contrast to the Si(111) surface, pretreatment of the SiO_2 surface has little effect on the crystallographic texture of the TiN films.

Deposition with the N_2 plasma on Si(111) also shows a difference with pretreatment gas, as shown in Fig. 2. Whereas the film that is pretreated with the N_2 plasma exhibits exclusively (200) texture, pretreatment with NH_3 leads to the appearance of the (111) reflection. The N_2-pretreated sample on SiO_2 was damaged prior to analysis,

Experiments were also carried out using a H_2 plasma to pretreat the substrates, followed by depositions with NH_3 or N_2. XRD patterns of the NH_3 films were similar to Figs. 1a,b and exhibited strong (111) reflections and extremely weak (200) reflections. XRD patterns of films pretreated with H_2 and deposited with N_2 exhibited (111) and (200) reflections of equal intensity on the single crystal substrates. The XRD pattern for the film deposited on SiO_2 resembled Fig. 1c, with a slightly more intense (111) reflection.

Finally, we ran a series of experiments where the substrates were pretreated with NH_3 for 10 min. and the deposition gas was changed halfway through the deposition. Depositions with N_2 for 20 min. followed by NH_3 for 20 min. produced a film exhibiting strong (200) texture on Si(111) (Fig. 3a). Comparing this film with one that was pre-treated in NH_3 and deposited with N_2 (Figs. 2b), we see a decrease in the relative intensity of the (111) reflection. The preferred (200) orientation is evident on both Si and oxide, although in the latter, the (111) reflection is more evident (Fig. 3b). The width of the peaks and, therefore, the grain size of the films, appear to be comparable in the alternated and single gas depositions.

When the sequence of plasma gases for deposition is reversed, that is, we start the deposition with NH_3 for 20 min. followed by N_2 for 20 min., only the film on SiO_2 exhibits the strong (111) texture that we previously saw in the single gas experiments (Fig. 3e). There is a increase in the width of the peak, however, that may reveal a disruptive effect on the grain growth by the N_2 plasma, although the preferred orientation is retained. The films deposited on Si(111) exhibit random orientation (Fig. 3d), more closely resembling the film deposited on SiO_2 with N_2 after NH_3 pretreatment (Fig. 2c).

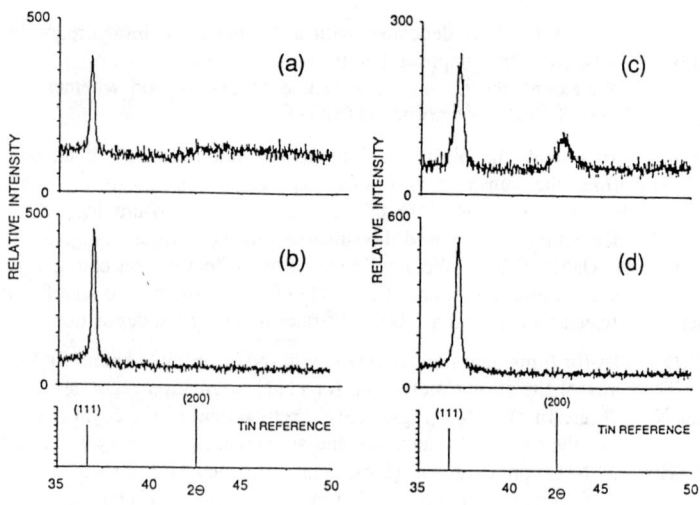

Fig. 1. XRD patterns of TiN films deposited with downstream NH_3 plasma on Si(111) (a,c) and SiO_2 (b,d) after pretreatment with NH_3 plasma (a,b) or N_2 plasma (c,d). Film thicknesses: 800 Å (a,b) and 700 Å (c,d); deposition time: 40 min.

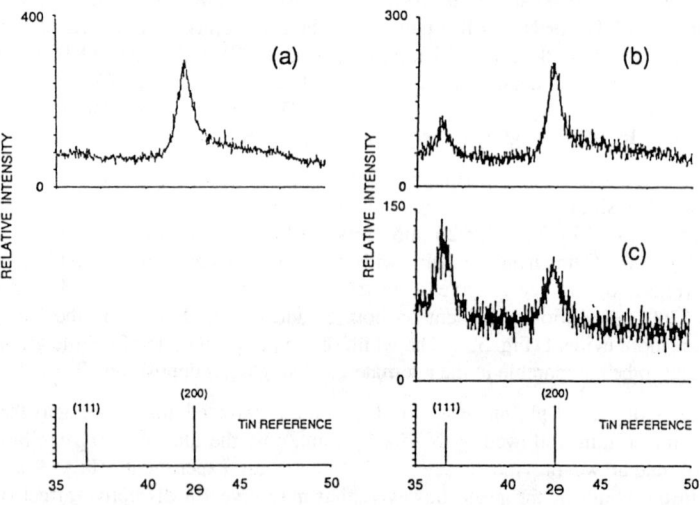

Fig. 2. XRD patterns of TiN films deposited with downstream N_2 plasma on Si(111) (a,b) and SiO_2 (c) after pretreatment with N_2 plasma (a) or NH_3 plasma (b,c). Film thicknesses: 600 Å (a) and 700 Å (b,c); deposition time: 40 min.

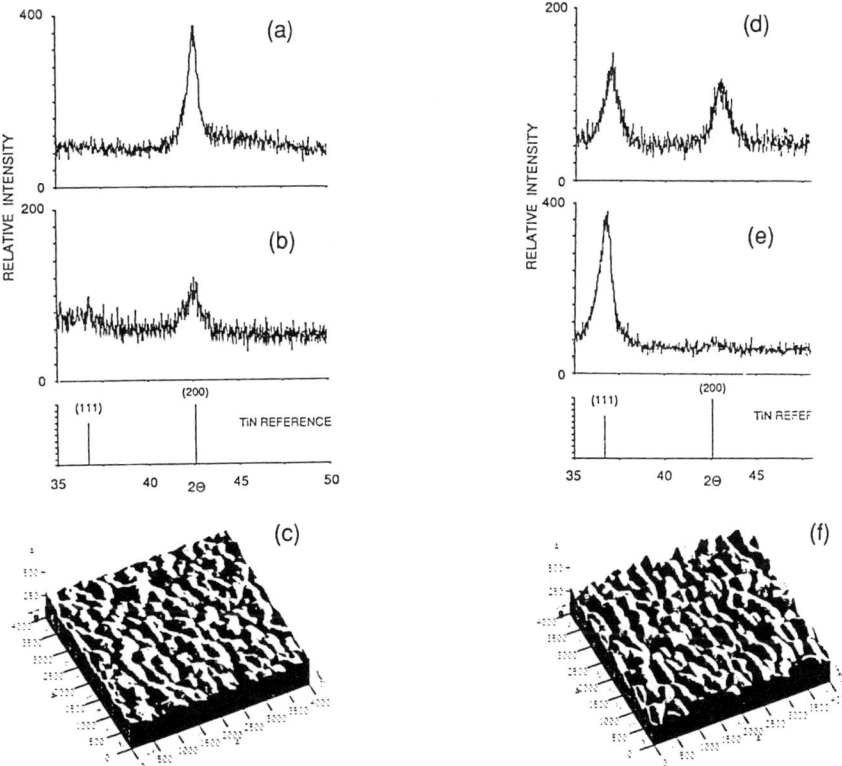

Fig. 3. XRD patterns of TiN films deposited on Si(111) and SiO$_2$ and AFM images of films deposited on Si(111) with (a,b,c) N$_2$ (20 min.) followed by NH$_3$ (20 min.) and (d,e,f) NH$_3$ (20 min.) followed by N$_2$. Pretreatment: NH$_3$ (10 min.); film thicknesses: 600Å.

Atomic force micrographs of the films deposited with the alternating plasma gases reveal the difference in grain size and roughness as suggested by the widths of the XRD reflections. The NH$_3$/N$_2$ film has an average grain size of about 230Å and surface roughness of about 35Å. The N$_2$/NH$_3$ film has an average grain size of 160Å and a surface roughness of 20Å.

The differences in film textures that we found, predominantly on the Si surfaces, suggest a surface modification caused by the plasma pretreatment. Bolmont, et al., reported the nitridation of Si(100) in a downstream N$_2$ ECR plasma at a microwave power of 100 W with the substrate at room temperature.[12] Under those conditions, the Si$_3$N$_4$ layer saturates at a thickness of 20Å. We propose that the effects of pretreatment that we observe on single crystal Si substrates is due to nitridation of the Si.

It is interesting to note that TiN films deposited from a very different metal-organic Ti complex, cyclopentadienyl cycloheptatrienyl Ti, using a downstream rf plasma exhibit textures similar to those observed with TDMAT depending on the N-containing gas that is used.[13] Films deposited with NH_3 and N_2 plasmas exhibit preferred (111) and (200) orientations, respectively. Further experiments are underway to further elucidate the similarities and differences in these very different chemical systems.

CONCLUSIONS

In this report of our continuing work on the deposition of TiN from TDMAT and downstream N_2 or NH_3 plasmas, we have explored the effect of plasma pretreatment of the substrate on the film texture and morphology. Pretreatment of single crystal Si ((111) or (100)) with N_2 plasma followed by deposition with NH_3 plasma leads to the development of some texture associated with deposition using N_2 and vice versa. NH_3 plasma pretreatment of SiO_2 also has an effect on the film texture, whereas little effect is observed for N_2 plasma pretreatment of the oxide.

REFERENCES

[1] J.-E. Sundgren, B.O. Johansson, and S.-E. Karlson, Thin Solid Films **105**, 353 (1983).

[2] R.V. Joshi and S. Brodsky, Proc. Ninth Intl. IEEE VLSI Multilevel Interconnection Conf., 1992, p. 253.

[3] A.E. van Arkel and J.H. de Boer, Z. Anorg. allgem. Chem. **148**, 345 (1925).

[4] O. Ruff and F. Eisner, Ber. **41**, 2250 (1908).

[5] J.B. Price, P.J. Tobin, F. Pintchovski, and C.A. Seelbach, U.S. Patent No. 4,570,328 (18 February 1986).

[6] S.R. Kurtz and R.G. Gordon, Thin Solid Films **140**, 277 (1986).

[7] R.M. Fix, R.G. Gordon, and D.M. Hoffman, Mater. Res. Soc. Symp. Proc. **168**, 357 (1990). R.M. Fix, R.G. Gordon, and D.M. Hoffman, Chem. Mater. **3**, 1138 (1991).

[8] A. Weber, R. Nikulski, and C.-P. Klages, Appl. Phys. Lett. **63**, 325 (1993); A. Weber, Proc. 5th Annual Scumacher Symp.,pp. 362-400 (Feb. 1993, San Diego, CA).

[9] A. Intemann, Proc. 5th Annual Scumacher Symp.,pp. 159-182 (Feb. 1993, San Diego, CA).

[10] A. Weber, U. Bringmann, R. Nikulski, R. Pockelmann, C.-P. Klages, M.E. Gross, W.L. Brown, R.M. Charatan, E. Dons, and D.J. Eaglesham, "Advanced Metallization for ULSI Applications in 1993," edited by D. Favreau and Y. Shacham-Diamand, in press.

[11] A. Weber, R. Nikulski, and C.-P. Klages, M.E. Gross, E. Dons, R.M. Charatan, and W.L. Brown, J. Electrochem. Soc., in press.

[12] D. Bolmont, J.L. Bischoff, F. Lutz, and L. Kubler, Surf. Sci. **269/270**, 924 (1992).

[13] R.M. Charatan, M.E. Gross, and D.J. Eaglesham, this volume.

INVESTIGATIONS OF TIN AND TI FILM DEPOSITION BY PLASMA ACTIVATED CVD USING CYCLOPENTADIENYL CYCLOHEPTATRIENYL TITANIUM, A LOW OXIDATION STATE PRECURSOR

Robert M. Charatan, Mihal E. Gross and David J. Eaglesham
AT&T Bell Laboratories, Murray Hill, NJ 07974

ABSTRACT

Sequential Ti and TiN thin film deposition by CVD is highly desirable for advanced Si integrated circuit applications. To date, most CVD TiN work has been performed using Ti(IV) compounds. We have investigated plasma assisted CVD using a lower oxidation state precursor, cyclopentadienyl cycloheptatrienyl titanium, $(C_5H_5)Ti(C_7H_7)$ (CPCHT), which might provide a more facile pathway to both Ti and TiN film formation. CPCHT was introduced with H_2 carrier gas into the downstream region of an NH_3, N_2 or H_2 plasma. Low resistivity (100-250 $\mu\Omega$-cm), nitrogen rich TiN films with little C or O incorporation were deposited at 300 to 600°C, inclusive with either activated N_2 or NH_3. Although the film texture was influenced by the chosen plasma gas, the average grain size of the N_2 and NH_3 plasma deposits was similar. Annealing studies showed that the CVD TiN was an effective diffusion barrier between aluminum and silicon to at least 575°C. TEM micrographs revealed that, in contrast to many CVD metal films, the growth of this TiN was not columnar. Film conformality was investigated by scanning electron microscopy (SEM). Experiments performed with activated H_2 resulted in deposits of Ti contaminated with C. No depositions were observed in the absence of plasma excitation.

INTRODUCTION

TiN is increasingly used as a barrier material in ultra large scale integrated (ULSI) devices to prevent interaction of the Al or W metallization with the Si.[1] To ensure low contact resistance at the Si/TiN interface, a thin film of Ti is deposited prior to the TiN and reacted with the Si substrate to form $TiSi_2$. Ti and TiN are currently deposited by sputtering or reactive sputtering with N_2, respectively, of a Ti target. The limitations of this line-of-sight physical vapor deposition (PVD) technique in achieving adequate bottom coverage in the high aspect ratio contact windows necessary for ULSI devices have led to a search for alternative chemical vapor deposition (CVD) and plasma enhanced CVD processes for depositing these materials.

The most studied CVD TiN precursors are titanium tetrachloride, $TiCl_4$,[2-4] and tetrakis(dimethylamido) titanium, $Ti(N(CH_3)_2)_4$, (TDMAT).[5-9] In both precursors, Ti is in the +4 oxidation state, requiring a one-electron reduction to the +3 oxidation state in TiN and a four-electron reduction to Ti metal. Akahori et al. deposited low resistivity TiN (40 $\mu\Omega$-cm) by reacting $TiCl_4$ in the downstream region of a N_2 electron cyclotron resonance (ECR) plasma.[3,4] The material was a succesful diffusion barrier between Si and Al to 650°C.[3] Ti-rich films were deposited in the same system using a H_2/Ar plasma.[4] Weber et al. have deposited low resistivity TiN (50-150 $\mu\Omega$-cm) at temperatures as low as 100°C by reacting TDMAT in the downstream region of either a N_2 or NH_3 ECR plasma.[9]

Although the two Ti(IV) compounds mentioned above are commercially available and, therefore, lend themselves to wider investigation, a potentially more attractive route for the deposition of TiN and Ti is to use a lower oxidation state Ti precursor such as cyclopentadienyl cycloheptatrienyl titanium, $(C_5H_5)Ti(C_7H_7)$, (CPCHT).[10] The Ti center in CPCHT was determined to have an oxidation state of +1.1 by X-ray photoelectron spectroscopy.[11] We have investigated the reactions of CPCHT with N_2, NH_3 and H_2, both thermally and with RF plasma activation of these gases.

EXPERIMENTAL

Experiments are conducted in a load-locked, diffusion pumped chamber with a wall temperature maintained at 150°C and a base pressure better than 8×10^{-7} torr (see Fig. 1). Si(100) substrates are cleaned in trichloroethane, acetone, methanol, and 4% HF (60 sec) followed by a DI water rinse. Patterned Si wafers are used for the conformality studies. The temperatures referred to in this paper are measured at the resistively heated stage; the wafer, which is not clamped to the stage, is undoubtedly at a lower temperature than the stage.

To minimize recombination of the active species, experiments are conducted at a low pressure (200 mTorr) with the inductively coupled RF helical coil resonator placed inside the chamber. The N_2, NH_3 or H_2 flow rate is 10 sccm. The discharge tube is made of alumina to minimize wall recombination effects,[12] and surrounded by a grounded aluminum shroud to reduce deposition on the outside of the tube. Solid CPCHT is dispersed with glass helices inside a glass tube that is heated to 100°C during deposition. CPCHT vapor is transported with 10 sccm H_2 carrier gas, to a gas dispersion ring in the chamber that is positioned downstream from the plasma cavity. Experiments are performed for 30 minutes at temperatures between 300°C and 600°C and at applied plasma powers between 0 and 80 W. The activated species are monitored with optical emission spectrometry (OES). During deposition, the chamber is pumped by a mechanical pump through a LN_2 cooled trap and the process pressure regulated with a capacitance manometer coupled to a throttle valve.

The film compositions are quantified by Rutherford backscattering (RBS) and forward recoil scattering (FRS) spectrometries. RBS analyis is carried out with 2 MeV $^4He^+$ at a scattering angle of 170° with the incident ion beam channeled along the <100> axis. Channeling reduces the signal from the Si substrate allowing better resolution of the C, N, and O signals. The hydrogen content of the films is determined using FRS of hydrogen at 30°, as previously described.[9] Film resistivities are calculated using four point probe sheet resistance measurements and thicknesses estimated from the RBS spectra, assuming a film density that is 80% of the bulk value of 5.39 g cm^{-3}. Film texture is examined using X-ray diffraction (XRD) and transmission electron microscopy (TEM). The barrier properties were explored by annealing Si / CVD TiN / Al structures, where the CVD TiN films were exposed to air prior to the Al deposition. RBS analysis of films before and after annealing was used to determine whether any diffusion had occurred. A lower energy 1.5 MeV $^4He^+$ beam, which increases the ion energy straggle, was used to increase the separation of the Al and Si signals.

RESULTS AND DISCUSSION

CPCHT is an extremely stable bright blue solid that sublimes at 125°C at 1 torr and is thermally stable under vacuum to 320°C.[10] In the absence of plasma activation, no films could be deposited with either N_2, NH_3, or H_2, as demonstrated in the RBS spectra shown in Fig. 2. Using plasma activated N_2 or NH_3, gold-colored TiN films could be deposited at rates of 2.5 - 4.0 nm min^{-1}. The ratio of N to Ti and the hydrogen content of films deposited with N_2 and NH_3 plasmas at 200 mTorr are plotted in Fig. 3. Note that all of the films tend to be nitrogen rich, but that a decrease in plasma power from 38 W to 20 W leads to a decrease in the N/Ti ratio for the N_2 plasma films. Except for the 300°C N_2 plasma deposited film, no C or O was detected by RBS in any of the films. The H content of the N_2 plasma films was more sensitive to the Ti:N ratio than those films deposited with activated NH_3. The estimated resistivities are presented in Fig. 4 for films grown with 38 W N_2 or NH_3 plasmas. The film stoichiometries and resistivities did not show a significant dependence on the substrate temperature.

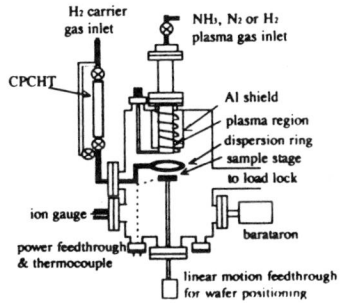

Figure 1. Experimental Apparatus.

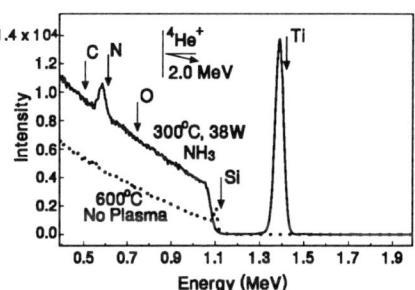

Figure 2. RBS spectra showing effect of plasma gas.

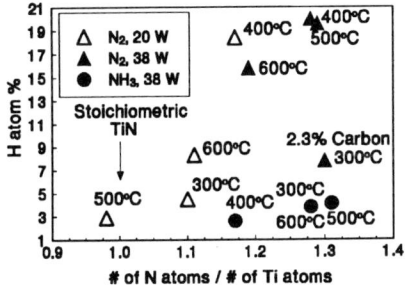

Figure 3. Plot of NH$_3$ and N$_2$ plasma deposited film compositions.

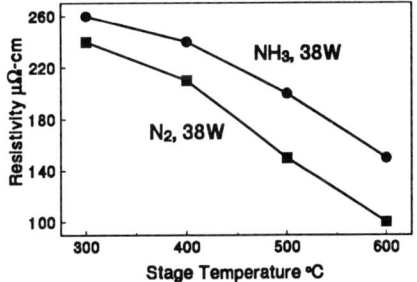

Figure 4. Film resistivity vs. substrate temperature for NH$_3$ and N$_2$ plasma CVD films.

Films deposited with the N$_2$ plasma exhibit preferred (200) orientation (Fig. 5), whereas those deposited with NH$_3$ plasmas are more randomly oriented. As seen in Fig. 5, increasing the applied plasma power leads to a decrease in the XRD signals when N$_2$ is used, whereas the reverse is seen with the NH$_3$ plasma. Although the XRD patterns reveal different textures for the N$_2$ and NH$_3$ processes, plan view TEM micrographs (Fig. 6) reveal similar grain size distributions (avg. ~100Å). The plasma gas dependence of the textures in these films is similar to that observed in TiN films deposited from TDMAT and downstream N$_2$ or NH$_3$ plasmas.[9] TEM analysis of the TDMAT films, however, reveals differences in grain size and distribution between the two plasma gases.

RBS analysis of a Si(100) / NH$_3$ plasma deposited TiN / Al sample before and after annealing in vacuum shows that the TiN is a good diffusion barrier to 575°C (Fig. 7). Barrier properties were similar for all of the activated N$_2$ and NH$_3$ processing conditions explored. Cross sectional TEM micrographs, presented in Fig. 8 show that in contrast to many CVD metal processes, neither the N$_2$ nor NH$_3$ plasma process leads to columnar structures. A SEM cross section of an NH$_3$ film deposited in a 0.35 µm window (aspect ratio 1.15) shows that the sidewall and bottom coverages are about 30% and 50%, respectively (Fig. 9).

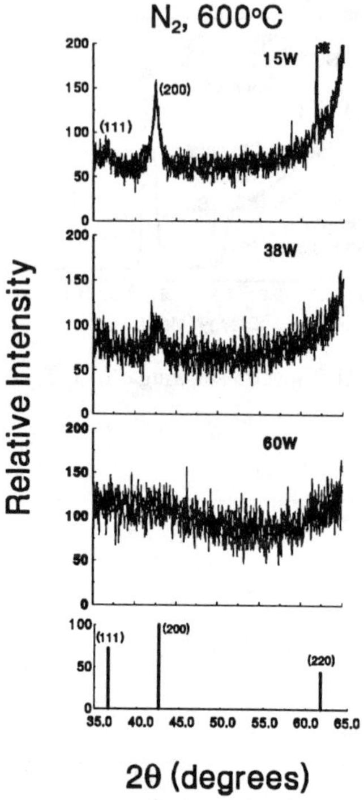

Figure 5. XRD patterns of TiN films deposited at 600°C with N$_2$ at different applied plasma powers. (*: Si reflections)

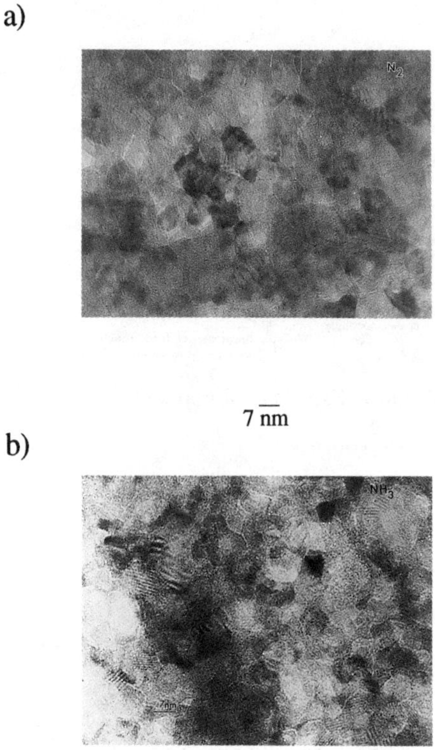

Figure 6. Plan view TEM micrographs for (a) N$_2$ and (b) NH$_3$ plasma films deposited at 600°C.

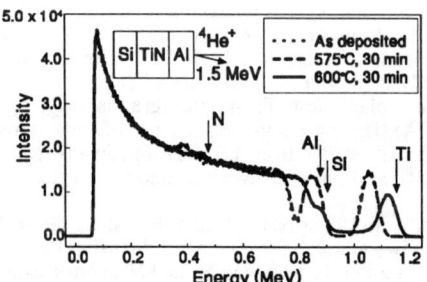

Figure 7. TiN diffusion barrier study. The Al was deposited on the CVD TiN after exposing it to air.

a) 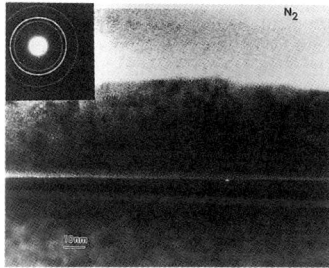 b)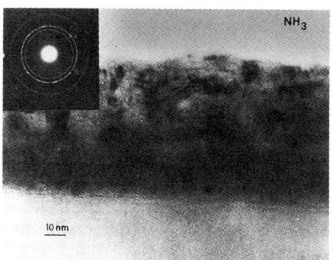

Figure 8. Cross sectional TEM micrographs for (a) N_2 and (b) NH_3 plasma films deposited at 600°C.

In addition to depositing TiN, we were interested in seeing whether Ti could be deposited from the low oxidation state CPCHT precursor for use as a low resistance contact to Si. Using a H_2 plasma, we did succeed in depositing films, but these films contained substantial amounts of C. The level of both C and H decreased with increasing plasma power, as shown in Fig. 10. Although all films had a C:Ti ratio well below the 12:1 ratio of the CPCHT precursor, we were not able to achieve stoichiometries of better than $TiC_{2.7}$. The estimated resistivities for these deposits were all greater than 800 μΩ-cm. No films could be deposited with H_2 in the absence of plasma excitation up to 600°C.

 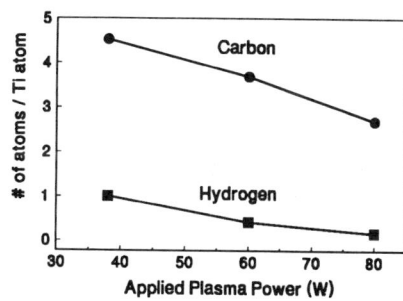

Figure 9. SEM cross section of a via hole with 38 W NH_3 plasma deposited TiN grown at 600°C.

Figure 10. Plot of hydrogen plasma deposition results as a function of applied plasma power. The films were all deposited at 600°C.

SUMMARY

We have successfully deposited near stoichiometric TiN films with little C or O incorporation from the reaction of CPCHT with either activated N_2 or NH_3 at temperatures as low as 300°C. These films are generally nitrogen rich. Although XRD patterns show different textures for the NH_3 and N_2 films, the grain size and structure as revealed by plan view TEM micrographs is similar. Annealing studies of Si(100) / CVD TiN / Al samples show that the CVD TiN is an effective diffusion barrier to at least 575°C. Cross sectional TEM micrographs show that, unlike many CVD metal processes, the growth of these films is not columnar. A SEM cross section of a NH_3 plasma film deposited in a 0.35 μm window reveals sidewall and bottom coverages of about 30% and 50%, respectively. The H_2 plasma experiments always produced deposits of Ti contaminated with C. No depositions were observed in the absence of plasma excitation.

ACKNOWLEDGMENTS

The authors are grateful to K. P. Cheung for many helpful discussions and to E. Dons for assistance with the FRS measurements.

REFERENCES

1. M. -A. Nicolet, Thin Solid Films **52**, 415 (1978).
2. For example, see R. Kurtz and R. G. Gordon, Thin Solid Films **140**, 277 (1986); N. J. Ianno, A. U. Ahmed and D. E. Englebert, J. Electrochem. Soc. **136**, 276 (1989); F. Pintchovski, T. White, E. Travis, P. J. Tobin and J. B. Price, "Workshop on Tungsten and Other Refractory Materials for VLSI Applications, IV", R. Blewer and C. McConica, eds., (Mat. Res. Soc., Pittsburgh, 1989) p. 323; A. Sherman, J. Electrochem. Soc. **137**, 1892 (1990); N. Yokoyama, K. Hinode and Y. Homma, J. Electrochem. Soc. **138**, 190 (1991); M. J. Buiting and A. F. Otterloo, J. Electrochem. Soc., **139**, 2580 (1992).
3. T. Akahori, A. Tanihara and M. Tano, Jap. J. Appl. Phys. **30**, 3558 (1991).
4. T. Akahori, Tech. Proc. Semicon./Japan, Dec. 2-4, 1992, p. 364.
5. R. M. Fix, R. G. Gordon and D. M. Hoffmann, Chem. Mater. **2**, 235 (1990)
6. G. S. Sandhu, S. G. Meikle and T.T. Doan, Appl. Phys.Lett. **62**, 240 (1993)
7. A. Weber, R. Nikulski and C.-P. Klages, Appl. Phys. Lett. **63**, 325 (1993)
8. A. Intemann, H. Koerner and F. Koch, J. Electrochem. Soc. **140**, 3215 (1993).
9. a) A. Weber, R. Nikulski, C. -P. Klages, M. E. Gross, W. L. Brown, E. Dons and R. M. Charatan, J. Electrochem. Soc., in press.
   b) M. E. Gross, A. Weber, R. Nikulski, C. -P. Klages, R. M. Charatan, W. L. Brown, E. Dons and D. J. Eaglesham, this volume, in press.
10. H. O. Van Oven and H. J. De Liefde Meijer, J. Organometal. Chem. **23**, 159 (1970).
11. C. J. Groenenboom, G. Sawatzky, H. J. De Liefde Meijer and F. Jellinek, J. Organometal. Chem. **76** C4 (1974).
12. N. M. Johnson, J. Walker and K. S. Stevens, J. Appl. Phys. **69**, 2361 (1991).

VAPORIZATION CHEMISTRY OF ORGANOALUMINUM PRECURSORS TO AlN

PAUL D. CROCCO*, JOHN B. HUDSON*, YI WANG** AND LEONARD V. INTERRANTE**,
*Materials Engineering Department and
**Chemistry Department, Rensselaer Polytechnic Institute, Troy, NY 12180.

ABSTRACT

We have used the technique of molecular beam mass spectrometry, with time of flight measurement, to study the vaporization of two organoaluminum precursors to AlN. The compounds studied were the isomers $(CH_3)_3Al \cdot NH_3$ (I) and $(CH_3)_3N \cdot AlH_3$ (II). Compound I was synthesized in house by reaction of $(CH_3)_3Al$ with NH_3. Compound II was obtained commercially. Mass spectrometric measurements of Compound I indicated a room temperature vapor pressure below 1 Torr, and slow decomposition at room temperature to yield CH_4 and a solid product, identified as the trimer, $[(CH_3)_2AlNH_2]_3$, as well as possible concentrations of the corresponding monomer and dimer. Similar measurements on Compound II indicated partial decomposition to metallic aluminum and gaseous trimethylamine and hydrogen.

INTRODUCTION

In recent years there has been increasing use of specially designed single component molecular precursors for the chemical vapor deposition of binary compounds. One aim of this work has been to develop precursors containing the two desired elements in the desired stoichiometric ratio as an aid to growing stoichiometric films. A critical assumption in this effort is that the desired stoichiometry is maintained in the vaporization and transport processes that take place prior to decomposition of the precursor on the growth surface. In this work, we report measurements of the vaporization process for two potential aluminum nitride precursors, namely the adduct formed by reaction of ammonia and trimethylaluminum, $(CH_3)_3Al \cdot NH_3$ (hereafter, Compound I) and its isomer, trimethylamine alane, $(CH_3)_3N \cdot AlH_3$ (hereafter, Compound II).

EXPERIMENTAL

The principal measurement technique used in this study is mass spectrometric measurement of the vapor diffusing from a valved Knudsen cell, using time of flight molecular beam techniques to differentiate between decomposition in the vapor source and fragmentation in the mass spectrometer ion source. Preliminary measurements of the stability of the two compounds studied were also made by measuring the pressure over the condensed phase of the precursor as a function of time.

Compound I was produced in house through the reaction of trimethylaluminum with NH_3, as first described by Interrante *et al.* [1]. Compound II was obtained commercially.

The apparatus used in this study is a modified version of the system used in previous studies of the vaporization of aluminum-nitrogen compounds [2], modified to improve the sensitivity and the time resolution as shown in Figure 1. The sample to be studied is contained in a reservoir connected to a solenoid operated pulsed valve (General Valve Model 9) which can be operated in either a pulsed or dc mode. Typical pulse width is 1 msec. A 0.05 mm valve aperture is

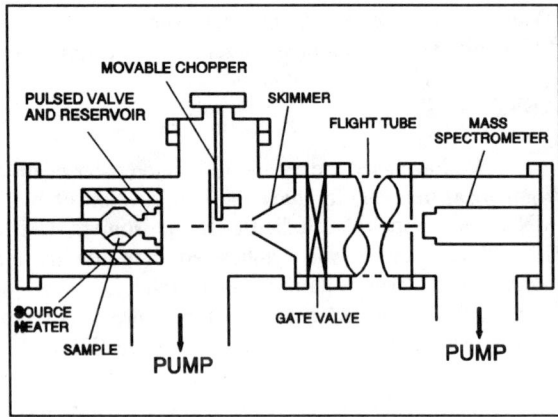

Figure 1. Schematic diagram of apparatus.

used to insure molecular flow into the vacuum system. (For preliminary measurements, in which the total pressure in the source was measured, this arrangement was replaced with a source external to the vacuum system, provided with a Baratron pressure gauge and connected to the pulsed valve by a stainless steel tube). Source pressure with the source at ambient temperature was typically in the one to ten Torr range. The reservoir and pulsed valve are surrounded by a heater assembly that can be used to control the precursor temperature between ambient and 150°C. Just downstream from the pulsed valve is a mechanical chopper wheel, which can be moved normal to the beam direction to provide either an unimpeded flux from the pulsed valve, square wave modulation at 400 Hz, or pulse modulation with a 7 μ-sec pulse width. The operation of the pulsed valve is electronically synchronized with the rotation of the chopper to provide high signal intensity while minimizing consumption of the source material. After modulation, the flux from the source passes through a skimmer with a 1.0 mm diameter orifice to form the molecular beam.

The source chamber is separated from the flight tube and mass spectrometer by a gate valve that can be closed to permit sample replenishment without loss of vacuum in the beam column. Two stages of differential pumping separate the skimmer from the mass spectrometer, a homemade quadrupole using Extranuclear electronics. The spectrometer mass range was calibrated using fluxes of helium and argon from the source region; the sensitivity was not calibrated, consequently all peak heights reported are referenced to the most intense peak in the spectrum. Signal acquisition is by pulse counting, using a Channeltron electron multiplier. The mass spectrometer tuning and data acquisition are controlled through a personal computer, using a multichannel scaling card from The Nucleus, Inc. The effective length of the flight path was determined using the method described by Wu and Hudson [3].

Two principal measurements are made for each compound at each temperature studied. In the first, the mass spectrum of the vapor effusing from the cell is collected by operating the valve in the dc mode, with the chopper wheel withdrawn from the beam. Spectra are taken over the mass range of interest with the valve open and with the valve closed. Subtraction of the background spectrum from the sample spectrum identifies the peaks to be studied by time of flight techniques, and provides information on the way in which the relative peak heights change with source temperature.

Once the peaks of interest have been identified, time of flight spectra of each peak are obtained by operating the pulsed valve and the chopper in synchronism, with the mass spectrometer tuned to the peak of interest. The resulting time of flight distributions are analyzed by using the program Jandel PeakFit to fit the experimentally observed waveform to the time of flight distribution expected for an effusive beam having a Maxwell-Boltzmann velocity distribution to determine the parent mass of the species giving rise to the peak, and to deconvolute effects arising when multiple parent species contribute to the peak. Measurements of this sort, over a range of source temperature, serve to characterize the vaporization process.

RESULTS AND DISCUSSION

Preliminary measurements were made on Compound I to determine whether spontaneous decomposition takes place at room temperature. Such decomposition was characterized by measuring the total pressure over the sample as a function of time using the capacitance manometer. Results of these measurements showed slow decomposition of Compound I with time at 25°C. The source pressure rose linearly with time over a period of several hours. Cooling the sample to 77 K resulted in a decrease in system pressures to 9.2 Torr, the equilibrium vapor pressure of methane at 77 K. This indicates that the reaction

$$(CH_3)_3AlNH_3 \rightarrow (CH_3)_2AlNH_2 + CH_4$$

has a significant rate even at room temperature. (This reaction has been observed previously in the production of the trimeric amide, $[(CH_3)_2AlNH_2]_3$, from Compound I at temperatures above 150°C [1]). Quantitative characterization of the reaction rate constant was not possible, due to the unknown total amount and total surface areas of the precursor in the source.

The spectrum obtained for Compound I in a dc measurement at ambient temperature is shown in Figure 2. In this spectrum, all peak heights are referenced to the largest peak, at m/e =43. Time of flight measurements of the principal peaks indicated that the parent species contributing to the spectrum are Compound I (m/e values of 84-86), a species with a molecular weight of 72-73 (m/e values of 27-73) and methane (m/e values of 15-16). The parent mass of 72-73 would be consistent with either trimethylaluminum (M =72) or with the monomer of dimethylaluminum amide, $(CH_3)_2AlNH_2$, (M = 73). The cracking pattern of Figure 2 is very different from the reference spectrum of trimethylaluminum [4], and all of the observed peaks are consistent with ion source fragmentation of the monomeric amide. This fact, along with the observation of the peak at m/e = 101, which was observed in a previous study of the trimeric amide $[(CH_3)_2AlNH_2]_3$ [2], suggest that the major species in the vapor phase is the monomeric amide, with smaller amounts of Compound I and higher amide oligomers.

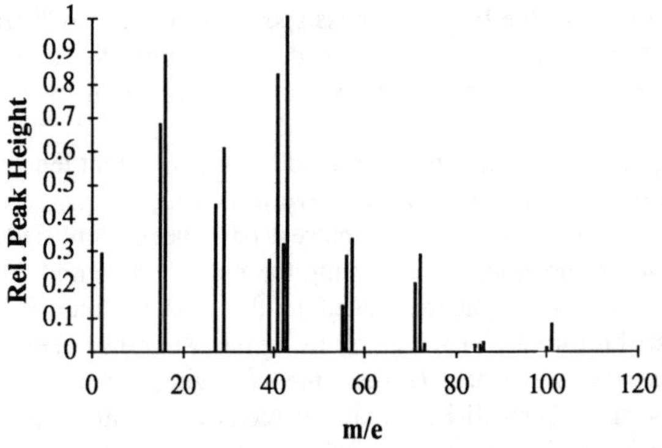

Spectra were also taken of the vapor over Compound I after holding

Figure 2 Mass spectrum of vapor phase over Compound I at 300 K.

for various times. Figure 3 is such a spectrum for the case of Compound I held for three weeks at ambient temperature. This spectrum shows primarily methane, with minor peaks consistent with amide monomer or oligomers. Spectra of the vapor over a sample of Compound I held for eight months at -40 °C were similar to those observed in a previous study of the trimeric amide, indicating that the reaction of Compound I to form the amide trimer takes place even at very low temperature.

The spectrum of Compound II obtained in a dc measurement at ambient temperature is shown in Figure 4. As before, the peak heights are referenced to the most intense peak, in this case the peak at m/e = 58. Time of flight measurements indicated that the species contributing to the spectrum are Compound II (m/e values of 60-88), trimethylamine (m/e values of 15-59) and hydrogen (m/e = 2). The region of the spectrum between m/e values of 15

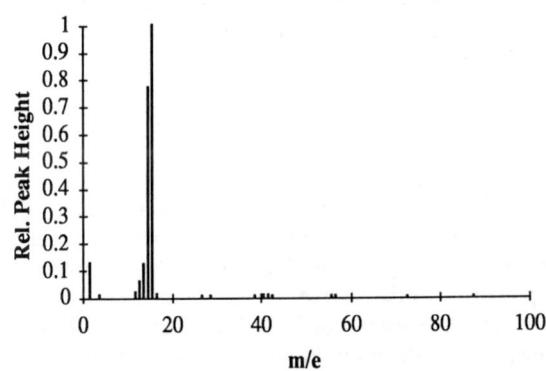

Figure 3 Mass spectrum of the vapor over Compound I after holding for three weeks at ambient temperature.

and 59 is very similar to the reference spectrum for trimethylamine [3]. This result is consistent with partial decomposition of Compound II in the source to deposit metallic aluminum and release trimethylamine and hydrogen, and indicates the presence of a surface that catalyses the decomposition somewhere in the source assembly. These results differ, however, from those obtained as part of a microwave spectroscopy experiment being carried out by other members of our research group [5], who observed decomposition of Compound II only above 60 °C. The difference may be related to differences in the materials to which the compound was exposed in the course of the experiments.

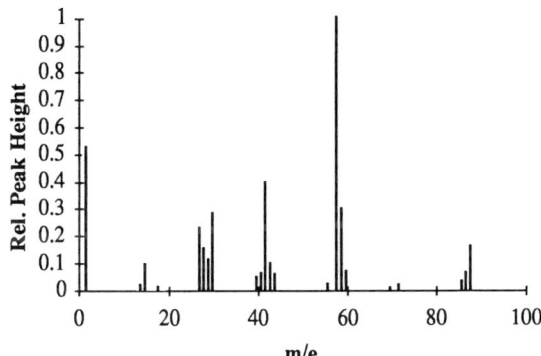

Figure 4 Mass spectrum of the vapor over Compound II at ambient temperature.

CONCLUSIONS

Both of the compounds studied showed significant decomposition in the vapor source at ambient temperature. In the case of Compound I, the process involved was an oligomerization reaction to produce first the monomer, dimethylaluminum amide and subsequently the trimer, tris-dimethylaluminum amide. In the case of Compound II, the primary reaction was the deposition of aluminum, accompanied by the release of the trimethylamine and the production of hydrogen. It is clear that the use of these compounds as CVD precursors must involve precautions to mitigate the effects of these reactions on the deposition process. In the case of Compound I, such precautions would be aimed at the prompt use of the precursor following synthesis, or alternatively allowing the reaction to produce the trimeric amide to go to completion. This latter strategy has been used to produce the trimeric amide, which has been shown to produce stoichiometric AlN films in both hot wall [1] and cold wall CVD [6] reactor conditions. In the case of Compound II, avoidance of potentially catalytic surfaces in the precursor source should suffice.

ACKNOWLEDGEMENT

This work was funded in part by the Chemistry Program of the National Science Foundation under Grant No. CHE-9202973.

REFERENCES

1. L. V. Interrante, L. E. Carpenter II, C. Whitmarsh and W. Lee, in Better

Ceramics Through Chemistry II, C. J. Brinker and D. R. Ulrich, Eds., Mat. Res Soc. Symp. Proc. **73**, 359, (1986).
2. C. C. Amato Wierda, PhD Thesis, Rensselaer Polytechnic Institute, 1993.
3. P. K. Wu and J. B. Hudson, J. Vac. Sci. Technol. **A11**, 2603, (1993).
4. S. R. Heller EPA/NIH Mass Spectral DataBase, US. Govt Printing Office, Washington, DC, 1978.
5. H. Warner and C. W. Gillies, Personal communication.
6. M. J. Cook, P. K. Wu, N. Patibandla, W. B. Hillig and J. B. Hudson, Mat. Res. Soc. Symp. Proc. (in press).

CHARACTERIZATION OF SILICON-NITRIDE FILM GROWTH BY REMOTE PLASMA-ENHANCED CHEMICAL-VAPOR DEPOSITION (RPECVD)

ZHONG LU, YI MA, SCOTT HABERMEHL AND GERRY LUCOVSKY, Departments of Physics, Materials Science and Engineering, and Electrical and Computer Engineering, North Carolina State University, Raleigh, NC 27695-8202

ABSTRACT

We have characterized RPECVD formation of Si-nitride films by relating the chemical bonding in the deposited films to the growth conditions. Gas flow rates for different N- and Si-atom source gases have been correlated with (i) the film stoichiometry, i.e., the Si/N ratio, and the (ii) the growth rate. N_2 and NH_3 were used as N-atom source gases, and were either delivered (i) up-stream through the plasma-generation tube, or (ii) down-stream. Different flow-rate ratios of NH_3/SiH_4 were found for deposition of stoichiometric Si-nitride films using up-stream or down-stream introduction of NH_3. This is explained in terms of competition between excitation and recombination processes for the N-atom precursor species. Stoichiometric nitride films could not be obtained using the N_2 source gas for (i) either up-stream or down-stream delivery, and (ii) for plasma powers up to 50 W. This is attributed to the higher relative binding energy of N-atoms in N_2 compared to NH_3, and to significant N-atom recombination at high N_2 flow rates through the plasma generation region.

1. INTRODUCTION

In addition to its application as a passivation layer for microelectronic devices, silicon nitride films have also been used as an active insulating gate layer for silicon solar cells [1], and particularly, carefully engineered silicon nitride film deposited by low temperature Remote R.F. Plasma Chemical Vapor Deposition (RPECVD) have shown improved performance in amorphous silicon based thin film transistor (a-Si:H TFTs) [2]. As a low temperature, and/or low thermal budget process, the RPECVD process is also of interest for use in ULSI processing technologies. [3] Si-nitride films, deposited by RPECVD under different conditions were previously studied by Infrared Absorption spectroscopy (IR), ellipsometry, Auger Electron Spectroscopy (AES), and mass spectrometry (MS). Issues such as the growth rate, index of refraction, bonded hydrogen content for films deposited from NH_3 or N_2 nitrogen source gases at different gas flow ratios and different deposition temperatures have been addressed. [4] This paper pays special attention to correlations between the chemical composition of the silicon nitride films for different flow conditions and methods of plasma excitation (up-stream and down-stream) using either NH_3 or N_2 as the nitrogen source gas. Post deposition film characterizations were carried out using infrared absorption spectroscopy (IR) and capacitance-voltage (C-V) measurements. This type of study then provides a database linking deposition process variations to film properties including (i) chemical composition (e.g., the Si/N ratio, and hydrogen bonding), and (ii) electrical performance as a gate dielectric material (e.g., the trapped charge and fixed charge densities). This data can provide parameters for empirical modeling of various aspects of the deposition process such as the dissociation rates of source gases and recombination rates of plasma-excited species. Insight into these deposition mechanisms can be used to optimize films for device applications.

2. EXPERIMENTAL

The experiments were performed in a multi-chamber integrated processing system; the particular chambers used in this study included an RPECVD chamber for Si-nitride deposition, and an analysis chamber for on-line Auger Electron Spectroscopy (AES).[5] The RPECVD chamber has base pressure in the mid 10^{-8} Torr range, whereas the analysis chamber routinely maintains a base pressure of $\sim 2 \times 10^{-10}$ Torr. Si (100) wafers were chemically cleaned by UV/O_3 for four minutes, rinsed in diluted 1:30 HF (49 wt% HF in deionized water) for 10 sec, and then blown dry in nitrogen. The wafer was immediately inserted into a load lock chamber of the multi-chamber integrated system. The silicon wafer could then be transferred in vacuum either to the deposition chamber for RPECVD silicon nitride deposition or to the analysis chamber for AES analysis for pre- or post-silicon nitride deposition.

There are two different types of RPECVD processes that can be used to produce silicon nitride films; these are up-stream and down-stream processes which are defined by the injection point of the N-atom source gas. In the up-stream process, the N-atom source gas, and He diluent are directly plasma excited, and silane, SiH$_4$, is introduced into the deposition chamber via a down-stream gas dispersal ring. In the down-stream process, only the He gas is subjected to direct plasma excitation, and the other two process gases are introduced into through the two down-stream gas dispersal rings. The overall deposition reactions for up-stream and down-stream processes are given in Equations (1) and (2) respectively.

$$[\text{He} + \text{NH}_3 \text{ (or N}_2\text{)}]^* + \text{SiH}_4 \rightarrow \text{Si}_3\text{N}_4 \qquad (1)$$

$$[\text{He}]^* + \text{NH}_3 \text{ (or N}_2\text{)} + \text{SiH}_4 \rightarrow \text{Si}_3\text{N}_4 \qquad (2)$$

where the gas species in the stared bracket, [ ]*, are subjected to direct plasma excitation in the plasma tube at the top of the deposition chamber. The other RPECVD conditions are listed in Table 1. Since both NH$_3$ and N$_2$ are used as N-atom source gases, and they can be fed either through plasma tube (up-stream) or gas-ring (down-stream), there are four different growth conditions for silicon nitride films denoted, respectively by I, II III, IV in Table 1. We define R as the nitrogen source gas flow ratio with respect to SiH$_4$, but since the effective SiH$_4$ flow is always 1 sccm (10 sccm flow of 10% SiH$_4$ diluted in He), it is also the flow rate of the N-atom source gas.

Table 1: Deposition Conditions for RPECVD Silicon Nitride Films

Process	Plasma tube (sccm)	Gas Ring 1 (sccm)	Gas Ring 2 (sccm)	Results shown in
I) NH$_3$ up	[He+NH$_3$] (200, R)	SiH$_4$ (10)		Fig. 2
II) NH$_3$ down	[He] (200)	SiH$_4$ (10)	NH$_3$, R	Fig. 3
III) N$_2$ up	[He, N$_2$] (200, R)	SiH$_4$ (10)		Fig. 4
IV) N$_2$ down	[He] (200)	SiH$_4$ (10)	N$_2$, R	Fig. 5

Note that the R.F. power is 50 W; the substrate temperature, T$_s$ = 300°C; the process pressure is 300 mTorr; and the SiH$_4$ is 10% diluted in He. For all the AES measurements shown in Figs 1-5, the deposition time is five minutes. The SiH$_4$ flow was fixed, and the nitrogen source gas flow was then characterized by R (see Table 1). The silicon sample sits about six inches below the bottom of the 10 inches long plasma tube, and gas ring is located half way in between bottom of the plasma tube and sample surface. For the C-V and ellipsometry measurements, the silicon nitride film were grown for 30 minutes. A two hour deposition time was used to deposit thick Si-nitride films for FTIR measurements to ensure strong signals from the Si-H and Si-NH bonding groups. For case I), R varies from 2.5 to 7 sccm; for II), R varies from 0.5 to 4; for III), R varies from 31 to 120, and for iv) R varies from 30 to 120.

Prior to the systematic study of silicon nitride depositions, AES spectra were taken for the Si substrate after chemical cleaning, and after a 5 min RPECVD silicon nitride deposition (see Fig. 1). Trace amounts of carbon (~ 270 eV) and oxygen (~510 eV) can be detected on a Si (100) substrate after a four minute UV/O$_3$ clean, followed by a 10 sec. dip in diluted HF (1:30) solution (see Fig. 1a). The dominant feature of the Si LVV AES is at 92 eV. After a five minute RPECVD nitride deposition using case i) conditions with R=8, there is an AES nitrogen KLL peak at ~379 eV, the Si-LVV peak was shifted toward lower energy, peaking at ~82 eV (Fig. 1b), the characteristic peak in stoichiometric Si-nitride. A five minutes Si-nitride deposition under these conditions typically grows a 120 Å thick film. The escape depths of the AES electrons are much less than the film thickness, so that all the AES signals comes mostly from the top surface of the silicon nitride film. The fact that Si Auger transition is at different peak positions for bare silicon (92 eV) and for silicon nitride (82 eV) provides the basis for the study of nitride composition for different gas flow conditions. For example, a detectable signal at the high energy side of the Si$_3$N$_4$ peak at 82 eV is an indication of Si-Si bonding in the film; i.e., a silicon-rich silicon nitride film. The absence of any signal at about 90-92 eV implies very little Si-Si bonding, but does not give any information relative to the bonded-H concentration, which is easily obtained from the IR measurements.

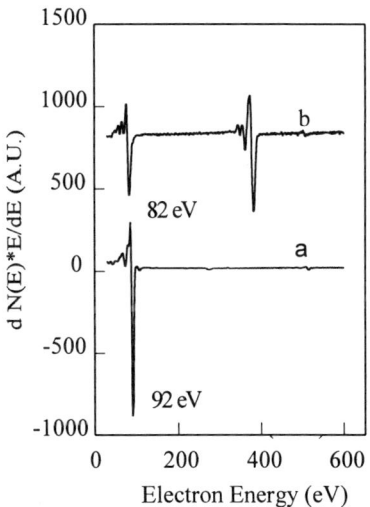

Fig. 1 AES spectra of Si (100) a) after UV/O3 and HF dip; b) after 5 min. RPECVD Si-nitride deposition, deposition condition is the same as case I in Table 1 with R=8.

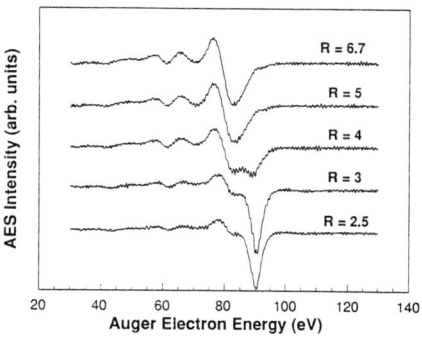

Fig. 2 AES spectra of RPECVD Si-nitride deposition by NH_3 up-stream process under various NH_3/SiH_4 ratio R, see Table 1, case I for deposition condition.

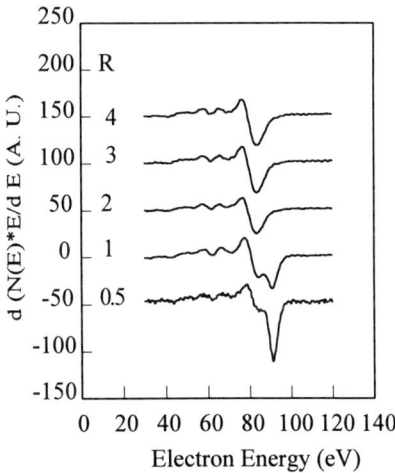

Fig. 3 AES spectra of RPECVD Si-nitride for NH_3 down-stream process under various NH_3/SiH_4 ratio R, see Table 1, case II for deposition condition.

3. RESULTS AND DISCUSSIONS

Based on the AES signatures for stoichiometric nitrides and silicon-rich nitrides, we carried out AES analyses on the RPECVD silicon nitride films. Four different sets of AES data (two up-stream, two down-stream, using either NH_3 or N_2, and labeled cases I, II, III, IV) are shown in Figs. 2, 3, 4, 5 respectively. The results are summarized below:

I) NH_3 up-stream (see Fig. 2): R > 6 is needed to obtain stoichmetric nitrides. NH_3 flow ratios larger than 6 (we tried up to 15) always produce stoichiometric nitrides. We typically use R = 8 for device applications such as ONO structures [6]. IR studies indicate for R ~6-15, Si-H is significantly reduced and Si-NH approaches a constant value. In this range, the total hydrogen concentration is a minimum, and in addition, the Si-N stretching frequency does not vary with R, and is equal to ~840 cm^{-1}.

II) NH_3 down-stream case (see Fig. 3): R > 2 is needed to obtain stoichmetric nitrides. We carried out experiments for R up to 8, and found that the AES spectra still showed stoichiometric nitride formation.

For cases I and II, an *insufficient* supply of NH_3 will result in a silicon-rich nitride; however, the minimum value of R is different for these two cases. The concentration of excited nitrogen at the growth surface is a result of a competition between the efficiency of plasma excitation and gas transport, and recombination. Plasma excitation is much greater in the up-stream case than the down-stream case, but the recombination is also much greater in the up-stream case than the down-stream case since the excited species has to travel much longer distance in the up-stream case than the down-stream case to reach the sample surface. The balance between delivery of excited species and recombination is a function of the plasma power and geometry of the particular RPECVD system. For up-stream case, although NH_3 is directly excited in the plasma tube with He, but the excited species have to travel a longer distance during which some NH_3 recombines. The net effect is R is smaller for down-stream as compared to up-stream NH_3 injection and excitation.

III) N_2 up-stream (see Fig. 4): First, we notice that it requires a significantly larger R (~ 60) to get to even close to stoichiometric nitride. This derives from the higher bonding energy in N_2 as compared to NH_3.[7] Second, further increases in the N_2 flow do not make the nitride more stoichiometric, on the contrary, the nitride film becomes more silicon-rich as we see by the increase of 92 eV Si-Si peak. This is a direct consequence of increased recombination of a higher concentration of short lived excited nitrogen species.

IV) N_2 down-stream (see Fig. 5): The results are essentially the same as for III) a ratio R=60 yields a nitride close to stoichiometry, but increasing R further yields silicon rich nitrides.

At an R.F. power of 50 W, we can't make stoichiometric Si-nitride film using N_2 as the N-atom source gas. We found that Si-nitride films move toward stoichiometry if we increase power. However, stoichiometry could not be obtained for powers up to 200 W. This is a manifestation of the larger bonding energy of N_2 relative to NH_3, and also to shorter lifetimes for N-atoms, than for the active species generated by direct or remote excitation of NH_3, so that it is difficult to use N_2 as the N-atom source gas for applications requiring stoichiometric nitride films [8], i.e., films that do not have AES detectable Si-Si bonding.

The film thickness and index of refraction of silicon nitride films deposited using the NH_3 up-stream process were measured using single wavelength ellipsometry (6328Å), with results summarized in Fig. 6 The film deposition conditions were identical to case I, except the deposition time is 30 minutes for all five samples instead of 5 minutes. These samples were also used for C-V measurements. As R increases, the film thickness monotonically increases with increasing R, and index of refraction monotonically decreases. The increase in film thickness comes from an increase of growth rate, while the decrease in index of refraction is another indication of hydrogen incorporation in the form of Si-NH as evidenced by IR measurements [6]. Each sample was split into two halves, one half was annealed at 900°C for thirty seconds, and other was studied without further post-deposition processing. The annealed sample showed a 10% thickness reduction for the R=10 and R=15 samples, indicating densification after the annealing, and suggesting the as-deposited film, which deposited a relatively fast rate, had an open local atomic structure. The IR spectra for the as-deposited films typically displayed two H-related features, e.g., Si-H and Si-NH bond-stretching absorptions peaking at approximately 2100 cm^{-1} and 3400 cm^{-1}, respectively. The IR absorbance in these two bands was monitored as function of the annealing

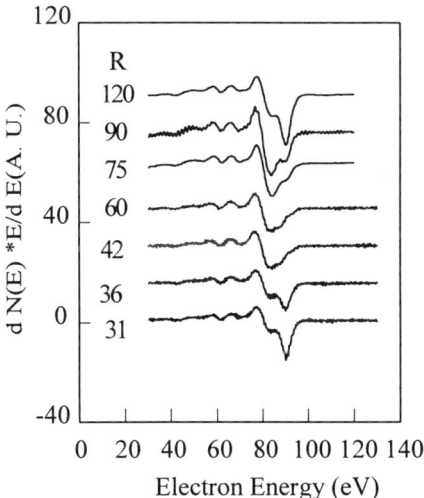

Fig. 4 AES spectra of RPECVD Si-nitride for N_2 up-stream process under various N_2/SiH_4 ratio R, see Table 1, case III for deposition condition.

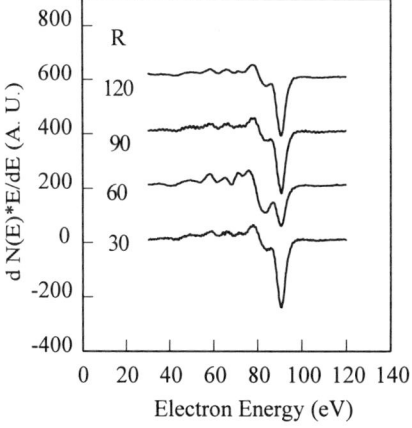

Fig. 5 AES spectra of RPECVD Si-nitride for N_2 down-stream process under various N_2/SiH_4 ratio R, see Table 1, case IV for deposition condition.

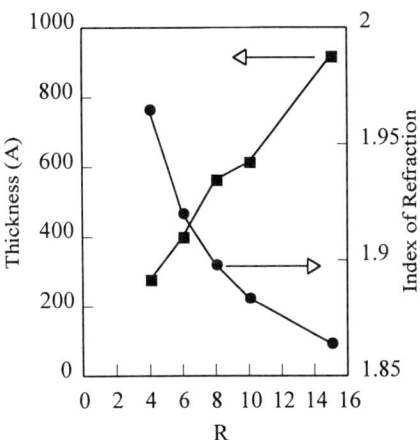

Fig. 6 Thickness and index of refraction measured by ellipsometry for Si-nitride films deposited by RPECVD, deposition is the same as case I in Table 1 with R=8 and 30 min. deposition time

temperature. As this temperature was increased, the absorbance in the Si-H band decreased and was not observable for annealing temperatures in excess of ~400°C. In contrast, the absorbance in the Si-NH band decreased initially and then remained relatively constant at the higher annealing temperatures. This difference in behavior is consistent with the approximately 33% higher bond energy of the N-H bond, as compared to the Si-H bond [7]. We will present a model for the temperature dependence of the relative absorbances in these two vibrational modes in a future publication.

Capacitance-voltage measurements were also carried out to characterize bulk electrical properties of the films. A trapped charge density was calculated from a high frequency (1 MHz) trace-retrace measurement, and a fixed nitride charge density was obtained from flat band voltage. This showed a systematic decrease as R increase from 4 to 8 and then flattened out as R increased to 15. Annealing at 900°C significantly reduces both trapped and fixed charges in the nitride films. However, both the fixed charge and interface state densities are still too high for any gate dielectric applications, even after annealing.

4. CONCLUSIONS

In conclusion: 1) Different flow-rate ratios of NH_3 to SiH_4 were obtained for deposition of stoichiometric Si-nitride films using up-stream or down-stream introduction of NH_3. This is explained in terms of competition between (i) excitation and transport, and (ii) recombination processes for the N-atom precursors. 2) Stoichiometric nitride films could not be obtained using the N_2 source gas for (i) either up-stream or down-stream delivery, and (ii) for plasma powers up to about 200 W. This is attributed to the higher relative binding energy of N-atoms in N_2 compared to NH_3, and to significant N-atom recombination at high N_2 flow rates through the plasma generation region. 3) IR absorption measurements were used to determine the relative bond cooncentrations of Si-H and Si-NH groups in films deposited from both N-atom source gases, and to track the changes in these bond strengths as a function of the post-deposition annealing for films deposited from the NH_3 source gas. 4) C-V measurements were used to characterize trapped and fixed charge in silicon nitride films deposited using the NH_3 source gas.

ACKNOWLEDGMENTS

Z. Lu acknowledges helpful discussions with S.S. He and David Lee at NC State. This work is supported by ONR.

References

1. R. Hezel and R. Schorner, J. Appl. Phys. **52**, 3076 (1981).
2. S.S. He, D.J. Stephens, G. Lucovsky and R.W. Hamaker, MRS Symp.Proc. **284**, 414 (1992).
3. G. Lucovsky D.V. Tsu, R.A. Rudder, and R.J. Markunas, in Thin Film Processes II, edited by J. Vossen and W. Kern, (NY: Academic Press, 1991) p. 565.
4. D. V Tsu, Ph. D. thesis, North Carolina State University, 1989, p. 77-86; S. S. Kim, Ph. D. thesis, North Carolina State University, 1990. p. 135-149; J. A. Theil, Ph. D. thesis, North Carolina State University, 1992, p. 136-157
5. J. T. Fitch, J. J. Sumarkeris, and G. Lucovsky, SPIE **1188**, 39 (1990).
6. Y. Ma, T. Yasuda, S. Habermehl and G. Lucovsky, J. Vac. Sci. Technol. **A11**, 952 (1993).
7. The bonding energy for Si-H is 3.09 eV, N-N is 9.80 eV, and for H-NH_2, it is 4.77 eV: CRC Handbook of Chemistry & Physics, 62nd edition, ppF-181, pp. F-183 and F-192 respectively.
8. G. Lucovsky, Y. Ma, S. S. He, T. Yasuda, D. J. Stephens and S. Habermehl, MRS Symp. Proc. **284**, 34 (1993).

SPECTROSCOPIC INVESTIGATIONS OF LASER ABLATED GERMANIUM OXIDE

Paul J. Wolf*, Brian M. Patterson*, and Sarath Witanachchi**
*Frank J. Seiler Research Laboratory, Materials Physics Division, USAF Academy, CO 80840
**Department of Physics, University of South Florida, Tampa, FL 33620

ABSTRACT

Laser produced plumes from a GeO_2 target were investigated using both optical emission spectroscopy and laser induced fluorescence. Elemental neutral and ionic Ge and diatomic GeO were monitored as a function of O_2 pressure at 4 cm from the target surface. Data obtained from laser fluorescence experiments showed no indication of chemical production of GeO in these plumes under any conditions studied; however, the atomic emission intensities of several ionic and neutral Ge lines appeared to increase with O_2 pressure. This signified plume interactions with the background gas. The trends in the species concentrations were determined and the results are interpreted with respect to mechanisms for thin film growth.

INTRODUCTION

Our recent work in growing thin films of germanium dioxide (GeO_2) using laser ablation deposition showed that stoichiometric thin films can be fabricated under two different deposition conditions, and the results of these studies suggest the possibility of two separate kinetic mechanisms for growing these films. First, stoichiometric films could be produced by ablating a GeO_2 target in 150 mTorr of ambient O_2 [1]. At these relatively high pressures, elemental Ge in the ablation plume can react gas kinetically [2] with O_2 to form GeO, which can subsequently recombine with excess O_2 at the substrate surface in a disproportionation reaction. Generally, the oxygen content in the films increased with O_2 pressure from 10^{-5} to 0.15 Torr.

Performing depositions at 100's mTorr of an ambient gas causes the plume to become very highly forward directed [3] and leads to films of nonuniform thickness. For that reason, we explored film depositions at low pressures to produce plumes that are spatially more extended and allow a better opportunity to produce flat films for optical waveguides. Stoichiometric films were prepared at lower background pressures when the ablation plume was passed through a dc discharge in a 30 mTorr oxygen background [4]. In these studies, we found that the combination of the substrate temperature and the dc discharge enabled us to incorporate the correct oxygen content in the film. At these lower pressures, chemical reactions would appear unlikely because of the low collision rates between Ge and any ambient gas. However, the discharge could activate the material in the plume or the ambient gas providing additional energetic species to promote surface recombination reactions.

Reactions between elemental species in laser produced plumes and a reactive background gas are known to be important for producing good quality high T_c superconducting thin films [5]. An understanding of the potential chemical processes occurring between the plume material and any reactive gas in between the target and the substrate is needed to elucidate the important thin film growth mechanisms. Thus, the aim of this work is to explore the microscopic mechanisms responsible for forming high quality dielectric thin films under both growth conditions discussed above.

EXPERIMENT

Figure 1 shows a schematic diagram of the experimental apparatus. A pulsed ArF excimer laser [λ = 193 nm, pulse width of 28 ns (FWHM)] was focused onto the surface of a GeO_2 target at normal incidence with an energy density of 1 J cm^{-2}. The target was housed in a chamber that routinely achieved a base pressure of 5 x 10^{-6} Torr. Oxygen was metered into the chamber with a needle valve and a ring electrode biased at -1.5 kV was inserted in the plume path to determine the effect of a dc discharge on the constituents. The emission from the laser-produced plume (with and without the discharge) was imaged onto a spectrometer and the dispersed signal was detected using a linear diode array (OMA III). The time-integrated emissions were examined as a function of O_2 pressure at 4 cm from the target surface in spectral regions that covered emission features originating from electronically excited states of Ge I, Ge II, GeO, and O I.

The laser fluorescence experiments were performed using the frequency doubled output of a pulsed dye laser. The dye laser probe beam interrogated the plume perpendicular to the excimer laser beam, and the fluorescence was collected at 90° to the dye laser beam. The timing between the two pulsed lasers was varied between 1 and 50 μsec so that the plume constituents could be probed as a function of time from the ablation laser pulse. The timing of the OMA gate was maintained so that the probe beam sampled the plume material within the 500 ns gate width of the OMA. Ge atoms were probed by exciting from the ground $^3P_0$ state to the $^3P_1$ level at 265.16 nm and observing the fluorescence from this state with the gated OMA III system. GeO molecules were probed at 261.4 nm which promoted population from the $X^1\Sigma$ ($v'' = 0$) to the $A^1\Pi$ ($v' = 1$) state. The entire $v' = 1$ progression from GeO (A→X) was captured with the gated OMA in a single trace.

RESULTS

Figure 2 shows a sample of a time-integrated plasma emission spectrum. The plume luminescence was caused by electronic transitions in neutral Ge (Ge I), singly ionized Ge (Ge II), and neutral O (O I) atoms. At pressures below ≈ 50 mTorr and with no discharge, the emission emanated solely from excited Ge I. The emission intensities remained constant from vacuum to about 10 mTorr of added O_2 and increased by a factor of 10 between 10 and 100 mTorr. This behavior could be attributed to species confinement at higher backing gas pressures which increases the number of emitters per volume [6]. Application of the discharge at low pressures (i.e., $P(O_2) < 50$ mTorr) enhanced both the Ge I and O I atomic emissions as shown in Fig. 2. No GeO emission was observed under any conditions in these experiments.

The various channels that are open to creating these electronically excited states in the plume material make it difficult to monitor the behavior of any one species independent of the formation mechanism. In order to circumvent this limitation, the behavior of species in the plume was monitored as a function of ambient gas pressure using laser induced fluorescence (LIF). Figure 3 shows a sample LIF spectrum of Ge I. The ground state Ge I concentration appears to decrease with increasing O_2 pressures as illustrated in Fig. 4. The decay, when converted to a rate coefficient, is consistent with the known reaction rate coefficient for Ge + O_2 → GeO + O [2]. The Ge LIF intensities were also monitored as a function of N_2 pressure to test whether or not the ddecay is due to this reaction scheme. Collisions between N_2 and Ge are non-reactive and should yield little or no decrease in the Ge I concentration with pressure. We found, however, that the Ge concentration decreased with increasing N_2 pressures with a rate that was only 40% slower

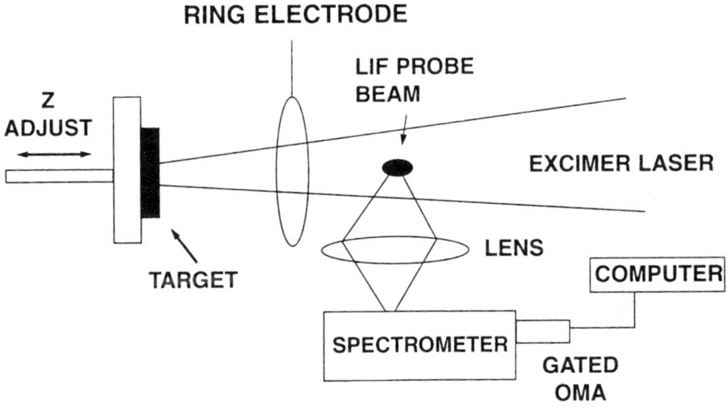

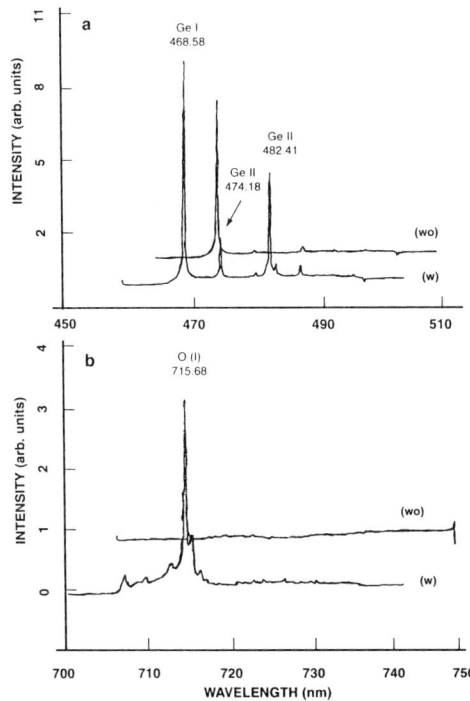

Figure 1. Diagram of the experimental system. The diagram shows the arrangement for the LIF studies. The set-up for the optical emission studies changes by removing the dye laser beam and operating the OMA in an ungated mode.

Figure 2. Time-integrated plasma emission spectrum recorded at an oxygen pressure of 30 mTorr and at 4.0 cm from the target surface. Trace (a) shows Ge emission with (w) and without (wo) the discharge while trace (b) shows the behavior of atomic oxygen.

than that for O_2. This may indicate that the decays are not caused by a chemical depletion of ground state atoms. Electronic quenching could account for these decays, but the radiative lifetimes of the states under study ($\approx$250 ns) prohibit such encounters [7]. For example, the time between collisions in an equilibrium mixture is about 12 µs so the electronic states will never see a collision. These pressure experiments were repeated with the dc discharge with no difference in the results.

Although the emission studies did not reveal the presence of GeO*, the removal of ground state Ge I by reaction is not totally ruled out. Reactions between Ge and O_2 produce GeO in the electronic ground state with $\Delta E = +1.2$ eV [7]. The exothermicity of the reaction is sufficient to populate GeO(X) up to $v" = 5$ so no visible emission would be produced. We searched for vibrational excitation in GeO(X), which would indicate chemical formation of GeO, by pumping from various X state vibrational levels ($0 \leq v" \leq 5$) to the A state using the strongest Frank - Condon allowed transitions. We discovered strong emission lines from the molecule upon excitation from only $v" = 0$. No LIF emission was observed after exciting from $v" \geq 1$ even in the presence of various pressures of O_2. In experiments similar to those performed with Ge I, the GeO LIF emission intensities decreased with increasing O_2 pressures. The GeO(A $\rightarrow$ X) LIF signals became undetectable at O_2 pressures near 40 mTorr. The radiative lifetime for GeO(A) is not known, but, assuming typical values of 3-5 µs, a significant number of collisions would not occur eliminating electronic quenching as the source of the intensity decrease. Although the mechanism for these intensity decreases at various O_2 pressures is currently not understood, the results do indicate that chemical production of GeO via interactions of the plume material with the background O_2 does not occur.

Since the chemical production of GeO via plume-ambient gas interactions does not appear to be important, we traced the origin of diatomic GeO by monitoring LIF signals as a function of distance from the target surface. These LIF experiments showed that GeO molecules were present in the plume at distances as near as 2.5 mm from the target surface. Molecular LIF emission was also observed in vacuum. These results indicate that GeO is either produced as a result of recombination of material in the plume near the target surface where the ion and neutral number densities are high or the molecule dissociated directly from the target.

DISCUSSION

These experiments showed that interactions between the plume material and the ambient O_2 gas directly affected the concentrations of the excited neutral and ionic states of Ge and O atoms as well as molecular GeO. Two possible mechanisms can be postulated to explain the observed emission and LIF behavior described above. The first mechanism involves non-reactive plume - ambient gas interactions. Plume interactions with the background gas, especially at the higher gas pressures (> 50 mTorr), spatially confine the plume and should cause both the atomic emission features and the LIF intensities to increase. However, we observe increases in only the atomic emission intensities and not in the LIF signals. Therefore, this mechanism may not be plausible. The second mechanism involves reactive encounters between the plasma and the ambient gas. Scattering events between electrons and atoms or molecules and high kinetic energy collisions between atoms could produce neutral electronic excitation as well as additional ionization. Therefore, atomic emission experiments would show an increase in emission intensities as was observed here. These interactions would also result in a concurrent decrease in ground state species which is consistent with the LIF results.

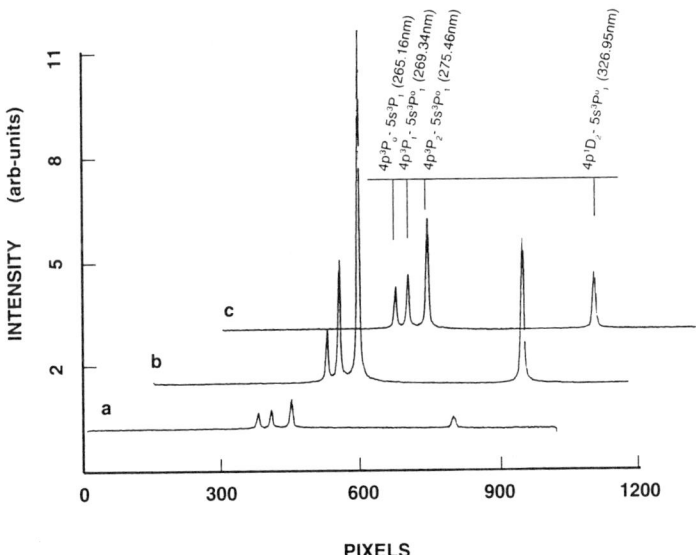

Figure 3. LIF emission spectra from Ge as a function of O_2 pressure: (a) $P(O_2) = 25$ mTorr, (b) $P(O_2) = 0$ mTorr, and (c) $P(O_2) = 10$ mTorr. These spectra were recorded 4 cm from the target surface with a 10.5 μs delay time between the excimer and the probe lasers.

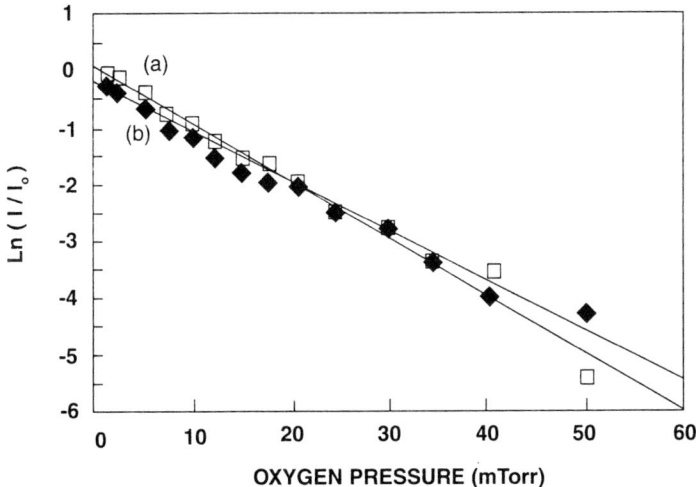

Figure 4. Plot of the natural log of the relative fluorescence intensities vs. O_2 pressure. Decays from two LIF emission lines are plotted: (a) λ = 326.9 nm and (b) λ = 275.5 nm. The cause of this decay may be due to a combination of reactive and non-reactive collisions with the ambient oxygen.

These results may have significance in determining the mechanisms for GeO_2 thin film growth. Since chemical reactions do not appear to dominate the plume-ambient gas processes, surface reactions must control the thin film growth. At high ambient gas pressures (i.e., $P(O_2)$ = 150 mTorr), excess O_2 may be required to complete recombination reactions on the substrate surface, whereas at lower pressures, the discharge produces excited O atoms which can then interact more energetically with the substrate. These processes produce the same result: stoichiometric thin films. The importance of GeO in the film formation process is not completely understood. Diatomic GeO is abundant in vacuum but disappears as the O_2 pressure is increased. Since we cannot detect population in other vibrational levels of the ground state, GeO may either be scattered from the excitation region or the molecule could be dissociated as a result of plasma-O_2 interactions.

SUMMARY

Constituents of a laser-produced plume from a GeO_2 target have been spectroscopically examined. The results indicate complicated processes that occur when the plume interacts with the ambient O_2. High kinetic energy species are produced and the behavior of individual species in the plume may be governed by several reactive and non-reactive encounters. Chemical production of diatomic species, which are critical for growing good quality high T_c thin films, does not occur for laser ablated GeO_2 plumes in O_2, and thus does not appear to be important for growing stoichiometric GeO_2 thin films.

REFERENCES

1. P.J. Wolf, T.M. Christensen, N.A. Coit, and R.W. Swinford, J. Vac. Sci. Technol. A, **11**, 2725 (1993).
2. P.M. Swearengen, S.J. Davis, and T.M. Niemczyk, Chem. Phys. Lett., **55**, 274 (1978).
3. D.J. Lichtenwalner, O. Auciello, R. Dat, and A.I. Kingon, J. Appl. Phys., **74**, 7497 (1993).
4. S. Witanachchi and P.J. Wolf, submitted to J. Appl. Phys.
4. C.E. Otis and R.W. Dreyfus, Phys. Rev. Lett., **67**, 2102 (1991) and references therein.
5. P.J. Wolf, in preparation.
6. M.H. Miller and R.A. Roig, Phys. Rev. A, **7**, 1208 (1973).
7. G.A Capelle and J.M.Brom Jr, J. Chem. Phys., **63**, 5168 (1975).

INDIUM TIN OXIDE FILMS FORMATION BY LASER ABLATION

J.P. Zheng* and H.S. Kwok
Department of Electrical and Computer Engineering
State University of New York ar Buffalo
Amherst, New York 14260

*Present Address:
Army Research Laboratory
Electronics and Power Sources Directorate
Fort Monmouth, New Jersey 07703-5601

ABSTRACT

The electrical and optical properties of room temperature laser deposited indium tin oxide films were studied. It was found that the resistivity of the film was quite sensitive to the deposition conditions. At the optimized conditions, films with a bulk resistivity value of 2.8×10^{-4} Ω-cm and optical transmission of greater than 90% could be obtained. By using an in situ resistance measurement, it was shown that the initial growth mode was via island formation. Additionally, a classic transition from two- to three-dimensional behavior for the resistance was observed.

INTRODUCTION

Indium tin oxide (ITO) films are found in a wide range of applications in optoelectronic devices[1-6], because of their optical transparency and electrical conducting properties. Recently, very large optical nonlinearities have been obtained from this material[7]. The lowest resistivity of 7×10^{-5} Ω-cm was obtained by an activated reactive evaporation technique at a substrate temperature of 350 °C[8]. However, for some devices, the deposition temperature is an important issue and heating is not desirable. For example, high temperatures are known to lead to deleterious interface effects in ITO/InP solar cells[9], and in ITO/polymer opto-ferroelectric memory devices[10], the polymer dissociates at high temperatures. Therefore, it is necessary to grow low resistivity ITO films at low temperatures without post-annealing. In this paper, we demonstrated that the technique of pulsed laser deposition (PLD) was used to successfully deposit ITO films at room temperature without post-annealing.

FILM PREPARATION

ITO films were prepared in a clean vacuum chamber with a base pressure of 5×10^{-6} Torr. Details of the deposition system were discussed previously[11]. An excimer laser beam at a wavelength of 193 nm and pulse duration of 15 ns was focused onto a rotating target at a oblique angle of 30°. The laser was operated at 10 Hz. An energy of 50 mJ per pulse was delivered to the target with a focal size of 3×1.5 mm^2. This corresponded to a fluence of about 1 J/cm^2 which was sufficient to produce a laser plume. The target was a 2.5 cm diameter In_2O_3(90%) + SnO(10% by weight) sintered ceramic. The substrates were glass of size 1.5×1.5 cm^2. The substrates were cleaned with trichlorethylene, acetone, methanol and

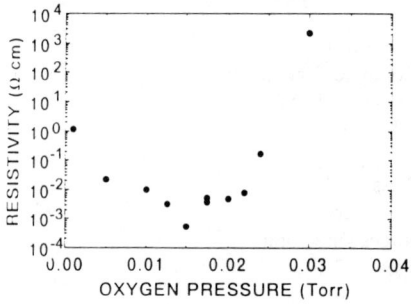
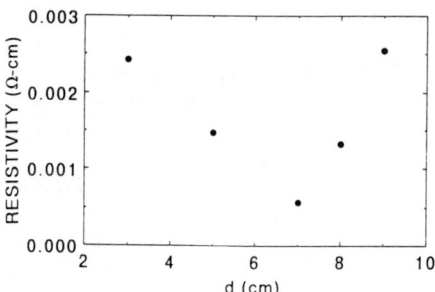

FIG. 1. Resistivity as a function of oxygen pressure. The target-substrate distance is 7 cm.

FIG. 2. Resistivity as a function of target-substrate distance. The oxygen pressure is 15 mTorr.

deionized water, and they were then thermally contacted with silverpaste on a copper block which acted as a heat sink to stabilize the substrate temperature. The substrates were not heated or cooled. The thermocouple was used to monitor the surface temperature of the substrate at all times. The temperature was found to remain around 20 °C during deposition. Oxygen gas (98% purity) was introduced into the chamber to the desired pressure. The film thickness was measured by a profilometer, allowing the deposition rate to be calculated. It was found that the deposition rate was strongly dependent on the oxygen background pressure and decreased with increasing oxygen pressure. For example, the deposition rate was 0.1 Å /pulse at 3×10^{-5} Torr and was reduced to 0.03 Å /pulse at 5×10^{-2} Torr. Incidentally, the absolute deposition rate is dependent on the laser fluence and can be increased significantly.

ELECTRICAL PROPERTIES

The resistivity of the film was calculated based on the resistance measurement by a standard four-probe technique[17]. The errors of the measured resistivity were mainly contributed from the variations of the film thickness. The maximum error of the resistivity was estimated to be less than 10% of this value. Fig. 1 shows the resistivity of ITO films as a function of oxygen pressure. It can be seen that the resistivity is very sensitive to oxygen pressure. Low resistivity films could be obtained only in a narrow pressure range. Films with the lowest measured resistivity of 5.6×10^{-4} Ω-cm were deposited at 15 mTorr. The sample thickness in these measurement was 1500 Å. It will be shown later that there is a significant surface scattering contribution to the measured resistance. The actual bulk resistivity for the ITO film is 2.8×10^{-4} Ω-cm after some theoretical fitting to the data by a classic model of the two- to three-dimensional transition[12]. This resistivity value is the lowest ever reported for films deposited by any method at this low temperature without post-annealing.

On the another hand, it was also found that the resistivity is quite sensitive to the target-substrate distance at a fixed oxygen pressure as shown in Fig. 2. It can be seen that the minimum resistivity was obtained for films deposited at the optimal target-substrate distance of 7 cm for the oxygen pressure of 15 mTorr.

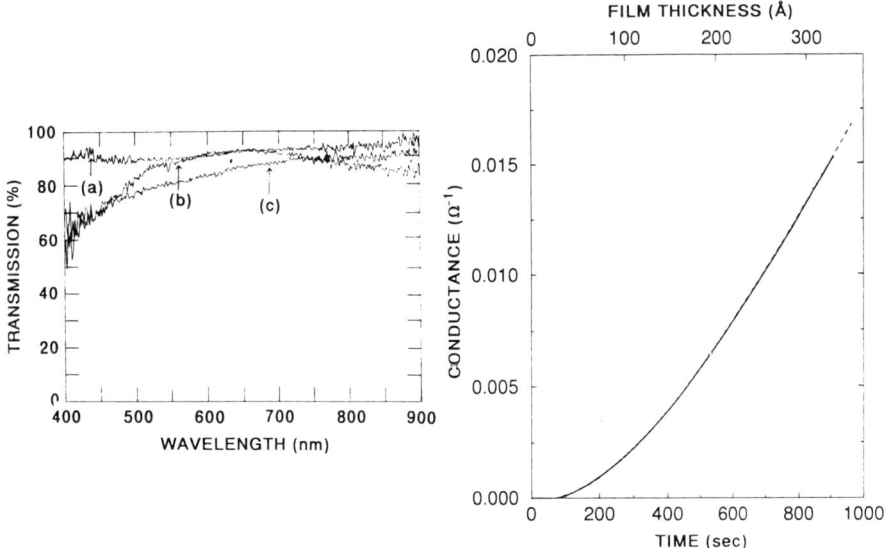

FIG. 3. Optical transmission spectra for films deposited at oxygen pressures of (a) 30, (b) 15 and (c) 1 mTorr. The film thicknesses are 500, 1500 and 600 Å respectively.

FIG. 4. Conductance of an ITO film during deposition. The dotted lines (barely distinguishable from the data) are theoretical fits.

OPTICAL PROPERTIES

Optical transmission spectra of samples were measured with a 360 W halogen lamp light passing through samples. The transmitted light was analyzed by an optical multichannel analyzer in the wavelength range 400-900 nm. This wavelength range is limited by the sensitivity of the silicon photodiode detector. Fig. 3 shows optical transmission spectra for films deposited at different oxygen pressures. It should be pointed out that these transmission spectra include the loss from the surface reflection of the glass substrate itself. The transmission increased with increasing O_2 pressure during the deposition. For films deposited at the optimal conditions, the transmission is greater than 90% between 600 nm and 900 nm. This transmission is comparable with that of the best films deposited at high temperatures.

IN-SITU RESISTANCE MEASUREMENT

It is well known that the resistivity of thin films changes from two-dimensional to three-dimensional behavior as the thickness is increased[12]. Because of surface scattering, the resistivity of thin films is enhanced for films thinner than λ, the mean free path of the free carriers. It can be shown that the conductivity is given by[12]

$$\sigma = \sigma_o \left[1 + \frac{3}{2K} \int_1^\infty \left(\frac{1}{z^3} - \frac{1}{z^5} \right) \exp(-Kz) dz \right] \quad (1)$$

where σ_0 is the bulk (three-dimensional) conductivity, $K=l/\lambda$ and l is the film thickness. In situ measurement of resistance has been used to measure the resistance change during ITO film formation as follows. Four silver strips were thermally evaporated onto a glass substrate. Each strip was separated by a distance of 1 mm. During the deposition, a standard four-probe method was used to measure the resistance change. A computer was used to record the resistance (conductance) values with a time resolution of 0.3 sec.

Fig. 4 shows the conductance change during film growth. It was found experimentally that the resistance remained very large (greater than 100 MΩ) for a period of time corresponding to a film thickness of 22 Å. This is direct evidence that the film forms via an island growth mechanism. After this initial thickness of 22 Å, the observed resistance decreased, indicating coalescence of the islands. After coalescence, the increase in conductance is not linear in l until very large l, because of contributions from surface scattering. Equation (1) was used to fit the data in Fig. 4 with σ_0 and λ as adjustable parameters. The result is plotted as a dotted line in the same figure. The fit and the data overlap very well and are almost indistinguishable. The values of λ and ρ_0 obtained are 550 Å and 2.8×10^{-4} Ω-cm respectively. For $l < \lambda$, the electronic transport behavior can be described as a two-dimensional system with surface scattering dominating. When the film thickness is larger than the electron mean free path, normal bulk three-dimensional behavior becomes dominant. Fig. 4 is an almost textbook example of the two-dimensional to three-dimensional transition in σ.

DISCUSSION

The strong dependence of the resistivity of ITO films on oxygen pressure as shown in Fig. 1 can be explained by the nature of PLD. From an optical time-of-flight study[13], it was found that laser-generated atoms and ions travel towards the substrate with time duration of about 10 µs and peak velocity of approximately 10^6 cm/s. The lighter atoms travel faster initially. However, these velocities eventually equilibrate owing to the many collisions between fast and slow atoms and ions. The most important collision is between the energetic atoms and the ambient O_2. From our previous work[13,14], it was also found that the quality of the deposited film was directly related to the velocity distributions of the different atoms. There exists a distinct pressure-distance scaling law for the deposition of high quality films. The reason is that at low O_2 pressures, the velocity distribution of various constituents will not be uniform on the surface of the substrate. This situation is not conducive to epitaxial films formation and is actually similar to sequential evaporation of various constituent metals. However, if the O_2 pressure is too high, the velocities of the various species in the laser plume slow down as a result of collisions. Surface activation by moderately energetic ions and atoms will not be possible. Therefore, at the optimal O_2 pressure, energetic ions and atoms with a uniform velocity distribution of various species combine to produce good quality films. Essentially, the quality of films deposited by PLD is equivalent to that deposited by other methods at high substrate temperatures.

It is known that the resistivity of ITO films is dependent on the concentration of oxygen, because the free carriers are due to oxygen vacancies. Therefore, the O_2 pressure may affect both the deposition dynamics and the O-vacancy. The results shown in Fig. 2 indicate that the velocity distributions of different species in the laser plume play a major role in determining the resistivity of the ITO film. Furthermore, from our previous work[15] it was found that oxygen content in the film was mainly contributed by the oxygen coming from the target itself. This is especially true for films deposited at low temperatures, because the oxygen ejected from the target is in the form of atoms or ions which are much more chemically active than the ambient O_2. Therefore, when they impinge on the substrate they are easily incorporated with other constituents to form the stoichiometric film.

Finally, these results are contrary to those for films grown at high temperatures[16]. For those cases, the resistivity of the film is not sensitive to the oxygen pressure. This is reasonable because even if the different species arrive at the substrate at different times or have low kinetic energies, they can still migrate on the surface of the substrate because of their thermal energy and can rearrange themselves to form a stoichiometric film.

CONCLUSION

The dependence of electrical and optical properties on deposition conditions for ITO films grown at room temperature has been studied. The resistivity of the films was strongly dependent on the deposition pressure and the target-substrate distance at room temperature. The energetic atoms and ions in the laser plume are believed to be responsible for improving the quality of films. By carefully controlling deposition conditions, a bulk resistivity value as low as 2.8×10^{-4} Ω-cm and an optical transmission of greater than 90% for a 1500 Å thick film have been achieved. We have also shown by in situ measurement that the film grows via an island formation mechanism. The coalescence thickness is approximately 20 Å. A classic two-dimensional to three-dimensional behavior of the film resistivity was also observed for films thinner than 1000 Å.

REFERENCES

1. V.K. Jain and P.A. Kulshreshtha, Sol. Engery Mater., 4, 151 (1981).
2. X. Li, M.W. Wanlass, T.A. Gessert, K.A. Emery and T.J. Coutts, Appl. Phys. Lett., 54, 2674 (1989).
3. I. Hamberg and C.G. Granqvist, Appl. Opt., 24, 1815 (1985).
4. M. Green, W.C. Smith and J.A. Weiner, Thin Solid Films, 38, 89 (1976).
5. M. Hagerott, H. Jeon, A.V. Nurmikko, W. Xie, D.C. Grillo, M. Kobayashi and R.L. Gunshor, Appl. Phys. Lett., 60, 2825 (1992).
6. L.A. Goodman, RCA rev., 35, 613 (1974).
7. D.H. Kim and H.S. Kwok, to be published.
8. R. Nath, R.F. Bunshah, B.M. Basol and O.M. Staffsud, Thin Soild Films, 72, 463 (1980).
9. Q.X. Jia, J.P. Zheng, H.S. Kwok and W.A. Anderson, to be published.
10. M. Date, T. Furukawa, T. Yamaguchi, A. Kojima and I. Shibata, IEEE Trans. Electr. Ins., Vol. 24, 537 (1989).
11. H.S. Kwok, J.P. Zheng, S. Witanachchi, P. Mattocks, L. Shi, Q.Y. Ying, X.W. Wang and D.T. Shaw, Appl. Phys. Lett., 52, 1095 (1988).
12. C.R. Tellier and A.J. Tosser, Size Effects in Thin Films, Elsevier, Amsterdam, 1982.
13. J.P. Zheng, Q.Y. Ying, S. Witanachchi, Z.Q. Huang, D.T. Shaw and H.S. Kwok, App;. Phys. Lett., 54, 954 (1989).

14. H.S. Kim and H.S. Kwok, Appl. Phys. Lett., 61, 2234 (1992).
15. J.P. Zheng, D.H. Kim, S.Y. Dong, C. Lehance, W.P. Shen and H.S. Kwok, AIP conf. Proc., 273, 437 (1992).
16. J.P. Zheng and H.S. Kwok, Appl. Phys. Lett., 63, 1 (1993).
17. Q.Y. Ying and H.S. Kwok, Appl. Phys. Lett., 56, 1478 (1990).

EXCIMER LASER INTERACTIONS WITH PTFE RELEVANT TO THIN FILM GROWTH

J.T. DICKINSON, M.G. NORTON, J.-J. SHIN, W. JIANG, AND S.C. LANGFORD
Washington State University, Pullman, WA 99164-2814

ABSTRACT

Recently, thin films of polytetrafluroethylene (PTFE) have been grown using pulsed laser ablation of Teflon™ at 266 nm.[1,2] To provide further insight into the growth mechanisms we have examined the neutral and charged particle emissions generated in vacuum by 0 - 3 J/cm² pulses of 248 nm radiation incident on solid PTFE. Measurements include quadrupole and time-of-flight mass spectroscopy. We find in addition to the neutral monomer (C_2F_4), copious emissions of highly reactive neutral and charged radicals, e.g., CF_2, CF_3, CF, F, and C_x. A careful analysis of the fluence dependence of these products provides definitive evidence that their precursors are generated by a thermally driven unzipping reaction. Models for the production of the radical species with the observed energies (several eV) involving gas phase processes are presented. Implications for improving PTFE thin film growth will be discussed.

INTRODUCTION

Polytetrafluoroethylene (PTFE) has many valuable properties, including low dielectric constant, high resistivity, high thermal and chemical stability, low surface energy, low coefficient of friction, potential biocompatibility, and low molecular diffusivities for a number of species. Thin PTFE films are desirable for a number of applications, and a number of techniques have been developed to fabricate such films. A plasma environment is often employed to provide the reactive species necessary for polymerization. Blanchet and Shah have recently fabricated PTFE films by pulsed-laser deposition (PLD) in Ar and CF_4 atmospheres (50-250 mTorr) on heated substrates;[1,2] the PTFE targets were irradiated with 10 ns pulses of 266 nm radiation at fluences of 0.5-2 J/cm². They propose that the primary ablation product is the monomer (C_2F_4) produced by pyrolytic decomposition of the target. PTFE film formation is attributed to repolymerization of the monomer on the substrate surface.

The polymerization of PTFE films by plasma-assisted polymerization (PAP) has been shown to require radical and ionic species.[3,4] In particular, CF, CF_2, CF_3, CF_3^+, $C_3F_5^+$, F, and C have been proposed as important reactive species. Several groups have identified these and other species during UV laser irradiation with PTFE.[5-8] Thus laser irradiation produces all the essential species for polymerization and film growth, although not necessarily in optimum proportions. For example, polymerization requires a high C/F ratio in the incident flux of particles; a low C/F ratio can actually result in etching.[4]

In this work, we show that the fluence dependence and velocity distributions of the neutral products produced by pulsed 248-nm irradiation of PTFE are consistent with a thermally activated emission mechanism. The low activation energy is consistent with known activation energies of un-zipping reactions, which produce predominately monomers, but also other neutral species. Ionic species are also observed, presumably due to the collisional ionization of neutral molecules in the weak plasma generated at the target surface.

EXPERIMENT

The PTFE targets used in this study were 0.5-mm-thick films supplied by Goodfellow Corporation. Experiments were conducted in vacuum at background pressures of 10^{-7}-10^{-5} Pa.[9] 20-ns pulses of 248 nm radiation from a Lambda Physik EMG203 excimer laser (KrF) were directed at the sample through a 114-cm focal length lens and mask. The fluence at the sample was varied by adjusting the lens-to-sample distance. A low pulse repetition rate (1 Hz) was used to avoid accumulative heating. The total energy of the laser was measured with a GenTec ED500 Joule meter.

Neutral molecule mass spectra were obtained with a UTI 100C quadrupole mass spectrometer (QMS) with an electron impact ionizer operating at 70 eV. Time resolved signals at single masses provided time-of-flight (TOF) distributions of the emitted particles. After correcting the TOF distribution for the velocity dependence of the ionization probability and the time required for ions to pass through the mass filter, a least squares fit to a Maxwell-Boltzmann distribution was used to determine the effective temperature of the neutral emission products. The resulting equation for the measured intensity vs time, $I(t)$, is of the form:

$$I(t) \sim N_o \left(\frac{\alpha}{t^2}\right)^2 \int_D^{D+S} x^3 \exp\left(\frac{-a x^2}{t^2}\right) dx \quad , \quad (1)$$

where D is the distance from the sample to the tip of the ionizer (17 cm), S is the length of the ionizer (2 cm), and $\alpha = m/kT$ where m is the molecular mass, k is Boltzmann's constant, and T is the temperature. N_o is an adjustable parameter which simply adjusts the amplitude of the curve, accounting for the overall ionization efficiency and peak emission intensity. The only other adjustable parameter is T, which determines the shape of the TOF distribution (including the position of the peak).

Ion mass spectra were measured with both the QMS (ionizer removed) and a Reflectron time-of-flight mass spectrometer.

RESULTS AND DISCUSSION

Neutral Emission

PTFE is normally transparent at most excimer laser wavelengths.[10] The absorption constant, α, of nonirradiated samples is typically about 158 cm^{-1} at 248 nm.[11] Despite this weak absorption, irradiation at 248 nm (and 308 nm) at fluences of 0.5–3 J/cm^2 produces strong particle emission on the first few laser pulses; the emission intensities at these fluences subsequently drop to roughly constant values.

In agreement with several other groups, we find that the monomer, C_2F_4, is the dominant neutral species at fluences of about 2 J/cm^2. Spectra are shown in reference 12. Using cracking fractions from the EPA/NIH Mass Spectral Data Base,[13] we conclude that (in order of decreasing intensity) the parent molecules include C_2F_4, C_3F_6, and C_4F_8. In addition, we observe the neutral radicals CF_2, CF_3, CF, C, C_2, C_3, F, and other species of the form C_2F_x, C_3F_y, and C_4F_z. The relative intensities of the major neutral species changes with laser fluence and wavelength. Preliminary examination of the neutral emission during irradiation at 308 nm shows relatively high intensities of CF, CF_2, and CF_3. These high radical concentrations may benefit thin film growth and polymerization.

Intense C_2F_4 emission and relatively strong C_3F_6 emission are consistent with thermal decomposition, as suggested elsewhere [e.g., Refs. 1 and 9]. However, as shown below, the route to decomposition does not appear to be direct thermal bond breaking; rather a free radical mechanism is proposed. We can calculate an effective surface temperature by modeling the neutral particle TOF with a nonlinear least squares routine.[14] Figure 1(a) shows a TOF signal for mass 100 (C_2F_4) at a laser fluence of 2 J/cm^2 averaged over 50 pulses. The curve fit was generated assuming the signal was due to a single mass (100 amu, corresponding to the parent C_2F_4) at a temperature of 987 K. Figure 1(b) shows the corresponding signal at the same fluence with the QMS tuned to mass 31. Here, four parent molecules were needed to obtain a reasonable fit: 31 amu (CF), 50 amu (CF_2), 100 amu (C_2F_4—the most intense), and 200 amu (C_4F_8), all at a temperature of 988 K. The agreement with the mass 100 temperature is somewhat fortuitous. Similar unconstrained curve fits to TOF data taken at the same laser fluence at mass 50 [assuming two parent masses: 50 amu (CF_2) and 100 amu (C_2F_4)] and mass 69 [assuming two parent masses: 69 amu (CF_3) and 100 amu (C_2F_4)] yielded good fits with temperatures of 930 K and 1050 K, respectively. We are therefore confident that these fits yield reasonable temperatures of the material yielding the emission. SEM images of the PTFE surface indicate that decomposition at these fluences is highly localized probably associated with defects; thus these temperature estimates do not apply to the irradiated surface as a whole but to localized regions.

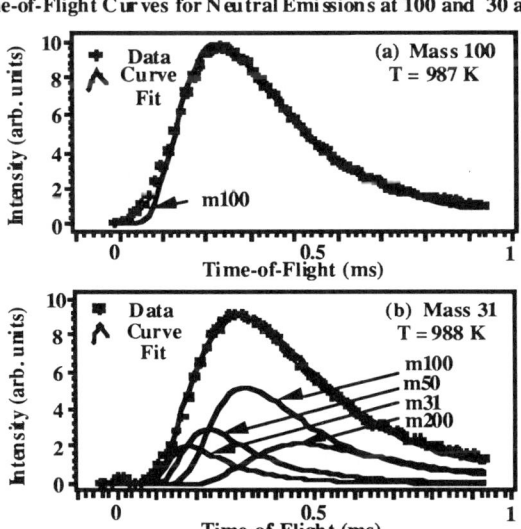

Fig. 1. Time of flight curves for the neutral emission detected with the QMS tuned to (a) 100 amu (the monomer) and (b) mass 31. The curves drawn through the data points are derived from Maxwell-Boltzmann distributions at the temperatures shown.

Negative ion emission measurements show that, at a fluence of 1.2 J/cm^2, F^- is the dominant negative ion; large quantities of C_3^- and CF_3^- are also detected, consistent with the work of Gardella et al..[8] Copious electron emission is also observed, with an intensity increasing approximately linearly with laser fluence. Laser-stimulated electrons are easily discriminated from negative ions by their time-of-flight in weak electric fields.[15]

Fluence Dependence and Kinetics Model

Figure 2 shows the fluence dependence of the emission intensities for neutral C_2F_4 and CF under 248 nm irradiation; also shown are the fluence dependence of the ionic species CF^+ and F^-/F, where the curves are normalized to a single point. The F^- intensity has been divided by a factor of F, the laser fluence, consistent with the formation of negative ions by electron reattachment with neutral species in the plume, where we assume that the electron density is proportional to fluence. With this adjustment, the behavior of all these species is essentially the same. The similar fluence dependence of the ionic and neutral species (including the monomer) suggests that they are both controlled by the rate of polymer decomposition.

TOF data for the neutral monomer at various fluences were used to determine the average temperature vs fluence. The resulting temperatures over a range of fluences are shown in Fig. 3(a), where we have fit the data to a straight line. As expected, the onset of significant monomer emission is close to the isothermal decomposition temperature for PTFE ($T_d \sim 700$ K).[16]

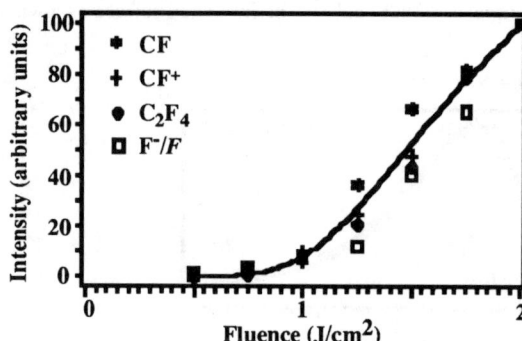

Fig. 2. Emission intensities (at 248 nm) as a function of fluence for neutral C_2F_4 and CF, normalized to a single point. Also shown are the emission intensities of the ionic species CF^+ and F^-, where the F^- intensity has been divided by a factor of F (the laser fluence).

We take the simplest form for the temperature increase, ΔT, vs F, namely:

$$\Delta T = \frac{(1-R)\alpha F}{\rho C_v} \quad (2)$$

where R is the reflectivity of the PTFE surface (~ 0.15), α is the absorption constant, F is the incident laser fluence, ρ is the density of PTFE (2.1 g/cm^3) and C_v is the heat capacity, which we take to be constant = 1.05 J/(g·K). The slope of the data in Fig. 3(a) thus determines α, which here is about 1770 cm^{-1}—reasonable for "damaged" PTFE. It is important to note that the resulting value of α depends on the incubation treatment (i.e., the fluence and number of laser pulses) before the TOF curve is acquired. Nevertheless, after incubation, provided that only a few pulses are used to determine T, α does not change with fluence.

The decomposition kinetics (if thermal) should be reflected in the dependence of the monomer emission intensity on the gas temperatures as determined by TOF analysis. Figure 3(b) shows an Arrhenius plot of $\log_e(I)$ vs $1/T$ for two data sets (symbols * and + , respectively). This data does not obey a simple Arrhenius equation (i.e., a single, straight line). At low fluences (low T), where the slope is the largest, the apparent activation energy, E_a, is ~ 0.9 eV, well below the activation energy for isothermal decomposition (3.5 eV/molecule).[17] Furthermore, as T increases, the slope of the Arrhenius plot (apparent E_a) decreases to ~ 0.2 eV.

We propose that the kinetics of monomer production is rate limited by a low energy process, namely a depropagation or unzipping reaction,[18] where photochemically produced free radicals sequentially catalyze bond scissions between monomer units as they move along the polymer chain. Unzipping reactions account for the high monomer yields in isothermal decomposition of PTFE, and would explain the high monomer yields accompanying ablation. It has been suggested that at elevated temperatures, a single radical can produce 10^5 monomer units during a period equal to the thermal time constant for the cooling of laser irradiated PTFE (10^{-7}-10^{-8} s).[1]

To model our data we assume that free radicals are created photochemically at a concentration R_0 proportional to the laser fluence F. From Fig. 2(a), F is proportional to T; therefore, R_0 is also proportional to T. Under equilibrium thermal conditions, R_0 is small and termination typically occurs by disproportionation.[19] Under ablation conditions, R_0 is much larger and termination by the "collision" of radicals on the same chain is more likely. This would occur at a rate $\sim R^2$; if we assume that the temperature is constant over the effective reaction time Δt, R will be given by second order decay kinetics of the form:

$$R = \frac{R_0}{1 + R_0 k_t t} \quad (3)$$

where k_t is the rate constant for the termination reaction, $k_t \sim \exp(E_t/kT)$. The emission intensity will be proportional to the product of R and the rate constant for the depropagation reaction, $k_d \sim \exp(E_d/kT)$. E_d and E_t are the activation energies for depropagation and termination, respectively; for the proposed termination mechanism, both E_d and E_t equal the activation energy for the propagation of the radical along the chain ($E_d = E_t$). The total emission $I(T)$ is therefore proportional to the integral of $k_d \cdot R$ over time, yielding

$$I(T) = A \cdot \ln\left[1 + B \cdot T \cdot \exp\left(-\frac{E_t}{kT}\right)\right] \quad (4)$$

where A and B are constants. Equation 4 was fit the data in Fig. 3(b), yielding $E_t (= E_d) = 0.9 \pm 0.2$ eV. This energy is appropriate for the activated transport of radicals along the polymer backbones, as opposed to the much more energetic process of forming radicals by thermally induced bond scissions.

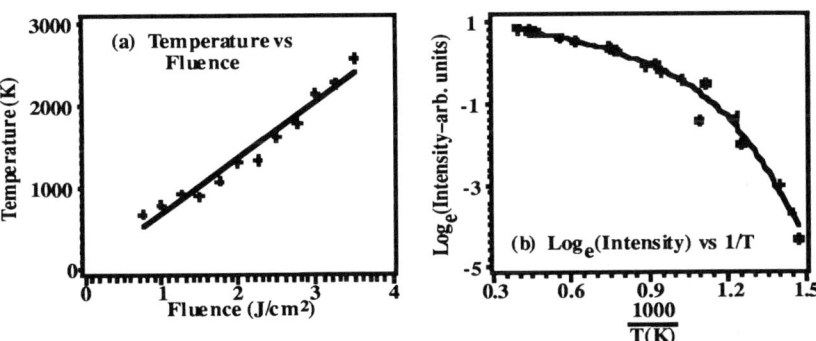

Fig. 3. (a) Temperature estimates (derived from the TOF data) versus fluence for the monomer emission. (b) Log_e(monomer intensity) vs (1/T). The symbols * and + represent measurements taken on different spots.

At low fluences and temperatures, the termination reaction can be neglected and emission is rate limited by the initial radical density and the thermally stimulated "diffusion" of these radicals along the chain. This yields a rapidly rising (superexponential) emission with fluence, consistent with the low-fluence portion of Fig. 2. The rapidly rising emission intensities between 0.5 and 1.0 J/cm^2 in Fig. 2 are significant and cannot be realistically described in terms of threshold behavior. This strong dependence of the emission intensities on temperature implies that the hottest regions of the sample dominate the emission process. At the lower temperatures before and after the material reaches its peak temperature, the emissions are much reduced; this would account for the very nearly Maxwellian velocity distribution of the monomer (where the signal is dominated by a single parent species) despite the necessary heating and cooling.

At higher fluences and temperatures, the emission intensities are limited by the termination reaction, resulting in saturation behavior. Only a hint of this saturation appears in Fig. 2 due to the low fluences employed. However, this saturation has been previously reported (e.g., in the etch rate measurements of Küper and Stuke)[11] and attributed to other factors. Despite the limited evidence for saturation in Fig. 2, the fact that the model parameters derived from this data show saturation behavior in the correct fluence range is promising. We also note that this model does not employ an intensity-dependent absorption coefficient.

CONCLUSIONS

Substantial quantities of radical and reactive ion components accompany the ablation of PTFE at 248 nm. These species, along with the intense C_2F_4 emission, are important for the repolymerization of PTFE at the deposition substrate. The kinetics of monomer emission can be explained by the photochemical production of free radicals, resulting in a thermally activated unzipping reaction. Unzipping reactions would also account for the high monomer yields accompanying laser ablation of PTFE. Termination reactions at higher fluences would account for the saturation in etch rates with increasing fluence observed from UV ablation of PTFE.[14] We are currently extending the kinetic investigations to the well studied PMMA ablation at 248 nm and 308 nm and finding evidence for a low activation energy process. It would also be interesting to look at the kinetics of PTFE decomposition at 157 nm (the F_2 excimer laser) irradiation where strong single photon absorption and clean etching are observed.[10,20]

We thank Les Jensen, and Richard Webb for their assistance. This work was supported by Department of Energy-BES (DE-FG06-92ER14252), the National Science Foundation (DMR-9201767), the Washington Technology Center, and Washington State University.

REFERENCES

1. G.B. Blanchet and I. Shah, Appl. Phys. Lett. **62**, 1026 (1993).
2. G. B. Blanchet, C. R. Fincher Jr., C. L. Jackson, S.I. Shah, and K. H. Garner, Science **262**, 719 (1993).
3. F. Fracassi, E. Occhiello, J. W. Coburn, J. Appl. Phys. **62**, 3980 (1987).
4. R. d'Agostino, F. Fracassi, and F. Illuzzi, J. Appl. Poly. Sci. Appl. Poly. Symp. **46**, 17 (1990).
5. M. Stuke and Y. Zhang, Proc. Electrochem. Soc. **88** (10), 70-81 (1987).
6. P.M. Goodwin and C.E. Otis, J. Appl. Phys. **69**, 2584 (1991).
7. K. Kurosaki, J. Appl. Polym. Sci. Applied Polym. Symp. **48**, 401 (1991).
8. J A. Gardella, Jr., D.M. Hercules, and H.J. Heinen, Spectros. Lett. **13**, 347 (1980).
9. P.A. Eschbach, J.T. Dickinson, S.C. Langford, and L.R. Pederson, J. Vac. Sci. Technol. A **7**, 2943 (1989).
10. D. Basting, U. Sowada, F. Voss, and P. Oesterlin, in: Gas and Metal Vapor Laser and Applications, SPIE Proceedings **1412**, 80 (1991).
11. S. Kuper and M. Stuke, Appl. Phys. Lett. **54**, 5 (1989).
12. J. T. Dickinson, J-J. Shin, W. Jiang, and M. G. Norton, J. Appl. Phys. **74**, 4729 (1993).
13. S.R. Heller and G.W.A. Milne, EPA/NIH Mass Spectral Data Base (NBS, U.S. Dept. of Commerce, Washington DC, 1978).
14. J.T. Dickinson, L.C. Jensen, D.L. Doering, and R. Yee, J. Appl. Phys. **67**, 3641 (1990).
15. S.C. Langford, L.C. Jensen, J.T. Dickinson, and L.R. Pederson, J. Appl. Phys. **68**, 4253 (1990).
16. D.W. Van Krevelen, Properties of Polymers, (Elsevier, Amsterdam, 1976), p. 461.
17. S.L. Madorsky, Thermal Degradation of Organic Polymers, (Interscience, New York, 1964), pp. 135-137.
18. N. Grassie and G. Scott, Polymer Degradation and Stabilization, (Cambridge University Press, Cambridge, 1985), pp. 23-24.
19. J.C. Siegle, L.T. Muus, T.-P. Lin, and H.A. Larsen, J. Polym. Sci. A **2**, 391 (1964).
20. P.R. Herman, B. Chen, and D.J. Moore, to be published.

PHOTOCHEMICAL MODIFICATION OF FLUOROCARBON RESIN SURFACE TO ADHERE WITH EPOXY RESIN

M. OKOSHI*, T. MIYOKAWA*, H. KASHIURA* and M. MURAHARA**
* Graduate Student of Tokai University
** Department of Electrical Engineering, Tokai University, 1117 Kitakaname, Hiratsuka, Kanagawa 259-12, JAPAN

ABSTRACT

An epoxy-compatible layer was produced in the near-surface region of a fluorocarbon resin by irradiating with an ArF excimer laser in a gaseous B(CH3)3 or a B(OH)3 water solution atmosphere. The pure photochemical reaction was employed in the modification process, the defluorination of the surface was performed with boron atoms which were photo-dissociated from B(CH3)3 or B(OH)3. The CH3 or OH radicals, also photo-dissociated, replaced the fluorine atoms of the surface. As a result, chemical bonding of the surface with the epoxy was performed. The adhesive strength was evaluated by the shearing tensile test, and the epoxy break value of 130 kgf/cm2 was sucessfully achieved.

INTRODUCTION

Fluorocarbon resin (TEFLON) is superior in chemical, physical, electrical and mechanical properties. For these reasons, the surface is chemically stable and is deficient in hydrophilic, oleophilic and adhesive properties. Thus, a modification technique of the resin surface has been required.
As conventional methods, the plasma treatment[1,2] and the chemical treatment [3,4] have been carried out. Plasma treatment, however, produces cracks on the resin surface. Although the adhesive property of surface is improved by the anchor effect of the rough surface with an adhesive agent, the overall adhesion is weak compared with the case of chemical treatment. On the contrary, in chemical treatment, environmental pollution arises from the waste chemicals. In the future, the chemicals used will be limited; a new modification method would be needed.
Therefore, we have proposed a photochemical modification method using an excimer laser light[5]. By this method, the surface can be modified to hydrophilic[6,7], oleophilic[8,9] and to metallic[10] states at will. In this paper, the resin surface was photochemically modified to an epoxy-compatible layer; chemical bonding of the resin surface with epoxy was performed without a primer to obtain an adhesive strength which is equal to the epoxy cohesive strength.

PRINCIPLE OF PHOTOCHEMICAL MODIFICATION

Surface Defluorination

A property of material surface is determined by an

functional group. In case of the fluorocarbon resin, fluorine atoms at the surface determine the hydrophobic and oleophobic properties. Therefore, defluorination is needed to modify the fluorocarbon resin surface. We have reported a photochemical defluorination method using ultraviolet light. In this method, an ArF excimer laser which has the photon energy of 6.4 eV(147 kcal) higher than the C-F bonding energy of 5.6 eV(128 kcal/mol) is used; therefore, the C-F bonds can be cut directly by the ArF laser photons. Then, boron atoms[5], aluminum atoms[7] or hydrogen atoms[11] are employed to prevent the recombination reaction of fluorine atoms with carbon atoms of the surface; these three atoms are possible to pull out the fluorine atoms. Reason why these atoms are employed for is that the B-F, the Al-F and the H-F bonding energies are higher than the C-F bonding energy. Especially, the boron atoms and the aluminum atoms have the bonding energy with fluorine atoms of 8.1 eV(184 kcal/mol) and 6.9 eV (159 kcal/mol), respectively; the B-F and the Al-F bonding energies are higher than the photon energy of ArF laser. Accordingly, B-F and Al-F bonds are not photo-dissociated under the irradiation of ArF laser, which means the defluorination reaction effectively takes place. On the contrary, the hydrogen atoms have the bonding energy with fluorine atoms of 5.9 eV(135 kcal/mol), which is lower than the photon energy of ArF laser. Fortunately, the absorption band of the HF is shorter than 161nm wavelength. As a result, the H-F bonds are not photo-dissociated by ArF laser light; the defluorination can be performed. In consequence, by using these atoms, the surface defluorination is effectively performed. In this paper, the boron atom which is the highest in efficiency for defluorination is employed.

Substitution of CH3 and OH groups

In general, epoxy resin is widely used as an adhesive agent. The resin is composed of C-H and O-H bonds. If the fluorine atoms of the surface are replaced by CH3 or OH groups, the fluorocarbon resin can be modified to an epoxy-compatible layer. Thus, chemical bonding of the surface with the epoxy is possible without a primer.

In order to modify to the epoxy-compatible layer, B(CH3)3 and B(OH)3 which are compounds of boron with CH3 or OH groups are used. B(CH3)3 and B(OH)3 are photo-decomposed by the ArF laser light as follows[12]:

$$B(CH3)3 + h\nu(193nm) \longrightarrow B + 3CH3$$

$$B(OH)3 + h\nu(193nm) \longrightarrow B + 3OH$$

As a result, the surface is defluorinated by the boron atoms which are photo-dissociated; the substitution reaction of the functional group such as CH3 or OH takes place[12].

EXPERIMENTAL METHOD

Schematic diagram of the experimental setup is displayed in Figure 1. (a) is the case of using the gaseous B(CH3)3.

The fluorocarbon resin film was set in the reaction chamber. After evacuation, gaseous $B(CH3)3$ was sealed in the chamber at the gas pressure of 40 Torr. Then, the ArF laser irradiated to the surface at the laser fluence of 0-50 mJ/cm2 and at the laser shot number of 0-10000.
(b) shows the case of using the $B(OH)3$ water solution. $B(OH)3$ solution was dropped on the surface of fluorocarbon resin; the solution dropped surface was covered on the fused silica window. And the ArF laser irradiated to the sample surface. In this setup, a liquid layer of about 50 μm was formed between the sample surface and the window by the capillary phenomenon. For this reason, the laser light irradiated effectively to the sample surface; production of bubbles in the solution were prevented.

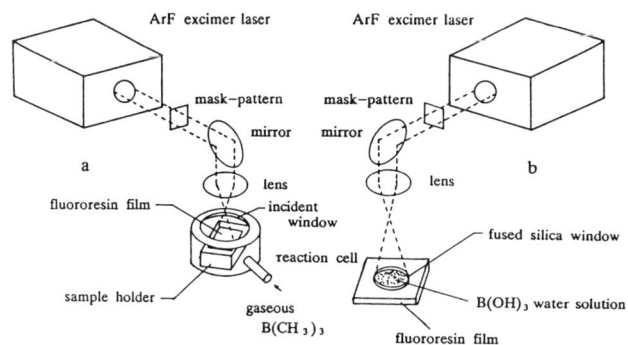

Fig.1 Schematic diagram of the experimental setup.
(a) is the case of using the gaseous $B(CH3)3$,
(b), the case of using the $B(OH)3$ water solution.

RESULTS AND DISCUSSION

Substitution of CH3 groups

Fluorocarbon resin in an atmosphere of $B(CH3)3$ gained the affinity for oil after being irradiated with ArF laser. In order to evaluate the oleophilic property of the modified surface, the contact angle with benzene was inspected[9] as a function of the laser fluence, as shown in Figure 2. The contact angle of the non-irradiated surface was about 57 degrees. With increasing the laser fluence, the modification density of the surface became high; the contact angle gradually became small. In case of 25 mJ/cm2 fluence, the angle became about zero degree. However, the contact angle became larger at the fluence of over 25 mJ/cm2 because the substituted CH3 groups were broken.

Therefore, the epoxy was applied to the modified sample, and a strip of metal was placed there; the shearing tensile test was examined. The strength as a function of the laser fluence is also exhibited in Figure 2. The shearing tensile strength became stronger with increasing the laser fluence. In case of 25 mJ/cm2 fluence, the strength was 130 kgf/cm2 of the epoxy break value. However, the strength became weak at the fluence of over 25 mJ/cm2. The strength dependence on the laser fluence precisely agreed with the results of contact angle measurement.

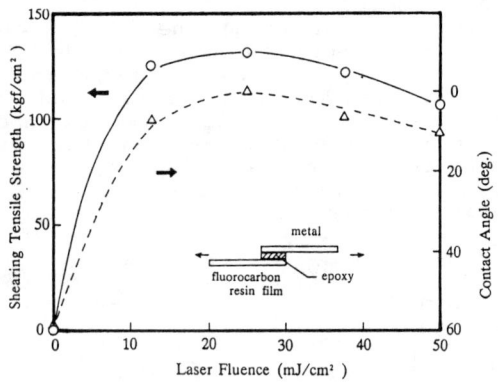

Fig.2 Shearing tensile strength and contact angle with benzene as a function of the laser fluence. The specimens were modified in the B(CH3)3 gas ambient of 40 Torr.

Then, the shearing tensile strength as a function of the laser shot number was investigated, as shown in Figure 3. The laser fluence was kept at 25 mJ/cm2. With increasing the laser shot number from 0 to 500 and 1000, the strength gradually became strong; the epoxy break value of 130 kgf/cm2 was obtained at the shot number of 3000. However, the strength became weak at the shot number of over 3000. The strength dependence on the laser shot number also corresponded to the results of contact angle measurement.

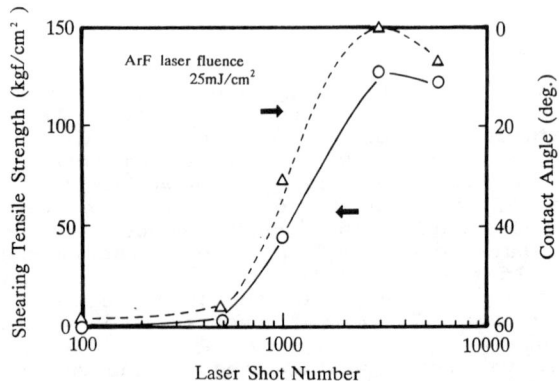

Fig.3 Shearing tensile strength and contact angle with benzene as a function of the laser shot number. The specimens were modified in the B(CH3)3 gas ambient of 40 Torr.

In order to investigate the relation between the contact angle and the shearing tensile strength, X-ray photoelectron spectroscopy (XPS) analyses of the modified surfaces were conducted. Figure 4 shows F 1s (690eV) spectra. The peak of the non-irradiated surface was due to CF2-CF2 fluorocarbon resin structure. With increasing the laser shot number from 0 to 500 and 1000, the fluorine of surface gradually decreased. In case of 3000 laser shot number, the F 1s peak almost disapeared. By these defluorinations, the substitution density of CH3 groups became high [9]. As a result, the chemical bond density of the CH3 groups with the epoxy becomes higher; therefore, the shearing tensile strength became high.

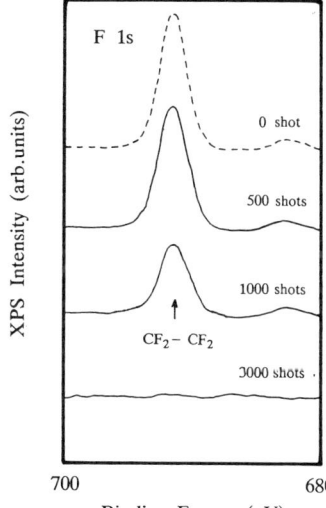

Fig.4 XPS spectra of the modified surfaces at four different laser shot number.

Substitution of OH groups

B(OH)3 water solution is very stable compared with the case of using the gaseous B(CH)3. Moreover, the fluorocarbon resin can be modified in air atmosphere by using the B(OH)3 water solution. Therefore, the sample surface was irradiated with the ArF laser in the way as illustrated in Figure 1 (b). Thus, the irradiated surface remarkably became strong in the affinity for water. Figure 5 displays the contact angle with water as a function of the laser shot number. The laser fluence was kept at 25 mJ/cm2. The contact angle of the non-irradiated surface was about 114 degrees. When the concentration of B(OH)3 water solution was zero, that is H2O only, the ArF laser induced the photo-decomposition as follows:

$$H_2O + h\nu(193nm) \longrightarrow H + OH$$

As a result, the surface was defluorinated with the hydrogen atoms. Since the hydrogen atoms were difficult to defluorinate compared with the case of using the boron atoms, the contact angle of less than 30 degrees was not obtained by irradiating the shot number of 6000. In case of 0.3 Normal concentration, the contact angle of about zero degree was obtained at the shot number of 4000. Then, the concentration of B(OH)3 solution was changed from 0.3 to 2 Normal. As the absorption of B(OH)3 solution became strong, the laser shot number of 10000 was needed to obtain the contact angle of about zero degree.

Thus, the shearing tensile stength of the modified sample at the shot number of 4000 is examined, as exhibited in Figure 6. The shearing tensile strength of the non-irradiated sample was less than 0.2 kgf/cm2[12]. After modification, the strengthbecame over 550 times as strong as before and was 110 f/cm2.

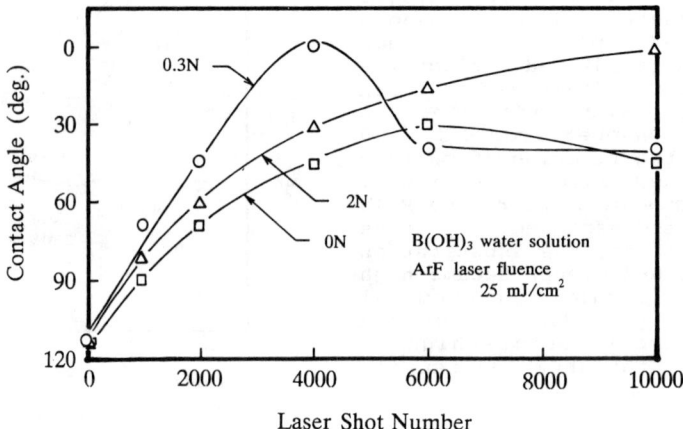

Fig.5 Contact angle with water as a function of the laser shot number at three different concentrations of B(OH)3 water solution.

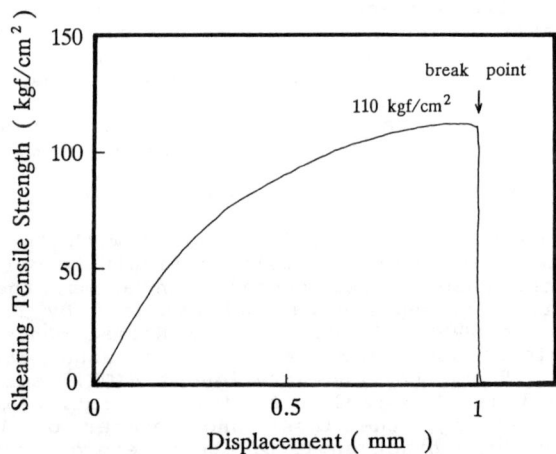

Fig.6 Break-point chart of the shearing tensile test. The specimens were modified in the B(OH)3 water solution of 0.3 N concentration.

CONCLUSION

   A fluorocarbon resin surface was photochemically modified by using the ArF excimer laser light and gaseous B(CH3)3 or B(OH)3 water solution. Chemical bonding of the surface with epoxy was carried out. The adhesive strength corresponded to the results of the contact angle measurement. And the epoxy break value was achieved at the optimum laser fluence and laser shot number. Moreover, the surface morphology of the sample was observed smooth and similar to the non-irradiated surface, which was by the scanning electron microscope (SEM). Thus, chemical bonding of the fluorocarbon resin and the epoxy agent was demonstrated. As a result, the photomodification method can now be applied to the development of a high electric insulation material or high strength structural materials.

REFERENCES

1. J.R.Hall, C.A.L.Westerdahl, et al., J.Appl.Polym.Sci., 16, 1465 (1972)
2. R.H.Hasen and H.Schonhorn, J.Polym.Sci., B-4, 203 (1966)
3. E.R.Nelson, T.J.Kilduff and A.A.Benderly, Ind.Eng.Chem., 50, 329 (1958)
4. L.Soukup, Int.Polym.Sci.Technol., 5, No.9, 19 (1978)
5. M.Okoshi, N.Sugahara, T.Shigeeda, M.Murahara and K.Toyoda, Digest of Technical Papers, Ninth Annual Meeting of the Laser Society of Japan, 173 (1989)
6. M.Okoshi, M.Murahara and K.Toyoda, Mat.Res.Soc.Symp.Proc., Vol.201, 451 (1991)
7. M.Okoshi, H.Kashiura, T.Miyokawa, K.Toyoda and M.Murahara, Mat.Res.Soc.Symp.Proc., Vol.279, 737 (1993)
8. M.Okoshi, M.Murahara and K.Toyoda, Mat.Res.Soc.Symp.Proc., Vol.158, 33 (1990)
9. M.Okoshi, M.Murahara and K.Toyoda, J.Mater.Res., Vol.7, No.7, 1912 (1992)
10. M.Okoshi, M.Murahara and K.Toyoda, Mat.Res.Soc.Symp.Proc., Vol.236, 377 (1992)
11. M.Okoshi, Y.Hayashi, T.Miyokawa, K.Toyoda and M.Murahara, Extended Abstract (The 54th Fall Meeting, 1993) The Japan Society of Applied Physics, No.3, 953 (1993)
12. M.Murahara and M.Okoshi, J.Poly.Sci.Technol., Vol.6, No.3, 379(1993)

THE EFFECTS OF DEPOSITION PARAMETERS ON THE PROPERTIES OF SiO$_2$ FILMS DEPOSITED BY MICROWAVE ECR PLASMAS

T. T. Chau, P. M. Lam, and K. C. Kao, *Electrical and Computer Engineering, University of Manitoba, 15 Gillson Street, Winnipeg, Manitoba, Canada, R3T 5V6.*

Abstract

Electronic and physical properties of SiO$_2$ films deposited by microwave ECR plasmas of the mixtures of SiH$_4$ and N$_2$O have been measured as functions of the pressure and the gas-flow ratio of N$_2$O to SiH$_4$ gases in the processing chamber. Experimental results show that the film deposition rate increases with increasing SiH$_4$ concentration, that is, with decreasing gas-flow ratio. The films deposited at N$_2$O/SiH$_4$ gas-flow ratios smaller than 10 tend to have a refractive index higher than the thermally grown oxide. However, for the N$_2$O/SiH$_4$ gas-flow ratios between 10 and 20, the films have the refractive index close to that of thermally grown oxide, which is about 1.45-1.46. The film deposition rate increases linearly with increasing pressure. In general, the films deposited at high pressures (>100 mTorr) have a higher refractive index as compared with the thermally grown oxide; also films deposited at high pressures have more electron traps. Good quality SiO$_2$ films can be deposited at pressures with the range of 20-50 mTorr and the N$_2$O/SiH$_4$ gas-flow ratio of 10.

Introduction

Plasma enhanced chemical vapor deposition (PECVD) is an important fabrication process for microelectronics. It has been used for the deposition of metal and insulating films for many years[1,2]. Its advantages are: a) it operates at low pressures so that the contamination of foreign particles can be easily controlled and b) it operates at low temperatures, low thermal budget, which is essential for sub-micron device fabrication[3]. This is why silicon dioxide (SiO$_2$) films fabricated by PECVD have attracted the interest of microelectronic industry.

Many investigators[4-6] have reported that for SiO$_2$ films fabricated by PECVD the deposition rate has to be low in order to achieve high quality SiO$_2$ films. In practice, the low deposition rate may be acceptable for the gate oxide fabrication because the thickness of the films can be easily controlled precisely. However, for other applications such as SiO$_2$ films used as the inter-layers for metal routing, thick films (>1000 Å) are required. In this case a high deposition rate is necessary. In this paper we present our experimental study on the effects of deposition parameters on the quality of the SiO_2 films deposited by PECVD, and discuss the trade-off between the quality of the films and the deposition rate.

Experimental

The microwave electron cyclotron resonance (ECR) plasma system used in the present investigation has been reported elsewhere[7]. The substrates were n-type, 1-2 Ωcm, <100> oriented silicon wafers. All of the substrates were cleaned using RCA method, and then dipped in $HF:H_2O$ (1:100) solution for 2 min. The deposition temperature was 300°C, and the gases used were N_2O and 5 % of SiH_4 in Ar (for simplicity the latter gas is referred to as SiH_4 from here on). SiO_2 films were deposited as functions of: a) gas-flow ratio of $N_2O:SiH_4$ (apparent ratio with the understanding that SiH_4 is in fact the mixture of SiH_4 and Ar) between 1 and 20 at a constant pressure of 27 mTorr and b) the pressure of the mixed gas between 10 mTorr and 220 mTorr at a constant gas-flow ratio of 10. After the SiO_2 films were deposited, each substrate was cut into two portions. One portion of the substrate was used for the thickness measurement by ellipsometry and kept for later use. The remaining portion was immediately loaded into another vacuum system for metalization. Aluminum was deposited on the top of the SiO_2 through a shadow mask to form the gate electrodes, the size of each electrode being 5×10^{-3} cm^2 in area. After this, aluminum was also deposited on the silicon side to form the back electrode. The capacitance-voltage (C-V) characteristics were measured before and after the MOS devices being subjected to thermal annealing at 400 °C in forming gas (10% of H_2 in N_2) for 30 min.

Results and Discussion

Figure 1(a) shows the deposition rate and the refractive index of the SiO_2 films as functions of the gas-flow ratio. The deposition rate is about 65 Å/min at the gas-flow ratio of 1. The films deposited at gas-flow ratio of 1 have a refractive index of 1.477 indicating that the films are silicon rich. However, as the N_2O gas-flow ratio increases, the deposition rate tends to decrease and so does the refractive index. But when the gas-flow ratio reaches 10, the refractive index and the deposition rate tend to be saturated if the gas-flow ratio is further increased. The films deposited at the gas-flow ratio of 10 have a refractive index of 1.46, which is close to the refractive index of the thermally grown oxide. Figure 1(b) shows the deposition rate and the refractive index of the deposited films as functions of the gas pressure. The deposition rate increases linearly with the gas pressure. Thus much higher deposition rate can be achieved. However, a high gas pressure also results in a high

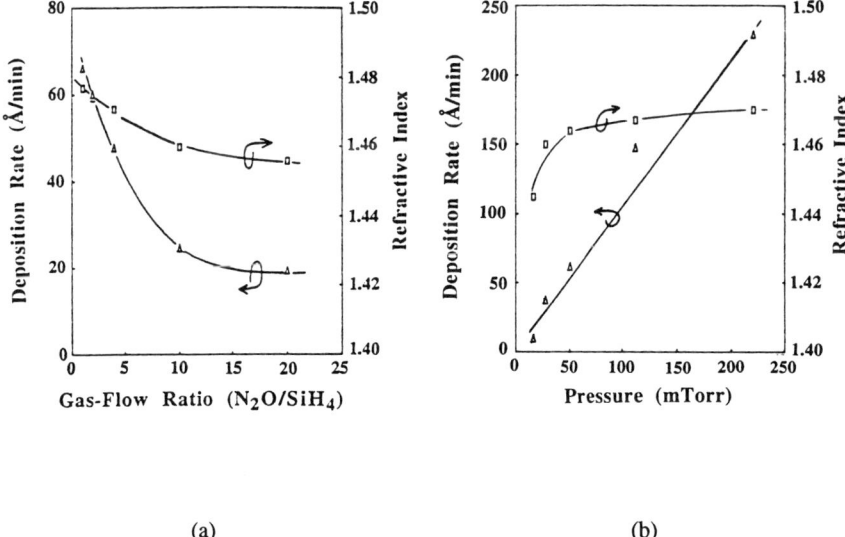

(a) (b)

Figure 1: (a) Deposition rate and refractive index of SiO_2 films as functions of gas-flow ratio at constant pressure of 27 mTorr and (b) deposition rate and refractive index of SiO_2 films as functions gas pressure at constant gas-flow ratio of 10.

refractive index. Figure 2(a) shows the breakdown strength as a function of the gas-flow ratio. The breakdown strength increases with increasing gas-flow ratio for the gas-flow ratios less than 10. But for gas-flow ratios between 10 and 20, the breakdown strength becomes saturated. For the pressure dependence, the breakdown voltage decreases with increasing pressure in the range between 10 and 50 mTorr as shown in figure 2(b). However, the final breakdown strength is around 8-9 MV/cm for films deposited at gas pressures larger than 100 mTorr. This can be explained as due to the increase in electron trapping in the high field region, which is caused by the non-stoichiometry of SiO_2 or silicon rich films[8]. The presence of trapped electrons due to high field injection creates an internal field that opposes to the applied field as indicated by the ledge in the I-V characteristics. Figure 3(a) shows that the oxide fixed charge and the interface state density (D_{it}) as functions of gas-flow ratio. Films deposited at a low gas-flow ratio have a higher oxide fixed charge than those deposited at a high gas-flow ratio. An increasing in oxide

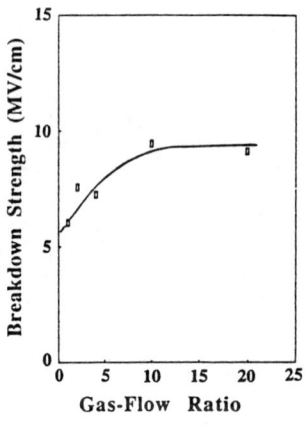 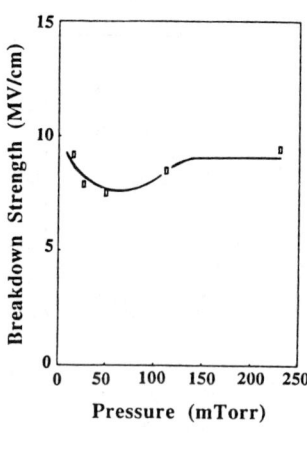

(a)   (b)

Figure 2: (a) Breakdown strength of SiO_2 films as a function of gas-flow ratio at a constant pressure of 27 mTorr and (b) breakdown strength of SiO_2 films as a function of gas pressure at a constant gas-flow ratio of 10.

fixed charge is also observed with increasing gas pressure as shown in figure 3(b). For the PECVD films fabricated at low substrate temperatures, the adsorbed gas molecules on the substrate surface have a low mobility and hence they need a certain time to move around to reach a stable state. When the deposition rate is high, the adsorbed gas molecules could not move around to search for the minimum energy state. This incomplete chemical reaction may lead to creating more defects or silicon rich films. Figure 3(a) and 3(b) show that after PMA (post metalization annealing at 400 °C) process, the value of D_{it} can reach as low as 10^{-11} eV^{-1} cm^{-2} for films deposited at gas-flow ratios between 10 and 20. The value of D_{it} follows the same trend of the oxide fixed charge. However, the value of D_{it} is not dependent on pressure. This may suggest that the value of D_{it} may not depend on the growth process. This may be due to the fact that our films were fabricated using our new system[7], and therefore they were not subjected to either ionic particle or photon bombardment, and that prior to the deposition process a thin native oxide film was already present on the silicon surface. Of course, more experiments are needed to verify this speculation. It is also worth to mention that before the PMA process, the D_{it} level is quite

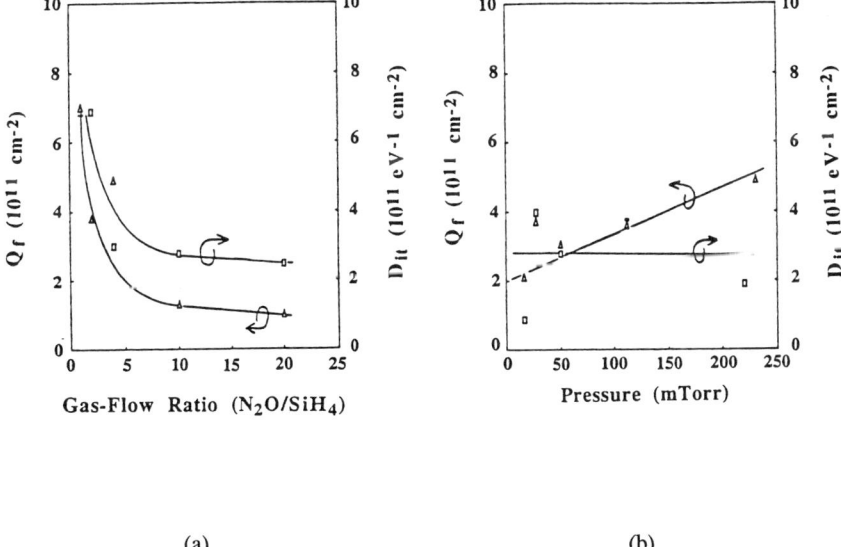

Figure 3: (a) Oxide fixed charge (Q_f) and interface state density (D_{it}) as functions of gas-flow ratio at constant pressure of 27 mTorr and (b) oxide fixed charge (Q_f) and interface state density (D_{it}) as functions of gas pressure at constant gas-flow ratio of 10.

high (>10^{13} ev^{-1} cm^{-2}). However, this high value of D_{it} can be annealed out by the PMA process. The high value of D_{it} before annealing may be due to the difference in the formation of an interface transition layer between Si and SiO_2 film. For PECVD films the transition layer is grown at low temperatures; thus this layer may be similar to the native oxide that had already been present prior to the deposition process. This may be why the quality may not be as good as the transition layer of the thermally grown oxide before annealing. Nevertheless, the treatment of the deposited SiO_2 in the H_2 environment at 400 °C can anneal out the excess D_{it} as we have mentioned. This indicates that the role of the H_2 in the annealing of PECVD films is as important as in the thermally grown oxide.

Conclusion

The deposition rate can be increased by either increasing the SiH_4 concentration or by increasing the gas pressure. The increase in deposition rate is more sensible to the increase in gas pressure than to the increase in gas-flow ratio. However, the trade off is that a high deposition rate is always accompanied with a high electron trap concentration and less stoichiometric for SiO_2 films. Good quality SiO_2 films can be deposited at a relative high deposition rate by keeping the gas-flow ratio of 10 and the gas pressure within 20 - 50 mTorr.

Acknowledgment

We thank the Natural Sciences and Engineering Research Council (NSERC) of Canada for supporting this research.

References

1. S. V. Nguyen, J. Vac. Sci. Technol. **B4**, 1159(1986).
2. J. L. Vossen and W. Kern, Physic Today, may 1990.
3. R. B. Fair and G. A. Ruggles, Solid State Technol., **33**, 31(1990).
4. J. Batey, E. Tierney, and T. N. Nguyen, IEEE Electron Device Lett., **EDL-8**, 148 (1987)
5. G. G. Foutain, R. A. Rudder, S. V. Hattangady, R. J. Markunas, and P. S. Lindorme, J. Appl. Phys., **63**, 4744(1988).
6. T. T. Chau, S. R. Mejia, and K. C. Kao, J. Vac. Sci. Technol., **B9**, 50(1991).
7. T. T. Chau, S. R. Mejia, and K. C. Kao, J. Vac. Sci. Technol., **B10**, 2170(1992).
8. D. J. Maria, R. Ghez, and D. Wong, J. Appl. Phys., **51**, 4830(1980).

FLOW TUBE KINETICS OF GAS-PHASE CVD REACTIONS

BRUCE H. WEILLER
The Aerospace Corporation, Mechanics and Materials Technology Center,
PO Box 92957/M5-753, Los Angeles, CA 90009-2957

ABSTRACT

This paper explores the use of a flow-tube reactor coupled to an FTIR spectrometer to study gas-phase chemical reactions in CVD systems. We show that our apparatus can generate reliable kinetics data by reproducing the literature rate constant for the reaction between O_3 and isobutene. We present data from this apparatus on two technologically important systems: TiN from $Ti(NMe_2)_4$ (TDMAT) and NH_3 and SiO_2 from tetraethoxysilane (TEOS) and O_3. The results presented include kinetics data for the reaction of $Ti(NMe_2)_4$ with NH_3 and ND_3 at room temperature and the IR spectra of the products from the reaction of TEOS with O_3 at 175 °C.

INTRODUCTION

Gas-phase chemical reactions often occur in CVD systems but relatively little effort has been made to understand them. These reactions often determine what species reach the surface of the growing film and can affect the growth rates, purity, morphology and step coverage. Therefore, in order to control the properties of the resultant film, it is important to understand this chemistry. Information on the identity of reactive intermediates and the rates of chemical reactions is needed in order to develop quantitative kinetics models of deposition processes.

We have developed a useful apparatus for this purpose, a flow tube reactor coupled to an FTIR spectrometer. The flow tube reactor is a relatively simple technique in which two reactive flow streams are mixed via sliding injector that provides control over the distance between where the reagents are mixed and the products detected. Under constant laminar flow conditions, the distance along the flow tube is simply related to the reaction time. This technique has proven itself to be an extremely useful kinetics tool over the years.[1] However, it remains largely unexplored for the study of CVD chemistry. Here we demonstrate the use of the flow-tube reactor to study two technologically important CVD processes: 1) SiO_2 from TEOS and O_3 and 2) TiN from $Ti(NMe_2)_4$ and NH_3.

EXPERIMENTAL

The experimental apparatus is shown in Figure 1. The flow-tube reactor is a 1-m long, 1.37" i.d. teflon-coated stainless-steel tube equipped with a sliding injector port that provides a variable distance from the focus of the IR beam. The injector is a 1/4" tube ending in a pyrex loop with many equally spaced holes for gas injection counter-current to the main flow for good mixing. The observation region and sliding injector are shown in Figure 2. The observation region is a standard cross (NW-40) equipped with purged windows, purged capacitance manometers, and a throttle valve controller to maintain constant pressure. A stainless-steel tubular insert with 0.75" diam. holes aids in separating the reactants and products from the KRS-5 windows. Both the flow tube and the observation region are wrapped with heat tape for temperature regulation when desired. The IR beam from the FTIR spectrometer (Nicolet 800) exits the spectrometer on the right-hand side and is focused in the middle of the observation region using a combination of flat and off-axis parabolic mirrors. A Nicolet detector module with focusing mirror is mounted on the other side of the flow tube. The focusing optics and the detector module are enclosed in plexiglass boxes and the entire beam path is purged with dehumidified, CO_2-free air. Mass flow meters measure the separate flows of buffer (Ar or He), bubbler, and purge gases, and the flow of a dilute mixture of NH_3 in buffer (6.0%). The flow meters are calibrated by monitoring the pressure rise in a standard volume as a function of time. The buffer gas and NH_3 flows are mixed and fed into the side arm of the flow tube while the $Ti(NMe_2)_4$ mixture flows into the sliding injector. A mechanical pump (Sergeant Welch 1397) was equipped with a liquid nitrogen trap for pumping. For the O_3 experiment, a mixture of He and O_2 flows into an ozone generator and then into the sliding injector of the flow tube. It was possible to generate a mixture of 1.4 % O_3 in O_2 and He. This was confirmed by IR absorption using the integrated absorbance of the 1042 cm^{-1} band and the accepted integrated band intensity.[2] Isobutene flowed into the side arm of the flow tube along with the buffer gas. The following chemicals were used as received from the following suppliers: Ti(NMe$_2$)$_4$ (Strem); Ar (Matheson, UHP grade); NH$_3$ (Matheson, electronic grade), He (Spectra Gases, UHP grade), O_2 (Airco), isobutene, (Aldrich). TEOS was obtained from Aldrich and was distilled prior to use. The spectrometer was operated at 8 cm^{-1} resolution and 256 scans were averaged. For kinetics measurements integrated intensities were used.

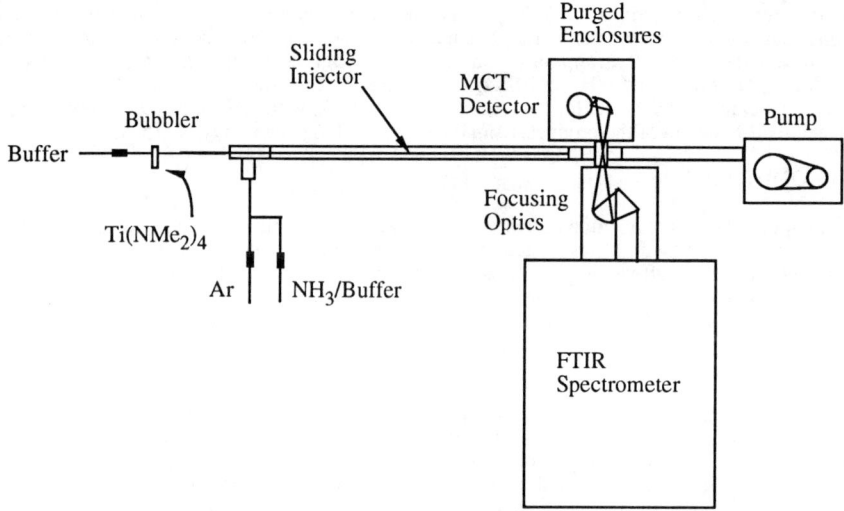

Figure 1. Experimental apparatus.

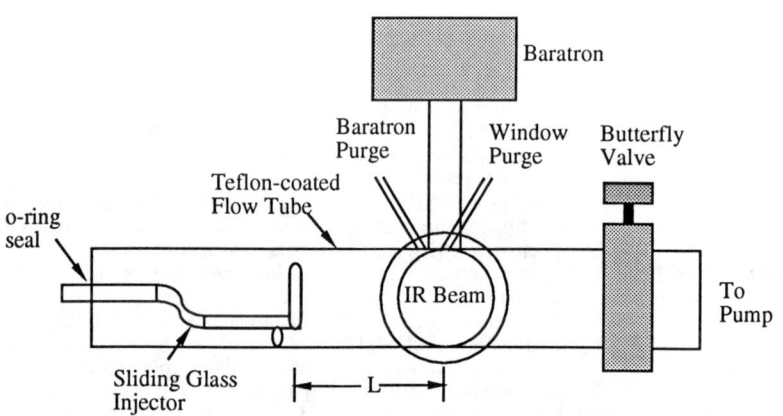

Figure 2. Diagram of the observation region and sliding injector.

KINETICS OF A KNOWN REACTION: O_3 + ISOBUTENE

In order to confirm that this flow tube reactor produces reliable kinetics data, we have measured the rate constant for the reaction of O_3 with isobutene. This reaction was selected for several reasons: 1) a reliable value is available since this reaction is important in atmospheric chemistry and has been measured numerous times over the years by several different techniques,[3] 2) the rate constant is close in magnitude to the value for $Ti(NMe_2)_4$ + NH_3, and 3) the IR absorption bands of isobutene are sufficiently removed from the O_3 band at 1042 cm^{-1} to prevent interference.

The disappearance of O_3 was measured as a function of reaction time as shown in Figure 3. The isobutene pressure ranged from 0.5 to 2.3 torr and is far in excess over the $[O_3]$. Therefore pseudo-first order conditions apply and we expect an exponential decay of O_3: $[O_3] = [O_3]_0 \exp(-k_{bi}[C_4H_8]t)$, where k_{bi} is the bimolecular rate constant. Furthermore, we expect a linear relationship between the logarithm of the IR absorbance and the reaction time: $\ln(A/A_0) = -k_{obs}t$, $k_{obs} = k_{bi}[C_4H_8]$. Here A is the integrated absorbance of the O_3 band, A_0 is the average integrated absorbance of the O_3 band before and after isobutene addition, and k_{obs} is the observed, decay constant. Figure 3 shows such a linear dependence for two isobutene pressures. The slopes of the lines in Figure 3 (k_{obs}) should show a linear dependence on isobutene partial pressure. Figure 4 shows this expected result and provides the bimolecular rate constant, $k_{bi} = (13.8 \pm 0.1) \times 10^{-16}$ cm^3(molec.-s)$^{-1}$ at 25 °C and 10 torr total pressure. We have also measured the rate constant at 5 torr total pressure and find no significant difference.

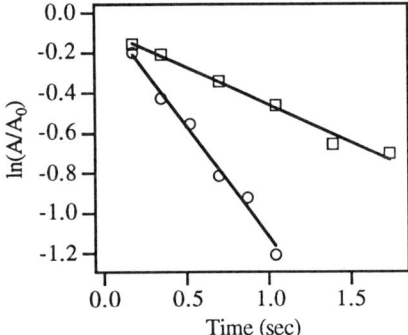 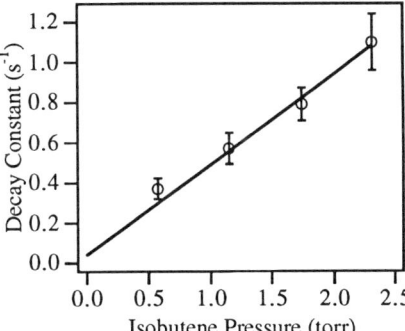

Figure 3. Plot of $\ln(A/A_0)$ vs. time for the reaction of O_3 with isobutene. The squares and circles are for 1.15 and 2.32 torr isobutene respectively.

Figure 4. Plot of the observed decay constants (k_{obs}) vs. isobutene pressure for the reaction of O_3 with isobutene.

The best literature value for this reaction is $(13.6 \pm 0.2) \times 10^{-16}$ cm^3(molec.-s)$^{-1}$ which is in excellent agreement with our result.[3] This is particularly significant since very different conditions and techniques were used, a static sample at a total pressure of 760 torr and ex-situ measurement of O_3. The fact that we are able to reproduce this rate constant is important. A serious concern in flow tube studies is the rate of mixing of the two reagents to produce a homogeneous mixture. For very fast reactions this is the limiting rate and can prevent measurement of the reaction rate constant. The lack of dependence on total pressure is also significant since the diffusional mixing time will be substantially shorter at the lower pressure. This observation coupled with the good agreement with the literature clearly demonstrates that mixing is not important under these conditions and validates the reliability of our kinetics data.

KINETICS OF LOW TEMPERATURE CVD OF TiN: Ti(NMe$_2$)$_4$ + NH$_3$

TiN is a material with a unique combination of properties including high hardness, high melting point, chemical inertness and good electrical conductivity. The potential applications of this material include wear resistant coatings, optical or thermal control coatings for spacecraft components that would be less susceptible to erosion by small particles and as a diffusion barrier for metallization layers in integrated circuits.

This last application is of much current interest and one of the best methods of depositing TiN diffusion barriers is the reaction of Ti(NMe$_2$)$_4$ with NH$_3$ since it proceeds at low temperatures and gives high quality films with good conformal coverage.[4] Other workers have shown that a gas-phase pre-reaction between Ti(NMe$_2$)$_4$ and NH$_3$ is required in order to deposit low-carbon films of TiN.[5] Quantitative kinetics data on this gas-phase reaction is required in order to optimize this process and design improved CVD reactors. We have presented preliminary results on the use of the flow-tube reactor to study this reaction.[6] Here we present refined data for this reaction.

When Ti(NMe$_2$)$_4$ is reacted with NH$_3$ in the flow tube reactor, the IR spectra show the formation of HNMe$_2$ concomitant with the removal of Ti(NMe$_2$)$_4$.[7] As the reaction time is increased from 0.7 to 5.5 s at a fixed NH$_3$ partial pressure (0.245 torr), we observe the disappearance of the Ti(NMe$_2$)$_4$ bands at 2779, 1254, 949 and 594 cm^{-1} and formation of bands assigned to HNMe$_2$ (1450, 1158, and 735 cm^{-1}).[8] This is expected for a transamination reaction and similar results were reported by Dubois et al. although, as those authors pointed out, their rates were limited by mixing.[5]

Figure 5 shows the disappearance of Ti(NMe$_2$)$_4$ as a function of time and NH$_3$ pressure. The integrated intensity of the NC$_2$ stretch (949 cm^{-1}) was used to determine the number density of Ti(NMe$_2$)$_4$. Using the available vapor pressure data, we estimate the partial pressure of Ti(NMe$_2$)$_4$ to be ~0.01 torr.[9] The NH$_3$ pressure ranged from 0.1 to 0.4 torr and therefore pseudo-first order conditions hold and we expect an exponential decay for [Ti(NMe$_2$)$_4$] (as discussed above for O$_3$ above). The data were fit to a straight line using a weighted least squares routine and the slopes of these lines give the observed rate constant (k_{obs}) at each NH$_3$ pressure. For this data He buffer gas was used but we have obtained similar data using Ar and find no significant difference. Since the diffusion rate is about three times slower in Ar, this result demonstrates that mixing is not rate limiting under these conditions.

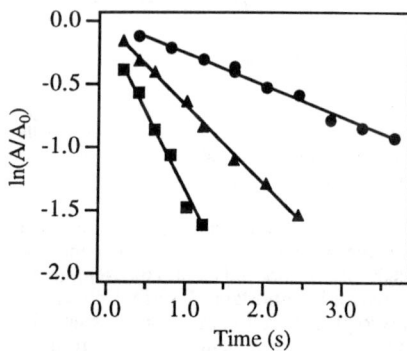 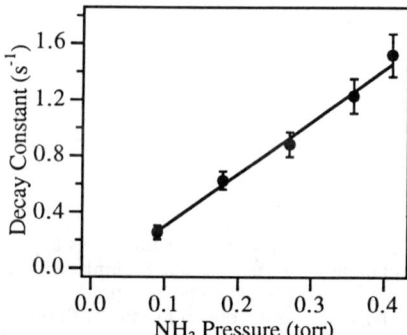

Figure 5. Plot of ln(A/A$_0$) vs. time for the reaction of Ti(NMe$_2$)$_4$ with NH$_3$. The NH$_3$ pressures are 0.090 (circles), 0.179 (triangles) and 0.357 torr (squares).

Figure 6. Plot of the observed decay constants (k_{obs}) vs. NH$_3$ pressure for the reaction of Ti(NMe$_2$)$_4$ with NH$_3$.

Figure 6 shows that a plot of the observed decay constants (k_{obs}) vs. NH_3 pressure is linear with a zero intercept as expected for pseudo-first order conditions. The slope of Figure 6 gives the bimolecular rate constant $k_{bi} = (1.1 \pm 0.1) \times 10^{-16}$ cm^3(molec.-s)$^{-1}$. To our knowledge, this is the first measurement of the rate constant for such a reaction. The rate constant is fast, especially considering that the temperature is only 26 °C and that both $Ti(NMe_2)_4$ and NH_3 are closed-shell species.

We have also obtained kinetics data for this reaction using ND_3 instead of NH_3. The rate constant is reduced by a factor of $k_h/k_d = 2.4 \pm 0.4$ from value for NH_3 indicating a primary isotope effect and that H-atom transfer is involved in the rate limiting step of this reaction.[7] Our observation of a kinetic isotope effect is consistent with labeling studies that showed the amine proton in the product $HNMe_2$ originates from NH_3.[10] This result is important for two reasons. It serves as a good confirmation that the disappearance of $Ti(NMe_2)_4$ is due to a chemical reaction with the added reagent and it also provides some mechanistic insight. One likely scenario for the mechanism of this reaction is the formation of a weakly bound adduct between NH_3 and $Ti(NMe_2)_4$ followed by rate limiting H-atom transfer to the dimethylamido moiety and elimination of amine. We have shown that a simple mechanism involving transamination and the production of $Ti(NH_2)_4$ can quantitatively simulate our data.[7] Work in progress includes direct spectroscopic identification of the intermediates from this reaction, temperature dependent kinetics measurements as well as related experiments with $Ti(NEt_2)_4$.

CVD OF SiO_2 : THE REACTION OF O_3 WITH TETRAETHOXYSILANE (TEOS)

The deposition of SiO_2 from TEOS and O_3 is an important process since it provides thin films of this important dielectric material at relatively low temperatures (< 400 °C) and with good conformal coverage for electronic applications.[11] In addition, SiO_2 for multilayer optical components could be deposited on temperature sensitive substrates with this approach. Because O_3 is a reactive, unstable molecule, the possibility of gas-phase reactions with TEOS is significant. Therefore, we have made a preliminary examination of the gas phase chemistry in this system with our flow tube reactor.

Figure 7 shows the IR spectra obtained when we react TEOS with O_3 in our flow tube reactor. Using a bubbler, TEOS was injected via the sliding injector and O_3 was mixed via the side arm. At room temperature the reaction rate is too slow to measure with this apparatus and therefore we heated the flow tube to 175 °C with heat tape. Even at this temperature the reaction is relatively slow and in order to observe reaction we raised the total pressure to increase the residence time and the partial pressures of the reactants. The top spectrum is taken at 2.0 s while the middle and bottom spectra are taken at 10.0 and 18.1 s respectively. We observe the decay of the TEOS bands at 2982, 1394, 1119, 962 and 794 cm^{-1} and the O_3 band at 1042 cm^{-1}.[3,8] We also see the formation of products at 3591, 2349, 1786, and 648 cm^{-1}. These bands can be assigned to CO_2, CH_3CHO, and possibly CH_3COOH.[8] These results are consistent with those of Kawahara et al.[12] who made similar measurements in a static cell and with recent results of Mucha.[13] Currently we are working on determining if the products result from the reactions of O atoms or O_3 and to what extent surface reactions may be involved.

ACKNOWLEDGMENTS

This work was supported by The Aerospace Sponsored Research Program. The laboratory assistance of B. V. Partido is gratefully acknowledged. We thank Professor R. G. Gordon and Dr. L. H. Dubois for helpful discussions and communication of results prior to publication. We also thank Dr. N. Cohen for helpful discussions regarding kinetics issues and Dr. R. L. Martin for the loan of the O_3 generator.

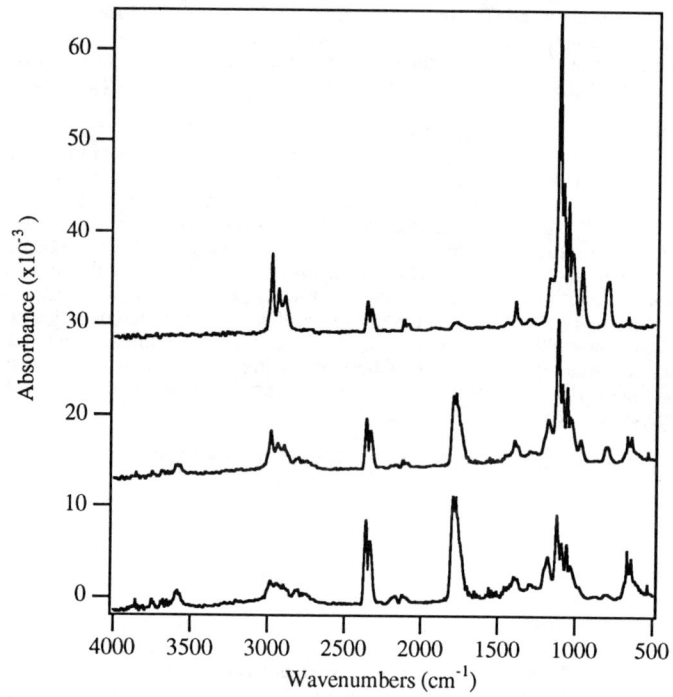

Figure 7. IR spectra for the reaction of TEOS with O_3 as a function of reaction time. The temperature is 175 °C and the reaction times are 2.0, 10.0 and 18.1 s from top to bottom.

REFERENCES

1. L. F. Keyser, J. Phys. Chem. **88,** 4750 - 4758 (1984) and references therein.
2. L. A. Pugh and K. N. Rao, "Intensities from Infrared Spectra" in Molecular Spectroscopy: Modern Research; Academic: New York, 1976, Vol. II, pp. 165f.
3. S. M. Japar, C. H. Wu, and H. Niki, J. Phys. Chem. **78,** 2318 (1974).
4. a) R. M. Fix, R. G. Gordon, and D. M. Hoffman, Mat. Res. Soc. Symp. Proc. **168,** 357 (1990), b) R. M. Fix, R. G. Gordon, and D. M. Hoffman, Chem. Mater. **3,** 1138 (1991).
5. L. H. Dubois, B. R. Zegarski, and G. S. Girolami J. Electrochem. Soc. **139,** 3603 (1992).
6. B. H. Weiller, MRS Symp. Proc. **282,** 605 (1993)
7. B. H. Weiller, MRS Symp. Proc., in press.
8. The Aldrich Library of FTIR Spectra, Edition I, Vol. 3, Vapor Phase; C. J. Pouchert, ed.; Aldrich Chemical Company, Inc.
9. The vapor pressure at 25 °C is 0.11 torr, D. Roberts, The Schumacher Corporation, personal communication.
10. J. A. Prybyla, C.-M. Chiang, and L. H. Dubois J. Electrochem. Soc. **140,** 2695 (1993).
11. K. Fujino, Y. Nishimoto, N. Tokumasu, and K. Maeda, J. Electrochem. Soc. **137,** 2883 (1990).
12. T. Kawahara, A. Yuuki and Y. Matsui, Jpn. J. Appl. Phys. **31,** 2925 (1992).
13. J. Mucha, personal communication.

HIGH RATE DEPOSITION OF SiO$_2$ BY THE REMOTE PECVD TECHNIQUE

ARICHIKA ISHIDA, MASATO HIRAMATSU and YOSHITO KAWAKYU
R&D Center, Toshiba Corporation, 1, Komukai Toshiba-cho, Saiwai-ku, Kawasaki 210, Japan.

ABSTRACT

The deposition of SiO$_2$ by remote PECVD using microwave activated oxygen species and SiH$_4$ were examined. The deposition rate has been found to be greatly enhanced by reducing the distance between the substrate surface and the end of the supplying nozzle for the plasma excited oxygen species.
The deposition rate decreased with increasing substrate temperature under the condition that the activated oxygen species and SiH$_4$ molecules collide with each other many times. Its activation energy was -0.27 eV.
It has been found that the gas phase reaction caused by the collision is important to obtain high quality films. A SiO$_2$ film with a resistivity larger than 1×10^{16} Ω·cm was able be obtained with high deposition rates beyond 200 nm/min, at low temperatures below 400°C. The minimum interface state density was 4.3×10^{10} cm^{-2}eV^{-1} for a film deposited at 230 nm/min, and 2.0×10^{10} cm^{-2}eV^{-1} for a film deposited at 30 nm/min.

INTRODUCTION

A low temperature deposition technique for high quality surface passivation films is strongly required for Si-ULSIs and thin film transistors (TFTs) used for active matrix liquid crystal displays (AMLCDs). In remote PECVD, plasma excitation and film deposition were spatially separated. Therefore, the ion bombardment effect, which occurs in conventional PECVD, is expected to be eliminated. Some groups have reported the deposition of insulator films by the remote PECVD technique using RF (13.56 MHz) plasma [1,2]. Fuyuki reported that high quality SiO$_2$ films could be obtained by the remote PECVD technique using activated oxygen species generated by a microwave (2.45 GHz) plasma and silane (SiH$_4$) as the source gas [3].
A high deposition rate is necessary to obtain a high throughput in order to apply this technique to the deposition of surface passivation films for TFTs used in AMLCDs. However, the deposition rates for SiO$_2$ films previously reported were below 10 nm/min, and were insufficient to apply this technique to the deposition of surface passivation films in actual devices.
In this study, the authors have discussed the deposition mechanism and the effect of the deposition conditions on film quality, varying the gas pressure, substrate temperature, and the distance between the substrate surface and the end of the supplying gas nozzle for the plasma excited oxygen species. The deposition rate was greatly enhanced by reducing the distance. It has been found that high quality SiO$_2$ films could be obtained at high deposition rates beyond 200 nm/min.

EXPERIMENTS

Figure 1 shows a schematic diagram of the remote PECVD apparatus. SiH_4 gas was introduced into a deposition chamber through a ring shaped nozzle made by quartz located at a distance of 30 mm from the substrate surface. Oxygen was excited in the plasma generated by a microwave (2.45 GHz, 100 W) in a quartz tube (12.7 mm diameter). The activated oxygen species were introduced into the deposition chamber through the quartz nozzle. The distance between the substrate surface and the end of the gas nozzle for the activated oxygen species was varied from 10 mm to 90 mm. The gas pressure was changed from 0.05 Torr to 0.5 Torr, by adjusting the conductance of the evacuation system. A typical O_2 gas flow rate was 10 cm^3/min, and the SiH_4 gas flow rate was 1-10 cm^3/min. The substrates were treated with the activated oxygen species prior to deposition in order to improve the interface property [4].

The film thickness was measured by a surface roughness profiler. The etching rate was examined with a buffered HF solution ($HF:NH_4F:H_2O=6:30:64$ at 23°C). The capacitance-voltage (C-V) characteristic was measured using $Al/SiO_2/Si$ MIS capacitors to analyze the SiO_2/Si interface property. The substrates for the C-V measurement were (100) oriented n-type Si wafers. MIS capacitors were annealed prior to the C-V measurement in a forming gas at 400°C, for 1 h. The film resistivities were also measured.

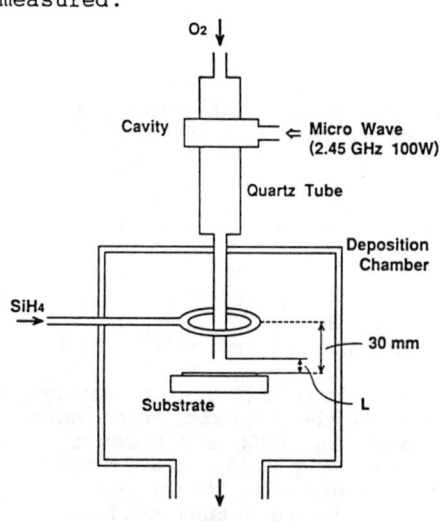

Fig.1 Remote PECVD apparatus

RESULTS and DISCUSSION

DEPOSITION RATE

Figure 2 shows the substrate temperature dependence of the deposition rate for depositions obtained by varying the gas pressure (P) and the distance between the substrate surface and the end of the supplying gas nozzle for the activated oxygen species (L). The flow rate of the O_2 gas and that of the SiH_4 gas were both 10 cm^3/min.

For the conditions of 10 mm L and 0.5 Torr P, the deposition rate decreased with increasing substrate temperature, and it saturated below 200°C. The deposition rate at 400°C was extremely high at 230 nm/min. The activation energy was -0.27 eV. The negative value of the activation energy shows that the deposition rate depends on the desorption rate of the reactive species from the substrate surface.

For 0.05 Torr P and 10 mm L, the deposition rate was independent of the substrate temperature. The mean free path is about 0.1 mm for 0.5 Torr P, and 1 mm for 0.05 Torr P. Therefore, the collision between the activated oxygen and SiH_4 molecules occurred only a few times for 0.05 Torr P. The amount of precursors generated by a gas phase reaction for 0.05 Torr P was much less than that for 0.5 Torr P, taking account of the number of collisions. The difference in the temperature dependence of the deposition rate seems to be caused by the difference in the amount of precursors.

On the other hand, for 90 mm L, the deposition rate decreased with increasing substrate temperature not only for the deposition for 0.5 Torr P but also for the deposition for the 0.05 Torr P. The activation energy was the same value as for the deposition for 10 mm L and 0.5 Torr P. The amount of precursors increased by elongating L due to the increase in the number of collisions between the activated oxygen species and the SiH_4 molecules. Consequently, even under 0.05 Torr P, it is considered that the deposition rate has the same temperature dependence as in the deposition under 0.5 Torr P.

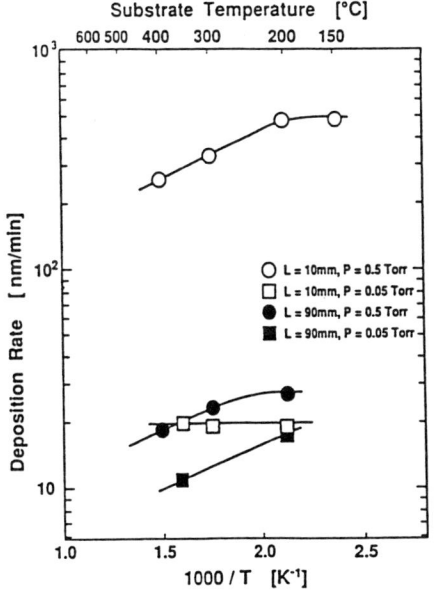

Fig.2
Deposition rate as a function of substrate temperature

FILM QUALITY

Figure 3 shows the etching rate of films with a buffered HF solution. The refractive indexes measured by elipsometry were in the range from 1.45 to 1.46. For all deposition conditions, the etching rate decreased as the substrate temperature increased. The etching rate for the film deposited under 10 mm L and 0.05 Torr P was much higher than that for other films. Except for the film deposited under 10 mm L and 0.05 Torr P, the etching rate tended to saturate beyond 300°C substrate temperature. The etching rate for films deposited beyond 300°C were about three times higher than that for thermal SiO_2 films.

Figure 4 shows the pressure dependence of the resistivity for films deposited under 10 mm L and 300°C substrate temperature. The applied electric field was 1 MV/cm. The resistivity increased with increasing gas pressure. A high resistivity beyond 10^{16} $\Omega\cdot$cm was obtained for the deposition at 0.5 Torr P. Although the deposition rate increased with increasing gas pressure, the film quality deposited at higher gas pressure was superior to that deposited at lower gas pressure.

These data indicate that precursors produced by the gas phase reaction played an important role in obtaining the high quality SiO_2 films.

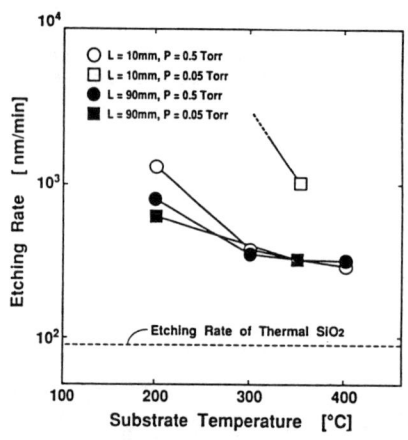
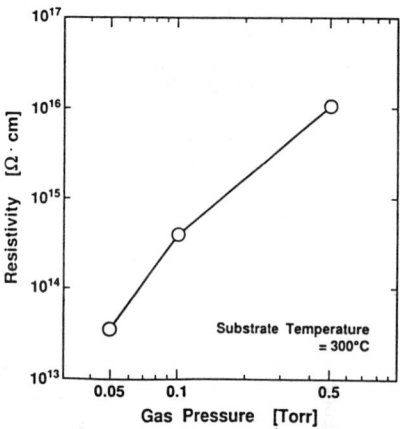

Fig.3
Etching rate of films with buffered HF solution
($HF:NH_4F:H_2O = 6:30:64$ at 23°C)

Fig.4
Resistivity versus gas pressure

C-V CHARACTERISTIC

Figure 5 shows the high frequency (1 MHz) C-V curve and quasi-static C-V curve for the SiO_2 film deposited under 10 mm L and 0.5 Torr P. The O_2 gas flow rate was 10 cm^3/min, and the SiH_4 gas flow rate was 10 cm^3/min. The substrate temperature was 400°C. The deposition rate was 230 nm/min under this condition.
The minimum interface state density derived by the Terman method was 4.3×10^{10} $cm^{-2}eV^{-1}$. The hysteresis width was below 0.05 V. It has been shown that excellent film can be obtain with a high deposition rate of 230 nm/min.
However, the fixed charge density derived from a flat band voltage shift was slightly high, $3.0 \times 10^{11} cm^{-2}$. It is speculated that Si rich layer might be formed by the activated oxygen prior to the deposition.
The interface property can be improved by reducing the SiH_4 gas flow rate. Figure 6 shows the C-V characteristic of the sample fabricated under SiH_4 gas flow rate of 1 cm^3/min. The deposition rate was 30 nm/min. The minimum interface state density was 2.0×10^{10} $cm^{-2}eV^{-1}$, comparable to thermal SiO_2 films.

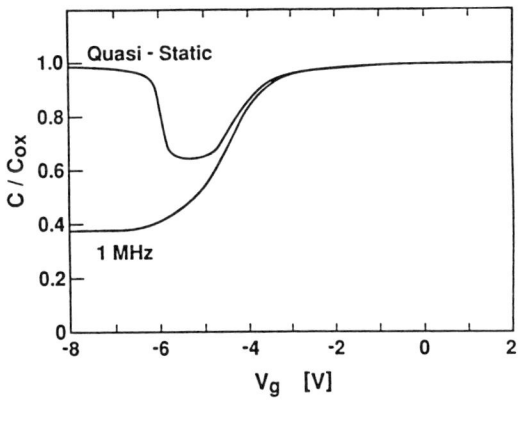

Fig.5
C-V characteristic for film deposited with deposition rate of 230 nm/min

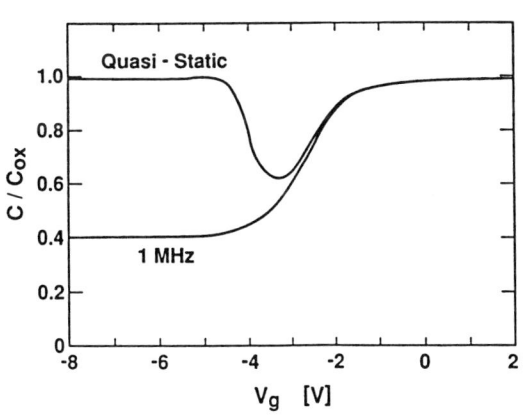

Fig.6
C-V characteristic for film deposited with deposition rate of 30 nm/min

CONCLUSIONS

The deposition of SiO_2 films by remote PECVD using microwave activated oxygen species and SiH_4 were examined, varying the gas pressure, substrate temperature, and the distance between the substrate surface and the end of the supplying nozzle for the plasma excited oxygen species. The deposition rate has been found to be greatly enhanced by reducing the distance.
The temperature dependencies of the deposition rate varied with the change in the gas pressure and the distance. The deposition rate decreased with increasing substrate temperature under the deposition condition that the activated oxygen species and the SiH_4 molecules collide with each other many times. Its activation energy was -0.27 eV. On the other hand, the deposition rate was constant under the condition that the number of collisions was small. The difference in the temperature dependencies seems to be caused by the difference in the amount of precursors generated by the gas phase reaction.
It has ben found that the gas phase reaction caused through the collision is important to obtain high quality films. Films with resistivities larger than 10^{16} $\Omega \cdot cm$ were able to be obtained with high deposition rate beyond 200 nm/min, and at low temperatures below 400°C. The minimum interface state density was 4.3×10^{10} $cm^{-2}eV^{-1}$ for the film deposited at 230 nm/min, and 2.0×10^{10} $cm^{-2}eV^{-1}$ for the film deposited at 30 nm/min.
The feasibility of the remote PECVD technique to be applied to the deposition of surface passivation films of practical devices has been confirmed.

REFERENCES

1. L.G.Meiners, J.Vac.Sci.Technol.,21,655(1982).

2. Lucovsky G.,Tsu D.V., Kim S.S.,Markunas R.J., and Fountain G.G., Appl.Surface Sci.,39,33(1989)

3. Fuyuki T.,Furukawa T., Oka T., and Mtsunami H., IEICE Trans.Electron.,E75-C,1013(1992)

4. Yasuda T., Ma Y.,Habermehl S., and Lucovsky G., Appl.Phys.Lett.,60,434(1992)

KrF EXCIMER LASER INDUCED PHOTOCHEMICAL MODIFICATION OF
POLYPHENYLENESULFIDE SURFACE INTO HYDROPHILIC PROPERTY

H.KASHIURA, M.OKOSHI and M.MURAHARA

Faculty of Engineering, Tokai University, 1117 Kitakaname,
Hiratsuka, Kanagawa 259-12, JAPAN

ABSTRACT

   Carboxyl groups were photochemically substituted for hydrogen atoms at a polyphenylenesulfide surface by using an KrF excimer laser light and vaporized formic acid. In the process, the polyphenylenesulfide film was placed in the formic acid vapor, and the surface was irradiated with the KrF excimer laser light. Irradiating with the laser, the surface was dehydrogenated by the hydrogen atoms which were photodissociated from the formic acid; the dangling bonds on the surface combined with the COOH radicals which were also photodissociated. The hydrophilic property of the photomodified surface was evaluated by the measurement of the contact angle with water. The dehydrogenation reaction and the substitution reaction of the COOH radicals were inspected by XPS analysis. Surface morphology of the sample was observed by SEM photograph.

INTRODUCTION

   Engineering plastics have been developed and widely used as structual materials. When extending the use of the material, the requirements for industrial use in various fields are difficult to satisfy by using a single type of engineering plastic. Thus, the polymer blending is carried out to improve the capacity of material; by this method, the properties of individual materials to be blended can not be maintained. Therefore, surface modification method is considered more useful and easier. The method is generally conducted by the plasma treatment and is essencially physical. Such a physical method, however, generally damages the material surface; the mechanical strength of material decreases. Consequently, we have proposed a photochemical modification method using an ArF excimer laser[1-4] at very low fluence.
   Of many polymers, polyphenylenesulfide (PPS), which has excellent chemical and thermal properties[5-7], has been employed as a structual material and expected for wider use[8,9]. However, adhesion characteristics of the surface are not sufficient to form composite materials with metals and to use for filtration devices. Thus, surface modification for adhesion is needed and is often carried out by the plasma treatment method. The method is physical and the resulting adhesion is not sufficient. To increase adhesion, the surface must be treated chemically. We propose the photochemical modification of the PPS surface by using KrF excimer laser

light and formic acid vapor; the surface was successfully modified to a hydrophilic state by substituting the COOH radicals for the hydrogen atoms of the PPS.

PHOTOCHEMICAL REACTION

Generally, a photochemical reaction requires two conditions as follows: first, the substrate must have an optical absorption at the wavelength of laser; second, the dissociation energy of molecule or atom must be lower than the photon energy of laser [1-4].
Both formic acid (HCOOH) and PPS used in this study have strong absorption at the 248nm wavelength of KrF laser as shown in Figure 1. And, the C-H bond of 80.6kcal/mol(3.5eV) is lower than the photon energy of KrF laser (equivalent to 114kcal/mol, 5.0 eV). Therefore, the above conditions are satisfied in order for the photochemical reaction to be effectively induced.

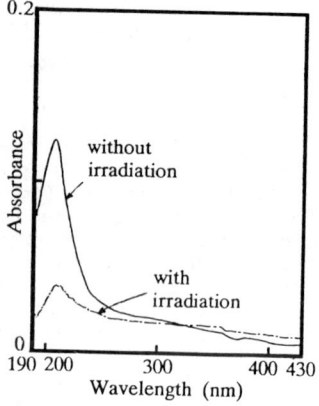

Figure 1. UV Spectra of HCOOH vapor

In order to modify the surface to a hydrophilic state, the surface must be dehydrogenated and be replaced with hydrophilic radicals. Figure 2 illustrates the principle of photochemical reaction in this method. The C-H bonds were cut by KrF laser photons; the hydrogen atoms of the surface were pulled out by the hydrogen atoms which are photodissociated from the HCOOH vapor. Thus, the surface was dehydrogenated to form H-H bonds. At the same time, the hydrophilic radicals combined with the dangling bond of carbon atoms on the surface. By these reactions, the PPS surface was photochemically modified to be hydrophilic only on the parts irradiated.

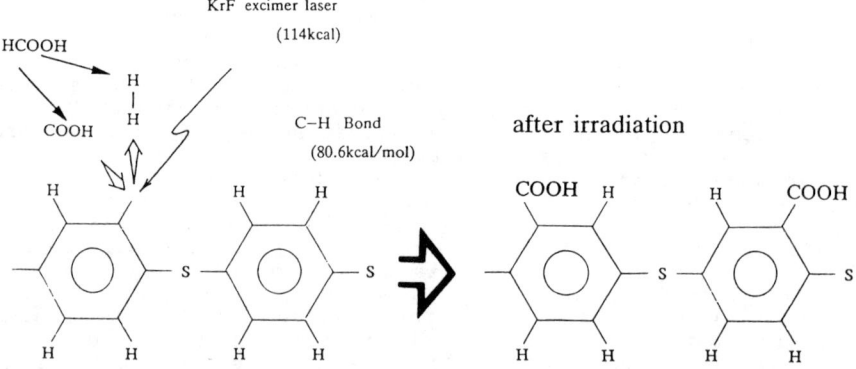

Figure 2. Principle of this photomodification method

EXPERIMENTAL PROCEDURE

Schematic diagram of the experimental setup is shown in Figure 3. A PPS film of 50μm thickness was put in a reaction chamber. After evacuating the chamber, the vaporized formic acid was sealed in the chamber at the gas pressure of 0-120Torr. And, the surface was irradiated with the KrF laser at the fluence of 0-200mJ/cm2 through a mask-pattern.

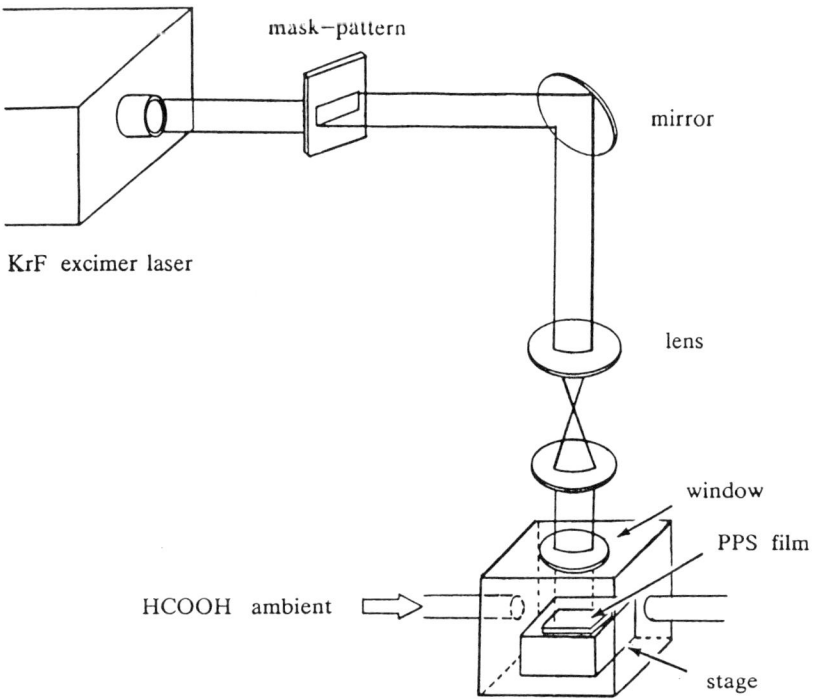

Figure 3. Schmatic diagram of the experimental setup

RESULTS AND DISCUSSION

Hydrophilic property

Generally, a measurement of contact angle with liquid is important to quantify the adhesive property of a surface. Stronger the hydrophilic property of surface becomes, smaller the contact angle becomes. Therefore, the hydrophilic property of a photomodified surface can be evaluated by measuring its contact angle with water. Figure 4 (a) shows the contact angle as a function of laser fluence. Non-treated PPS surface indicated about 70 degrees. In the range of 0-25mJ/cm2 fluence, the contact angle gradually became smaller with the increase of laser fluence. The contact angle was about 15 degrees at the 25mJ/cm2 fluence and stayed there in the range of 25-50mJ/cm2 fluence. However, the contact angle became larger at the laser fluence of 50mJ/cm2 and over due to the damege given to the substituted COOH radicals by the excess of laser irradiation. Figure 4 (b) indicates the dependence of contact angle on the gas pressure. Without HCOOH vapor, the contact angle of the surface did not change. With the increse of HCOOH gas pressure, the contact angle became lower; it was about 15 degrees at the pressure of 30 Torr. That is, the amount of hydrogen atoms increased when the amount of hydrogen atoms enlarged. And the dehydrogenation and substitution reactions easily took place. With the HCOOH gas pressure of 120 Torr, the contact angle exhibited little change. This resulted from the photon of KrF laser that could not reach to the surface because of the absorption of HCOOH vapor.

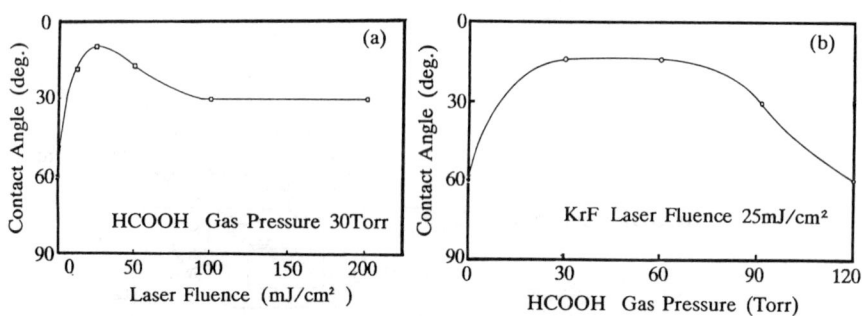

Figure.4 Dependence of the contact angle
(a) is on the laser fluence
(b) is on the HCOOH gas pressure

Substitution of COOH radicals

To investigate the incorporation of COOH radicals into the PPS surface, the carbon and oxygen bondings of the photomodified surface were analyzed by the X-ray photoelectron spectroscopy (XPS). Figure 5 (a) indicates the C1s XPS peak with and without laser irradiation. The C1s peak of the non-treated surface was at 285eV. The peak of the treated surface was shifted from 285eV to 287.6eV. Figure 5 (b) shows the O1s peak with and without irradiation. Without laser irradiation, the O1s peak was at 532eV. With laser irradiation, the peak increased in corresponding to the C1s peak shift. The curve-fitting analysis of the C1s peak was carried out, as indicated in Figure 6. The peak was separated to 286.1, 287.6, and 289.7eV. The first peak was identified as C-O bonding; the second, as C=O; the third, as COO. We propose that these peaks are derived from the substitution reaction of the hydrophilic radicals, reacted like Figure 2. Figure 7 displays the scanning electron microscope (SEM) photographs of the photomodified surface and the laser ablated surface. (a) shows the photomodified surface without ablation; its surface morphology was the same as the non-treated surface. (b) shows the surface ablated with the laser fluence of 400mJ/cm2. The modification was photochemically performed.

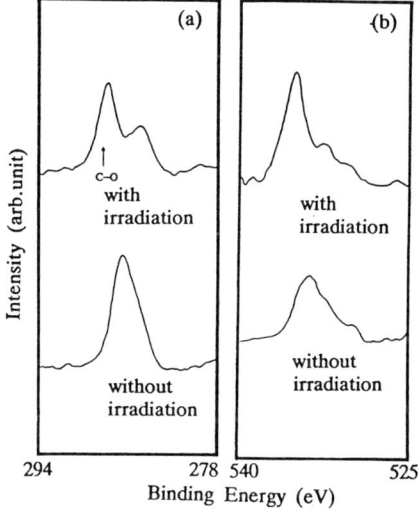

Figure.5 XPS spectra of the photomodified surface
(a) is C1s, (b) is O1s

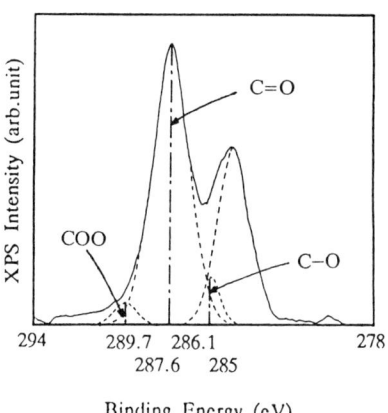

Figure.6 The results of the curve fitting analysis

CONCLUSION

A polyphenylenesulfide(PPS) surface is photochemicallymodified to be hydrophilic by using the formic acid vapor and KrF excimer laser light. The surface is dehydrogenated with the hydrogen atoms which are photodissociated from HCOOH, and the dangling bonds of the surface are occupied by the COOH radicals which were also produced by photodissociation. The contact angle with water of about 15 degrees is easily achieved with proper laser fluence and HCOOH gas pressure. From the results of XPS and SEM photographs, it is confirmed that this is an effective method for photochemical modification without damage on the surface. This photomodification method will be applied to other plastics which have C-H structures.

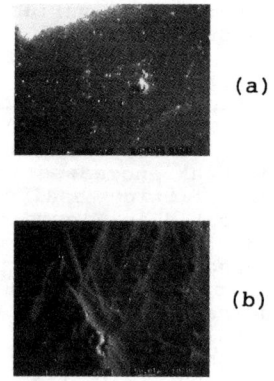

Figure.7 The SEM photograph of the surface
(a) is photomodified
(b) is ablated

REFERENCES

1) M.Okoshi, M.Murahara and K.Toyoda, Mat. Res. Soc. Symp. Proc., Vol.158, 33 (1990)

2) M.Okoshi, M.Murahara and K.Toyoda, Mat. Res .Soc. Symp. Proc., Vol.201, 451 (1991)

3) M.Okoshi, M.Murahara and K.Toyoda, Mat. Res. Soc. Symp. Proc., Vol.236, 377 (1992)

4) M.Okoshi, M.Murahara and K.Toyoda, J.Mater.Res., Vol.7, No.7,1912 (1992)

5) B.J.Tabor,et al, Eur.Polym., J., 7, 1127 (1971)

6) D.G.Brady, J.Appl.Polym.Sci., 20, 2541 (1976)

7) H.W.Hill, et al, Polym.Eng.Sci., 16 (12), 831 (1976)

8) Modern Plastics, 66[1], 19 (1989)

9) L.Broens, D.Bargeman and C.A.Smolders, Proc. 6th Int. Symp. Fresh Water from Sea, 3, 165 (1978)

PART V

Etching

NEW DRY-ETCH CHEMISTRIES FOR III-V SEMICONDUCTORS

S. J. PEARTON, U. K. CHAKRABARTI, F. REN, C. R. ABERNATHY, A. KATZ,
W. S. HOBSON AND C. CONSTANTINE*
AT&T Bell Laboratories, Murray Hill, New Jersey 07974
*Plasma Therm IP, St. Petersburg, FL

ABSTRACT

For some dry etching applications in III-V semiconductors, such as via hole formation in InP substrates, the currently used plasma chemistries have etch rates that are up to a factor of 30 too slow. We report on the development of 3 new classes of discharge chemistries, namely $Cl_2/CH_4/H_2/Ar$ at 150°C (yielding InP etch rates of >1 µm · min^{-1} at 1 mTorr and −80 V dc), HBr/H_2 for selective etching of InGaAs over AlInAs, and iodine-based plasmas (HI/H_2, CH_3I/H_2) that offer rapid anisotropic etching of all III-V materials at room temperature. In all cases, Electron Cyclotron Resonance sources (either multipolar or magnetic mirror) with additional rf biasing of the sample position are utilized to obtain low damage pattern transfer processes that generally use metal contacts on device structures as self-aligned etch masks. The temperature dependence of etch rates with these new chemistries display non-Arrhenius behavior in the range 50-250°C and a detailed study of the phenomenon are reported. Electrical, optical and chemical analysis of the etched surfaces show that it is possible to achieve essentially damage-free pattern transfer.

INTRODUCTION

Modern lightwave communication systems are based on transmission at 1.3 or 1.5 µm through optical fiber. The long wavelength lasers currently used for photonic signal generation are InP-InGaAsP solid state devices grown by epitaxial growth techniques like metal organic chemical vapor deposition (MOCVD). Processing of these structures involves several etching steps for grating or mesa formation.[1-7] Many of the common laser structures are relatively large in microelectronic terms, with stripe widths of more than 5 µm and stripe lengths of ≥ 250 µm. Wet chemical etching is generally employed for these structures because tolerances on active area dimensions are not stringent.[1]

Increasingly, however, there is a need for closer control of the pattern transfer fidelity, particularly in microphotonic arrays where the active elements may be only a few microns in diameter. Dry etching must be used for these structures, but a number of issues must be addressed to derive the full benefits of this technique. First, the choice of plasma chemistry determines the etch rate and surface stoichiometry of the InP-InGaAsP heterojunction. There is often a need epitaxially to regrow a current blocking layer on the etched surface,[1] and, therefore, the morphology must be smooth and defect-free. Second, there must be sufficient selectivity for etching the semiconductor relative to the masking material to achieve the required etch depth without loss of the mask. Finally, the ion sputtering component of the etching must be minimized in order to avoid introduction of non-radiative generation-recombination centers in the semiconductor materials, while still maintaining anisotropic pattern transfer.

The normal dry etching chemistries for III-V materials are based on chlorine-containing mixtures, and these work well for GaAs and related compounds. Etch rates of >1 μm $\cdot$ min^{-1} are easily obtainable for these compounds for applications such as via hole etching, while for most purposes much lower rates are desirable to improve control of the etch depth. Dry etching of InP and related compounds is more difficult because of the low volatiles of In chlorides at room temperature. The CH_4/H_2 chemistry provides slow smooth etching of all III-V materials,[8] but suffers from several drawbacks including the extensive polymer deposition that occurs on the mask and within the reactor chamber, the need for careful conditioning of the chamber and the slow rates (<500 Å $\cdot$ min^{-1}). In this paper we report on the development of 3 new dry etching chemistries with particular advantages for InP, namely

(i) $Cl_2/CH_4/H_2/Ar$, with the sample held at $\geq 150\,°C$

(ii) $HI/H_2/Ar$ or CH_3I/H_2

(iii) HBr/H_2

RESULTS AND DISCUSSION

Dry etching was performed in a PlasmaTherm SL720 shuttlelock system [10], employing a Wavemat Model 300 ECR source operating at 2.45 GHz of the type developed by Asmussen.[9] Gases were introduced into this source through electronic mass flow controllers at total flow rates of approximately 30 standard cubic centimeters per minute (sccm). The sample position was biased by application of 13.56 MHz radio-frequency power to produce dc biases of -100 to -250V. In this hybrid ECR-rf approach the plasma density and ion energies are controlled separately and, therefore, allow creation of high density ($\geq 5 \times 10^{11}$ cm^{-2}) plasmas with moderate ion energies at low pressure (1 mTorr). The flow rates of $Cl_2/CH_4/H_2/Ar$ were 10, 5, 17, 8 sccm respectively, while those of HI, HBr and CH_3I were all 10 sccm.

1. $Cl_2/CH_4/H_2/Ar$ Chemistry

Table 1 shows the relative volatilities of possible etch products for III-V materials. Note that $InCl_3$, which is the main etch product for removing In when using chlorine-based mixtures, is quite involatile. One can improve desorption of $InCl_3$ by increasing the sample temperature during etching.[10-13] At high pressures (>0.3 Torr) the apparent activation energy for etching InP is close to the latent heat of vaporization of $InCl_3$, even though the published vapor pressure would suggest this would not be a limiting factor.[14]

Table I. Volatilities of possible etch products for GaAs, AlGaAs and InP etched in HBr- or HI-based discharges. Chlorine related etch products are included for comparison.

Substance	Boiling Point (°C)	Melting Point (°C)	$\Delta H (kCal \cdot mole^{-1})$
$AlCl_3$	262	190	−139
$AlBr_3$	263	98	−102
AlI_3	360	191	−49.6
$GaCl_3$	201	78	−107
$GaBr_3$	278	121	−92
GaI_3	345	212	−34
$AsCl_3$	130	−8.5	−62.5
$AsBr_3$	221	33	−26
AsI_3	403	146	?
AsI_2	?	136	?
AsI_5	?	76	?
$InCl_3$	600	586	−128
$InCl_2$	550	235	?
$InCl$	608	225	?
$InBr$	662	220	−13.6
$InBr_2$	632	235	−67.4
InI	711	351	1.8
InI_2	?	212	?
InI_3	?	210	−28.8
PCl_3	75	−112	−68.6
PBr_5	173	−40	−33
PI_3	decomposes	61	?
PH_3	−88	−132	+1.3
AsH_3	−55	−116	+15.9

We have investigated use of elevated temperature Cl_2 ECR etching, and found that while the etch rates are rapid (up to several microns per minute), the etched surfaces are rough due to preferential loss of In above $\sim 130°C$, and the features are undercut.[15] The surface morphology can be drastically improved by adding H_2 to the discharge to promote the removal of P (or other group V elements) as PH_3 (or other hydrides). This leads to a stoichiometric surface, with equi-rate removal of both In and P. The undercutting can be suppressed by adding a small amount of CH_4, which causes a sidewall passivation reaction with polymer deposition on the sidewall.[13] Ar is also added to facilitate ignition of the discharge at low pressure and improve the thermal properties of the plasma.

Figure 1 shows etch rate vs. time plots for different plasma chemistries. Under our conditions (1 mTorr, $-100V$ dc, 200W microwave power) the etch rates of InP, InGaAs, InGaAsP and InGaP are $\sim 300 Å \cdot min^{-1}$, requiring very long plasma exposures for deep structures such as laser mesa etching (~ 4 μm deep) when using $CH_4/H_2/Ar$. Substitution of C_2H_6 for CH_4 produces slightly faster ($\sim 20\%$) etch rates. The fastest etch rates were obtained with the $Cl_2/CH_4/H_2/Ar$ chemistry.

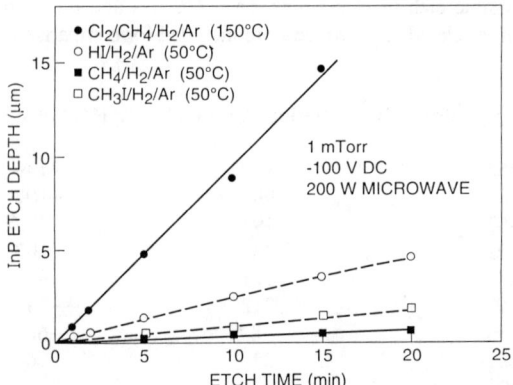

Fig. 1. InP etch depth versus time for four different discharge chemistries. The process pressure was 1 mTorr, the dc bias −100 V and the microwave power 200W in all cases.

Figure 2 shows the etch rate of InP in these mixtures as a function of CH_4 flow rate. As the CH_4 flow is increased, the semiconductor etch rate drops rapidly as expected because of the much slower rates achievable with CH_4/H_2. At low CH_4 additions it is possible to retain rates of ~1 $\mu m \cdot min^{-1}$ while having sufficient polymer deposition on the sidewall to maintain anisotropic profiles. We have used the $Cl_2/CH_4/H_2/Ar$ chemistry to produce laser mesas suitable for regrowth[16] and also through-wafer via holes for InP power device applications.[15]

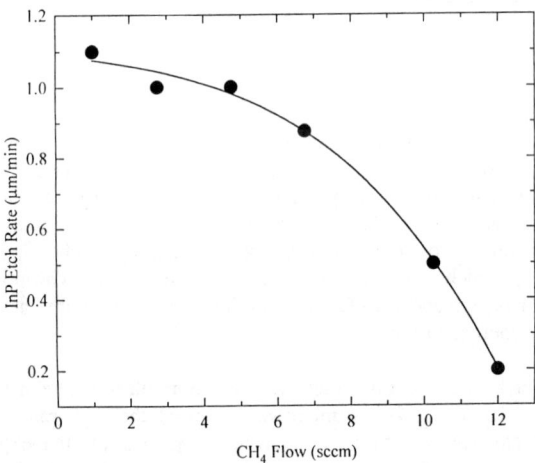

Fig. 2. InP etch rate as a function of CH_4 flow rate in $Cl_2/CH_4/H_2/Ar$ ECR discharges (2 mTorr pressure and −80 V dc).

2. Iodine (HI, CH₃I) Discharges

The iodine-related etch products are more volatile than those associated with CH_4/H_2, and therefore iodine-based chemistries provide faster etch rates with less polymer deposition.[17-19] Pure I_2 is a difficult material to handle and usually requires some form of heating to obtain a sufficient flow rate. By contrast, HI is a gas easily applicable to dry etching.

Figure 3 shows the etch rates of III-V materials as a function of de bias in $HI/H_2/Ar$ discharges. The etch rates for all these semiconductors increase monotonically with dc bias due to the increase in the sputtering component of the etching. The rates are 8-10 times faster than for $CH_4/H_2/Ar$ discharges under the same conditions in the same reactor. With $HI/H_2/Ar$ there is no polymer deposition and we did not observe any incubation time required for the onset of etching with any of the materials, indicating that the native oxides are rapidly removed in these, discharges.

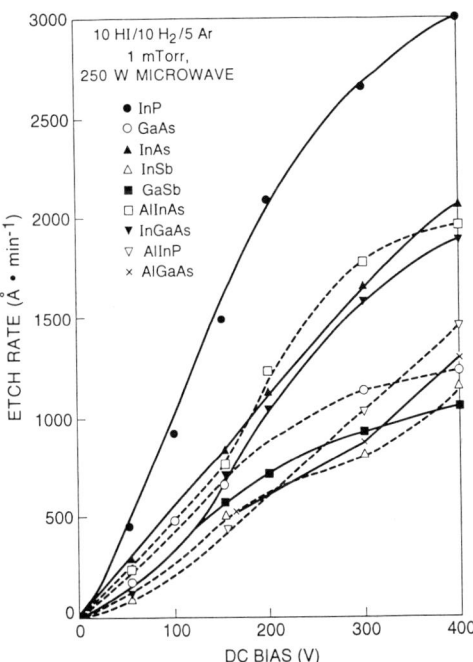

Fig. 3. Etch rates of III-V materials in 1 mTorr, 250W (microwave) $10HI/10H_2/5Ar$ discharges as a function of rf-induced dc bias on the samples.

Figure 4 shows removal rates of various mask materials during exposure to $HI/H_2/Ar$ discharges with different dc biases on the sample. The selectivity for etching InP over photoresist is ~100:1 at −100V dc bias, and higher values are obtained for dielectrics or metal masks. A compilation of data from the different discharge chemistries investigated is given in Table 2. High selectivities over resist and dielectrics are obtained in most cases, with the exception being use of resist with the $Cl_2/CH_4/H_2/Ar$ mixture. The need to hardbake the resist also leads to difficulties with its removal after the semiconductor etch is completed. For this reason, dielectric masks are preferred with this mixture.

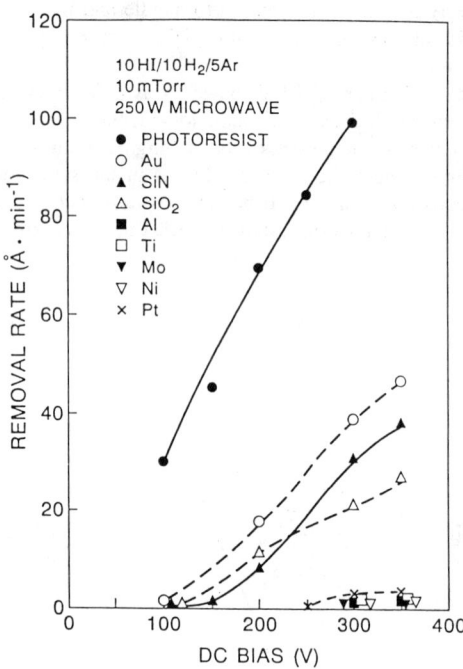

Fig. 4. Removal rate of various masking materials for InP in $HI/H_2/Ar$ discharges, as a function of dc bias.

Table 2. Selectivity for etching InP over various mask materials in different discharge chemistries. The plasma parameters were the same in each case — 1 mTorr pressure, −100V dc bias and 200W microwave power.

	Selectivity over InP		
Discharge Chemistry	Photoresist	SiO_2	Si_3N_4
$CH_4/H_2/Ar$	infinite[1]	80:1	60:1
$CH_3I/H_2/Ar$	infinite[1]	120:1	100:1
$HI/H_2/Ar$	100:1	>500:1	>500:1
$Cl_2/CH_4/H_2/Ar$	4:1[2]	10:1	12:1

[1] polymer deposition occurs on the resist with these mixtures.
[2] resist is hard-baked (150°C) prior to plasma exposure.

An Arrhenius plot of etch rates of InP, GaAs and AlGaAs is shown in Fig. 5. The plots are not described by a single activation energy, but the curves are similar for all three materials, with an initial rapid increase in etch rates between 50 and ~100°C, followed by less rapid increases thereafter.

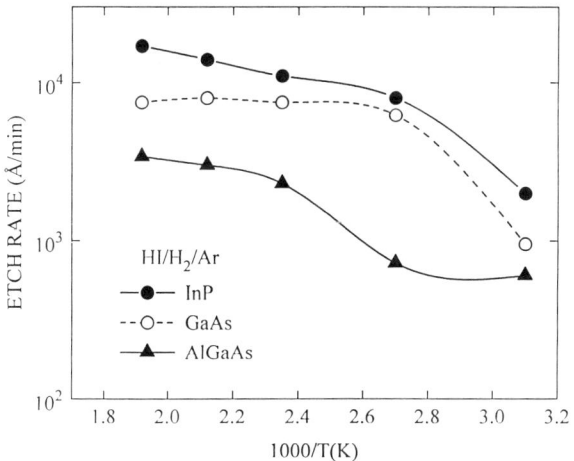

Fig. 5. Arrhenius plot of etch rates of InP, GaAs and AlGaAs in a 1 mTorr, 250W (microwave), −200V dc HI/H_2/Ar discharge.

Carrier density profiles in n-type GaAs samples were also measured as a function of the temperatures during HI/H_2/Ar dry etching. Figure 6 shows that a reduced carrier concentration in the surface region occurred for etch temperatures ≥ 100°C. There are two possible explanations — compensation of the Si donors in the material by deep acceptors caused by ion bombardment damage, or passivation of these donors by atomic hydrogen from the discharge. It is a common feature of ion-damaged semiconductors that the depths over which the effects of the lattice disorder are evident are much greater than the projected ranges of the most energetic ions in the plasma, and hence mechanisms such as recombination-enhanced defect diffusion or channelling are usually invoked to explain the discrepancy.[20–22] The other alternative, that of hydrogen passivation, is more amenable to direct measurement.[23] SIMS profiling of deuterium in a n-type GaAs sample after dry etching for 5 mins at 200°C in an HI/D_2/Ar discharge showed there is a substantial indiffusion of deuterium, with the concentration being above 10^{18} cm^{-3} for ~4000 Å, in good agreement with the electrical profile for this etch condition. The carrier profiles were restored to their initial state by annealing at 400°C for 5 min in a N_2 ambient — this is a signature of hydrogen passivation in GaAs, and implicates the hydrogen as being the cause of the carrier reduction in as-etched samples. Dopant passivation during plasma etching is also a problem when using other hydrogen-containing mixtures, such as CH_4/H_2.[24]

In summary for HI/H_2/Ar chemistries, Electron Cyclotron Resonance discharges at elevated sample temperatures were examined using InP, GaAs and AlGaAs. The etch rates of these materials increase rapidly with temperature, but in a non-Arrhenius fashion. InP surfaces show a smooth-to-rough transition at ~150°C in HBr/H_2/Ar discharges, with degraded anisotropy of etched features. Both GaAs and AlGaAs are more resistant to this surface roughening, but preferential loss of As occurs at elevated temperatures. Similar results are obtained with $HI/H_2/Ar$ discharges, with loss of the group V elements above ~100°C. The electrical properties of all three materials are affected by the etch treatments, with creation of a highly conducting surface layer on InP and dopant passivation occurring in GaAs (and AlGaAs). Use of elevated temperatures does not appear useful for dry etching of InP in either of the two chemistries investigated here because of the surface roughening that occurs, but for GaAs and AlGaAs the increased etch rates may be a useful trade-off against the much less obvious morphology changes.

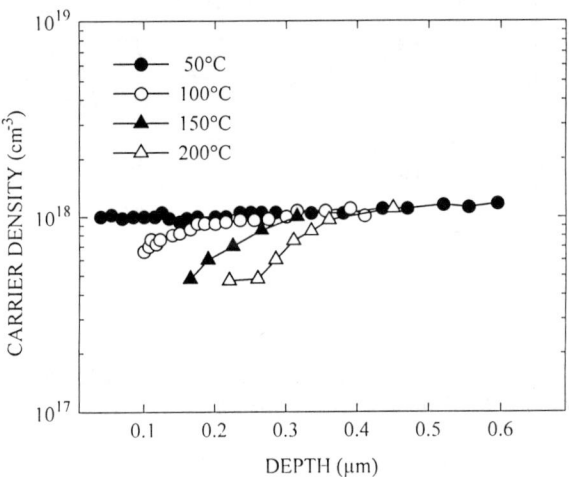

Fig. 6. Carrier profiles in n-type GaAs after dry etching in $HI/H_2/Ar$ discharges at different temperatures.

Since HI is very corrosive and will attack the gas delivery lines, mass flow controllers and the mechanical pumps, and has a shelf life of approximately six months before decomposing to hydrogen and iodine, there are clearly drawbacks to its use. Alternative chemistries involving iodine include methyl, ethyl- and propyl-iodide (CH_3I, C_2H_5I, C_3H_7I). These iodine-containing liquids are stable at room temperature but have sufficiently high vapor pressures that practical gas flows can be obtained even without heating the liquid source, and moreover they are less corrosive than HI.[25]

Figure 7 shows the etch rates of different III-V materials in methyl- or ethyl-iodide discharges as a function of dc bias. The etch rates are always faster for CH_3I than for C_2H_5I, and similarly the rates were slower still for C_3H_7I. These results can be understood if the presence of CH_X species reduce the availability of active iodine to the semiconductor surface, either by blocking surface sites, or by recombining with iodine in the gas phase. We found previously that addition of CH_4 to $HI/H_2/Ar$ discharges reduces the resulting etch rates. The fact that increasing dc bias leads to more rapid etch rates and that thresholds for the commencement of etching ($\geq -100V$) are observed are consistent with a mechanism of sputter-induced desorption of the etch products. The etch rates of InP, InAs, InSb, GaAs, AlGaAs and GaSb are ~50–100% faster with $CH_3I/H_2/Ar$ than with $CH_4/H_2/Ar$ at the same flow rates, pressures and microwave power for dc biases of ≥ -150 V. Under low bias conditions (≤ -150 V dc) the halocarbon mixtures do not offer any advantage over $CH_4/H_2/Ar$ and have significantly slower rates than $HI/H_2/Ar$. The fact that the etch rates increase rapidly with bias is similar to what we and others have observed with methyl chloride dry etching of InP and related compounds. The surface morphologies were smooth for all III-V materials over a wide range of discharge conditions with the halocarbon iodides, and this window was wider than for $CH_4/H_2/Ar$. The etched surfaces are also chemically clean, with no evidence for polymeric contamination. While the etch rates with CH_3I, C_2H_5I and C_3H_7I discharges are still considerably slower than those obtained with HI, the former offer advantages with respect to their stability and less corrosive nature.

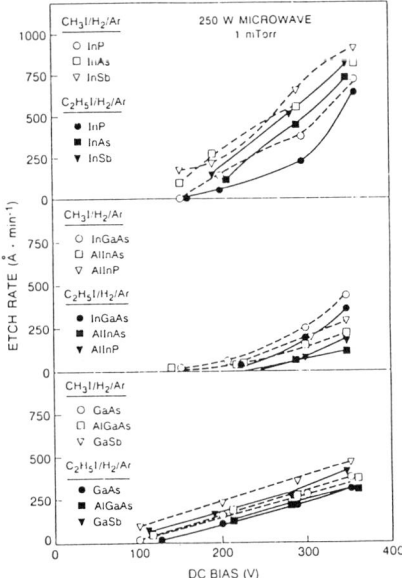

Fig. 7. Etch rates of various III-V semiconductors in 10 $CH_3I/10H_2/5$ Ar or $10C_2H_5/10H_2/5$ Ar, 1 mTorr, 250W (microwave) discharges as a function of the dc bias on the sample during the plasma exposure.

3. Bromine Chemistries

Little information is available on the etching characteristics of bromine-based discharges for III-V's, particularly under the ion-enhanced conditions necessary for the production of small feature sizes.[26-28] Ibbotson et al.[27] reported extremely fast etch rates for GaAs (60 μm · min^{-1}) in high pressure (0.3 Torr) Br_2 plasmas. Temperatures in excess of 200°C were required to achieve etching of InP. Takimoto et al.[28] fabricated etch-facet lasers using Br_2/N_2 and Br_2/Ar reactive ion etching to obtain rates of 2 μm · min^{-1} for power densities of 0.8 W · cm^{-2} at a pressure of ~4 mTorr. Also, HBr plasma etching has recently attracted attention for deep feature fabrication in Si technology.[29]

Figure 8 shows the etching rate of different III-V materials as a function of the dc bias at fixed pressure (1 mTorr) and microwave power for 15 HBr/5 H$_2$, 15 HBr/5 CH$_4$ or 15 HBr/5 Ar discharges. The etch rates are quite slow and have the same trend for each different semiconductor in that CH$_4$ addition produces faster rates than H$_2$, which in turn is faster than Ar addition. The surface morphologies were smooth, and the features anisotropic over a wide range of discharge conditions with HBr-based chemistries.[30]

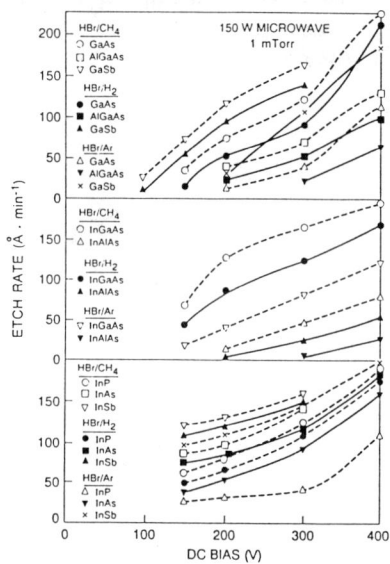

Fig. 8. Etch rate of different III-V materials in 1 mTorr, 150W (microwave) discharges of 15 HBr/5 CH$_4$ 15 HBr/5 H$_2$ or 15 HBr/5 Ar, as a function of dc bias on the sample.

Figure 9 shows an Arrhenius plot of the etch rates of InP, GaAs and AlGaAs in the HBr/H$_2$/Ar mixture. The etching is not characterized by a single activation energy, which would imply there are several different product desorption mechanisms present. One might assume that ion-assisted desorption of group III (and possibly group V) bromide species and reaction of hydrogen with P and As to form PH$_3$ and AsH$_3$ are occurring. It is generally observed that dry etching at low pressures does not occur with a single activation energy because of the importance of ion-enhanced reactions.[31-33] By contrast, at higher pressures (≥ 0.3 Torr) the apparent activation energies for etching of InP and GaAs are close to the latent heats of vaporization for InCl$_3$ and GaCl$_3$, respectively, even though their published vapor pressures would suggest that these would not be a limiting factor. For each of the three different semiconductors there is a break in the etch rates above ~150°C, with more rapid removal of the material. Chemical reactions, such as evaporation of the etch products from the surface without need for ion-enhancement, may become more important at elevated temperatures.

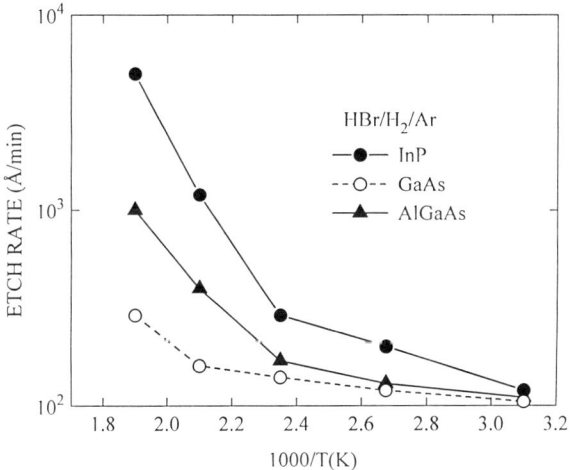

Fig. 9. Arrhenius plot of etch rates of InP, GaAs and AlGaAs in a 10 mTorr, 250W (microwave), −200 V dc HBr/H_2/Ar discharge.

The etch rates of III-V materials in HBr-based discharges are quite slow under the conditions needed for smooth, anisotropic etching. Relatively high self-biases are needed to initiate etching with any of the three mixtures we investigated (HBr/H_2, HBr/CH_4, HBr/Ar). At high dc biases, the binary In-based materials show surface roughening due to preferential sputtering of the group V element and at high microwave power levels the surfaces of all the In-based semiconductors except InAlAs become rough. For these materials the HBr/CH_4 mixture provides smooth etching over the widest range of conditions. Changes in the electrical and optical properties of the semiconductors are generally not as severe as might be expected at the high (−250 V) dc biases used in many of our experiments because of the passivation of deep levels by hydrogen from the discharges.

CONCLUSIONS

A variety of dry etch chemistries was examined from the viewpoint of patterning III-V semiconductors. Conventional CH_4/H_2 discharges offer smooth, controlled etching with high selectivities to standard masking materials, but the etch rates are too slow for practical fabrication of deep structures. Iodine-based mixtures (HI/H_2, CH_3I/H_2) offer faster etch rates, but the HI has a relatively short shelf-life and the CH_3I/H_2 chemistry requires careful seasoning of the reactor. Fast, smooth etching of all In-containing III-V semiconductors is obtained with Cl_2/CH_4/H_2/Ar discharges with the sample held at ~150 °C. This requires the use of dielectric masks. Changes to the electrical and structural properties of the InP are minimal with this etch mixture, and it has been used to fabricate novel microdisk lasers that may form the basis of photonic arrays.

ACKNOWLEDGMENTS

The authors acknowledge the technical support of J. R. Lothian, T. R. Fullowan, B. Tseng and R. A. Keane.

REFERENCES

1. A. Katz, S. J. Pearton, and M. Geva, Appl. Phys. Lett. *59* 286 (1991).
2. J. Yuba, K. Gamo, X. G. He, J. S. Zhang, and S. Namba, Jap. J. Appl. Phys. *22* 1211 (1983).
3. M. Schilling and K. Wunstel, Appl. Phys. Lett. *49* 710 (1986).
4. T. R. Hayes, InP and Related Materials ed. A. Katz (Boston, MA: Artech House 1991), Chapter 8.
5. G. P. Agrawal and N. K. Dutta, Long Wavelength Semiconductor Lasers (New York, Van Nostrand 1986).
6. S. J. Pearton, F. Ren, W. S. Hobson, C. A. Green, and U. K. Chakrabarti, Semicond. Sci. Technol. *7*, 1217 (1992).
7. D. T. C. Huo, M. F. Yan, and J. D. Wynn, J. Electrochem. Soc. *137*, 3639 (1990).
8. U. Niggebrugge, M. Klug and G. Garus, Inst. Phys. Conf. Ser. *79* 367 (1985).
9. J. Asmussen, J. Vac. Sci. Technol. A*7* 883 (1989).
10. S. C. McNevin, J. Vac. Sci. Technol. B*4* 1216 (1986).
11. G. J. Van Gurp, J. M. Jacobs, J. J. M. Binsma and C. F. Tiemeijer, Jap. J. Appl. Phys. *28* 1236 (1989).
12. R. Van Roijen, M. Kemp, C. Bulle-Lieuwma, L. van Izjendoorn and T. Thijssen, J. Appl. Phys. *70* 3983 (1991).
13. C. Constantine, C. Barratt, S. J. Pearton, F. Ren and J. R. Lothian, Appl. Phys. Lett. *61* 2899 (1992).
14. D. Flamm, in Plasma Etching-An Introduction, ed. D. Manos and D. E. Flaman (Academic Press, San Diego, 1989).
15. C. Constantine, C. Barratt, S. J. Pearton, F. Ren and J. R. Lothian, Electron. Lett. *28* 1749 (1992).
16. F. Ren, S. J. Pearton, B. Tseng, J. R. Lothian and B. P. Segner, J. Electrochem. Soc. Nov. 1993 (in press).
17. D. C. Flanders, L. D. Pressman and G. Pinelli, J. Vac. Sci. Technol. B*8* 1990 (1990).
18. S. J. Pearton, U. K. Chakrabarti, A. Katz, F. Ren and T. R. Fullowan, Appl. Phys. Lett. *60* 838 (1992).

19. S. J. Pearton, U. K. Chakrabarti, W. S. Hobson, C. R. Abernathy, A. Katz, F. Ren, T. R. Fullowan and A. Perley, J. Electrochem. Soc. *139* 1763 (1992).
20. H. F. Wong, D. L. Green, T. Liu, D. G. Lishan, M. Bellis, E. L. Hu, P. M. Petroff, P. O. Holtz and J. L. Merz, J. Vac. Sci. Technol B*6*, 1906 (1988).
21. H. Hidaka, R. Akita, M. Taneya and Y. Sugimoto, Electron. Lett. *26* 1112 (1990).
22. V. Swaminathan, M. T. Asom, U. K. Chakrabarti and S. J. Pearton, Appl. Phys. Lett. *58*, 1256 (1991).
23. S. J. Pearton, J. W. Corbett and T. S. Shi, Appl. Phys. A*43*, 153 (1987).
24. M. Moehrle, Appl. Phys. Lett. *56* 542 (1990).
25. U. K. Chakrabarti, S. J. Pearton, A. Katz, W. S. Hobson and C. R. Abernathy, J. Vac. Sci. Technol. B*10* 2378 (1992).
26. V. M. Donnelly, D. L. Flamm, C. W. Tu and D. E. Ibbotson, J. Electrochem. Soc. *129* 2533 (1982).
27. D. E. Ibbotson, D. L. Flamm, C. W. Tu and D. E. Ibbotson, J. Electrochem. Soc. *129* 2533 (1982).
28. D. E. Ibbotson, D. L. Flamm and V. M. Donnelly, J. Appl. Phys. *54* 5974 (1985).
29. K. Takimoto, K. Ohnaka and J. Shibata, Appl. Phys. Lett. *54* 5974 (1985).
30. G. S. Oehrlein, Y. Zhang, G. M. Kroesen, E. de Fresurt and T. D. Bestwick, Appl. Phys. Lett. *58* 2252 (1991).
31. S. J. Pearton, U. K. Chakrabarti, E. Lane, A. Perley, C. R. Abernathy, W. S. Hobson and K. S. Jones, J. Electrochem. Soc. *139* 856 (1992).
32. R. J. Contolini, J. Electrochem. Soc. *135* 929 (1988).
33. N. Furuhata, H. Miyamoto, F. Okamoto and K. Ohata, J. Appl. Phys. *65* 168 (1989).
34. S. J. Pearton, A. B. Emerson, U. K. Chakrabarti, E. Lane, K. S. Jones, K. T. Short, A. E. White and T. R. Fullowan, J. Appl. Phys. *66*, 3839 (1989).

GaAs ETCHING BY Cl_2 and HCl: Ga- vs. As- LIMITED ETCHING

CHAOCHIN SU, ZI-GUO DAI, HUI-QI HOU, MING XI, MATTHEW F. VERNON, AND BRIAN E. BENT
DEPARTMENT OF CHEMISTRY, COLUMBIA UNIVERSITY, NEW YORK, NY 10027

ABSTRACT

Results in the literature indicate that Cl_2 etches GaAs at room temperature but HCl etches GaAs at a measurable rate only at temperatures above ~ 670 K. In this work, molecular beam scattering and surface analysis techniques have been applied to address the fundamental kinetic differences between these two systems. The results indicate that the onset of GaAs etching by Cl_2 is determined by the kinetics of Ga-removal as $GaCl_3$ while etching by HCl is limited by As evaporation as As_2. The results also suggest that HCl selectively etches gallium from GaAs at temperatures between 600 and 650 K.

INTRODUCTION

Cl_2 and HCl are two halogen compounds widely used in various GaAs etching processes because of the high reactivity of chlorine and the high volatility of the generated products. Since the etching presumably involves the dissociative adsorption of the etchant molecules on the surface, it is of interest to compare GaAs etching by Cl_2 and HCl, Cl-containing molecules with dramatically different bond energies (59 kcal/mol for Cl_2 and 103 kcal/mol for HCl). Previous studies of Cl_2 and HCl etching of GaAs at steady state have shown a dramatic difference in the onset of etching: Cl_2 etches GaAs from room temperature while HCl etches at a measurable rate only at temperatures above 670 K [1-5]. The experiments reported in this paper address the rate-limiting step in these etching systems. In the case of GaAs etching by Cl_2, the reaction products and kinetics suggest that the onset of etching is controlled by the rate of $GaCl_3$ formation/desorption. By contrast, the onset of GaAs etching by HCl is limited by the rate of arsenic evaporation as As_2.

EXPERIMENTAL

Studies of the etching products, etching rate at steady state, and etching kinetics for the Cl_2 and HCl reactions with GaAs were carried out in a molecular beam scattering apparatus which has been described elsewhere [3]. To measure the surface Cl coverage, Auger electron spectroscopy experiments were performed in an ultra-high vacuum (UHV) chamber as described in reference 4.

RESULTS AND DISCUSSION

Reaction of GaAs and Cl_2

Previous molecular beam scattering studies of the $GaAs/Cl_2$ etching reaction have reported the product formation rates as well as the total etching rate as a function of surface temperature under steady state conditions [3]. Five product species distributed over the temperature range of 330 - 930 K are found as summarized in Table 1. The etching rate profile obtained from summing up all the Ga- or As- products formation rates is given by the open circles in Fig. 1. An interesting feature of the

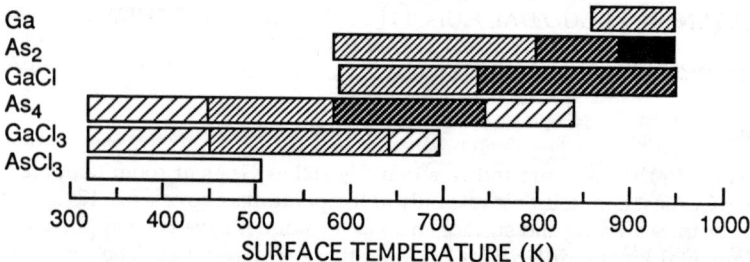

Table 1. The steady state product distribution over the temperature range of 330 - 900 K for the etching of GaAs by Cl_2. The shading represents the relative yields of different products at the indicated temperature ranges, with the yield increasing in the order ▫, ▨, ▨, ▨, ■.

temperature dependence of the etching rate is that there are two regions in which the etching rate increases with surface temperature (330 - 460 K and 600 - 700 K), consistent with the ranges in which the formation rates of $GaCl_3$ and GaCl increase. This result suggests that the steady state etching of GaAs by Cl_2 is limited by Ga removal as $GaCl_3$ and GaCl. The additional increase in the etching rate above 850 K is due to the evaporation of GaAs.

The temperature dependence of the steady state etching rate can be correlated with the surface chlorine coverage. Fig. 1 also shows the temperature dependence of the surface Cl coverage as measured by Auger electron spectroscopy. Since the Cl_2 flux in this experiment was two orders of magnitude lower than that used in measuring the steady state etching rate, the dashed curve indicates the behavior expected for the higher Cl_2 flux (which is not experimentally accessible in our UHV apparatus). The counter profiles for the steady state rate and the surface Cl coverage are striking; as the surface Cl coverage decreases, the etching rate increases by a proportionate amount. This relationship has been quantified in previous work which shows that the surface Cl coverage during steady state etching is in the monolayer regime and that the etching rate is proportional to $(1-\theta_{Cl})$ [4], the fractional coverage of empty sites on the surface. Such a functional form has also been found for the Cl_2 dissociative adsorption rate on GaAs(100) and GaAs(110) by Sullivan et al. [6].

The correlation between the number of Cl-vacancies in the surface monolayer and the steady state etching rate suggests a Langmuir adsorption model for $GaAs/Cl_2$ etching. At steady state, the Langmuir formulation can be written as [4]:

$$\phi s(1-\theta) = k\theta^n \quad (1)$$

where ϕ is the flux of reactive species to the surface, s is the sticking probability on vacant surface sites, θ is the fractional occupation of surface sites, n is the reaction order, and k is the rate constant for monolayer removal. Equation (1) can be used to derive the steady state surface Cl coverage as follows [4]:

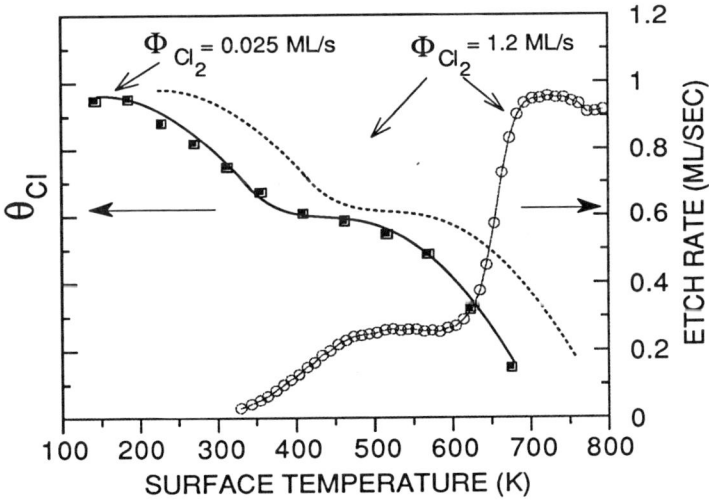

Figure 1. Surface temperature dependence of the absolute etching rate (open circles) and surface Cl coverage (solid squares) for GaAs etching by Cl_2. The dashed line indicates the behavior expected for the higher Cl_2 flux used in the etching rate study.

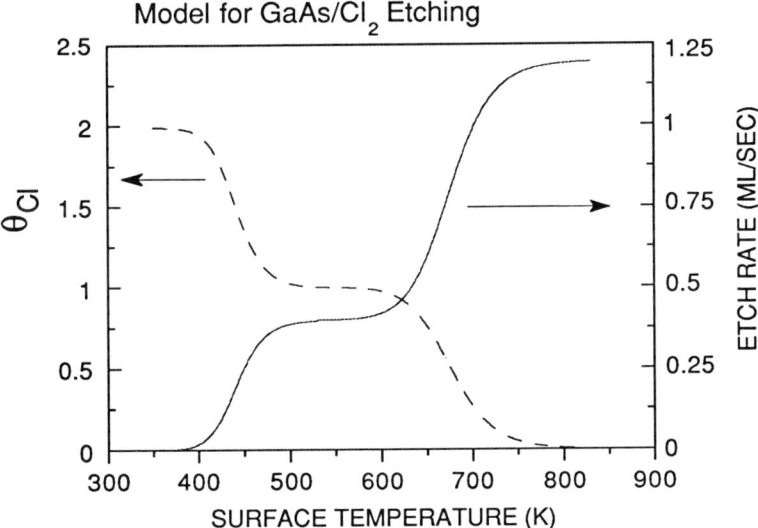

Figure 2. Simulation of the steady state GaAs/Cl_2 etching rate (solid line) and surface chlorine coverage (dashed line) as a function of surface temperature.

$$\theta_{Cl}^{s.s.} = \frac{1}{1 + \frac{\alpha k}{2\Phi s}} \qquad (2)$$

where α is the Cl/Ga stoichiometry of the etch product (3 for $GaCl_3$ and 1 for GaCl) and the factor 2 accounts for the two Cl atoms per Cl_2 molecule.

In the case of $GaAs/Cl_2$ etching, two Langmuir models (one for low temperature and one for high temperature) are required to explain the two plateau structure in the etching rate as a function of surface temperature and to account for the formation of two different gallium chloride products, $GaCl_3$ and GaCl. Having previously studied the desorption kinetics of $GaCl_3$ and GaCl by temperature-programmed reaction (TPR) experiments, the rate constants k for chlorinated monolayer removal have been determined by analyzing the TPR peaks as pseudo first order processes with pre-exponential factors of 10^{13} s^{-1}: the activation energies for $GaCl_3$ and GaCl evolution are 25.5 and 39.0 kcal/mol, respectively. Since the incident Cl_2 flux is known to be 1.2 monolayers (ML)/sec [4], and the reaction probability is ~1 [6] for Cl_2 incident on vacant surface sites, the surface Cl coverage and steady state etching rate as a function of surface temperature can be calculated. The results in fig. 2 show the same qualitative behavior as the experimental results (compare to fig. 1). Note, in particular, that without any adjustment of the kinetic parameters, the calculated etching rate increase over the same temperature ranges and the absolute rates are within 30% of those measured experimentally. This agreement provides strong evidence that a Langmuir model with kinetic parameters from monolayer adsorption experiments captures the dominant features of the steady state $GaAs/Cl_2$ etching reaction.

Three important points emerge from the results above. First, the major chloride products, $GaCl_3$ and GaCl, are both Ga-containing species. The only $AsCl_x$ species is $AsCl_3$, and it is formed in small yield (see Table 1). Second, the temperature dependent changes in the steady state etching rate correlate with the changes of the surface Cl coverage as well as with the evolution of $GaCl_3$ and GaCl in TPR studies. Third, a Langmuir model, which includes desorption kinetic parameters for only the Ga chloride species, semiquantitatively describes the steady state etching rate.

It should be emphasized that while the kinetics of arsenic removal have been neglected in predicting the steady state GaAs rate, the observation of stoichiometric etching [4] necessarily implies that the Ga and As removal kinetics are interconnected. What the results here imply is that the onset of etching is dominated by the rate of Ga removal. The interplay between the Ga and As removal rates becomes significant as etching approaches the flux-limited regime. We note also that the fact that the vast majority of Cl leaves the surface as Ga chlorides does not necessarily imply that Cl exits on the surface primarily as $GaCl_x$ species. Arsenic to gallium transfer of surface Cl is possible, and recent photoemission studies of Shuh et al. have documented the presence of AsCl, $AsCl_2$, GaCl, and $GaCl_2$ on a chlorinated GaAs(110) surface [7].

Reaction of GaAs and HCl

Fig. 3 shows the three major etching products (GaCl, As_2 and Ga) detected in the reaction of GaAs with HCl over the surface temperature range of 670 to 930 K. No gallium hydrides or arsenic hydrides are formed. Below 850 K, etching occurs through the reaction of HCl with GaAs, while above 850 K etching also occurs by GaAs evaporation. The total etching rate can be represented by the As$_2$ formation rate since this is the only As-containing product and the reaction of GaAs/HCl is stoichiometric in Ga and As [2]. Two important results emerge from this experiment. First, the onset of

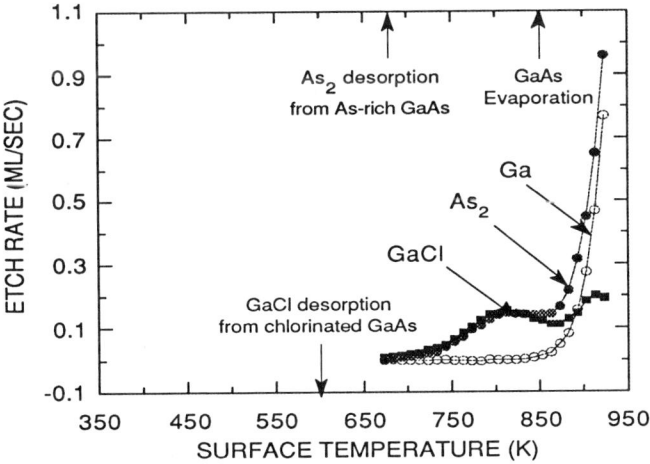

Figure 3. Surface temperature dependence of the indicated product formation rates for GaAs etching by HCl. The vertical arrows indicate the approximate temperatures at which the indicated processes achieve measurable rates.

etching by HCl (670 K) is significantly higher than the TPR peak temperature for GaCl (600 K) which is the sole gallium product at the onset of the GaAs/HCl reaction. Second, Auger measurements show no surface Cl at temperatures above 650 K for GaAs/HCl system. Since GaCl is the only Cl containing product in this reaction and the surface Cl coverage remains below detection limits, it is unlikely that GaAs/HCl etching is rate-limited by Ga removal.

To see whether the etching is controlled by arsenic removal, modulated molecular beam scattering measurements of the evolution kinetics of GaCl and As$_2$ were performed. Fig. 4 shows the modulation waveforms for GaCl (A) and As$_2$ (B) acquired at 770 K. Also plotted in fig. 4(B) with a dashed line is the modulation waveform of the reflected HCl. Clearly, the rate of arsenic evolution responds slowly to the changes in the HCl flux compared with the GaCl evolution rate. Also notice that Ga is only removed when the HCl beam is on while As$_2$ continues to desorb after the HCl beam is off. These results suggest that during the steady state etching by HCl the surface is As-rich. The role of HCl is apparently to remove Ga atoms as GaCl to generate a sufficiently As-rich surface so that evaporation as As$_2$ can occur. This conclusion is strongly supported by the report of Banse et al. that the onset of As$_2$ evaporation from an As-rich surface occurs around 670 K [8], which is same temperature as the onset of GaAs etching by HCl. The inability of HCl to etch As-rich GaAs at lower temperatures may be related to the recombinative desorption of HCl reported by Nooney et al. for such surfaces [9].

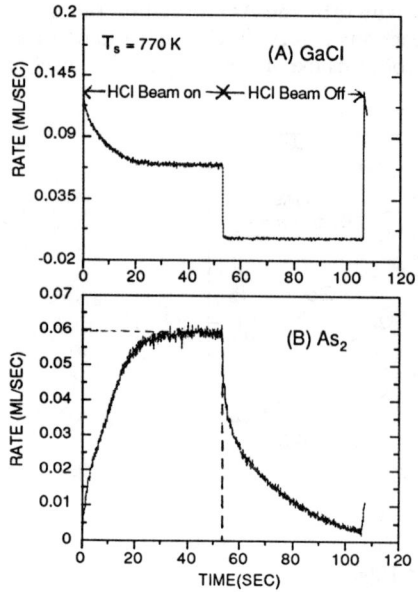

Figure 4. Modulated molecular beam scattering waveforms at 770 K for GaCl and As$_2$ evolution during GaAs etching by HCl.

Importantly, the results above indicate that HCl selectively etches gallium from GaAs at temperature between 600 and 670 K. 600 K is the onset for GaCl formation while 670 K is the onset for As$_2$ evolution from an As-rich surface. Consequently, HCl does not etch GaAs between 600 and 670 K because only Ga atoms can be removed, leaving an As rich surface which only evaporates at a significant rate above ~ 670 K. Studies are in progress to develop a chemical process for selectively removing this arsenic layer at lower temperature.

CONCLUSIONS

The rate limiting steps for the GaAs/Cl$_2$ and GaAs/HCl etching reactions has been addressed by measuring the temperature dependence of the steady state etching rate, the steady state surface Cl coverage, and product desorption kinetics. In the GaAs/Cl$_2$ reaction, the etching is controlled by Ga removal as GaCl$_3$ at low temperature and as GaCl at high temperature. In the GaAs/HCl reaction, the onset of etching is controlled by the removal of arsenic as As$_2$. The results provide indirect evidence for selective Ga removal from GaAs by HCl at 600 - 670 K.

ACKNOWLEDGMENTS

Financial support from the NSF (#DMR-89-57236) as part of the Presidential Young Investigator Program and from the Joint Services Electronics Program (DAAL03-91-C-0061) is gratefully acknowledged. The authors thank Dr. Phillip Kash for useful comments.

REFERENCES

[1] J. Saito and K. Kondo, J. Appl. Phys. **67**, 6274 (1990).
[2] C. Su, Z.-G. Dai, W. Luo, D.-H. Sun, M.F. Vernon and B.E. Bent, (Submitted).
[3] C. Su, H.Q. Hou, Z.G. Dai, G.H. Lee, W. Luo, M.F. Vernon and B.E. Bent, J. Vac. Sci. Technol. B **11**, 1222 (1993).
[4] C. Su, M. Xi, Z.G. Dai, M.F. Vernon and B.E. Bent, Surf. Sci. **282**, 357 (1993).
[5] N. Furuhata, H. Miyamoto, A. Okamoto and K. Ohata, J. Appl. Phys. **65**, 168 (1989).
[6] D.J.D. Sullivan, H.C. Flaum and A.C. Kummel, Surf. Sci. (Submitted).
[7] D.K. Shuh, C.W. Lo, J.A. Yarmoff, A. Santoni, L.J. Terminello and F.R. McFeely, Surf. Sci. (in press).
[8] B.A. Banse and J.R. Creighton, Appl. Phys. Lett. **60**, 856 (1992).
[9] M. Nooney, V. Liberman, M. Xu, A. Ludviksson and R.M. Martin, (submitted).

SCANNING TUNNELING MICROSCOPY OBSERVATION OF THE REACTION OF AlCl₃ ON Si(111)-7x7 SURFACE

Katsuhiro UESUGI, Takaharu TAKIGUCHI, Michiyoshi IZAWA, Masamichi YOSHIMURA and Takafumi YAO
Department of Electrical Engineering, Hiroshima University, Higashi-Hiroshima 724, Japan

ABSTRACT

The initial stage of the reaction of aluminum trichloride ($AlCl_3$) with the Si(111)-(7x7) surface and the annealing effects of the adsorbed surface are investigated with a scanning tunneling microscope (STM). Reacted and unreacted sites manifest in the contrast of adatom sites on the $AlCl_3$-exposed surface. An $AlCl_3$ molecule dissociatively adsorbs onto the Si(111)-(7x7) surface at room temperature, which yields Cl atoms. Preferential adsorption site is found to be the center adatom site rather than the corner adatom site, and Cl atom adsorbs onto the top site of the center Si adatom. The migration of adsorbed molecules ($AlCl_x$) in the (7x7) unit cell is observed at room temperature. The $SiCl_x$ species desorb from the surface by thermal annealing at 600 °C, leaving vacancies behind. After annealing at 1200 °C, Al deposition occurs in a limited area and the surface shows the $\sqrt{7} \times \sqrt{7}$ reconstruction, which is a characteristic Al-induced surface structure.

Introduction

Halogen gases such as chlorine, chlorosilane and dichlorosilane are widely used as source gases for chemical etching or chemical vapor deposition (CVD). The interaction of gas molecules with Si surfaces is both scientifically and technologically important. The reaction of gas molecules with clean Si surfaces has been intensively investigated using many techniques such as X-ray photoemission spectroscopy (XPS),[1] low-energy electron diffraction (LEED),[1,2] temperature programmed desorption (TPD),[3,4] and so forth. However, few attention has been paid to the atomistic and electronic studies of the reacted sites and resulting structures.

Schnell et al. reported Cl exposed Si(111)-(7x7) surface by XPS.[1] SiCl, $SiCl_2$, and $SiCl_3$ were found on the surface at room temperature. After annealing at 400 °C, only SiCl remained on the surface. Recently, Boland et al. reported the adsorption and desorption kinetics for Cl on Si(111)-(7x7) surfaces by scanning tunneling microscopy (STM).[5,6] Reacted and unreacted sites were distinguishable in both current-voltage curve and topographic image. Chlorine was observed to adsorb on the top site of Si adatom. At saturation coverage, the Cl atoms eliminated the Si adatom layer in the (7x7) reconstruction by thermal annealing, so that the underlying Si rest-atom layer appears.

In this paper, we investigate the initial stage of the reaction of $AlCl_3$ with the Si(111)-(7x7) surfaces by STM. It is found that the $AlCl_3$ molecules adsorb dissociatively onto the Si(111) surface at room temperature. The annealing effects of the $AlCl_3$ exposed surfaces was also investigated to enhance the surface reaction. Thermal annealing resulted in the desorption of $SiCl_x$ species from the surface at 600 °C, while at 1200 °C the deposition of Al atoms were observed.

Experimental

An ultra-high vacuum (UHV) STM-system was used in the present experiments, which consists of an STM chamber, a preparation chamber, and a load-lock chamber. The STM chamber was evacuated by an ion pump and the base pressure was $\sim 5 \times 10^{-11}$ Torr. The specimen was cut from a Si(111) wafer of 0.1–1 Ω·cm. It was prebaked at ~500 °C for 12 h, and was cleaned by repeated flash heating to ~1250 °C to obtain a clean Si(111) surface. The STM observation confirmed that the surface had a large flat area of (7x7) reconstruction. Then, the surface was exposed to 0.1–20 L (1 Langmuir (L) = 1×10^{-6} Torr sec) $AlCl_3$ gas at room temperature in the preparation chamber. After exposure, $AlCl_3$ was evacuated and the sample was transferred again to the STM chamber for the observation. The tip used was a tungsten wire with a diameter of 0.3 mm which was sharpened by electrochemical etching.

Results and Discussion

(a) $AlCl_3$-exposed Si(111)-(7x7) surfaces

Figures 1 show topographic STM images of a 0.1 L-exposed Si(111) surface. The tunneling current is 1.1 nA and the sample bias is (a) 1.15 V and (b) 1.96 V. In both images one can see the contrast of adatom sites which reflects reacted and unreacted sites. We observed many kinds of protrusions which showed different contrast with the bias voltage. In another part of the surface, the as exposed surfaces show essentially the same features. These protrusions can be classified into three patterns: A site is considered to be a vacancy because no protrusion was found at any bias voltage. B sites are invisible at 1.15 V, while with increase of the bias voltage up to 1.96 V, we can observe a protrusion at B sites though they are still darker than the surrounding adatoms. This fact suggests that the B sites are not vacancies but reacted sites. Since this behavior is similar to that observed by Boland and Villarrubia for Cl adsorbed Si(111)-(7x7) surfaces,[5,6] we can assign the B site to a Cl atom being bonded to the underlying Si adatom. Namely, $AlCl_3$ molecules dissociatively adsorb onto the Si(111) surface at room temperature, which yields Cl atoms. It is noted that the site B is preferentially distributed on the center-adatom site. In this area, the reactivity ratio of the center adatom to corner adatom is 22:0. The other protrusions C through E are also observed on the surface. The brightest protrusions (E) are located either at adatom sites (left E) or between adatoms (right E). These are considered to be undissociative $AlCl_3$ molecules or its fragments ($AlCl_x$) since they have not been observed on the Cl-adsorbed Si(111) surfaces.[5,6]

On the other area of this surface, we have observed migration of adsorbed molecules in the (7x7) unit cell at room temperature. Figures 2 show successive STM images of 7.5x6.7 nm^2 area of a 0.1 L-exposed surface taken every 60 sec. The unit cell at the same position is shown by a solid line. The protrusion (A) is located near a rest atom site. The protrusion indicated by arrows moves from A site to B site within the (7x7) unit cell, and it eventually sits on the center adatom site

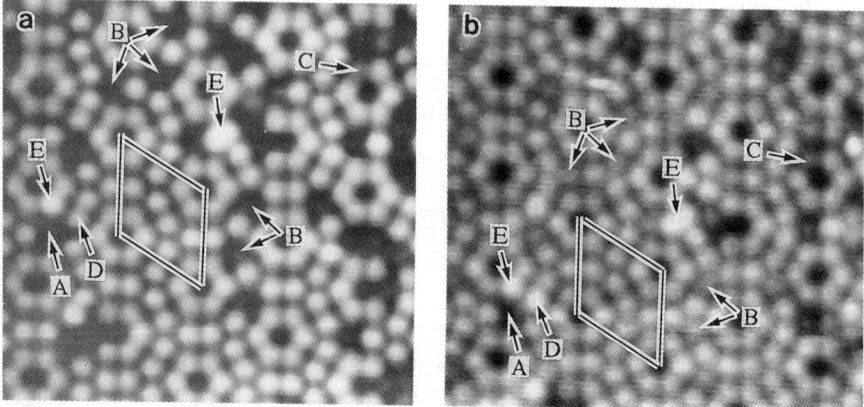

Fig. 1. STM topographic images of a partially reacted surface after 0.1 L $AlCl_3$ exposure (12x11 nm^2). The tunneling current is 1.1 nA and the sample voltage is (a) 1.15 V and (b) 1.96 V. The unit cell at the same position is shown with a solid line.

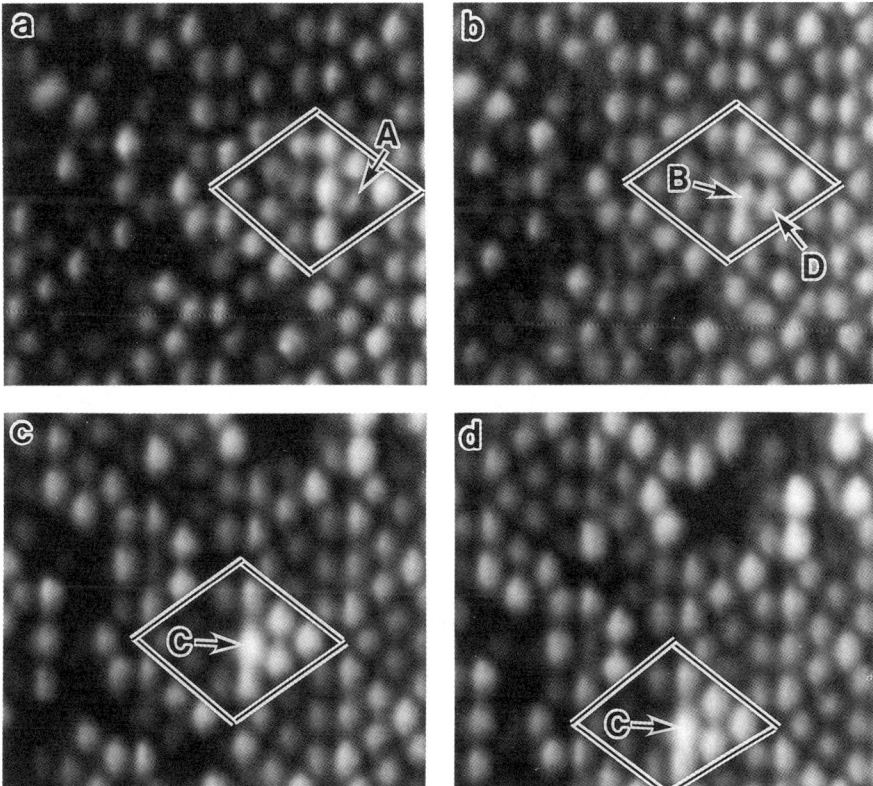

Fig. 2. Successive STM images of 7.5x6.7 nm² area of a 0.1 L exposed surface at room temperature taken every 60 sec. The tunneling current is 1 nA and the sample voltage is −1.15 V. The unit cell at the same position is shown with a solid line.

(C). Although an adatom is not observed at the center adatom site (D) in Fig. 2(a), it becomes visible when the protrusion moves to B site as shown in Fig. 2(b). This is because the electronic state of the center adatom is affected by weak chemisorption of the molecule onto the proximal rest atom. We note that not Cl atom but $AlCl_x$ species dissociates and adsorbs on the center adatom site (C) because the site shows bright contrast. This fact suggests also the important role of the rest atom at the initial stage of the reaction of $AlCl_3$ with Si(111) surface.

Figure 3 shows the dependence of density of Cl-adsorbed sites and reactivity ratio of center adatom to corner adatom for various $AlCl_3$-gas exposure. The density of Cl-adsorbed site becomes larger as the exposure increases. Although Cl atom is preferentially adsorbed onto the center-adatom site rather than the corner adatom site at lower exposure, the reactivity ratio is reduced as the exposure increases.

The mechanism of the initial stage of the reaction of $AlCl_3$ with the Si(111) surfaces is proposed as follows. It is well known that the Si(111)-(7x7) reconstruction contains different surface dangling bonds with different electronic structure through charge transfer, where the

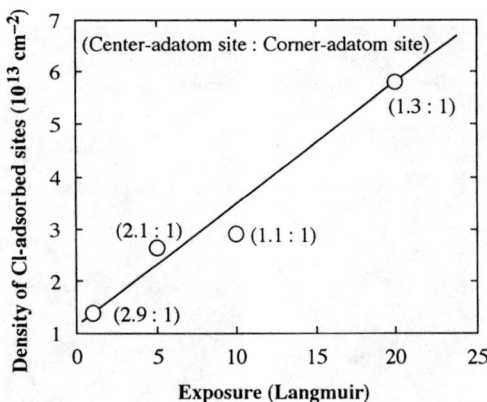

Fig. 3. The relationship between the $AlCl_3$-gas exposure and the surface density of Cl-adsorbed sites. The reactivity ratio of the center adatom to corner adatom is dependent on the $AlCl_3$-gas exposure.

adatom has much smaller density of state in the occupied state near the Fermi level than that of the rest atom.[7,8] As a result, the rest-atom can act as a Lewis base, while the adatom plays as an electron acceptor. Since $AlCl_3$ is a Lewis acid, the rest atom plays a crucial role in the decomposition process of $AlCl_3$. $AlCl_3$ gas molecule exists as Al_2Cl_6 below 440 °C, where Al atoms and Cl atoms are positively and negatively charged, respectively. Hence, as the molecule approaches the surface, one of the Al atoms in the molecule is located at the rest-atom site. We note that the distance between an Al atom and a furthermost Cl atom in Al_2Cl_6 is ~0.46 nm, which is close to the distance between the rest atom and proximal adatom (~0.44 nm). As a result, the negatively charged Cl atom can easily bond to the proximal Si adatom to break the Al–Cl bond. This process produces the Si–Cl species observed at site B. Owing to both the geometric difference around the rest atom and the distortion effect caused by adjacent dimer rows, the center adatom has more unoccupied character than the corner adatoms.[7,8] Therefore, it is considered that the center adatom is more reactive to the Cl atoms. Based on the above discussion, it is considered that the migration of adsorbed molecules, observed in Fig. 2 shows the process that the molecule of Al_2Cl_6 migrates around the rest atom until the Cl atom reaches to its react site.

(b) Annealing effects on the $AlCl_3$ exposed surfaces

Although the surface density of Cl adsorbed Si adatom sites on the 0.2 L-exposed surface was 4.4×10^{12} cm^{-2} before annealing, we could not observe any Cl-adsorbed sites after annealing at 600 °C. In contrast, the number of vacancies increases to 3.7×10^{12} cm^{-2} by annealing, which is about three times larger than that of vacancy before annealing. These facts suggest that the desorption of $SiCl_x$ species is promoted by annealing, leaving vacancies behind. This result is consistent with the annealing experiments on Cl exposed surface,[5,6] where annealing at 470 °C results in desorption of $SiCl_x$ from the surface.

Figure 4 shows an STM image of a 0.2 L-exposed surface after annealing at 1200 °C. The protrusions form the $\sqrt{7} \times \sqrt{7}$ reconstruction with each protrusion 1 nm apart. The $\sqrt{7} \times \sqrt{7}$ reconstruction is observed on the Si(111)-Al surface, which appears at the coverage of 3/7 ML.[9] Since Si(111)-Al structure was observed on the annealed surface, it suggests the deposition of Al atoms through annealing.

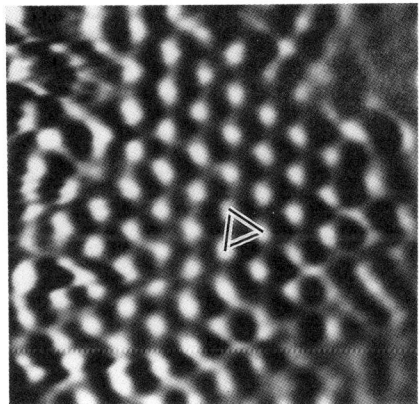

Fig. 4. STM topographic image of a 0.2 L-exposed surface (11x10 nm^2) after annealing at 1200 °C. The protrusions form $\sqrt{7}\times\sqrt{7}$ reconstruction. The sample voltage is −1.14 V and tunneling current is 1.1 nA.

Conclusions

We have presented the STM study on the initial reaction of $AlCl_3$ with the Si(111)-(7x7) surfaces. It is found that $AlCl_3$ molecules dissociatively adsorb onto the Si surface at room temperature, resulting in Cl atoms and $AlCl_x$ fragments. The Cl atom preferentially adsorbs onto the center adatom site rather than the corner adatom site. The migration of adsorbed molecules in the (7x7) unit cell is observed real time at room temperature. The desorption of $SiCl_x$ species from the surface by thermal annealing at 600 °C, and the Al deposition with the $\sqrt{7}\times\sqrt{7}$ reconstruction in a limited area occurs by thermal annealing at 1200 °C.

References

[1] R.D. Schnell, D. Rieger, A. Bogen, F.J. Himpsel, K. Wandeltand and W. Steinmann, Phys. Rev. B **32** (1985) 8057.
[2] W. Sesselmann and T.J. Chuang, J. Vac. Sci. Technol. B **3** (1985) 1507.
[3] P. Gupta, P.A. Coon, B.G. Koehler and S.M. George, Surf. Sci. **249** (1991) 92.
[4] J. Matsuo, F. Yannick and K. Karahashi, Surf. Sci. **283** (1993) 52.
[5] J.S. Villarrubia and J.J. Boland, Phys. Rev. Lett. **63** (1989) 306.
[6] J.J. Boland and J.S. Villarrubia, Phys. Rev. B **41** (1990) 9865.
[7] R. Wolkow and Ph. Avouris, Phys. Rev. Lett. **60** (1988) 1049.
[8] Ph. Avouris and R. Wolkow, Phys. Rev. B **39** (1989) 5091.
[9] R.J. Hamers, Phys. Rev. B **40** (1989) 1657.

CHEMICAL TOPOGRAPHY OF Si ETCHING IN A Cl_2 PLASMA, STUDIED BY X-RAY PHOTOELECTRON SPECTROSCOPY AND LASER-INDUCED THERMAL DESORPTION

V. M. DONNELLY, K. V. GUINN, C. C. CHENG, AND I. P. HERMAN[a]
AT&T Bell Laboratories, 600 Mountain Ave., Murray Hill NJ, 07974

ABSTRACT

This paper describes x-ray photoelectron spectroscopy (XPS) studies of etching of Si in high-density Cl_2 plasmas. Polycrystalline Si films, masked with photoresist stripes, are etched and then transferred in vacuum to the XPS analysis chamber. Shadowing of photoelectrons by adjacent stripes and differential charging of the photoresist and poly-Si were used to separate contributions from the top of the mask, the side of the mask, the etched poly-Si sidewall, and the bottom of the etched trench. In pure Cl_2 plasmas, surfaces are covered with about one monolayer of Cl. If oxygen is introduced into the plasma, either by addition of O_2 or by erosion of the glass discharge tube, then a thin Si-oxide layer forms on the sides of both the poly-Si and the photoresist. Laser-induced thermal desorption (LITD) was used to study etching in real time. LITD of SiCl was detected by laser-induced fluorescence. These studies show that the Si-chloride layer formed during plasma etching is stable after the plasma is extinguished, so the XPS measurements are representative of the surface during etching. LITD measurements as a function of pressure and discharge power show that the etching rate is limited by the positive ion flux to the surface, and not by the supply of Cl_2, at pressures above 0.5 mTorr and for ion fluxes of $\sim 4 \times 10^{16} cm^{-2} s^{-1}$.

INTRODUCTION

With the advent of high charge density ($10^{11} - 10^{12} cm^{-3}$), low ion energy (<100 eV), low pressure (0.3-10 mTorr) plasmas for anisotropic etching of microelectronics materials, mechanisms for etching need to be re-examined. At ion energies of several hundred eV traditionally used in higher pressure processes, etching is generally believed to occur by an ion-enhancement of surface reactions with species (e.g. Cl-atoms) formed from the etchant gas (e.g. Cl_2) in the plasma [1,2]. The chemically enhanced ion sputtering yields exceed one Si atom-per-ion, and the flux of neutrals is orders of magnitude larger than the ion flux. In low pressure, high density plasmas, however, the low ion energies cause the yields to be less than one, and the ion flux is comparable to the neutral flux.

In this paper, we describe *ex situ* x-ray photoelectron spectroscopy (XPS) studies of anisotropic etching of photoresist-masked polycrystalline Si in a high-density helical resonator Cl_2 plasma. The XPS analysis uses shadowing of photoelectrons by adjacent features to spatially resolve adsorbate coverages. This principle was first reported by Fadley [3,4], and was more recently adapted by Oehrlein and co-workers to study etched features [5-8]. We also use *in situ* laser induced desorption measurements performed in real time during etching to determine Cl-coverages as a function of time and plasma conditions. More detailed accounts of this work are reported elsewhere [9-11].

EXPERIMENTAL PROCEDURE

The apparatus described in detail previously (Figure 1) [9-12], consists of a plasma reactor and an ultrahigh vacuum chamber equipped with an X-ray photoelectron spectrometer (XPS). The plasma reactor contains a high-density helical resonator source with magnetic confinement.

Samples used for XPS analysis consist of SiO_2-covered Si(100) substrates on which are deposited 3000 Å of undoped polycrystalline Si. These samples are then masked with patterned photoresist (PR). Samples are clamped to a stainless steel holder so that after etching they can be moved under high vacuum to the analysis chamber. Etching was carried out at a pressure of 1 mTorr Cl_2. The plasma density was $1-2 \times 10^{11}$ positive ions/cm^3, and the ion current was 6.3 mA/cm^2 (3.9×10^{16} ions/cm^2/s) [11]. A radio frequency bias was applied to the sample stage to obtain a DC bias of -35 V in most experiments, or 0 to -100 V in experiments that investigated the effects of bias. The plasma potential was 50 V, so the energy of ions bombarding the substrate was 85 eV, when the DC bias voltage was -35V.

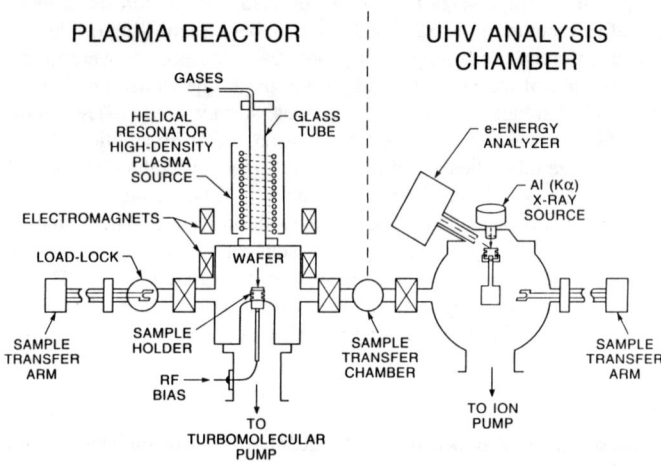

Figure 1. Schematic depiction of the plasma etching chamber and XPS analysis chamber.

After etching, samples were transferred under vacuum to the XPS chamber. In XPS analysis, the sample is irradiated with an X-ray beam that penetrates several microns into the material. Photoelectrons are detected at different take-off angles, θ. When θ=90°, the top of the mask and the bottom of the trench are mainly detected. When θ=30°, the top of the mask and the side of the features are seen. Also, since PR is an insulator, the XPS peaks are shifted, allowing adsorbates on PR and Si to be resolved. We can also flood the surface with low energy electrons, neutralizing the charge and shifting the peaks back to their normal binding energies to aid in making assignments.

Laser-induced desorption measurements were performed *in situ*, both during and after etching of bare Si(100) substrates. The excimer laser (XeCl, 308 nm $0.2-0.6 J/cm^2$, 15 ns pulses at 1-80 Hz) irradiated the surface at normal incidence and caused transient heating of the surface. Products are thermally desorbed and are excited both by the tail of the laser pulse and by electron impact excitation in the plasma. The former, laser-induced fluorescence (LIF) method was used to detect laser-induced thermal desorption of SiCl, and is described elsewhere [11,12]. The latter method was not used in the study reported here and is discussed in another report [12].

RESULTS AND DISCUSSION

X-Ray Photoelectron Spectroscopy

Figure 2 shows an example of Si(2p) spectra for 0.75 µm lines and spaces after etching in a Cl_2 plasma. Etching was stopped before reaching the SiO_2 layer. At $\theta = 90°$, we see elemental Si (Si(2p) peak at 99.5 eV) at the bottom of the poly-Si trench. Since the poly-Si does not charge up, the flood gun has no effect on the electron kinetic energy and hence apparent binding energy. At $\theta = 30°$, photoelectrons from the top and side of the mask and a small amount from the poly-Si sidewall can reach the electron energy analyzer. The small Si signal at 99.6 eV comes from the poly-Si sidewall. In addition, with the charge neutralizer off, a broad feature is observed between 100 and 114 eV. This apparent binding energy is too high to be due to a chemical shift. When the flood gun is turned on, the broad feature shifts to the position expected for SiO_2. Therefore this feature originates from oxidized Si (SiO_xCl_y, see below) on PR. Since the broad feature is not observed at $\theta = 90°$, this SiO_xCl_y adsorbate is only on the side and not on the top of the PR.

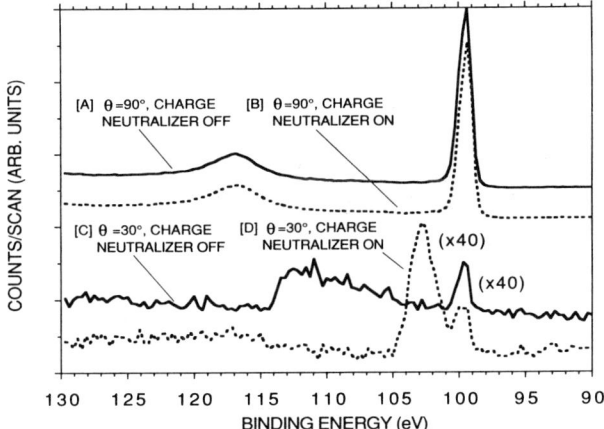

Figure 2. Si(2p) spectra for the region of 0.75 µm lines and spaces, recorded after etching. (a): Charge neutralizer off, $\theta = 90°$. (b): Charge neutralizer on, $\theta = 90°$. (c): Charge neutralizer off, $\theta = 30°$. (d): Charge neutralizer on, $\theta = 30°$.

We can perform similar analyses for the other elements to derive separate XPS signals for elements on poly-Si and PR. We then use measurements at $\theta = 90°$ to determine atomic concentrations for species on top of the PR and at the bottom of the poly-Si trenches. This information, in conjunction with measured θ dependences and the angle response function of the analyzer, and modelling of geometric shadowing of photoelectrons and X-ray absorption, is used to determine the atomic concentrations of adsorbates on the PR and poly-Si sidewalls. These concentrations are then converted into a two-dimensional bar graph or "chemical topography" of the etched surface, as shown in the example in Figure 3. In this method of presenting the results, the atomic concentration of each surface species is ratioed to that of the main bulk component (i.e. Si for poly-Si regions, and C for PR regions). In this example, the bottoms of etched poly-Si features are covered with roughly a monolayer of Cl and not much else. There appears to be somewhat less Cl on the poly-Si sidewall, in addition to some O. In contrast, PR picks up much more Cl on the side and less on top. The PR sidewalls are also covered with a thin layer of oxidized Si (SiO_xCl_y). The O on the poly-Si sidewall and SiO_xCl_y on the side of the PR come from deposition of products from erosion of the glass tube where the helical resonator plasma is formed. [9] When the reactor was modified to minimize this erosion, etching in a pure Cl_2 plasma produced about a monolayer of Cl on etched surfaces, and much less O (on poly-Si) and Si (on PR). One of the more striking conclusions is the lack of C on the Si over a wide range of conditions, even though the mask was also etching. It is often speculated that C eroding from PR produces a thin film that protects poly-Si sidewalls and prevents undercutting. We find no evidence for this process under the etching conditions used in this study.

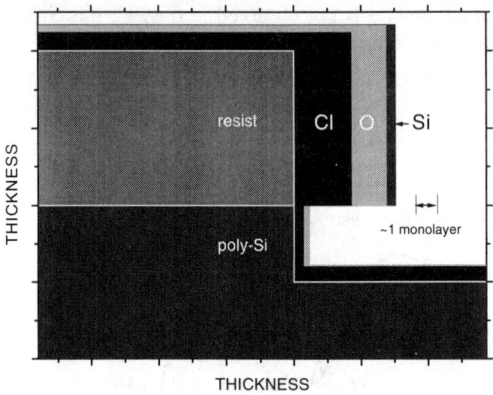

Figure 3. Chemical topography plot of the photoresist/poly-Si structure after etching with a pure Cl_2 plasma, under conditions where the glass tube containing the high-density plasma was eroding at an accelerated rate. The thickness of each of the bands is proportional to the atomic concentration of adsorbates on that surface, ratioed to the major component for the bulk material (i.e. Si for the poly-Si regions and C for the photoresist regions).

Laser-Induced Thermal Desorption

One question frequently asked in post-processing analysis is whether the analyzed surface is representative of the one present during etching, i.e. do adsorbates desorb in the time that it takes to move the sample to the XPS chamber, or is the surface further chlorinated during the pump down and transfer period. To address these issues, we have developed an *in situ* analysis technique. We use laser pulses to thermally desorb and detect adsorbates present during etching. In the present case, the same pulse excites LIF of SiCl [11-14], unambiguously identified from the observed fluorescence spectrum corresponding to the transition ($B^2\Sigma_o^+ \rightarrow X^2\Pi$). The signal observed as a function of time (Figure 4) at a fixed wavelength (2930Å) is a real-time, relative measure of the coverage of Cl during exposure to Cl_2 with the plasma either on or off [11]. We see that the Cl coverage from exposure to Cl_2 immediately doubles and then remains nearly constant when the plasma is turned on. When the plasma is turned off with the laser blocked, the Cl_2 is pumped away, and laser irradiation is resumed (~5min later), the signal observed on the first laser pulse is nearly equal to (about 90% of) the steady-state signal during etching, indicating that the surface has changed little during this 5 min period. Consequently, the chemical topographies derived form XPS measurements are a good reflection of the surfaces during etching.

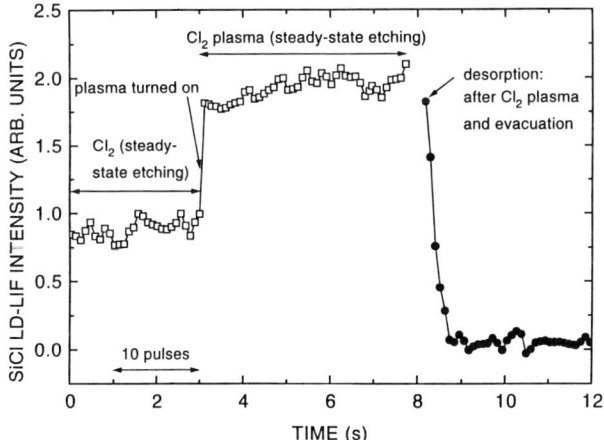

Figure 4. LIF intensity of laser-desorbed SiCl (2930Å) as a function of time. The open squares show steady-state laser-induced etching of Si by Cl_2, and then etching with the plasma suddenly turned on (at 3 s). (●): Time dependence of desorbed SiCl after chlorination with the plasma and subsequent pump down. For comparison, this trace has been placed immediately after the steady-state measurement. Laser repetition rate = 5 Hz, laser fluence=0.5J/cm^2.

We have also used the laser-desorption method to determine Cl coverage in real time as a function of discharge power, sample bias, and pressure. These results are reported elsewhere [11]. We show the power and bias dependences in Figures 5 and 6. Also plotted in Figures 5 and 6 are the poly-Si etch rates and (in Figure 6) the saturated ion current density. The Cl coverage increases rapidly with power and saturates at ~100W. The etch rate rises less steeply with power, and saturates at the same power (~200W) at which the ion current density saturates. These results indicate that the etching rate is limited by the flux of ions to the surface, and not by the neutral chlorine flux, under typical etching conditions (a pressure of 1-10 mTorr, and a power density of ~1W/cm^3 (~400W in our plasma source)). This is also substantiated by an etching rate that is nearly independent of pressure. From etch rate measurements as a function of DC bias voltage (Figure 6) and saturated ion current density measurements (3.9×10^{16} cm^{-2}s^{-1}, independent of bias), we determined that the chemically enhanced ion sputtering yield (Si atoms-per-positive ion) is 0.38 at an ion energy of 50 eV and 0.60 at 120 eV.

CONCLUSIONS

We have used *ex situ* x-ray photoelectron spectroscopy, with vacuum sample transfer, to study etching of Si in a high-charge-density ($1-2 \times 10^{11}$ ions/cm^3), low pressure (0.3-10 mTorr) Cl_2 plasma. After etching in a pure Cl_2 plasma, polycrystalline Si films, masked with photoresist stripes, are covered with about one monolayer of Cl on the top of the mask, the side of the mask, the etched poly-Si sidewall, and the bottom of the etched trench. When oxygen is

Si (100) ETCHING IN A Cl$_2$ HELICAL RESONATOR PLASMA

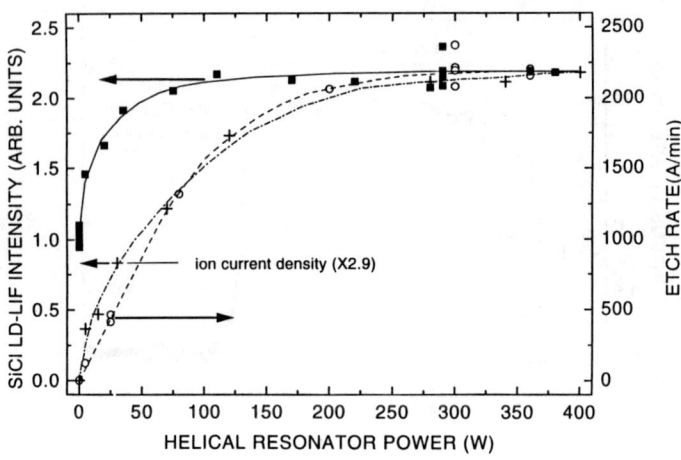

Figure 5. SiCl LD-LIF signal intensity (—■—), etch rate (--O--), and I_{sat}, the saturated ion current density (-.-+-.-) vs. helical resonator power. I_{sat} can be obtained by multiplying the left y-axis scale by 2.9 (i.e. $I_{sat} = 6.3 \text{mA/cm}^2$ at 400 W). Other conditions are the same as for Figure 4.

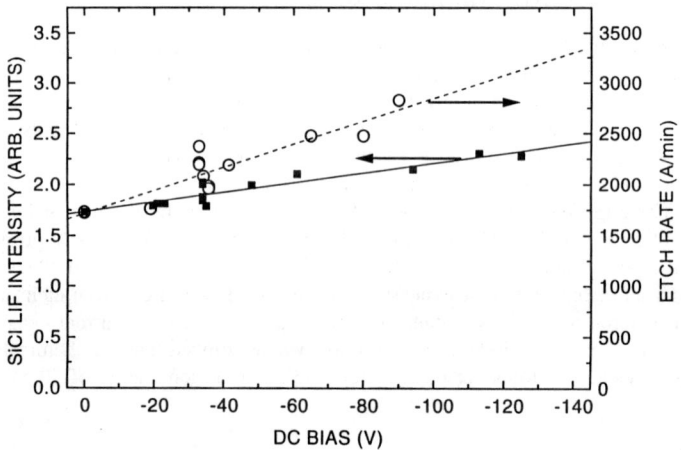

Figure 6. SiCl LD-LIF signal intensity (■) and etch rate (O) vs. substrate bias voltage. Other conditions are the same as for Figure 4.

present in the plasma, either by addition of O_2 or by erosion of the glass discharge tube, then a thin Si-oxide layer forms on the sides of both the poly-Si and the photoresist.

Using *in situ* laser-induced thermal desorption, combined with laser-induced fluorescence detection of SiCl, we have shown that the Si-chloride layer formed during etching is stable after the plasma is extinguished, so the XPS measurements are representative of the surface during etching. Laser-desorption measurements as a function of pressure and discharge power also show that the etching rate is limited by the positive ion flux to the surface, and not by the supply of Cl_2 at pressures above 0.5 mTorr, and ion fluxes of $\sim 4 \times 10^{16} \text{cm}^{-2}\text{s}^{-1}$. The chemically enhanced sputtering yield is 0.38 Si atoms-per-ion at 50 eV ion energy and 0.60 at 120 eV.

ACKNOWLEDGMENTS

Acknowledgment is made to the donors of the Petroleum Research Fund, administered by the ACS, for partial support of I. P. Herman's involvement in this research.

REFERENCES

(a) Also with the Department of Applied Physics and the Columbia Radiation Laboratory, Columbia University, New York, NY, 10027.

1. D. L. Flamm, V. M. Donnelly, and D. E. Ibbotson, in *VLSI Electronics Microstructure Science*, eds. N. G. Einspruch and D. M. Brown, (Academic Press, Inc., 1984), pp. 189-251.

2. H. F. Winters and J. W. Coburn, Surface Science Reports, *14*, 161 (1992).

3. C. S. Fadley, Prog. Surf. Sci. *16*, 275 (1984).

4. C. S. Fadley, Prog. in Solid State Chem. *11*, 265 (1976).

5. G. S. Oehrlein, K. K. Chan, and M. A. Jaso, J. Appl. Phys. *64*, 2399 (1988).

6. G. S. Oehrlein, A. A. Bright, and S. W. Robey, J. Vac. Sci. Technol. *A6*, 1989 (1988).

7. G. S. Oehrlein, J. F. Rembetski, and E. H. Payne, J. Vac. Sci. Technol. *B8*, 1199 (1990).

8. G. S. Oehrlein, K. K. Chan, M. A. Jaso, and G. W. Rubloff, J. Vac. Sci. Technol. *A7*, 1030 (1989).

9. K. V. Guinn and V. M. Donnelly, J. Appl. Phys., in press (1993).

10. K. V. Guinn, C. C. Cheng, and V. M. Donnelly, J. Vac. Sci. Technol., submitted for publication.

11. C. C. Cheng, K. V. Guinn, V. M. Donnelly, and I. P. Herman, J. Vac. Sci. Technol. submitted for publication (1994).

12. I. P. Herman, K. V. Guinn, C. C. Cheng, and V. M. Donnelly, to be published (1994).

13. *Spectroscopic Data, Vol. 1, Heteronuclear Diatomic Molecules*, ed. S. N. Suchard (IFI Plenum, NY, 1975), part B, p.992.

14. A. Aliouchouche, J. Boulmer, B. Bourguignon, J.-P. Budin, D. Debarre, and A. Desmur, Appl. Surf. Sci. *69*, 52 (1993).

KINETICS OF REACTIVE ION ETCHING OF POLYMERS IN AN OXYGEN PLASMA:
THE IMPORTANCE OF DIRECT REACTIVE ION ETCHING

SANDRA W. GRAHAM and CHRISTOPH STEINBRÜCHEL
Materials Engineering Department and Center for Integrated Electronics
Rensselaer Polytechnic Institute
Troy, NY 12180

ABSTRACT

The etching of polymer films in oxygen-based plasmas has been studied between 5 and 100 mTorr in a reactive ion etch reactor using Langmuir probe and optical actinometry measurements. Results for the etch yield (the number of carbon atoms removed per incident ion) are analyzed in terms of a surface-chemical model for ion-enhanced etching proposed by Joubert et al. (J. Appl. Phys. 65, 5096 (1989)). A proper description of the results requires that this model be modified by including a term due to direct reactive ion etching and physical sputtering. The contribution by direct reactive ion etching to the overall etching turns out to be significant under all conditions and even dominant at the lowest pressures. The modified model should be applicable to the etching of polymers in other types of reactors, especially high-plasma-density reactors. The relationship between these results and the anisotropic patterning of polymer films is also discussed.

INTRODUCTION

Dry etching of polymers in oxygen-based plasmas has been an essential step in microelectronics fabrication for many years. The earliest application of the process was for the stripping of photoresist by isotropic etching [1]. More recently, fine-line patterning of polymers by anisotropic etching has become important [2-4]. The general features of the process are fairly well understood: oxygen atoms are the main neutral reactants, and in the presence of ion bombardment the etch reaction is ion-enhanced. However, a detailed picture of all the components contributing to the overall process is still lacking. Specifically, it is unclear which of the four possible mechanisms for ion-enhanced etching, (1) surface-damage-promoted etching, (2) chemical sputtering, (3) chemically enhanced physical sputtering [5], or (4) direct reactive ion etching [6], are dominant. Joubert et al. have proposed a surface-chemical model for the ion-enhanced etching of polymers [7], without specifying which of the mechanisms (1) to (3) are really involved. It is the purpose of this paper to demonstrate that in order to obtain good agreement with experiments, it is necessary to include mechanism (4), direct reactive ion etching. Our results show, in fact, that direct reactive ion etching is important for the reactive ion etching of polymers in the entire pressure range from 5 to 100 mTorr, and is even dominant at the lowest pressures.

THEORETICAL BACKGROUND

In the surface kinetics model of Joubert et al [7], the surface concentration of oxygen atoms is assumed to be given by the following mass balance:

$$aF\Theta = b[O](1-\Theta) \tag{1}$$

where F is the ion flux to the substrate (ions/square cm and s), [O] is the oxygen atom concentration in the plasma, Θ is the oxygen surface coverage, a is an ion-induced desorption factor containing the etch yield, and b is an adsorption factor containing the sticking coefficient of O atoms. The left side of Eq. 1 represents removal of oxygen by ion-enhanced etching, and the right side of Eq. 1 represents adsorption of oxygen. From Eq. 1, the following expression follows for the etch rate R, assuming a steady state for O:

$$R = \frac{abnF[O]}{aF + b[O]} \tag{2}$$

where R is the number of C atoms removed per square cm and s, and n is the number of C atoms removed per O atom removed (n = 0.5 to 1).

In order to incorporate into the model the effects of direct reactive ion etching and sputtering, we propose to add another term cF in Eq. 2 [8]:

$$R = \frac{abnF[O]}{aF + b[O]} + cF \tag{3}$$

The constant c in Eq. 3 denotes the sum of the direct ion etch and the sputtering yield, and cF represents the etch rate in the limit of [O] = 0. This limit, which is almost realized in ion beam etch experiments [9], is not described properly by Joubert's model since it gives R = 0 when [O] = 0.

Now, if one divides Eq. 3 by the ion flux F, and if one defines the total etch yield Y by Y = R/F, so that Y stands for the total number of C atoms removed per oxygen ion,, then Eq. 3 can be rewritten in the form

$$1/Y = F/R = \frac{1 + AF/[O]}{B + CF/[O]} \tag{4}$$

where A = a/b, B = an + c, and C = ca/b. If c = 0 (thus C = 0) as in Joubert's model, then Eq. 4 simplifies to

$$1/Y = F/R = (1 + AF/[O])/B \tag{5}$$

An expression essentially equivalent to Eq. 5 has been used by Joubert et al. [7] and Carl et al. [10]. One way of comparing Eqs. 4 and 5 is to note that 1/Y is a non-linear function of F/[O] in Eq. 4, but a linear function of F/[O] in Eq. 5. The goal of this work is to show which of these equations provides a better description of experimental results.

EXPERIMENTAL DETAILS

The details of the present experiments are the same as described before [8]. The etch reactor was an Applied Materials 8130 hexode reactor. Ion densities and electron temperatures were measured as a function of position above a wafer with a Langmuir probe and analyzed as described previously [11,12]. From the probe data, the ion flux to the substrate was determined according to the Bohm criterion [12]. Optical emission actinometry was performed using the 844.5 nm oxygen line and the 750.4 nm argon line. Emission from O atoms represents an average across the discharge.

The polymer films investigated were: (1) Shipley 1400-27 positive photoresist, (2) Ciba-Geigy Probimide 285 polyimide, and (3) amorphous carbon [8]. Experiments were performed at pressures between 5 and 100 mTorr, either in pure oxygen or in oxygen with varying amounts of carbon tetrafluoride and nitrogen. The rf power was held constant at 1000 Watts.

Various loading conditions were also examined. No loading refers to the situation where all bare silicon wafers and exposed polymer (cloctrode-covering) surfaces were removed except for the wafer being etched. Medium loading was achieved by covering all polymer surfaces but leaving all bare dummy Si wafers exposed to the plasma. High loading was obtained by exposing all Si wafers and polymer surfaces to the plasma. Further details can be found in [13].

The total etch yield Y was deduced from the formula [14]

$$Y = \frac{R}{F} = r \frac{d(N_c - N_o)}{MF} N \qquad (6)$$

where r is the etch rate in cm/s (i.e. the rate of change in polymer film thickness), d is the density of the polymer film, N_c is the number of C atoms per monomer, N_o is the number of O atoms per monomer, M is the molecular weight of the monomer, and N is Avogadro's number (see Eqs. 2 and 4). An assumption implicit in Eq. 6 is that oxygen atoms in the polymer are used up to volatilize carbon atoms of the film. The etch rate r was measured interferometrically.

RESULTS AND DISCUSSION

Data were collected for 30 different combinations of sample material, gas composition, and reactor loading configuration. Regression analysis was performed to fit both Eq. 4 and Eq. 5 to the data. In about half the cases, Eq. 4 clearly provided the better fit (see Figs. 1 and 2), whereas in the other half of the cases neither equation was better than the other. (Figs. 1 and 2 also show an additional fitting curve for which c, the yield for direct ion etching and sputtering, was fixed at a value of 2, see below, but this case is seen to be almost equivalent to the linear Eq. 5). The non-linear relationship between 1/Y and F/[O] is particularly evident for those sets of data which had been taken over the widest possible range of pressures (5 - 100 mTorr). The same trends were also observed at different loadings [13]. We note that the lowest values for 1/Y (i.e. the largest yields) in Figs. 1 and 2, at the low end of the F/[O] scale, correspond to the highest pressures, and vice versa.

It is worth pointing out that the shape of the curves in Figs. 1 and 2 is independent of the accuracy of determining the ion flux F to the substrate, since F enters as a factor common to both the x- and the y-scales (F/[O] and 1/Y = F/R). The effects of other sources of uncertainty have been discussed elsewhere [13] and do not affect the conclusions reached here either.

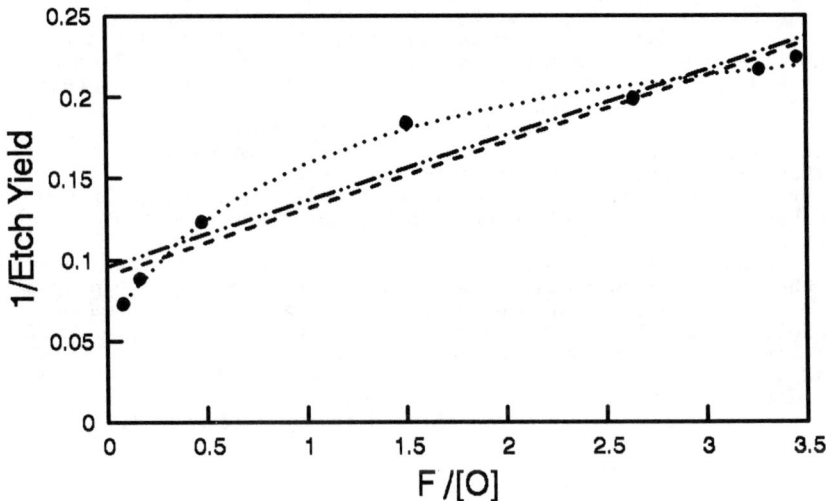

Fig. 1: Plot of data for 1/Y vs F/[O] of photoresist in pure oxygen plasma (low loading), with various model fits.
... : Eq. 4;   --- : Eq. 5;   -..- : Eq. 4 with c = 2.

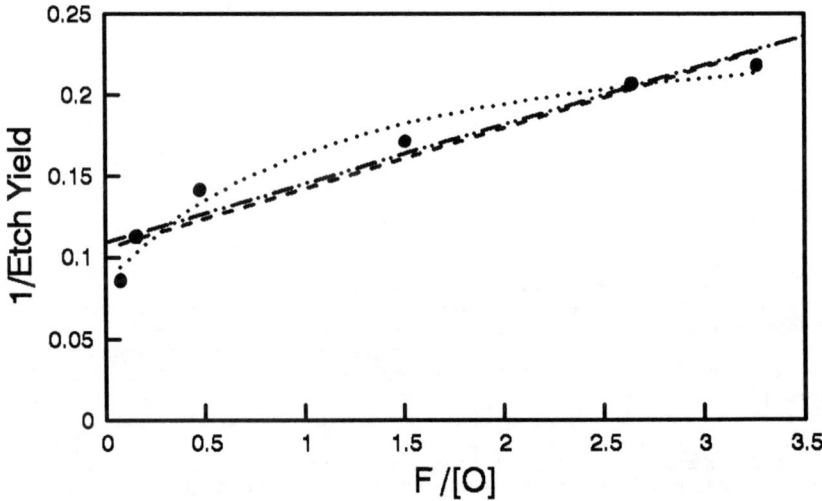

Fig. 2: Plot of data for 1/Y vs F/[O] of polyimide in pure oxygen plasma (low loading), with various model fits.
... : Eq. 4;   --- : Eq. 5;   -..- : Eq. 4 with c = 2.

Sets of fitting parameters corresponding to the data of Fig.1 are given in the tables below:

Table I: Fitting parameters for data of Fig. 1

	A	B	C
Eq. 4	4.0	16.5	14.7
Eq. 5	0.46	11.1	-

Table II: Contribution of direct reactive ion etching and sputtering to total etching for data of Fig. 1

pressure (mTorr)	Y	c	c/Y
100	13.6	3.7	0.27
60	11.4	3.7	0.32
20	8.1	3.7	0.46
5	4.5	3.7	0.81

The fitting parameters corresponding to the data of Fig. 2 are similar to those given above in Tables I and II. We note that the yield c for direct reactive ion etching plus sputtering is obtained from the fitting parameters A, B, and C by $c = C/A$. The constants a and b in the surface kinetic model (Eq. 3) can be deduced in a similar fashion. The value chosen for n (n between 0.5 and 1) turns out not to affect the fit in a significant manner.

It is apparent from these numbers that direct reactive ion etching plus sputtering are the dominant etch mechanism at the lowest pressures (5-10 mTorr) and still contribute almost 30 % of the overall etching at 100 mTorr. If the sputtering yield under our conditions is estimated to be about 0.3, from ion beam experiments [9], then the direct reactive ion etch yield is of the order of 3 C atoms per ion. This number is somewhat larger than a yield of about 2 expected on the basis that molecular oxygen ions react completely to give CO as a product. However, SIMS experiments, bombarding a polymer with oxygen ions, indicate that it is possible to produce species containing more than one C atom per O atom [15]. Furthermore, the experimental total yields reported here are in good agreement with previous results from the reactive ion etching and the magnetron ion etching of polymers in different reactors [2].

CONCLUSIONS

The etching of polymers in oxygen-based plasmas has been shown to require inclusion of direct reactive ion etching and sputtering for a reasonably complete description of the etch yields. In a reactive ion etch process, direct reactive ion etching accounts for anywhere between 80 % and 30 % of the total etch yield in a pressure range from about 5 to 100 mTorr. This mechanism should clearly be taken into account also in corresponding experiments involving high-density, magnetically enhanced plasmas. Future work in our laboratory will incorporate isotropic etching into the surface-kinetic model of the process and will examine to what extent such a model can form the basis for a description of the evolution of etched profiles.

REFERENCES

[1] G. N. Taylor and T. M. Wolf, Polymer Sci. Eng. 20, 1087 (1980).
[2] Ch. Steinbrüchel, B. J. Curtis, H. W. Lehmann, and R. Widmer, IEEE Trans. Plasma Sci. PS-14, 137 (1986).
[3] J. E. Heidenreich, J. R. Paraszczak, M. Moisan, and G. Sauve, Microel. Eng. 5, 363 (1986).
[4] M. A. Hartney, D. W. Hess, and D. S. Soane, J. Vac. Sci. Technol. B7, 1 (1989), and references therein.
[5] H. F. Winters, J. W. Coburn, and T. J. Chuang, J. Vac. Sci. Technol. B1, 469 (1983).
[6] Ch. Steinbrüchel, H. W. Lehmann, and K. Frick, J. Electrochem. Soc. 132, 180 (1985).
[7] O. Joubert, J. Pelletier, and Y. Arnal, J. Appl. Phys. 65, 5096 (1989).
[8] S. W. Graham and Ch. Steinbrüchel, Mat. Res. Soc. Symp. Proc. 282, 617 (1993).
[9] H. Gokan and S. Esho, J. Electrochem. Soc. 131, 1105 (1984).
[10] D. A. Carl, D. W. Hess, and M. A. Lieberman, J. Appl. Phys. 68, 1859 (1990).
[11] Ch. Steinbrüchel, J. Electrochem. Soc. 130, 648 (1983).
[12] Ch. Steinbrüchel, J. Vac. Sci. Technol. A8, 1663 (1990).
[13] S. W. Graham, Ph.D. Thesis, Rensselaer Polytechnic Institute (1993).
[14] H. Gokan, S. Esho, and Y. Onishi, J. Electrochem. Soc. 130, 143 (1983).
[15] M. R. Lorenz, V. J. Novotny, and V. R. Deline, Surf. Sci. 250, 112 (1991).

HIGH SELECTIVE ETCHING OF SiO2/Si
BY ArF EXCIMER LASER

K. KITAMURA* and M. MURAHARA**
*Graduate student of Tokai University, Faculty of Engineering
**Faculty of Engineering, Tokai University,
1117, Kitakaname, Hiratsuka, Kanagawa 259-12, JAPAN

ABSTRACT

Dry etching of SiO2 insulation layer has been required in the Si semiconductor manufacturing process. The etching of SiO2/Si is chemically carried out by using HF solution. We successfully demonstrated a new method for exclusive etching of SiO2 using the nitrosyle fluoride (NFO) gas which was produced from the mixed gas of NF3 and O2 with an ArF excimer laser irradiation.

SiO2 and Si substrates were placed side by side in a reaction cell which was filled with 3% O2 gas in NF3 at the gas pressure of 380 Torr. ArF excimer laser beam was irradiated parallel to the substrates. The laser fluence was kept at 100mJ/cm2. As soon as the mixed gas of NF3 and O2 was irradiated with the ArF laser beam, an intermediate product of NFO was produced. The chemical behavior of NFO was confirmed from the UV absorption spectrum with absorption in the 310 to 330nm wavelength region. In the presence of SiO2, the absorption of NFO diminished. The absorption of NO2, instead of NFO, appeared at 350nm. This indicates that the oxygen atoms of SiO2 were pulled out by NFO.

The etching reactions continued for 3 minutes after irradiation when the SiO2 and Si substrates were kept in an atmosphere of the reactant gases. As a result, not the Si but SiO2 substrate was etched with the depth of 2000Å.

INTRODUCTION

In the LSI process, the etching of SiO2 insulation layers is important. The HF solution has been used for wet etching; however, there is a resolution limit with this method. A dry etching such as a plasma bombarding method has thus been widely employed. The plasma method, unfortunately, induces etching not only on the SiO2 layer but also on the Si semiconductor layer. Thus, an exclusive SiO2 etching is required and has been extensively studied [1,2]. In this work, we successfully demonstrated the selective etching of SiO2 using the NFO gas, which was produced from a mixed gas of NF3 and O2 by an ArF excimer laser irradiation [3,4,5]. In this novel etching method, the NFO molecules were produced by the photochemical reaction of the mixed gas of NF3 and O2, which is very important [6]. The NFO gas pulled the oxygen atoms out of the SiO2 insulation layer; the isolated Si atoms of the etched SiO2 layers were then captured by fluorine atoms [7,8,9]. The exclusive SiO2 etching method was found, and its etching mechanism was clarified by the spectroscopic measurement.

PHOTOCHEMICAL REACTION

Production of NFO gas

NF3 gas has an absorption band at the 193nm wavelength of ArF excimer laser; the N-F dissociation energy of 72kcal/mol is lower than the laser photon energy of 147kcal. Accordingly, ArF laser decomposes NF3 into NFx and F, as given in the following equation:

$$NF_3 + h\nu(193nm) \longrightarrow NF_x + F \quad (x=0\sim2) \quad \text{---------(1)}$$

In order to clarify this photo decomposition mechanism, the ultraviolet absorption spectra of the mixed gas of NF3 and O2 were measured by the spectroscopic measurement system as Fig.1. Fig.2 shows the UV spectra of the NF3 and O2 mixed gas with and without the ArF laser irradiation. When the mixed gas was irradiated with ArF laser light, an absorption appeared at 330nm wavelength, and the spectrum was identified as NFO. The absorption gradually increased with the laser shot repetition. Thereupon, O2 concentration in the mixture ratio was changed from 1 to 6% in order to produce as many NFO molecules as possible. In the case of 1 to 3% O2 concentration, as shows in Fig.3, the NFO UV absorption gradually increased. The NFO absorption remarkably diminished at the O2 concentration of 6%. As a result, an absorption of 350nm appeared, along with F2 IR absorption [10], and was identified as NO2. This is caused by the NFO molecules to easily combine with oxgen atoms in the gas phase. The amount of NO2 thus becomes more than that of NFO.

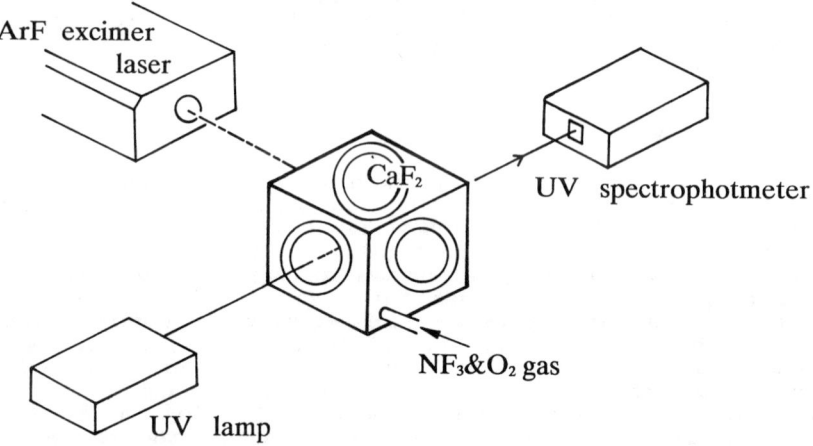

Fig.1 Measurement system of UV spectra with and without ArF laser irradiation in an atmosphere of NF3 and O2 mixed gas.

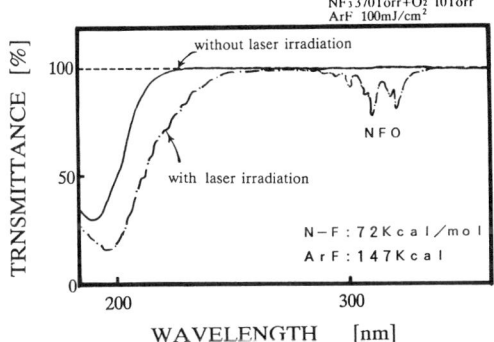

Fig.2 UV spectra of the NF3 and O2 mixed gas with and without ArF laser irradiation.

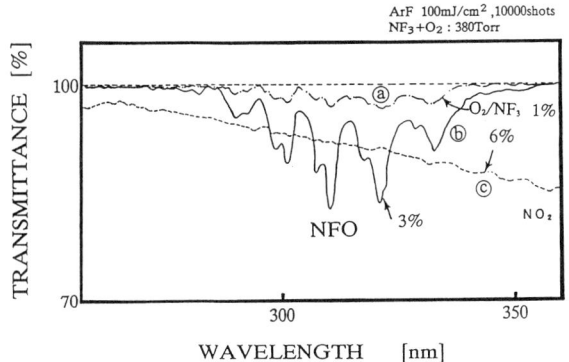

Fig.3 UV spectra of intermediate products for three different percentage of O2 in NF3.

Etching mechanism of SiO2

SiO2 substrate was put in an atmosphere of the mixed gas of NF3 and 3% O2 ambient. When the mixed gas was irradiated with ArF laser beam, NO2 absorption increased and NFO decreased. This means that the oxygen atoms of the SiO2 insulation layer were pulled out by the NFO molecules.

As mentioned above, NFO was produced by the ArF laser irradiation in an optimum mixture ratio of 3% O2 in the NF3 gas. The NFO have high chemical activities and can etch the SiO2 layers. In the reduction of SiO2, the oxygen atoms from SiO2 were pulled by the F atoms in the gas phase and the Si atoms became isolated. The etching mechanism of SiO2 was given by the following equations:

$$2NF_3 + O_2 + h\nu(193nm) \longrightarrow 2NFO + 2F_2 \quad \text{---------------(2)}$$
$$SiO_2 + 2NFO + 2F_2 \longrightarrow 2NO_2 + SiF_4 \quad \text{---------------(3)}$$

Experimental method

As illustrated in Fig.4, the SiO2 and Si substrates which were coated with PMMA resist were placed in a reaction cell with the CaF2 window. A thermaloxide Si wafer was employed as the SiO2 sample. The cell was evacuated and then filled with NF3 and O2 gases. The ArF laser was irradiated parrarel to the substrates, which were kept in an atmosphere of the reactant gases for 3 minutes after irradiation. In the experiment, the mixture ratio of NF3 and O2 gases and the ArF laser fluence were changed. The SiO2 etching rate was compared with that of Si under each condition. The ArF laser shot number was 10000 at a pulse repetition of 20pps; the total gas pressure was kept at 380 Torr.

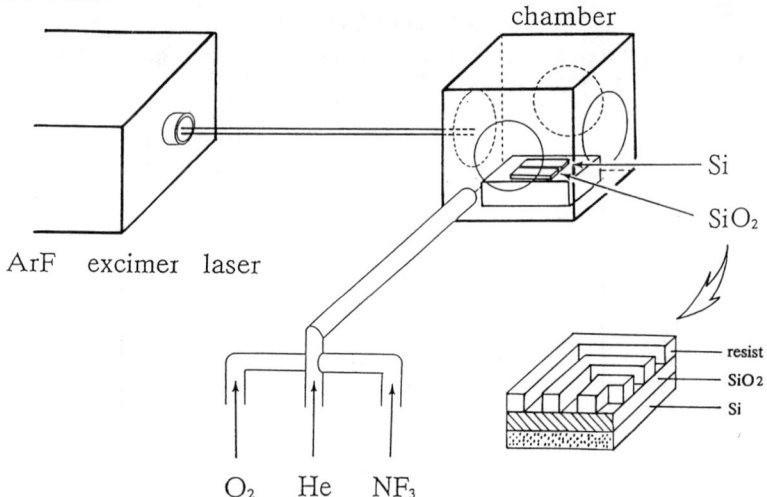

Fig.4 Shematic diagram of SiO2/Si selective etching system by ArF excimer laser

Results and discussion

In order to find the optimum conditions for an exclusive SiO2 etching, experiments with different mixture ratios of NF3 and O2 gases as well as with different fluences of ArF laser were carried out. The etching rates of SiO2 and Si substrates were then measured.

First, displayed in Fig.5 is the etching rate as a function of the gas mixture ratio. The laser fluence and the shot number were kept at 100mJ/cm2 and 10000 shots, respectively. The respective etching depths on the SiO2 and Si layers were 2000Å and 150Å when the partial pressure of NF3 and O2 was 100:3; this was the optimum condition. When the partial pressure became either lower or higher than 100:3, the Si layerwas also etched.

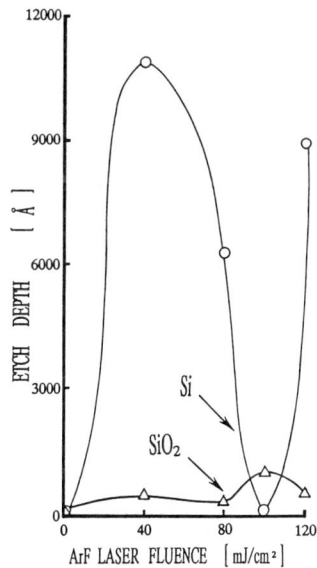
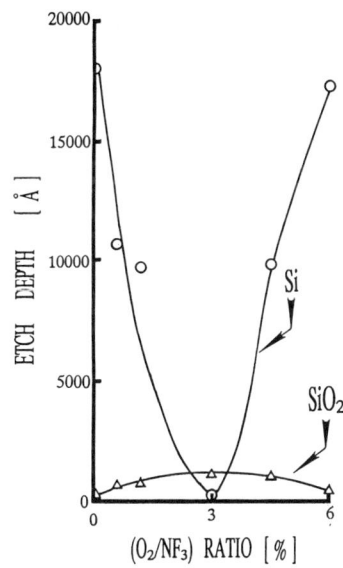

Fig.5 Dependence of the etching depth on O2 mixing percentage against NF3 for SiO2 and Si substrates.

Fig.6 Dependence of the etching depth on laser fluence for SiO2 Si substrates.

Second, indicated in Fig.6 is the etching depth as a function of the laser fluence when the mixture ratio was set at 3 %. When the laser fluence was lower than 40mJ/cm2, the Si layer was etched deeper than the SiO2. As the fluence was increased, the etch depth of Si had a peak at 40mJ/cm2 and then decreased to about zero at a fluence of 100mJ/cm2. When the fluence increased to 120mJ/cm2, Si was etched again. It appears that the most effective SiO2 etching was at a fluence of 100mJ/cm2, of which condition described above.

The following were found from the results of the two experiments. When the laser fluence was higher than 100mJ/cm2, the Si was etched by the F radicals because the oxygen content was insufficient for photo-decomposition of NF3 gas to NFO. On the other hand, when the laser fluence was lower than 100mJ/cm2, the Si was etched by the F radicals because the oxygen content was excessive and oxygen reacted with the NFO to form NO2 and F radicals. That is, the large amount of NFO is required in the reaction cell for exclusive etching of SiO2. The SEM photo of the etched feature obtained under the optimum O2/NF3 condition was shown as in Fig.7. The space between lines was 5μm, and the etching depth was 2000Å.

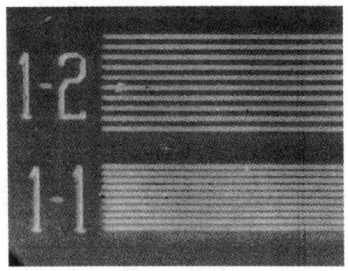 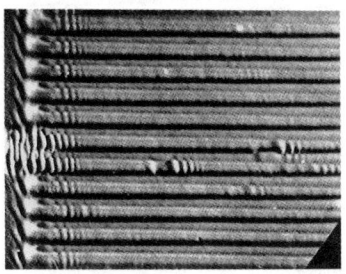

Fig.7 Etched feature of the thermal oxide Si wafer

Conclusion

The selective etching of SiO2 was successfuly demonstrated by using the NFO gas and the ArF excimer laser beam. The production amount of the NFO gas was closely related to the selectivity of SiO2 etching. The amount of NFO produced critically depended on the laser fluence and the partial pressure of NF3 and O2. It was substantiated that the oxide insulation film which covered the semiconductor surface was exclusively etched. It was presumed that the mechanism of the SiO2/Si selective etching was the deoxidization reactions by NFO molecules. This new method may be applied for improvement of high integration and high efficiency semiconductor devices.

Reference

1) S.Yokoyama, Y.Ymakage and M.Hirose, APL, 47, 389 (1985)
2) T.Akimoto, K.Harasima and K.Kasama,Extended Abstracts (The 40th Spring Meeting, 1993); The Japan Society of Applied Physics and Related Societies, 614, 2
3) K.Yamazaki, K.Tsunokuni and K.Nojiri,Extended Abstracts (The 40th Spring Meeting, 1993); The Japan Society of Applied Physics and Related Societies, 517, 2
4) J.H.Brannon, J. Phys. Chem. 90, 1784 (1986)
5) G.L.Loper and M.D.Tabat, APL, 46,654 (1985)
6) A.L.Flores and B. deB. Darwent, J. Phys. chem, 73, 7 (1969)
7) T.Obara and M.Murahara, Mater, Res, Soc, Symp. Proc. 26, 65 (1992)
8) M.Murahara, M.Yonekawa and K.Sirakawa, CLEO'90, CWB6 (1990)
9) M.Murahara, S.Aomori and M.Yonekawa, CLEO'91, CWF55 (1991)
10) A.Aomori and M.Murahara, Mater, Res, Soc, Symp. Proc. Vol.236,9(1992)

Plasma Etching and Surface Analysis of a-SiC:H Films Deposited by Low Temperature Plasma Enhanced Chemical Vapor Deposition

J. H. Thomas, III,* David Sarnoff Research Center, Princeton NJ 08543
and M. J. Loboda, and J. A. Seifferly, Dow Corning Corp. Midland MI 48686

Abstract

Organosilicon precursors were used to deposit thin films of highly stable amorphous hydrogenated silicon carbide at low temperatures by plasma enhanced chemical vapor deposition. X-ray photoemission and Auger electron spectroscopy were employed to characterize the surface chemistry of air exposed and plasma etched films. Auger analysis of the as-deposited material shows its composition to be constant throughout its depth. RF plasma etching was performed in CF_4/O_2 and SF_6/O_2 gas mixtures. Etch rates in these atmospheres are similar to those reported in the literature. After plasma etching, the surface was converted to a highly fluorinated state. In addition to the expected SiC bond, Si-OF bonds were found after plasma etching. Fluorocarbon residue was not produced in this process. Chlorine etching of this low temperature PECVD film is described and is shown to be compatible with standard integrated circuit manufacture processing as an hermetic-like sealant.

Introduction

Hydrogenated silicon carbide derived from plasma enhanced chemical vapor deposition (PECVD) has been investigated for many uses. In this study, we have investigated the compatibility to silicon device processing of reactive ion etching (RIE) gases, such as, pure chlorine, CF_4/O_2 (1-3) and SF_6/O_2 (3,4). The surface of the etched a-SiC:H was investigated using x-ray photoelectron spectroscopy. The material system of interest is that used to form ceramic-based hermetic sealed integrated circuits as proposed by Chandra (5). This requires that a-SiC:H be applied at low temperatures on SiO_2 coatings and adhere to gold. It is preferred that the a-SiC:H etch stop at the SiO_2 surface. Chlorine was chosen as the preferred etchant. Chlorine and the other etchants of a-SiC:H will be described in this paper.

Experimental

The details of film growth parameters and composition are shown in table 1. Films were deposited using standard PECVD techniques (6) from trimethylsilane and silacyclobutane source gasses. The carrier gas used was argon at 2 Torr, deposition temperature =250 - 350°C. Layer stoichiometry and uniformity was studied using Auger electron spectroscopy. a-SiC:H was found to be uniform throughout the deposited layer and has a thin oxide layer at the surface. Within experimental error, the C/Si ratio was held in the range of 1 - 1.2. The oxygen content was less than 3 at. % and the hydrogen content was 20-30 at. %.

Fluorine based etchants have been shown to be effective in removing a-SiC:H, if the gas contains an appreciable amount of oxygen. This results in SiC etch selectivity over silicon and silicon dioxide (1-4). The selectivity is not large and requires significant ion bombardment, that is, etching does not occur spontaneously. To demonstrate etching with fluorine based etchants, the CF_4/O_2 and SF_6/O_2 systems were chosen. Published data describing the gas mixture that provides the highest selectivity and etch rate were chosen as operating points.

In the CF_4/O_2 etchant system the ratio of CF_4 to O_2 was set at 0.44. The operating pressure was set at 100 millitorr and the plasma was RF generated at 13.56 MHz, 30 watts to develop a self bias of - 240 volts. The etching system is described in detail in reference 7 and is operated in a diode configuration.

In the SF_6/O_2 etchant system (ratio of $SF_6/O_2 = 0.25$), measurements were performed in an Innotec Vertical In-line system. This system has large electrodes but the power was adjusted to roughly match the first experiment (-240 volts bias). The operating pressure was set at 200 millitorr. With a-SiC:H on SiO_2, the process could not be controlled to chemically stop at the interface since both materials are etched. The measured etch rate on a-SiC:H was ~ 100 nm/min. This etchant was also shown to sputter gold (SF_5^+ -> Au).

Experiments using chlorine plasma etching were performed in two reactors, a PlasmaTherm In-Line RIE system and an MRC Aries MERIE (magnetically enhanced RIE) reactor. Table 2 shows the etch rates as a function of operating power and pressure for the two materials deposited and for the two reactors. All etching was performed at 40 sccm chlorine flow rate.

X-ray photoelectron spectroscopy (XPS) measurements were performed on a system built around a Leybold Heraeus EA-10 hemispherical analyzer. This system is controlled using an HP-1000/E minicomputer and photoelectrons are excited using Mg K_a non-monochromated radiation. Survey spectra were obtained at a 200 eV pass energy and high resolution spectra at 50 eV pass energy. Data were obtained digitally at 1 eV increments for the survey spectra and 0.12 eV increments for the high resolution spectra.

Results and Discussion

Figure 1 shows a typical XPS survey spectra taken before and after exposing the a-SiC:H sample to the plasma without etching through the film to SiO_2. The sample was transferred into the analysis chamber without exposure to air. Prior to exposure, the surface shows peaks due to silicon, oxygen and carbon. Figure 2 shows the C 1s, O 1s, Si 2p and F 1s peaks before and after plasma exposure. High resolution spectra show that the surface is oxidized (also shown by Auger electron spectroscopy). After exposure, shifts are observed in the oxygen and carbon and structure developed in the silicon and fluorine peaks. After plasma exposure, the surface is fluorinated. However, the fluorine is not bound to the carbide surface. Chemically shifted peaks at binding energies of >289 eV may be expected in the C 1s spectrum (8) for bound fluorine to carbon. This structure is not observed. The Si 2p and F 1s spectra show a two peak structure. In the Si spectrum, one component at 100 eV is due to carbide bonding and the other component is due to $Si(O_xF_y)$ bonding. The components of the F 1s spectrum are due to Si-F bonding (685 eV) and Si-O-F bonding (688 eV). Exposing the carbide surface to the oxygen containing plasma results in oxidation of the surface and consequently the formation of the $Si(O_xF_y)$. The C 1s peak also shows the presence of the dominant Si-C bonding. Similar results have been reported in exposure of the surface of silicon to a CF_4/O_2 plasma (9).

Table 1
Characteristics of Organosilicon Source on PECVD a-SiC:H Films

Source Gases:	Silacyclobutane or $H_2SiCH_2CH_2CH_2$	Trimethylsilane $(CH_3)_3SiH$
Carrier Gas:	Argon	
Pressure:	2 Torr	
C/Si Ratio:*	1 < C/Si < 1.2	
Oxygen Content:	< 3 at. %	
Hydrogen Content:	20-30 at. %	

* from Auger Electron Spectroscopy

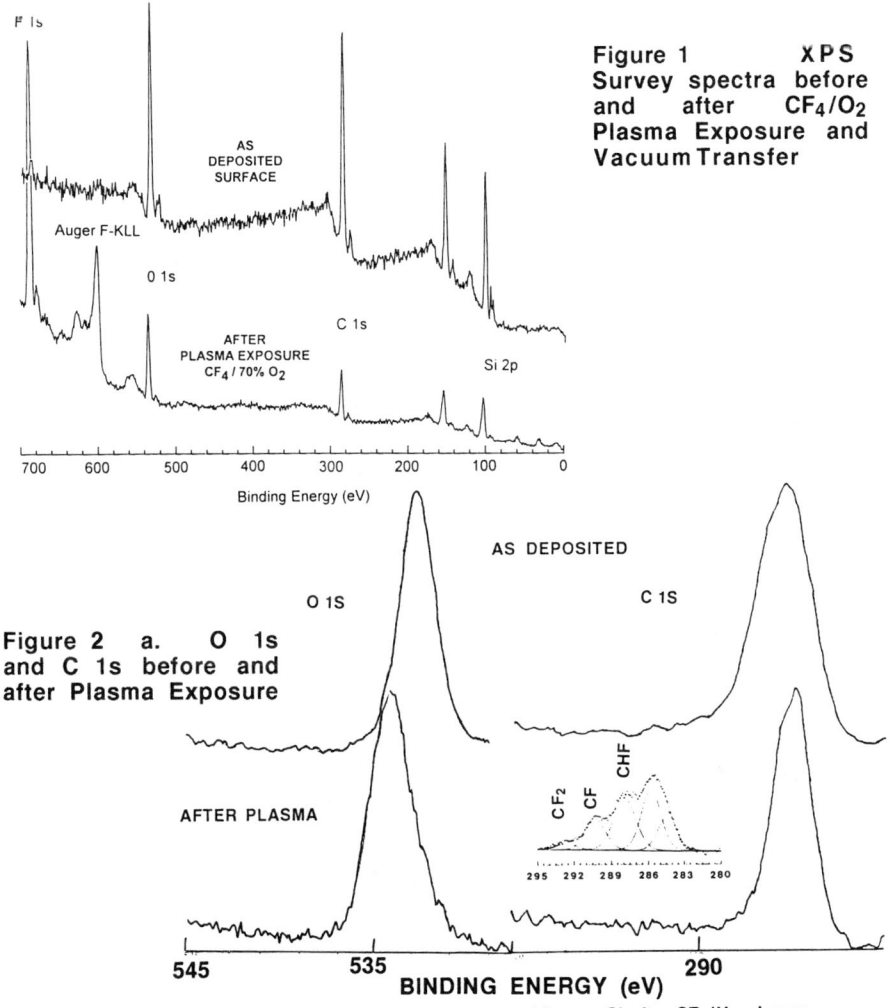

Figure 1 XPS Survey spectra before and after CF_4/O_2 Plasma Exposure and Vacuum Transfer

Figure 2 a. O 1s and C 1s before and after Plasma Exposure

Inset: C 1s - Fluorocarbon residue on Si after CF_4/H_2 plasma

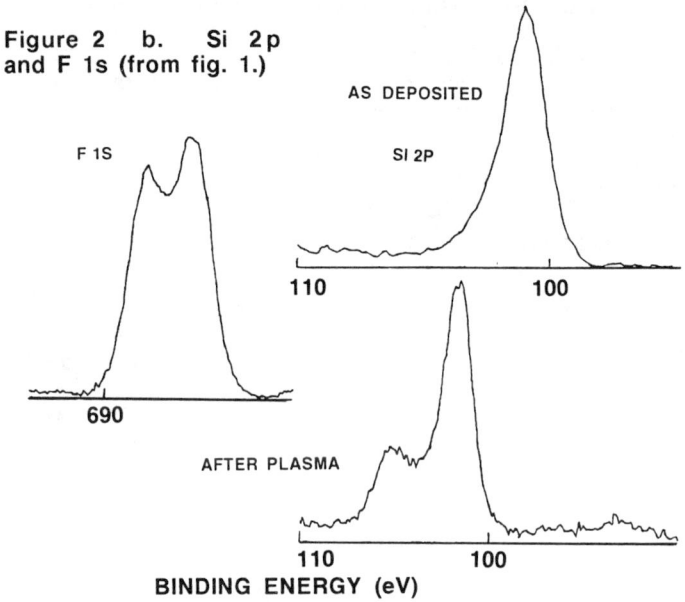

Figure 2 b. Si 2p and F 1s (from fig. 1.)

Figure 3   SEM micrograph of a-SiC:H / SiO_2 on Si after Cl Plasma etch and after HF etch

A similar effect is observed when etching a-SiC:H with the SF_6/O_2. Results are not shown here for brevity. However, there is a difference with respect to the CF_4 system in ion bombardment (10). The ion fragments in the SF_6/O_2 etchant system are heavier and hence sputtering yields may be higher than with the CF_4/O_2 system. When etching a-SiC:H/SiO$_2$/Au, the Innotec reactor etched through the SiO_2 and gold layer to the substrate. Without in-situ thickness monitoring in the etching chamber, this and the CF_4 etchant system would be difficult to control. Selectivity for these etchants have been published by other authors (1-4) and will not be reported here.

The etching mechanism appears similar to fluorine etching of silicon and silicon dioxide. Atomic fluorine and oxygen in the plasma reacts with the silicon and carbon to form SiF_x and CO_x species, respectively. With ion bombardment these species are removed at an accelerated rate over the rate found without ion bombardment.

A better approach to etching a-SiC:H selectively over SiO_2 is to use pure chlorine reactive ion etching. It is well known that chlorine RIE has a good selectivity of Si to SiO_2 (11,12). It is expected that the chlorine plasma will react with C, Si and H to form the volatile species CCl_4, $SiCl_4$ and HCl. Spontaneous etching is expected. The selectivity with respect to silicon dioxide is good at 10 and 300 milliTorr (see table 2). Photoresist is shown for completeness and etches at a comparable rate to a-SiC:H at 300 milliTorr. The source material does not appear to strongly affect the etch rate, that is, the use of Me$_3$SiH and SCB. Clearly, the chlorine etch is superior to the fluorine based etchants for etching a-SiC:H and should be effective in etching single crystal SiC. An example of chlorine etching is shown in figure 3 after chlorine etching and after wet etching the oxide layer with HF on a typical bond pad window on an integrated circuit.

X-ray photoelectron spectroscopy shows that the surface of the a-SiC:H exposed to a chlorine plasma is similar to that observed in fluorine etching where Cl is observed in place of F. The SiC peak of the Si2p is observed along with a peak due to Si-ClO$_x$ bonding (101.6 eV) where the ClOx is due to air exposure of the sample. Chlorine is bound to the carbon peak and shows a chemical shift due to C-Cl bonding (13). It is reasonable to project that the etching mechanism is the same as fluorine except the roles of Cl and F are switched making it possible for more volatile species to be formed.

Table 2

Chlorine Reactive Ion Etching

Film	Pressure (mTorr)	Power (watts)	Etch Rate nm/min
a-SiC:H from SCB (250C)	10	500	60
a-SiC:H from SCB (250C)	300	150	100
SiO$_2$ (Thermal)	10	500	26
SiO$_2$ (Thermal)	300	150	9.8
Photoresist	10	500	190
Photoresist	300	150	94
a-SiC:H from Me$_3$SiH (350C)*	10	500	60
a-SiC:H from Me$_3$SiH (350C)	300	150	97

Summary

In summary, we have observed that CF_4/O_2 based reactive ion etching leaves a typical SiO_yF_x type of layer behind and is not a good etchant for selectively etching SiC with respect to SiO_2. This reaction chemistry is similar to that found on other oxygen containing silicon compounds or silicon. SF_6/O_2 produces similar results to the CF_4/O_2 etchant. Etch rates of a-SiC:H using fluorocarbon and sulfur hexafluoride are in good agreement with results published for etching high temperature SiC films. However, the etching appears to depend more on ion bombardment rather than significant chemically induced etching of the a-SiC:H.

Chlorine etching at ~ 300 milliTorr is not sensitive to the starting material or composition (SCB or Me_3SiH) and can be used in hermetically sealing bond pads on integrated circuits. In using a pure chlorine etchant at moderate pressures (~ 300 mTorr), the a-SiC:H is removed with good selectivity against etching SiO_2. Chlorine appears to be the preferred etchant system that is compatible with integrated circuit manufacture.

Acknowledgments

The authors wish to acknowledge A. Levine, J.-S. Maa, V. Patel, J. Shaw, G. Chandra and K. Michael for their assistance and support throughout this project.

*Present Address: 3M CRL, 3M Center 201-2S-16, St. Paul MN 55144

References

1. J. Sugiura, W. J. Lu, K. C. Cadien and A. J. Steckl, J. Vac. Sci. Technol. B4, 349 (1986).
2. W. S. Pan and A. J. Steckl, Mat. Res. Soc. Symp. Proc. 76, 157 (1987).
3. W. S. Pan and A. J. Steckl, J. Electrochem. Soc. 137, 212 (1990).
4. J. W. Palmour, R. F. Davis, P. Astell-Burt, and P. Blackborow, Ceramic Trans. 2, 491 (1989).
5. Grish Chandra, Mat. Res. Soc. Symp. Proc. 203, 97 (1991).
6. M. J. Loboda, J. A. Seifferly and F. C. Dall, J. Vac. Sci. Technol. A 12(1), (1994), in print.
7. J. L. Vossen, J. H. Thomas, III, J.-S. Maa, O. R. Mesker and G. O. Fowler, J. Vac. Sci. Technol. A 1, 1452-1455(1983).
8. J. H. Thomas, III, L. K. White and N. Miszkowski, Journal of Vacuum Science and Technology A 8, 1706-1711 (1990).
9. J. H. Thomas III and J-S. Maa, Applied Physics Letters 43, 859-861 (1983).
10. J. H. Thomas, III and L. H. Hammer, Journal of Vacuum Science and Technology B 5, 1617-1621 (1987).
11. T. M. Mayer, R. A. Barker and L. J. Whitman, J. Vac. Sci. Technol. 18, 349 (1981).
12. C. J. Mogab and H. J. Levinstein, J. Vac. Sci. Technol. 17, 721 (1980).
13. H-P. Chang and J. H. Thomas, III, Journal of Electron Spectroscopy and related phenomenon 26, 203-212 (1982).

CONFINEMENT AND LOW-ENERGY EXTRACTION OF PHOTO-FRAGMENT IONS USING RF ION TRAPPING

Seiji Yamamoto, Kozo Mochiji, Isao Ochiai, and Naohiko Mikami*
Central Research Laboratory, Hitachi Ltd., Higashi-koigakubo 1-280, Kokubunji, Tokyo 185, Japan
*Department of Chemistry, Faculty of Science, Tohoku University, Aoba-ku, Sendai 980, Japan

ABSTRACT

Spatial confinement of mass-selected ions and extracting them to a solid surface at low kinetic energy is achieved by using rf ion trapping which is combined with multi-photon ionization by a KrF laser. SF_5^+ fragment ions of SF_6 generated by photo-ionization are separated from other ionic species and then confined in the cylindrical ion trap cell for at least 100 ms. The stored ions are extracted onto the seniconductor surface. This technique is promising for clarifying surface reactions of semiconductors, which is the key to controlling surface structure on an atomic scale.

INTRODUCTION

In the near future microfabrication on an atomic scale will be needed in processes such as dry etching. For future devices that have a scale of nanometer order, it is necessary to greatly improve material selectivity, damage effects and dimension accuracy. So far plasma etching using microwaves has been the main method for dry etching processes. However, it is difficult to make conventional plasma etching satisfy the above requierments because of the two main drawbacks of plasma etching. First, the selectivity for material is not very high because of many kinds of chemical species in plasma. Second, the kinetic energy of the etchant particles in the plasma is so large that the etching is mainly due to collision induced spattering. This in turn induces damage in the semiconductor. To prevent damage, we need to lower the kinetic energy of the particles. However, this lowers their reactivity. Instead of kinetic energy we must use the inner state potential of the particles. Dependence of chemical species on internal state for chemical properties was reported in the case of Si etching with Cl [1]. Nanometer scale etching is very likely to be achieved by using these chemical species whose internal states are specified and by uniformly irradiating them onto the semiconductor surfaces, that is, their kinetic energy and direction of motion are made uniform.

On the basis of the above discussion, we used ions as etchant species in this study because ions are easy to control and are advantageous for fine pattern fabrication. In the first step we selected ion species and gave them low kinetic energy. The next step will be to control their internal state. Conventionally the mass selection is achieved by using a quadrupole mass filter. However, this separates ions while they are being transmitted. Therefore it is not suitable for controlling the internal state of ions by laser because the intersection of the two beams is so small. To control the internal state by laser, it is necessary to confine the ions. The rf ion trapping method is a promising key to the confinement of ions for laser excitation. In this paper we report on the confinement and low-energy extraction of photo-fragment ions by using rf ion trapping.

EXPERIMENTAL

The ion trapping method and its general characteristics have been described in detail elsewhere [2-4].   Figure 1 shows a schematic of the apparatus and the timing chart of the experimental procedure. The cylindrical ion trap (CIT) cell consists of three electrodes: a cylindrical electrode and two endcap electrodes. The inner dimensions of the cell are 40 mm in diameter and 38 mm in height. The cylindrical electrode was biased by an rf voltage whose frequency was in the range of 100-200 kHz and whose amplitude (Vac) was less than 300 V. The end cap electrodes were kept grounded while ions were being trapped. A dc voltage (Vdc) could also be superimposed on the rf potential in order to separate the trapped ions depending on their charge to mass ratio (m/z). Moreover the initial phase (θ) of the sine wave of the rf is also adjusted in order to make the confinement easier. The phase corresponds to the initial value of Vac including its magnitude and polarity.

A supersonic free jet of a 1.5-atm gaseous mixture of 20% SF_6 and 80% He was generated by a pulsed nozzle system with an orifice of 0.8 mm diameter. The free jet was skimmed by a 1-mm-diameter skimmer. The pulsed molecular beam was brought into the CIT cell through one of the inlet holes. The SF_6 molecules in the beam were photo-ionized by 248-nm light of a KrF excimer laser at the center of the cell where the molecular beam and the laser beam were crossed. The generated ions were trapped easily in the CIT cell because the translational energy of SF_6 molecules is of the order of several tens of millielectronvolts. After trapping the desired ions for the required time, they were extracted downwards by applying pulsed positive voltage to the upper end cap electrode. First extracted ions were directly detected by a quadrupole mass analyzer to evaluate ion trapping characteristic. Next extracted ions were thus irradiated onto the semiconductor surface. The ion species from the surface were analyzed by a quadrupole mass analyzer.

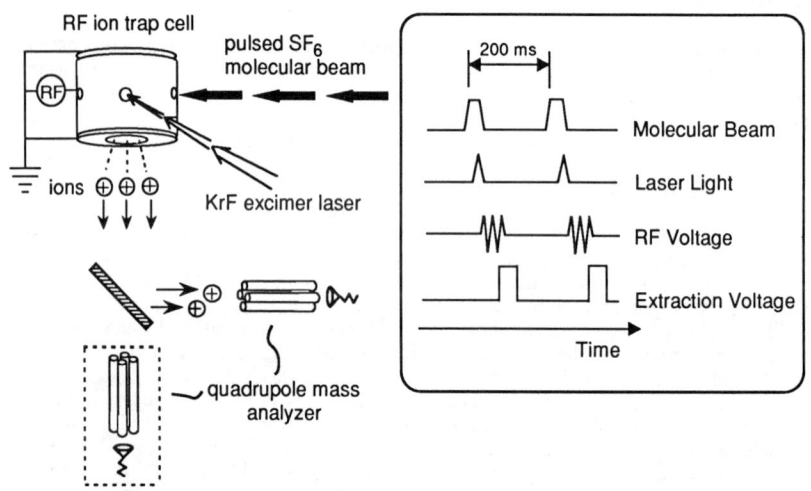

Figure 1. Experimental apparatus and timing chart for this experiment.

RESULTS AND DISCUSSION

Figure 2 shows the mass spectrum obtained by extracting directly without trapping the ions produced from SF_6 at 1 μs after laser irradiation. In this case the quadrupole mass analyzer was positioned below the ion trap cell without the semiconductor sample. Since the ionization potential of SF_6 is 15.6 eV and the photon energy of KrF laser light is about 5 eV, SF_6 is ionized by absorption of more than three photons. The parent SF_6^+ ion was not be observed. SF_6^+ ion is unstable due to Jahn-Teller distortion and dissociates into the smaller ions immediately [5]. The observed singly charged fragment ions were SF_x^+ (x=0-5), F^+ and He^+ ions. Fragment ions with an odd number of fluorine atoms were produced more abundantly. The doubly charged ionic species such as SF_x^{2+} (x=1-4) were also observed. In this case fragment ions having an even number of fluorine atoms were produced more abundantly. The doubly charged ions were produced by ionization of the singly charged ions. Probably an F atom was dissociated when the singly charged ions were ionized, that is,

$$SF_n^+ + h\nu \rightarrow SF_n^{2+} + F + e^-.$$

The ionic fragmentation of SF_6 via S2p inner shell excitation has already been investigated by electron impact [6]. The observed ionic species were the same, but the intensity distribution was different from the results of this study, that is, in the case of high-energy electron impact the intensity of the smaller fragment ions are high but not in the case of multi-photon ionization. So this comparison shows the multi-photon ionization is a rather soft one.

Figure 3 shows the mass spectrum obtained after trapping the ions stored in the CIT cell for 1 ms. The trapping condition of rf bias and dc bias which was adjusted for SF_5^+ ions is described in the figure. SF_5^+ ions were separately confined and the other ion species were excluded. The trapped ion number is estimated to be about 10^4 ions/pulse from the resulting ion current and the amplification factor of the secondary electron multiplier of the mass analyzer (10^7).

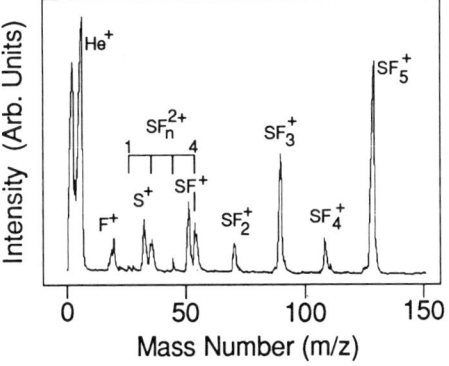
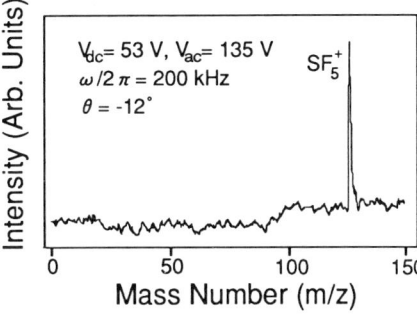
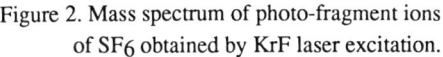

Figure 2. Mass spectrum of photo-fragment ions of SF_6 obtained by KrF laser excitation.

Figure 3. Mass spectrum of trapped ions selected from photo-fragment ions.

The change in intensity of trapped SF_5^+ ions as a function of trapping time is shown in Figure 4. The SF_5^+ ions survive for at least 100 ms. This property suggests that it is possible to control the internal state of the ions by another laser excitation during this trapping time. On the other hand, no SF_5^+ ions were observed even at 3 μs after multiphoton ionization when ion trapping was switched off. This is because the ions continue with their initial velocity, which is estimated to be about several tens of millielectronvolts.

The dependence of the SF_5^+ ion intensity on extraction voltage is shown in Figure 5. This voltage is applied to the upper end cap electrode. The ion intensity is almost constant at extraction voltages down to 10 V from 300 V. This is probably because trapped ions with low kinetic energies of a few electronvolts are confined in the small volume of about several tens of cubic millimeters in the center of the cell.

Next, we show the tentative results obtained when the extracted ions of about 100 eV were irradiated onto the semiconductor surfaces. Figure 6 shows the mass spectra obtained by analyzing the ions from Si(100) surfaces. Figure 6 (a) shows the mass spectrum for the irradiation of ions as shown in Figure 2. Figure 6 (b) shows in the case of irradiation of ions as shown in Figure 3. The observed ion species are almost the same as the irrcident ion species. They are the elastically scattered ions. However, no product ions were observed. The reason is probably that the reactivity of incident ions are not enough to cause detectable amounts of product ions. Deposition of sulfer may block the surface etching [7]. We will study the surface reactions with more reactive ions in the trap cell such as F^+ ions.

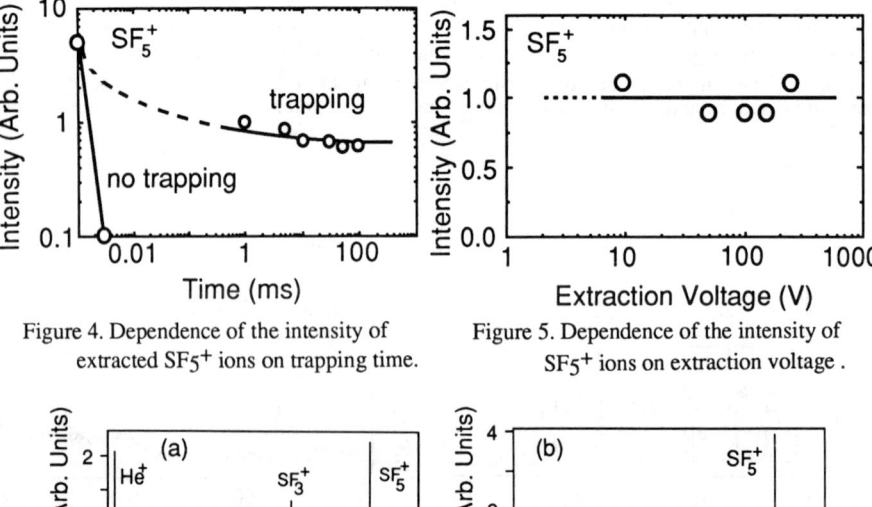

Figure 4. Dependence of the intensity of extracted SF_5^+ ions on trapping time.

Figure 5. Dependence of the intensity of SF_5^+ ions on extraction voltage.

Figure 6. Mass spectra of the ions from the Si surfaces after irradiation with (a) all ions and (b) SF_5^+ ions.

CONCLUSION

Spatial confinement of photo-dissociated ions and their extraction with low kinetic energy have been demonstrated. We were able to select, confine, and extract specific reactive ion species out of various kinds of photo-fragment species by using ion trapping and multiphoton ionization. By using this method for the study of surface chemical reactions with ions, etching phenomena will be greatly simplified. Moreover, from the viewpoint of industrial needs, the potential of this ion trapping method is promising for achieving dry etching using state selective ions.

REFERENCES

[1] W. Muller-Markgraf and M. J. Rossi, Rev. Sci. Instrum., **61**, 1217 (1990).
[2] R. F. Bonner, J. E. Fulford, and R. E. March, Int. J. Mass Spectrum. Ion Phys., **24**, 255 (1977).
[3] N. Mikami, T. Sasaki, and S. Sato, Chem. Phys. Lett., **180**, 431 (1991).
[4] R. E. Mather, R. M. Waldren, and J. F. J. Todd, Int. J. Mass Spectrum. Ion Phys., **33**, 201 (1980).
[5] V. H. Dibeler and J. A. Walker, J. Chem. Phys., **44**, 4405 (1966).
[6] A. P. Hitchcock, C. E. Brion and M. J. Van der Wiel, J. Phys. B. Atom. Molec. Phys., **11**, 3245 (1978).
[7] T. Ogawa, K. Mochiji, I. Ochiai, S. Yamamoto, and K. Tanaka, submitted to J. Appl. Phys.

SURFACE PERTURBATION BY ENERGETIC PARTICLE BEAMS

Che-Chen Chang, Jung-Yen Yang, Jaw-Chang Shieh
Department of Chemistry, National Taiwan University, Taipei, Taiwan, R.O.C.

Abstract

The state of the surface after energetic keV particle bombardment is investigated using molecular dynamics. The model utilizes a Ag{110} microcrystallite which is statically bombarded by Ar particles at normal incidence. After being bombarded at incident energy of 1 keV, the relocation probability is <0.3 for all the surface atoms initially residing within four lattice spacings from the target. The probability decreases exponentially as the initial distance of the substrate atom from the target is increased. The most probable distances of displacement from the lattice site also vary with the initial atomic distance from the target atom. The probable displacement of the surface atom, except the target atom, is less than one twentieth of the surface lattice spacing. An analytical formulae for the initial-distance dependence of the relocation probability is also proposed. The formula has three adjustable parameters which are determined by the least-squares method.

I INTRODUCTION

When the impinging energy is above the bond strength of the solid, the series of collisional events that are initiated by particle bombardment may result in changes of the surface morphology [1] and ultimately may alter the surface property [2]. This knowledge of the structural perturbation that impinging particles introduce to a solid surface during collision [4-6] is essential to the fabrication of nanometer-scale structures in electronic and optoelectronic devices. Collision-induced damage, such as that caused by a focused-ion beam [7] during etching, affects the electrical and optical properties of materials. Schottky barrier characteristics of metal to ion-etched semiconductor surfaces may be degraded [8] and the luminescence efficiency in optical materials processed using energetic ion beam may be reduced [9]. On the other hand, the ion-induced damage process may produce conducting channels under the patterned areas in the heterostructure material. The regions outside these areas are nonconducting such that the electrons are confined in the channel as the two-dimensional gas [10].

Characterization of the state of the crystal surfaces that have been perturbed by an energetic ion beam is clearly central to our understanding of the properties of the material surface. In this work, we perform a series of molecular dynamics calculations to study the collision-induced structural damage in the crystal surface. We first focus on the influence of energetic incident particles on surface structural integrity by examining the displacement distribution and the

Surface Perturbation by Energetic Particle

probability that the surface atoms in the impact region are displaced from their lattice sites. The ejection behavior of the underlayer atoms which leave late in the collision cascade when the surface is damaged is then investigated. Our calculations of keV Ar atoms incident normal to Ag{110} show that under the static bombardment condition the surface retains to a significant extent the initial crystal order. The most probable displacement, i.e. the distance by which most substrate atoms are displaced from their initial locations, is less than one twentieth of the crystal lattice spacing (LS, 1 LS = 4.0862 Å). In addition, the relocation probability, i.e., the probability that the substrate atom is relocated to a position away from the initial lattice site by more than half of the substrate Ag-Ag bond-length, is found to be <0.3 for all atoms (except the target atom) remaining in the surface after the collision. This probability decreases exponentially as the distance of the initial site from the target atom is increased. It approaches zero at an initial atomic distance of ~4.5 lattice spacings from the target.

II DESCRIPTION OF THE CALCULATION

A full three-dimensional molecular dynamics calculation procedure [11] is employed to simulate the particle collision process in the surface. In the calculation, the positions and momenta of all the particles in the particle-substrate system are determined as a function of time by integrating Hamilton's equations of motion. The primary particle used for the bombardment is Ar atom at normal incidence. The substrate consists of nine layers of Ag atoms with about 200 atoms per layer and with the {110} crystal face exposed. To represent the interaction potential, we assume that a sum of pairwise additive potentials exists for all of the atoms. A Thomas-Fermi Moliere potential is used to describe the interaction between the primary particle and the Ag atom. Our previous study [11] showed that this potential, with a scaling factor of 1.0 for the Firsov screening length [12], was adequate in describing Ar atoms scattering from the Ag{110} surface. Further, a potential consisting of three parts was utilized to describe the interactions among Ag atoms: an attractive Morse function at long range, a repulsive Moliere potential for small internuclear separations, and a cubic spline to connect the two.

For each set of initial conditions, approximately 2000 collision sequences are computed. The points of impact are uniformly distributed over a zone of irreducible surface symmetry. For each collision sequence, the final positions and velocities of all the atoms with five lattice spacings from the target atom are stored for subsequent analysis. Specific sequences of interest may be recalculated in order to trace the important microscopic collision events of atoms in solids.

III RESULTS AND DISCUSSION

The surface damage created by bombardment of energetic particles are simulated by a series of full molecular dynamics calculations of the collision

Surface Perturbation by Energetic Particle

cascade. The extent of the damage may be explored by examining the displacement of the surface atoms from their initial lattice sites after the cascade subsides. Fig. 1 shows the displacement distribution of the surface atoms which

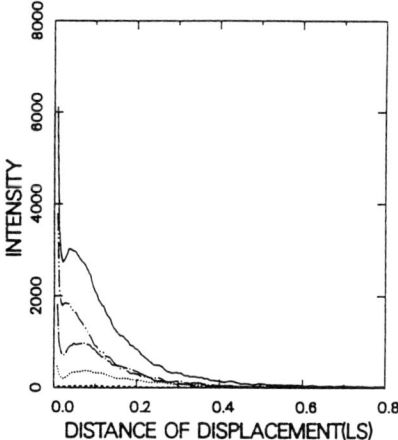

Fig. 1. Displacement distributions of the atoms remaining in the Ag{110} surface after the surface is bombarded by Ar atoms of 1 keV incident energy. The solid curve represents the distribution of the displacement from the lattice site for all the atoms initially residing within four lattice spacings from the target atom. The dashed, dotted, dot-dashed, and dot-dot-dashed curves represent the displacement distribution for the atoms initially located within the first , the second, the third, and the fourth lattice spacing, respectively.

are initially located within four lattice spacings from the target atom and left in the Ag{110} surface after the surface is bombarded by Ar atoms of 1 keV incident energy. As shown in the figure, most surface atoms remain at their lattice sites after the bombardment. For those atoms displaced from their initial sites, the most probable distance (i.e. the distance to which most atoms are displaced) of displacement decreases with increasing initial distance from the target atom. The overall distribution shows that these displaced residual atoms have a most probable displacement of about 0.04 LS (or 0.12 Å).

Since the most probable displacement of the atoms near the target atom is much less than half of the Ag-Ag bond length in the bulk, the surface thus retains to a significant extent the initial crystal order during static particle bombardment. This is quite interesting since the impinging energy is significantly higher than the strength of the chemical bond in the solid. The retention of the surface structure may also be studied by numerically following the ejecting trajectories of the underlayer atoms which leave the surface late in the collision cascade. It has been suggested that when particles leave the surface late in the cascade much of the surface order is no longer present [13]. The take-off points of these atoms are expected to be randomly distributed in the surface

Surface Perturbation by Energetic Particle

plane of the top-layer atoms if the substrate lattice structure is completely destroyed during the collision. Shown in Fig. 2 is the take-off point distribution

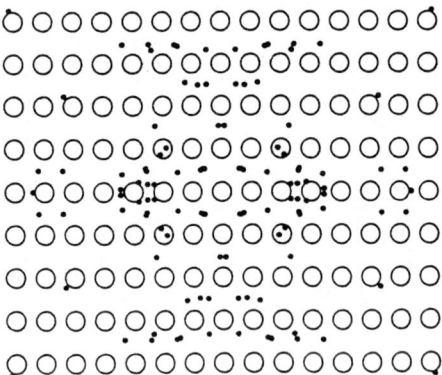

Fig. 2. The calculated distribution of the take-off points, in the plane of the top-layer atoms, for the atoms originating from the second layer of Ag{110} and with long collision time of >150 fsec. The open circles represent surface atoms in the topmost layer and the small closed circles the take-off points. The incident energy is 2 keV.

of the second-layer atoms which eject with collision time, i.e. the time interval between the moment of the primary impact and the instant the ejecting atom is out of the interaction range with the surface, greater than the most probable collision time. As shown in the figure, the path of these ejected atoms are mostly constrained to the spacings on the {110} face between the first layer atoms. Although the processes that lead to the ejection of the underlayer atoms are very complex, the constraint of the ejection path reveals that the first-layer atoms are present close to their lattice sites and that these atoms exert strong directional force on the ejecting atoms. The brevity of the interaction between the keV incident particle and the surface atom as well as the presence of the weak chemical bond between surface atoms may account for the retention of crystal order in the surface during particle bombardment.

It is of interest to also understand the dependence, on the initial atomic distance from the target atom, of the probability that the surface atom is relocated to a position more than half of the Ag-Ag bond length away from the initial lattice site. This information is very useful in determining adequate etching conditions for precise control of the generation and propagation of damage in fabricating nanostructure devices. In Fig. 3, the relocation probability of the atom remaining in the surface after the collision is shown as a function of the initial distance of the atom from the target atom. As shown in Fig. 3, about 60% of the incident trajectories results in a significant displacement of the target atom. The probability decreases rapidly as the distance is increased. The relocation probability decreases to ~0.3 for the first-nearest-neighboring atoms of the target, to ~0.2 for atoms within the second lattice spacing from the target, to ~0.1 for the ones within the third, and to ~0.01 for the atoms located at 4 LS away from the

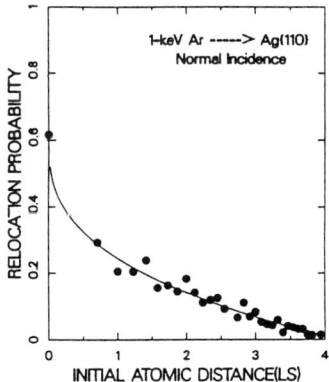

Fig. 3. Dependence of the relocation probability on the initial distance of the residual surface atom from the target for Ar atoms of 1 keV energy incident on Ag{110}. The solid line represents the least-squares best-fit curve.

target. Fig. 3 also gives calculated best-fit curves, using the least-squares method, for the relation between the relocation probability and the initial atomic distance from the target. The corresponding analytical expression in the exponential form is

$$\Pi(r) = -0.37\, r^{0.35} + 0.61$$

where r is the distance, in LS, between the initial location of the target atom and that of the residual substrate atom and Π is the relocation probability. In addition to the prediction of the extent of surface damage during particle bombardment, this equation may also be used to select an adequate size of the model crystallite for the computer simulation of sputtering. The equation predicts that the relocation probability falls to zero for the atoms originally located at a distance of greater than ~4.5 LS. The crystal size used in this work is 7.5 LS, which is sufficiently large compared with the distance at which the relocation probability is zero.

IV CONCLUSION

In this work, molecular dynamics calculations are employed to study the surface damage induced by energetic particle collision. A high degree of crystal order is found to be retained under the static keV particle bombardment. For the residual atoms in the impact region of within four lattice spacings from the target atom, the most probable displacement is less than one twentieth of the lattice spacing. The relocation probability of the atoms remaining in the impact region is less than ~0.3. Further, the probability decreases rapidly as the initial distance

Surface Perturbation by Energetic Particle

of these atoms from the target increases. Because of the retention of crystal order during the collision, the ejection path of the underlayer atoms are mostly confined within the spacings between the top-layer atoms.

The retention of crystal order under static keV particle bombardment makes dry etching by a focused-ion beam [a] a useful technique for realizing nanostructure formation and for precisely introducing damage to the substrate in producing conducting channels of desired dimensions. Further studies to explore the damage mechanism on the atomic scale and the transport properties of the ion-damaged, heterostructure material at various temperature are underway. These studies are critical for better controlling the electrical potential created by the damage in the conducting channel of the heterostructure material such that the sheet resistance of the nanowires does not exceed a certain value to pinch off the wire.

ACKNOWLEDGMENT

The financial support of ROC Naional Science Council is gratefully acknowledged.

REFERENCE

1. I. H. Wilson, N. J. Zheng, U. Knipping, I. S. T. Tsong, Appl. Phys. Lett. 53 (1988) 2039.
2. a) E. Collart, R. J. Visser, Surf. Sci. 218 (1989) L497; b) G. E. McGuire, Surf. Sci. 76 (1978) 131.
3. a) H. I. Smith, H. G. Craighead, Physics Today, February (1990) 24; b) A. H. Al-Bayati, K. G. Orrman-Rossiter, R. Badheka, D. G. Armour, Surf. Sci. 249 (1991) 293; 237 (1990) 213.
4. a) B. D. Weaver, D. R. Frankl, R. Blumenthal, N. Winograd, Surf. Sci. 222 (1989) 464; b) Y. Kido, I. Konomi, M. Kakeno, K. Yamada, K. Dohmae, J. Kawamoto, Nucl. Instr. Meth. B15 (1986) 42.
5. a) G. S. Oehrlein, R. M. Tromp, J. C. Tsang, Y. H. Lee, E. J. Petrillo, J. Electrochem. Soc. 132 (1985) 1441; b) K. L. Seaward, N. J. Moll, W. F. Stickle, J. Electron. Mat. 19 (1990) 385.
6. M. Posselt, J. P. Biersack, Nucl. Instru. Meth. B15 (1986) 20.
7. H. Sakaki, Jpn. J. Appl. Phys. 19 (1980) L735.
8. T. Hara, H. Suzuki, A. Suga, J. Appl. Phys. 62 (1987) 4109.
9. H. F. Wong, D. L. Green, T. Y. Liu, D. G. Lishan, M. Bellis, J. Vac. Sci. Technol. B6 (1988) 1906.
10. T. L. Cheeks, M. L. Roukes, A. Scherer, H. G. Craighead, Appl. Phys. Lett. 53 (1988) 1964.A. Scherer, M. L. Roukes, Appl. Phys. Lett. 55 (1989) 377.
11. C.-C. Chang, Surf. Interface Anal. 15 (1990) 79.
12. O. B. Firsov, JETP (Sov. Phys.) 6 (1958) 534.
13. N. Winograd, B. J. Garrison, D. E. Harrison, Phys. Rev. Lett. 41 (1978) 1120.

CLEANING DURING INITIAL STAGES OF EPITAXIAL GROWTH IN AN ULTRA-HIGH VACUUM RAPID THERMAL CHEMICAL VAPOR DEPOSITION REACTOR

MAHESH K. SANGANERIA*, KATHERINE E. VIOLETTE*, MEHMET C. ÖZTÜRK*, GARI HARRIS**, C. ARCHIE LEE** AND DENNIS. M. MAHER**
*North Carolina State University, Department of Electrical and Computer Engineering, Box 7911, Raleigh, NC 27695-7911,
**North Carolina State University, Department of Materials Science and Engineering, Box 7907, Raleigh, NC 27695-7907

ABSTRACT

In this paper, we report our results on surface preparation for the growth of epitaxial Si films. Hydrogen passivated surfaces are currently being investigated for application in Si epitaxy to eliminate the high temperature in-situ bake necessary to remove the native oxide. Hydrogen passivation is obtained by a dilute HF dip before the substrate is loaded in the process chamber. However the passivation is partially lost when the HF dip is followed by a water rinse which results in oxygen absorption on the substrate. It was found that the peak oxygen concentration at the epitaxy substrate interface increase by an order of magnitude due to a five minute water rinse. We report here that oxygen and carbon at the epitaxy substrate interface can be desorbed during initial stage of epitaxial growth by reducing epitaxial growth rate. In this work, epitaxial Si films were deposited over a wide range of growth rates obtained by varying Si_2H_6 flow rates. The peak oxygen concentration decreases by an order of magnitude by changing the growth rate from 3000 to 700Å/minute for a deposition temperature of 800°C. We believe that at higher growth rates Si overgrows on absorbed oxygen maintaining epitaxial alignment reflected in the good electrical quality of the epitaxial films. However, at low growth rates oxygen has sufficient time to desorb before overgrowth can take place, improving the epitaxy substrate interface quality.

INTRODUCTION

Low temperature Si epitaxy for application in deep submicron devices has been the subject of numerous investigations [1-6]. Good quality Si epitaxy has been demonstrated at low temperatures by several different techniques [1-6]. An atomically clean substrate surface is essential for the successful growth of device quality epitaxial Si. In conventional reactors, a clean surface is obtained by baking the substrate in a hydrogen ambient at high temperatures for long times which is unacceptable in future submicron technologies. A variety of methods have been proposed to obtain clean surfaces at low temperatures. These include sputter cleaning [7], thermal desorption in ultra-high vacuum (UHV) environment [8], hydrogen plasma clean [9], reaction with reactive gases like Si_2H_6 [10], SiH_4 [11], GeH_4 [12] and reduction by molecular beam of Ga [13], Si [14], Ge [15]. Another important issue in surface preparation for epitaxy is that an atomically clean Si surface is very reactive and hence it can recontaminate if the surface is not passivated or the growth is not commenced immediately after cleaning.

HF cleaning provides a potential method by which the high temperature in-situ clean can be eliminated completely. In this method the substrate is dipped in a dilute HF solution before introduction to the process chamber. The HF dip leaves a predominantly hydrogen covered surface which is stable in air for several minutes. Meyerson et. al. have shown that in a UHV environment, if the Si deposition is initiated on a hydrogen passivated surface during ramp up, before the hydrogen desorption begins (<400°C), a good quality film can be deposited without a high temperature bake [3]. Similar results have also been reported by Iyer and co-workers [16]. They have reported good quality films on HF treated surfaces using UHV/CVD as well as molecular beam epitaxy (MBE). In their work, epitaxy was accomplished without a high temperature clean by commencing deposition at temperatures below the hydrogen desorption temperature at extremely low rates, and gradually increasing both the growth rate and temperature to conventional MBE values. One disadvantage of this technique is that the film quality can degrade if the HF treated surface is rinsed in water before introduction to the process chamber. Gräf et. al. have reported that water rinse results in oxygen absorption on the HF treated surface [17]. Therefore, although the technique developed by Meyerson et. al. is very successful for blanket Si substrate, it may not be acceptable for patterned substrate with oxides as any left over HF will result in nonuniform and uncontrolled etching of the oxide.

Rapid thermal chemical vapor deposition (RTCVD) is one of the promising techniques for application in single wafer in situ processing in multichamber clustertools. In RTCVD, the deposition process can be initiated and terminated by rapid changes in the substrate temperature. Therefore, films can be deposited at higher temperatures for short times while maintaining a low thermal budget. Good quality Si epitaxy has been demonstrated using RTCVD [4, 18, 19].

In this study, our objective was to develop a low thermal budget surface preparation process for the growth of epitaxial films using RTCVD. The approach presented in this report combines an ex-situ HF dip with cleaning during initial stages of growth for removal of any residual oxygen on the wafer surface.

We have previously reported epitaxial films with generation times exceeding 100 µs in our UHV-RTCVD reactor, in the temperature range from 700°C to 800°C using Si_2H_6 as the reactive gas, without an in-situ clean [5]. This was achieved even though the integrated dose of oxygen (O) at the epitaxy-substrate surface was on the order of 10% of a monolayer. In RTCVD, since the substrate is rapidly heated to the deposition temperature, the deposition starts immediately as the hydrogen desorbs and no contamination of the interface occurs in the deposition chamber. High interfacial O was obtained due to a water rinse used in this previous work. In this report, we present our results on controlling the interfacial O and carbon (C) by changing the deposition parameters, Si_2H_6 partial pressure and temperature. Gated diodes fabricated in epitaxial Si have been used to evaluate the electrical quality of the substrate epitaxy interface.

EXPERIMENTAL

In this work, the depositions were carried out in an UHV-RTCVD reactor. The detailed description of the UHV-RTCVD system can be found elsewhere [5]. In brief, the system consists of a load-lock (sample entry chamber), an intermediate chamber and a reaction chamber. The intermediate chamber serves as a vacuum buffer. The load-lock is pumped with a dry molecular drag pump to a base pressure of 10^{-5} Torr. The intermediate chamber is pumped by a cryopump to a base pressure of 10^{-9} Torr. The reaction chamber is pumped by a cryopump to a base pressure of 10^{-9} Torr and the process gases are pumped by a turbomolecular pump. The reaction chamber is water cooled stainless steel with a quartz window on top. The wafer is heated through the window by a 35 kW PEAK^{TM} Systems LXU-35 arc lamp. The wafer sits on a quartz wafer holder which can rotate during film growth.

In this study, we have used 100mm, p (100) oriented Si wafers. The wafers were cleaned using a standard RCA solution and stored in a nitrogen purged box. Immediately prior to loading, each wafer was individually dipped in a dilute 5% HF solution for 20 seconds and rinsed for 10 seconds in deionized (DI) water and finally blow dried in N_2 obtained from a liquid source. Upon insertion into the reactor each wafer underwent the following procedure: the entry chamber was pumped down to ~ $8X10^{-5}$ Torr in ~ 10 minutes before transfer to the intermediate chamber. An intermediate chamber base pressure of high 10^{-9} Torr was routinely achieved within 10 minutes. The wafer was then transferred into the reaction chamber and was cryopumped to a base pressure of about $3X10^{-9}$ Torr. Then, the chamber was switched to the process pumps and gas flows were initiated. Gases used in this work were 10% Si_2H_6 in H_2 and 500ppm B_2H_6 in H_2. After establishing steady gas flows, the arc lamp was turned on to heat the substrate to the deposition temperature at a ramp rate of approximately 150°C/s. Oxygen and carbon concentrations were measured using Secondary Ion Mass Spectroscopy (SIMS).

RESULTS AND DISCUSSIONS

In the first experiment, we have studied the effect of water rinse on the O and C at the epitaxy substrate interface. Si substrates were individually dipped in 5% HF and subsequently rinsed in deionized (DI) water for 0, 10, 30, 90 and 300 seconds. Each wafer was then inserted into the epitaxy reactor. Silicon films were deposited at 650°C with 250 sccm of 10% Si_2H_6/H_2 (25 sccm Si_2H_6) gas mixture and at a pressure of about 90 mTorr. The deposition temperature was selected to be 650°C because there is no desorption of O and C at or below this temperature. The O and C profiles obtained in these films are shown in Figure 1. As shown in Figure 1, a 10 second water rinse results in a about 4 to 5 time increase in the O at the substrate epitaxy interface compared to a no water rinse condition. Further increase in the rinse time to 300 seconds i. e., a 30 fold increase in the water rinse time results in only 2 to 3 times increase in the O at the interface. These results are similar to the result of Gräf et. al. on the reaction of water with a Si surface treated with 40% HF [17]. According to Gräf et. al. reaction of water with an HF treated surface proceeds in two steps. When HF treated Si is dipped in water there is an instantaneous increase in Si-OH bonds and then the number of Si-OH bonds increase progressively following a logarithmic variation with time for about 3-5 hours. The Si-OH bonds on the Si surface are responsible for the increased O at epi-substrate interface in water rinsed substrates. It is interesting to note here that C at the epitaxy substrate interface was almost unaffected by the water rinse. All the other experiments in this work were conducted with a 10 second water rinse. Since there is only a logarithmic increase in the O at the interface with the rinse time, all the data obtained in this work can be extended for higher rinse times.

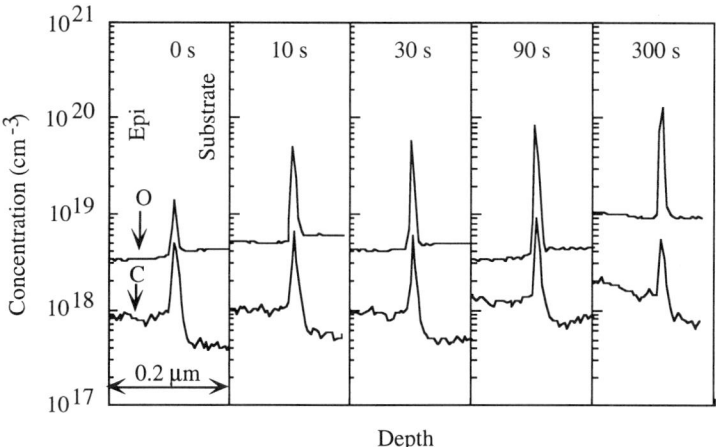

Fig. 1 Effect of water rinse time on the O and C absorption on HF treated Si.

To study the effect of deposition temperature on the O and C at the interface, films were deposited at 650, 700, 750 and 800°C with 250 sccm of 10% Si_2H_6/H_2 gas mixture (25 sccm Si_2H_6) and a pressure of 90 mTorr. Figure 2 shows the integrated dose of O and C at epitaxy substrate interface as a function of the deposition temperature. As seen clearly, O and C decreases as the deposition temperature is increased. We propose that this is achieved because there is a desorption of O and C during initial stages of epitaxial growth. As the temperature is increased, the desorption becomes more efficient resulting in decreased O and C at higher deposition temperatures. We have previously reported that good quality Si films with generation time exceeding 100 µs are obtained with these deposition conditions which requires a good epitaxial alignment [5]. We argue here that during the initial stages of epitaxial growth, the film is nucleated only on silicon sites and not on O and C absorbed sites on the substrate. At high growth rates Si overgrows on the O and C sites while maintaining a good epitaxial alignment. A fraction of absorbed O and C is desorbed before overgrowth takes place.

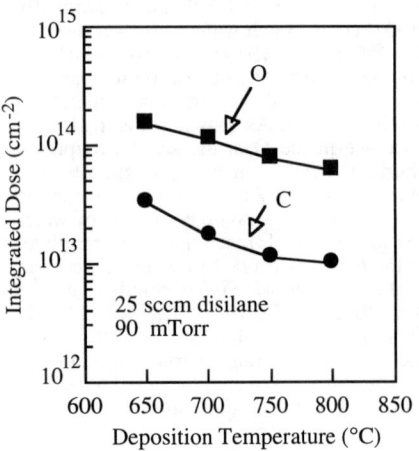

Fig. 2 Integrated dose of C and O at the epitaxy substrate interface as a function of the deposition temperature.

It appears that if the hypothesis presented above is true, more O and C will be desorbed if the film growth rate is reduced during the initial stages of epitaxial growth as there is more time for the desorption of O and C before overgrowth can take place. To investigate this, films were deposited with different growth rates at 800°C. The growth rate was changed by varying the Si_2H_6 flow rate during the first 10 seconds of the film deposition. The Si_2H_6 flow rate and temperature profiles used in this experiment for different wafer are summarized in Figure 3. The wafer was ramped to 800°C in a Si_2H_6 flow of 0, 1, 2.5, 5, 10, 20 or 40 sccm. At the deposition temperature, the Si_2H_6 flow rate was changed to 20 sccm after 10 seconds and the deposition was continued for additional 30 seconds to obtain a cap layer of about 2000Å for SIMS measurements. It should be noted that since a 10% Si_2H_6/H_2 gas mixture was used in this experiment the amount of H_2 and pressure changed proportionately as Si_2H_6 flow rate was changed. Figure 4 shows the integrated dose of O and C at the epitaxy substrate interface as a function of the Si_2H_6 flow rate. It is observed that the O and C decrease monotonically as the Si_2H_6 flow rate decreases. For flow rates of 0 (vacuum) and 1 sccm O and C were below the detection limit of SIMS instrument which were 1×10^{18} and 5×10^{17} cm^{-3} respectively.

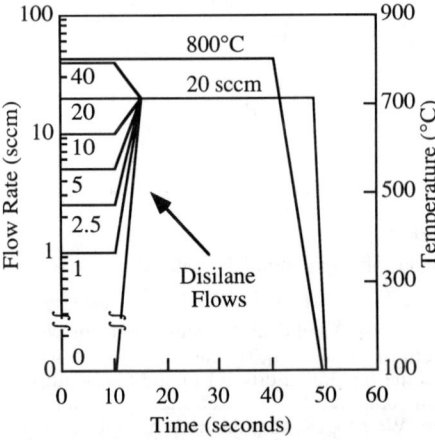

Fig. 3 Temperature and flow rate profiles used to study the effect of growth rate on interfacial O and C

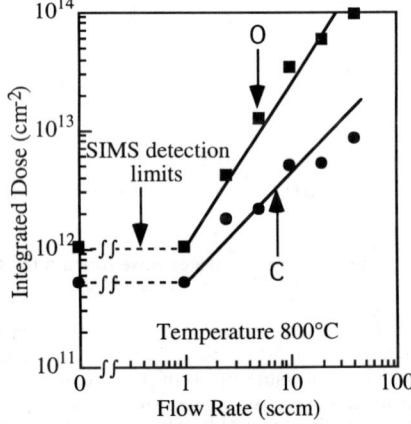

Fig. 4 Integrated O and C dose as a function of disilane flow rate during initial stage of the epitaxial growth.

Electrical quality of the epitaxial film and the substrate epitaxy interface were evaluated using gated diodes fabricated in the epitaxial films. For this purpose, approximately 4000Å insitu boron doped ($\approx 5 \times 10^{16}$ cm^{-3}) epitaxial films were deposited at 750°C with 200 sccm of 10% Si$_2$H$_6$/H$_2$ (20 sccm Si$_2$H$_6$) and 0.15 sccm of 500 ppm B$_2$H$_6$/H$_2$ gas mixture. Gated diodes (n$^+$p) were fabricated in these films by diffusing phosphorus from in-situ doped Si$_{0.7}$Ge$_{0.3}$ deposited selectively in 600μm X 600μm active areas defined in isolation oxide. The junction depth was approximately 400 Å. A detailed description of the fabrication and characterization of gated diodes have been reported elsewhere [20]. It should be noted here that low temperature oxides and shallow junction were employed in this fabrication to minimize the thermal budget received by epitaxial films. A typical current-voltage (I-V) characteristics obtained for the diodes is shown in Figure 5. In the forward bias an ideality factor (n) of 1.04 was calculated over at least 5 orders of magnitude of current. The same ideality factor was obtained for diodes fabricated in bulk Si. The reverse leakage current at a reverse bias voltage of 3 volts was on the order of 5 pA which compares favorably with what is typically obtained in industry for devices of this size. It is interesting to note that high quality epitaxial films can be obtained even with high levels of O on the order of 10% of a monolayer, at the epitaxy substrate interface.

Even though good quality Si films are obtained with high interfacial O, the heavily contaminated interface may result in excessive generation if the interface is included in the active region. To investigate the effect of interfacial O on the electrical quality, about 200 Å of epitaxial film was deposited at 5 sccm Si$_2$H$_6$ flow at 750°C for about 20 seconds. This was followed by deposition of 4000Å film according to the condition described in the previous paragraph. We expect that the film with a buffer layer deposited with 5 sccm Si$_2$H$_6$ will have a substantially lower oxygen at the interface. Figure 6 shows reverse leakage current for film deposited with and without the buffer layer. It is seen that in the film without the buffer layer a kink is observed in the I-V curve. This kink corresponds to the depletion region reaching the epitaxy substrate interface. The location of the interface was identified from the change in the doping concentration associated with the interface from capacitance voltage characteristics. This kink is not observed in the film deposited with a low growth rate buffer layer which has lower interfacial O and C. This suggests that interfacial O can contribute to degradation of electrical quality and therefore deposition conditions must be optimized to reduce O and C contamination at the epitaxy substrate interface.

It is clear that films deposited with lower growth rates results in superior interface properties. However, in a single wafer reactor, growth rate can not be reduced below a certain

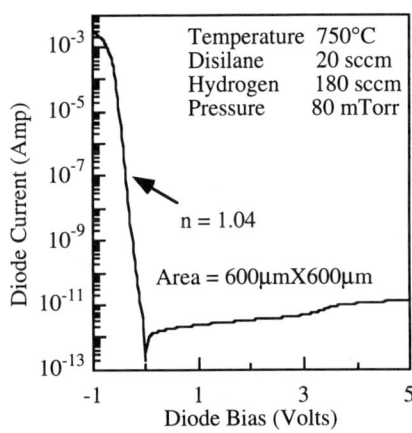

Fig. 5. A typical I-V characteristic of diode fabricated in UHV-RTCVD Si epitaxy

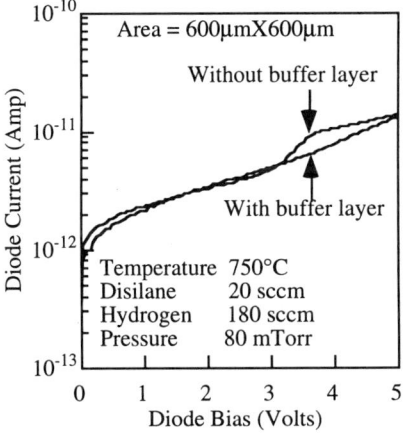

Fig. 6. Reverse I-V characteristic of diodes in films with and without a buffer layer

limit to meet the throughput requirements. In this work, we have observed that a Si$_2$H$_6$ flow rate of 5 sccm gives an interface with good electrical properties. We have previously reported the growth rate of epitaxial Si as a function of Si$_2$H$_6$ flow rate for the deposition temperatures of

750°C and 800°C [21]. It was found that with Si_2H_6 flow rate of 5 sccm the growth rate was about 700 Å/minute. This is an acceptable growth rate for single wafer manufacturing especially for applications that require films thinner than 1000Å. Therefore, we conclude that epitaxial films can be deposited without a separate high temperature in-situ clean by optimizing the growth conditions to remove any residual oxygen via desorption during initial stages of growth following an ex-situ HF dip which removes bulk of the oxygen on the surface.

CONCLUSIONS

In conclusion, we have shown that a hydrogen passivated surface can be recontaminated by O during a water rinse. However, the O and C can be removed in a UHV-RTCVD reactor during initial stages of epitaxial growth by reducing the growth rate by allowing sufficient time for desorption. Good quality films can be deposited even with a high interfacial O, although the interface does contribute to excessive generation if it is included in the active region of the device. The process however can be optimized to remove the interfacial components below the SIMS detection limits while maintaining growth rates compatible with single wafer manufacturing.

ACKNOWLEDGMENTS

The authors are grateful to S. P. Ashburn, D. T. Grider, X. W. Ren and S. M. Celik for many helpful discussions and K. Seastrand for his help in measurements. This research was partially supported by NSF Engineering Research Centers Program through the center for Advanced Electronic Materials Processing (grant CDR-8721545).

REFERENCES

1. J. Murota, N. Nakamura, M. Kato, N. Mikoshiba and T. Ohmi, Appl. Phys. Lett. **54**, 1007 (1989).
2. T. O. Sedgwick, M. Berkenblit and T.S. Kuan, Appl. Phys. Lett. **54**, 2689 (1989).
3. B. S. Meyerson, F. J. Himpsel and K. J. Uram, Appl. Phys. Lett. **57**, 1034 (1990).
4. M. L. Green, D. Brasen, H. Luftman and V. C. Kannan, Appl. Phys. Lett. **65**, 2558 (1989).
5. M. K. Sanganeria, K. E. Violette and M. C. Öztürk, Appl. Phys. Lett. **63**, 1225 (1993).
6. I. Rahat, J. Shappir, D. Fraser, J. Wei, J. Borland and I. Beinglass, J. Electrochem. Soc. **138**, 2370 (1991).
7. W. R. Burger and R. Reif, J. Appl. Phys. **62**, 4255 (1987).
8. A. Ishizaka and Y. Shiraki, J. Electrochem. Soc. **133**, 666 (1986).
9. T. Hsu, B. Anthony, R. Qian, J. Irby, S. Banerjee, A. Tasch, S. Lin, H. Marcus and C. Magee, J. Elect. Mats. **20**, 279 (1990).
10. Y. Kunii and Y. Sakakibara, Jap. J. Appl. Phys. **26**, 1816 (1987).
11. H. Hirayama, T. Tatsumi, A. Ogura and N. Aizaki, Appl. Phys. Lett. **51**, 2213 (1987).
12. M. M. Moslehi, Proceedings of the SPIE symposium on Rapid Thermal and Related Processing Techniques, (SPIE, 1990), Vol. 1393, p. 90.
13. S. Wright and H. Kroemer, Appl. Phys. Lett. **36**, 210 (1980).
14. T. Tatsumi, N. Aizaki and H. Tsuya, Jap. J. Appl. Phys. **24**, L227 (1985).
15. J. F. Morar, et. al., Appl. Phys. Lett. **50**, 463 (1987).
16. S. S. Iyer, M. Arienzo and E. Fresart, Appl. Phys. Lett. **57**, 893 (1990).
17. D. Gräf, M. Grunder and R. Schulz, J. Vac. Sci. Technol. A **7**, 808 (1989).
18. J. C. Sturm, C. M. Gronet and J. F. Gibbons, J. Appl. Phys. **59**, 4180 (1986).
19. T. Y. Hsieh, K. H. Jung, D. L. Kwong and S. K. Lee, J. Electrochem. Soc. **138**, 1188 (1991).
20. D. T. Grider, Ph. D. Thesis, North Carolina State University, (1993).
21. M. K. Sanganeria, K. E. Violette and M. C. Öztürk, Proceedings of the MRS Symposium on Rapid Thermal and Integrated Processing, (Materials Research Society, 1993), Vol. 303, p. 25.

PART VI

Special Topics: Deposition on Patterned Substrates, Reactor Design, Heteroepitaxy, Selective Epitaxy

THE IMPACT OF GAS PHASE AND SURFACE CHEMICAL REACTIONS ON STEP COVERAGE IN LPCVD

GREGORY B. RAUPP AND TIMOTHY S. CALE
Department of Chemical, Bio and Materials Engineering
Center for Solid State Electronics Research
Arizona State University, Tempe, Arizona 85287-6006

ABSTRACT

The characteristic step coverage behavior which a given LPCVD process exhibits depends on the nature of the controlling gas phase and/or surface chemical reactions. Physically-based ballistic transport and reaction film profile evolution simulation has provided a structure wherein the origins of step coverage limitations can be understood in the context of the interaction of transport and the controlling chemistry. Based on comparisons of the simulations to literature and in-house experimental data, we have categorized LPCVD mechanisms into one of three types. In *heterogeneous deposition*, conformal step coverage can usually be found under at least some process conditions. Step coverage typically degrades with increasing deposition temperature. In *homogeneous precursor-mediated deposition*, a reactive intermediate is formed in the gas-phase above the wafer surface, resulting in poor to moderate step coverage. Step coverage may or may not degrade with increasing temperature. In *byproduct-inhibited deposition*, a gas-phase byproduct generated via a surface reaction readsorbs on the growing film surface and slows the deposition rate, yielding a poor to moderate, relatively temperature-insensitive step coverage. Poor step coverage is manifested in a marked film thickness discontinuity at the feature mouth, with a relatively uniform film down the feature sidewalls.

INTRODUCTION

Step coverage is a critical production-level constraint in low pressure chemical vapor deposition (LPCVD) processes employed to deposit conducting, semiconducting and insulating films over patterned wafers in microelectronics fabrication. This constraint must be met while simultaneously meeting film intensive property constraints, as well as intrawafer and interwafer or run-to-run film thickness uniformity constraints. The process must also be "production-worthy", *i.e.*, films must be deposited at acceptable deposition rates with acceptable source gas utilization so that reasonable throughputs and costs are realized. In this context, an understanding of the origins of step coverage limitations for a given process chemistry may be instrumental in identifying operating windows, in guiding experiments to attempt to optimize the process, or in guiding the search for alternative CVD source gases.

Our understanding of step coverage behavior in LPCVD has dramatically improved in recent years due largely to the development of first-principles, physically-based descriptions of transport, reaction and film profile evolution in features on patterned wafers [1-6]. Because transport is rigorously predicted, these models have allowed researchers to test alternative reaction mechanistic schemes and associated kinetics for a given process chemistry by comparing feature-scale simulations to experimental film profiles. With these new tools our database of chemistries has now grown to the point where we can begin to categorize the various process chemistries of technological interest according to the controlling or dominant chemical reaction mechanism as follows:

- Class I: Heterogeneous Deposition
- Class II: Homogenous Precursor-mediated Deposition
- Class III: Byproduct-inhibited Deposition

In this paper we first review the ballistic transport and reaction model (BTRM) valid for LPCVD processes. We then describe the characteristics of each chemistry class and in particular the origins of step coverage limitations unique to that class. Selected examples are provided based on our current knowledge.

BALLISTIC TRANSPORT AND REACTION MODEL

Full descriptions of the BTRM can be found in publications by Cale and coworkers [1-3], Islam-Raja *et al.*[4], Singh *et al.* [5], and Hsieh and Joshi [6]. The model is based on material balances for the various species which exist in the deposition system, both in the vacuum and on the surface of the evolving film. The basic principles are briefly reviewed here with a focus on LPCVD processes. Assumptions appropriate for LPCVD processes are:
1. The frequency of particle-particle collisions is negligible relative to particle-wall collisions; *i.e.*, intra-feature transport is by free molecular flow.
2. The film grows slowly relative to the redistribution of fluxes to the walls due to film growth.
3. Deposition occurs by heterogeneous reactions between gas phase species which are transported ballistically within the feature and the evolving film surface.
4. The open end of the feature is exposed to a low pressure, ideal gas with a Maxwellian distribution of velocities.
5. Molecules are emitted from surfaces with cosine flux distributions.
6. Reactive sticking factors do not depend on the incident angle or collision history of the impinging molecules.

The second assumption is supported by comparing molecular speeds with the speed of an evolving film profile. The third assumption is a direct consequence of the first assumption, and does not imply that homogeneous reactions do not occur in the reactor volume above the wafer. As discussed below, such homogeneous reactions can indeed lead to precursors which then react on the surface of the evolving surface. The fourth assumption is reasonable for most LPCVD processes, because it implies only local thermal equilibrium in the gas phase above the wafer. There is little information regarding the last two assumptions; however, all of our CVD process simulations to date have included them and we have achieved good agreement between simulated and experimental film profiles. To complete the model for a specific LPCVD chemistry one must specify the intrinsic kinetics of the deposition reaction(s), the surface diffusivities of adsorbed species, and the local deposition conditions (partial pressures at the feature mouth and wafer temperature).

EVOLVE [7], a low pressure deposition process simulator based on the BTRM, is used to simulate topography evolution during deposition processes. Instantaneous deposition rates as a function of position on the interior surface of a feature are obtained by solving the integrodifferential equations which represent the BTRM. Film profile evolution is performed using the method of characteristics [8] to move surfaces in a manner which guarantees that conservation laws are satisfied.

CLASS I: HETEROGENEOUS DEPOSITION

In this chemistry class, source gases are fairly stable and do not react in the gas phase. Instead, the source gas molecules adsorb on the growing film surface and undergo unimolecular or bimolecular decomposition reactions to deposit new film and release byproducts into the gas phase. This reaction scheme may be simply represented as follows:

$$R_{(g)} + * \xleftrightarrow{K_a} R*$$
$$R* \xrightarrow{k_s} F_{(s)} + * + B_{(g)}$$

The first reaction represents reversible adsorption of the source gas molecule R on a bare surface site $*$ to produce a molecularly-chemisorbed $R*$ surface intermediate, which subsequently decomposes heterogeneously to add to the growing solid film F, regenerate a surface site and release byproduct gas(es) B. The dependence of the individual reaction rates on temperature enters through the adsorption/desorption equilibrium parameter K_a and the surface reaction rate parameter k_s. If Langmuirian behavior governs adsorption/desorption and one further assumes that the surface decomposition step is rate limiting, then the film deposition rate G is given by

$$G = \frac{k_s K_a P_R}{1 + K_a P_R}$$

where P_R is the partial pressure of the source gas. Depending on conditions of local partial pressure and surface temperature, the reaction rate may exhibit a range of partial pressure dependence from zeroth to first order. High partial pressures and low temperature yield a zeroth order or low fractional order dependence, corresponding to a surface which is saturated or nearly saturated with the surface adsorbate. Low partial pressure and high temperature yield a first order or near first order partial pressure dependence, corresponding to a surface which is largely free of adsorbate molecules; under these conditions the reaction becomes adsorption rate limited.

For the heterogeneous chemistries, step coverage for a given feature is determined by the reactive sticking factor(s) for the source gas precursor(s). The local reactive sticking coefficient σ is the fraction of molecules striking the surface which react, and can therefore be estimated by dividing the specific reaction rate G by the local flux η of source gas molecules to the surface. For the generic, single source gas mechanism summarized above, the reactive sticking coefficient can also be written

$$\sigma = \frac{G}{\eta} = S_o \theta_*$$

where S_o is the "initial" or sticking coefficient at zero surface coverage and θ_* is the fraction of surface sites which are unoccupied. At high to moderate fractional surface coverage, the reactive sticking factor is less than the zero coverage by a factor of θ_*, and is a function of reactant gas flux. For low coverages, the reactive sticking factor approaches S_o and is independent of reactant flux.

These ideas can be used to understand the step coverage behavior of blanket tungsten LPCVD through the hydrogen reduction of tungsten hexafluoride. Although the reaction mechanism has not been unequivocally established, it is generally agreed that homogeneous reactions are unimportant and that deposition proceeds through surface reactions between adsorbed WF_x species and adsorbed hydrogen atoms or impinging H_2 molecules. For conditions typical of low pressure commercial operation, the reaction rate exhibits a fractional order dependence on H_2 partial pressure and a near-zeroth order dependence on WF_6 partial pressure [9]. Figure 1(a) shows typical step coverage behavior in trenches for the hydrogen reduction chemistry at low total pressure and moderate temperature [10]. EVOLVE simulations employing a heterogeneous reaction rate expression, shown in Figure 1(b), are in excellent agreement with the experimental profiles. A key-hole shaped void is formed, with the void increasing in size as aspect ratio of the trench increases. Under these process conditions poor step coverage is largely caused by high reactive sticking coefficients for WF_6 [10], with local values increasing spatially down the length of the trench, and increasing temporally as the instantaneous feature aspect ratio increases as the feature fill proceeds.

Consideration of the reaction rate partial pressure dependencies suggests that lower sticking coefficients can be achieved *while simultaneously increasing the deposition rate* by increasing the total pressure and holding the $WF_6:H_2$ partial pressure ratio and temperature constant. Indeed, this new operating window has been found to yield excellent step coverage even for high aspect ratio trenches [11], as can be seen in Figure 2.

In the Langmuir chemisorption model employed above, site exclusion prohibits an impinging molecule from adsorbing if the site is occupied by an adsorbate. Alternatively, one can envision an adsorption process in which a reactant molecule condenses, or physisorbs on the surface without making a strong chemical bond; this weakly-bound, mobile molecular precursor then diffuses on the surface until it either desorbs or locates a site for strong, dissociative chemisorption. There exists mounting evidence that this "precursor-mediated adsorption" mechanism [12] is important in the chemisorption of a variety of gases on semiconductor surfaces, including O_2 [13], $Si(C_2H_5)_2H_2$ [14], $SiCl_4$ [15] and $SiCl_2H_2$ [16] on silicon. An obvious question to ask is: "To what extent does the precursor mobility effect film conformality?"

To focus the discussion, consider deposition of silicon through heterogeneous decomposition of silane. For this chemistry, the reaction sequence can be written as follows:

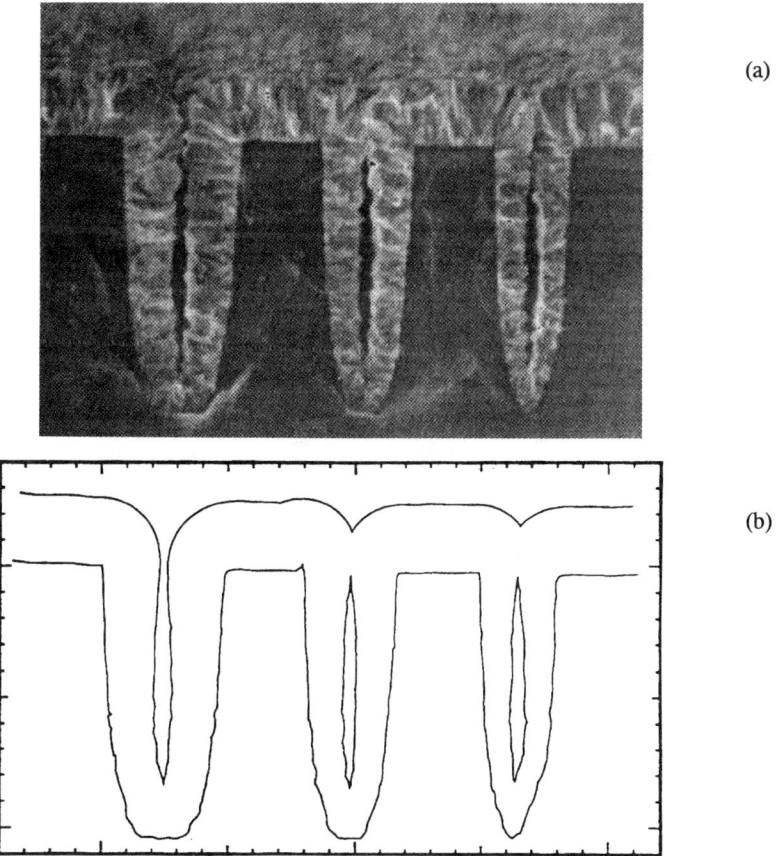

Fig. 1 (a) Experimental and (b) EVOLVE-simulated tungsten film profiles deposited in tapered trenches at 753 K, 0.88 Torr H_2 partial pressure, and 0.0078 Torr WF_6 partial pressure [10].

$$SiH_{4\,(g)} \underset{k_d}{\overset{\alpha}{\longleftrightarrow}} SiH_{4\,(p)}$$

$$SiH_{4\,(p)} + 2* \xrightarrow{k_a} SiH_2* + H_2*$$

$$SiH_2* \xrightarrow{k_f} Si_{(s)} + H_{2(g)} + *$$

Physisorbed (p) silane may be thought of as a weakly-bound entity, not associated with any particular surface site and free to move about on the surface in search of a site at which strong dissociative chemisorption can occur. Once dissociatively chemisorbed, the silylene

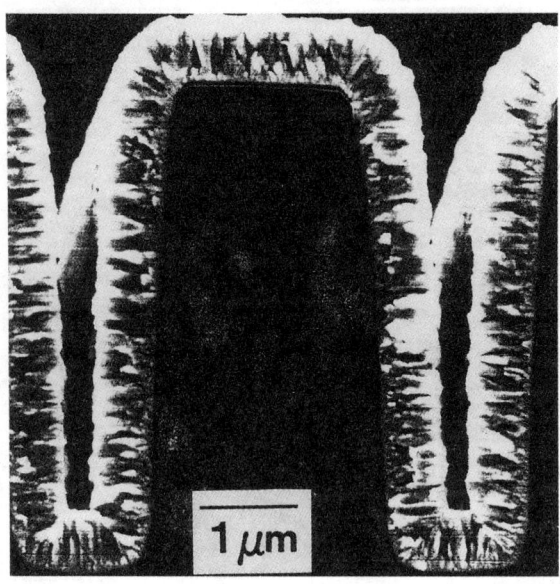

Fig. 2 Conformal tungsten film profile deposited in a high aspect ratio trench at 60 Torr total pressure [11].

and hydrogen intermediates are assumed to be immobile. Thus we distinguish between two types of surface diffusion: in the first type, the mobile precursor behaves like a two-dimensional gas and the precursor need not surmount a significant activation barrier for diffusion to take place; in the second type, the chemisorbed intermediate must surmount a significant energy barrier (chemical bonds must be broken and reformed) in moving from one specific site to an adjacent site. We presume that this latter diffusion type is unimportant in this example. This assumption is supported in the case of adsorbed hydrogen by the work of Koehler et al. [17], who used laser-induced thermal desorption to show that hydrogen surface mobility on Si(111) 7x7 is negligible up to at least 740 K.

A relatively simple rate expression for the mechanistic model presented above can be developed by assuming that (i) the condensation coefficient α for the precursor is constant, independent of flux, temperature and surface concentration; (ii) the binding energy of the precursor is the same above both filled and empty sites; (iii) the chemisorbed state follows usual Langmuirian behavior, i.e., a maximum of one monolayer adsorption at specific, uniform surface sites; and (iv) adsorbed silylene and adsorbed H_2 are kinetically indistinguishable. The steady rate of film formation G is

$$G = \frac{\alpha}{1+\frac{k_d}{k_a \theta_*^2}} \eta_{SiH_4} = k_f \theta_{SiH_2}$$

where θ_{SiH_2} is the fractional coverage of sites by the silylene surface intermediate. The term preceding the silane flux is the *reactive* sticking factor of SiH_4, *i.e.*, the fraction of impinging silane molecules which react. A site balance allows the steady surface coverages and the deposition rate to be predicted once all rate parameters are set. We have used the precursor model to fit the low pressure data of Buss *et al.* [18]. At a fixed silane flux, the data can be characterized by two regimes: a low temperature, highly-activated regime and a high temperature, mildly activated regime. At low temperatures the reaction rate is controlled by the surface decomposition reaction. At high temperature, the process becomes adsorption rate limited.

At low temperatures silane's *adsorptive* sticking coefficient approaches the value of the condensation coefficient. In physical terms, desorption from the precursor state is the rate-limiting step in the overall adsorption process, precursor molecules remain on the surface for relatively long periods, encountering many potential chemisorption sites as they migrate across the surface. The fact that the precursor molecule can "seek out" vacant active sites leads to a coverage-independent, high sticking factor as the surface sites fill. At higher temperatures, the desorption rate becomes higher, the mean residence time decreases, the average diffusion length of the precursor molecules therefore decreases, and the likelihood that a precursor reacts with an active site before it desorbs decreases.

An order of magnitude estimate of the residence time and the number of diffusive hops a precursor can make in that time can be obtained from the Frenkel equation and transition state theory. Assuming a desorption activation energy of 40 kJ/mol (the high end of values for physisorption), we estimate a residence time at 300 K of *ca.* 10^{-6} s and *ca.* 10^7 diffusive hops. The situation is markedly different at the elevated temperatures employed in deposition. At 850 K these values drop to 2 x 10^{-11} s and 370 hops, respectively. On the basis of this estimate, and considering the fact that atomic separation distances are on the order of tenths of nanometers and that typical feature dimensions are on the order of microns, it seems unlikely that precursor mobility will have a major effect on film profiles. On the other hand, mobility may have a significant affect on microscopic film properties (*e.g.*, crystallinity), and the most general model should therefore account for surface mobility.

CLASS II: HOMOGENEOUS PRECURSOR-MEDIATED DEPOSITION

In this chemistry class the source gases are relatively unstable and react in the gas phase to form highly reactive precursors to deposition, which then adsorb and ultimately react on the growing film surface to deposit new film. This reaction scheme may be generically written as follows:

$$R_{(g)} + M_{(g)} \xrightarrow{k_g} I_{(g)} + M_{(g)} + B_{(g)}$$
$$I_{(g)} + * \xleftrightarrow{k_I} I*$$
$$I* \xrightarrow{k_s} F_{(s)} + * + B_{(g)}$$

The first reaction represents collisionally-activated gas-phase decomposition of the source gas R to produce a reactive gas-phase precursor I. This reactive precursor adsorbs reversibly to produce a molecularly-chemisorbed I* surface intermediate, which subsequently decomposes heterogeneously to add to the growing crystalline film F, regenerate a surface site and release byproduct gas(es) B. The film deposition rate G is given by

$$G = \frac{k_s K_I P_I}{1 + K_I P_I}$$

where P_I is the local partial pressure of the reactive intermediate. The deposition rate and the step coverage depend on the partial pressure and reactive sticking coefficient of the intermediate. The precursor partial pressure is in turn a complex function of the interaction between the fluid mechanics and the gas-phase and surface chemical kinetics, so that in general a complex relationship results between apparent step coverage phenomena and the partial pressure of the feed gas R.

Potential gas-phase reactions in LPCVD systems include isomerization, abstraction, insertion, and various free-radical reactions. For chemistries in which significant free-radical concentrations are produced, a complex chain network of homogeneous reactions, not unlike those found in combustion chemistry, may control the production of a large slate of reactive intermediates. Examples include silicon deposition from silane [19], and silicon dioxide deposition from silane and N_2O or oxygen [20].

For accurate prediction of film conformality, the local reactive precursor concentrations above the wafer surface and their heterogeneous reaction kinetics must be known. To demonstrate the impact of homogeneous reactions on step coverage, we have used the reactor-scale model predictions of Coltrin et al. [19] for silicon deposition from silane in a rotating disk reactor as input to the feature scale simulator EVOLVE. Coltrin and coworkers employed 27 homogeneous reactions and a detailed description of the reactor fluid dynamics to predict the spatial distribution of 17 gas-phase species at steady state over a rotating, heated wafer. The data of Buss et al. [18] were used for the reactive sticking factor for silane; unity sticking was assumed for all gas-phase intermediates. Figure 3 summarizes EVOLVE-predicted step coverages at feature closure for a silicon film deposited from a 0.1% silane in hydrogen carrier feed in long rectangular trenches. The reactor scale model predicts that homogeneous reactions make an insignificant contribution to the deposition rate at 900 K and below; conformal step coverage is obtained under these conditions. As the temperature is increased, the homogeneous silane decomposition rate increases. Silylene and to a lesser extent disilylene become primarily responsible for the overall deposition rate. Because these precursors react with unit probability, the predicted step coverage degrades.

In general, several strategies can be employed to reduce the extent of homogeneous reactions when they have a deleterious effect on step coverage. These strategies include: (i) a reduction in temperature to slow the rates of the (usually) highly-activated homogeneous reactions; (ii) reduction in pressure to slow the rate of collisional activation; (iii) judicious choice of carrier gas to promote reverse reactions; and (iv) careful design of the reactor and choice of process conditions to minimize gas residence time, and in particular gas recirculation loops, in the heated or active zone of the reactor.

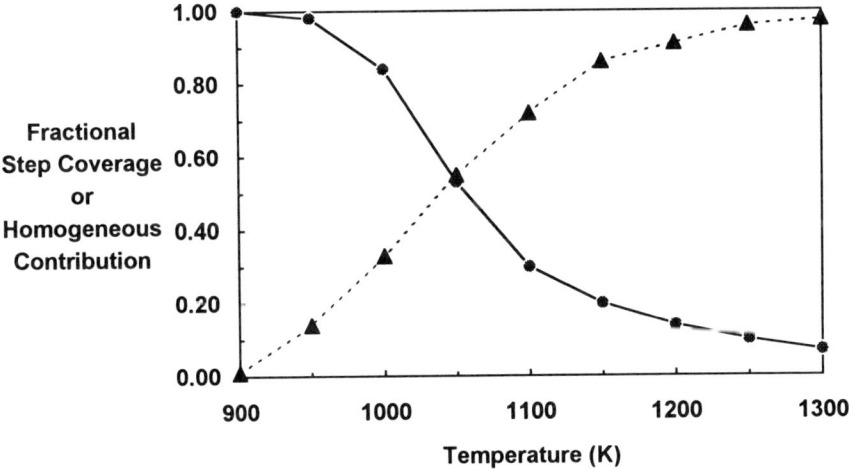

Fig. 3 Step coverage (solid line) vs. temperature for Si deposition from silane pyrolysis in a trench of aspect ratio equal to 2. Also plotted is the fractional contribution of the total deposition rate from homogeneous precursors (dashed line).

CLASS III: BYPRODUCT-INHIBITED DEPOSITION

In this chemistry class, byproducts released into the gas phase during the course of film deposition readsorb on the growing film, blocking sites and slowing the local deposition rate according to:

$$R_{(g)} + * \xrightleftharpoons{K_a} R*$$
$$R* \xrightarrow{k_s} F_{(s)} + B*$$
$$B* \xrightleftharpoons{K_B} B_{(g)} + *$$

The first two reactions are nearly identical to those presented for the Class I chemistries, except that the byproduct B does not desorb in the film formation step. The last reaction represents desorption/readsorption of the byproduct. The film deposition rate G is

$$G = \frac{k_s K_a P_R}{1 + K_a P_R + K_B P_B}$$

where P_B is the local partial pressure of the byproduct gas. The deposition rate expression clearly shows that the greater the byproduct partial pressure, the lower the deposition rate, and in turn the lower the reactive sticking factor of the source gas. We speculate that the byproduct inhibition mechanism may be most commonplace in metallorganic CVD (MOCVD) processes, since release of the last metallorganic ligand, perhaps through radical desorption, is thought to be the slow step in several such processes. If the byproduct released is indeed a hydrocarbon radical, then one would expect that readsorption would occur with a high probability. Moreover, free radical reactions in the gas phase over the wafer and the sweeping action of the convective gas flow over the wafer surface should serve to give a significantly lower level of the byproduct in the bulk gas space than within microfeatures.

Tetraethoxysilane (TEOS) sourced thermal LPCVD silicon dioxide film thickness in deep trenches drops dramatically in the region of the feature mouth, but is of fairly uniform thickness down the walls of the trench [21]. Figure 4 shows that this characteristic behavior is not reproduced if a homogeneous precursor-mediated deposition model is employed, but can be accurately predicted with the byproduct inhibition model and an appropriate choice of K_B [10,22,23]. The lower film thicknesses in the interior of the feature are caused not by "depletion" of source gas down the length of the trench (a decreasing source gas flux profile with increasing depth), but rather by a greater flux of byproduct to the interior walls of the feature relative to the external wafer surface, as graphically illustrated in Figure 5(a) for the initial stages of film deposition. Figure 5(b) shows that the byproduct flux builds up as the film deposition proceeds and the instantaneous feature aspect ratio increases. The feature "traps" the byproduct species in the feature in the sense that byproducts molecules or radicals desorbed from the growing film must undergo a relatively greater number of randomizing collisions with the feature surfaces to escape the feature as the aspect ratio increases.

We have undertaken preliminary studies which suggest that several MOCVD TiN chemistries may be governed by byproduct inhibition [24,25]. This conclusion does not preclude the existence of homogeneous reaction, but rather implies that byproduct inhibition is a controlling factor in film profile evolution. We suspect that additional chemistries will be shown to belong to this class of processes in the years to come as new MOCVD chemistries are developed and as this relatively newly-discovered phenomenon is tested more broadly.

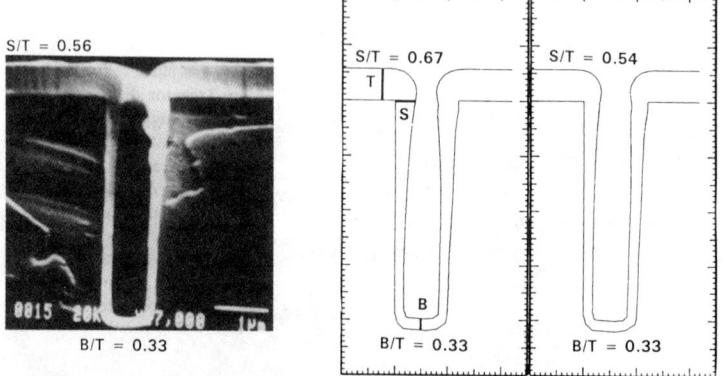

Fig. 4 Comparison of experimental TEOS-sourced silicon dioxide film in a high aspect ratio trench with EVOLVE simulations assuming a homogeneous precursor-mediated kinetic model (left) and a byproduct-inhibition model (right) [23].

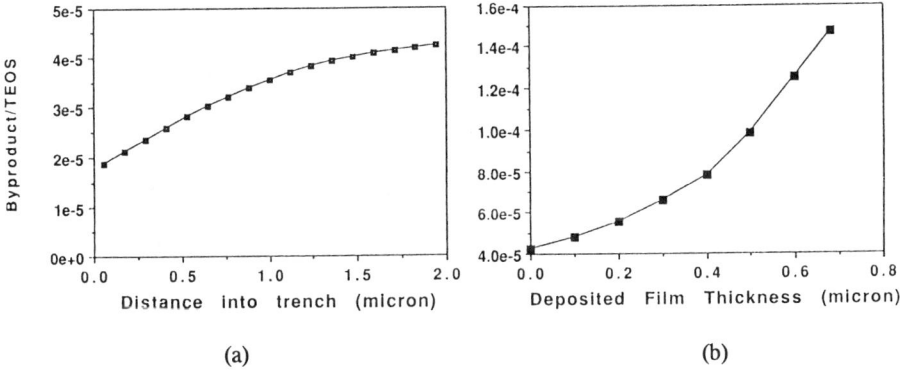

Fig. 5 EVOLVE predictions of the byproduct to TEOS flux ratio for (a) distance down the depth of the trench at initiation of film deposition, and (b) at the bottom of the trench as a function time (amount of film deposited).

SUMMARY

Table I summarizes a number of practical and developmental LPCVD chemistries by class based on current experimental evidence and current understanding.

TABLE I. Summary of LPCVD Process Chemistries by Class

Heterogeneous Deposition	Homogeneous Precursor-mediated Deposition	Byproduct-inhibited Deposition
$SiH_4 \rightarrow Si$ (low T, low P)	$SiH_4 \rightarrow Si$ (high T, high P)	$TEOS \rightarrow SiO_2$
$WF_6 + H_2 \rightarrow W$	$SiH_4 + O_2 \rightarrow SiO_2$	$SiH_4 + PH_3 \rightarrow$ P-doped Si
$TIBA \rightarrow Al$	$SiH_4 + N_2O \rightarrow SiO_2$	t-DMAT + $NH_3 \rightarrow TiN$
$Cu(hfac)_2 + H_2 \rightarrow Cu$		t-DEAT + $NH_3 \rightarrow TiN$
$TiCl_4 + NH_3 \rightarrow TiN$		

ACKNOWLEDGMENTS

We gratefully acknowledge the financial support of the National Science Foundation (NSF), the Semiconductor Research Corporation (SRC), and the Advanced Research Projects Agency (ARPA).

REFERENCES

[1] T. S. Cale and G. B. Raupp, *J. Vac. Sci. Tech. B* **8**, 649 (1990).
[2] T. S. Cale and G. B. Raupp, *J. Vac. Sci. Tech. B* **8**, 1242 (1990).
[3] T. S. Cale, G. B. Raupp and T. H. Gandy, *J. Vac. Sci. Tech. A* **10**, 1128 (1992).
[4] M. M. Islamraja, M. A. Capelli, J. P. McVittie and K. C. Saraswat, *J. Appl. Phys.* **70**, 7137 (1991).
[5] V. K. Singh, E. S. G. Shaqfeh and J. P. McVittie, *J. Vac. Sci. Tech. B* **10**, 1091 (1992).
[6] J. J. Hsieh and R. V. Joshi, in <u>Advanced Metallization for ULSI Applications</u>, V. V. S. Rana, R. V. Joshi and I. Ohdomari, eds., MRS, 1992, p. 77.
[7] EVOLVE was developed by T. S. Cale at Arizona State University with funding from the Semiconductor Research Corporation and the National Science Foundation.
[8] D. S. Ross, *J. Electrochem. Soc.* **139**,1714 (1993).
[9] C. M. McConica and K. Krishnamani, *J. Electrochem. Soc.* **133**, 2542 (1986).
[10] T. S. Cale, G. B. Raupp, M. B. Chaara, and F. A. Shemansky, *Thin Solid Films* **220**, 66 (1992).
[11] G. B. Raupp and T. S. Cale, *J. Inst. Electron. Telecomm. Engin. (India)* **37**, 206 (1991).
[12] J. B. Taylor and I. Langmuir, *Phys. Rev.* **33**, 423 (1933).
[13] M. P. D'Evelyn, M. M. Nelson and T. Engel, *Surface Sci.* **186**, 75 (1987).
[14] S. M. George, private communication.
[15] P. Gupta, P. A. Coon, B. G. Koehler and S. M. George, *J. Chem. Phys.* **93**, 2827 (1990).
[16] P. Gupta, P. A. Coon, B. G. Koehler, M. L. Wise and S. M. George, *Mat. Res. Soc. Symp. Proc.* **204**, 311 (1991).
[17] B. G. Koehler, C. H. Mak, D. A. Arthur, P. A. Coon, and S. M. George, *J. Chem. Phys.* **89**, 1709 (1988).
[18] R. J. Buss, P. Ho, W. G. Breiland and M. E. Coltrin, *J. Appl. Phys.* **63**, 2808 (1988).
[19] M. E. Coltrin, R. J. Kee and G. H. Evans, *J. Electrochem. Soc.* **136**(3), 819 (1989).
[20] C. J. Giunta, J. D. Chapple-Sokol and R. G. Gordon, *J. Electrochem. Soc.* **137**, 3237 (1989).
[21] F. S. Becker, D. Pawlik, H. Anzinger and A. Spitzer, *J. Vac. Sci. Tech. B* **5**, 1555 (1987).
[22] J. Schlote, K.-W. Schröder and K. Drescher, *J. Electrochem. Soc.* **138**, 2393 (1991).
[23] G. B. Raupp, F. A. Shemansky and T. S. Cale, *J. Vac. Sci. Tech. B* **10**, 2422 (1992).
[24] T. S. Cale, G. B. Raupp, J. T. Hillman and M. J. Rice, Jr., in <u>Advanced Metallization for ULSI Applications 1992</u>, T. S. Cale and F. Pintchovski, eds., MRS, 1993, p. 195.
[25] T. S. Cale, M. B. Chaara, G. B. Raupp and I. J. Raaijmakers, *Thin Solid Films* **236**(2), 294 (1993).

THE INFLUENCE OF TEMPERATURE GRADIENTS ON PARTIAL PRESSURES IN A CVD REACTOR.

T.G.M. OOSTERLAKEN, G.J. LEUSINK, G.C.A.M. JANSSEN, S. RADELAAR
DIMES/ Section Submicron Technology, Delft University of Technology, P.O. Box 5046, 2600 GA Delft, The Netherlands
K.J. KUIJLAARS, C.R. KLEIJN, H.E.A. VAN DEN AKKER
Kramers Laboratorium voor Fysische Technologie, Delft University of Technology, Prins Bernhardlaan 6, 2628 BW Delft, The Netherlands

ABSTRACT

The influence of temperature gradients on the partial pressures of a binary mixture in a cold wall low pressure chemical vapor deposition reactor was determined by Raman spectroscopy of the gaseous species in the reactor. It is demonstrated for the first time that the partial pressure of the heavy constituent in the hot region of a low pressure reactor is reduced by 35 % due to the Soret effect. Model calculations that included the Soret effect are in agreement with the experimental data.

INTRODUCTION

Due to the decreasing size of the smallest features in integrated, circuits chemical vapor deposition (CVD) is more and more used as deposition technique for films in microelectronics [1]. The properties of CVD processes and films are dependent on the partial pressures of the reactants at the wafer surface. The CVD films have to meet ever more demanding requirements in terms of uniformity, step coverage and stress. The transport processes responsible for supplying the reactants to and removal of reaction products from the wafer surface are, therefore, becoming increasingly important. For instance in W-CVD the step coverage of H_2 reduced tungsten films depends both on the H_2 and the WF_6 partial pressures [3]. In addition, the mechanical stress in tungsten films depends on the partial pressures of the reactants [4].

In a cold wall reactor (the type of reactor most frequently used for tungsten CVD) the gas mixture is subject to large temperature gradients. These gradients have a separating effect on the mixture due to the Soret effect [5,6]. The Soret effect or thermal diffusion forces the heavy and large species to concentrate in the colder regions of the accessible volume. The Soret effect therefore effectively reduces the partial pressures of the heavy constituents in the hot area.

In this paper the results of experiments with a model system are given. The gas mixture consists of hydrogen and nitrogen. This system allows to study the influence of the Soret effect on the partial pressures of the species without any interference of chemical reactions. The experiments are performed in a cold wall reactor under conditions resembling those of actual LPCVD processes. Raman scattering allows the determination of the partial pressures and local temperatures without disturbing the heat or gas flow profile in the reactor. The experimentally observed values for the partial pressures are compared to the values calculated with a numerical reactor model [7,8].

RAMAN SCATTERING

The partial pressures and temperatures of the gas were determined by Raman scattering. When light is scattered by a gas from an intense monochromatic light beam, spectral lines with a low intensity can be observed on either side of the elastically scattered Rayleigh line [9]. These Raman lines are due to photons which have lost or gained energy in the scattering process. The photons that emerge as red-shifted photons are called Stokes shifted photons while the photons that are shifted towards higher energies are called anti-Stokes photons. The Raman lines are due to transitions between the various rotational and vibrational energy levels [10] under influence of light. The use of Raman scattering for determination of the composition of a gas mixture has become feasible with the introduction of high power laser systems. Nevertheless, when the determination of local partial pressures is required under the combined conditions of low total pressures, elevated temperatures, and short sampling times the technique is still very demanding [11]. The Raman lines can also be used for temperature measurements. Most methods used for determination of the temperature in gases are based on the relative intensity of Stokes and anti-Stokes Raman scattered radiation. Since for most Raman wavelength displacements the anti-Stokes lines are at least one order of magnitude less intense than the Stokes lines, we relied exclusively on the Stokes lines for determination of the partial pressures and temperatures. The intensity of the Stokes vibrational Raman line at 2331 cm^{-1} was used to determine the density of the nitrogen. The rotational lines at 587 cm^{-1} and 1034 cm^{-1} were used to determine the density as well as the temperature of hydrogen.

NUMERICAL REACTOR MODEL

The numerical reactor model is described more extensively elsewhere [7,8]. The model is based on the assumption that the gas in the reactor behaves as a continuum. The density of the species and the temperature distribution were calculated using the fundamental equations for mass, momentum, heat, and chemical species. These equations were solved numerically with a finite volume discretization technique on a numerical grid that was refined near the wafer surface and other places in the reactor where steep gradients are likely to occur. The model includes a detailed description of thermal diffusion, based on the kinetic theory of gases.

EXPERIMENT

The experiments were performed in a cold-wall LPCVD single wafer reactor suitable for the processing of 100 mm Si wafers. The gases are introduced into the reactor at the top of the rotational symmetrical vessel and pumped at the bottom of the vessel. During the experiments the flows of the gases were controlled by mass flow controllers. The 100 mm wafer in the reactor was heated by a resistance heater. The temperature was varied from 290 K to 800 K. The pressure in the process chamber during the experiments was 533 Pa. A nitrogen flow of 0.130 slm and a hydrogen flow of 0.310 slm were used for the experiments.

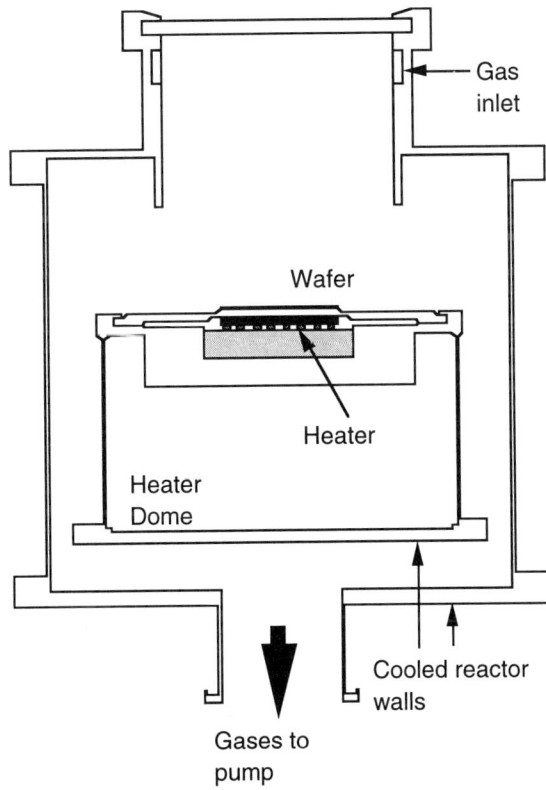

An excimer pumped dye laser (Lambda Physik 315i, 3002 fl, 390 nm)[12] was used as a light source for the Raman experiments. The dye laser generates 30 mJ pulses with 25 nanoseconds duration at repetition rates up to 100 hertz. The probing volume was viewed at 90° with respect to the propagation direction and the polarization direction of the laser beam. The collected light is filtered by a double monochromator (Spex 1403). The photons were detected by an array detector. The elements of the array were grouped and acted as a single light sensitive channel. Due to the small cross section for Raman scattering integrating times of 200 seconds were necessary to achieve sufficient signal to noise ratios for accurate determination of the partial pressures and temperatures at low pressures. The intensity of each measured line was determined with the monochromator tuned to the specific wavenumber of the line. The background signal and the relative sensitivity of the experimental set-up to each line were determined at room

Figure 1 Cross section of the reactor used in this study. The wafer is placed on top of the heater dome, and is heated by the graphite chuck in the heater dome. The diameter of the reactor is 0.40 m.

temperature at various pressures for each specific species. The power of the laser during the experiments was monitored at the exit window of the reactor. The temperature as well as the composition of the mixture is probed at 17 mm above the center of the surface of the 100 mm-wafer. The number of counts that was measured during the integration time was divided by the total energy delivered by the laser to the energy monitor. The total energy measured by the energy monitor during such a measurement amounted to approximately 60 J.

RESULTS AND DISCUSSION

In the figures 2, 3 and 4 the results of a study on the composition of a gas mixture consisting of H_2 and N_2 are shown as a function of the wafer temperature. The intensity of the measured

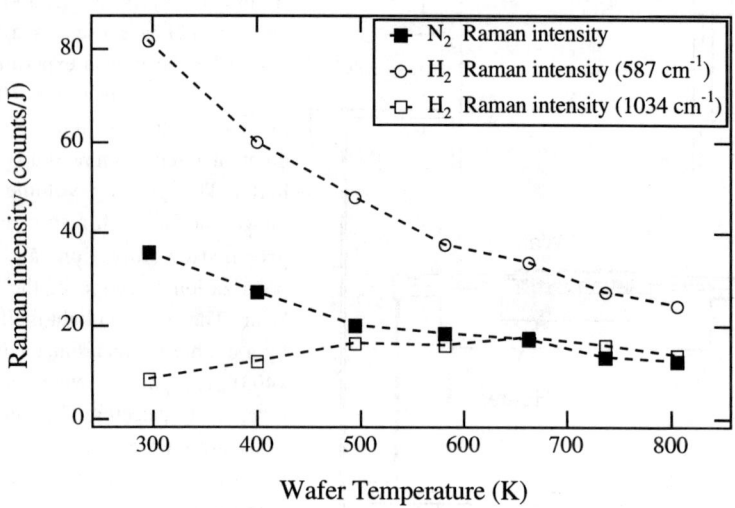

Figure 2 *The intensity of the measured Raman lines. The measurements were taken at 17 mm above the center of the wafer. The intensity is shown as a function of the wafer temperature. The dashed lines connect the points of a single Raman line.*

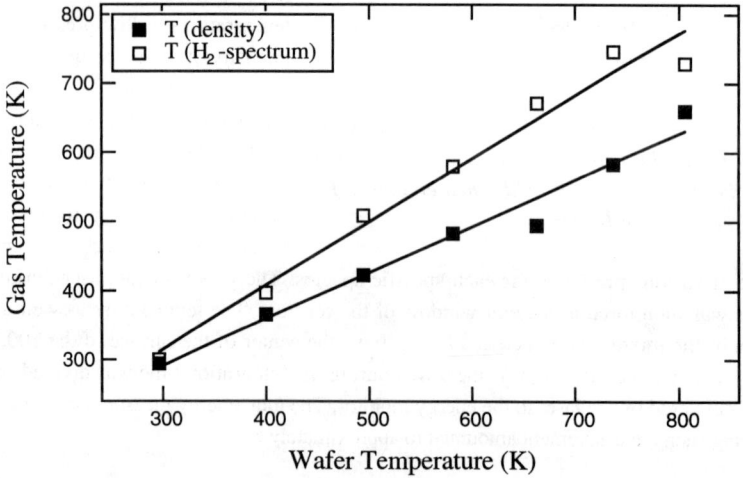

Figure 3 *The temperatures of the gas at 17 mm above the center of the wafer, as calculated from the total density (black squares) and as calculated from the hydrogen Raman spectrum (open squares). The solid lines show the best linear fit to the measured values.*

Raman lines is shown as a function of the wafer temperature in figure 2. As the temperature increases in the probing volume the density decreases, this effect is directly observed in the decreasing intensity of the N_2 line in figure 2. The behaviour of the lines of the hydrogen spectrum is more complicated since the occupancy of the energy levels changes with the temperature. The intensity of the line at 587 cm^{-1} decreases due to the decrease in density and due to a decrease in occupancy of the low rotational levels at higher temperatures. The occupancy of higher rotational levels increases and causes an increase in the intensity of the line at 1034 cm^{-1}.

In figure 3 the temperature of the hydrogen is shown as it was deduced from the hydrogen Raman spectrum. The temperature was also determined from the ratio between the total density and pressure in the reactor. Both values of temperature are in agreement. The rotational temperature from the hydrogen is not very accurate especially at higher temperatures. The error is systematic and due to the weakness of the hydrogen line at 1034 cm^{-1} at the calibration temperature of 290 K. The temperatures during these experiments are such that the rotational and translational temperatures are in equilibrium.

In figure 4 the partial pressure of nitrogen and hydrogen are shown as a function of the wafer temperature. From the intensity of the Raman lines the density of both hydrogen and nitrogen were determined. The partial pressure is then easily calculated from the partial density, the total density, and the total pressure. The partial pressure of nitrogen is significantly reduced in the hot area of the reactor. This reduction is due to the Soret effect. The results of the numerical calculations are also shown in figure 4. The solid lines are the partial pressures of both species when a Lennard-

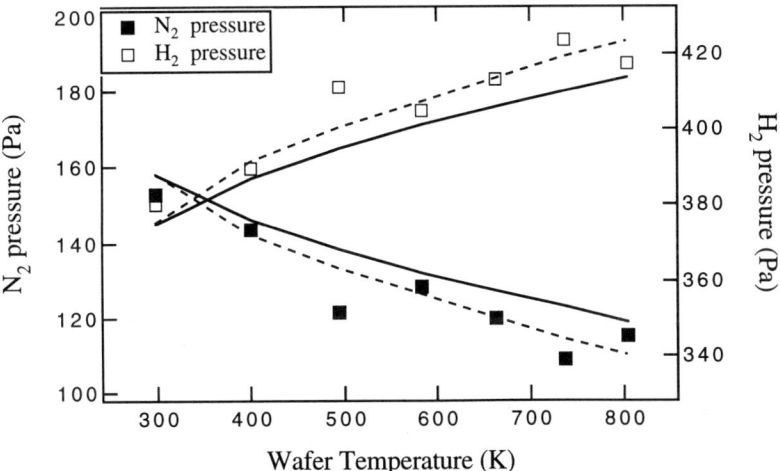

Figure 4 The partial pressures of hydrogen and nitrogen at 17 mm above the center of the wafer. The reduction in the partial pressure of nitrogen is due to thermal diffusion. The solid lines are the partial pressures calculated with the assumption of Lennard Jones interactions between the molecules. The dashed lines were calculated with the rigid spheres approximation.

Jones(6-12) type intermolecular potential function was used in the calculations. The dashed lines are calculated assuming rigid spheres interactions between the molecules. In addition to the boundary conditions of the reactor the gas flows and the wafer temperature were used as input for these calculations. The calculated values are in agreement with the observed values for the partial pressures.

CONCLUSIONS

In this paper it is shown that the influence of temperature gradients on partial pressures can be observed by Raman scattering. This was done for the first time in a low pressure CVD system under conditions comparable to actual process pressures and temperatures. The partial pressure of the heavy species (N_2) was reduced from 155 Pa at the inlet to 110 Pa near the wafer surface when the wafer temperature is increased to 800 K. The calculated values are in agreement with the observed values.

ACKNOWLEDGEMENTS

This work is part of the research program of the "Stichting voor Fundamenteel Onderzoek der Materie", which is financially supported by the "Nederlandse Organisatie voor Wetenschappelijk Onderzoek (NWO)". We gratefully acknowledge financial support by the "Innovatie gericht Onderzoeksprogramma voor IC-Technologie (IOP-IC)". The help of J. Tóth and G.J. Kuiper in performing the experiments was appreciated. Prof. E.H.A. Granneman is acknowledged for valuable discussions about the experiments.

REFERENCES

1. E.H.A. Granneman, Thin Solid films, **228**, 1 (1993).
2. A. Hasper, C.R. Kleijn, J. Holleman, J. Middelhoek, and C.J. Hoogendoorn, J. Electrochem. Soc. **138**, 1729 (1991).
3. J.E.J. Schmitz, W.L.N. van der Sluys, and A.H. Montree, Proceedings of "Tungsten and other advanced metals for VLSI/ULSI Applications V (1989), p 117, ed. S.S. Wong, S. Furukawa, Materials Research Society, Pittsburgh, Pennsylvania (1990).
4. G.J. Leusink, T.G.M. Oosterlaken, G.C.A.M. Janssen, and S.Radelaar, Thin Solid Films **228**, 125-128 (1993).
5. J.O. Hirschfelder, C.F. Curtiss and R.B. Bird, "Molecular theory of Gases and Liquids", John Wiley and Sons Inc., New York, 1967.
6. S. Chapman, F.W. Dootson, Phil. Mag. **33**, p 268 (1917).
7. C.R. Kleijn, C.J. Hoogendoorn, A. Hasper, J. Holleman, and J. Middelhoek, J. Electrochem. Soc. **138**(2), 509 (1991).
8. C.R. Kleijn, J. Electro. Chem. Soc. **138** (7), 2190 (1991).
9. R.M. Measures, Analytical Laser Spectroscopy, Chemical Analysis **50** (1979), ed. Nicolò Omenetto, John Wiley & Sons, New York, USA.
10. G. Herzberg, Molecular Spectra and Molecular structure **I,II**, second edition, (1950), Van Nostrand Reinhold Company, New York USA.
11. P.J. Hargis jr., Applied Optics **20**, (1), 149 (1981).
12. T.G.M. Oosterlaken et al, to be submitted to J. Appl. Phys.

SURFACE REACTION INTERMEDIATES IN Ge CHEMICAL VAPOR DEPOSITION ON SILICON

C. MICHAEL GREENLIEF AND LORI A. KEELING
University of Missouri-Columbia, Department of Chemistry, Columbia, MO 65211

ABSTRACT

Computational programs are often used for estimating molecular orbital energies and geometries. The results of these calculations can also provide information needed for investigation of adsorption and decomposition mechanisms on surfaces. In this study, *ab initio* methods are used to calculate properties of gas phase diethylgermane and GeH_x-substituted silanes (x=1-3). The silanes are used as models for surface Ge hydrides. The calculated gas phase molecular orbital energies are then compared with experimentally determined molecular orbital energies for the adsorbed species. The surface species are prepared by germane, digermane, ethylbromide, or diethylgermane adsorption on the Si(100)-(2×1) surface and the molecular orbital energies are measured by photoelectron spectroscopy.

INTRODUCTION

The growth of $Si_{1-x}Ge_x$ alloys and heterostructures on Si(100) has received attention recently because of the potential electronic devices that can be made with these materials. Digermane, Ge_2H_6, and germane, GeH_4, are the two most common germanium containing molecular precursors that are used in the chemical vapor deposition (CVD) of Ge [1-5]. However, these two gases are pyrophoric and their handling and disposal are becoming important safety concerns. With these concerns in mind, there is interest in the development of alternative precursors for germanium. Alkylgermanes, such as diethylgermane (GeH_2Et_2), are possible substitutes because they are liquids at room temperature with suitable vapor pressures for CVD, are less reactive with air, and are easier to handle than conventional hydride sources.

As part of an effort to understand the surface chemistry of possible alternative precursors, the adsorption and decomposition of GeH_2Et_2 on Si(100) at low temperatures is investigated by ultraviolet photoelectron spectroscopy (UPS). The position and identification of molecular orbitals is aided by *ab initio* calculations. The eventual thermal decomposition products are molecular hydrogen, ethylene, and adsorbed germanium. During the decomposition of GeH_2Et_2, ethyl groups are deposited onto the surface and desorb at higher temperatures as C_2H_4. The proposed mechanistic step for the decomposition of surface ethyl groups to form C_2H_4 is a β-hydride elimination [6-8]. Ethylbromide (EtBr) is also used as a precursor to deposit ethyl groups onto Si(100) [9]. *Ab initio* calculations are used to help distinguish adsorption on Si sites and these results are compared with those obtained for GeH_2Et_2 on Si(100).

β-Hydride and β-alkyl elimination reactions and their role in chemical vapor deposition have been reported by Bent and co-workers [10,11]. The thermal decomposition of trisisobutylaluminum (TIBA) on Al(111) and Al(100) were investigated. This molecule was an effective precursor for the deposition of Al with the evolution of isobutylene and hydrogen. The formation of isobutylene was possible only through a β-hydride reaction. However, when the same molecule is exposed to Si(111) or Si(100), reversible desorption is observed at low surface temperatures [11]. This observation suggests that TIBA decomposes differently on Si than on Al and makes the comparison with diethylgermane studies difficult.

EXPERIMENTAL

The TPD and UPS experiments are performed in a stainless steel ultra-high vacuum (UHV) chamber that is described in detail elsewhere [12]. Diethylgermane (Gelest, purity > 98%), germane (Solkatronic, electronics grade, min purity 99.99%), digermane [Voltaix, ultrahigh purity (UHP) grade, min purity 99.999%], and ethylbromide (Aldrich, 99+% purity) are further purified by several freeze-pump-thaw degassing cycles and the purity of the gases is checked *in-situ* by mass spectrometry. The gas is admitted to the chamber through an effusive doser and directed onto the front face of the crystal at an apparent pressure of 3×10^{-10} torr above the base pressure for various periods of time.

Sample preparation and cleaning procedures have been described previously [13]. All the reported UPS binding energies are referenced to the Si valence band edge, which is assigned as 0 eV binding energy.

Ab initio molecular orbital calculations are performed with the Gaussian 92 system [14] using a modified double zeta effective core potential basis set (LANL1DZ [15]). All equilibrium geometries and transition state geometries are fully optimized at the Hartree-Fock level. Vibrational frequencies and zero point energies are obtained from the analytical second derivatives [16] calculated at the HF/LANL1DZ level using the HF/LANL1DZ optimized geometries.

RESULTS AND DISCUSSION

Figure 1 shows three substituted silanes used as models for adsorbed Ge hydrides (GeH, GeH_2, and GeH_3) on Si(100). These structures reflect the calculated equilibrium geometries for each molecule. The Ge–H bonding molecular orbital energies for each of these molecules is also determined. These energies are listed in Table I and compared with the experimentally determined binding energies for each of the Ge hydrides on Si(100) [13] and Si(111) [17]. The Ge hydrides are generated by the thermal decomposition of either GeH_4 or Ge_2H_6. The calculated molecular orbitals are in good agreement with the experimentally-measured results.

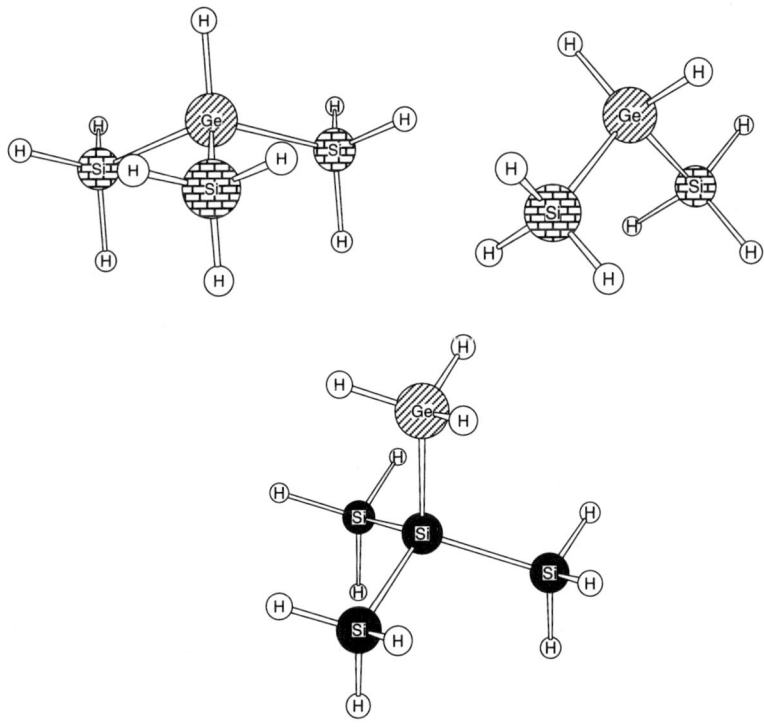

Figure 1: Ge-hydride substituted silane and germanes used in *ab initio* calculations to examine the molecular orbital energies of Ge-H bonds.

Table I: Summary of GeH$_x$ Measurements and Calculations

Hydride	Bond Lengths, Å		[a]σ_{GeH}	Orbital Energies (- eV) [b]σ_{GeH}	[c]σ_{GeH}
GeH				5.1	4.9
GeH(SiH$_3$)$_3$	R$_{Ge-H}$	1.564	5.372		
GeH$_2$				5.6	5.4
GeH$_2$(SiH$_3$)$_2$	R$_{Ge-H}$	1.566	5.930		
GeH$_3$				6.1	5.9
GeH$_3$Si(SiH$_3$)$_3$	R$_{Ge-H}$	1.535	6.389		

[a] Calculated energy level
[b] Reference 13, Si(100)
[c] Reference 17, Si(111)

The room temperature adsorption of GeH$_2$Et$_2$ on Si(100) is dissociative [8] with both ethyl groups transferring to the Si surface. This can be summarized in the following reaction:

$$\text{GeH}_2\text{Et}_2 \text{ (g)} + 3 \text{ Si}^* \rightarrow 2 \text{ Et (ad)} + \text{GeH}_2 \text{ (ad)} \qquad (1)$$

where Si* represents a surface Si atom. There is also the possibility that the ethyl groups remain attached to the Ge atom. In order to help distinguish between these two possibilities, EtBr was thermally dissociated to generate surface ethyl groups [9]. This process generates a surface with all the ethyl groups adsorbed on Si sites. Figure 2a shows the UPS spectrum obtained from 2.7×10^{14} ethyl groups cm^{-2} by the thermal dissociation of EtBr. This spectrum can be compared with the spectrum shown as Fig. 2b, which obtained after the room temperature adsorption of GeH$_2$Et$_2$. The peaks observed near 6.5, 9, 14, and 18 eV are assigned (using the symmetry notation of Jorgensen and Salem [18]) as follows: $\pi(\text{CH}_3)$, 6.5 eV; $\pi'(\text{CH}_3)$, 9 eV; $\sigma(\text{CH}_3,\text{CC})$, 14 eV; and C(2s), 18 eV. The close agreement between the two spectra is interpreted as due to the ethyl groups generated from the decomposition of GeH$_2$Et$_2$ being adsorbed at Si sites rather than being bound to Ge.

This interpretation is also consistent with the calculated results. The bar graph at the top of Figure 2 indicates the calculated molecular orbital energies for gas phase C$_2$H$_5$Si(SiH$_3$)$_3$. This substituted silane is a model molecule for adsorbed ethyl groups on the Si surface. There is reasonable agreement between the calculated levels and the measured spectra. The calculated level near 16 eV is due to the $\sigma(\text{CH}_3,\text{CC})$ molecular orbital. The almost 2 eV difference between the measured and calculated spectra is due to the effective core approximation that is used in the basis set and this approximation is starting to break down for these more tightly bound molecular orbitals. Calculations were also completed for a substituted germane, C$_2$H$_5$Ge(SiH$_3$)$_3$, to use as a model for ethyl groups adsorbed on a Ge atom. The molecular orbital energies obtained were in poorer agreement with the measured spectra.

It is also interesting to note that the ethyl-substituted silane is more stable than the ethyl-substituted germane. This result shows that there is a thermodynamic driving force for the ethyl groups to reside on Si sites over the Ge sites. The calculation is also consistent with measured carbon bond strengths to Si and Ge. The typical Si-C bond strength is 90 kcal mol^{-1} [19], while the Ge-C bond strength averages about 58 kcal mol^{-1} [20].

SUMMARY AND CONCLUSIONS

Several different reaction intermediates involved in the CVD of Ge on Si have been isolated and identified by photoelectron spectroscopy. The identification of the reaction intermediates is aided by the use of calculations. The model systems used in the calculations are simple substituted silanes and germanes. The molecular orbital energies calculated are in good agreement with the experimentally-measured results. While this approach works well for molecular orbital energies, it is not expected to predict reaction energetics reliably.

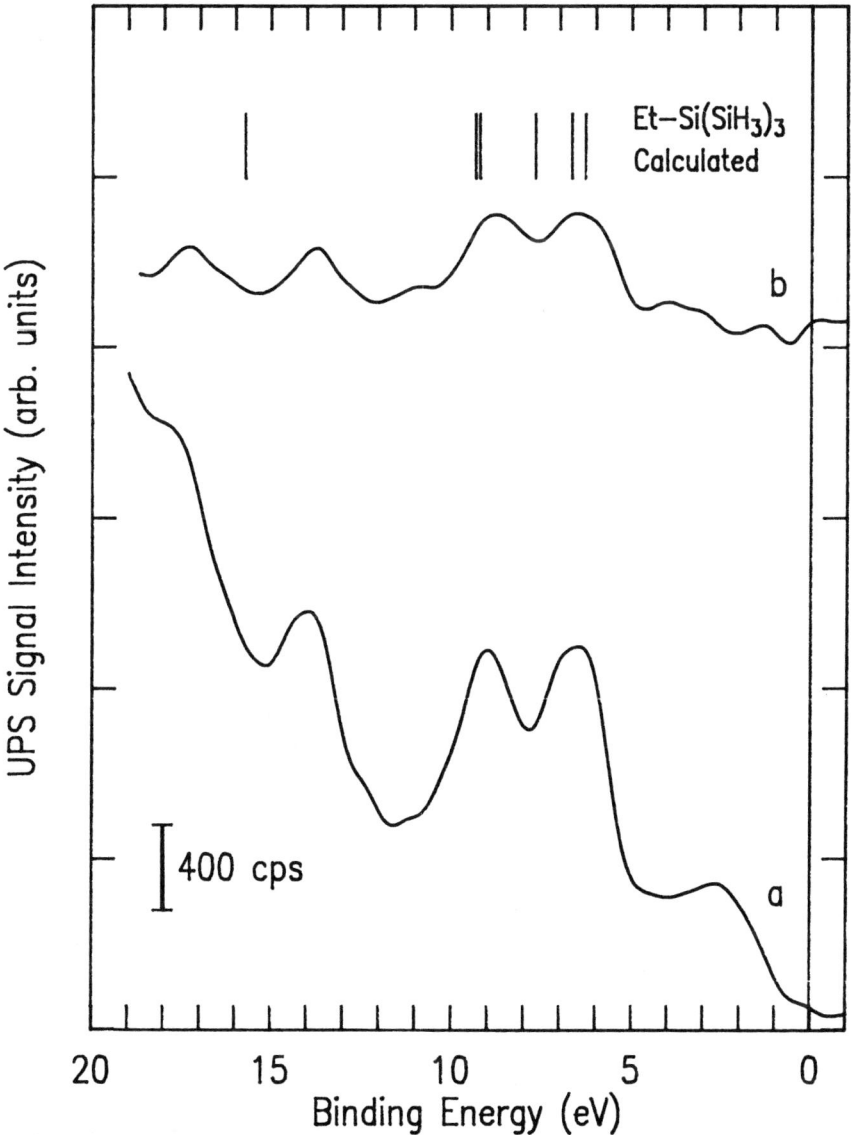

Figure 2: a) He II spectrum for a coverage of 2.7×10^{14} cm^{-2} ethyl groups on Si(100) generated by the room temperature adsorption of EtBr. b) He II difference spectrum obtained after the room temperature adsorption of GeH$_2$Et$_2$ on Si(100). The coverage of ethyl groups is 1.4×10^{14} cm^{-2}. The bar graph at the top of the figure is the calculated photoelectron spectrum of C$_2$H$_5$Si(SiH$_3$)$_3$.

Acknowledgements

The authors gratefully acknowledge the Donors of The Petroleum Research Fund, administered by the American Chemical Society, and the National Science Foundation (Grant CHE-9100429) for support of this research. CMG also acknowledges the National Science Foundation for a Young Investigator Award.

REFERENCES

1. B.S. Meyerson, K.J. Uram, and F.K. LeGoues, Appl. Phys. Lett. **53**, 2555 (1988).
2. P.M. Garone, J.C. Strum, and P.V. Schwartz, Appl. Phys. Lett. **56**, 1275 (1990).
3. M. Racaneli and D.W. Greve, Appl. Phys. Lett. **56**, 2524 (1990).
4. D.J. Robbins, J.L. Glasper, A.G. Cullis, and W.Y. Leong, J. Appl. Phys. **69**, 3729 (1991).
5. S.-M. Jang and R. Reif, Appl. Phys. Lett. **59**, 3162 (1991).
6. P.A. Coon, M.L. Wise, Z.H. Walker, S.M. George, and D.A. Roberts, Appl. Phys. Lett. **60**, 2002 (1992).
7. A.C. Dillon, M.B. Robinson, S.M. George, and D.A. Roberts, Surf. Sci. **286**, L535 (1993).
8. W. Du, L.A. Keeling, and C.M. Greenlief, J. Vac. Sci. Technol. A, accepted.
9. L.A. Keeling, L. Chen, C.M. Greenlief, A. Mahajan, and D. Bonser, Chem. Phys. Lett., in press.
10. B.E. Bent, R.G. Nuzzo, and L.H. Dubois, J. Amer. Chem. Soc. **111**, 1634 (1989).
11. B.E. Bent, R.G. Nuzzo, and L.H. Dubois, J. Vac. Sci. Technol. A **6**, 1920 (1988).
12. C.M. Greenlief and D.A. Klug, J. Phys. Chem. **96**, 5424 (1992).
13. D.A. Klug, W. Du, and C.M. Greenlief, J. Vac. Sci. Technol. A **11**, 2067 (1993).
14. Gaussian 92, Revision C, M.J. Frisch, G.W. Trucks, M. Head-Gordon, P.M.W. Gil, M.W. Wong, J.B. Foresman, B.G. Johnson, H.B. Schlegel, M.A. Robb, E.S. Replogle, R. Gomperts, J.L. Andres, K. Raghavachari, J.S. Binkley, C. Gonzalez, R.L. Martin, D.J. Fox, D.J. Defrees, J. Baker, J.J.P. Stewart, and J.A. Pople, Gaussian, Inc., Pittsburgh PA, 1992.
15. P.J. Hay and W.R. Wadt, J. Chem. Phys. **82**, 270 (1985); W.R. Wadt and P.J. Hay, J. Chem. Phys. **82**, 284 (1985); P.J. Hay and W.R. Wadt, J. Chem. Phys. **82**, 299 (1985).
16. J.A. Pople, R. Krishnan, H.B. Schlegel, and J.S. Binkley, Int. J. Quantum Chem. Symp. **13**, 225 (1979).
17. S. Van, D. Steinmetz, F. Ringeisen, D. Bolmont, and J.J. Koulmann, Phys. Rev. B **44**, 13807 (1991).
18. W. L. Jorgensen and L. Salem, The Organic Chemist's Book of Orbitals (Academic Press, New York, 1973).
19. R. Walsh, in: The Chemistry of Organic Silicon Compounds, edited by S. Patai and Z. Rappoport (Wiley, New York, 1988), chap. 5.
20. F. Glockling, The Chemistry of Germanium (Academic Press, New York, 1969), p. 10.

Growth and Characterization of Si-GaP and Si-GaP-Si Heterostructures

N. Dietz, S. Habermehl, J. T. Kelliher, G. Lucovsky and K. J. Bachmann
Department of Material Science and Engineering and Department of Physics
North Carolina State University, Raleigh, NC 27695

ABSTRACT

The low temperature epitaxial growth of Si / GaP / Si heterostructures is investigated with the aim using GaP as a dielectric isolation layer for Si circuits. GaP layers have been deposited on Si(100) surfaces by chemical beam epitaxy (CBE) using tertiarybutyl phosphine (TBP) and triethylgallium (TEG) as source materials. The influence of the cleaning and passivation of the GaP surface has been studied in-situ by AES and LEED, with high quality epitaxial growth proceeding on vicinal GaP(100) substrates. Si / GaP / Si heterostructures have been investigated by cross sectional high resolution transmission electron microscope (HRTEM) and secondary ion mass spectroscope (SIMS). These methods reveal the formation of an amorphous SiC interlayer between the Si substrate and GaP film due to diffusion of carbon generated in the decomposition of the metalorganic precursors at the surface to the GaP/Si interface upon prolonged growth (layer thickness > 300Å). The formation of twins parallel to {111} variants in the GaP epilayer are extended into the subsequently grown Si film with minor generation of new twins.

I. INTRODUCTION

The dielectric isolation of Si epilayers by epitaxial interlayers of GaP in vertical epitaxial interlayers of integrated microelectronic circuits and the possible use of epitaxial GaP on Si for optoelectronic applications is of interest in the context of optical interconnection[1,2]. The potential use in microelectronic applications requires low temperature processing to avoid interdiffusion. A low temperature processing also is favorable to decrease impurity segregation and to reduce thermal induced stress due to the mismatch of thermal expansion coefficients of GaP and Si[1]. In this paper, we report the growth of GaP on Si(100) substrates using metalorganic sources and the epitaxial growth of Si from 10% silane in hydrogen on GaP surfaces using remote plasma CVD. Epitaxial growth at low substrate temperature was achieved by both techniques.

II. EXPERIMENTAL

Thin GaP films deposited on Si substrates have been demonstrated in a CBE system which has been described elsewhere[3]. The Si substrate received an ex-situ RCA clean followed by a buffered HF etch and a final DI water rinse. The substrate is then loaded immediately into the CBE growth chamber via a load lock chamber. The TBP and TEG flows are first established into a separated pumped bypass chamber and switched over into the growth chamber when the desired temperatures are reached. The TBP is flowing into the growth chamber as the subtrate is heated up to the GaP deposition temperature.

The Si thin film deposition and the initial surface treatment are done in an ultra-high vacuum-compatible integrated processing system, described previously[4]. An analysis chamber attached to the remote plasma enhanced chemical vapor deposition (RPECVD) system allows to monitor the surface chemical composition by Auger electron spectroscopy (AES) while surface crystallinity was determined with a four grid reverse-view LEED system. The ex-situ cleaning procedure involves a 60 sec etch of $NH_4OH : H_2O_2 : H_2O$ (1:1:10) followed by 5 min DI water rinse, both at room temperature. The deposition of Si on GaP surfaces was achieved with a remote plasma process, where the deposition is accomplished by exciting a downstream H_2-SiH_4 injected mixture by active species extracted from an upstream He plasma source. The process parameter are RF power of 50 W, H_2 flow of 25 sccm, He flow of 200 sccm and a dilute SiH_4:He (10:1) flow of 10 sccm. The substrate temperature and the process pressure, are maintained in the range from 300-400°C and from 50-500 mtorr, respectively.

III RESULT AND DISCUSSION

a) Si deposition on GaP surfaces

After the ex-situ wet chemical treatment the AES spectra reveals significant amounts of residual carbon and oxygen on the GaP(111) surface, as shown in fig. 1a. The GaP surface was then heated to 530°C and annealed in a 10 mtorr H_2 ambient (50 sccm H_2 flow). After a two minute H-plasma exposure (fig. 1b) the O KLL peak is significantly reduced, which is due to the chemical reduction of native Ga-O and P-O bonds by activated hydrogen species[5]. After 6 min H-plasma exposure no oxygen or carbon traces are found in the AES spectra (fig. 1c) within the detection limits, which indicates a clean GaP surface. The vicinal GaP(100) surface was successfully cleaned using the same parameters except for a longer H-plasma exposure time of 15 - 20 min. The longer H-plasma exposure time can be understood assuming a more complex bonding situation at steps on the vicinal (100) surface[4,5]. Si films are deposited on (111) and vicinal (100) GaP surfaces in the temperature range between 300-400°C with process pressures from 50-500 mtorr. For the GaP(111) surface, LEED analysis of the deposited Si films for various temperature and pressure conditions revealed surface disorder after only a few monolayers of Si growth. Compared to a clean GaP(111) LEED pattern, a broadening of the diffraction spots and a loss of contrast is observed. Better growth results are obtained for stepped GaP(100) surfaces with a substrate temperature of 400°C and a process pressure of 500 mtorr. LEED patterns were observed for film thicknesses up to 1000 Å.

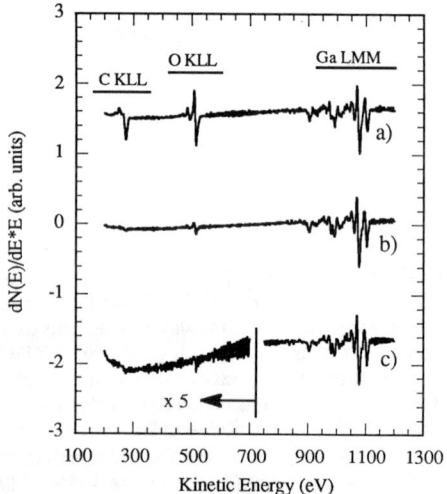

Fig. 1
Auger electron spectra of GaP(111) after:
a) wet chemical cleaning,
b) 20 min annealing at 530°C and 2 min H plasma exposure
and
c) 6 min H plasma exposure at 530°C.

Figure 2 shows a LEED pattern for a 530Å Si film deposited at 400°C and 500 mtorr process pressure. A fairly well ordered Si(100) - 2x1 reconstructed surface can be observed, which is slightly streaked along the <01> direction, as a result of surface steps.

Fig. 2

2x1 LEED pattern of 530Å Si on vicinal GaP(100) surface.

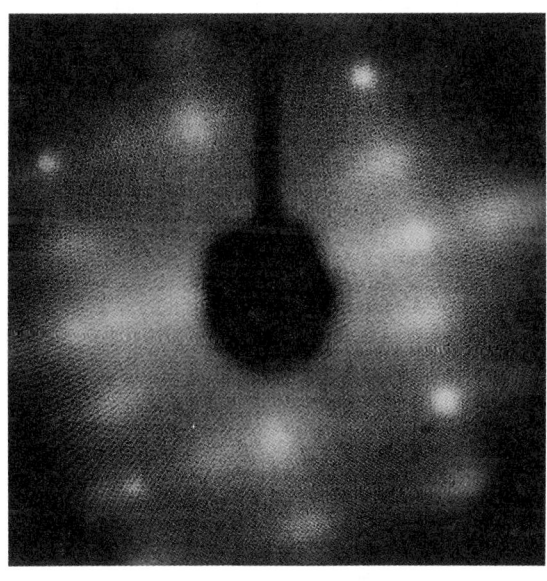

Fig. 3

High resolution cross sectional TEM image of a 775Å Si epilayer grown on a stepped GaP(100) substrate at 400°C.

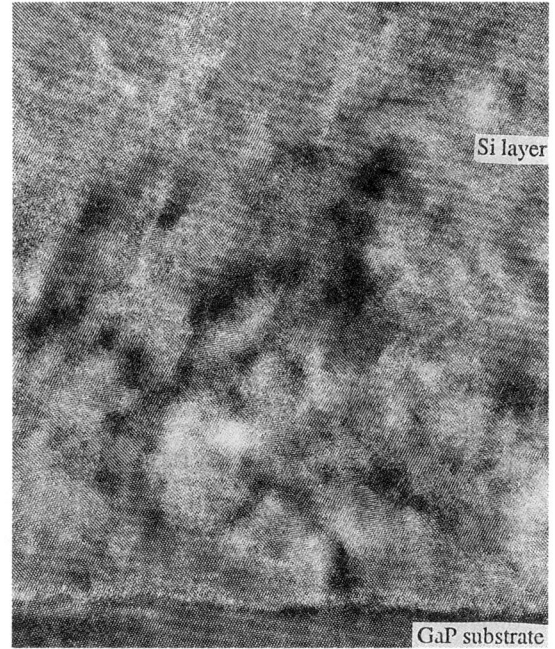

The high-resolution cross-sectional transmission electron micrograph in fig. 3 shows high quality single crystal growth can be maintained upwards to at least 775Å film thickness. The image indicates a heteroepitaxial growth with no evidence of misfit slip dislocations propagating in the bulk surface. Closer examination of the interface revealed terraces in the range of 50 to 60 Å. Depostion on stepped (100) surfaces results in a better epitaxial Si layer than similar deposition experiments on unstepped surfaces, which indicates an improved surface mobility of deposition species at the stepped surface.

b) GaP deposition on Si

SIMS analysis on GaP films grown at 310°C with a TBP:TEG flow ratio of 25 for longer growth times (>30min) revealed an increase in the signals of mass over charge (m/e) ratio 12 for carbon and m/e=16 for oxygen in the interfacial region[3,6]. Figure 4a shows a plot of the signals m/e=12 and m/e=40 versus film thickness for a GaP layer deposited at 310°C for 3 hours. The m/e=40 signal increases sharply at the GaP/Si interfacial region marked by arrows. The m/e=12 signal decreases inside the GaP film and shows also a maximum at the GaP/Si interface.

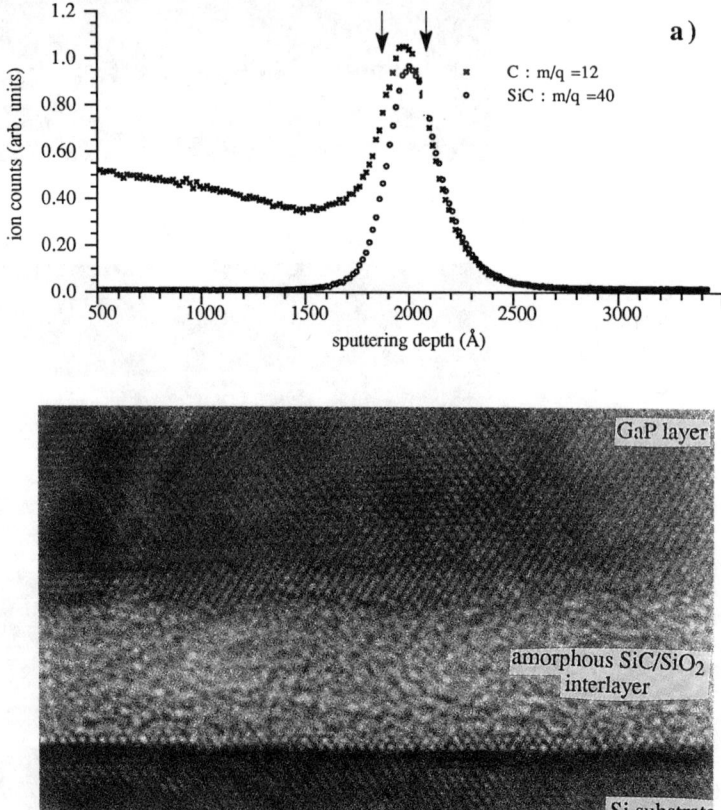

Fig. 4 a) SIMS plot for the carbon and silicon carbide signals for a GaP layer deposited at 310°C for 180min. b) HRTEM image of a 800Å GaP film deposited at 310°C.

The results give evidence to the formation of a subcutaneous SiC/SiO_2 interlayer for film thicknesses above 300Å[3], while for layers thinner than 300Å no significant carbon accumulation is revealed at the interface.

Figure 4b shows a HRTEM image of a 800Å GaP film deposited on Si(100) at 310°C for 1 hour growth time revealing the formation of an amorphous layer in the interfacial region. The thickness of this layer varies from 40 to 50Å. The formation of the amorphous SiC interlayer is understood as a diffusion-controlled steady state growth due to carbon diffusion through the GaP film with an estimated diffusion constant at 310°C of $\approx 2 \times 10^{-14}$ cm^2s^{-1}. (The GaP grows highly selective to silicon with regard to SiO_2, the selectivity of GaP on SiC is still maintained[3]). The delay in the formation can be explained as an induction period which is needed to get a sufficient supersaturation of carbon at the silicon surface to start the nucleation of a SiC interlayer. This is consistent with the epitaxial growth of the overlaying GaP film. However, as seen in fig 4b, several twins along the (111) plane are formed at the interface as well as during the growth phase. This twin formation at the interface may be significantly reduced by using an in-situ atomic-H plasma cleaning procedure, as shown on grown silicon layers on in-situ cleaned GaP surfaces.

c) Si-GaP-Si heterostructures

Figure 5 shows a HRTEM image of a Si-GaP-Si heterostructure. The 800Å GaP layer is deposited on an ex-situ cleaned Si(100) surface at 310°C. Before the silicon has been deposited, the surface of the GaP film was cleaned in-situ using 5 min annealing at 530°C in 10 mtorr H_2 ambient followed by a 15 min H-plasma exposure.

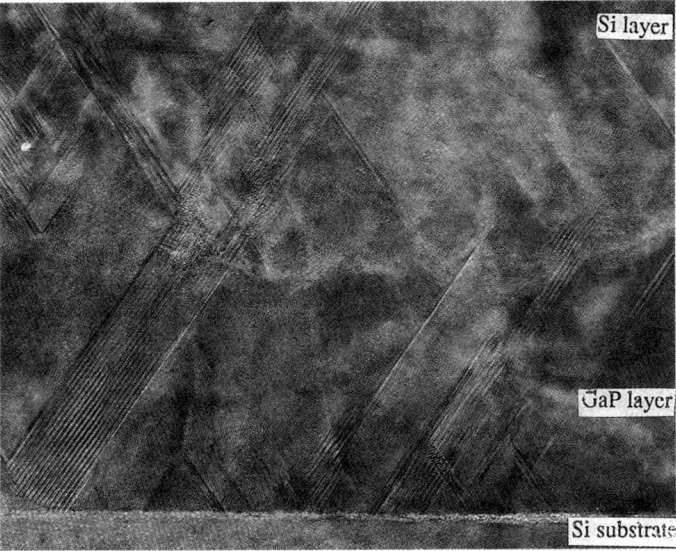

Fig. 5 Cross sectional HRTEM image of a (100)Si-GaP-Si heterostructure, deposited at 310°C and 400°C for GaP and Si, respectively.

The AES spectrum reveals a clean GaP surface, the LEED pattern however, shows already fairly broadened diffraction spots. This is consistent with the HRTEM image in fig. 5, which shows a fairly high amount of twin formation starting at the Si-GaP interface along the (111) plane with a relatively rough GaP-Si interface. The twin formation can also be seen at the GaP-Si interface and during the growth of the Si layer. Here however, the amount of twin formation is significantly reduced compared to the amount generated at the Si-GaP interface. This reduction can be explained by an improved surface quality due to the in-situ surface cleaning procedure on the second interface. A further improvement of the film morphology and crystallinity can be obtained by interrupted cycle chemical beam epitaxy with H_2 flowing[6].

IV CONCLUSION

We have shown the epitaxial growth of Si-GaP-Si heterostructures at low substrate temperatures. High-resolution cross sectional transmission micrograph images show the existence of an amorphous SiC interlayer and the formation of twins on the Si-GaP interface. It is shown, that the in-situ H-plasma exposure is an effective cleaning procedure of GaP surfaces, which allows the growth of high quality Si epilayer on GaP surfaces.

ACKNOWLEDGMENT

The authors would like to thank Dr. Y.L. Chen for the HRTEM microscopy. This research is supported by NSF grants DMR 9282201 and CDR 8721505.

REFERENCES

1 J.T. Kelliher, J. Thornton, N. Dietz, G. Lucovsky and K.J. Bachmann, Material Science and Engineering, B22 (1993) 97-102.

2 M. Yoshimoto, K. Ozasa and H. Matsunami, J. Appl. Phys., **70** (1991) 5708.

3 J.T. Kelliher, N. Dietz and K.J. Bachmann, Proc. Electroch. Soc. Symp., R. Enstrom, S.N.G. Chu, T. Kamijoh and O. Oeda, eds., Honolulu, Hawaii 1993, in print.

4 S. Habermehl, N. Dietz, Z. Lu, K.J. Bachmann and G. Lucovsky, presented at the 39th Annual AVS Symposium, Nov. 1993 (to be published in the symposium proceedings).

5 Z. Lu, S. Habermehl, G. Lucovsky, N. Dietz and K.J. Bachmann, presented at the Third International Symposium on Cleaning Technology in Semiconductor Device Manufacturing, The Electroch. Society, Oct.10-15, 1993 New Orleans, LA (to be published in the symposium proceedings).

6 J.T. Kelliher, J. Thornton, P.E. Russell, and K.J. Bachmann, in Mechanism of thin film Evolution; edited by S.M. Yalisove, C.V. Thompson and D.J. Eaglesham, (Mat. Res. Soc. Symp. Proc. 317, Pittsburg, PA, 1993).

THE ROLE OF BARIUM IN THE HETEROEPITAXIAL GROWTH OF INSULATOR AND SEMICONDUCTORS ON SILICON

T. K. CHU, F. SANTIAGO, M. STUMBORG[†], AND C. A. HUBER
Naval Surface Warfare Center Dahlgren Division, White Oak Laboratory, 10901 New Hampshire Ave., Silver Spring, MD 20903, USA
[†]Also The Catholic University, Washington, DC 20064, USA

ABSTRACT

The epitaxial growth of an insulator, BaF_2, and semiconductors of the II-VI and the IV-VI families on Si substrates were carried out. In-situ XPS analyses during the growth of the first monolayers were used to study the surface chemical reactions involved. The results point to a common ingredient in these growths: that the Ba atoms are involved in forming interfacial compounds that would facilitate the heteroepitaxies. In the case of BaF_2/Si, a $BaSi_2$ compound has been identified previously. In the case of PbTe and CdTe, the heteroepitaxies on Si are made possible with the $BaSi_2$ buffer. As a result, the impinging semiconductor molecules are broken up, and the metallic elements are ejected from the $BaSi_2$ surface. A new surface chemical, BaTe, is thereby formed. These surface Ba compounds appear to be the dominant factors as the crystal orientations of the BaF_2, CdTe, and PbTe layers are independent of those of the Si substrates.

INTRODUCTION

The heteroepitaxial growth of materials is a subject of not only of scientific interest, but has practical technological applications also. The majority of efforts in the past have dealt with materials that are similar in terms of crystal structures and lattice constants. This is a natural outcome of the atomic stacking model of the heteroepitaxial interface, in which atoms above the interface are perceived to perch either directly on atoms below the interface, or on some interstitial site between atoms below the interface. The successful growth of heteroepitaxies between semiconductors that are lattice-matched and nearly lattice-matched has resulted in practical and commercial devices; examples include diode lasers and the high electron mobility transistor. The growth of heteroepitaxies between insulators and semiconductors, on the other hand, is not as successful; even though the technological pay-off may also be high. An important example is the development of better insulating layers for micro-electronic circuitry. Another is substrate engineering whereby semiconductors of different structures and different electronic and/or optical properties can be combined into a single integrated chip.

A group of insulators that has received attention is the (cubic) Group IIa fluorides. This is because of their superior dielectric properties, and their potential replacement for SiO_2 in MOS and other devices. Of special interest[1] is CaF_2, since its lattice constant is close to that of Si. In addition, BaF_2 is also of interest as a substrate material for the IV-VI semiconductors[2].

Early attempts[3,4] to grow BaF_2 on Si by the MBE process indicated that [111] is the preferred growth direction irrespective of the orientation of the Si substrate. A detailed investigation[5] of the growth process revealed that there is a strong chemical interaction between the impinging BaF_2 molecules and the Si substrate so that a new Ba-Si compound is formed at the interface between the substrate and the deposited BaF_2. In fact, by a proper heat-treatment of the deposited BaF_2/Si film, a "pure" Ba-Si (i.e. fluorine-free) surface layer can be formed. This surface layer is very thin - on the order of 10 Å - and can serve as a buffer for the epitaxial growth of PbTe[6]. In this paper, we will report on our on-going investigation on the properties of the Ba-Si layer and the successful growth of epitaxial CdTe thereon. We will present evidence to show that the growth of BaF_2, PbTe, and CdTe are facilitated by surface chemical reactions in which the Ba occupies a crucial role.

FILM DEPOSITION

Our present investigations are facilitated by two important ingredients: (1) the growth rates are very low, on the order of one or two molecular layers per minute; and (2) the chemistry of the deposited layers were analyzed by XPS in an interlocking chamber. The growth conditions for BaF_2 and PbTe have been described previously[2]. The condition for CdTe growth were the same as for PbTe, except that a CdTe molecular source and atomic Cd and Te sources were used instead. The epitaxial quality was monitored in-situ with RHEED, and after growth by x-ray diffraction. XPS analyses were carried out after various times of growth. These time periods are on the order of a few minutes.

BARIUM FLUORIDE/BARIUM SILICIDE

In the MBE deposition of BaF_2 at 700°C, as soon as BaF_2 molecules impinge upon the Si surface, a chemical reaction takes place and a new compound $BaSi_2$ is formed[5]. This is illustrated in Figure 1, where the spectra of the Ba $4f_{5/2}$ and Ba $4f_{3/2}$ levels and the Si $2p_{3/2,1/2}$ level are shown in the same XPS window. As the deposition time increases (at deposition rates of ~ 6 Å/min.) the Ba levels displayed a shift to higher binding energies. Eventually it becomes that of Ba in BaF_2, obtained after approximately 15 minutes of deposition. Concurrently, the Si

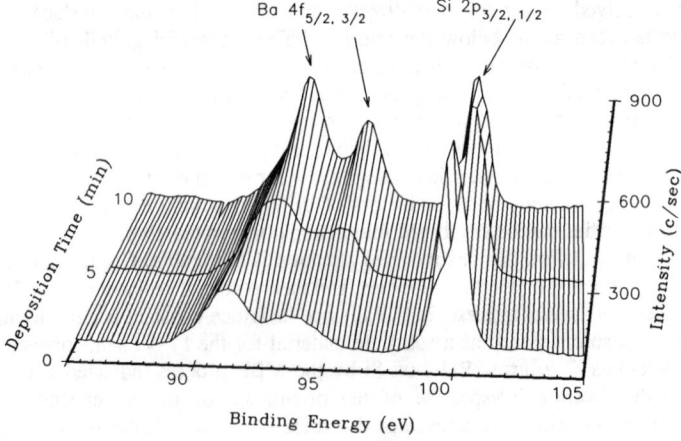

Figure 1. Variation of the XPS Spectra of the Ba $4f_{5/2}$, Ba $4f_{3/2}$ and the Si $2p_{3/2,1/2}$ levels with time of BaF_2 deposition.

levels showed a shift toward lower energies. Notice that the magnitude of the Si signal stayed more or less the same while the Ba magnitudes increased sharply after 2 minutes of deposition. A BaF_2 film thus deposited can be converted to a $BaSi_2$ film[7] by a prolonged (~ 1 hour) annealing at the deposition temperature. The position of the Ba spectra of the $BaSi_2$ surface is essentially those of the 1 minute film. In fact, it had been shown that a Ba spectrum can be deconvoluted into two distinct components: pertaining to the BaF_2 and the $BaSi_2$ states respectively[5]. The behavior of the Si signal indicates that the initial growth of BaF_2 proceeds in an island mode, mostly likely on a $BaSi_2$ surface.[6] (The thickness of the $BaSi_2$ layer is on the order of 10 Å.) As a result, a certain portion of the $BaSi_2$ molecules are exposed and the Si signal persists even after about 20 Å of BaF_2 coverage. After about twenty minutes of deposition, the BaF_2 converts to 2-D growth. This has been confirmed by RHEED patterns,[7] and by the disappearance of the Si in the XPS spectrum.[5]

The crystal structure of the $BaSi_2$ layer can not be determined from the present results, because the $BaSi_2$ layers obtained are too thin to yield any signal with our x-ray diffractometer. Furthermore, the strength of the XPS signal from the $BaSi_2$ layers is independent of the thickness of the BaF_2. It appears that the $BaSi_2$ layer may not be effective against any further reaction between BaF_2 and Si. In bulk form, $BaSi_2$ has been reported to be either orthorhombic[8] or hexagonal (AlB_2 type)[9]. An examination of the RHEED pattern, given in Figure 2, indicates that this thin BaS_2 layer is similar to that of a Si(111) surface, and hence may be hexagonal.

 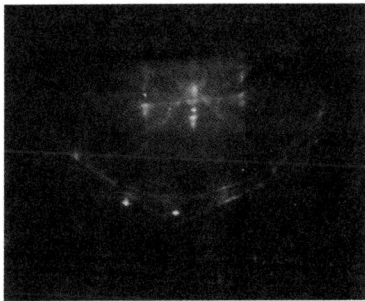

Figure 2. RHEED Pattern of a $BaSi_2$ Film on a Si(100) Substrate (left). For comparison, the Pattern of a Si(100) Substrate, Heat-cleaned at 950°C for 3¾ hrs., is also shown (right).

LEAD TELLURIDE AND CADMIUN TELLURIDE

The deposition of epitaxial PbTe on the $BaSi_2$/Si substrate has been reported previously[6]. It was shown that at 500°C, PbTe essentially decomposed at the $BaSi_2$ surface, with most of Pb atoms ejected from the surface. Te was retained in some sort of chemical bond with Ba and Si. The orientation of the PbTe film was (100). This orientation was independent of the Si substrate with the thin layer of $BaSi_2$ prepared prior to the PbTe deposition. Because of this decomposition, the deposition rate at 500°C was extremely low. At lower temperatures < ~450°C, Pb was retained. The XPS spectra of the Te $3d_{5/2}$ level for deposited PbTe films can be deconvoluted into two components: one pertaining to the PbTe compound, and the other apparently due to some surface compound involving Te, Ba, and perhaps Si. The nature of the data precludes any definitive conclusion on this interfacial compound. It was suggested that since PbTe and CdTe have the same anion, and that their lattice constants are almost identical, CdTe could grow epitaxially on the $BaSi_2$/Si substrate[2].

XPS analyses of the surfaces of CdTe deposited on the $BaSi_2$/Si substrate revealed behaviors that were essentially the same as that of PbTe. At a temperature of 500°C or higher, CdTe largely decomposed, leaving Te in a reacted state with atoms on the substrate surface. As the substrate temperature is lowered, more Cd is retained. Figure 3(b) shows the wide scan XPS spectrum of a CdTe surface after 30 sec. of deposition at 400°C. Figure 3(c) shows the spectrum for a film after 5 min. deposition. (For comparison, the spectrum for a $BaSi_2$ surface is also shown in Figure 3(a).) Except for their magnitude, the spectra at deposition times longer than 5 min. are unchanged from that of the 5 min. deposition, indicating that the surface is CdTe. It is obvious that after a 30 sec. deposition time, there is a deficiency of Cd at the surface. A closer examination of the XPS spectra suggests that the chemical state of Cd is essentially unchanged once it adheres to the surface. On the other hand, the Te peaks positions vary with deposition time. A detailed monitoring of the Te 3d levels showed that they shifted to lower energies as the deposition time increased. This is shown in Figure 4. Concurrently, there were shifts in the Ba 3d levels to higher energies. As the Ba signal was rather small, and became progressively so with increasing coverage of CdTe, no quantitative analysis was pursued. It thus appears that the deposition of CdTe at 400°C resulted in a partial decomposition of CdTe. The retention of Te is accompanied by a chemical reaction with Ba. The atomic Cd is ejected from the surface. Subsequent growth of CdTe occurs thereafter on a surface at least partially covered by the Ba-Te compound. In view of the observation that the chemical state of Cd is unchanged, this compound is tentatively identified as BaTe.

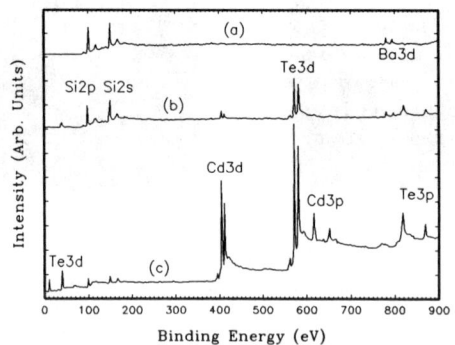

Figure 3. The Wide Scan XPS Spectra of (a) A $BaSi_2$ Surface, (b) A CdTe Film of 30 sec. Deposition at 400°C, and (c) A CdTe Film of 3 min. Deposition. The Vertical Scales of These Curves Are Displaced with Respect to Each Order to Show Their Differences.

The epitaxial quality of the CdTe film was monitored in-situ with RHEED. Figure 5 shows the pattern obtained with an electron beam at an energy of 25kV. A surprising result of the CdTe deposition is that the film is (111) oriented. And, similar to the cases of PbTe and BaF_2, the orientation is independent of the Si wafer orientation. Figure 6 displays the evidence on a two-theta plot of the x-ray diffraction result, where the (111) and the (333) CdTe peaks are evident.

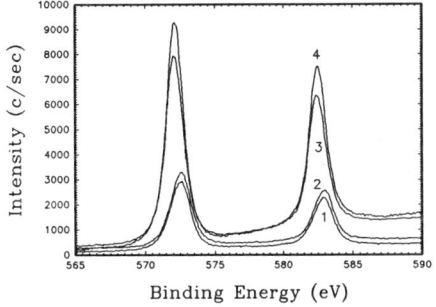

Figure 4. XPS Spectra of the Te 3d Levels at CdTe Deposition Times of: 1 - 30 sec., 2 - 60 sec., 3 - 3.5 min., and 4 - 5 min. at 400°C on a $BaSi_2$/Si(100) substrate.

Figure 5. RHEED Pattern of a CdTe Film on a $BaSi_2$/Si Substrate.

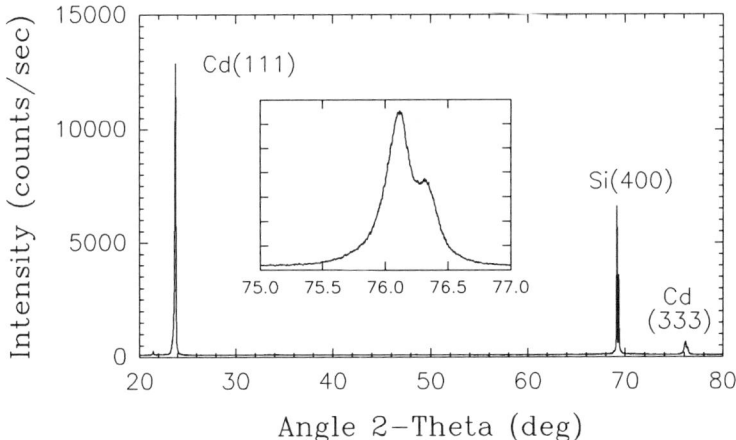

Figure 6. Diffraction of the Cu Kα Lines by an CdTe Film Deposited on a $BaSi_2$/Si(100) Substrate. Inset shows details of the Cd(333) peak.

DISCUSSION AND CONCLUSION

In a series of growth studies involving an insulator (BaF_2) and two semiconductors (PbTe and CdTe) on Si substrates, reported here and previously,[2,5,7] we have demonstrated that there are important chemical reactions at the interfaces between atoms on the substrate surface and the incoming molecules. Exploitation of these chemical reactions have allowed the epitaxial growth of BaF_2, PbTe, and CdTe on Si substrates. All of these materials have lattice constants that are considered quite large compared to that of Si by conventional standards. (6.21 Å for BaF_2, and

6.46 Å for PbTe and CdTe, versus 5.43 Å for Si.) Their crystal structures are also different from that of Si, and from each other: BaF_2 - fluorite, PbTe - NaCl, and CdTe- zincblende.

Even though the occurrence of chemical shifts and the formation of interfacial compound have been reported previously, e.g. for CaF_2 on Si[10] and PbSe[11] on BaF_2, our results demonstrate that the chemical reactions play crucial roles in the growth of these materials. The atomic stacking model for heteroepitaxial growth is therefore quite inadequate, if not entirely inapplicable. In any case, any atomic stacking model should also include an interfacial layer, even if only a few angstroms in thickness, in order to account for the large mismatch in lattice constants of the present examples.

Specifically, our series of studies shows that in the heteroexpitaxies of BaF_2, PbTe, and CdTe on Si, the interfacial chemicals all involve a common element, Ba. A rather surprising property of Ba is that it has relatively large crystal ionic radii[12] - 1.34 Å for Ba^+ and 1.53 Å for Ba^{++}. Furthermore, the Ba atom in the $BaSi_2$ compound occupies a very large volume: in the orthorhombic form, the Ba-Ba distance varies between 4.38 and 4.44 Å, and in the hexagonal form 4.39 and 4.83 Å. In BaTe, which has the NaCl structure, this distance is 4.95 Å. It may be conjectured that the large size of the Ba atom allows it to bond to other atoms in a more flexible manner. For example, $BaSi_2$ has been reported[13] to be tri-morphic. In addition, the Si-Si bond length in $BaSi_2$ varies between 2.34 and 2.48 Å, which is very close to that of the Si-Si bond, 2.34 Å in a Si lattice. In other words, Ba can bond to Si atoms without large disruptions to the Si-Si bonds. In view of the fact that the Ba-Ba distance in BaF_2 is 4.38 Å, and the Te-Te distance in PbTe and in CdTe is 4.53 Å; it may not be surprising that these three compounds can be grown epitaxially on $BaSi_2$ as we have described. It may be further argued that successful manipulation of the Ba chemistry at interfaces can result in heteroepitaxies that are heretofore considered unlikely.

ACKNOWLEDGEMENT

This work was supported by the Naval Surface Warfare Center Independent Research and Independent Exploratory Development Funds.

REFERENCES

1. L. J. Schowalter, CRC Critical Rev. Sol. State and Mat. Sci. 15(4) 367 (1989).
2. T. K. Chu et.al, Mat. Res. Soc. Symp. Proc. 221 483 (1991), and
   F. Santiago, D. Woody, T. K. Chu, and C. Huber, Mat. Res. Soc. Symp. Proc. 281 603 (1993).
3. T. Asano, H. Ishiwara, and N. Kaifu, Jpn. J. Appl. Phys. 22 1474 (1983).
4. A. P. Taylor, W. Li, Q.-F. Xiao, and L. J. Schowater, Mat. Res. Soc. Symp. Proc. 220 537 (1991).
5. F. Santiago, Ph. D. Thesis, the American University, 1992.
6. The identification of this interface compound was done by an analysis of the Si XPS signal by the same method used by L. G. Feldman and J. W. Mayer, Fundamentals of Surface and Thin Film Analysis (Elsevier, New York, 1986), Chapter 5, P. 125. Details of this analysis and of the determination of the $BaF_2/BaSi_2/Si$ layer structure will be published elsewhere.
7. Chu, T. K., Santiago, F., and Huber, C. A., Naval Surface Warfare Center Dahlgren Division Technical Digest Vol. 4, Materials, 98 (1993).
8. H. Schaefer, K. H. Janzon, and A. Weiss, Angew. Chem. Internat. Edit. 2 393 (1963).
9. M. Hansen and K. Anderko, Constituion of Binary Alloys, (McGraw Hill, New York, 1958).
10. for example, F. J. Himpsel et.al., Appl. Phys. Lett. 48 596 (1986).
11. P. J. McCann, Mat. Res. Soc. Symp. Proc. 221 289 (1991).
12. CRC Handbook of Chemistry and Physics, 46th Ed., P. F-117, (The Chemical Rubber Co., 1965).
13. J. Evers, G. Oehlinger, and A. Weiss, Angew. Chem. Internt. Edit. Engl. 16 659 (1977).

TOPOGRAPHICAL EFFECTS REGARDING TRENCH STRUCTURES COVERED WITH RTCVD SIGE THIN FILMS

G. RITTER, J. SCHLOTE, AND B. TILLACK
Institute of Semiconductor Physics, W. Korsing Str. 2, D-15230 Frankfurt (Oder), Germany

ABSTRACT

The paper presented discusses certain topographical effects being significant for the coverage of nonplanar structures with undoped and doped thin SiGe films. They are essential for new integrated heterojunction Si/SiGe devices. To investigate the coverage of topographical surfaces SiGe layers have been deposited on different trench structures in a Rapid Thermal Low Pressure single wafer CVD reactor (RTCVD) from the system SiH_4, GeH_4, and H_2 using B_2H_6/H_2 and PH_3/H_2 for the in-situ doping, respectively. Various deposition conditions and different film compositions have been used. The results especially the differences of the thickness distribution within trenches and on the surface have been discussed with regard to different CVD models taking into account distinguished reaction mechanisms.

INTRODUCTION

Recently, the silicon- germanium (SiGe) material system has become increasingly important because of applications in a wide range of electronic devices. These applications extend from the heterojunction bipolar transistor to quantum well based devices and have considerably widened the opportunities of devices which can be fabricated on silicon substrates with only minor deviations from the well-established silicon technology. Recent research has shown that chemical vapor deposition (CVD) techniques can successfully grow SiGe-structures produced first by molecular beam epitaxy (MBE).

Nowadays a multitude of thermally as well as plasmachemically activated CVD processes are used to coat workpieces with good homogeneity of both the thickness and properties of the deposited layers. Nevertheless, difficulties will arise if the workpieces have a more complicated surface shape containing small structural details such as edges, steps, strips or cavities and if the final thickness of the deposited layer is comparable with the geometrical dimensions of such structural details. A typical example contributed by the microelectronics industry is a silicon wafer patterned with narrow trenches to be coated with undoped and in-situ doped SiGe- thin films.

In order to describe the layer thickness distribution within trenches, previous pu-

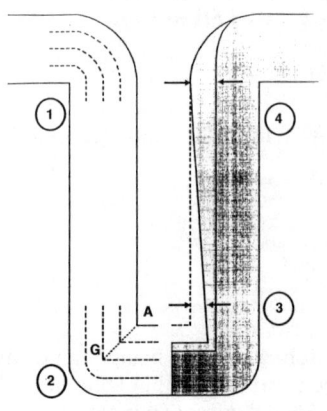

Fig. 1: Schematic diagram of topographical phenomena in an idealized trench structure:
left-hand side: edge phenomena (1,2)
right-hand side: cavity phenomena (3,4)

blications of several authors have preferred line-of-sight models [1,2] as well as "continuum-like" diffusion-reaction models [3-6] because of the relatively high computing time consumption of direct Monte Carlo simulations. In these models the reactive sticking coefficient (or the reaction rate constant) of the layer-forming precursor serves as the disposable fitting parameter. On the other hand, some deposition processes characterized by an abrupt decrease in the growth rate at the trench orifice and a nearly constant but smaller rate everywhere within the trench seem to require another model type for an adequate description of the observed thickness distribution [7].

Various topographical effects are significant for displacement of the surface and the developement of its final shape due to deposition (see Fig. 1). Some of them (edge phenomena) are of fundamental nature [8-10] and will not be discussed in this paper while others (cavity phenomena) depend on the chemistry of the deposition reaction as well as on the geometry of the cavity. For the latter it is assumed that the mean free path length is much greater than the characteristic measure of the cavity and that the reactants are transported into the cavity via particle-wall collisions with a low (reactive) sticking coefficient. "Shadowing" effects connected with a high sticking coefficient (typical for physical vapor deposition processes) will not be considered here.

Cavity phenomena

The restricted efficiency of reactant transportation through the gas phase within a narrow trench results in a modified non-uniform thickness distribution which overlaps the edge phenomena. The cavity phenomena are defined by the kinetics of the reaction controlling the deposition rate and either the geometry of the cavity (depletion) or the existence of the cavity in itself (rate jump).

Depletion
If the consumption of reactants by the heterogeneous deposition reaction on the trench walls becomes comparable with the supply from the gas phase, the film thickness will decrease with increasing trench depth, because surface elements in the depth are discriminated against surface elements near the orifice. The degree of non-uniformity is controlled by the Thiele modulus a, which is given considering Knudsen diffusion (mean free path length very large compared with trench width w) as [7]

$$a^2 = \frac{6q}{v}(d/w)^2$$

Uniform thickness is to be expected for small values of the heterogeneous rate constant q and the aspect ratio of depth to width, d/w, and for higher values of mean molecular velocity v, i.e. if the Thiele modulus a is not too large.

Rate jump

For certain deposition processes which are controlled by the so-called byproduct mechanism the deposition rate diminishes abruptly at the top edge of the trench from its surface value to a smaller value which is nearly constant everywhere within the trench. This "rate jump" is caused by the strong and almost uniform adsorption of a byproduct not taking part in layer formation but retarding the real deposition reaction. The level of byproduct adsorption is always higher within the trenches than outside and is nearly independent on the trench width (but it depends on the reaction-adsorption kinetics of the byproduct, of course) [7].

Unfortunately, the different topographical cavity effects frequently do not appear clearly isolated from each other. Thus the reliable distinction between depletion and rate jump needs special attention and can often be done only using special process parameters and trenches of different widths simultaneously. For this purpose in our investigations the common gas phase transport induced depletion effect has been suppressed by using low deposition rates.

EXPERIMENTAL

The trench structures have been etched by reactive ion etching (RIE) in the silicon substrate. The aspect ratio of depth to width d/w is about 5 µm/ 2 µm. The Si-surface has been covered by 80 nm thick Si_3N_4 film.

On these patterned substrates the film stack containing $Si_{0.8}Ge_{0.2}$ (undoped, B-doped, P-doped), SiO_2 or Si_3N_4, undoped $Si_{0.8}Ge_{0.2}$, and amorphous Si has been deposited by various LPCVD techniques. The layer structure is shown schematically in Fig.2. The different variants of the corresponding samples are described in Table I.

SiO_2 has been deposited at 420°C from SiH_4/ O_2, Si_3N_4 has been deposited plasma enhanced at 350 °C from SiH_4/ NH_3, a-Si at 650 °C from SiH_4.

The Si- and SiGe-films have been deposited by rapid thermal CVD (RTCVD) in a single wafer radiation heated stainless steel reactor at a temperature of 600°C and a pressure of 200 Pa from the mixture of SiH_4, H_2, and GeH_4. The films have been in-situ doped using B_2H_6/ H_2 and PH_3/ H_2. The Ge-content (20%) in the SiGe-films has been measured by X-ray analysis, the dopand concentration (about 10^{20} cm^{-3}) has been determined by secondary ion mass spectroscopy (SIMS).

After the deposition processes the structures have been studied by cross-sectional transmission electron microscopy (TEM).

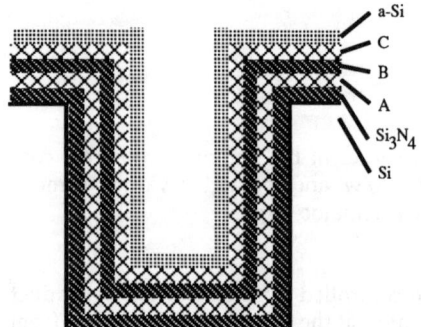

Fig.2 Schematical layer stack deposited in trench-structures in silicon.

TABLE I. Variation of layers A, B, C

sample	layer A	layer B	layer C
1	Si, undoped	SiO$_2$	Si, undoped
2	Si, B-doped	Si$_3$N$_4$	Si, undoped
3	Si, P-doped	Si$_3$N$_4$	Si, undoped
4	SiGe, undoped	SiO$_2$	SiGe, undoped
5	SiGe, B-doped	Si$_3$N$_4$	SiGe, undoped
6	SiGe, P-doped	Si$_3$N$_4$	SiGe, undoped

RESULTS AND DISCUSSION

The coverage of trenches by undoped and doped SiGe layers are demonstrated in the following TEM micrographs (Fig. 3a-c). One can see that both undoped and B-doped SiGe-layers (the outer SiGe-layers of all three samples and the inner SiGe-layers in figs. 3a and b) has been deposited uniformely on the wafer surface and in the whole trench. There is neither depletion nor rate jump effect. The corresponding Si-layers show the same distribution (not demonstrated here).

On the other hand both the P-doped Si- and SiGe- films exhibit a remarkable rate jump of the deposition rate inside the trench structures. This is demonstrated for the

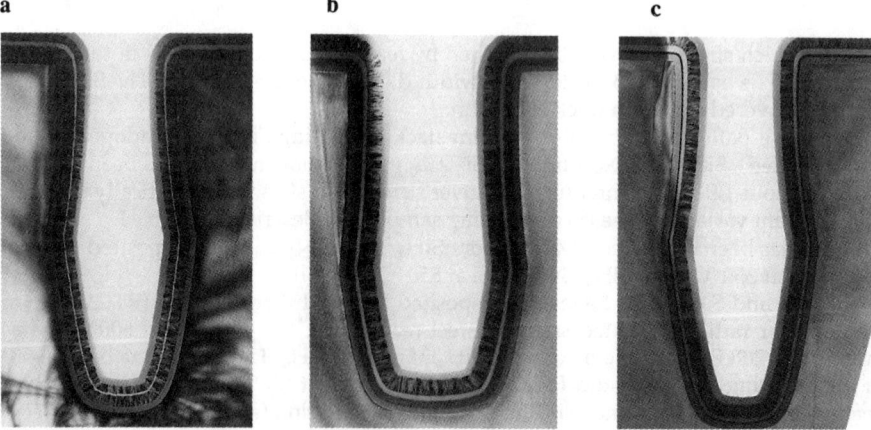

Fig. 3: XTEM micrographs of trenches coated by an outer undoped SiGe-films and an inner undoped (a), B- doped (b), and P- doped (c) SiGe- film. (⊢――⊣ 1µm)

inner SiGe-layer in fig. 3 c and in detail for the P-doped Si and SiGe-layers in figs. 4 and 5, respectively.

 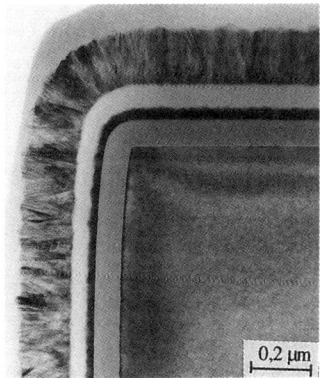

Fig. 4: XTEM detail of the top trench edge covered by P- doped and undoped Si films

Fig. 5: XTEM detail of the top trench edge covered by P- doped and undoped SiGe films

For the sake of better clarity, the normalized (versus top-layer) thickness distribution of different Si and SiGe-films measured from the top across the sidewall to the trench bottom are shown in figs. 6 and 7, respectively.

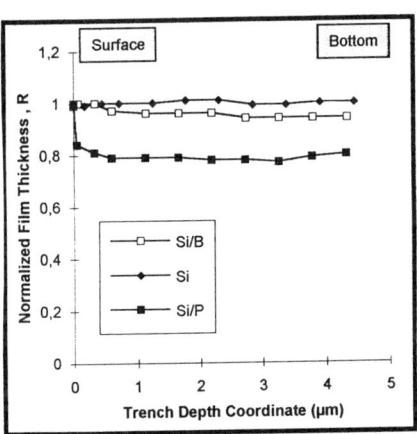 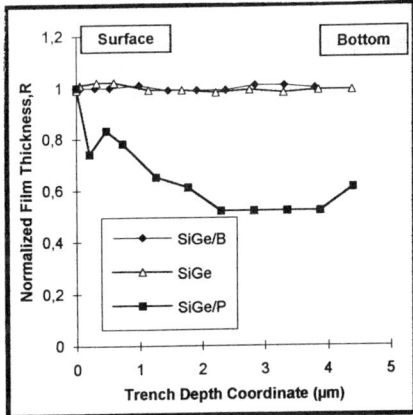

Fig. 6: Thickness distribution of different Si-films in trenches.

Fig. 7: Thickness distribution of different SiGe films in trenches.

It has been demonstrated that the deposition processes in trench structures of corresponding Si and SiGe films are of the same kind for LPCVD conditions with low deposition rates.

The absence of any nonuniformities and depletion effect allows the conclusion that these processes are not controlled by a byproduct adsorption mechanism.

The significant rate jump of the P-doped (by PH_3) films within trenches correlates with a strong decrease of the average deposition rate by adding PH_3 to the other gases. This fact emphasizes the evidence of the byproduct mechanism for explanation of the experimental results.

CONCLUSIONS

Concerning the thickness distribution of thin films of undoped and doped Si and SiGe on trench structures deposited by low pressure RTCVD-processes from the system SiH_4/ GeH_4/ H_2 and B_2H_6 or PH_3, respectively we can conclude:

1. The thickness distribution of corresponding Si and SiGe films is of the same kind.

2. Undoped as well B-doped films cover trench structures with uniform thickness. For these processes a byproduct mechanism is not adequate.

3. Both Si- and SiGe- films doped by PH_3 show a distinct rate jump inside the trench, but no depletion effect. Therefore, the process can be described with help of a byproduct model.

ACKNOWLEDGEMENT

The authors would like to thank B. Quick for XTEM micrographs and A. Wolff for reactive ion etching of the samples.

REFERENCES

1 T.S. Cale and G.B. Raupp, J. Vac. Sci. Technol. B, **8**(6), 1242 (1990).
2 H. Wille, E. Burte, and H. Ryssel, J. Appl. Phys., **71**(7), 3532 (1992).
3 K. Watanabe and H. Komiyama, J. Electrochem. Soc., **137**(4), 1222 (1990).
4 G.B. Raupp and T.S. Cale, Chem. Mater., **1**, 207 (1989).
5 S. Chatterjee and C.M. McConica, J. Electrochem. Soc. **137**(1), 328 (1990).
6 K. Fujino, Y. Egashira, Y. Shimogaki, and H. Komiyama, J. Electrochem. Soc. **140**(8),2309 (1993).
7 J. Schlote, K.W. Schröder, and K. Drescher, J. Electrochem Soc. **138**(8), 2393 (1991).
8 J. Schlote, K.W. Schröder, P. Koy, and G. Morgenstern, J. Electrochem. Soc. **137**(6), 1939 (1990).
9 K.W. Schröder, J. Schlote, and S. Hinrich, J. Electrochem. Soc. **138**(8), 2466 (1991).
10 J. Schlote, S. Hinrich, B. Kuck, and K.W. Schröder, Surf. Coat. Technol, **59**, 316 (1993).

NEW DAMAGE-LESS PATTERNING METHOD OF A GaAs OXIDE MASK AND ITS APPLICATION TO SELECTIVE GROWTH BY MOMBE

Seikoh Yoshida and Masahiro Sasaki
Optoelectronics Technology Research Laboratory
5-5 Tohkodai, Tsukuba, Ibaraki 300-26, Japan

ABSTRACT

A new damage-less patterning method of the photo-oxidized GaAs mask used for the selective-area growth of GaAs has been developed. We have found a new characteristic of the GaAs oxide: a metal Ga deposition onto the GaAs oxide surface lowers the desorption temperature of the oxide. The patterning method employed is based upon this characteristic. The GaAs oxide where 15 atomic layers (ALs) of Ga is deposited is locally removed at 540°C to form an opening area in the oxide mask. After forming this opening area, GaAs is selectively grown there by metal-organic molecular beam epitaxy (MOMBE).

INTRODUCTION

Photo-oxidized GaAs [1], SiO_2 [2], and SiN_x [3] have been used as mask materials for the selective-area epitaxy of GaAs. Among these mask materials, photo-oxidized GaAs has been used for the *in-situ* selective-area epitaxy of GaAs using a metal-organic molecular beam epitaxy (MOMBE) method [1,4]. Here, the *in-situ* process means that all of the processes concerning selective-area epitaxy, including mask formation, patterning, regrowth, and mask removal, are carried out in an ultrahigh-vacuum (UHV) system [1]. Patterning methods of the mask in the *in-situ* process using a focused ion beam (FIB) [5] and an electron beam (EB) [6-8] have been developed. In the case of patterning using FIB, defects induced by ion irradiation were observed to extend to a depth of more than 2000Å [9]. On the other hand, in the case of patterning using EB, the EB-induced damage was reported to be sufficiently small, compared with that by FIB [6,10].

In this paper we propose a completely damage-free patterning method of a photo-oxidized GaAs mask using selective Ga deposition in which we do not use an energetic charged beam, such as EB and/or FIB. We have recently found a new characteristic of the GaAs oxide mask: the desorption temperature of the oxide can be controlled by varying the amount of Ga deposited onto the oxide surface [11-13]. The oxides on which Ga molecules from a Knudsen effusion cell were deposited can be selectively removed at lower temperatures than that in the case of no Ga deposition. This characteristic of the oxide can also be utilized for patterning the mask. In this paper we describe the selective-area epitaxy of GaAs using the above-mentioned newly proposed patterning method.

PATTERNING PRINCIPLE OF GaAs OXIDE BY Ga DEPOSITION

We have reported that Ga-deposited GaAs oxide desorbs at lower temperatures than that in the case of no Ga deposition [11-13]. Figure 1(a) shows the temperature-programmed desorption (TPD) spectra of Ga_2O^+ and As_2^+ from the oxide when Ga was not deposited onto the surface. Sharp peaks of Ga_2O^+ and As_2^+ desorption were observed at 660°C. The origin of an As_2^+ peak was proposed to be a reaction between the oxide and GaAs substrate [14,15]. That is, when the oxide changes to volatile Ga_2O by reacting with the substrate, excess arsenic is produced as a reaction product from the substrate and, as a result, both Ga_2O and As desorb thermally at the same temperature.

On the other hand, when 15 atomic layers (ALs) of Ga-deposited oxide was desorbed, only the Ga_2O^+ signal was observed at 540 °C and that of As_2^+ was not observed, as shown in Fig.1(b). It is considered that the oxide was completely deoxidized by 15 ALs of Ga, and a

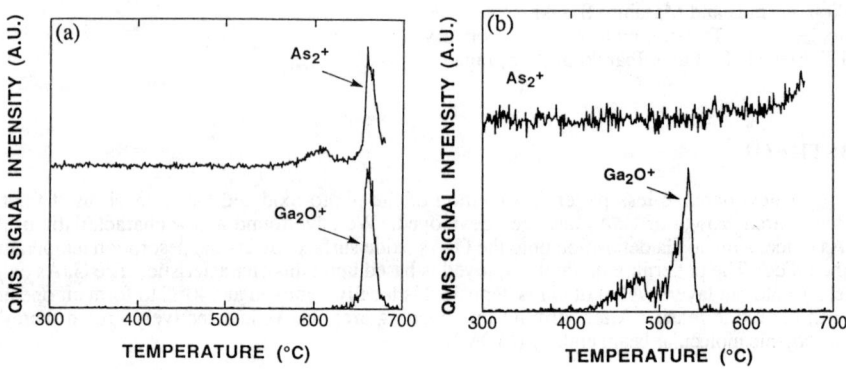

Fig.1 Thermal desorption spectra of Ga_2O^+ and As_2^+ when the amount of Ga deposited onto the photo-oxidized GaAs surface was (a) 0 ALs of Ga and (b) 15 ALs of Ga, respectively.

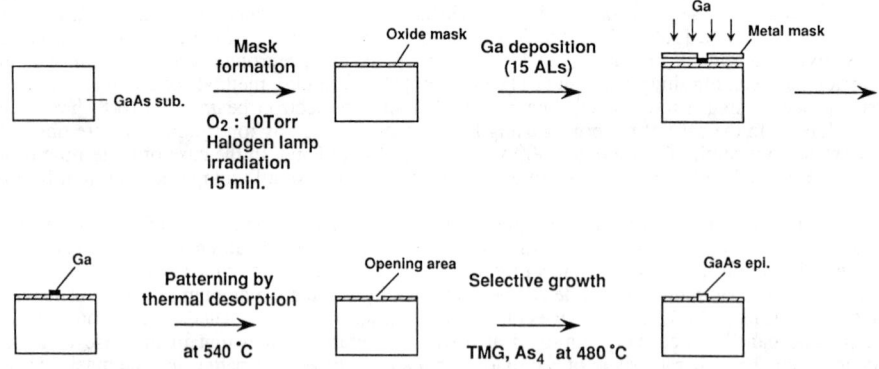

Fig.2 Illustration of the patterning procedure of GaAs selective growth using the newly proposed method.

volatile Ga_2O as a reaction product desorbed at lower temperatures than that in the case of no Ga deposition.

We used this characteristic of GaAs oxide for patterning the oxide mask. That is, metal Ga was locally deposited onto the oxide mask and only the Ga-deposited oxide desorbed at a specific temperature. The procedure concerning the selective growth using this patterning method is shown in Fig.2. That is, at first, a GaAs surface is photo-oxidized. After that, metal Ga is locally deposited onto the oxide surface by using a metal mask. The Ga-deposited oxide is locally removed at 540°C. After forming an opening area, a GaAs is selectively grown there.

SiO_2 is often used as a mask for selective-area epitaxy. It is well known that SiO_2 is deoxidized to volatile SiO by Ga deposited onto the surface [16]. Therefore, it is expected that this patterning method using Ga deposition can be applied for other mask materials, such as SiO_2, by controlling the amount of Ga.

EXPERIMENTAL

The MOMBE apparatus used in this experiment is shown in Fig.3. This system comprises a growth chamber, a surface-treatment chamber and a patterning chamber. A cryoshrouded quadruple mass spectrometer (QMS) and reflection high-energy electron diffraction (RHEED) equipment are set in the growth chamber. Metal-organic gas nozzles for introducing trimethylgallium (TMG) (or triethylgallium (TEG)) and a Knudsen effusion cell of As are also positioned in the growth chamber. A halogen lamp used for the photo-oxidation of GaAs is placed in the surface-treatment chamber. A metal mask for local Ga deposition is positioned in the patterning chamber. This metal mask is made of a 150 μm-thick tantalum plate and has five opening areas of 110×1000 μm^2 in size.

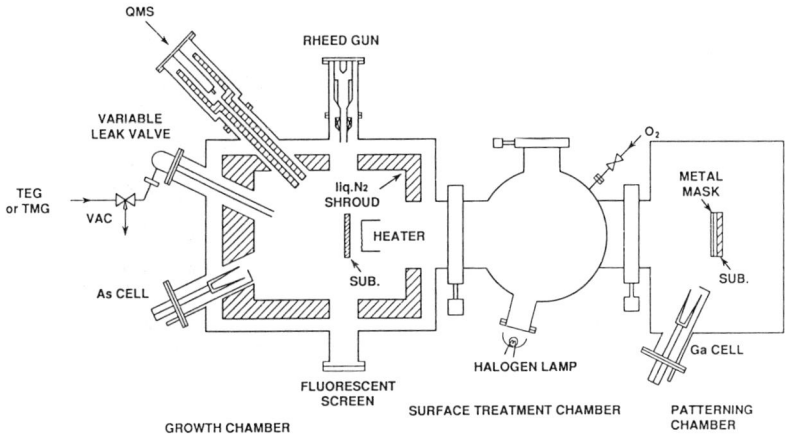

Fig.3   Schematic illustration of the MOMBE system used in this experiment.

At first, a GaAs buffer layer was grown at 550°C on a GaAs (100) substrate by using TEG and As$_4$ in order to obtain a well-defined surface with a (2×4) reconstruction. After that, the sample was transferred into the surface-treatment chamber and the GaAs surface was photooxidized by light irradiation from the halogen lamp (1.2 W/cm^2) under an oxygen ambient of 10 Torr. After forming the oxide mask, the substrate was transferred into the patterning chamber. 15 ALs of Ga were locally deposited at room temperature onto a particular area of the mask in the patterning chamber by using the metal mask positioned close to the substrate surface. The separation distance between the metal mask and the substrate was about 300 μm. After deposition of Ga, the substrate was transferred back into the growth chamber. The substrate was then heated up to 540°C in order to desorb the Ga-deposited oxide. After forming an opening area, selective-area epitaxy of the GaAs was performed by simultaneously supplying TMG and As$_4$ at 480°C for 90 minutes. The beam-equivalent pressure of TMG and As$_4$ were 2×10^{-7} Torr and 1.5×10^{-6} Torr, respectively

The selectively grown layer of GaAs was observed using a Nomarski microscope. The thickness profile of the selectively grown layer was measured using a contact-type surface profiler (TENCOR ALPHA-STEP). The residual oxygen on the substrate after removing the GaAs oxide was measured by secondary ion mass spectrometry (SIMS).

RESULTS AND DISCUSSION

Figure 4(b) shows a Nomarski microphotograph of selectively grown GaAs. Its shape was exactly the same as an opening area of the metal mask (Fig.4(a)). No polycrystalline GaAs was observed on the oxide mask. The thickness of the GaAs epitaxial layer was about 500Å, as shown in Fig.4(c). The growth rate of the selectively grown GaAs was almost the same as that on the GaAs substrate using no mask. The boundary between the GaAs epitaxial region and the oxide mask became abrupt compared with our previous report [13]. This is because the metal mask was closer to the substrate until 300 μm in distance, while the separation distance of the metal mask and substrate was about 1mm in our previous report [13]. The increase in the growth rate at the edge of the epitaxial layer, which generally appeared during selective-area growth by metal-organic chemical vapor deposition (MOCVD) [17,18], did not occur. However, the top shape of the selectively grown GaAs was not flat, as shown in Fig.4(c).

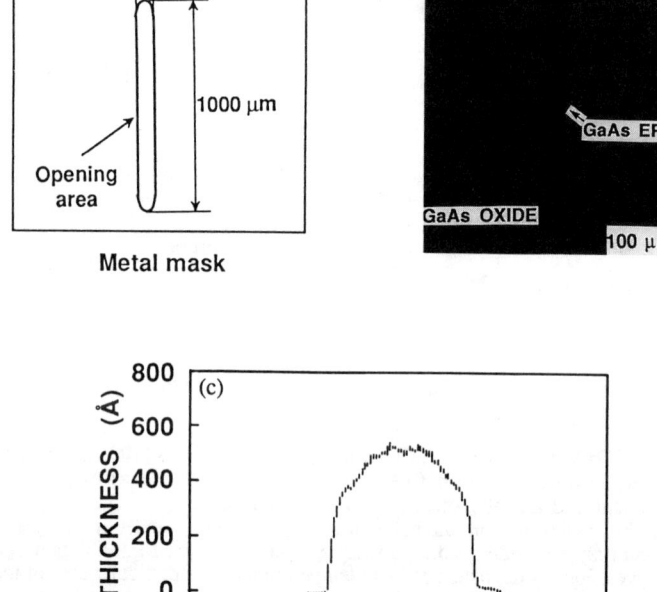

Fig.4   (a) Schematic illustration showing an opening area of a metal mask. (b) Nomarski microphotograph of a region of selectively grown GaAs. (c) Thickness profile to the lateral direction of the selectively grown GaAs in Fig.4(b).

It is conjectured that this inhomogeneity of the thickness was caused by the distribution of the amount of Ga deposited through the metal mask onto the oxide surface. This was probably due to a shadowing effect of the thick metal mask and the long distance between the mask and the substrate compared with the mask size. In this patterning size, Ga diffusion was not seen, since the size of a selectively grown GaAs was the same as an opening area of the metal mask. The subject of a future study is whether the Ga diffusion becomes a serious problem or not in the case of nano-meter scale patterning.

We next measured the depth profile of the residual oxygen concentration from the GaAs regrown layer towards the substrate by SIMS in order to investigate whether the oxide mask was completely removed or not. As a result, no residual oxygen was observed at the interface between the regrowth layer and the substrate. Residual oxygen-free patterning was therefore realized by a comparatively low-temperature desorption process of the oxide mask.

In this patterning method, it is expected that the surface after removing the oxide mask using Ga deposition remains flat although it is known[15] that the surface after removing the oxide is usually rough in the case of no Ga deposition. The reason is that no reaction occurs between the oxide and the GaAs substrate, since the entire oxide is deoxidized to volatile Ga_2O by deposited-Ga. In fact, we investigated the entire flateness after desorption of the Ga-deposited oxide using another sample. That is, when 15 ALs Ga-deposited oxide was desorbed, the RHEED pattern showed a streaky (1×6) structure. This streaky pattern immediately changed to a streaky (2×4) pattern when As_4 was supplied onto the surface. On the other hand, the RHEED pattern of the surface after the oxide desorption using no Ga deposition showed a three-dimensional spotty pattern of the GaAs.

From the above-mentioned results, the possibility of an *in-situ* patterning method for a GaAs oxide mask using selective Ga deposition was confirmed. This patterning method has some excellent advantages: damage-free patterning, no residual oxygen and a flat surface after patterning.

Furthermore, in order to realize the formation of a nanometer-scale structure, such as a quantum box or dot, as a future process, it is necessary to develop a method for direct Ga deposition by using a close-spaced liquid Ga-ion source [18-20] without the use of a metal mask. By this method, Ga can be finaly deposited onto a substrate. That is, the tip shape used to emit Ga-ions can be sufficiently sharp by supplying an electric field, and Ga can be deposited onto the substrate surface region of less than 100Å in diameter. The voltage used to initiate the emission of Ga-ions can be sufficiently lowered (20~30V) by reducing the distance of the ion source to the substrate (<50Å) [18]. The damage after Ga deposition is expected to be remarkably small due to the sufficiently low acceleration voltage. A very limited area of the thus Ga-deposited oxide mask can be thermally removed, as mentioned above. As a result, the fabrication of a nanometer-scale structure can be achieved by the patterning method of a GaAs oxide mask using the new patterning principle proposed here.

SUMMARY

A new damage-free patterning method of a GaAs oxide mask for the selective-area growth of GaAs was developed. The GaAs oxide mask where Ga molecules had been selectively deposited was thermally removed at 540°C, resulting in the formation of an opening area in the oxide mask. After forming the opened region, GaAs was selectively grown there by MOMBE. No residual oxygen was observed by SIMS at the interface between the GaAs selectively grown layer and the GaAs substrate.

ACKNOWLEDGEMENT

The authors would like to thank Dr. M.Tamura, Dr. Y.Katayama, and Dr. I.Hayashi for their valuable discussions and continuous encouragement.

REFERENCES

[1] Y.Hiratani, Y.Ohki, and M.Sasaki, J. Crystal Growth **115**, 74 (1991).
[2] E.Tokumitsu, K.Kudou, M.Konagai, and K.Takahashi, J. Appl. Phys. **55**, 3163 (1984).
[3] K.Kamon, S.Takagishi, and H.Mori, J. Crystal Growth **73**, 73 (1985).
[4] Y.Hiratani, Y.Ohki, Y.Sugimoto, K.Akita, M.Taneya, and H.Hidaka, Japan. J. Appl. Phys. Lett. **29**, 1360 (1990).
[5] H.Temkin, L.R.Harriott, R.A.Hamm, J.Weiner, and M.B.Panish, Appl. Phys. Lett. **54**, 1463 (1989).
[6] M.Taneya, Y.Sugimoto, H.Hidaka, and K.Akita, Japan. J. Appl. Phys. Lett. **28**, 515 (1989).
[7] K.Akita, M.Taneya, Y.Sugimoto, H.Hidaka, and Y.Katayama, J. Vac. Sci. Technol. **7**, 1471 (1989).
[8] M.Taneya, Y.Sugimoto, H.Hidaka, and K.Akita, J. Appl. Phys. **67**, 4297 (1990).
[9] T.Kosugi, R.Mimura, R.Aihara, K.Gamo, and S.Namba, Japan. J. Appl. Phys. **29**, 2295 (1990).
[10] N.Tanaka, H.Kawanishi, and T.Ishikawa, Japan. J. Appl. Phys. **32**, 540 (1993).
[11] S.Yoshida, Y.Hiratani, Y.Ohki, and M.Sasaki, in : Extended Abstracts 52th Autumn Meeting Japan. Soc. Applied Physics, No.1, 288 (1991) (in Japanese).
[12] M.Sasaki and S.Yoshida, submitted to J. Appl. Phys.
[13] S.Yoshida and M.Sasaki, J. Crystal Growth. **133**, 201 (1993).
[14] K.Tone, M.Yamada, Y.Ide, and Y.Katayama, Japan. J. Appl. Phys. **31**, 721 (1992).
[15] T.Van Buuren, M.K.Weilmeier, I.Athwal, K.M.Mackenzie, and T.Tiedje, Appl. Phys. Lett. **59**, 464 (1991).
[16] S,Wright and H.Kroemer, Appl. Phys. Lett. **36**, 210 (1980).
[17] K.Yamaguchi and K.Okamoto, Jpn. J. Appl. Phys. **29**, 1408 (1990).
[18] O.Kayser, J.Crystal Growth **107**, 989 (1991).
[19] G.Benassayag, P.Sudraud and L.W.Swanson, Surface Science **191**, 362 (1987).
[20] R.Clampitt, K.L.Aitken and D.K.Jefferies, J. Vac. Sci. Technol. **12**, 1204 (1975).
[21] V.E.Krohn and G.R.Ringo, Appl. Phys. Lett. **27**, 479 (1975).

SILICON NUCLEATION ON SILICON DIOXIDE AND SELECTIVE EPITAXY IN AN ULTRA-HIGH VACUUM RAPID THERMAL CHEMICAL VAPOR DEPOSITION REACTOR USING DISILANE IN HYDROGEN

Katherine E. Violette[1], Mahesh K. Sanganeria[1], Mehmet C. Öztürk[1], Gari Harris[2], and Dennis M. Maher[2],
[1]North Carolina State University, Department of Electrical and Computer Engineering, Box 7911, Raleigh, NC 27695-7911, [2]North Carolina State University, Department of Materials Science and Engineering, Box 7916, Raleigh, NC 27695-7916

ABSTRACT

Silicon nucleation on silicon dioxide and selective silicon epitaxial growth (SEG) were studied in an ultra high vacuum rapid thermal chemical vapor deposition (UHV-RTCVD) reactor. Experiments were performed using 10% Si_2H_6 in H_2 in a pressure range of 10 - 100 mTorr at 760°C. Under these conditions, the growth rate ranged from 75 to 330 nm/minute. Loss of selectivity via Si island formation on SiO_2 was studied using scanning electron microscopy (SEM) and atomic force microscopy (AFM) revealing a strong dependence on deposition pressure. Cross sectional transmission electron microscopy (XTEM) was employed to study the vertical oxide/epitaxy interface where faceting can occur. The incubation time for nucleation was found to increase from 10s to 70s as pressure is reduced from 100 mTorr to 10 mTorr, allowing thicker selective epitaxial film growth in spite of the reduced growth rates. This was attributed to the reduction in gas phase supersaturation of the Si containing species resulting in a lower density of adsorbed atoms on the SiO_2 surface. This process shows a potential for chlorine free selective epitaxial growth and provides insight to the surface morphology of polycrystalline films deposited at low pressures.

INTRODUCTION

Low temperature selective epitaxial growth (SEG) is currently being considered for various novel applications in Si integrated circuit processing. Current SEG processes typically use dichlorosilane with H_2 and HCl [1-5]. Reducing the use of chlorinated gases would be desirable because of existing concerns for chlorine based processing, including the corrosive nature of HCl leading to contamination issues, thin oxide degradation due to etching or pinhole formation, and difficulty in pumping chlorinated species potentially leading to cross-contamination in multi-chamber cluster tools. Also, the role of chlorine at reduced temperatures (<850°C) is poorly understood: while Cl provides a Si etching mechanism, the efficiency of this mechanism is questionable at temperatures below 850°C [6].

SEG using non-chlorinated chemistries has been reported before where SiH_4 [7] or Si_2H_6[8] were used in hot wall, low pressure chemical vapor deposition or gas source molecular beam epitaxy systems. Tatsumi and his colleagues at the NEC corporation have shown that pure Si_2H_6 in a gas source MBE reactor can give a critical selective epitaxial thickness of approximately 100 nm before loss of selectivity by Si nucleation on the SiO_2 surface[8]. Ohmi et. al. demonstrate that SiH_4 can give approximately a 50 nm critical layer thickness before loss of selectivity in their ultra clean low pressure CVD system [7]. These reports demonstrate that SiH_4 and Si_2H_6 both exhibit an intrinsic selectivity to oxide via an incubation time seen under ultra clean conditions, reinforcing the already well established cleanliness requirement for good quality epitaxial growth at low temperatures [7, 9].

In recent years, single wafer manufacturing using multi-chamber cluster tools has gained in popularity [10]. As a potential technique in such systems, rapid thermal chemical vapor deposition (RTCVD) has been considered for processes including Si epitaxy with successful results [11]. In a typical RTCVD reactor, the wafer is heated optically through a quartz window, providing thermal switching to initiate and terminate chemical reactions on the wafer surface in a matter of seconds. Thus, thin films with abrupt interfaces can be grown at high growth rates which is an important concern in single wafer processing. The technique aims at minimizing the thermal budget instead of temperature. Satisfactory growth rates for single wafer processing can be achieved at elevated temperatures while the process time can be controlled accurately down to a few seconds by means of temperature switching and computer control.

In this work, Si nucleation on SiO_2 from Si_2H_6 was studied using ultra high vacuum rapid thermal chemical vapor deposition (UHV-RTCVD). The method combines temperature switching capability of RTCVD with the clean growth environment of UHV-CVD essential for high quality epitaxial growth. The process temperature/pressure window was chosen to obtain an acceptable throughput for single wafer processing. For the same reason, Si_2H_6 was chosen over SiH_4 to maximize the growth rate at lower pressures[12]. At lower pressures, the amount of impurities such as oxygen and water vapor can be minimized providing a cleaner growth environment for low temperature epitaxy.

EXPERIMENTAL

Figure 1 shows the top-view of the UHV-RTCVD reactor used in this study. The system consists of a load-lock (sample entry chamber), an intermediate chamber, and a reaction chamber. The load lock is pumped with a dry molecular drag pump to a base pressure of 10^{-5} torr. The intermediate chamber is pumped by a cryopump to a base pressure of 10^{-9} Torr. This chamber serves as a vacuum buffer between the sample entry chamber and the reaction chamber. The reaction chamber is pumped to its base pressure with a dedicated UHV cryopump. O-ring seals are completely eliminated in this chamber in order to achieve UHV conditions. During film growth, the UHV cryopump is isolated with a gate valve and the process gases are pumped through a turbomolecular / molecular drag combination pump. This

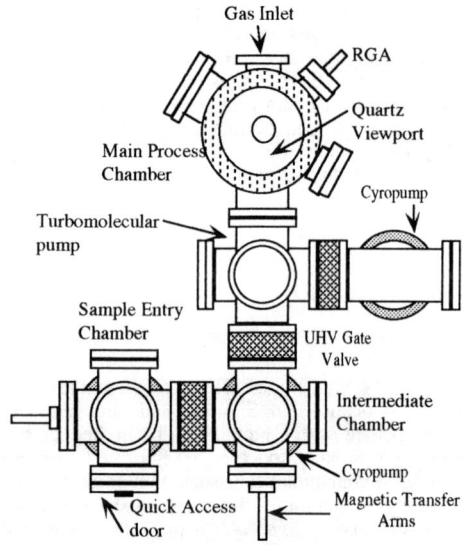

Figure 1 A schematic of UHV/RTCVD reactor.

pump features magnetic levitation for oil free processing and is backed by an oil free foreline mechanical pump. Sample transfer between chambers is achieved using magnetically coupled transfer arms. Point-of-use gas purifiers are used to remove contaminants such as oxygen and water vapor from the process gases. The reaction chamber is water-cooled, stainless-steel with a quartz window on top. The wafer is heated through the window with a 35 kW Peak Systems[TM] LXU-35 arc lamp. An optical pyrometer ($\lambda = 2.2$ µm) measures temperature in a closed-loop feedback control system. The wafer sits on a quartz wafer holder which is able to rotate during the process. We have recently reported the generation lifetime of epitaxial films grown in the same system using the same chemistry[13].

In this study, 4" Si (100) wafers with 100 nm thick thermal SiO_2 were patterned by photolithography and wet chemical etching. Seven sets of 100µm x 2 mm windows, spaced 300µm apart, were aligned in the [110] crystal direction and distributed along the center of the wafer. The samples were cleaned in a batch Huang clean: 5 min. in a 1:1:5 $NH_4OH:H_2O_2$:DI water bath at 80°C, 5 min. DI rinse, 5 min. in a 1:1:5 $HCl:H_2O_2$:DI water bath at 80°C followed by a second 5 min. DI rinse. The wafers were spin-dried in nitrogen and stored in a nitrogen purged box until use. Immediately before insertion into the UHV system, a 20 s, 5% HF dip, 30 s water rinse, and nitrogen blow dry were performed to obtain a hydrogen passivated native oxide-free Si surface[14]. It should be noted that the final water rinse results in a partial loss of the hydrogen coverage. The correlation between the rinse time and the hydrogen coverage is currently under investigation. Upon insertion into the reactor each wafer underwent the following procedure: the entry chamber was pumped down to ~ 8x 10^{-5} Torr in 10 minutes before transfer into the intermediate chamber. An intermediate chamber base pressure of high 10^{-9} Torr was routinely achieved within 10 minutes. The wafer was then transferred into the reaction chamber and cryopumped to a base pressure of about 3x10^{-9} Torr. Then, the chamber was switched to the process pumps and gas flows were initiated. The gas used in this work was ULSI grade 10% Si_2H_6 in H_2 further purified at point-of-use for oxygen and water vapor levels below 50 ppb. After establishing steady gas flows, the arc lamp was turned on to heat the substrate to the deposition temperature at a ramp rate of approximately 150°C/s. The deposition process was initiated and terminated by temperature switching. When the deposition was complete, the lamp was extinguished, the gas flow stopped and the wafer allowed to cool.

RESULTS

We have previously reported the temperature dependence of the Si growth rate from 10% Si_2H_6 in H_2 at a total pressure of 100 mTorr [13, 15]. Figure 2 shows the pressure dependence of silicon growth rate from 10% disilane in hydrogen. The deposition pressure was set by the input gas flow and the pumping speed of the process pump. Throttling was deliberately not used in order to maximize the pumping speed and hence minimize the background impurity levels. As shown in Figure 2, even in this very low pressure regime (10 - 100 mTorr) very high growth rates can be obtained with Si_2H_6. Because the total input gas flow at

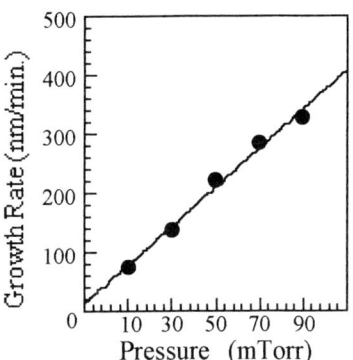

Figure 2: Pressure dependence of growth rate from 10% Si_2H_6 in H_2 at 760°C.

these pressures is of the order of <250 sccm, the amount of impurities introduced into the chamber by the gas itself is greatly reduced. Furthermore, in this low pressure regime, turbomolecular pumps operate with maximum efficiency, further reducing impurity levels. The growth rate data presented in Figure 2 exhibit a typical dependence on temperature with two distinct regions: a surface reaction limited regime governed by hydrogen desorption and a nearly temperature independent regime where growth is predominantly limited by mass transport of disilane from the gas phase to the Si surface.

To investigate the effects of pressure on Si nucleation on SiO_2, we have grown films on patterned wafers. The pressure dependence of Si nucleation on SiO_2 at 750°C is demonstrated in Figure 3. As shown, the incubation time, defined as the time before nucleation occurs on SiO_2, increases as pressure decreases. Also shown in Figure 3 is the incubation thickness, defined as the thickness of the SEG during the incubation time: in spite of lower growth rates, the incubation thickness increases at lower pressures, allowing growth of thicker selective epitaxial layers.

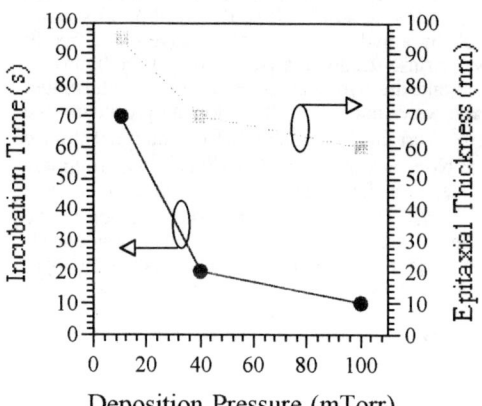

Figure 3: Incubation time and thickness as a function of pressure at 760°C

In their recent paper on selective Si epitaxy from Si_2H_6, Tatsumi et. al. report a critical supply volume for Si_2H_6 defined as the maximum gas volume which can be delivered into the growth chamber before loss of selectivity occurs. Their experimental results indicate that the critical supply volume of Si_2H_6 increases with decreasing flow rate. The data here reflects the same trend as the pressure in the reactor is proportional to the gas flow rate. Improved selectivity at reduced pressures has also been observed with SiH_2Cl_2 by Fitch and Denning[6]. This behavior can be explained by considering the requirements for nucleation[16]. In chemical vapor deposition, formation of a continuous film consists of an initial nucleation phase followed by coalescence and film growth. For homogeneous film growth; e.g., Si on Si, growth occurs readily because the cohesive forces between the adsorbed atoms (adatoms) and the substrate and those among the adatoms are identical. Heterogeneous film deposition; e.g., Si on SiO_2, begins via nucleation because the cohesive forces between the substrate and adatoms are weak compared to the relatively strong cohesive forces among the adatoms [16]. The nucleation process is governed in part by the adatom density and in part by the substrate surface structure. The adatom density is related to the available amount of the depositing component in the gas phase, in this case Si_2H_6, which is determined by the reactant species input pressure and the deposition temperature. At any given moment, there exists a random spatial distribution of Si adatoms on the SiO_2 surface. When the local density of adatoms reaches a critical value, a stable nucleus can form. In our experiments, the Si_2H_6 partial pressure directly affects the amount of available silicon in the gas phase and hence the adatom density. By decreasing the pressure, the average adatom density is reduced which is expected to reduce the probability of nucleation. Si nucleation on SiO_2 is also determined by the surface defect density on the oxide. Fitch et. al.[6]assert that, for the SiH_2Cl_2-HCl-H_2 system, at reduced pressures and in clean, leak-free environments, improvement in selectivity is due to an increase in passivation of the dangling bonds on the SiO_2 surface by atomic hydrogen[6]. Impurities such as oxygen in the growth ambient can also promote nucleation by providing reactive bonding sites for the adatoms. The level of such impurities will be reduced at lower pressures since the primary sources of these impurities are the reactor background and the reactant gases. This principle is the basis of epitaxy by UHV-CVD [17].

Figure 4 shows a cross sectional TEM of a film deposited at 750°C and 10 mTorr. The micrograph shows the crystalline structure of the film and facets typical of selective epitaxy.

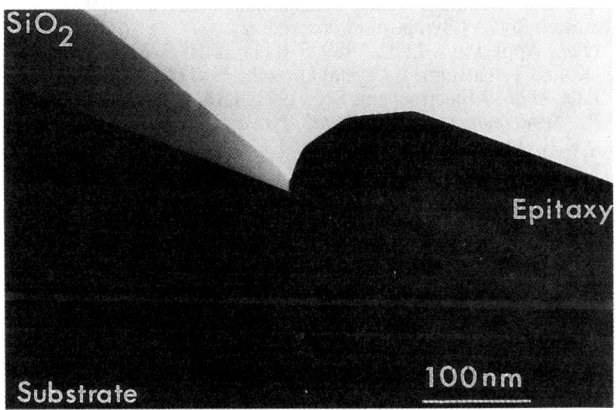

Figure 4: Cross sectional TEM of sample deposited at 760°C and 10 mTorr.

CONCLUSIONS

In this study, we have demonstrated that an inherent selectivity in the Si_2H_6/H_2 system exists with no apparent nuclei etching mechanism on SiO_2. The incubation time was found to decrease at higher pressures which was explained by considering the requirements for critical nuclei formation on SiO_2. Conversely, the incubation thickness was found to increase at low pressures due to the reduced adatom concentration. Epitaxial films as thick as 100 nm were grown on Si with no nucleation on SiO_2.

The use of Si_2H_6 provides a unique advantage over more commonly used silicon source gases since it provides relatively high growth rates even at pressures as low as 10 mTorr. Thus, growth rates compatible with single wafer manufacturing can be obtained at very low pressures. Low pressures not only reduce the probability for critical nuclei formation essential for selective growth but also reduce the impurity background in the growth ambient which is essential for good quality film growth.

ACKNOWLEDGMENTS

The authors are grateful to G. Xing of NCSU for many helpful discussions and D. Batchelor for his help in SEM microscopy. Help received from S. P. Ashburn, D. T. Grider and F. S. Johnson in the system construction is gratefully acknowledged. The authors would like to thank Henry Taylor, Joan O'Sullivan, Harold Morton and Richard Kuehn of NCSU Microelectronics Laboratory and Arthur Illingsworth of NCSU Precision Engineering Shop for their invaluable assistance. This research was partially supported by NSF Engineering Research Centers Program through the Center for Advanced Electronic Materials Processing (grant CDR-8721545).

1. Borland, J.O. *10th International Conference on CVD, CVD-X, 87-8.* 1987.
2. Jastrzebski, L., J.F. Corboy, and R. Soydan, J Electrochem. Soc., 1989. **136**(11): p. 3506-3513.
3. Rahat, I., *et al.*, J Electrochem. Soc., 1991. **138**(8): p. 2370-2374.
4. Klaasen, W.A. and G.W. Neudeck, IEEE Trans. Electron Devices, 1990. **ED-37**(1): p. 273-279.
5. Friedrich, J.A., J. Appl. Phys., 1989. **65**(4): p. 1713-1716.

6. Fitch, J.T. and D.J. Denning. *MRS Symposium. Proceedings* 1992. San Francisco, CA: Materials Research Society Symposium Proceedings.
7. Murota, J., *et al.*, Appl. Phys. Lett., 1989. **54**(11): p. 1007-1009.
8. Aketagawa, K. and T. Tatsumi, J. Crystal Growth, 1991. **11**: p. 860-863.
9. Sedgwick, T.O., *et al.*, J Electrochem. Soc. 1991. **138**(10): p. 3042-3046.
10. Burggraaf, P., *Semiconductor International.*, November, 1992, p. 71-74.
11. Hsieh, T.Y., K.H. Jung, and D.L. Kwong, J Electrochem. Soc., 1991. **138**(4): p. 1188-1207.
12. Sadamoto, M., J.H. Comfort, and R. Reif, J Elect. Mat., 1990. **19**(12): p. 1395.
13. Sanganeria, M.K., K.E. Violette, and M.C. Öztürk, Appl. Phys. Lett., 1993. **63**(9): p. 1225-127.
14. Meyerson, B.S., F.J. Himpsel, and K.J. Uram, Appl. Phys. Lett. 1990. **57**(10): p. 1034-1036.
15. Sanganeria, M.K., K.E. Violette, and M.C. Öztürk. *MRS Symposium Proceedings*. 1993. San Francisco, CA, USA:
16. Bloem, J., J. Crystal Growth, 1980. **50**: p. 581-604.
17. Meyerson, B.S., *et al.*, J Electrochem. Soc., 1986. **133**(6): p. 1232-1235.

Perfect selective Si epitaxial growth realized by synchrotron radiation irradiation during disilane molecular beam epitaxy

Yuichi Utsumi, Housei Akazawa and Masao Nagase
NTT LSI Laboratories, 3-1, Morinosato Wakamiya, Atsugi-shi, Kanagawa pref., 243-01, JAPAN

ABSTRACT

Perfect selectivity in Si epitaxial growth on a Si/SiO_2 substrate has been achieved by irradiating SR during molecular beam epitaxy (MBE) using disilane. This differs from conventional selective growth by MBE using disilane in that no Si nucleation occurs on SiO_2 irrespective of growth time. Temperature above 700 °C is necessary for SR-induced selective growth. This perfect selectivity might result from the elimination of adsorbed Si atoms induced by photo-stimulated evaporation of SiO_2 which results in the perfect suppression of Si nucleation.

INTRODUCTION

Selective epitaxial growth (SEG) of Si which enables Si deposition on Si but not SiO_2, is one of the most promising techniques for the fabrication of fine self-aligned structures for high-speed Si devices, and SEG by gas source molecular beam epitaxy (GSMBE) using silane and disilane has been widely investigated [1-3]. In these works, however, SEG has so far been unsatisfactory in the sense that it finally breaks down after an incubation period (IP) that ends in Si nucleation on the SiO_2 surface. This problem is caused mainly by the fact that Si nucleation on SiO_2 cannot be completely inhibited in conventional GSMBE where selectivity relies solely on the finite difference in the probability of silane or disilane decomposition on either the Si or SiO_2 surface. For perfect selectivity to be attained, the Si atoms adsorbed on the SiO_2 surface must be removed to prevent Si nucleation. In the present work synchrotron-radiation (SR) induced chemical reaction was adopted for this purpose. SR has recently been used as a high intensity source of vacuum ultraviolet light for photo-excited processes [4-7]. The photon energies of SR are very suitable for inducing electron excitation in materials that leads to the photo-stimulated desorption of atoms on solid surfaces or the breaking of atomic bonds in bulk materials. This work demonstrated that perfect selectivity can be achieved by combining GSMBE with SR irradiation and it investigated the conditions required for SR-induced selective epitaxial growth (SRSEG). The mechanism of this perfect SEG is also discussed here.

EXPERIMENT

The experiments used the BL1C beam line of the 2.5-GeV Photon Factory storage ring at the National Laboratory for High Energy Physics. The beam line and the reaction apparatus are described in detail elsewhere [3,6]. The reaction apparatus (Fig. 1) consisted of four chambers: one for load lock, one for cleaning sample surfaces, one for photochemical reaction and one for analysis. Silicon (100)

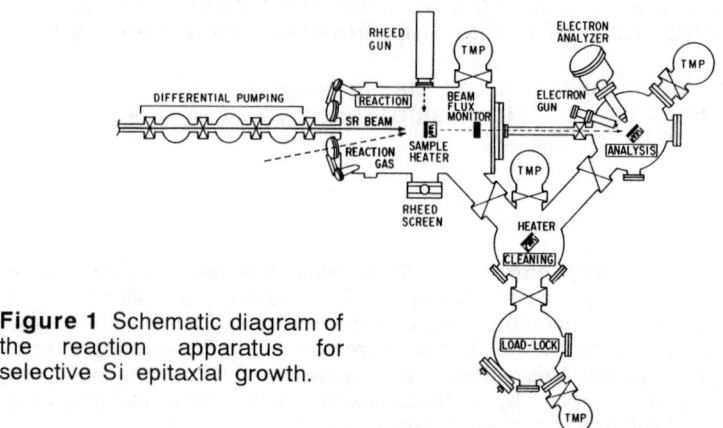

Figure 1 Schematic diagram of the reaction apparatus for selective Si epitaxial growth.

substrates covered with thermally-grown SiO_2 with Si windows were used for selective growth. Before growing the Si films we removed the native oxide in the Si windows by heating the substrate at 980°C for 10 min. Disilane gas with a purity of 99.99 % was fed into the reaction chamber while the SR beam with wavelengths of 1 to 100 nm and with a peak at 10 nm, irradiated the sample surface perpendicularly. The average storage ring current during experiment was 310 mA, which provide 3×10^{17} photons/sec^{-1}cm^{-2}. SEG was distinguished from non-SEG by observations made using reflective high energy electron diffraction (RHEED), secondary ion mass spectroscopy (SIMS), scanning electron microscopy (SEM), or Auger electron spectroscopy (AES).

RESULT AND DISCUSSION

A. Perfect selective epitaxial growth of Si under SR irradiation

Distinctions between SEG and non-SEG in a series of experiments were inferred mainly from RHEED patterns observed immediately after growth in the reaction chamber. When the growth was nonselective, the RHEED pattern for the SiO_2 region showed a ring structure because a polycrystalline Si layer grew on SiO_2. When growth was selective, however, no polycrystalline Si grew on SiO_2 and RHEED yielded a halo pattern characteristic of the SiO_2 surface. Regardless of selectivity, the Si windows yielded a 2x1 RHEED pattern reflecting two-dimensional epitaxial growth. Also, irrespective of whether growth was selective or nonselective, a SR-induced gas phase chemical reaction resulted in polycrystalline Si being deposited around the irradiated area.

Cross section SEM images of a substrate before and after SRSEG are shown in Fig. 2. No Si deposition occurred on the SiO_2 patterns, whereas epitaxial Si films with [113] facets on their edges grew on the Si windows. Comparing Fig. 2 (a) and 2 (b) reveals that the SiO_2 pattern edges have became narrower and rounded implying that SiO_2 patterns were etched during the SRSEG process. In fact, the thickness of the SiO_2 films decreased during the SRSEG process. We also found the etching rate increased with substrate temperature. This phenomenon is known as

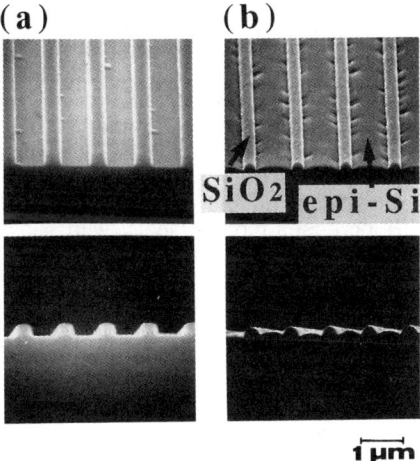

Figure 2 Cross-section SEM images of the substrate before (a) and after (b) the 20min of SRSEG. The thickness of the SiO2 patterns before SEG was 300 nm, substrate temperature was 750°C, and disilane pressure was 1.5×10^{-4} Torr.

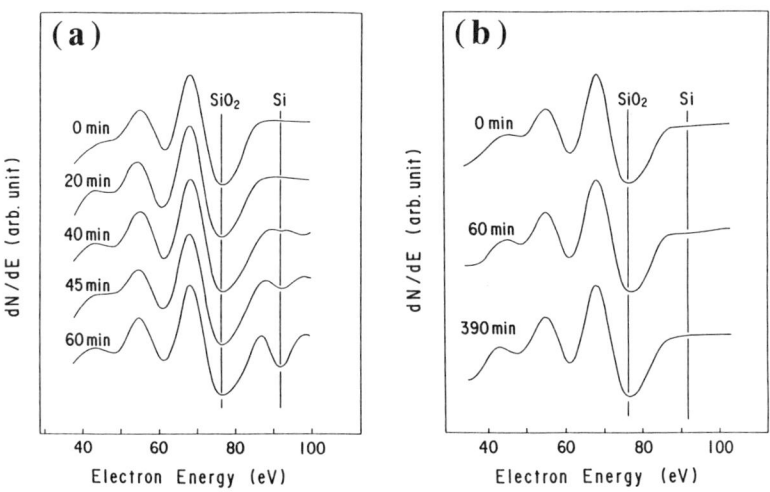

Figure 3 Time dependence of Auger spectra of SiO2 surfaces obtained after SEG without (a) and with (b) SR irradiation. Substrate temperature was 750°C and disilane pressure was 3×10^{-5} Torr.

SR-stimulated evaporation of SiO_2, first reported by Akazawa et. al.. The underlying mechanism is oxygen atom displacement into an interstitial site due to electron excitations followed by thermal desorption of SiO molecules [7].

AES spectra of SiO_2 surfaces (Fig. 3) show that without SR irradiation, the Si (LVV) peak at 92 eV appears as a result of Si nucleation on SiO_2 when the growth time exceeds 40 min. The emergence of the Si peak corresponds to the RHEED pattern change from a halo to a ring. With SR irradiation, the Si peak does not appear even after 390 min. We have found (data not shown) that the Si peak does not appear even at growth time up to 800 min.

With SR irradiation, IP increases monotonically with substrate temperature and becomes infinitely long above 700°C as shown in Fig.4. This is a distinctive feature of SRSEG. Without SR irradiation, in contrast, the IP vs temperature curve has a minimum near 700°C, which is characteristic of a SEG by GSMBE [3]. But perfect selectivity in this temperature range was not obtained without SR irradiation. The growth modes (SEG or non-SEG) at various substrate temperatures and disilane pressures are plotted in Fig. 5 for Si growth with and without SR irradiation [8]. Here the growth mode is defined as selective if no Si nucleation on SiO_2 is detected by RHEED within 40 min after the onset of the gas supply. At temperatures below 700 C, however even in the SEG mode, Si nucleation on SiO_2 starts after a certain period. At temperature above 700°C, on the other hand, it is confirmed SEG is maintained at least 800 min; that is; selectivity is perfect. In a separate experiment, it was shown that SR-irradiated SiO_2 surface became Si-rich below 700°C due to accumulation of oxygen -vacancy defects as evaporation of SiO_2 proceeded, whereas above 700°C, the surface stoichiometry was maintained as SiO_2 [9]. This was also confirmed for the present reaction system by using of AES measurements. Thus, temperature above 700°C are necessary for perfect SRSEG, since the formation of Si-rich surface cause Si nucleation.

From the results shown in Figs. 4 and 5, we conclude that this distinct difference between SRSEG and SEG without SR irradiation is caused by SR-induced chemical reaction.

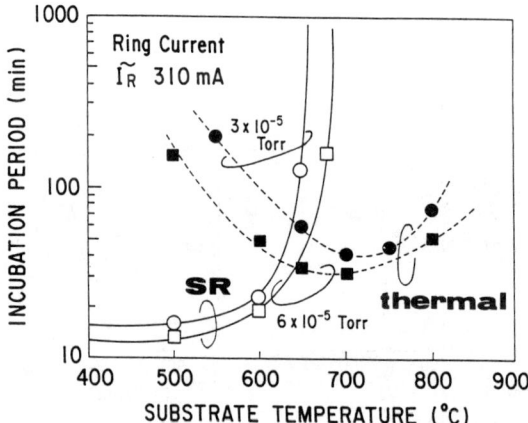

Figure 4 Incubation period as a function of substrate temperature for the cases that SR was irradiated (open symbols) and not irradiated (close symbols) during SEG. The disilane pressure was set up at 3×10^{-5} an 6×10^{-5} Torr.

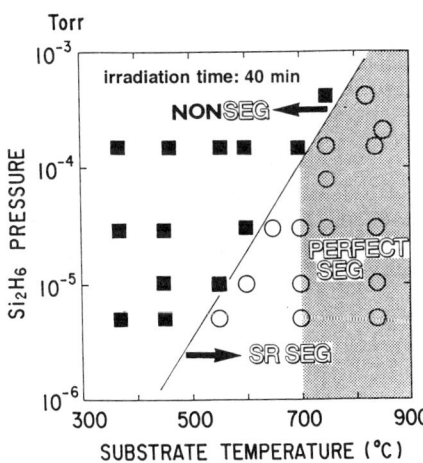
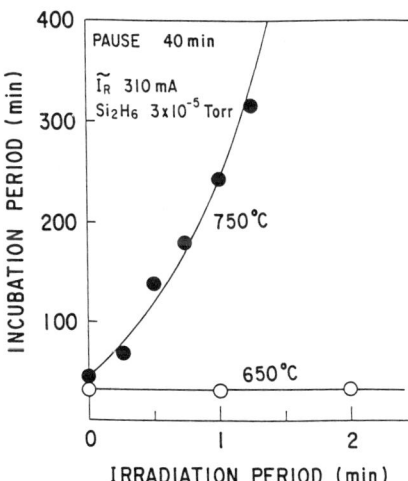

Figure 5 Selective growth condition in SRSEG as a function of disilane pressure and substrate temperature. Open circles indicate SEG and solid circles indicate non SEG.

Figure 6 Irradiation period dependence of incubation period under intermittent SR irradiation with the substrate as a parameter. The disilane pressure was set at 3×10^{-5} Torr.

B. Mechanism of perfect selective epitaxial growth

For perfect selectivity, adsorbed Si atoms have to be removed from the SiO_2 surface; otherwise they would become Si nucleation sites for polycrystalline Si deposition. We investigated the effects of SR irradiation on the annihilation of Si nucleation sites by irradiating the substrate intermittently. One cycle in intermittent SR irradiation consists of a 40 min pause period followed by an irradiation period ranging from 0 to 2 min. During the pause between exposures, the dissociative adsorption of disilane on the SiO_2 surface proceeds. If adsorbed Si atoms are desorbed completely during the SR irradiation period, perfect selectivity can also be obtained under intermittent irradiation. The IP for Si nucleation on SiO_2 is shown in Fig. 6 as a function of the SR irradiation period in one cycle. IP, here including both the pause and irradiation periods, increases with irradiation period at 750°C and we confirmed selectivity was maintained at least 800 min for irradiation period longer than 2 min. This result can be explained qualitatively on a model in which the Si atoms adsorbed during the pause are completely removed by SR irradiation. At 650°C, a 15-nm polycrystalline Si film grew on SiO_2 surface after a 40-min pause period. Once a Si film is formed on SiO_2 it cannot be removed by SR irradiation. The result also indicates that SR-induced surface chemical reaction contribute to the removal of Si from SiO_2 surface.

Next probable mechanism for perfect suppression of Si nucleation on the SiO_2 surface attained above 700°C is proposed based on the experimental results. Considering the fact that the substrate temperature above which the stoichiometry of the SiO_2 surface is maintained under photo-stimulated evaporation coincides with that above which perfect SRSEG is achieved, SRSEG is probably closely related to the photo-stimulated evaporation of SiO_2. The substrate SiO_2 film becomes chemically active in the photo-stimulated evaporation process because SiO_2 molecules are electronically excited and the molecules are broken into some active species. The adsorbed Si atoms assumed to change into volatile SiO through the reaction with the SiO_2. As a result, the adsorbed Si atoms that would otherwise become Si nucleation sites are removed from the SiO_2 surface. A substrate temperatures below 700 °C, however, the SiO_2 surface becomes Si-rich : the number of Si nucleation sites is greatly increased and more Si nucleation by adsorbed Si atoms occurs to a greater extent than does Si atom desorption. According to these mechanism, the principal features of SRSEG can be qualitatively explained.

CONCLUSIONS

Perfect selective growth has been achieved by irradiating SR during GSMBE using disilane at substrate temperatures above 700°C: No Si nucleation occurs on SiO_2 irrespective of growth time. This perfect selectivity has never been attained in conventional SEG by GSMBE using silane or disilane. It was also found temperature above 700°C is necessary for perfect SRSEG. Our experimental results on SEG with intermittent SR irradiation indicate that perfect selectivity results from the elimination of Si atoms from SiO_2 surface due to a SR stimulated-chemical reaction. This reaction is assumed to be the Si atom elimination induced by photo-stimulated evaporation of SiO_2.

ACKNOWLEDGMENT

The author would like to thank Dr. Katsutosh Izumi , Dr. Katsumi Murase and Dr. Hiroo Kinoshita for their continuous encouragement and useful suggestions throughout this work. The author also thank Hideo Namats and Toshihiko Ishiyama for the preparation of samples and the staff of the Photon Factory for their collaboration.

References

[1] H. Hirayama, T. Tatsumi and N. Aizaki, Appl. Phys. Lett. 5 224 (1988).
[2] F. Hirose, M. Suemitsu, and N Miyamoto, Jpn. J. Appl. Phys. L2003 (1989).
[3] K. Aketagawa, T. Tatsumi and J. Sakai , Appl. Phys. Lett. 5 173 (1991).
[4] H. Kyuragi and T. Urisu, J. Appl. Phys. 61 203 (1986).
[5] J.Takahashi, Y . Utsumi , H.Akazawa, I. Kawashima, and T. Urisu,
    Appl. Phys. Lett. 58 2776 (1991).
[6] Y. Utsumi , J. Takahashi, H. Akazawa, and T. Urisu,
    Jpn. J. Appl. Phys. 30 319 (1991).
[7] H. Akazawa, J. Takahashi , Y. Utsumi, I. Kawashiam, and T. Urisu,
    J. Vac. Sci, Techgnol. A9 2653 (1991).
[8] Y . Utsumi , H. Akazawa, M. Nagase, T. Urisu and I. Kawashimo,
    Appl. Phys. Lett. 6 164 (1993).
[9] H. Akazawa, M. Nagase, and Y. Utsumi to be published in the
    Proceedings of 7th International Conference on Radiation Effects i n
    Insulators

GATE QUALITY OXIDES PREPARED BY RAPID THERMAL CHEMICAL VAPOR DEPOSITION

R.T. KUEHN*, X. XU*, D.J. HOLCOMBE*, V. MISRA*, J.J. WORTMAN*, J.R. HAUSER, Q.-F. WANG**, D.M. MAHER***
*North Carolina State University, Department of Electrical and Computer Engineering, Raleigh, N.C. 27695
**IMEC, Kapeldreef 75, B-3001 Leuven, Belgium
***D.M. Maher, North Carolina State University, Department of Electrical and Computer Engineering, Raleigh, N.C. 27695

ABSTRACT

As the feature size of MOSFET devices shrink, issues such as thermal budget associated with controlling channel doping profiles and oxide growth kinetics raise concerns about using thermally grown furnace oxides for deep-submicron device applications. To address these concerns, we have developed a new RTCVD oxide process using a gas system of silane and nitrous oxide. The RTCVD oxides are deposited in a lamp-heated, cold wall, RTP system. Deposition rates ranging from 55 Å/min. to 624 Å/min. can be achieved at 800°C with silane nitrous oxide flow rate ratio of 2% and total pressure ranging from 3 to 10 Torr. The results indicate that this RTCVD process can be used to deposit both thin gate and thick isolation insulators for single wafer processing. Deposition rates of the RTCVD oxides exhibit a non-linear dependence on the total deposition pressure. Electrical characterization of the as-deposited RTCVD oxides shows a mid-gap interface trap density of $< 5 \times 10^{10}$ $eV^{-1}cm^{-2}$ and an average breakdown field of 13MV/cm. AES, RBS and TEM analyses have been used to study surface cleaning effects on the silicon-silicon dioxide interface quality and to determine the chemical composition of the RTCVD oxides. The results show that RTCVD oxides with stoichiometric composition and atomic flat silicon-silicon dioxide interface can be achieved using silane nitrous oxide flow rate ratio of <2%. I-V characteristics and transconductance degradation under hot carrier stress for MOSFET's using as-deposited RTCVD gate oxides have been found to be comparable to those of MOSFET's using thermal gate oxides.

INTRODUCTION

Thermally grown oxides are commonly used as gate dielectrics in MOSFET structures. As the feature size of MOSFET devices shrink, issues such as high oxidation temperatures, low growth rates, silicon consumption, and difficulty in controlling thin oxide growth kinetics raise concerns about the application of thermally grown oxides for advanced deep submicron devices. [1] We have investigated low temperature (≤800 °C) rapid thermal chemical vapor deposited (RTCVD) gate quality oxides and compared these films to thermal oxides used as controls. The RTCVD oxide films were deposited using silane nitrous oxide gas mixtures [2]. Deposition rates were compared to thermal oxidation rates and also were studied for a range of total system pressures from 1 -10 Torr. Boron concentration in silicon substrates have been compared after the formation of the deposited oxide and the control furnace oxide. Electrical performance and reliability of MOSFET devices and MOS capacitors has been studied for the deposited and control thermal oxides. Interface trapping and bulk charging have also been investigated for these RTCVD oxides and thermally grown oxides. Breakdown fields for RTCVD deposited oxides and rapid thermal oxides (RTO) were studied as a function of oxide thickness.

DEVICE FABRICATION

NMOS devices were fabricated with a simple four mask process. This mask set had various W/L ratios of devices plus enclosed polysilicon gate capacitors. The starting substrates were boron doped 0.3 ohm-cm silicon wafers with (100) orientation. After a RCA clean, followed by a field oxidation, the active areas were defined. The deposited gate oxides were formed in a RTCVD reactor with a cylindrical quartz reaction chamber, cold walls and oil-free pumps (10^{-8} Torr base pressure). The RTCVD oxide films were deposited by reacting silane (10% diluted in Argon) and nitrous oxide at a temperature of 800°C. The RTO oxides were formed in the same rapid thermal reactor at a temperature of 1050°C and a pressure of 760 Torr with a constant oxygen flow rate of 2 standard liter per minute (slpm). The furnace oxides were formed in a standard (Tylan) horizontal furnace system at 900°C using dry oxygen (no HCL). Both the deposited and the RTO oxides had 2000Å of polysilicon deposited in this RTP reactor. The remaining fabrication steps included polysilicon pattern plus etch, POCL3, LTO 410 oxide, contact masking and etch, and metalization plus pattern definition followed by forming gas anneal.

RESULTS

Figure 1 shows the deposition rate for the RTCVD deposited films and the growth rate for thermally oxidized films verses inverse temperature. The RTCVD deposited films were formed with flow ratios of 0.5% and 2% silane/nitrous oxide at 800°C with a system pressure of 3 Torr. As shown in Figure 1, the deposited films have a factor 40 to 90 higher deposition rate than the thermal oxide film growth rate [3]. Figure 1 also shows a dramatic reduction in thermal budget for deposited insulators verses thermally grown oxide films.

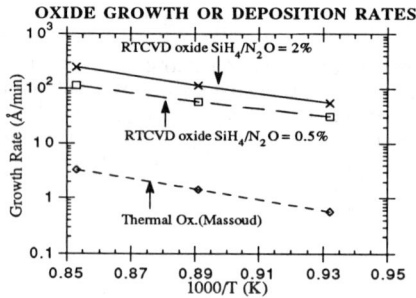 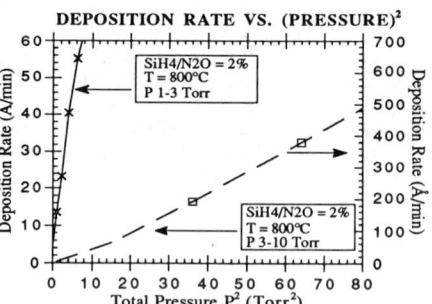

Figure 1. Deposition Rates for RTCVD Oxides vs. Temperature

Figure 2. Deposition Rates of RTCVD Oxides as a Function of Pressure

Figure 2 shows the deposition rates for deposited oxides verses pressure. The deposition rate for low pressures (1 - 3 Torr) is in the 10Å/min. to 50Å/min. range while for higher pressures (3 - 10 Torr) the deposition rate is 50Å/min. to 625Å/min. The deposition rates exhibit a linear dependence on the square of total system pressure. The square pressure dependence holds for the range of pressures studied. As shown this RTCVD process has an excellent deposition rate range for use in both very thin gate insulators and also for thick isolation insulators.

Figure 3 compares the boron concentration profile in silicon on which a 500Å thick RTCVD oxide has been deposited and one with a 500Å thermally grown oxide to the as implanted profile. As shown, the boron doping in the silicon, for the RTCVD deposited oxide, retains the same shape as the original implanted profile while the surface concentration has been reduced by an order of magnitude for the sample with the thermally grown oxide. This shows the advantage of the low thermal budget RTCVD process.

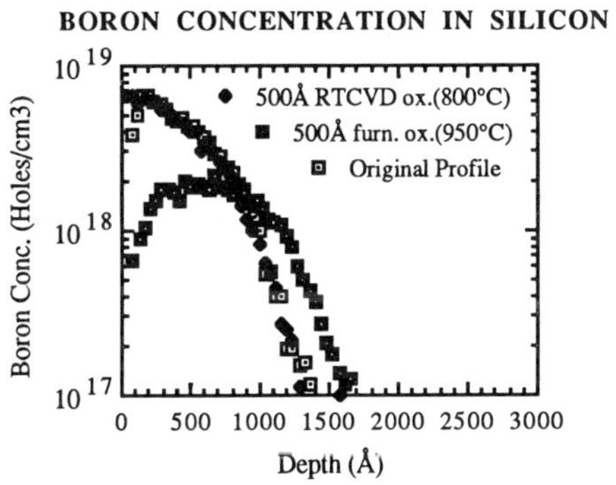

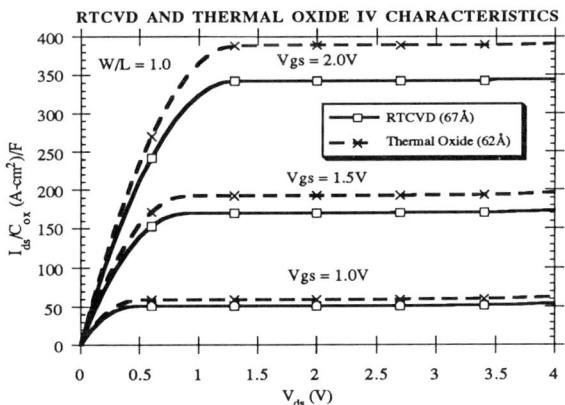

Figure 3. Boron Concentration as a Function of Thermal and Deposited Oxides

Figure 4 show the standard MOSFET Ids vs Vds curves with the vertical axis normalized to the gate capacitance per unit area. For gate drive voltage equal or below 2V the characteristic curves are comparable. As the gate drive voltage is increased, the thermal oxide has higher drive capability. However, for submicron devices with gate insulator thicknesses in the 50Å range operating voltages will likely be reduced to lower voltages.

Figure 4. MOSFET IV Characteristics of RTCVD and Thermal Oxides

Figure 5 is a plot of the field effect mobility verses the gate drive normalized to the oxide thicknesses for RTO and RTCVD oxides. The field effect mobility is derived from the transconductance ($\partial I_{ds}/\partial V_{gs}$ with constant V_{ds}). For higher gate drives the mobility falloff is slightly more for the deposited oxides. This falloff is possibly due to different substrate doping or scattering at the surface. This result is consistent with the results shown in Figure 4.

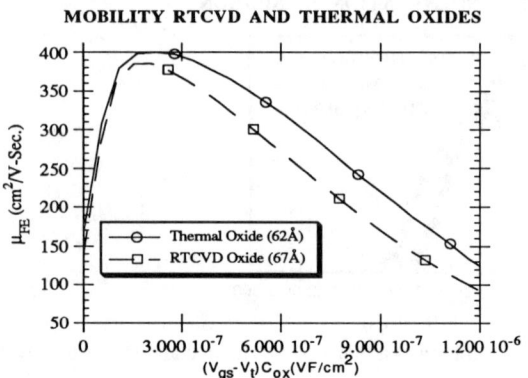

Figure 5. Field Effect Mobility vs Gate Voltage for Thermal and RTCVD Oxides

Figure 6 shows data for RTCVD oxides which were subjected to an in-situ post-deposition rapid thermal anneal in a nitrous oxide gas ambient at different annealing temperatures [4]. The three curves shown in Figure 6 are for MOSFET devices with RTCVD gate oxides. As shown post-deposition annealing in nitrous oxide at 800°C improves the mobility falloff characteristics at high gate fields. As the temperature of the rapid thermal anneal in nitrous oxide is increased, the high field mobility falloff is decreased.

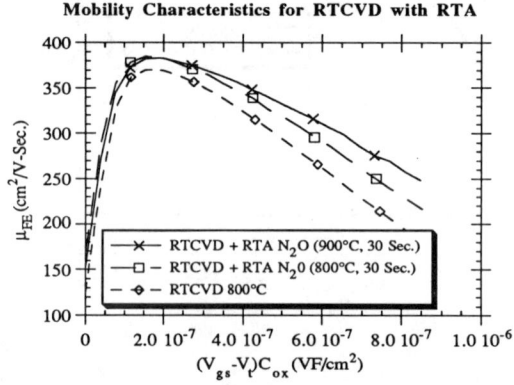

Figure 6. Field Effect Mobility for Different Post-Deposition Annealing Conditions

Figure 7 shows the effect of capacitors stressed under Fowler-Nordheim constant current densities of 10^{-4} A/cm^2 for 200 sec.(see Figure 7). Capacitors using RTCVD deposited films and standard 900°C furnace thermal oxides are compared with electron injection both from the gate and the substrate as a function of film thickness. Gate injection gives larger flatband voltage shifts as the oxide thickness is increased for both deposited and thermal oxides. The flatband shift approaches zero as the oxide thicknesses for both RTCVD and thermal oxides are reduced to 50Å.

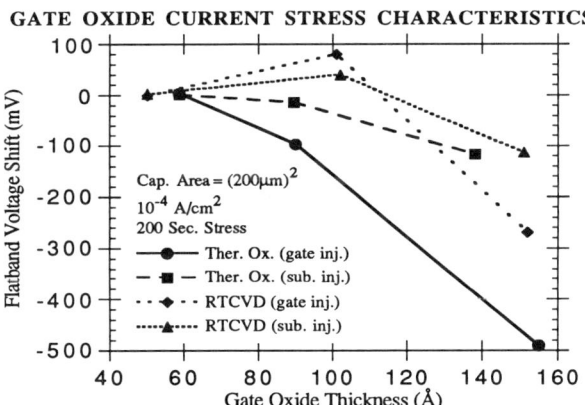

Figure 7. Flatband Voltage Shift vs Gate Oxide Thickness Under Gate Current Stress

Figure 8 shows the catastrophic breakdown field of RTCVD and RTO oxides as a function of gate oxide thickness. The breakdown fields of RTCVD deposited oxides are comparable to those of RTO in the gate oxide thickness range of 50Å. The breakdown fields of RTCVD oxides increases as the oxide thickness is reduced. This may be attributed to the reduced interface and bulk trapping of the ultra-thin RTCVD oxides (Figure 7).

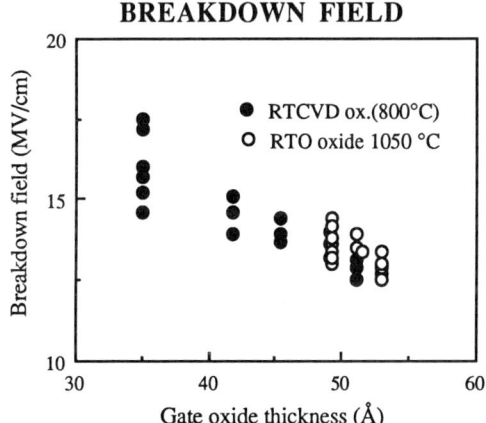

Figure 8. Breakdown Field of RTCVD and Thermal Oxides vs. Oxide Thickness

Figure 9 shows channel hot electron (CHE) effects on MOSFET devices with channel lengths of 1 μm(Vds = 5V, Vgs = 1.8V). The reduction of transconductance for both RTCVD deposited and thermal oxides is equivalent.

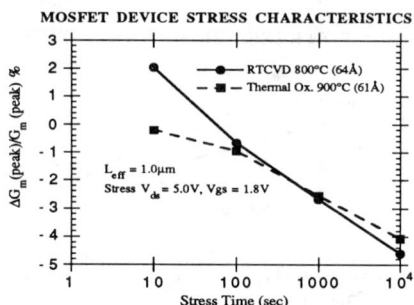

Figure 9. Peak Gm Change for RTCVD and Thermal Oxides Under Channel Hot Electron Stress

SUMMARY/CONCLUSIONS

We have found that silane/nitrous oxide RTCVD deposited oxides can be used to reduce the thermal budget by a factor of 40 or more over standard furnace thermal oxides. RTCVD deposition rates follow a linear dependence on the square of the total system pressure. These deposition rates are suitable for thin (30Å - 50Å) gate films (1-3 Torr pressure) and thick isolation type films (3-10 Torr pressure). Boron depletion in the silicon near the silicon dioxide/silicon interface is eliminated with RTCVD deposited films confirming the low thermal budget characteristics of these silane/nitrous oxide deposited films. MOSFET device IV indicate a slight reduction in current drive with the RTCVD films. This reduced current drive is consistent with a lower mobility for the RTCVD films especially at higher fields. The mobility characteristics are improved after the RTCVD films are post deposition rapid thermal annealed in nitrous oxide at temperatures ranging from 800°C to 900°C. Silane/nitrous oxide deposited films exhibit similar flatband voltage shifts for capacitor stress tests. The flatband shifts also approach zero as the insulator thickness is reduced to 50Å. Preliminary channel hot electron stress show similar peak transconductance reductions for the deposited and the thermally grown films. Breakdown fields for the RTO and RTCVD insulators are similar in the 50Å range. Breakdown fields of RTCVD films increase as the deposited oxide thickness is reduced below 50Å. Silane/nitrous oxide RTCVD deposited films show great promise for ultra-thin (30Å - 50Å) gate films required for deep submicron devices. In addition these deposited films are promising for sidewall oxides plus other potential thicker isolation films.

REFERENCES

1. R. Fair and G. Ruggles, "Thermal budget issues for deep submicron ulsi," Solid State Technology, p. 107, May 1990.
2. X. Xu, R. Kuehn, J. Wortman, and M. Ozturk, "Gate quality SiO_2 film deposited by rapid thermal low pressure chemical vapor deposition," Applied Physics Letters, Vol. 60, No.24, p. 3063 May 1992.
3..H. Massoud, J. Plummer, and E. Irene, "Thermal oxidation of silicon in dry oxygen: growth-rate enhancement in the thin regime," Journal Electrochemical Society: Solid State Science and Technology, Vol. 132, No. 11, p. 2685, Nov. 1985.
4. Z. Liu, H. Wann, P. Ko, C. Hu, Y.Cheng, "Improvement of charge trapping characteristics of N_2O-annealed and reoxidized N_2O-annealed thin oxides," IEEE Electron Device Letters, Vol. 13, No. 10, p. 519, Oct. 1992.

GROWTH KINETICS AND CORRESPONDING LUMINESCENCE CHARACTERISTICS OF AMORPHOUS C:H DEPOSITED FROM CH_4 BY DC SADDLE-FIELD PLASMA-ASSISTED CVD

Roman V. Kruzelecky, Chun Wang and Stefan Zukotynski
University of Toronto, Department of Electrical and Computer Engineering, Toronto, Ontario, Canada, M5S 1A4

ABSTRACT

Hydrogenated amorphous carbon (a-C:H) thin films were deposited onto various substrates at 200 °C by DC saddle-field glow-discharge dissociation of CH_4. In situ plasma probe mass and energy spectroscopy was employed to systematically study the effects of the discharge parameters on the formation of reactive precursors. The ion current at the substrate holder consists mainly of CH_3^+ for pressures < 75 mTorr; higher pressures and lower discharge currents encourage the formation of $C_2H_n^+$ and $C_3H_n^+$ radicals in the discharge. The growth rate was largely independent of the total pressure, increasing approximately linearly with the discharge current. Under 476 nm photo-excitation, the a-C:H films exhibited room-temperature photoluminescence in the visible, near 1.9 eV, with an intensity that was strongly dependant on the discharge parameters.

INTRODUCTION

Diamond-like hydrogenated amorphous carbon (a-C:H) films have gained technological importance due to their extreme hardness and chemical resistance [1], a wide band gap exceeding 2.0 eV [2], and a high dielectric strength [3]. Applications of a-C:H include protective coatings, dielectric layers, and more recently, electroluminescent devices [4]. The diamond-like a-C:H films are generally prepared by glow discharge decomposition of organic vapours such as CH_4 [5] and C_2H_2 [3] using DC [1] or RF [2,4] excitation. The microstructure, optical and electrical properties of a-C:H films are strongly dependent on the growth conditions [3].

In this paper, the preparation of a-C:H thin films by glow-discharge dissociation of methane (CH_4) ignited using a DC saddle-field electrode configuration [6] is discussed, focusing on the effects of the discharge parameters on the formation of reactive precursors and the resulting film growth rate, hydrogen content and opto-electronic characteristics. In the saddle-field electrode configuration, electrons can oscillate along the axis of the plasma chamber, resulting in a relatively large path length for ionizing collisions [6]. This allows discharge formation at current densities >200 $\mu A/cm^2$ over a wide range in pressures from about 20 to 150 mTorr, avoiding the tuning problems associated with RF techniques, while providing more direct control of the ion energies, ion densities and direction of ion motion.

EXPERIMENTAL PROCEDURE

The DC saddle-field deposition apparatus has been previously described in detail [6]. The saddle-field electrodes were in the form of discs constructed of stainless steel mesh, each about 15 cm in diameter, comprising a central anode sandwiched between two cathodes, spaced 15 cm apart. A cylindrical stainless steel tube, kept at ground potential, confined the glow discharge within the saddle-field cavity and defined the gas flow pattern from the gas inlet near the anode symmetrically outwards towards the two cathodes. Substrates were mounted on a heated holder, about 15 cm in diameter, positioned about 3.5 cm behind one of the cathodic screens. A second unheated substrate holder was symmetrically mounted 3.5 cm behind the second cathodic screen. It contained a crystal thickness monitor and a 6 mm orifice at the centre. A differentially-pumped Vacuum Generators SXP 300 plasma probe was located 5 mm directly behind the orifice. Due to the symmetry of the electrode configuration, measurements by the plasma probe reflect the mass and energy distributions of radicals impinging on the substrates during film growth.

The growth chamber was cryopumped to a base pressure of about 5×10^{-8} Torr. Just prior to processing, the cryopump was isolated and pumping was provided by an Alcatel molecular drag pump. CH_4 was introduced into the growth chamber through a mass flow controller. The resulting pressure, P_T, as measured by a capacitance manometer, was controlled by throttling the pumping speed. Film growth was initiated by applying a positive DC potential, V_{dc}, to the central anode, keeping the cathodic screens at ground potential, resulting in a discharge current, I_{dc}. The substrate holder was grounded via a 10 Ω resistor to enable determination of the substrate bias current, I_b. The a-C:H films were deposited onto glass slides, resistive ($\rho > 10$ Ωcm) crystalline silicon (c-Si), and indium-tin-oxide (ITO) coated glass held at 200 °C. Several sequences of samples were prepared to systematically study the effects of I_{dc} and P_T on the formation of reactive precursors and the resulting film growth.

The morphology of the a-C:H films was examined using a Leitz DMR optical microscope. An Alpha-step 200 profilometer was used to determine film thickness. The incorporation of bonded hydrogen into the films was investigated using far infrared vibrational spectroscopy. The FIR transmittance of a-C:H films on c-Si was measured at room temperature between 4000 and 500 cm^{-1} relative to an uncoated c-Si substrate using a N_2-purged Perkin-Elmer 621 double-beam spectrophotometer. The optical transmittance of a-C:H films on glass was measured between 1.5 and 3.5 eV relative to the glass substrate using a Perkin-Elmer 451 double-beam spectrophotometer. The corresponding photoluminescence (PL) was recorded between 1.2 and 2.5 eV using 476 nm photo-excitation provided by an Ar laser operating at 100 mW.

RESULTS AND DISCUSSION

Figure 1 shows the mass spectra of neutrals in the gas-phase as obtained using the plasma probe for 3 sccm of CH_4 at $P_T = 60$ mTorr with the discharge "off" ($I_{dc} = 0.0$ mA), corresponding to the cracking pattern of CH_4, and with the discharge ignited ($I_{dc} = 20$ mA). After discharge ignition, the dominant component in the gas-phase is H_2, resulting from the dissociation of the CH_4. The relative partial pressures over the mass range 12-16 amu, corresponding to CH_n (n=0,4), are similar to those observed for the cracking pattern of CH_4, suggesting that these peaks are largely due to residual CH_4. The partial pressures in the mass range 25-28 amu, attributed to the formation of C_2H_n (n=1,4), are enhanced by the glow discharge. The most prominent peaks are at masses 26 and 28, corresponding to C_2H_2 and C_2H_4, respectively. Increasing P_T, enhances the partial pressures associated with C_2H_n, indicating greater gas-phase polymerization. Increasing I_{dc}, for a given value of

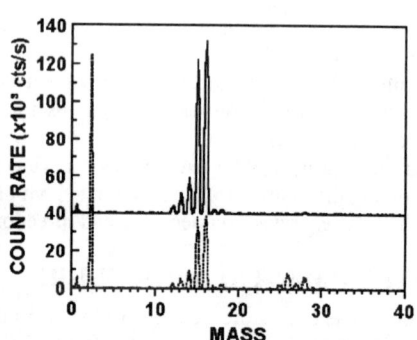

Fig. 1 Mass spectra of neutrals in the gas-phase for CH_4 with $I_{dc} = 0.0$ mA (solid line) and $I_{dc} = 20$ mA (dotted line).

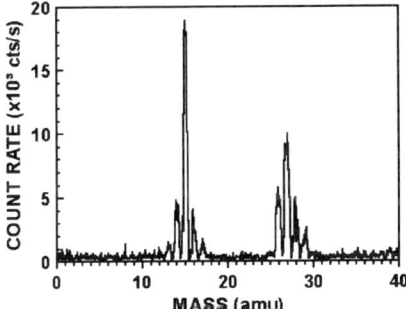 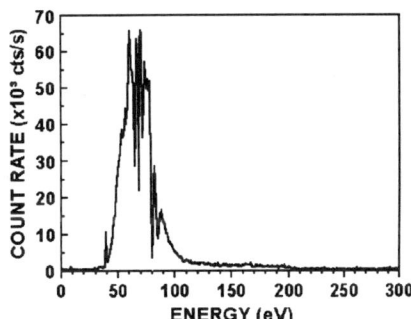

Fig. 2 (a) The mass spectrum of 70±10 eV positive ions and (b) the energy distribution of CH_3^+ for a discharge in CH_4 at P_T= 30 mTorr (F(CH_4)=5 sccm) and I_{dc}=20 mA (V_{dc}= 410 V).

gas flow, F(CH_4), reduces the partial pressures associated with CH_n and C_2H_n, reflecting greater activation of input CH_4 and subsequent deposition on the walls of the growth chamber.

The composition and energy distribution of positive ions impinging onto the substrate holder during discharge formation in CH_4 was examined using the plasma probe in the external ion mode. Figure 2(a) shows the mass spectrum of 70±10 eV positive ions for P_T= 30 mTorr and I_{dc}=20 mA. The two main components of the ion current at the substrate holder are positive ions at mass 15 and 27, associated with CH_3^+ and $C_2H_3^+$, respectively. Although P_T is largely due to H_2 during discharge formation in CH_4, the plasma probe results indicate that relatively little of the background H_2 is activated. Figure 2(b) shows the energy distribution of positive ions at mass 15. The distribution is centred about 70 eV with a tail that extends above 200 eV. The peak in the energy distribution of positive ions at other mass numbers was also at about 70 eV.

Figure 3(a) shows the mass spectrum of 40±10 eV positive ions impinging onto the substrate holder for a discharge in CH_4 at P_T=130 mTorr and I_{dc}=20 mA. The plasma probe measurements indicate that operating the discharge at higher pressures enhances the relative

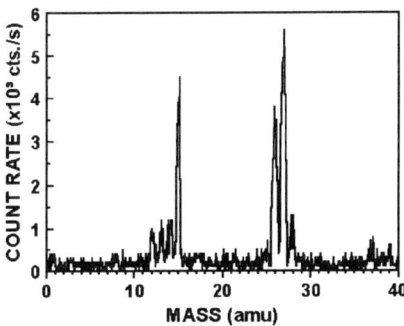 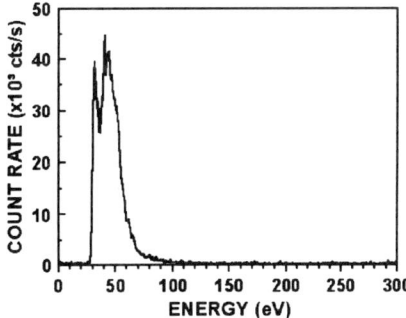

Fig. 3 (a) The mass spectrum of 40±10 eV positive ions and (b) the energy distribution of CH_3^+ for a discharge in CH_4 at P_T= 130 mTorr (F(CH_4)=5 sccm) and I_{dc}=20 mA (V_{dc}= 400 V).

contribution to the ion current by heavier hydrocarbon radicals at mass numbers 27 and 26, associated with $C_2H_3^+$ and $C_2H_2^+$, respectively. Moreover, increasing P_T shifts the peak in the energy distribution of positive ions from about 70 eV at 30 mTorr (see Fig. 2(b)), downwards in energy to about 40 eV at 130 mTorr (see Fig. 3(b)). This is accompanied by a narrower high-energy tail in the distribution that is likely due to a shorter mean free path at higher P_T.

The growth of a-C:H exhibited a dependence on the nature of the substrate and on the substrate bias conditions. For a-C:H depositions with the substrate holder at virtual ground potential, the growth rate on glass was about 1/3 of the growth rate on c-Si and ITO. The morphology of a-C:H films on glass ($\approx$1500 Å thick) was relatively smooth and free of pinholes for $P_T <$ 100 mTorr. Higher values of P_T encouraged the formation of greyish speckles within the largely transparent a-C:H films that were most likely graphitic in structure. Thicker a-C:H films on c-Si and ITO ($\approx$4000 Å thick) exhibited some cracking due to stress that was enhanced for films prepared at higher I_{dc} and P_T. This will be discussed in a subsequent paper.

Figure 4(a) shows the variation in the current I(x) impinging onto the substrate holder due to positive ions at mass numbers x=27 ($\blacklozenge$), associated with $C_2H_3^+$, and x=39 ($\square$), associated with $C_3H_3^+$, relative to I(15) (CH_3^+) as a function of I_{dc} for F(CH_4)= 3 sccm and P_T=60 mTorr. The corresponding variation in the growth rate ($\blacksquare$), as measured for a-C:H on c-Si, is also shown. Increasing I_{dc} by increasing V_{dc}, for fixed P_T, reduces the relative fraction of $C_2H_n^+$ and $C_3H_n^+$ contributing to the ion current. This may be due to a higher dissociation rate of the larger hydrocarbon radicals in the plasma at higher discharge current densities. The growth rate increases approximately linearly with I_{dc}, saturating at higher currents due to insufficient CH_4 flow.

Figure 4(b) shows the effect of P_T on the formation of reactive precursors, as measured by I(x)/I(15), and on the resulting growth rate for I_{dc}= 40 mA and F(CH_4)= 5 sccm. I(27)/I(15) and I(39)/I(15) tend to increase with P_T, reflecting an increase in the relative formation rate of $C_2H_3^+$ and $C_3H_3^+$, respectively. Increasing P_T reduces the mean free path of radicals, encouraging additional gas-phase reactions between CH_3^+ and neutrals that reduce the relative fraction of CH_3^+ contributing to I_b. The growth rate did not vary significantly with P_T despite the changes in the composition of the ion current at the substrate holder.

The FIR transmittance spectra of 4000 Å a-C:H/c-Si prepared at 200 °C exhibited two significant peaks; a broad peak near 2900 cm^{-1} associated with C-H bond-stretching vibrational

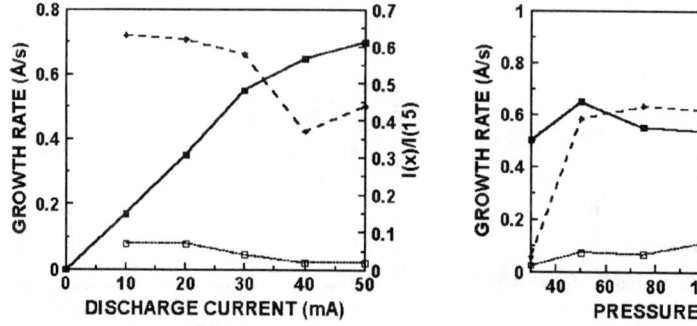

Fig. 4 Effect of (a) I_{dc} and (b) P_T on the relative positive ion current at the substrate holder (($\blacklozenge$)-I(27)/I(15), ($\square$)- I(39)/I(15)) and the resulting growth rate ($\blacksquare$).

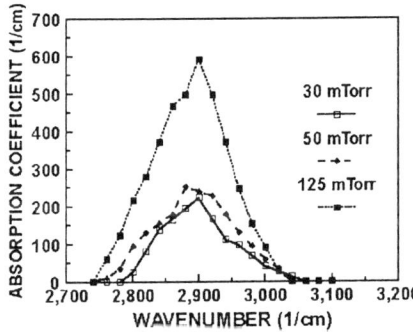 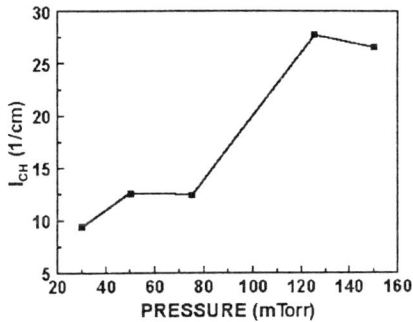

Fig. 5 Variation of (a) α(ω) and (b) I_{CH} with P_T (F(CH$_4$)=5 sccm, I_{dc}=40 mA, T_s = 200°C).

modes [7] and a weaker peak near 650 cm^{-1} that is attributed to C-H wagging modes [7]. Figure 5(a) shows the effect of P_T on the optical absorption coefficient α(ω) associated with the C-H stretching-mode absorption band near 2900 cm^{-1}. A number of C-H$_n$ bond-stretching modes contribute to this absorption band [7], including C-H$_3$ near 2960 cm^{-1} (asymmetric), C-H$_2$ near 2910 cm^{-1} (asymmetric) and C-H near 2800 cm^{-1}. The strong optical absorption peak near 2900 cm^{-1} suggests that hydrogen was incorporated mainly into C-H$_2$ sites. The total bonded hydrogen content is proportional to the integrated absorption,

$$I_{CH} = \int \frac{\alpha(\omega)}{\omega} d\omega \qquad (1)$$

taken over the C-H stretching-mode absorption band. As shown in Fig. 5(b), I_{CH}, and hence, the total bonded hydrogen content, increases with P_T.

Figure 6(a) shows the effect of P_T on the optical transmittance characteristics of 1500 Å thick a-C:H films deposited onto glass slides at 200 °C, keeping F(CH$_4$) = 5 sccm and I_{dc}= 40 mA. Figure 6(b) shows the corresponding PL spectra as measured at 300 K using 476 nm photo-excitation. Films prepared at lower P_T (30 mTorr) exhibit significant optical absorption below 2.9 eV that extends to about 2 eV. Indeed, the films have a slight yellowish tinge. This suggests the formation of fairly broad tail-state distributions at the conduction and valence band edges. Samples prepared at lower pressures (30-50 mTorr) exhibited relatively strong PL with the peak in the intensity at about 1.9 eV. The a-C:H films prepared at intermediate pressures (75 mTorr) were largely transparent below 2.9 eV. The corresponding reduced PL intensity is likely due to weak absorption of the 2.6 eV photo-excitation by the a-C:H films. The preparation of a-C:H films at relatively high P_T (>120 mTorr) again results in appreciable optical absorption below 2.9 eV. Despite the relatively strong absorption of the laser light, a-C:H films prepared at high P_T exhibit weak PL intensity. Moreover, the PL peak is shifted to about 1.75 eV.

These results can be explained in terms of the plasma probe spectra. At lower P_T, positive ions such as CH$_3^+$ have a pronounced high-energy tail in their energy distribution that extends above 200 eV (see Fig. 2(b)). Damage due to bombardment by energetic ions during film growth at low P_T may result in defect structures that contribute to the broad tail-state distributions indicated by the transmittance measurements. The extent of the high-energy tail as well as the average energy of the ions is reduced as P_T is increased, resulting in less defective film growth. However, higher P_T also increases gas-phase polymerization, resulting in the formation of larger radicals such as C$_3$H$_n^+$ ions (see Fig. 4(b)). The corresponding films have a pronounced two-phase

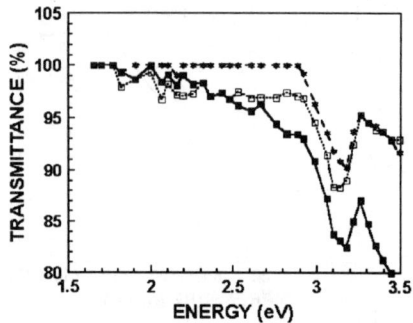 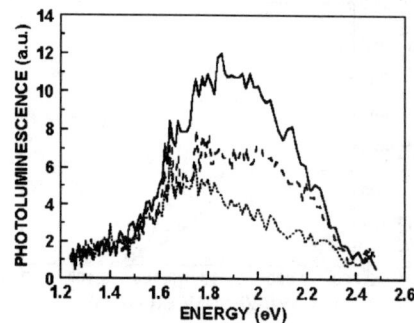

Fig. 6 (a) Optical transmittance and (b) 300 K photoluminescence characteristics for different P_T. Solid lines correspond to 30, broken lines to 75 and dotted lines to 150 mTorr respectively.

structure consisting of diamond-like, four-fold coordinated a-C:H surrounding inclusions of greyish, three-fold coordinated graphitic a-C:H. This is associated with greater incorporation of bonded hydrogen and additional gap states in the a-C:H that degrade the PL intensity.

CONCLUSIONS

The morphology and optoelectronic characteristics of the a-C:H films prepared by glow-discharge excitation of CH_4 using a DC saddle-field cavity are correlated with the discharge parameters and their influence on the formation and energy distribution of reactive precursors. Transparent a-C:H films with relatively good PL characteristics at 300 K were obtained at intermediate discharge pressures of about 75 mTorr. Film preparation at lower pressures results in broad tail-state distributions due to damage arising from bombardment by energetic ions, predominantly CH_3^+. Moreover, relatively high P_T (>100 mTorr) encourages additional gas-phase interactions to form heavier hydrocarbons (i.e. $C_3H_3^+$) that enhance the growth of graphitic a-C:H.

ACKNOWLEDGEMENTS

The authors wish to acknowledge the financial support extended by the Natural Sciences and Engineering Research Council of Canada, The University Research Incentive Fund of Ontario and Ontario Hydro.

REFERENCES

1. D.S. Whitmell and R. Williamson, Thin Solid Films, **35**, 255 (1976).
2. D.A. Anderson, Phil. Mag., **35**, 17 (1977).
3. B. Meyerson and F.W. Smith, J. Non-Cryst. Solids, **35&36**, 435 (1980).
4. Y. Hamakawa, T. Toyama and H. Okamoto, J. Non-Cryst. Solids, **115**, 180 (1989).
5. M. Shimozuma, G. Tochitani, H. Ohno, H. Tagashira and J. Nakahara, J. Appl. Phys., **66**, 447 (1989).
6. R.V. Kruzelecky and S. Zukotynski, "DC Saddle-Field Plasma-Enhanced Vapour Deposition" in Plasma Processing of Materials, editors J.J. Pouch and S. Alterovitz (Trans Tech Publications, Ltd., Aedermannsdorf, Switzerland) in press.
7. N.B. Colthup, L.H. Daly and S.E. Wiberly, Introduction to Infrared and Raman Spectroscopy, 2nd ed. (Academic Press, New York, 1975) pp. 190-340.

MOMBE GROWTH OF YBCO SUPERCONDUCTING FILMS WITH *IN-SITU* MONITORING OF RHEED OSCILLATIONS

K. Endo, F. Hosseini Teherani, S. Yoshida and K. Kajimura.
Electrotechnical Laboratory, Tsukuba, Ibaraki 305, Japan.
Y. Moriyasu.
Asahi Chemical Ind., Fuji, Shizuoka 416, Japan.

ABSTRACT

We have succeeded in in-situ growth of YBCO superconducting films by developing a new MOMBE technique which employs in-situ monitoring of RHEED intensity. The X-ray diffraction pattern showed peaks corresponding to those calculated from the Laue function which indicates the high quality of films grown using MOMBE. AFM and RHEED oscillations show 2-dimensional unit cell by unit cell growth.

INTRODUCTION

Among the various deposition techniques used for the *in situ* deposition of High Tc materials, such as sputtering or laser ablation, chemical vapor deposition (CVD) has become increasingly important. Molecular beam epitaxy (MBE), however, is superior to CVD in control of composition and regulation of film thickness, allowing monolayer resolution through *in situ* monitoring and shutter control. Problems such as oxidation of sources and requirements for a higher evaporating temperature, however, limit the use of MBE systems. In order to overcome these problems we have developed a metalorganic molecular beam epitaxy (MOMBE) system which combines the advantages of both CVD and MBE. Through the use of metal chelates as source materials we can obtain a higher oxygen pressure than with metal sources, because the volatilization temperatures of the former are much lower than those of metals. The new technique also allows us to characterize the growth *in situ* through reflection high energy electron diffraction (RHEED). Furthermore the use of metal chelate sources offers the possibility of atomic layer epitaxy with a self-limiting growth mechanism, which is one of the more promising techniques for fabricating future electronic devices.

In our previous studies, we initially used $Ba(PPM)_2$ (PPM: pentafloro propanoylpivaloyl methane) as a source material for Ba, where PPM ligand contains fluorine[1]. Since the as-grown films were amorphous, insulating and contained fluorine, a post-annealing step was necessary in order to obtain a superconducting $YBa_2Cu_3O_x$ (YBCO) thin film. The presence of fluorine inclusions, however, was reported to be responsible for a substantial decrease in superconducting phase[2]. We then demonstrated that we could obtain superconducting YBCO thin films (T_C=45K) on MgO(100) substrates, using high quality, fluorine-free, β-diketonate chelate of Ba: $Ba(DPM)_2$ (DPM: dipivaloyl methane) and ex-situ post-annealing[3]. In this paper, we report on the successful *in situ* growth of superconducting YBCO thin films (T_C=85 K) using liquid ozone.

EXPERIMENTAL

Figure 1 is a schematic drawing of our MOMBE apparatus. The main chamber was evacuated to less than 10^{-8} Torr using a turbo molecular pump with a liquid nitrogen shroud. The pressure in the chamber during growth varied between 5×10^{-5} and 6×10^{-5} Torr. The $Y(DPM)_3$, $Ba(DPM)_2$ and $Cu(DPM)_2$ source materials were loaded into PBN crucibles and heated to 77~78, 143~150 and 60~64 °C, respectively. MgO(100), Nd:YAlO3(100)

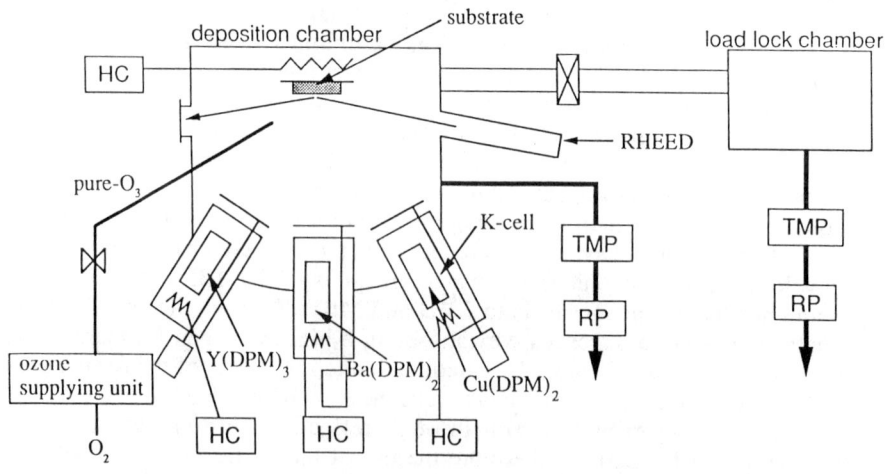

Fig.1: Schematic drawing of MOMBE apparatus. TMP: Turbo Molecular Pump, RP: Rotary Pump, HC: Heater Controller, VLV: Variable Leak Valve.

and SrTiO$_3$(100) substrates were fixed to an Inconel block using silver paste. The substrate temperature was varied from 536 to 615 °C, using a graphite heater coated with silicon carbide. The substrate temperature was measured using a pyrometer. Ozone, initially liquid in the supplying unit, was introduced during growth through a nozzle near the substrates with a flow rate of 0.24 SCCM. The distance between substrate and nozzle was about 3 cm. Under these conditions, the growth rate varied between 70 and 100 Å/hr.

The deposition was monitored *in situ* using RHEED with an accelerating voltage of 25 kV. Following the deposition, the compositions and the structures of the films were examined by inductively coupled plasma spectroscopy (ICP) and X-ray diffraction (XRD), respectively. The surface morphology was observed using an atomic force microscope (AFM). Finally, electrical characterizations were carried out using a standard four probe DC method, where Cu wires were attached with Ag paste to Au electrodes deposited onto the films.

FILM CHARACTERIZATION

Figure 2 shows the curve of resistive transition versus temperature for YBCO thin films deposited on SrTiO$_3$(100) under the conditions described in Table 1. This curve shows a sharp superconducting transition and exhibits one of the highest zero resistance temperatures $T_C(0) = 84$ K. Longer deposition time, for a given substrate temperature and ozone pressure, is observed to cause an increase in T_C. This table also shows the importance of the ozone pressure in controlling the T_C value.

Table 1 Growth conditions for YBCO on SrTiO$_3$.

Sample	Deposition Conditions		Thickness	Tc(zero)
	Substrate Temperature (°C)	Ozone Pressure (Torr)	(Å)	(K)
1	607	5.0×10^{-5}	320	85
2	615	5.2×10^{-5}	80	74
3	605	5.0×10^{-6}	210	76
4	548	5.0×10^{-5}	150	78

Figure 3 shows the RHEED intensity oscillation during YBCO thin film growth on a SrTiO$_3$(100) substrate. The electron beam was directed parallel to SrTiO$_3$[100]. During the deposition, clear changes in RHEED intensity and a streaky RHEED pattern were observed. This result shows that two dimensional layer-by-layer growth is continuous throughout film

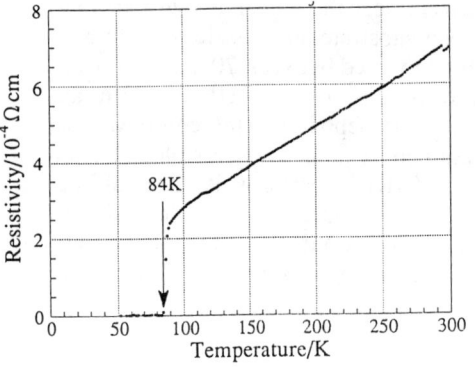

Fig.2: Resistance versus temperature for a 150 Å thick YBCO film deposited on SrTiO$_3$(100).

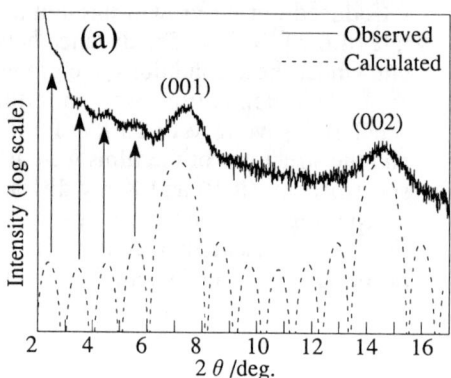

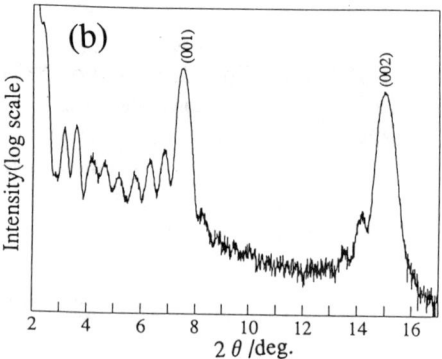

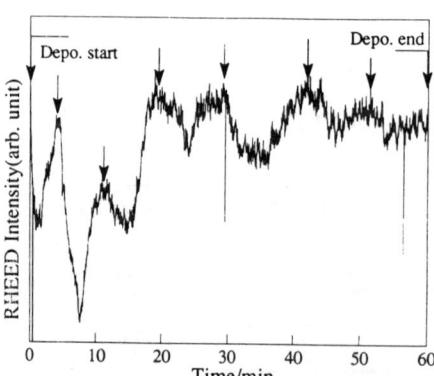

Fig.3: The change in RHEED intensity through the formation of YBCO on SrTiO$_3$(100). The signal was measured for a specular diffraction spot. Arrows indicate each period of oscillation.

Fig.4: X-ray diffraction pattern measured around the YBCO(001) peak for (a) the same sample as in Fig.3, showing a pattern which corresponds to 7 YBCO atomic layers. The continuous line corresponds to the experimental measurement and the broken line corresponds to the theoretically calculated Laue function. (b) the same sample as in Fig. 2 showing a pattern which corresponds to 14 YBCO atomic layers.

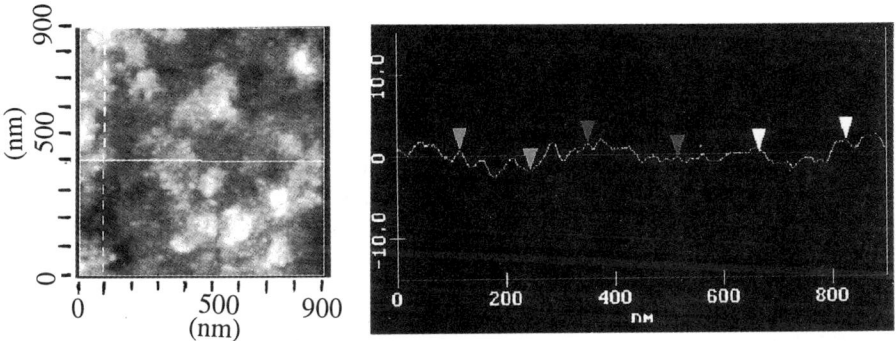

Fig.5: AFM image of YBCO on SrTiO₃

growth. The deposition was stopped after seven oscillations. The development of surface morphology is different from that observed with MgO(100) or Nd:YAlO$_3$(100) substrates. Indeed, for YBCO thin films grown on MgO(100), after 5 minutes the streaky RHEED pattern, which characterizes the substrate surface, gradually becomes diffuse, revealing surface roughness. However, the streaky pattern can be observed again as the film thickness increases further. The results obtained with Nd:YAlO$_3$(100) substrates are better than those for MgO(100), but still not as good as for SrTiO$_3$(100).

The film thickness was deduced from the Laue function. Figure 4(a). shows the X-ray diffraction pattern measured around the YBCO(001) peak. The observed oscillatory peaks are attributed to a Laue function calculated for 7 YBCO atomic layers (about 80Å). The reliability of this method was confirmed by measuring a YBCO thin film with 14 layers (Fig. 4(b)). It can clearly be seen that there is a critical interdependence of thickness and Tc. Indeed, the measured Tc for the 7 atomic layer thick YBCO thin film was 74 K, and that for the 14 atomic layer thick film was 84 K.

Finally, AFM observations for a 200Å thick film (Fig. 5) showed that the surfaces of these films have steps with a height of either 1 or 2 monolayers. Moreover, one can notice the absence of screw dislocations. These 2 points are consistent with RHEED measurements and confirm the 2-dimensional growth mechanism.

CONCLUSIONS

In this paper, we successfully used MOMBE to grow YBCO thin films. Through use of RHEED intensity oscillations we could control the 2 dimensional growth. In this way, YBCO thin films could be grown with a roughness of the order of 1 to 2 atomic layers without the use of miscut substrates. Finally, our samples were screw-dislocation free, which represents the first step towards realization of viable devices.

REFERENCES

[1] K. Endo, Y. Ikedo, S. Hayashida, J. Ishiai, N. Nakatsuka, S. Misawa and S. Yoshida, in T. Ishiguro and K. Kajimura (eds.), "Advances in Superconductivity II" Springer, Tokyo (1990) p. 805.
[2] Y. Yokoyama, T. Okumura, R. Sugise, T. Matsubara, N. Koshizuka, M. Hirabayashi, N. Terada and H. Ihara, J. Ceram. Soc. Jpn. Int., 96 (1988) p. 481.
[3] K. Endo, S. Saya, S. Misawa and S. Yoshida, Thin Solid Films, 206 (1991) p.143.

Author Index

Abernathy, C.R., 399
Akazawa, Housei, 525
Allendorf, Mark D., 105

Bachmann, K.J., 495
Bartram, Michael E., 57
Bent, Brian E., 413
Blaisten-Barojas, Estela, 75
Bohling, David A., 169
Bonnell, David W., 305
Breiland, William G., 3, 51
Brewer, P.D., 245
Brough, Lawrence F., 213
Brown, W.L., 323
Buss, R.J., 51

Cale, Timothy S., 471
Carrier, Michael J., 75
Chakrabarti, U.K., 399
Chang, Che-Chen, 457
Chang, Y.J., 251
Charatan, Robert M., 323, 329
Chau, T.T., 373
Chen, Wei, 317
Cheng, C.C., 425
Chu, T.K., 501
Colacot, Thomas J., 213
Coltrin, Michael E., 3
Conner, W.T., 123
Constantine, C., 399
Crocco, Paul D., 335

Dai, Zi-Guo, 413
Danek, M., 231
Dhote, A.M., 299
Dickinson, J.T., 359
Dietz, N., 495
Dillon, Anne C., 37
Donnelly, V.M., 425
Dons, E., 323

Eaglesham, David J., 323, 329
Endo, K., 543
Eneva, J., 265
Ernie, Douglas W., 141
Evans, John F., 273

Filipov, D., 93
Fisher, E.R., 51

Gang, Liu, 213
George, Steven M., 25, 37
Gladfelter, Wayne L., 273
Gleason, K.K., 123
Goedken, Virgil L., 213
Gordon, D.C., 231
Graff, I.B., 129
Graham, Sandra W., 433
Greenlief, C. Michael, 489

Greenwald, Anton C., 207
Greuel, Peter G., 141
Gross, Mihal E., 323, 329
Guinn, K.V., 425

Ha, Jeong Sook, 183
Habermehl, Scott, 341, 495
Haefke, H., 265
Harris, Gari, 463, 519
Harris, Stephen J., 117
Hastie, John W., 305
Hauser, J.R., 531
Herman, I.P., 425
Hiramatsu, Masato, 385
Ho, Pauline, 3, 51
Hobson, W.S., 399
Holcombe, D.J., 531
Hou, Hui-Qi, 413
Huang, Chen, 99
Huang, J.-W., 189
Huber, C.A., 501
Hudson, John B., 335
Huh, J-S., 231

Ingle, N.K., 177
Interrante, Leonard V., 335
Ishida, Arichika, 385
Ivanov, B., 93
Izawa, Michiyoshi, 419
Izumi, Hirohiko, 45

Janssen, G.C.A.M., 483
Jasinski, Joseph M., 11
Jensen, Klavs F., 169, 201, 231
Jeon, Yoo-Chan, 81
Jiang, W., 359
Jönsson, J., 155
Joo, Seung-Ki, 81
Just, Oliver, 219

Kajimura, K., 543
Kao, K.C., 373
Kashiura, H., 365, 391
Katz, A., 399
Kawakami, H., 135
Kawakyu, Yoshito, 385
Kawasaki, Atsushi, 45
Keeling, Lori A., 489
Kelliher, J.T., 495
Kitamura, K., 439
Kitova, S., 265
Klages, C.-P., 323
Kleijn, C.R., 483
Kosar, W.P., 231
Kruzelecky, Roman V., 537
Kuech, T.F., 189, 201
Kuehn, R.T., 531
Kuijlaars, K.J., 483
Kwok, H.S., 353

Lam, P.M., 373
Langford, S.C., 359
Lay, Uwe W., 207
Lee, C. Archie, 463
Lee, El-Hang, 183
Lee, Seok-Woon, 81
Leung, Denise L., 283
Leusink, G.J., 483
Lipkovich, Matthew A., 213
Liu, Wen-Shryang, 239
Loboda, M.J., 445
Lu, Zhong, 341
Lucovsky, Gerry, 341, 495

Ma, Yi, 341
Maher, Dennis M., 463, 519, 531
Mao, M.Y., 147
Martorell, J., 251
Matsuyama, H., 135
Melius, Carl F., 105
Mendicino, M.A., 63
Mikami, Naohiko, 451
Misra, V., 531
Miyokawa, T., 87, 365
Mochiji, Kozo, 451
Moffat, Harry K., 57
Moriyasu, Y., 543
Mountziaris, T.J., 177
Mucha, J.A., 31
Murahara, M., 87, 257, 365, 391, 439

Nagase, Masao, 525
Nayak, S., 189
Nikulski, R., 323
Norton, M.G., 359

Ochiai, Isao, 451
Ogale, S.B., 299
Ogi, Katsumi, 293
Ohtsuka, Nobuyuki, 225
Okada, Lynne A., 25, 37
Okoshi, M., 87, 365, 391
Oosterlaken, T.G.M., 483
Osterheld, Thomas H., 105
Ott, Andrew W., 25
Öztürk, Mehmet C., 463, 519

Panov, A., 265
Papasouliotis, George D., 111
Park, Seong-Ju, 183
Patterson, Brian M., 347
Paul, Albert J., 305
Pearton, S.J., 399
Ploska, K., 155
Popov, C., 93
Pugliese, Jr., R.A., 129

Radelaar, S., 483
Raupp, Gregory B., 239, 471
Redwing, J.M., 189, 201
Rees, Jr., William S., 207, 213, 219
Reinhardt, F., 155
Ren, F., 399

Richter, W., 155
Ritter, G., 507
Ro, Jeong-Rae, 183
Roberts, Jeffrey T., 141, 317
Rumberg, J., 155

Saitou, Norimichi, 293
Salim, Sateria, 169
Sanganeria, Mahesh K., 463, 519
Santiago, F., 501
Sasaki, Masahiro, 513
Sato, N., 135
Satou, Masamitu, 293
Sawin, H.H., 123
Schenck, Peter K., 305
Schlote, J., 507
Scott, T.W., 251
Seebauer, E.G., 63
Seifferly, J.A., 445
Shanov, V., 93
Shieh, Jaw-Chang, 457
Shin, J.-J., 359
Sim, Jae-Ki, 183
Simka, H., 201
Simmonds, Michael G., 273
Smith, F.T.J., 177
Sneh, Ofer, 25, 37
Sosa, Carlos, 289
Sotirchos, Stratis V., 111
Southwell, R.P., 63
Späth, M., 245
Speckman, Donna M., 283
Steinbrüchel, Christoph, 433
Stuke, M., 245
Stumborg, M., 501
Su, Chaochin, 413

Takeuchi, Hideki, 45
Takiguchi, Takaharu, 419
Tan, S.S., 147
Tebakari, Masayuki, 293
Teherani, F. Hosseini, 543
Theodoropoulos, C., 177
Thomas III, J.H., 445
Tillack, B., 507
Toyoda, K., 87
Tsang, Wing, 19

Uchida, Hiroto, 293
Ueda, Osamu, 225
Uesugi, Katsuhiro, 419
Utsumi, Yuichi, 525

Van Den Akker, H.E.A., 483
Vernon, Matthew F., 413
Violette, Katherine E., 463, 519

Wang, Chun, 537
Wang, Q.-F., 531
Wang, W.Y., 147
Wang, Yi, 335
Washington, J., 31
Weber, A., 323

Weiller, Bruce H., 379
Weinberg, W. Henry, 69, 99
Weiner, Anita M., 117
Wendt, Jerry P., 283
Westmoreland, P.R., 129
Wexler, R.M., 177
Widdra, Wolf, 69, 99
Wise, Michael L., 25, 37
Witanachchi, Sarath, 347
Wolden, C.A., 123
Wolf, Paul J., 347
Wortman, J.J., 531

Xi, Ming, 169
Xi, Ming, 413
Xu, X., 531

Yamamoto, Seiji, 451

Yamane, K., 257
Yang, Jung-Yen, 457
Yao, Takafumi, 419
Yoon, Hyun J., 141
Yoshida, S., 543
Yoshida, Seikoh, 513
Yoshimura, Masamichi, 419

Zachariah, Michael R., 19, 75
Zau, G., 123
Zazzera, Larry A., 273
Zhang, X.K., 147
Zheng, J.P., 353
Zorn, M., 155
Zukotynski, Stefan, 537

Subject Index

ab initio
   calculations, 489
   molecular orbital, 19
ablation, 245
acetylene, 69
adsorbates, 251
adsorption, 69, 317
   of ethylene on Si, 99
AFM, 543
Ag_2S films, 265
Al deposition, 273, 283, 289, 419
AlInAs, 399
alkane, 117
AlN, 335
aluminum trichloride ($AlCl_3$), 419
amorphous C:H, 537
angular flux distributions, 245
ArF excimer laser, 87, 365, 439
argon, 81
arsenic-containing electronic materials, 213
arsine, 177, 183, 213
a-SiC:H, 445
atomic
   hydrogen, 123
   layer epitaxy, 25, 225
Au wiring, 293

BaF_2, 501
ballistic transport and reaction model, 471
$BaSi_2$, 501
$BaTiO_3$, 305
$B(CH_3)_3$, 365
β-hydride elimination, 37
bias-enhanced microwave plasma chemical vapor deposition, 147
binary ABAB... sequence, 25
$B(OH)_3$, 365
boranol, 57
byproduct-inhibited deposition, 471

cadmium sulfide, 251
carbon incorporation, 231
CdTe, 239, 501
CdTe(100), 245
chemical
   beam epitaxy (CBE), 183, 495
   topography, 425
   vapor deposition (CVD), 37, 57, 63, 87, 111, 177, 273, 283, 323, 329, 379, 489
     - CVD
        models, 507
        reactor, 483
chlorine etching, 445
$CH_4/O_2/H_2$ gas mixtures, 135
Cl_2, 413, 425
collision-induced structural damage, 457
controlled SiO_2 film deposition, 25
COOH radicals, 391
copper bromide vapor laser (CBVL), 93

coverage, 507
crystallinity, 81
cyclopentadienyl cycloheptatrienyl titanium, 329

damage-less patterning, 513
DC saddle-field plasma-assisted CVD, 537
decomposition
   of ethylene on Si, 99
   reactions, 169
density
   functional methods, 289
   of state, 373
deposition rate, 373, 385, 531
desorption, 419, 513
   of ethylene from Si, 99
destruction probability, 117
diallylselenium, 231
diamond(-)
   deposition, 135
   film, 141, 147
   like carbon, 129
   surface chemistry, 123
dielectric thin films, 305, 385
diethyl aluminum ethoxide, 189
diethyldiethoxysilane, 37
diethylgermane, 489
dimethyl aluminum methoxide, 189
diemethylethylamine alane, 273
dimethylzinc triethylamine adduct, 231
direct reactive ion etching, 433
disilane, 519
distribution of film quality, 45
DMAu(hfap), 293
dopant(s), 57, 201
   incorporation, 189
dry etching, 399, 439

ECR PECVD, 81
ECR plasma, 323
electron
   cyclotron resonance, 399
   energy, 81
epitaxy, 201, 495
epoxy-like layer, 365
erbium doping, 207
etch yield, 433
etching, 413, 425, 439
excimer laser, 245, 293, 365

fiber, 399
film
   quality, 385
   stoichiometry, 341
flow-tube reactor, 379
fluorocarbon resin, 365
fluororesin, 87

GaAs, 177, 201, 213, 413
GaAs-to-$Al_xGa_{1-x}As$ interface, 189

gasdynamics, 305
gas-phase
  chemistry, 189
  pyrolysis, 231
  reactions, 169, 177
gate quality oxides, 531
gated diodes, 463
GeO_2, 347
gold, 273, 293
growth
  kinetics, 537
  mechanisms, 359
  oscillations, 155

H atoms, 117
HCl, 413
hermetic seals, 445
heteroepitaxy, 501
heterogeneous
  deposition, 471
  reaction kinetics, 239
high(-)
  resolution photoregistering materials, 265
  temperature flow reactor, 105
homogeneous precursor-mediated, 471
HRTEM, 495
hydrogen, 69, 81
  pairing, 69
  passivation, 463
hydrophilic property, 391

InAs, 225
$(InAs)_1(GaAs)_1$ short period superlattice, 225
indium tin oxide films, 353
infrared spectroscopy, 31, 169, 273
InGaAs, 399
InP, 399
in situ
  cleaning, 463
  growth, 543
  sources, 213
  techniques, 305
interconnects, 299
interfacial compounds, 501
ion-enhanced etching, 433
IR region, 265

keV particle bombardment, 457
kinetic(s), 3, 111, 177, 379
  analysis, 45
  model, 177
KrF excimer laser, 391, 451

laser(-), 347, 399
  ablation, 299
  chemical vapor deposition, 257, 293
  induced
    chemical vapor deposition (LCVD), 93
    thermal desorption (LITD), 425
low(-)
  pressure
    chemical vapor deposition (LPCVD), 471
    MOCVD, 177

resistance, 299
temperature
  deposition, 299
  PECVD, 445
luminescence, 537

mass
  spectrometry, 141
  spectrum, 45
metal
  alkyl decomposition, 239
  organic vapor phase epitaxy (MOVPE), 189
metalorganic chemical vapor deposition
  (MOCVD), 177, 239, 293, 323
methyl radical, 123
methyltrichlorosilane, 105, 111
microcrystalline silicon films, 81
microscope (STM), 419
microwave plasma, 373
mid-gap interface trap density, 531
modeling, 111
models, 3, 177
molecular(-)
  beam
    epitaxy (MBE), 525
    mass spectrometry, 129, 335
    scattering, 413
  dynamics, 75, 457
MOMBE, 513, 543
monoethylarsine, 183
morphology, 283
multi-photon ionization, 451

nanosecond UV laser ablation, 245
new RTCVD oxide process, 531
NH_3, 379
$NH(X^3\Sigma^-)$, 51
nucleation, 19, 257, 283, 519

$OH(X^2\pi)$, 51
OMVPE, 169, 207, 219
optical
  emission, 305
  transmission, 353
optoelectronic devices, 207
organoaluminum, 335
organometallic chemical vapor deposition, 231
organosilanes, 105
oxide fixed charge, 373
oxygen
  doping, 189
  effect on diamond deposition, 135
ozone, 31, 379

partial pressures, 483
patterning, 513
PbTe, 501
PECVD, 129, 373
phase diagram, 135
photoelectron spectroscopy, 489
photomodified surface, 391
photo-oxidized GaAs mask, 513
photoreduction, 251

554

photoregistering, 265
photo-stimulated evaporation, 525
picosecond, 251
plasma, 425
    assisted, 329
    emission spectroscopy, 135
    enhanced CVD, 445
    etching, 445
plumes, 305
polymers, 433
polyphenylenesulfide (PPS), 391
polytetrafluoroethylene, 359
precursors, 335
p-type ZnSe, 219
pulse shape, 93
pulsed laser
    ablation, 359
    deposition, 305, 353, 359

radical(s)(-), 141
    chain mechanism, 105
Raman spectroscopy, 483
rapid thermal chemical vapor deposition
    (RTCVD), 507, 519, 531
reaction(-), 317, 419
    kinetics, 117
    mechanism, 45, 177
    rate theory, 19
    transport models, 177
reactive ion etch, 433
real-time control, 155
reflectance anisotropy spectroscopy, 155
remove PECVD, 385
resistivity, 353
RF ion trapping, 451
RHEED oscillation, 543
room temperature, 81
rotating disk reactor, 3
RPECVD, 341, 495
RTCVD oxides, 531

scanning tunneling microscope (STM), 419
second harmonic generation, 251
selective
    epitaxy, 519
    etching, 439
    growth, 147, 257
    Si epitaxial growth, 525
self-limiting growth, 225
semiconductors, 207
sequential deposition, 25
SF_5^+, 451
SF_6, 451
Si, 425
    atoms, 3
    epitaxy, 463
Si(100), 69, 289
Si(111)-(7×7), 419
Si/GaP/Si heterostructures, 495
$Si_{1-x}Ge_x$ alloys, 489
SiGe films, 507
SiH, 11
SiH_2, 11
SiH_3, 11
SiH_4, 87
Si_2H_6, 463
SiH_4/NH_3 plasma, 51
silane, 3
silanol, 57, 273
silicon, 37, 87, 93, 201, 519
    carbide, 105, 111
    cluster growth, 75
    CVD, 11
    nitride, 51, 341
SiO_2, 25, 31, 37, 273, 373, 379, 385, 439
site-selective dopant, 219
Soret effect, 483
spectroscopy, 347
step coverage behavior, 471
sub-picosecond, 245
superlattice structures, 189
surface, 169, 289
    analysis, 445
    kinetics, 123, 177
    loss coefficient, 11
    modification, 391
    morphology, 391
    reactions, 51, 177
    reconstructions, 155
synchrotron, 525

TDMAT, 323
Teflon, 359
temperature(-)
    distribution, 93
    gradients, 483
    independent optical devices, 207
    programmed
        desorption, 69
        reaction, 169
TEOS, 45, 379
tertiary-butyl(allyl)selenium, 231
tertiarybutylarsine, 201
tetraethoxysilane, 31, 37
tetraethyl orthosilicate (TEOS), 57
tetrakis-(alkyl) diarsines, 213
tetrakis(dimethylamido) titanium, 323
tetramethylsilane, 31
TFT's, 81
thermal
    CVD, 45
    desorption, 239
thin films, 347, 359
time-of-flight, 245
TiN, 323, 329, 379
$Ti(NMe_2)_4$, 379
$TiSi_2$, 63
titanium
    nitride, 323
    tetrachloride, 317
transition from two- to three-dimensional
    behavior, 353
trench structures, 507
triethyl-gallium, 177, 183
trimethylamine alane, 283
trimethyl borate (TMB), 57

trimethylgallium, 183
trimethylindium-dimethylethylamine adduct (TMIDMEA), 225
TRIS-dimethylamino arsenic, 169

UHVCVD, 183
UHV-RTCVD, 463
unzipping reaction, 359
UV laser ablation, 245